Lecture Notes in Electrical Engineering

Volume 875

The book series *Lecture Notes in Electrical Engineering* (LNEE) publishes the latest developments in Electrical Engineering - quickly, informally and in high quality. While original research reported in proceedings and monographs has traditionally formed the core of LNEE, we also encourage authors to submit books devoted to supporting student education and professional training in the various fields and applications areas of electrical engineering. The series cover classical and emerging topics concerning:

- Communication Engineering, Information Theory and Networks
- Electronics Engineering and Microelectronics
- Signal, Image and Speech Processing
- Wireless and Mobile Communication
- Circuits and Systems
- Energy Systems, Power Electronics and Electrical Machines
- Electro-optical Engineering
- Instrumentation Engineering
- Avionics Engineering
- Control Systems
- Internet-of-Things and Cybersecurity
- Biomedical Devices, MEMS and NEMS

For general information about this book series, comments or suggestions, please contact leontina.dicecco@springer.com.

To submit a proposal or request further information, please contact the Publishing Editor in your country:

China

Jasmine Dou, Editor (jasmine.dou@springer.com)

India, Japan, Rest of Asia

Swati Meherishi, Editorial Director (Swati.Meherishi@springer.com)

Southeast Asia, Australia, New Zealand

Ramesh Nath Premnath, Editor (ramesh.premnath@springernature.com)

USA, Canada:

Michael Luby, Senior Editor (michael.luby@springer.com)

All other Countries:

Leontina Di Cecco, Senior Editor (leontina.dicecco@springer.com)

** **This series is indexed by EI Compendex and Scopus databases.** **

More information about this series at https://link.springer.com/bookseries/7818

Pradeep Kumar Singh · Maheshkumar H. Kolekar ·
Sudeep Tanwar · Sławomir T. Wierzchoń ·
Raj K. Bhatnagar
Editors

Emerging Technologies for Computing, Communication and Smart Cities

Proceedings of ETCCS 2021

Springer

Editors
Pradeep Kumar Singh
Department of Computer Science
KIET Group of Institutions
Ghaziabad, Uttar Pradesh, India

Sudeep Tanwar
Department of Computer Engineering
Nirma University
Ahmedabad, Gujarat, India

Raj K. Bhatnagar
Department of Electrical Engineering
and Computer Science
University of Cincinnati
Cincinnati, OH, USA

Maheshkumar H. Kolekar
Department of Electrical Engineering
IIT Patna
Patna, Bihar, India

Sławomir T. Wierzchoń
Institute of Computer Science
Polish Academy of Sciences
Warsaw, Poland

ISSN 1876-1100 ISSN 1876-1119 (electronic)
Lecture Notes in Electrical Engineering
ISBN 978-981-19-0283-3 ISBN 978-981-19-0284-0 (eBook)
https://doi.org/10.1007/978-981-19-0284-0

This Springer imprint is published by the registered company Springer Nature Singapore Pte Ltd.
The registered company address is: 152 Beach Road, #21-01/04 Gateway East, Singapore 189721, Singapore

Preface

This volume constitutes the referred proceedings of the second International Conference on Emerging Technologies for Computing, Communications, and Smart-World (ETCCS-2021), held in Baba Farid College of Engineering and Technology, Bathinda, Punjab, India, in August 21–22, 2021.

The conference was organized jointly with the technical support from Maharaja Ranjit Singh Punjab Technical University, India, University of Arad, Romania, ISTE India, IAC Education, and with the academic partnership of Southern Federal University Russia, SD College of Engineering and Technology, India, and UJW Poland, as well. The conference was supported by the All India Council for Technical Education (AICTE), Council of Scientific & Industrial Research (CSIR), and Ministry of Electronics and Information Technology (MeitY). Technical support was provided by the Digital India, National e-Governance Division.

Thematically, the works submitted to the conference set out five main paths. These are as follows: (i) emerging computing technologies, (ii) network and computing technologies, (iii) wireless networks and Internet of Everything (IoE), (iv) communication technologies, security, and privacy, and (v) next-generation computing technologies. The attractiveness of this topic has attracted many researchers and scientists, and its potential is reflected in the large number of submitted papers. We hope that the reports collected in this volume will be met with wide interest.

The inaugural speech was delivered by Professor Anil D. Sahasrabudhe, Chairman, AICTE, New Delhi, India. Welcome address was given by Dr. Pardeep Kaura, Dean Academics, followed by the introduction of conference speakers by Dr. Nimisha Singh, Dean Trainings, BFCET. Dr. Jayoti Bansal, Principal, BFCET, briefed about the conference information and contributions. The first keynote was delivered by Dr. Sachin Kumar Mangla, University of Plymouth, UK, and vote of thanks by Dr. Tejinderpal Singh Sarao, Dean R&D, BFCET. The keynote talks were delivered by Dr. Anand Nayyar, Duy Tan University, Da Nang, Vietnam; Prof. Pao Ann Hsuing, National Chung Cheng University, Taiwan; Dr. Sanjay Misra, Convent University, Nigeria; Dr. Pastor Jr. Arguelles, Dean, College of Computer Studies; Calabarzon, Philippines; and Dr. Zdzisław Pólkowski, from the UJW, Poland. These keynote speakers also took part in technical discussion organized for the authors.

We are highly thankful to our valuable authors for their contribution and our technical program committee for their immense support and motivation toward making the ETCC-2021 a grand success. We are thankful to various session chairs for chairing the session and giving their valuable suggestions; few of them include Dr. Chaman Verma, Eötvös Loránd University (ELTE), Hungary, Dr. Vivek Sehgal, Dr.YugalKumar, Dr. R.K. Saini, Dr. Rohit Tanwar, Dr.Anupan Singh, Dr. Sudhanshu Tyagi, Dr. Ashwani Kumar, and many more co-session chairs.

The volume editors are thankful to Patron Dr. Gurmeet Singh Dhaliwal, Chairman, Baba Farid Group of Institutions, and Co-Patron, Dr. Manish Goyal, Principal, BFCET, Bathinda, for extending their support during the conference. BFCET is thankful to the All India Council for Technical Education (AICTE), Council of Scientific & Industrial Research (CSIR), and Ministry of Electronics and Information Technology (MeitY) for providing grant to BFCET to support the conference.

Lastly, we express our sincere gratitude to our publication partner, LNEE Series, Springer, for believing in us.

Editors

Ghaziabad, India — Pradeep Kumar Singh
Patna, India — Maheshkumar H. Kolekar
Ahmedabad, India — Sudeep Tanwar
Warsaw, Poland — Sławomir T. Wierzchoń
Cincinnati, USA — Raj K. Bhatnagar
August 2021

Contents

About the Editors

Dr. Pradeep Kumar Singh is currently working as a Department of Computer Science, KIET Group of Institutions, Delhi-NCR Campus, Ghaziabad, India. Dr. Singh is having life membership of Computer Society of India (CSI), Life Member of IEI and promoted to Senior Member Grade from CSI and ACM. He is Associate Editor of *International Journal of Information Security and Cybercrime* (IJISC) a scientific peer reviewed journal from Romania. He has published more than 120 research papers in various International Journals and Conferences of repute. He has received three sponsored research projects grant from Govt. of India and Govt. of HP worth Rs 25 Lakhs. He has edited total 10 books from Springer and Elsevier and also edited several special issues for SCI and SCIE Journals from Elsevier and IGI Global. He has Google scholar citations 1451, H-index 20 and i-10 Index 49 in his account. He is section editor of *Discover Internet of Things*, Springer Journal.

Dr. Maheshkumar H. Kolekar is working as Associate Professor in Department of Electrical Engineering at Indian Institute of Technology Patna, India. He received the Ph.D. degree in Electronics and Electrical Communication Engineering from the Indian Institute of Technology Kharagpur in 2007. During 2008 to 2009, he was a Post-Doctoral Research Fellow with the Department of Computer Science, University of Missouri, Columbia, USA. During May to July 2017, he worked as DAAD fellow in Technical University Berlin where he worked in the research area EEG signal analysis using machine learning. His research interests are in the area of Digital Signal, Image and Video Processing, Video surveillance, Biomedical Signal Processing, Deep Learning. Recently, he has authored a book titled as *Intelligent Video Surveillance Systems: An Algorithmic Approach*, CRC Press, Taylor and Francis Group, 2018. He has successfully completed R&D Project Sponsored by Principal Scientific Advisor to Government of India on abnormal human activity recognition.

Dr. Sudeep Tanwar (Senior Member, IEEE) is currently working as a Professor with the Computer Science and Engineering Department, Institute of Technology, Nirma University, India. He is also a Visiting Professor with Jan Wyzykowski University,

Polkowice, Poland, and the University of Pitesti in Pitesti, Romania. He received B.Tech. in 2002 from Kurukshetra University, India, M.Tech. (Honor's) in 2009 from Guru Gobind Singh Indraprastha University, Delhi, India and Ph.D. in 2016 with specialization in Wireless Sensor Network. He has authored two books and edited 13 books, more than 250 technical articles, including top journals and top conferences, such as *IEEE Transactions on Network Science and Engineering*, *IEEE Transactions on Vehicular Technology*, *IEEE Transactions on Industrial Informatics*, *IEEE Wireless Communications*, *IEEE Networks*, *ICC*, *GLOBECOM*, and *INFOCOM*. He initiated the research field of blockchain technology adoption in various verticals, in 2017. His H-index is 44. He actively serves his research communities in various roles. His research interests include blockchain technology, wireless sensor networks, fog computing, smart grid, and the IoT. He is a Final Voting Member of the IEEE ComSoc Tactile Internet Committee, in 2020. He is a Senior Member of IEEE, Member of CSI, IAENG, ISTE, and CSTA, and a member of the Technical Committee on Tactile Internet of IEEE Communication Society. He has been awarded the Best Research Paper Awards from IEEE IWCMC-2021, IEEE GLOBECOM 2018, IEEE ICC 2019, and Springer ICRIC-2019. He has served many international conferences as a member of the Organizing Committee, such as the Publication Chair for FTNCT-2020, ICCIC 2020, and WiMob2019, a member of the Advisory Board for ICACCT-2021 and ICACI 2020, a Workshop Co-Chair for CIS 2021, and a General Chair for IC4S 2019, 2020, and ICCSDF 2020. He is also serving the editorial boards of Computer Communications, International Journal of Communication System, and Security and Privacy. He is also leading the ST Research Laboratory, where group members are working on the latest cutting-edge technologies.

Prof. Sławomir T. Wierzchoń received M.Sc. and Ph.D. degrees in Computer Science from Technical University of Warsaw, Poland. He holds Habilitation (D.Sc.) in Uncertainty Management from Polish Academy of Sciences. In 2003 he received the title of Professor from the President of Poland. Currently he is full professor at the Institute of Computer Science of Polish Academy of Sciences. His research interests include Computational Intelligence, Uncertainty Management, Information Retrieval, Machine Learning and Data Mining. He is an author/co-author of over 100 peer reviewed papers in international journals and international conferences. He published, as author/co-author, 11 monographs from the field of Machine Learning. In the period 2000–2013 he co organized 13 international conferences on intelligent information systems. Co-authored proceedings from these conferences were published by Springer. He co-edited two volumes of proceedings of the international conference on Computer Information Systems and Industrial Management and he has served as a guest co-editor of three special issues of *Information & Control* journal. Currently, he is member of the editorial board for some international journals, as well as member of many program committees for international conferences. He cooperated with medical centers in the area of statistical data analysis and knowledge discovery in data bases.

Prof. (Dr.) Raj K. Bhatnagar is currently working as a Professor of Computer Science, Department of Electrical Engineering and Computing Systems, University of Cincinnati, USA. He completed his B.Tech. in IIT Delhi, India (1979) and Ph.D. in University of Maryland, USA (1989). He worked as a Visiting Professor in many institutions and he has taken many administrative responsibilities along with teaching profession at University of Cincinnati such as Graduate Program Director of CS Department and ECECS Department, Director of UG Computer Science Program, Co-Director of Graduate Certificate in BioInformatics, Assoc. School Director and Graduate Program Director School of Computing Sciences and Informatics, Interim Department Head, Computer Science Department. His research interests includes Data mining, AI and pattern recognition problems, knowledge discovery and big data and cloud computing environments.

Emerging Computing Technologies

Healthcare Transformation Traditional to Telemedicine: A Portrayal

Babita Yadav and Sachin Gupta

Abstract The biggest challenge nowadays is to cope with massive amount of data generated digitally which is being originated from various healthcare entities and sources. The use of this data in developing various healthcare models is to predict diseases as well as creating services that can be benefitted to humanity. The business of redesign through information technology practices which are broadening their spectrum to made possible that each and every health seeker should be benefitted through each and every ICT practices. As well as the data generated from various sources should be analyzed properly so that it can be taken out from healthcare information systems for prediction of future diseases and help medical practioners to advise patient in more advance ways. This is only possible through various change that took place in information technology. This merging of healthcare experts from medical domain with big data and predicting value points from this data will create a total difference in Medical industry. The prediction not only help health seekers but also helps in reducing the cost that a patient spend on health advises. The innovation in healthcare system storing information will not only improve service quality but also reduce delivery cost. Big data in clinical practices will also help in better management of data which will prove an effecting way of various data of patient as well as doctors in their expert field. In this chapter, the transformation of healthcare sector is amazingly changing because of taking perspective of big data into account.

Keywords Telemedicine · Health care · Big data · Clinical practices

B. Yadav (✉) · S. Gupta
MVN University, Haryana, India
e-mail: Babita.yadav@mvn.edu.in

S. Gupta
e-mail: Sachin.gupta@mvn.edu.in

P. K. Singh et al. (eds.), *Emerging Technologies for Computing, Communication and Smart Cities*, Lecture Notes in Electrical Engineering 875,
https://doi.org/10.1007/978-981-19-0284-0_1

1 Introduction

Analytics in healthcare system including taking account analysis of health data for patient. The patient records, cost of disease and diagnosis of data as well as in both the states micro- and macro-level are being considered. There are various visualization tool that help health managers by providing real-time insight about health data. Systematic analysis of data helps various professional working in healthcare system to get the findings and various opportunities that can improve health management system. Various trends and patterns can be generated from health data along with real-time data [1]. With these trends, various actionable insights from data can be taken out to improve existing health systems and services. When there is system of improved quality care, there will be improved diagnosis system and methods of improved treatment which in turn provide better health services to health system.

The health analytics also play an important role in initiating awareness about disease. It also helps public care units in predicting outbreak of certain disease [2]. Forecasting arrangements of healthcare units as well as arranging control of disease spread in healthy population. Finding trends will also uncover the certain hidden facts about disease that normal population should know to protect themselves from getting infection.

The various steps adopted to obtain, review and management of data related to health for entire population are carried out by public health organizations in an effort to maintain health of population [3]. The data related to health of public may include reports of mortality, demographic data, socioeconomic data and data of diagnosis of patient, and patient claiming medical claims. Health analytics monitor various disease using various patterns in population helping various public health organizations that may utilize analytics to monitor trends of disease and deriving patterns in certain category of populations which can be used to guide various programs that control disease and priorities setting for assigning resources to populations for various health need [4].

Predictive analysis plays an important role in healthcare system helping health providers as well as health seekers, medical claim companies applying machine learning to health care.

This type of analysis results in designing a model for prediction which can help in taking decision in a better way which may help in investing planning in healthcare system, helping health ministry to invest funds on health care and serving population in a much better way.

1.1 Telemedicine

According to World Health Organization (WHO), this term is said to be delivery of various services related to healthcare services.

The areas where distance is a crucial factor by which professional healthcare providers utilize various kind of information and technologies to communicate for exchanging various information related to patient diagnosis, treating them and how they can prevent disease and injuries. An electronic use of telehealth information and technologies to support how patient at a long distance can be cured without visiting them physically. There is a system which is required to be an interface between two remote locations and exchange information to avail consultancy regarding health of patient [5].

There is rapid change in the field of technology which has made the existence of telemedicine really a need. It's really became a technology to be used in health care to provide services related to health easily reach patients to locations far away. This technology has now transformed into an integrated service which is very much utilized in hospitals, homes, private physician offices, and in providing various healthcare facilities. The history of Telemedicine justified it use in the form of electromagnetic signals. A video consultation between patient and doctor made possible using this telemedicine technology [6].

In its original state Telemedicine was only used for treating patients which were located in places of remote areas and in case of local health facilities are shorter as well as less professional medical health caretaker [7]. Now the current state of Telemedicine is used as a very convenient tool in medical care. Everyone in this life is too much busy the health seekers really to save their time waiting in traffic or at hospital queue. Existing Telemedicine framework can be adopted to be untilled and implemented during pandemic times via existing Tele devices as shown in Fig. 1.

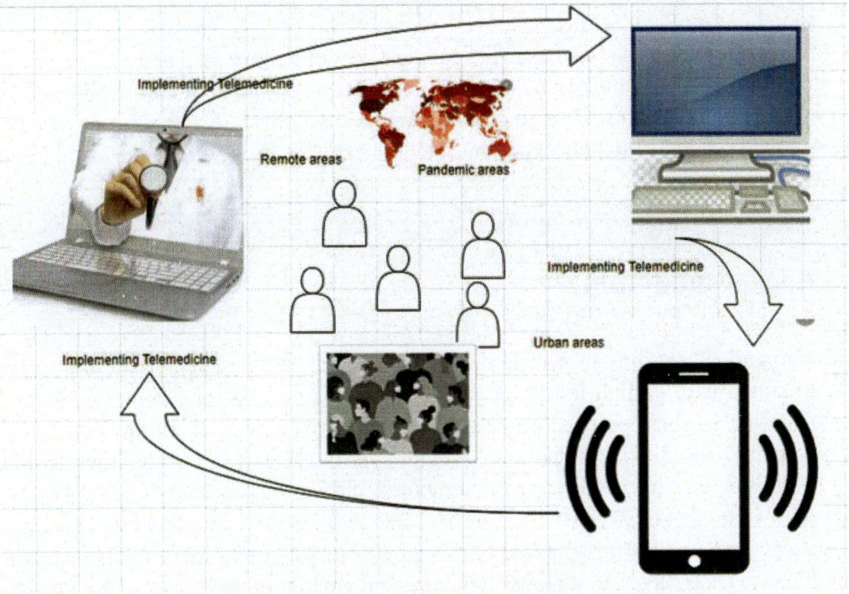

Fig. 1 Telemedicine adoption in pandemic

1.2 Current State: Telemedicine Today

Now a days the face of telemedicine has changed so fast because of technology advancement at exponential levels. The low cost of mobile and availability of Internet has given birth to live video telemedicine thorough face time, skype and many telemedicine applications ready to use at time of extreme cases [8].

1.3 Major Contributions

- The current state of art related to telemedicine has been identified.
- The medical experts are overburdened and sometime unavailable and it's expected by various patients to have individual expectation [9]. This expectation is more related to convenient care, individual exclusive attention. Sometime due to absence of some expert hospital units transfer and outsource the cases to some big extensive care units.
- In this age of mobile health various options given to patient in the form of mobile health applications, portals to take appointment etc. [10]. The patients have already started to use these technologies to monitor and using tracking their health continuously.
- Evaluation of the user acceptance, diagnostic accuracy and educational value of telemedicine. (Setting up hypothesis and research questionnaires to conduct a survey to retrieve real picture of Telemedicine in India from the perspective of medical experts) in future.
- After review study in this paper proposing a model of approach and techniques to that how Telemedicine can be implemented in remote and urban areas addressing various challenges based on above study including field study in healthcare sector.
- Analysis of telehealth data of patients generated by the proposed model to compute its efficiency.

2 Background Study

The demand of patients is prioritizing convenience and less expensive care. The author stated work which focus specifically described on various important use of telemedicine in disasters and emergency situation in health of public in various emergencies. Telemedicine has a huge challenge in creation and implemented but US health systems that are already implemented innovation in telemedicine [11]. The spread of virus globally rapidly become a global public health pandemic. The testing of patients because of this disease is started rising. The symptoms are not known completely. This has posed various challenges with managing health of such a vulnerable population. This has pointed out various fields where this technology can help a lot [12]. COVID-19 outbreak has impacted complete global healthcare systems.

The challenge imposed to many fields has also impacted clinical disciplines. The outbreak is so severe that adoption of telecommuting is an effective working option, others raise concerns about various potential downsides [13]. Persons with disabilities in this pandemic unable to visit hospital and clinic. Traditional healthcare was depending upon relationship between patient–physician and always been centered around contacting face-to-face. The challenge has been addressed by applicability, convenience and cost effectiveness of telemedicine [14]. The use of mobile phone technologies for promoting health care and preventing disease came into existence so rapidly. The purpose of this pilot evaluation study is to find out the efficiency of telemedicine and its importance. The proposed systematic review will evaluate the current evidence regarding finding out effectiveness of telemedicine and its role in situation like pandemic [15]. There are various patients in areas which are underserved and are at remote location. They are facing scarcity of specialists. The popularity about has increased dramatically in the situation like COVID-19. The experts related to medical are getting proper training regarding this technology including demonstration of the telemedicine programs. The telemedicine practice is flourishing with support from government agencies and various other sources to implement this technology. The technology like telemedicine despite of its popularity is not universally adopted in many regions and many people. The objective of this research paper is to show and review that how much technology has been in progress that this technology can prove a very emergency tool if being utilized properly in specialty care services [16]. Developed country like USA accepted telemedicine so widely. The programs related to telemedicine have been in existence since the 1960s, but only in the last two or three years, have they begun to proliferate and can be taken as miracle in situation like pandemic.

2.1 Gaps Identified

There is a lot of study and research work done by researchers where it has been found that despite having strategies, approaches existing regarding telemedicine, there is still a lack of adoption of proposed solutions. There are challenges and gaps still left in adoption of telemedicine that are extracted from the research article from survey of literature. Some of the gaps identified from literature are mentioned below.

1. There is a general lack of awareness among patients in most urban and rural areas.
2. Professional training of medical experts is an area of concern.
3. The various methods of medical information exchange are still developing. Standardization in information format to suit telemedicine is another gap area.
4. Technologically challenged practitioners and patients face resistance in adoption of telehealth care.
5. The cost factors of establishing e-medicine and telehealth care are currently high, due to less permeation in the market.

3 Current Trends in Telemedicine

Telemedicine can be considered to be the most effective diagnosing of treatment of patients by means of technology related to telecommunication. If this can be added and supported in government funding agencies, it can be further thereby providing substantial healthcare to low income regions also. Earliest records of telemedicine are in the first half of the twentieth century when ECG was transmitted over telephone lines. Today to then, telemedicine has been a long way in terms of both delivery of healthcare and technology. There are different types of telemedicine services including store and forward, real-time and remote or self-monitoring services, which provide different levels of educational, healthcare delivery and management options for screening of diseases. These services can be scaled to spread all over the world in a pandemic like situation for best benefits to humankind [10]. Even though telemedicine cannot be a solution to all the problems, it can surely help decrease the burden of the healthcare system to a large extent.

In various research studies, doctors utilizing telemedicine have acknowledged that they have successfully treated patients virtually. Sometimes, many patients visit doctor with minor medical conditions or questions then it does not require in-person diagnosis or treatment. It has been seen that 75% of all regular doctor as well as visits are either unnecessary and can be handled very effectively via phone or by video [5]. As well like patients with sinusitis, cold or flu will avoid coming out and will not spread the virus further. According to research sources, 60% of telemedicine visits ended with the doctor writing a prescription.

The latest trends in telemedicine today exploring benefits, the attitudes of patients and doctors, and telemedicine's potential cost-savings. The billion global market percentage adopting telemedicine technologies is valued $17.8 billion in 2014 as well as there will be about 800,000 online consultations in developing countries in 2015.

4 Reason of Adoption

The improvement in patient care as well healthcare delivery by remote methods is the first reason for rapid adoption of this technology. Telemedicine improves care delivery and patient communication. The consultation provided to health seekers by continuing multispecialty consultation via various mode of telemedicine. There can be several reasons of its adoption because of rise in digital revolution as well as digital literacy as shown in Fig. 2.

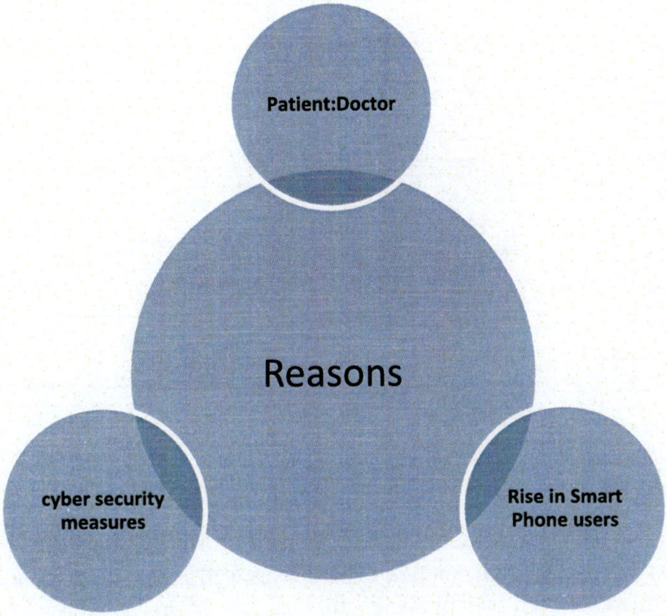

Fig. 2 Reasons of adoption

4.1 Patient Per Doctor Unavailability (Patient:Doctor)

The availability of doctors in developing countries like India has been estimated as per population according to registration comes to be low as compared to population growth. This has been seen in literature that 4.8 practicing doctors per 10,000 population available in India in 2014, in contrast to the belief of having seven doctors per 10,000 people. It is also been estimated that the country would also be able to achieve a ratio of about 6.9 practicing doctors per 10,000 people only by 2030 as shown in Fig. 3.

The findings also been shown that the current availability of doctors per 10,000 people and their growth prospects over the next 15-year period will be same, and also, it looks like an impossible task to achieve accepted doctor-population ratio of 1:1000 by 2030. Need of telemedicine is much needed to be implemented as widely accepted method for patient care.

- As well as adding it to more our 5.5 million km road network transports 64.5% or as well as 90% of India's total passenger traffic uses this road network to commute. The government itself has acknowledged that congestion on roads has been increasing. Telemedicine will help people and more and more doctor will be available for treatment which were later unavailable because of shortage of medical practioners in country and around the world. Implementation and adoption of telemedicine will help patient to get timely treatment and medical care.

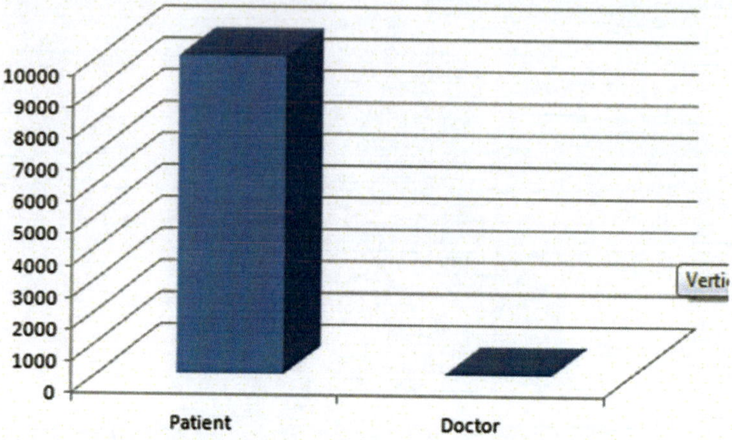

Fig. 3 Patient to doctor ration 7:10,000, need of telemedicine

Doctors can be in contact with patients frequently. The outcome of telemedicine adoption will be improved outcome of patients as well as better treatment. The gaps will be filled between patient and doctors because of this adoption. The doctors will available for more patient in one day. The health seekers will be saved from no less shows as well as late cancelations. The patients can keep regular contact with their doctors. The doctors can extend access to more specialists will save time and money that patients have to spend on travel for appointment especially will help rural patients. Discussing with their doctor from the comfort of their own home will result in shorter wait time; the care will be more convenient and on-demand. The involvement in decisions regarding treatment will be more.

4.2 Increase in Smart Phone Users

The number of smart phone users in India was estimated to reach over 760 million in 2021, with the number of smart phone users worldwide forecasted to exceed to 3.8 billion users in 2021.

In the world of 5G, India is also moving very rapidly at the path of digital revolution which is powered by increase in broadband and Internet availability at a very low cost; the data exponentially has been up taken [17]. The government's initiatives also focus on digitalization which has also increased trend of adopting technology across industries. Internet statistic says that due to digital revolution, the use of smart phone in India has much increased from 2015 to 2025. As well the estimation about smart phone users globally increased and to be exceed to 3.8 billion users in 2020 as well the number of smart phone users worldwide is almost projected to a figure nearly 2.7 billion by 2019 [18].

The rise in increase of smart phone users in country has been estimated to reach over 760 million by 2021. The statistic has been shown in Fig. 4.

The popularity of telemedicine has been shown growth from 2010, 35% to 76% in 2017 as shown in Fig. 5. Still, there is a pattern of visiting hospital then to consult doctor on Internet or phone. There are reluctance that has been still observed adopting Telemedicine [19].

As more people start using video chat apps and smart phones, telemedicine will become even more intuitive and accessible to doctors and patients. Literature shows that the number of video callers increased up to an estimated 380 million this year as compare to 2010 which was up 63 million as shown in Fig. 6.

In developing countries like India, more than 85% of population has Internet access. So partial infrastructure for adopting telemedicine is already existing, waiting to be utilized. About 30% of patients already use computers or mobile devices to

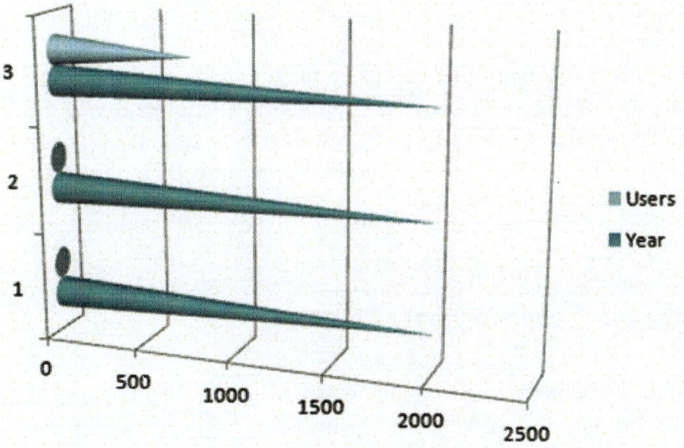

Fig. 4 Users increased by the year from 2029 to 2021

Fig. 5 35% to 76% increase from 2010 to 2017

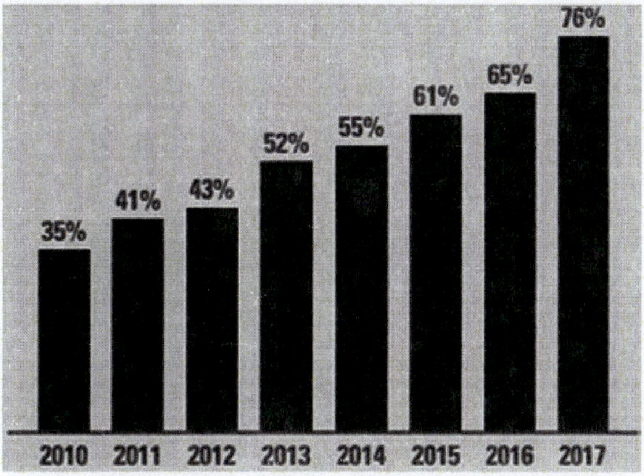

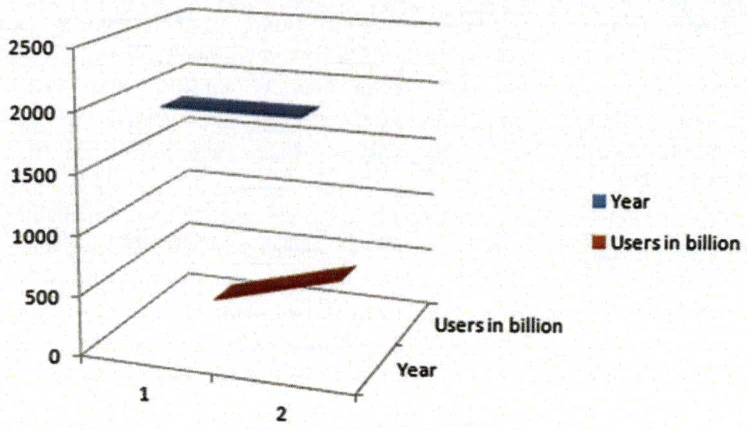

Fig. 6 Rise of users

check for medical or diagnostic information. It can be said that the adoption of telemedicine is been seen globally and population is equipped with basic requirements to adopt telemedicine completely as well as Nearly, half of healthcare professionals already planned to integrate mobile apps into their practices in coming years.

4.3 Stringent Cyber-Security Measures

The adoption of telemedicine is relying totally on technology because of this cyber-security has become a necessity. The contribution of cyber-security methods has made that the industry of health care is no more prone to cyber-attacks but with telemedicine, need of information protection has become priority [19]. The support of cyber-security industry in recent years has shown a backbone to health care which can be prove adoption of telemedicine with open arms.

4.4 Modeling Telemedicine Using Regression

To tackle the difficulties of telemedicine as well as defining relationship between patient, doctors and other dependent and independent variables, linear regression can be modeled to fit linear equation between variables [11]. The model supports relationship between the variables of interest like, time, cost, on time support and personal support, traffic, mobile users, digital literacy. The strength between patient and doctor variable is shown with the help of a scatter plot [14]. A valuable numerical measure of association between patient and doctor variables is the correlation coefficient, which is a value between −1 and 1, indicating the strength of the association of the observed data for the variables patient and doctor.

A linear regression line has an equation of the form $Y = a + bX$, where X is the explanatory variable and Y is the dependent variable. The slope of the line is b, and a is the intercept (the value of y when $x = 0$).

Hypothesis

In case of various variables assumed to be fit in regression equation, the multiple regression equation will take form: $y = b1x1 + \mathrm{b}2\,x2 + \ldots + bnxn + c$. Here, bi's ($i = 1,2\ldots n$) are the regression coefficients, which represent the value at which the criterion variable changes when the predictor variable changes.

$$y = b1x1 + b2x2 + \ldots + bnxn + c \quad (1)$$

time = $b1$, cost $b2$, on time support $b3$ and personal support $b4$, traffic $b5$, mobile users $b6$, digital literacy $b7$ it comes out to be

$$y = b1x1 + b2x2 + b3x3 + b4x4 + b5x5 + b6x6 + b7x7 + c \quad (2)$$

where x is the value of the independent variable (x)

5 Conclusion

There is no doubt that telemedicine will prove very effective in situations like pandemic. The need for transforming to an era of telemedicine is actually employing to its maximum utility which should be realized. The purpose of this research work is to showcase actual scenario of telemedicine as well as portraying why telemedicine needs to be implemented country like India where roads are full of traffic as well as availability of doctors is terribly low as compare to population. The support of government and healthcare organizations must produce required strategies to encourage fourfold commitment: to support and provide funding for telemedicine.

References

1. Mishra SK, Kapoor L, Singh IP (2009) Telemedicine in India: current scenario and the future. Telemed e-Health 15(6):568–575
2. Ganapathy K, Ravindra A (2009) Telemedicine in India: the Apollo story. Telemed e-Health 15(6):576–585
3. Chandwani RK, Dwivedi YK (2015) Telemedicine in India: current state, challenges and opportunities. In: Transforming government: people, process and policy
4. Bagchi S (2006) Telemedicine in rural India. PLoS Med 3(3): e82

5. Pal A, Mbarika VWA, Cobb-Payton F, Datta P, McCoy S (2005) Telemedicine diffusion in a developing country: the case of India (March 2004). IEEE Trans Inf Technol Biomed 9(1):59–65
6. Mishra SK, Gupta SD, Kaur J (2007) Telemedicine in India: initiatives and vision. In: 2007 9th international conference on e-health networking, application and services. IEEE, pp 81–83
7. Dasgupta A, Deb S (2008) Telemedicine: a new horizon in public health in India. Indian J Community Med: Official Publ Indian Assoc Preventive Soc Med 33(1):3
8. Myers MB (2003) Telemedicine: an emerging health care technology. Health Prog 22(3):219–223
9. Ebad R, Jazan KSA (2013) Telemedicine: current and future perspectives
10. Huston TL, Huston JL (2000) Is telemedicine a practical reality? Commun ACM 43(6):91–95
11. Hollander JE, Carr BG (2020) Virtually perfect? Telemedicine for Covid-19. New England J Med
12. Korupolu R, Stampas A, Gibbons C, Jimenez IH, Skelton F, Verduzco-Gutierrez M (2020) COVID-19: Screening and triage challenges in people with disability due to spinal cord injury. Spinal Cord Ser Cases 6(1):1–4
13. Lincoln MH, Khan R, Cai J (2020) Telecommuting: a viable option for medical physicists amid the COVID-19 outbreak and beyond. In: Medical physics
14. Ananthakrishnan AN, Singh S (2020) The doctor will call you now! telemedicine in the midst of a pandemic. In: Clinical gastroenterology and hepatology
15. Burnham JP, Fritz SA, Yaeger LH, Colditz GA (2019) Telemedicine infectious diseases consultations and clinical outcomes: a systematic review and meta-analysis protocol. Syst Rev 8(1):135
16. Wootton R (2001) Telemed BMJ 323(7312):557–560
17. Mishra S (2015) Drivers of mobile-commerce acceptance in India: an empirical study. In: XIV international business and economy conference, Bangkok, Thailand
18. Bhave K, Jain V, Roy S (2013) Understanding the orientation of gen Y toward mobile applications and in-app advertising in India. Int J Mobile Marketing 8(1)
19. Black PE, Scarfone K, Souppaya M (2008) Cyber security metrics and measures. In: Wiley handbook of science and technology for homeland security, 1–15

Comparative Analysis of Semantic Similarity Word Embedding Techniques for Paraphrase Detection

Shrutika Chawla, Preeti Aggarwal, and Ravreet Kaur

Abstract Most of the state-of-the-art plagiarism detection tools focus on verbatim reproduction of a document for plagiarism identification (PI), not keeping into account its semantic properties. Recently, deep learning models have shown considerable performance in identifying paraphrases using word embedding approach. This paper gives an overview and comparison of the performances of five word-embedding based deep learning models in the field of semantic similarity detection, such as TF-IDF, Word2Vec, Doc2Vec, FastText, and BERT on two publicly available corpora: Quora Question Pairs (QQP) and Plagiarized Short Answers (PSA). After extensive literature review and experiments, the most appropriate text preprocessing approaches, distance measures, and the thresholds have been settled on for detecting semantic similarity/paraphrasing. The paper concludes on FastText being the most efficient model out of the five, both in terms of evaluation metrics, i.e., accuracy, precision, recall, F1-score, receiver operating characteristic (ROC) curve, and resource consumption. It also compares all the models with each other based on the above-mentioned metrics.

Keywords Paraphrasing · Plagiarism · Paraphrase detection · Plagiarism detection · Paraphrase

1 Introduction

Paraphrasing is the procedure of rewriting a statement to change its form without changing the meaning of the original text. Many deep learning models have shown a promising performance in detecting semantic similarity between documents using word embedding technique. Word embedding is a representation of document vocabulary which is capable of capturing the semantic context of the document as a whole, or the individual words in a document. The idea behind word embedding is assigning

S. Chawla (✉) · P. Aggarwal · R. Kaur
UIET, CSE Department, Panjab University, Chandigarh, India

P. K. Singh et al. (eds.), *Emerging Technologies for Computing, Communication and Smart Cities*, Lecture Notes in Electrical Engineering 875,
https://doi.org/10.1007/978-981-19-0284-0_2

vector values to text, as the technique opens doors to various text evaluation and analysis approaches in the field of linear algebra.

The main objective of this paper is to review and compare the performance of various word-embedding based models (including some deep learning models) for semantic similarity detection which leads to the following contributions: (a) Systematic outline of corpus-based word embedding models, and (b) Elucidation of the best models, preprocessing techniques and thresholds for plagiarism detection and the kind of corpus they work well with.

The motivation behind this research is to in turn develop a framework which uses machine learning and statistics to measure semantic similarity between given documents and detect paraphrase plagiarism. The systematic study helps in cherry-picking the mechanisms and hyperparameters best suited for a new framework.

This research compares five models such as Word2Vec [1], Doc2Vec [2], BERT [3], TF-IDF [4], FastText [5] on two publicly available corpora, namely Quora Question Pairs and Plagiarized Short Answers (PSA), and in turn, these models are evaluated in terms of accuracy, precision, recall and F1-score. Before applying the models, data preprocessing techniques, distance measures, and a threshold value for each model were set after careful experimentation on each corpus.

The paper is organized as follows: Section 2 briefly discusses existing work in the related field as well as specific work in the paraphrase detection area. Section 3 discusses approach followed in this paper, models implemented and the results derived from the experiments. Section 4 draws conclusion and discusses future scope.

2 Related Work

The use of computers in the field of Natural Language Processing (NLP) is a challenge due to the equivocacy of texts and passages. For example, the term 'mango' can be referred to as a fruit as well as a clothing brand. Hence, the semantic features of a text play a major role over and above its linguistic features.

A large number of established researchers have contributed to various plagiarism detection stages [6]. For handling plagiarism and paraphrasing cases of higher complexities, [7] suggests that researchers should focus on linguistic, syntactic, and most importantly semantic information of text rather than just on verbatim reproduction.

Table 1 summarizes various works and approaches in the field of paraphrase detection.

Inspired by the recent successes of neural networks (NNs) in the fields of information retrieval and natural language processing, this paper experiments with various deep learning models to test semantic similarity detection power of the same and find out various factors affecting their performance. The succeeding section discusses the approach adopted in this research paper.

Table 1 Related work in paraphrase detection field

Research work	Technique/Algorithm	Dataset	Observations	Performance
[8]	Graph subsumption and isomorphism	Recognizing Textual Entailment (RTE)	Introduces a graph subsumption-based approach for paraphrase identification. In this, input sentences are mapped to corresponding graph structures and paraphrasing is evaluated based on graph isomorphism	Accuracy: 65.90 Precision: 64.70 Recall: 70.00 F1-score: 67.24
[9]	Neural Networks (NN)	Microsoft Research Paraphrase (MSRP)	Lexical features in association with syntactic, semantic, and composite features are used to train a back-propagation network, wherein extracted features are fed as input to the NN.	Accuracy: 75 and above
[10]	Combination of deep neural network with keywords	MSRP; Sentences Involving Compositional Knowledge (SICK) [16]	Detects paraphrasing based on lexical and semantic similarity, measured by combining neural networks and keywords, and hence identifies the relation between the inputs in the vector space.	MSRP- Accuracy: 80.2 F1-score: 86.4 SICK- Accuracy: 87.1
[11]	Textual Entailment	RTE	Paraphrase detection is viewed as a bidirectional textual entailment task. Word Sense Disambiguation is performed and using distant supervised learning technique, indirect facts are mined from the predicates.	Accuracy: 78.6 F1-score: 86.3
[12]	Logistic Regression (LR); SVM; Neural Networks	Quora Duplicate Question set	Compares various machine learning models after preprocessing and encoding the input and concludes recurrent neural network (RNN) to be the most efficient algorithm for paraphrase identification task.	Accuracies: LR—59.9 SVM—62.5 Siamese NN—63 Two-layer NN—74 RNN—80

(continued)

Table 1 (continued)

Research work	Technique/Algorithm	Dataset	Observations	Performance
[13]	Word Alignment Information (WAI)	Quora Question Pairs (QQP)	Introduces a novel approach leveraging WAI to improve deep PI baseline model RE2 [17]. Employs two major schemes to test the performance, proving 'pre-training' the unlabeled in-domain data majorly improving the performance of baseline model.	Accuracy on various schemes: Embedding—80.7 Multi-task—80.9 Pre-train—82.1 Hybrid (Pre-train + Multi-task)—82.9
[14]	Curriculum Learning	QQP; Large-scale Chinese Question Matching Corpus (LCQMC)	Estimates the effect of label noise on PI task and introduces an approach based on curriculum learning where: (a) a *loss-based noise metric* is developed to compute noise complexity of a sample, and (b) *similarity-based noise metric* classifies the paraphrase.	Accuracy: QQP—80.29 LCQMC—80.93
[15]	PIDG (Program Interaction Dependency Graph)	12 Undergraduate programming assignments diversified with the help of SPPlagiarise [18]	Introduces a novel behavior-based *Source Code Plagiarism Detection Tool*, BPlag. Each program's behavior is represented with the help of PIDG, and source code plagiarism is detected based on the similarity between the PID graphs.	Avg. Error Count at various % transformations: 4.3

3 Proposed Methodology

The goal of this paper is to measure and compare the efficiency of five word embedding models (Word2Vec, Doc2Vec, BERT, TF-IDF, and FastText) in the task of plagiarism/paraphrase detection, on two publicly available corpora: Quora Question Pairs and Plagiarized Short Answers.

Appropriate preprocessing techniques are applied on both the corpora before starting with model testing. The models are then evaluated based on the most appropriate threshold values and distance measure, and the results are produced based on standard metrics—accuracy, precision, recall, and F1-score. For each model tested against each corpus, ROC curve is generated for a deeper understanding of its performance.

3.1 Problem Definition

To formulate the task of paraphrase detection, the problem is defined as: Given two text inputs T_1 and T_2 such that $T_1 = \{T_1{}^1, T_1{}^2, T_1{}^3,\ldots.,T_1{}^n\}$ and $T_2 = \{T_2{}^1, T_2{}^2, T_2{}^3,\ldots.,T_2{}^n\}$. The task of paraphrase detection between the given inputs is formalized as a binary task where $L = \{0,1\}$ are the target labels such that if T_1 and T_2 are duplicate $L = 1$ otherwise $L = 0$, i.e., if not duplicate.

3.2 Corpora

The paper aims at comparing the listed models based on their semantic similarity detection. The following corpora, after taking into consideration their varied natures are tested on the models:

Quora Question Pairs.

In 2017, Quora Question Pairs,[1] an international competition by Quora, was released to identify plagiarism and paraphrasing to be able to group similar/duplicate questions. The dataset contains genuine examples from the Web site with over 400,000 records.

Plagiarized Short Answers.

Plagiarized Short Answers[2] (2009) is a corpus developed in a typical academic setting for the task of plagiarism identification in which four levels of plagiarism were committed. Since the input is classified based on a binary approach, four levels of paraphrasing are encoded to only two, i.e., 0 (non-plagiarized) and 1 (plagiarized). The dataset contained 95 documents plagiarized at different levels against 5 original documents sourced from Wikipedia.

The two corpora differ a lot from each other in various aspects which is partially the reason why they were chosen. (a) While PSA is a clean dataset containing only textual information, QQP contains mathematical equations, abbreviations, slang, typos, etc., all of which can be accounted for as noise. (b) PSA is a smaller dataset as compared to QQP. (c) PSA also differs from QQP in the context that the constituent documents at question are larger in length in PSA as compared to QQP.

The different natures of both the datasets helped in evaluating and reviewing the word embedding models more precisely.

Before experimenting with the models, the datasets need to be transformed into a standard form to maximize model performance. The following section discusses the various text preprocessing steps applied to both the datasets to achieve the best results.

---

[1] https://www.kaggle.com/c/quora-question-pairs/.

[2] https://ir.shef.ac.uk/cloughie/resources/plagiarism_corpus.html.

3.3 Text Preprocessing

Before feeding the input to the models, the text corpora need to be preprocessed to eliminate noise and maximize the processing effect. The major preprocessing steps are briefly discussed below and represented in the flowchart (Fig. 1) as well.

1. Each text input was transitioned from sensitive data to non-sensitive data by translating it to lower case and tokenizing it.
2. Each token obtained was checked against a dictionary of stop-words obtained from *nltk* library in python. If the word matched against the dictionary, it was dropped to avoid unnecessary processing, as stop-words rarely have any role in plagiarism.
3. Three alternatives of number manipulation were tested,

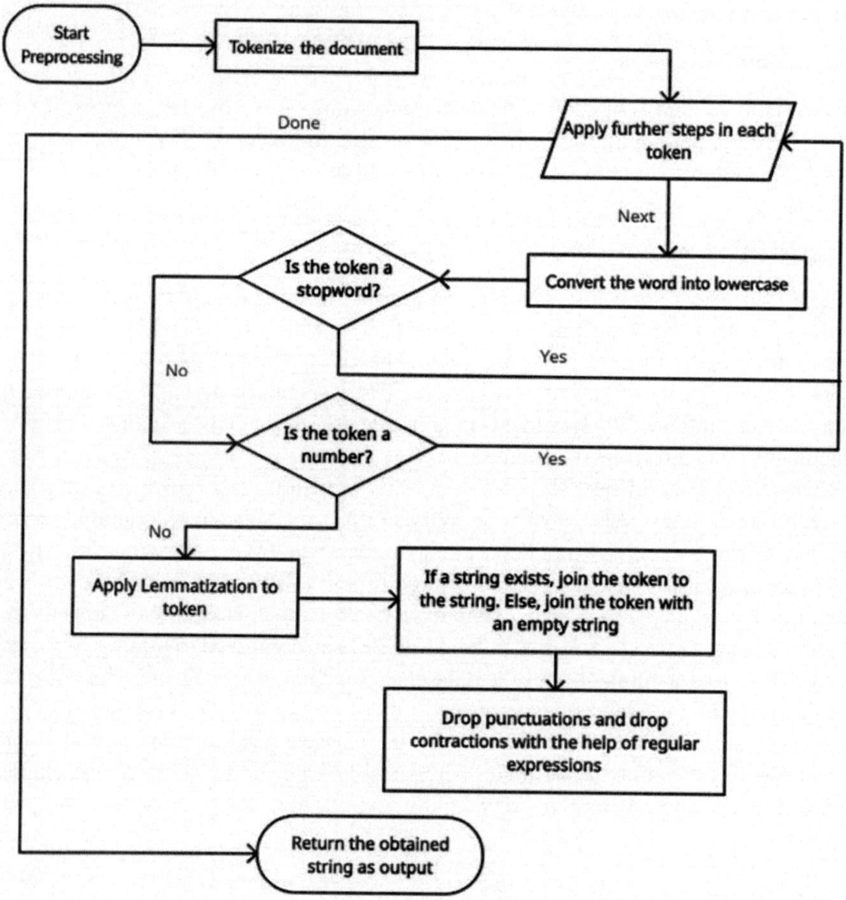

Fig. 1 Flowchart of text preprocessing steps applied to each corpus

a. Leaving them
b. Replacing them with <NAN> and
c. Deleting them from the text.

Best results were achieved once they were removed altogether.

4. Stemming [19] was applied to original tokens to reduce them to their word stem or word root, but the results proved to be better without this step, hence was not carried forward.
5. Lemmatization [20] was applied by using *WordNet* Lemmatizer [21]. Lemmatization helps in retaining the context of text or achieving its base form. For example, New and York are individual tokens which when lemmatized is treated as a single entity, i.e., New York, and hence, the context of the text is retained. The results proved to be slightly better when lemmatization was applied.
6. The tokens were then combined to form original text.
7. The original text was checked for punctuation marks and contractions. Punctuation marks were straight away eliminated while contractions, like can't, she'll, I've, etc., were expanded to their original form—here, cannot, she will, I have, respectively.
8. The whole corpus is then searched for duplicate entries which are subsequently removed to avoid overhead. The total number of duplicates were 353 which accounted for only about 0.9% of the whole dataset and hence didn't lead to any unbalancing.

The above steps are applied to each dataset. No additional preprocessing is done on any particular dataset to draw a fair result when tested with the models.

Following section discusses the various models which are applied on this preprocessed data to test semantic similarity of the documents through word embeddings.

3.4 Models Referred

Machine learning and almost all deep learning models are incapable of processing text in its raw form. One of the first and most popular approach to measure the semantic similarity between texts was vector space model [22], which introduced space density computations on text such that each entity (such as characters in words, words in sentences, sentences in documents and documents in dataset) could be represented in an n-dimensional space as a vector. Hence, the proximity of two entities in the space infers their semantic similarity.

Different models employ different strategies to utilize VSM for word embedding of input. A brief description of the five models taken on for research is given below:

TF-IDF

The paper proceeds with the oldest word embedding model, i.e., Term Frequency-Inverse Document Frequency which is a statistical frequency-based embedding strategy developed by [23], and measures the significance of a term among the corpus of documents. The words in a document are given different weights through the following equation:

$$w_{ij} = tf_{ij} \times \log(N/df_i') \quad (1)$$

where tf_{ij} = number of occurrences of term i in document j, df_i' = number of documents containing the term i, and N = number of documents.

Word2Vec

Word2Vec is a deep learning model which is actually a combination of two strategies—continuous bag of words (CBOW) and skip-grams. Both of these techniques learn weights of each input word which are thus taken as the final word vector representations.

The model creates vectors which are distributed numerical representations of word features such as the context of individual words. With enough data, usage, and contexts, Word2Vec can make accurate guesses about a word's synonyms based on its previous appearances (Fig. 2).

Figure 3 depicts the distribution of a small set of computed vectors in an n-dimensional space.

BERT

Sent2Vec is a sentence to vector strategy where distributed representations of sentences are achieved. Rather than extracting the semantic meaning of a word, the whole sentence is taken into consideration. It can be thought of as an extension of Word2Vec to sentences whereby, vectors are the average of source word embeddings.

BERTSimilarity library from pandas is used as sentence embedding technique in this research which employs forward pass of BERT model to compute vectors for sentences.

```
model = Word2Vec(questions_labeled, min_count=2, size=50, workers=4 )
model.wv.most_similar('book')
```

```
[('books', 0.8578062653541565),
 ('novel', 0.7103374004364014),
 ('novels', 0.6823523044586182),
 ('poem', 0.6142452359199524),
 ('ebook', 0.6063246130943298),
 ('textbook', 0.6029399633407593),
 ('journal', 0.5949798822402954),
 ('biography', 0.5928425788879395),
 ('pandey', 0.5822772979736328),
 ('congenitally', 0.5763487219810486)]
```

Fig. 2 Using Word2Vec to extract the most similar value of a given word based on nearest vector values plotted in n-dimensional space

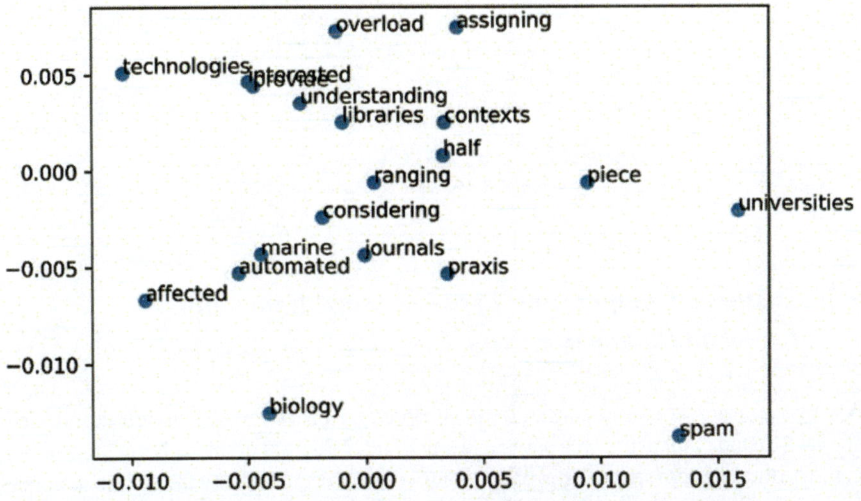

Fig. 3 Distribution of vectors computed by Word2Vec model in an n-dimensional space

Doc2Vec

Using this strategy, the document can be represented as vectors by using paragraph vector algorithm introduced by Mikolov et al. [2]. The model basically remembers the context of words encountered, and hence, the whole document can be plotted as a vector based on its semantic meaning.

FastText

FastText is a powerful word embedding model introduced by Facebook where in a word is assumed to be formed by n-grams of characters. For example, rainy can be represented as [rainy, rain, ainy], [rai, ain, iny]. It is particularly efficient over traditional approaches in the sense that it accounts for rarer words and can give vector representations to even the words absent from the dictionary.

All the above models have something in common: generating vector values for textual data. Raw textual data is incapable of extensive experimentation as majority of the existing algorithms work on numerical data. The advantage with extracted vector values is that it allows putting the data through various concepts of linear algebra, which opens doors to great level of experimentation and findings. The following section discusses how the generated word embeddings can be used to analyze semantic similarity between documents.

Table 2 Threshold values to classify plagiarized records in each model corresponding to the corpus

	Quora question pairs	Plagiarized short answers
TF-IDF	0.52	0.54
Word2Vec	0.90	0.98
BERT	0.89	0.96
Doc2Vec	0.94	0.93
FastText	0.92	0.92

3.5 *Distance Computation Between Word Embeddings to Detect Plagiarism*

After generating word embedding for the corpora based on various models mentioned above, a distance measure was chosen to calculate the actual distance between vectors in the n-dimensional space. This paper compares the vectors using the cosine distance measure which can be obtained from cosine similarity. The cosine similarity between two entities T1 and T2 can be calculated as:

$$\text{similarity}(T_1, T_2) = \cos(\theta) = (T_1 \cdot T_2)/(\|T_1\| \times \|T_2\|) \tag{2}$$

The cosine distance can be calculated from similarity such that,

$$\text{distance}(T_1, T_2) = 1 - \text{similarity}(T_1, T_2) = 1 - \cos(\theta) \tag{3}$$

The distance between the two entities is checked against a threshold value. If the score is greater than or equal to threshold, the entities are reported as plagiarized.

Since cosine distance is used as a metric, it is a given that the result of comparison will always be in the range [0,1]. Thus, the threshold always lies in the range [0,1].

The threshold value chosen varies according to the model into consideration and context of the given dataset. Hence, it is not a given that the same threshold value will produce the most accurate result for a model over any corpus and should be carefully chosen after experimentation.

The different threshold values which produced the most efficient results for chosen datasets are presented in Table 2.

3.6 *Evaluation and Results*

Standard measures are used in this paper to calculate the performance of each model. The measures include accuracy, precision, recall, and F1-score and can be calculated with the help of *sklearn* library in python. The library gives the scores under two major categories: macro-avg and weighted-avg. We have considered weighted-avg in our evaluation to accommodate the size positive and negative samples.

Table 3 Performance of models at their best thresholds

	Accuracy		Precision		Recall		F1-score	
	QQP	PSA	QQP	PSA	QQP	PSA	QQP	PSA
TF-IDF	0.64	0.79	0.65	0.86	0.64	0.79	0.64	0.80
Word2Vec	0.64	0.73	0.63	0.74	0.64	0.74	0.64	0.73
BERT	0.66	0.77	0.66	0.83	0.66	0.77	0.66	0.77
Doc2Vec	0.65	0.74	0.66	0.72	0.65	0.79	0.65	0.753
FastText	0.69	0.93	0.70	0.92	0.69	0.92	0.69	0.92

Table 3 summarizes the values obtained of mentioned metrics against each model and corpus.

Table 2 reveals that the threshold values for both the corpora against a particular model vary diversely. Table 3 gives the performance summary of all the tested models. The behavior is dramatically different when the evaluation metrics for both the corpora are compared against each other for all the models.

Both QQP and PSA datasets were made to undergo the same preprocessing steps for a fair comparison but QQP is a much larger dataset compared to PSA. Hence, it is lead to believe that the noise factor (like unaccounted abbreviations, orthographical errors, etc.) plays a major role in the performance of a model for a particular corpus. The derivation comes from the following two reasons:

- Deep learning models should ideally perform better when the size of dataset is large to train their neural network. Therefore, QQP should have worked well with all the models because of its size. But the performance of models, in contrast to the expectation has greatly worsened in QQP dataset, accounting that noise has a role to play.
- Since FastText is noise-tolerant to some extent as explained in Sect. 3.4, it should be able to account for some faults in the dataset. The evaluation metrics prove this point, as FastText performs significantly better on both corpora.

Another probable but indefinite reason for the poor performance of DL models on QQP dataset can be the presence of label noise. Label noise is extensively noticed in paraphrase identification datasets due to automatic labeling or non-expert labeling, and it can severely impact the outcome of deep learning models as mentioned in [24]. QQP has undergone human labeling to classify a pair of questions as paraphrased or non-paraphrased.

The presence of label noise, however, could not be confirmed in datasets due to the shortage of time during implementation.

The ROC curves for various models on each dataset are depicted in Figs. 4 and 5 to help assess the performance of the models further. As mentioned in [25], a perfect diagnostic test would constitute an ROC curve that is almost vertical from (0,0) to (0,1) and then horizontal to (1,1).

It is evident that FastText exhibits the best performance among all the listed models, with reference to Table 3 and Fig. 5.

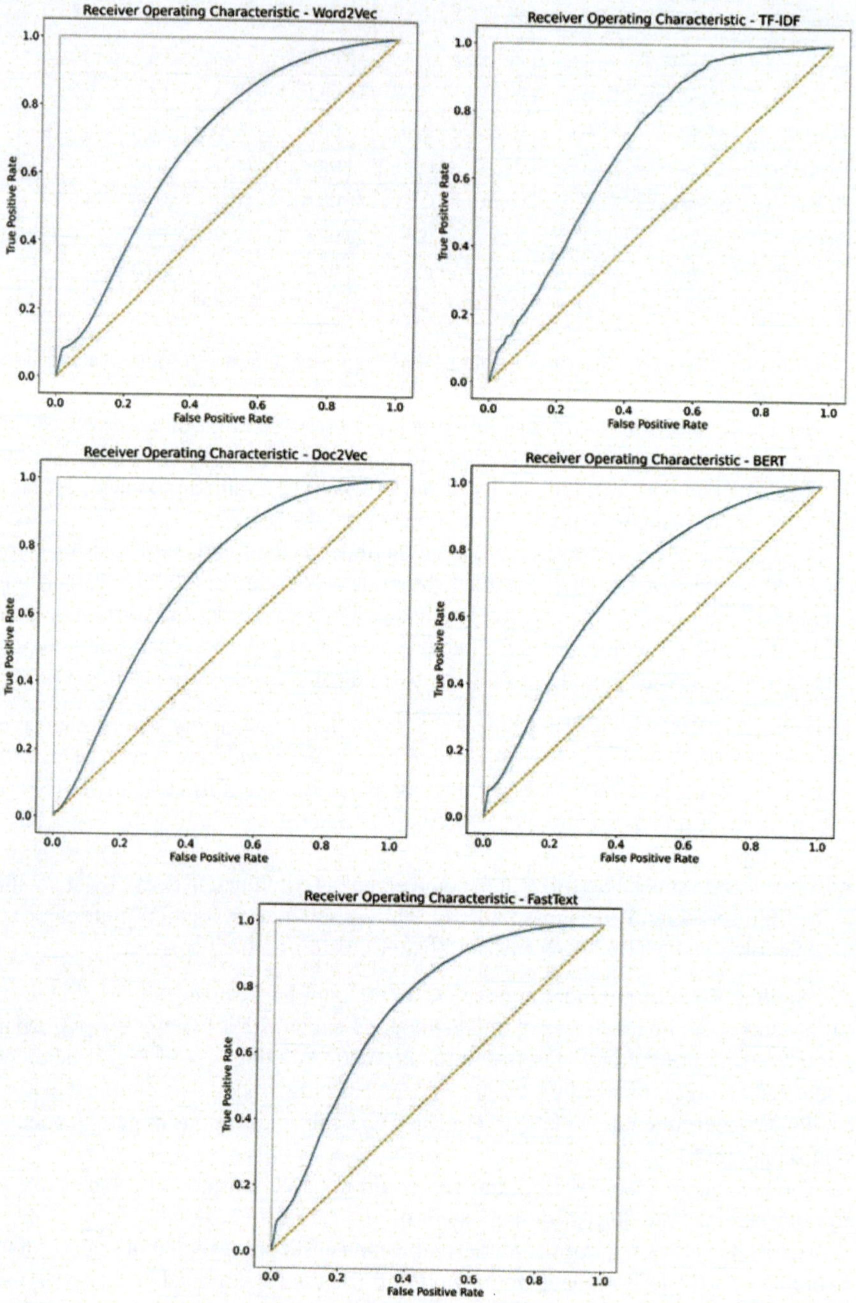

Fig. 4 ROC Curves for various models on QQP dataset (left to right, top to bottom: Word2Vec, TF-IDF, Doc2Vec, BERT, FastText)

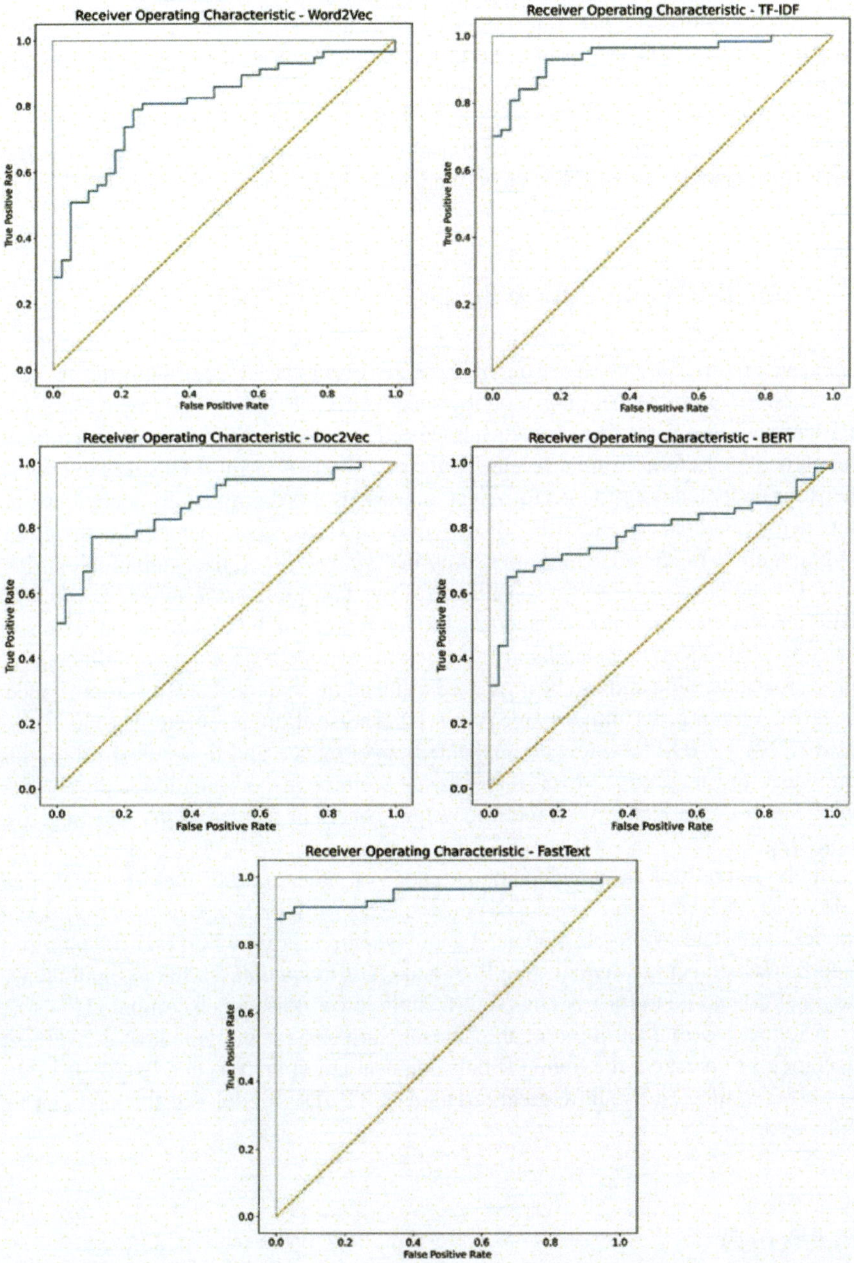

Fig. 5 ROC Curves for various models on PSA dataset (left to right, top to bottom: Word2Vec, TF-IDF, Doc2Vec, BERT, FastText)

Even though BERT performs slightly better than Word2Vec, it has considerably higher hardware and processing requirements, i.e., consumes more resources and time.

Doc2Vec has better performance over BERT as well as Word2Vec, in terms of both evaluation metrics and resource consumption. Contrary to expectations, TF-IDF performs at par with Doc2Vec, Word2Vec, and BERT.

4 Conclusion and Future Scope

This paper proposes comparison of various corpus-based deep learning as well as statistical word embedding models, specifically TF-IDF, Word2Vec, Doc2Vec, BERT, and FastText. The models are tested on two publicly available corpora namely—Quora Question Pairs and Plagiarized Short Answers. This research work helps in understanding the performance of various word embedding techniques on the mentioned corpora and effect of variations of various parameters like threshold, hyperparameters, distance measures, preprocessing steps, on the evaluation metrics.

It is concluded that FastText produces the most accurate and efficient results, and BERT is the most resource-demanding model. Based on the experimental results, it can also be deduced that standalone deep learning models are not quite sufficient for plagiarism detection and can be improved using additional mechanisms, for instance, negative sampling, dissimilarity measures, label noise elimination, etc. The paper also proves that TF-IDF has shown unexpected performance and hence probably holds the capability to produce even better results, either in combination with another well-suited DL model or with modifications to account for variations, like negative sampling.

In the future, this research can be extended on other state-of-the-art models like GloVe, ELMO, USE, etc. LSI can be considered as well for comparison of traditional models with state-of-the-art models. Another gap in this research is lack of testing on a corpus which is midway between PSA and QQP, i.e., a dataset that is significantly larger as compared to PSA but has significantly lower noise when compared to QQP.

Another possible variation of the dataset can consider the length of constituent documents to measure the trade-off between evaluation metrics and length. In addition to this, the results show great strength in TF-IDF model and the said can be improved.

References

1. Mikolov T et al (2013) Distributed representations of words and phrases and their compositionality. Adv Neural Inf Process Syst 1–9
2. Le Q, Mikolov T (2014) Distributed representations of sentences and documents. In: 31st Int Conf Mach Learn ICML 2014, vol 4, pp 2931–2939

3. Devlin J et al (2019) BERT: pre-training of deep bidirectional transformers for language understanding. In: NAACL HLT 2019, vol 1, pp 4171–4186
4. Ramos J (2003) Using TF-IDF to determine word relevance in document queries. Proc First Instr Conf Mach Learn 242(1):29–48
5. Joulin A, Grave E et al (2017) Bag of tricks for efficient text classification. In: 15th Conference Eur Chapter Association Computer Linguist EACL 2017—Proceedings Conference, vol 2, pp 427–431
6. Salim AN et al (2012) Understanding plagiarism linguistic patterns, textual features, and detection methods. IEEE Trans Syst Man Cybern 42(2):133–149
7. VMK and KK (2016) A survey on similarity measures in text mining. Mach Learn Appl Int J 3(1):19–28
8. Rus V, Mccarthy PM, Lintean MC, Mcnamara DS, Graesser AC (2008) Paraphrase identification with lexico-syntactic graph subsumption, pp 201–206
9. Rajkumar (2010) Paraphrase recognition using neural network classification 1(29):43–48
10. Wang X, Xu B (2018) Paraphrase recognition via combination of neural classifier and keywords
11. Seethamol S, Manju K (2018) Paraphrase identification using textual entailment recognition. ICICICT 2017, vol 2018-Janua, pp 1071–1074
12. Hunt E, Janamsetty R et al (2019) Machine learning models for paraphrase identification and its applications on plagiarism detection. In: 2019 IEEE international conference big knowledge, pp 97–104
13. Wang B, Wang L (2021) Enhancing deep paraphrase identification via leveraging word alignment information. CNCERT/CC, pp 7843–7847
14. Li B, Liu T, Wang B et al (2020) Label noise robust curriculum for deep paraphrase identification. In: 2020 international joint conference on neural networks (IJCNN), pp 1–8
15. Cheers H, Lin Y, Smith SP (2021) Academic source code plagiarism detection by measuring program behavioral similarity. IEEE Access 9:50391–50412
16. Marelli M et al (2014) A SICK cure for the evaluation of compositional distributional semantic models. In: Proceedings of 9th international conference language resource evaluation learning 2014, no May, pp 216–223
17. Yang R et al (2019) Simple and effective text matching with richer alignment features. In: Proceedings of the 57th annual meeting of the ACL, pp 4699–4709
18. Cheers H, Lin Y, Smith SP (2019) Spplagiarise: a tool for generating simulated semantics-preserving plagiarism of java source code. In: 2019 IEEE 10th ICSESS, pp 617–622
19. Porter MF (1980) An algorithm for suffix stripping. Program 14(3):130–137
20. Callison-burch C (2008) Syntactic constraints on paraphrases extracted from parallel corpora, no October, pp 196–205
21. Fellbaum C (2010) WordNet. In: Theory and applications of ontology: computer applications. Springer, Dordrecht, pp 231–243
22. Salton G, Wong A, Yang CS (1975) A vector space model for automatic indexing. Commun ACM 18(11):613–620
23. Salton G, Buckley C (1988) Term-weighting approaches in automatic text retrieval. Inf Process Manage 24(5):513–523
24. Li B, Liu T et al (2020) Label noise robust curriculum for deep paraphrase identification. In: Proceedings of the International Joint Conference on Neural Networks (1)
25. NCSS Statistical Software (1997) Comparing two ROC curves—paired design, pp 1–14

Machine Learning Approach for Identifying Survival of Bone Marrow Transplant Patients

Varun Sapra, Luxmi Sapra, Yashi Bansal, Gunjan Chhabra, and Rohit Tanwar

Abstract Bone marrow transplant is also known as hematopoietic stem cell transplant which replaces unhealthy cells or marrow with healthy cells. This is useful for curing many diseases especially few types of cancer. The success rate of bone marrow transplant depends on many factors like genetic matching with donors and also on some of the features of recipients. In this paper, we have implemented different machine learning approaches like logistic regression, multilayer perceptron, random forest and Bayes network to find out the survival of bone marrow transplant patients. The same techniques have been applied on full dataset as well as on reduced dataset. The dimensionality of the dataset is reduced using correlation feature subset method. The results show that reduced dataset provides better accuracy with all the implemented methods. Multilayer perceptron and Bayes network outshined other methods with 94.1% accuracy.

Keywords Machine learning · Bone marrow transplant · Random forest · Hematopoietic cell transplantation · Multilayer perceptron

1 Introduction

Bone marrow is a spongy material inside the center of the bones, such as pelvis and thigh bones; stem cells in it are responsible for producing platelets, white blood cells (WBC) and red blood cells (RBC) [1]. Platelets prevent blood clotting, WBCs fight against any infection and RBCs carry oxygen though out body. Bone marrow also contains supporting fibrous tissues. It is of two types: red and yellow; one cannot survive without a bone marrow because it is a factory that stores all the blood cells and stem cells which are later transformed into RBC, WBC and platelets when required [2, 3]. In some diseases, the bone marrow does not produce enough healthy

V. Sapra · G. Chhabra · R. Tanwar (✉)
University of Petroleum and Energy Studies, Dehradun, India

L. Sapra · Y. Bansal
Dev Bhoomi Institute of Technology, Dehradun, India

P. K. Singh et al. (eds.), *Emerging Technologies for Computing, Communication and Smart Cities*, Lecture Notes in Electrical Engineering 875,
https://doi.org/10.1007/978-981-19-0284-0_3

blood cells, and in this case, patients can be aplastic anemiatic which can cause leukemia, myelodysplastic syndrome, myeloproliferative disorders and so forth. In these cases, transplantation is required. Some symptoms of bone marrow diseases are tiredness, easy bleeding in injuries, extreme thirst, dehydration, loss of appetite, frequent urination and many more [4]. It is also known as hematopoetic stem cell transplantation. It is a medical procedure that is performed to replace unhealthy bone marrow with the healthy one. The process does not involve any major surgery but a process of blood transfusion. The procedure includes in which blood cells are transplanted so that they can produce new blood cells and advances the development of new marrow [5]. In some of the cases, chemotherapy or radiation is done at initial stage to kill the diseased cells, and new or healthy cells are infused into the body through an intravenous, i.e., IV catheter or tube. It is the same process as a patient is given blood or medicine with the help of IV. There is no specific age for bone marrow transplantation. The patient should fulfill certain criterion as approved by medical practitioners. Yet young age donors are preferred for better and long-term survival chances [6].

Machine learning is a field of computer science with the amalgamation of different branches such as statistics, image processing feature engineering and artificial intelligence [7–10]. It is the study of algorithm that learns automatically with examples and generate predictive and descriptive models. Due to advancement in machine learning methods and availability of huge amount of healthcare data in the form of clinical reports, symptoms, imaging use of machine learning have tremendously benefited the society. The paper is organized as follows. Section 2 contains the dataset description. Section 3 discusses the proposed framework to predict the survival of bone marrow transplant. Section 4 presents the results of the study. Finally, Sect. 5 presents the conclusion of the work.

Rosati et al. compared four different classifiers for the detection of hematopoietically active BM. They have used CT images with 36 radiomic features for the identification of hematopoietically active BM. For experiment purpose, they have used a data of 25 patients which were affected with cell carcinoma of the anal canal. Initially, the authors reduced the feature subset and parameters used for classification by using genetic algorithms. The results obtained by applying the classifiers on the CT images were compared with the literature. The performance of the classifiers was found satisfactory for iliac bone marrow and lumbosacral bone marrow. It has also been observed that the performance was poor in case of lower pelvis bone marrow [11].

The success rate of hematopoietic cell transplantation especially in case of unrelated donors depends on strict genetic matching. To improve the choice of donor, Buturovic et al. proposed a multivariate classifier model that can assign a label of preferred or not preferred based on the survival rate of the recipients. They evaluate the survival likelihood to five years based on clinical parameters of the donor. For experimental purpose, the data has been collected for 1255 patients who underwent unrelated donor HCT from Center for International Blood and Marrow Transplant Research Center. The data was collected for a period of 2000–2010. Their attempt to predict HCT outcomes was not successful [12].

Sharma et al. present a framework for improving the performance of nuclear segmentation technique for histopathological datasets using an AdaBoost classifier. For experimental purpose, they have used a dataset of slides with varying degrees of H&E stains. The results show that the research carried out by the authors is valuable for extracting visual data from heterogeneous datasets [13].

Kimura et al. in their research work proposed a convolutional neural network support system for identification of blood cell image. They implemented a deep learning system with gradient boosting for better decision making. They applied their model on the dataset comprised of 695,030 images. The proposed model was able to achieve the sensitivity of 93.5% and specificity of 96.0% for leucocyte detection, and for aplstic anemia, the model achieved the accuracy of 96.2% [14].

2 Dataset Description

Bone marrow transplant dataset is explored for the experimental purpose. The benchmark dataset is collected from UCI machine repository [15] having 37 features and 187 instances. The dataset consists of features such as donor age, ABO, age of recipient, gender of recipient, body mass of recipient, types of disease for which transplantation required, disease group, stem cell source, serological compatibility, compatibility of antigens, relapse of the disease, neutrophils recovery time, etc. Table 1 shows the description of the bone marrow transplant dataset.

3 Proposed Framework

In this paper, we proposed machine learning framework in order to predict the survival chances of bone marrow transplant in children's. The framework consists of relevant data selection, data preprocessing and feature engineering to reduce the dimensionality of the dataset in order to choose only those features that contribute more to outcome. Dimensionality reduction algorithm speedup the training and testing time of the algorithm by eliminating the redundant and irrelevant feature from the data [16, 17]. Initially, the dataset contained 37 features, and after implementing feature subset selection method, the dimensionality of the dataset has been reduced to six features. The most influential features are selected by correlation-based feature subset algorithms which are disease, PLT recovery, extensive chronic GvHD, relapse and survival time. Machine learning models are trained and tested using tenfold cross-validation method. The algorithms used to train and test models are logistic regression, multilayer perceptron, random forest and Bayes network. The same methods have been applied on full dataset as well as reduced dataset to evaluate the performance of these methods. Figure 1 shows the proposed framework for bone marrow transplantation survival prediction.

Table 1 Bone marrow transplant dataset description

Features	Description	Min	Max	Mean	stddev
Donor age	Numeric	18.6	55.55	33.47	8.272
Done age < 35	Yes/no	Yes-104, No-83			
Donor ABO	A, B, AB, O	A-71, B-28, AB-15, 0-73			
CMV_Donor	Preset/absent	Present-72,Absent-113			
Age recipient	numeric	0.6	20.2	9.9	5.3
Age < 10 recipient	Yes/no	Yes-99, no-88			
Recipient age	Numeric (count)	5 to 10–51, 0 to 5–47, 10 to 20–89			
Gender recipient	Male/female	Male-112, Female-75			
Body mass (recipient)	Numeric	6	103.4	35.8	19.6
ABO (Recipient)	Categorical (A, B, AB, O)	A-75, B-50, AB-13, O-48			
Rh (Recipient)	Categorical (plus, minus)	Plus -158, Minus -27			
CMV (Recipient)	Categorical (present, absent)	Present-100, Absebt-73			
Disease	Categorical (ALL, AML, chronic, nonmalignant, lymphoma)	ALL-68, AML-33, Chronic-45, Nonmalignant-32, Lymphoma-9)			
Disease group	Categorical (malignant, nonmalignant)	Malignant-155, Nonmalignant-32			
Gender match	Categorical Others, female to male	Others-155, female to male-32			
ABO_Match	Categorical matched, mismatched	Matched-52, Mismatched-134			
CMV Status	Numeric	0	3	1.509	1.12 9
HLA Mismatch	Categorical (matched, mismatched)	Matched-159, Mismatched-28			
Antigen	Numeric	0	3	0.92	1
Allel	Numeric	0	4	0.75	0.89
HLA Group-1	Categorical				

(continued)

Table 1 (continued)

Features	Description	Min	Max	Mean	stddev
Risk group	Categorical High, low	High-69, low-118			
Stem Cell Source	Categorical peripheral blood/bone marrow	Peripheral blood-145 Bone marrow-42			
Tx post relapse	Binary (Yes/no)	Yes-164, no-23			
CD34_x1e6 per kg	Numeric	0.79	57.7	11.8	9.9
CD3_x1e8per kg	Numeric	0.04	20.02	4.7	3.8
CD3_to_CD34_ratio	Numeric	0.204	99.5	5.3	9.59
ANC_Recovery	Numeric	9	1,000,000	26,752.8	161,747.2
PLT_Recovery	Numeric	9	1,000,000	90,937.9	288,242.4
acute_GvHD_II_III_IV	Binary (Yes/no)	Yes-112, No-75			
acute_GvHD_III_IV	Binary (Yes/no)	Yes-40, no-147			
time_to_acute_GvHD_III_IV	Numeric	10	1,000,000	775,408.04	418,425.0
extensive_chronic_GvHD	Binary (Yes/no)	Yes- 128, no-28			
Relapse	Binary (Yes/no)	Yes-159,no-28			
survival_time	Numeric	6	3364	938.7	849.5
Survival status	Binary Yes/no	Yes-102, no-85			

4 Results

The performance of the machine learning algorithms such as logistic regression, multilayer perceptron, random forest and Bayes network is evaluated on the base of performance measures such as accuracy, error rate, kappa stats, mean absolute error, room mean square error, sensitivity, specificity, precision, F-measure and ROC. Tables 2 and 3 display the performance of machine learning algorithm by considering all the features.

In the next step, we applied feature engineering process where dimensionality of the dataset is reduced by using CFS-based feature selection method. Dimensionality of the feature space is reduced to five features. Tables 4 and 5 show the result of reduced dataset.

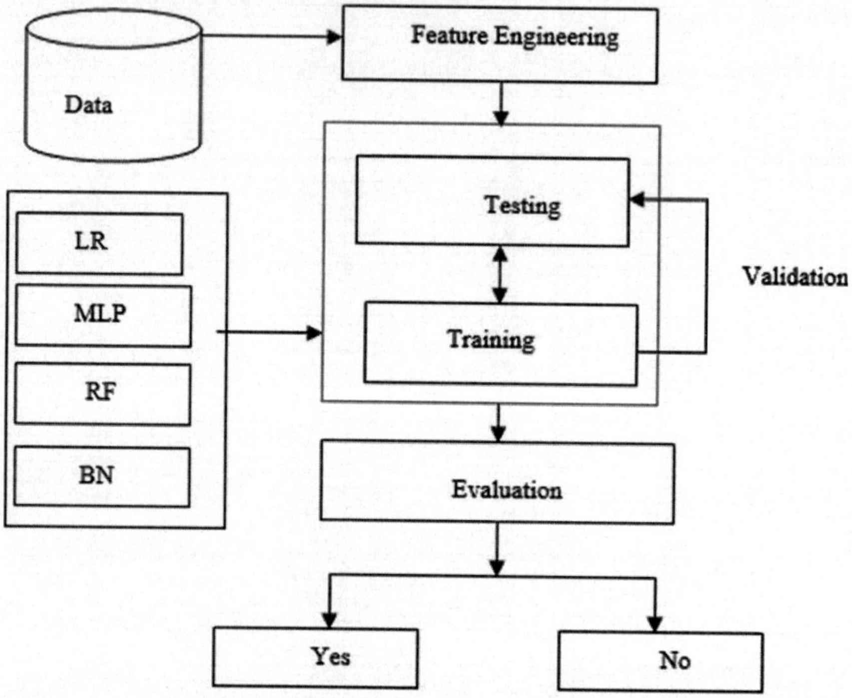

Fig. 1 Machine learning framework for bone marrow transplantation

Table 2 Accuracy/error/kappa/MAE and RMSE of models using all the features

Model	Accuracy (%)	Error rate (%)	Kappa	MAE	RMSE
Logistic regression	86.631	13.3	0.7301	0.134	0.3621
Multilayer perception	89.8396	10.1604	0.7953	0.1039	0.277
Random forest	93.5829	6.4171	0.8696	0.2495	0.2884
Bayes network	90.3743	9.6257	0.8051	0.0987	0.2354

Table 3 TPR/FPR/precision/recall/F-measure and ROC of models using all the features

Model	Class	TPR	FPR	Precision	Recall	F-measure	ROC
Logistic regression	Yes	0.882	0.153	0.86	0.882	0.73	0.936
	No	0.847	0.118	0.857	0.847	0.73	0.935
Multilayer Perception	Yes	0.902	0.106	0.911	0.902	0.906	0.968
	No	0.894	0.098	0.884	0.894	0.889	0.968
Random Forest	Yes	0.980	0.118	0.909	0.980	0.943	0.973
	No	0.882	0.020	0.974	0.882	0.926	0.973
Bayes network	Yes	0.931	0.129	0.896	0.931	0.806	0.977
	No	0.871	0.069	0.914	0.871	0.806	0.977

Table 4 Accuracy/error/kappa/MAE and RMSE of models after feature engineering

Model	Accuracy (%)	Error rate (%)	Kappa	MAE	RMSE
Logistic regression	93.04	6.9	0.8602	0.0885	0.2385
Multilayer perception	94.1	5.8	0.8815	0.0805	0.2437
Random forest	93.04	6.9	0.8594	0.084	0.2237
Bayes network	94.1	5.8	0.881	0.0869	0.2115

Table 5 TPR/FPR/precision/recall/F-measure and ROC of models after feature engineering

Model	Class	TPR	FPR	Precision	Recall	F-measure	ROC
Logistic regression	Yes	0.922	0.059	0.949	0.922	0.935	0.973
	No	0.941	0.078	0.909	0.941	0.925	0.973
Multilayer perception	Yes	0.941	0.059	0.950	0.941	0.946	0.957
	No	0.941	0.059	0.930	0.941	0.936	0.957
Random forest	Yes	0.951	0.094	0.924	0.951	0.937	0.986
	No	0.906	0.049	0.939	0.906	0.922	0.986
Bayes network	Yes	0.961	0.082	0.933	0.961	0.947	0.979
	No	0.918	0.039	0.951	0.918	0.934	0.979

5 Conclusion and Discussion

In this paper, we implemented machine learning framework to predict the survival of bone marrow transplant patients. Figures 2, 3, 4, 5 and 6 show the performance of different machine learning algorithm using parameters such as accuracy, error rate, true positive rate, false positive rate, F-measures and ROC. In case of full dataset, logistic regression achieves the accuracy of 86.6%, multilayer perceptron achieves the accuracy of 89.8%, random forest achieves the accuracy of 93.5% and in case of Bayes network, accuracy of 90.3% is reported. By reducing the dataset by using feature reduction algorithm, logistic regression achieves the accuracy of 93%, multilayer perceptron achieves the accuracy of 94.1%, random forest achieves

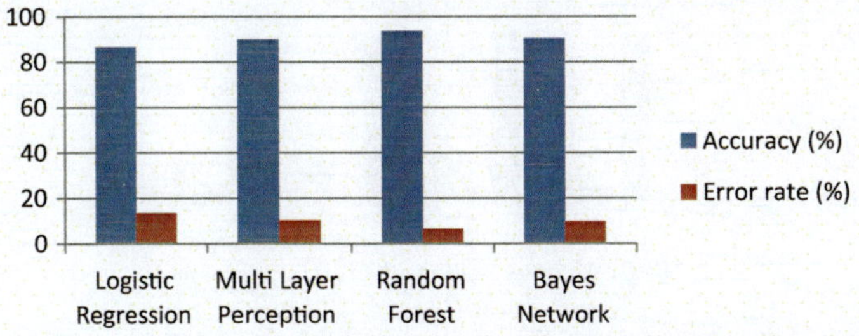

Fig. 2 Accuracy and error rate of models using all the features

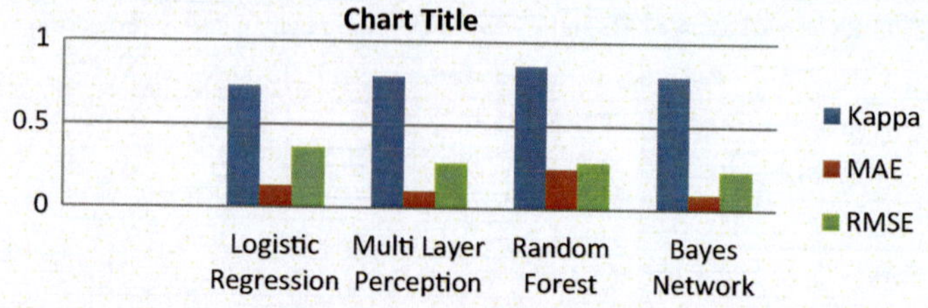

Fig. 3 Value of kappa/mean absolute and root mean square by considering all the features

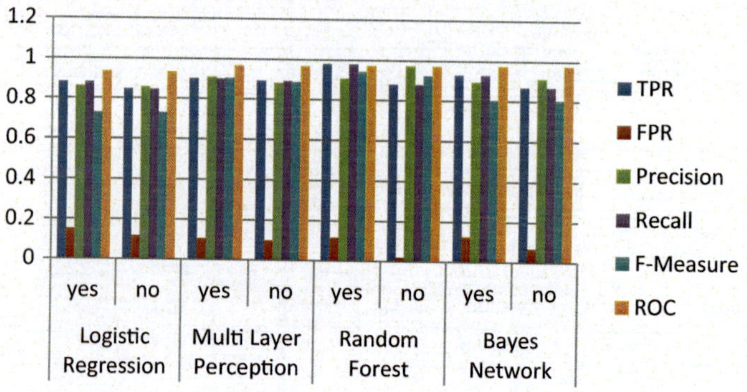

Fig. 4 Value of TPR/FPR/precision/recall/F-measure/ROC by considering all the features

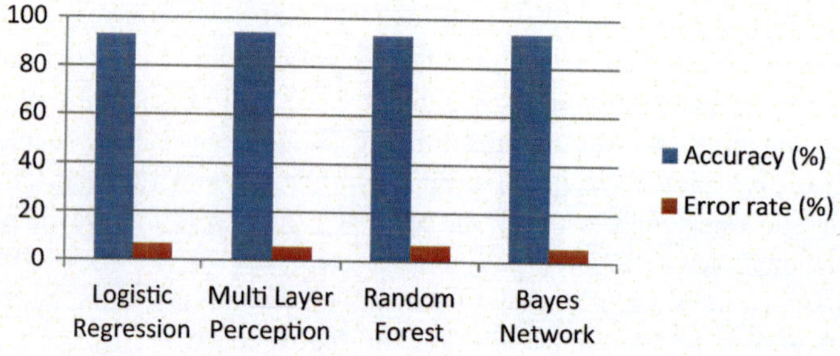

Fig. 5 Accuracy and error rate of models using feature engineering

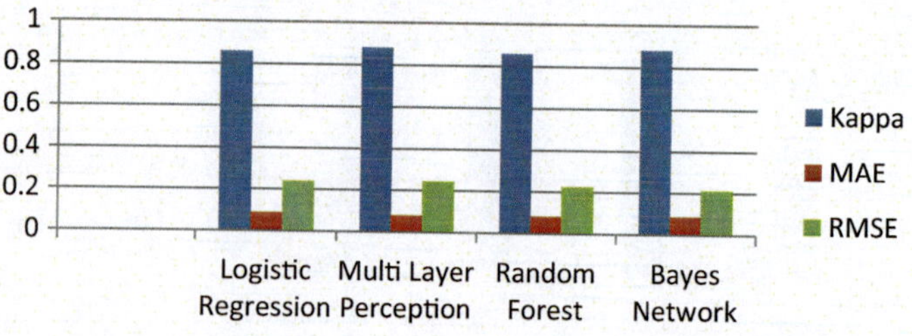

Fig. 6 Value of kappa/mean absolute and root mean square by using feature engineering

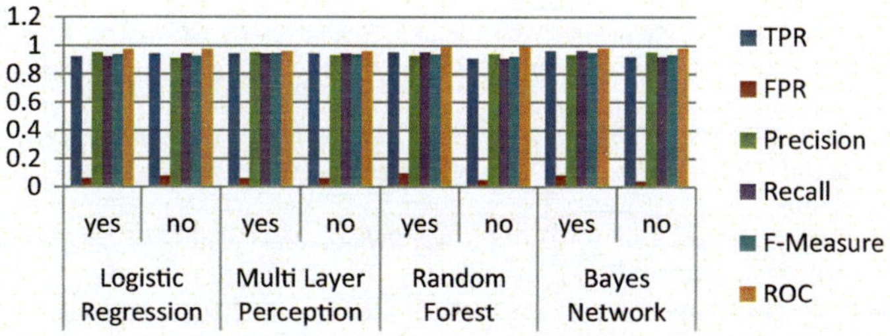

Fig. 7 Value of TPR/FPR/precision/recall/F-measure/ROC with feature engineering

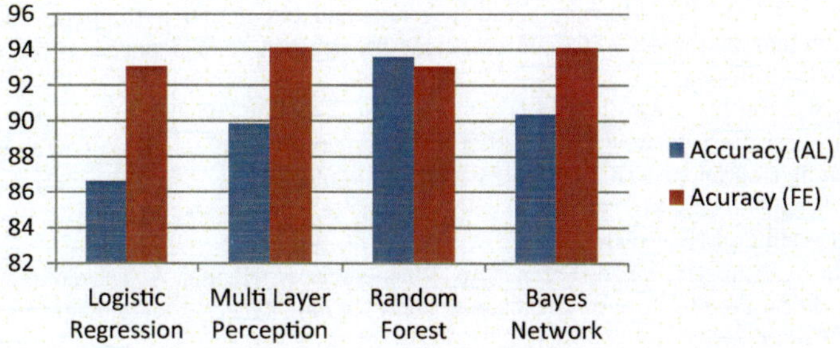

Fig. 8 Comparison of accuracy using complete features (AL) and feature engineering (FE)

the accuracy of 93% and Bayes network achieves the accuracy of 94.1%. The bone marrow transplant dataset consists of 37 features. We applied feature engineering for dimensionality reduction, and the improvement of accuracy has been reported in most of the cases. In case of logistic regression, the 6.44% of improvement is reported; in case of MLP, 4.27% improvement of accuracy has been reported as well as in case of Bayes network, 3.73% of improvement of accuracy is reported. But in case of random forest, 0.54% decrement of accuracy is reported. For this type of data, the performance of multilayer perceptron and Bayes network outperformed with accuracy of 94.1%. The experiment demonstrates that in most of the cases, improvement of accuracy is reported but in some case, accuracy can be decreased by using feature engineering (Figs. 7 and 8).

References

1. Henze K, Herten M, Haversath M, Busch A, Brandau S, Hackel A, Jäger M (2019) Surgical vacuum filter-derived stromal cells are superior in proliferation to human bone marrow aspirate. Stem Cell Res Ther 10(1):1–14
2. Larsson A, Smekal D, Lipcsey M (2019) Rapid testing of red blood cells, white blood cells and platelets in intensive care patients using the HemoScreen™ point-of-care analyzer. Platelets 30(8):1013–1016

3. Kawabata H, Isobe K, Watanabe T, Okudera T, Nakamura M, Suzuki M, Kawase T (2017) Quality assessment of platelet-rich fibrin-like matrix prepared from whole blood samples after extended storage. Biomedicines 5(3):57
4. Pranav Dheenathayalan S (2019) A descriptive study of Naadi Thervu and its clinical features based on the text "Sadhaga Naadi" (Doctoral dissertation, Government Siddha Medical College, Palayamkottai)
5. King W, Toler K, Woodell-May J (2018) Role of white blood cells in blood and bone marrow-based autologous therapies. BioMed Res Int
6. Kasamon YL, Bolaños-Meade J, Prince GT, Tsai HL, McCurdy SR, Kanakry JA, Jones RJ (2015) Outcomes of nonmyeloablative HLA-haploidentical blood or marrow transplantation with high-dose post-transplantation cyclophosphamide in older adults. J Clin Oncol 33(28):3152
7. Chandradevan R, Aljudi AA, Drumheller BR, Kunananthaseelan N, Amgad M, Gutman DA, Jaye DL (2020) Machine-based detection and classification for bone marrow aspirate differential counts: initial development focusing on nonneoplastic cells. Lab Invest 100(1):98–109
8. Ratley A, Minj J, Patre P (2020) Leukemia disease detection and classification using machine learning approaches: a review. In: 2020 first international conference on power, control and computing technologies (ICPC2T). IEEE, pp 161–165
9. Radakovich N, Nagy M, Nazha A (2020) Machine learning in haematologi cal malignancies. Lancet Haematol 7(7):e541–e550
10. ALzubi JA, Bharathikannan B, Tanwar S, Manikandan R, Khanna A, Thaventhiran C (2019) Boosted neural network ensemble classification for lung cancer disease diagnosis. Appl Soft Comput 80:579–591
11. Rosati S, Franco P, Fiandra C, Arcadipane F, Silvetti P, Gallio E, Balestra G (2020) Comparison of different classifiers to recognize active bone marrow from CT images. In: 2020 IEEE international symposium on Medical Measurements and Applications (MeMeA). IEEE, pp 1–5
12. Buturovic L, Shelton J, Spellman SR, Wang T, Friedman L, Loftus D, Lee SJ (2018) Evaluation of a machine learning-based prognostic model for unrelated hematopoietic cell transplantation donor selection. Biol Blood Marrow Transplant 24(6):1299–1306
13. Sharma H, Zerbe N, Heim D, Wienert S, Behrens HM, Hellwich O, Hufnagl P (2015) A multi-resolution approach for combining visual information using nuclei segmentation and classification in histopathological images. VISAPP (3):37–46
14. Kimura K, Tabe Y, Ai T, Takehara I, Fukuda H, Takahashi H, Ohsaka A (2019) A novel automated image analysis system using deep convolutional neural networks can assist to differentiate MDS and AA. Sci Rep 9(1):1–9
15. Dua D, Graff C (2017) UCI machine learning repository [Internet]. Available from: http://archive.ics.uci.edu/ml
16. El-Hasnony IM, Barakat SI, Elhoseny M, Mostafa RR (2020) Improved feature selection model for big data analytics. IEEE Access 8:66989–67004
17. ur Rehman MH, Liew CS, Abbas A, Jayaraman PP, Wah TY, Khan SU (2016) Big data reduction methods: a survey. Data Sci Eng 1(4):265–284

Human Disease Prognosis and Diagnosis Using Machine Learning

Sunil Kumar, Harish Kumar, Rashi Agarwal, and V. K. Pathak

Abstract The medical or healthcare sector is one of the trending and challenging fields of machine learning (ML). Machine learning is a branch of artificial intelligence (AI) that is concerned with machine training and learning. "It is the defining technology of this decade, though its impact on healthcare has been meagre"—stated James Collin of MIT in the USA. Healthcare-related machine learning encompasses a broad range of techniques that can be used to assess efficiency, scope, versatility, and ability in various machine learning applications. Many multinational corporations and research organizations do more research and development work in the medical sector, which can be associated with machine learning to find some fruitful solutions for human beings, as humans face a variety of health-related issues. Disease prognosis and diagnosis are new challenges of the medical or healthcare sector, in which computer technology can utilize. Disease prognosis and diagnosis are critical tasks in the medical or healthcare, in which the prediction and accuracy of the results must be up to the mark. The significance of machine learning in medical or healthcare systems has been discussed in this paper for further exploration.

Keywords Machine learning · Artificial intelligence · Healthcare · Disease prognosis · Disease diagnosis

1 Introduction

Evolution is a general form which occurs in most of the living and nonliving thing. Nature evolves living things, while nonliving or artificial things, such as technology,

S. Kumar (✉) · H. Kumar
J C Bose University of Science and Technology, YMCA, Faridabad, Haryana, India

S. Kumar · R. Agarwal
University Institute of Engineering and Technology, CSJM University, Kanpur, UP, India

V. K. Pathak
CSJM University, Kanpur, UP, India
e-mail: vinay@vpathak.in

P. K. Singh et al. (eds.), *Emerging Technologies for Computing, Communication and Smart Cities*, Lecture Notes in Electrical Engineering 875,
https://doi.org/10.1007/978-981-19-0284-0_4

evolve as well. First of all, we have to understand how evolution takes place in any terminology like biological or artificial. As far as we know, humans evolve biologically, learning from their experiences and failures. But with artificial things like computers, which follow instructions, they can provide some form of solution or understanding. But, can we say machines learn from past experiences as humans do? The answer is yes, the experience of the machine referred to as fed data into it; this concludes that the machines can learn from past data rather than experience as humans do. On this basis, we can say that the machines have started learning from past data results machine learning comes into the picture. Interdisciplinary fields are associated with each other to solve real-life problems. Machine learning is capable of solving many computational problems that can be used for solving medical-related issues in humans.

The prediction of the disease's severity is an important issue in the healthcare system, which may hurt the patient in future. The disease prognosis does this prediction through the observations, symptom identification, and signs related to the disease. With time limits, the disease prognosis keeps track of the disease and checks the impacts of the disease on patients, like its health is stable, good, bad, or for any complications. Using machine learning techniques, we can provide this functionality, which offers prediction, classification, and many more things by which we can predict or classify different types of diseases like diabetes, heart attack, cancer, etc., according to their characteristics. Medical diagnosis is usually made based on the doctor's expertise and experience. Based on observations, the doctor draws certain conclusions [1].

2 Machine Learning

Machine learning aims to improve the performance of an algorithm for a particular task based on experience. In machine learning, experience can be considered as training data. We take some sample data for observation from this training data. Machine learning algorithms may develop mathematical or computational models for making any decisions, forecasts, or classifications based on experience or training data [2].

In machine learning, data plays an important role because data is the basic building block of information as well as knowledge. If we have more data, then we can infer more knowledge by which machine can learn more. But as the data size increases, a large number of features also increases, which necessitates a large number of samples to be accurate. For providing accuracy, the machine must be capable of analysing, predicting, and making decisions on the past data. This functionality provides machine learning, in which the system learns automatically from the given past data.

Generally, machine learning is used for classification, prediction or making decisions, etc. It can achieve based on some patterns and relationships within the given datasets [3]. A concept of machine learning arises, which says from the given dataset, we can develop an automatic inference of patterns or relationships [4].

2.1 *Machine Learning Viewpoint*

Machine learning can be seen as a different-different point of views:

ML as learner's viewpoint—According to Tom Mitchell, machine learning is [5]—"… said to learn from experience with respect to some class of tasks, and a performance measure *P*, If (the learner's) performance at tasks in the class, as measured by *P*, improves with experience".

Humans or non-machines like biological agents or learners learn from past experiences, in which the learner repeats a set of tasks in order to assess their performance. As a result, it can be check that what the learner's performance is.

ML as Artificial Intelligence's viewpoint—Generally, it is an assumption that machine learning (ML) is a sub-domain of artificial intelligence (AI). This assumption can be true or false, both from the viewpoint of intelligence. All of the ML is not always performing intelligent behaviour, which is an aspect of artificial intelligence. Anyone using ML can think he is using AI; it is not always correct because not all of ML is AI, only some part of AI which belongs to ML and AI performing intelligent behaviour [6].

ML as a new programming paradigm's viewpoint—To understand the programming paradigm of machine learning, we need to understand traditional programming, which helps us in understanding the paradigm shift of programming. In traditional programming, we have the input data and a coded program, which will be run on a computer to produce the desired output.

In machine learning, we have the input and output data fed to the computer system and generate a program. In this scenario, we can check the performance of an algorithm. In machine learning, the general algorithms formulate the rules from the given input data as the programming paradigm shifts from traditional programming to machine learning that causes to deal with large datasets.

2.2 *Machine Learning Approaches*

Machine learning has different-different approaches to solve the different-different type of problems. Using one of the machine learning approaches, most of the problems can be solved. The division of approaches is important because it aids in understanding the essential nature of a problem that you can later work on. This will help you to provide a solution. Machine learning approaches have four broad categories as given below [2].

Supervised Learning—This type of learning is appropriate when the dataset contains labelled data; the labelled data have a tag or label and have some context or meaning based on some observation [2]. Supervised learning works on a mathematical or computational model in which a dataset contains both input and the desired output; it means you know the values of input and output [6, 7]. Supervised learning consists of classification and regression.

Unsupervised Learning—This type of learning is appropriate when the dataset contains no labelled data; data has no label given to the learning algorithm. This type of learning approach searched for undetected patterns and relationships in the given dataset. This approach generally draws inferences from the given dataset [2]. A medical or healthcare example of unsupervised learning methods is the prediction of heart diseases using clustering [8].

Semi-Supervised Learning—This type of learning is appropriate when the dataset contains both labelled and unlabelled data. This approach incorporates the benefits of both supervised and unsupervised learning methods so it has the potential to produce fruitful results in real-life applications. The real-life applications of semi-supervised learning are human speech analysis, content classification of the Internet, protein sequence classification and so on. In this approach, processing starts by clustering the data, which is unlabelled and dealt with by unsupervised learning. Then, next, for labelling the clustered unlabelled data, we use labelled data [2].

Reinforcement learning—This type of learning is appropriate in a dynamic environment (real or virtual) in which the algorithm has an objective to gain a particular goal. The algorithm or software agent acts as an action, and this action causes a reward or penalty. The agent's main objective is to learn how to perform perfect actions to gain maximum rewards [2].

3 Machine Learning in the Healthcare System

In the modern age, interdisciplinary domains are altogether to solve real-life related problems. ML and healthcare domains are also cooperating to solve real-life problems. The healthcare industry also uses ML to provide a better solution. The intersection of ML and healthcare is a new trending domain in which new possibilities of finding new solutions to real-life problems. As the demand for AI is increasing, the adoption of AI will be increasing soon. Over 50 executives from healthcare organizations that use or fund AI were surveyed by emerj [9]. This survey represented an industry's viewpoint for the adoption of AI in the USA. Emerj separated the companies that applying AI from those that do not. Companies consider return on investment (ROI) as a major factor that plays an important role in the adoption of AI. In the survey, the majority of surveyed companies are focusing on improving disease diagnostics and medical image analysis. As the field develops, disease diagnostic is a new trend to explore [10]. AI or ML is associated with the healthcare systems in which the disease prognosis and diagnosis is an important step to provide better solutions to patient-centric systems. To use AI in the healthcare systems is not so easy because of the complexities of both the systems [10].

Machine learning in the domain of medical or healthcare is an evolution of machine learning in which algorithms and models interpret data (like medical records, electronic health record, medical images, etc.), which can be used to identify and improve the disease prognosis and diagnosis of humans. Numerous machine learning applications will be used in the medical or healthcare systems [1], such as disease identification, medical imaging diagnosis, smart health records, proactive health management, drug discovery, clinical research and trial, customer complaint resolution and many more. In this process, we have to develop new ML methodologies or techniques that facilitate clinical decision-making and predictions. In clinical diagnoses, ML methods provide valuable insights [11, 12].

3.1 Sources of Healthcare Data

As we can see from the preceding section, no data sources are involved in supporting medical or healthcare systems. As a result, we must concentrate on medical or healthcare data sources in order to comprehend how data plays a significant role in the medical or healthcare system. Technology has progressed from the computer to the Internet. They work with a fundamental unit known as data. As data is a new fuel to the industry, the healthcare system is no exception. Due to the increasing volume of clinical data, operation performed on the clinical data is an important process. Patients' documents are now transformed to digital form in both the private and public healthcare systems. That digital data is generated, modified, and maintained by healthcare professionals. The healthcare data may have data from many sources like research data, wearable devices data, laboratory result data, medical images, smart phone data, environmental data, open access data and electronic health records, etc.

4 Disease Prognosis Using Machine Learning

The prediction of patients' health conditions is a critical task in clinical or medical decision-making. Machine learning is used not only for predicting the outcome or severity of diseases but also for post-operative events. According to neuraldesigner (a machine learning website/blog), the definition of the medical prognosis is as follows [8]:

> Medical prognosis consists of predicting the evolution of a patient so that the doctor can select the most appropriate treatment for him or her.

Medical prognosis may have one or more criteria like patients' life mortality, survivability of the patient, progression of disease, treatment, sensitivity, etc. These criteria are not the only factors that can help with prognosis; there are a number of other features or variables that can also help. A few of the most significant are mentioned below:

- Personal data of the patient like name, the age, sex, height, weight, etc.
- Data related to disease like symptoms of the patient, observations and test lab, etc.
- Environmental data like the patient is drinking or smoking, the location from which the patient belongs, etc.
- Additional information or condition which may help in disease prognosis.
- The benefits of disease prognosis using machine learning are that it provides:
- By the adjusting treatment level, we can improve the treatment structure.
- It prevents excessive treatment according to monitoring.
- It can prioritize patients according to their disease prognosis.

By using machine learning, predictive models also can be made. It helps doctors or healthcare professionals to find out the outcome of the existing disease of the patient. According to the outcome of the disease, appropriate treatment can be given to the patients [11].

5 Disease Diagnosis Using Machine Learning

Machine learning-assisted disease diagnosis is a great challenge that will be accomplished efficiently and more accurately. Machine learning is becoming increasingly popular for disease diagnosis (Fig. 1).

The medical or healthcare industry deals with a lot of data, including patient details, medical records, history, and so on, which makes it difficult to evaluate and maintain. To solve this issue, machine learning systems search for patterns in data and predict outcomes. The healthcare practitioners will draw certain conclusions based on the outcome, which helps in the patient's personalized care. As a result, we can see how machine learning systems can effectively manage huge amounts of medical data.

There are numerous machine learning techniques available, each with its own set of algorithms for data classification and prediction. As a result, we can say that it contributes to more accurate and efficient decision-making [1].

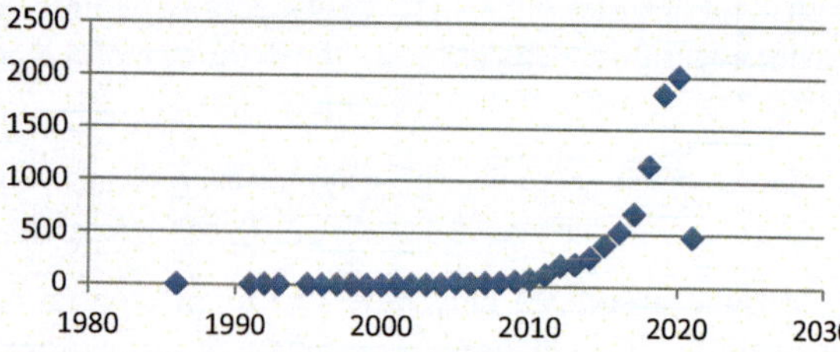

Fig. 1 Trend of machine learning in disease diagnosis. The comparison is obtained through searching disease diagnosis using machine learning on PubMed

6 Methodology

The methodology of the proposed system is to introduce a "human disease prognosis and diagnosis using machine learning". In general, machine learning begins with data acquisition and continues with pre-processing by the system. Find specific diseases and their relevant features in the form of numeric data that the machine learning algorithm can process.

The proposed approach for the prognosis and diagnosis of human diseases using machine learning (Fig. 2) in which some specific steps are presented are as follows:

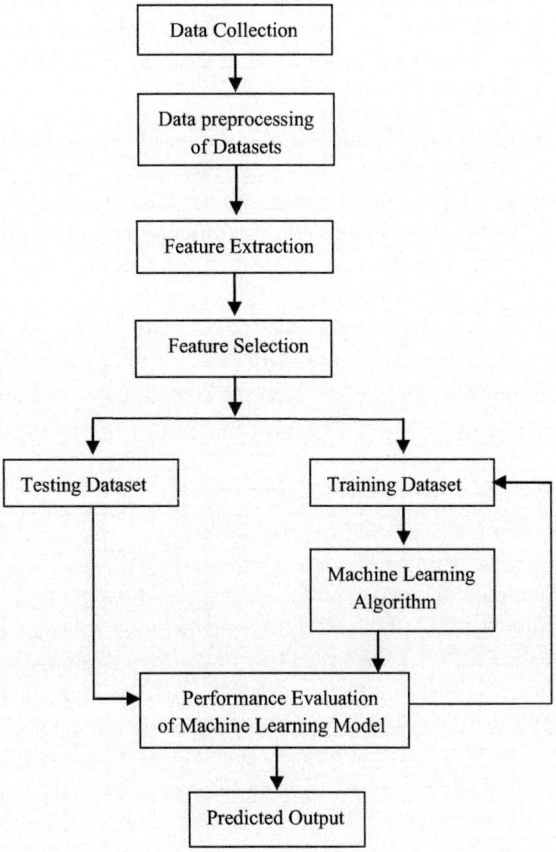

Fig. 2 Proposed approach for disease prognosis and diagnosis using machine learning

6.1 Datasets Collection

The first step after identifying the specific disease is to collect a dataset. Datasets can be obtained from the machine learning databases such as UCI, Kaggle, private datasets, and so on.

6.2 Data Pre-processing of Datasets

In this step, we process the data obtained in the first stage, which may or may not be perfect. This allows the data to be transformed into something useful and functional [1]. For example, in an X-ray image, the pre-processing of the dataset can be done as follows:

- On X-ray images, image resizing, image cropping, and dimension reduction are a few of the pre-processing tasks used [13].
- If you are working on an image classification problem, you will need to identify some issues before you can classify the images in the original dataset. Initially, image resizing to a certain dimension as the source, because the machine learning model demands that the size of the data being trained should be the same [13].
- The X-ray images can be in both two-dimensional images and in three-dimensional images. Dimension reduction was needed in order to convert three-dimensional images into two-dimensional images [13].

6.3 Feature Extraction

After selecting the region of interest (ROI), a mathematical calculation takes place for computation, which gives information about the ROI called a feature. For obtaining information from the ROI like the texture, a special kind of processing is performed called feature extraction [14]. In this step, features extracted from the image using machine learning techniques like grey level co-occurrence matrix (GLCM), LBP, DWT, etc. GLCM is a methodology of mathematics for the analysis of texture that provides the spatial ratio of pixels. Due to the large number of variables used in computation, GLCM reduces the variables using only needed variables [15]. By using feature extraction methods, all extracted features are passed to the next step.

6.4 Feature Selection

From extracted features, the only relevant features will choose for the study. Machine learning simply works on the rule which says, if you input surplus data, you will get surplus data. In machine learning, generally, algorithm prefers to feed only those features that are important. Some important reasons to use feature selection are like if the right subset of features is selected, it allows the machine learning algorithm to train faster, reduces model complexity, and increases model accuracy [16]. Wrappers, filters, and embedded methods are three major types of feature selection algorithms.

6.5 Division of the Datasets

Generally, in any machine learning models, the datasets are divided into two portions. The first portion is used for training the model, while the second portion is used to test the model, which may contain some biases in the test dataset, which affects the accuracy. To avoid this problem, cross-validation is used to estimate a model's ability on an unknown dataset. In k-fold cross-validation, the dataset is divided into k, equal-sized parts, such as 5, 10, and so on. Shuffle dataset with k-fold cross-validation to ensure that the order of input and output is absolutely random [17].

6.6 Selection of Machine Learning Algorithms

In this step, identification or classification of disease is done, which tells that the human has any ailment or not. Machine learning algorithms can perform this task. It is also difficult to choose the right machine learning algorithm, such as SVM, random forest, MLP, CNN, and so on. When we use machine learning algorithms, we can achieve the following two cases:

1. If the human does not contain any disease, the machine learning algorithm predicted that the human does not contain any disease, correctly.
2. If the human contains any disease, the machine learning algorithm predicted the disease by which the human is affected.

6.7 Performance Evaluation of a Machine Learning Model

A confusion matrix, also known as an error matrix, is typically represented as a table in machine learning (Fig. 3). A confusion matrix is used for visualization for the performance of a model. Each column in the matrix represents instances in a predicted class, but each row represents instances in an actual class (or vice versa) [18].

The following are the fundamental concepts used in the confusion matrix [13]:

True Positive (TP)—The positive class is correctly labelled. For instance, if a person has a disease, the model correctly classifies the disease existence.

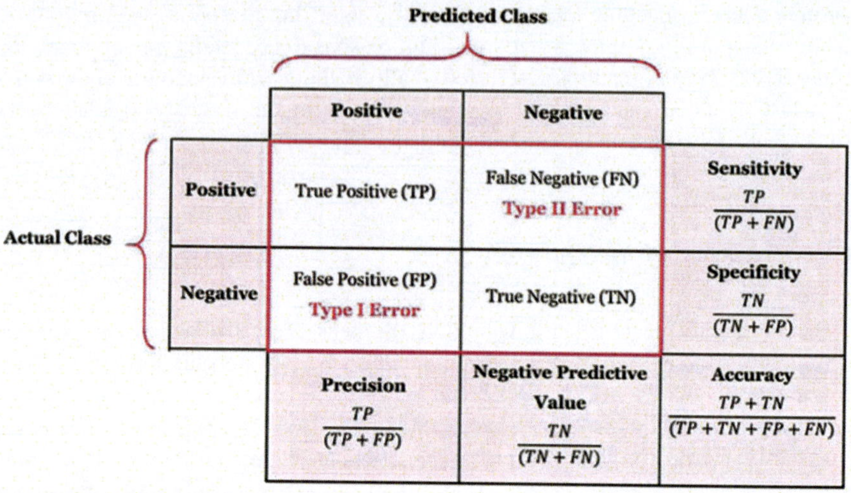

Fig. 3 Confusion matrix [19]

True Negative (TN)—The negative class is correctly labelled. For instance, if a person does not have a disease, the model correctly classifies the person as having no disease.

False Positive (FP)—Positives are predicted incorrectly. For instance, the person does not have any disease, but the system labels the person as having the disease.

False Negative (FN)—Negatives are predicted incorrectly. For instance, the person has a disease, but the model says the person does not have any disease [13].

For measuring the performance of a machine learning model, we can choose any performance metric from the performance metrics category. Some of the performance.

metrics discussed below are accuracy, precision, sensitivity or recall, specificity, and the F1 score:

Accuracy—This is a technique for determining the correct prediction value. To determine the accuracy, TP and TN are divided by the total amount of test data. This provides the correct classification results.

$$\text{Accuracy} = (\text{TP} + \text{TN})/(\text{TP} + \text{FP} + \text{FN} + \text{TN}) \quad (1)$$

Precision—Precision is evaluated by dividing the correct predictive value (TP) by the actual TP and FP values. This is used to calculate the accurate prediction outcomes of all positive class predictions. Precision can answer a question like how many of those who we labelled have disease from the total of those who had.

$$\text{Precision} = \text{TP}/(\text{TP} + \text{FP}) \quad (2)$$

Recall—Sensitivity is another name for recall. It is used to measure true positive predictions (TP) on all true predictions. The proportion of correctly positive-labelled disease to all disease cases in actual. Recall answers: Among all of the people who have the illness, people who have the disease have been correctly predicted [20].

$$\text{Sensitivity} = \text{TP}/(\text{TP} + \text{FN}) \tag{3}$$

Specificity—It is a metric for determining how many negatives are defined correctly. The test's specificity refers to its ability to correctly reject healthy people who do not have a disease's diagnosis, which means it indicates a healthy individual [21].

$$\text{Specificity} = \text{TN}/(\text{TN} + \text{FP}) \tag{4}$$

F1-Score—Precision and recall's harmonic mean or average can be used to calculate the F1-score. It is ideal if the model can find the right balance between precision and recall [20]. The F1-score is computed by multiplying the precision and recall values twice by the sum of the precision and recall values [22].

$$\text{F1 - Score} = 2 * \text{Precision} * \text{Recall}/\text{Precision} + \text{Recall} \tag{5}$$

6.8 Predicted Output

The machine learning model correctly predicted or classified the diseases that the patient is affected by. The algorithms expected performance is compared to the real output. The predicted output can be verified by experts or doctors.

7 Conclusion

More work is being done on disease detection using machine learning, as we have seen the trend for disease diagnosis using machine learning. The proposed methodology will help in diagnosing diseases using machine learning on which we will work on. In the current state of research and implementation, there was a lack of systematic disease detection. The investigation of how machine learning is used in disease detection which is essential for improving the efficiency of disease detection systems and ensuring that future research remains relevant.

References

1. Kalaiselvi K, Deepika M. Machine learning for healthcare diagnostics. In: Jain V, Chatterjee J (eds) Machine learning with health care perspective. Springer. ISBN: 978-3-030-40849-7. https://doi.org/10.1007/978-3-030-40850-3
2. Machine Learning Approaches that Every Data Scientist Should Know, Medium (2021) https://towardsdatascience.com/4-machine-learning-approaches-that-every-data-scientist-should-know-e3a9350ec0b9. Last accessed 2021/04/18
3. Yusuf M, Atal I, Li J et al (2020) Reporting quality of studies using machine learning models for medical diagnosis: a systematic review. BMJ Open 10:e034568. https://doi.org/10.1136/bmjopen-2019-34568
4. Hastie T et al (2005) The elements of statistical learning: data mining, inference and prediction. The Mathematical Intelligencer 27:83–85
5. Mitchell TM (1997) Machine learning, vol 45, issue 37. McGraw Hill
6. Machine learning, Wikipedia. https://en.wikipedia.org/wiki/Machine_learning. Last accessed: 2021/04/11
7. Russell SJ, Norvig P (2010) Artificial intelligence: a modern approach, 3rd edn. Prentice Hall. ISBN: 9780136042594
8. Medical prognosis using machine learning, Neuraldesigner.com (2021) https://www.neuraldesigner.com/solutions/medical-prognosis. Last accessed 2021/04/24
9. Faggella D. Machine learning in healthcare: expert consensus from 50+ executives. https://emerj.com/ai-market-research/machine-learning-in-healthcare-executive-consensus/. Last accessed 2021/04/15
10. How to Get Started with AI in Healthcare - A 3-Phase Approach, Emerj (2021) https://emerj.com/partner-content/get-started-ai-healthcare/. Last accessed 2021/04/20
11. Cleophas TJ, Zwinderman AH (2015) Machine learning in medicine—a complete overview. Springer International Publishing
12. Topol EJ (2019) High-performance medicine: the convergence of human and artificial intelligence. Nat Med 2019(25):44–56
13. Emhamed R et al (2020) Investigation of the performance of machine learning classifiers for pneumonia detection in chest X-ray images. In: 2020 IEEE international conference on electro information technology, EIT. IEEE, Chicago, USA, pp 98–104
14. Gupta et al (2020) Evolutionary algorithms for automatic lung disease detection. Measurement140:590–608
15. Boban, Megalingam (2020) Lung diseases classification based on machine learning algorithms and performance evaluation. In: International conference on communication and signal processing, ICCSP, 28–30 July 2020. IEEE, Chennai, India, pp 315–320
16. Kaushik S. Feature selection methods, machine learning, analytics Vidhya. https://www.analyticsvidhya.com/blog/2016/12/introduction-to-feature-selection-methods-with-an-example-or-how-to-select-the-right-variables/. Last accessed 2021/04/28
17. The importance of k-fold cross-validation for model prediction in machine learning. Medium. https://towardsdatascience.com/the-importance-of-k-fold-cross-validation-for-model-prediction-in-machine-learning-4709d3fed2ef. Last accessed 2021/04/24
18. Confusion matrix-Wikipedia. https://en.wikipedia.org/wiki/Confusion_matrix. Last accessed: 2021/05/01
19. Confusion Matrix. Image Source: https://2.bp.blogspot.com/-EvSXDotTOwc/XMfeOGZ-CVI/AAAAAAAAEiE/oePFfvhfOQM11dgRn9FkPxlegCXbgOF4QCLcBGAs/s1600/confusionMatrxiUpdated.jpg

20. Towards Data Science. https://towardsdatascience.com/accuracy-recall-precision-f-score, https://towardsdatascience.com/accuracy-recall-precision-f-score-specificity-which-to-optimize-on-867d3f11124-specificity-which-to-optimize-on-867d3f11124. Last accessed 2021/04/30
21. Sensitivity and Specificity—Wikipedia. https://en.wikipedia.org/wiki/Sensitivity_and_specificity. Last accessed 2021/05/02
22. Rachmad A et al (2020) Mycobacterium tuberculosis images classification based on combining of convolutional neural network and support vector machine. Commun Math Biol Neurosci 1–13. https://doi.org/10.28919/cmbn/5035

Incorporating Financial News Sentiments and MLP-Regressor with Feed-Forward for Stock Market Prediction

Junaid Maqbool, Preeti Aggarwal, and Ravreet Kaur

Abstract Stock market being very volatile depends on various political, environmental, and internal factors. The stock price prediction using news data is an interesting research topic. In this paper, an approach is proposed that represents textual news data as sentiment metrics using VADER sentiment analysis and price data scaled down between 0 and 1. The output predicted price of a stock on a particular day is fed forward to the next level of MLP*-Regressor to train as well predict the prices of following days. Experiments have been conducted on 10-year financial news as well price data of Reliance Company using the proposed model. The results show that the model because of feed-forward was able to learn the trend and the depths were followed more closely than the spikes. The model was tested on the news data of the same date as well as on the previous date separately. The model is an improvement made to MLP-Regressor whose results are also compared. The MLP-Regressor with feed-forward was able to learn long-term trends and also predict with an accuracy of 0.714 for the upcoming 7 days.

Keywords Stock prediction · News sentiment analysis · MLP-Regressor · Forecasting

*MLP Multi-level Perceptron

1 Introduction

The price of a single stock fluctuates more frequently compared to the stock indices and other metrics because of its dynamic and non-linear manner. The stock prices are very unstable and uncertain due to the dependence of its price on an uncountable number of parameters. The stock trade is done online dynamically. Traditionally, the prediction decisions are based on the historical trend as known by the trader

J. Maqbool (✉) · P. Aggarwal · R. Kaur
UIET, CSE Department, Panjab University, Chandigarh, India

P. K. Singh et al. (eds.), *Emerging Technologies for Computing, Communication and Smart Cities*, Lecture Notes in Electrical Engineering 875,
https://doi.org/10.1007/978-981-19-0284-0_5

and the current market knowledge [1]. The stock market is often influenced by political, financial, environmental factors, and the psychology of the investors [2]. The investors buy and sell their stocks based on all these events which let them predict possible up or down in the market. As the financial news is one of the most important factors which has the power to influence the psychology of traders to invest in a particular stock. The traders can gain profits if they will be able to predict correctly the stocks whose prices are going to increase in near future and buy them and sell the stocks whose prices are going to decrease [3]. The prediction of stock prices has attracted researchers from computer science, statistics, economic, and operations research [4].

Different models of machine learning like Bayesian Networks [5], Artificial Neural Networks [6], Support Vector Machines [7], Multi-Level Perceptron (MLP) [8], and Recurrent Neural Network especially Long Short-Term Memory (LSTM) [9] have already been utilized to predict the trends and the future stock prices. As the stock market exhibits natural chaotic behavior as different world markets react with different intensities to the period of crisis hence cannot be predicted by simple trading strategy [10].

The motivation for this study is to build a model which will be able to learn and predict the upcoming stock prices based on the sentiments calculated from financial news articles. The stock market being one of top most money investing destination, hence it will help investors to earn more profits and invest with more possibility of gains. This will help and encourage new investors to invest in stock market with limited information about market.

The main objective of this paper is to predict the future stock prices as well the trend of the direction of the stock price in which it will move based on sentiments [11] of news data provided for the same day and the previous day. This paper takes into consideration a new parameter called **label** which signifies if that particular date any news regarding the given stock is present. This paper will also check the effect of this new parameter "label" on the stock prediction. The machine learning models in [5–8] do not take into consideration the previous day prices thereby model learns only the fluctuations based on the change in sentiments while [9] do not take the sentiment of the news thereby predicting using historical trend and not taking in consideration the sentiment. Our model MLP-Regressor with feed-forward takes into consideration the previous day's price which given short-term historical support and news sentiments to predict the trend in the stock prices. Our model was able to follow the trend in long term.

This model is tested and trained on the stock data of reliance industries. The model is tested on the sentiments of the present day and the previous day separately. The model is compared with the MLP-Regressor without taking the previous day's price into consideration. The model was also tested with and without the label parameter and in turn, these models are evaluated in terms of Mean Absolute Percentage Error (**MAPE**), precision, recall, accuracy, and F1 measure.

The problem statement of this paper is to check the prediction accuracy of stock prices while incorporating financial news articles along with historical stock data using MLP-Regressor.

This paper is organized as follows: Section 2 briefly discusses existing work in the related field. Section 3 discusses the approach followed in this paper, models implemented, and the results derived from the experiments. Section 4 concludes and discusses future scope.

2 Related Work

There are two traditional approaches for stock prediction: Technical Analysis and Fundamental Analysis [12]. Most of the models made employ these two traditional approaches for prediction where fundamental analysis takes into consideration the overall financial conditions of the company, their management, and economy [1]. The models which employ a technical analysis approach generally take prediction as a classification problem where historic time series data is used to learn the market pattern. The models which work to predict exact stock price are termed as predictive regression in economic literature [13]. Though simple and naïve approaches mostly suffer from over-fitting when applied to real-world setups and are unable to learn the long-term trend. The Recurrent Neural Networks especially LSTM can work better on the long-term trend and prove to be superior to ARIMA (AR—autoregressive, I—integrated, MA—moving average) (Table 1) [9, 14].

The LSTM models are now being extensively used for stock prediction because of their ability to store the historical trend in architectural memory. LSTM models clubbed with other deep learning techniques show promising results in stock prediction.

3 Proposed Methodology

This section will introduce our methodology, feed-forward MLP-Regressor with sentiment analysis. The framework is shown in Fig. 1. The training of the model on time series data was done in a sequential manner without shuffling the data. The output of each day was fed as an input to the next day too.

During training, the output of the first prediction was fed into the input for the second prediction which helped the model to predict temporally for each day. The predictions have shown because of feed-forward the model was able to predict the trend for a longer period.

3.1 Problem Definition

To predict the stock price for an individual stock using the news sentiments, label, and historical stock price data. The task is to predict the movement of the stock price

Table 1 Related work in stock prediction

Research work	Technique/Algorithm	Dataset	Observation
Stock Prediction using LSTM [9]	RNN LSTM on historical time series data	NYSE [GOOGL][a]	The LSTM model can predict the opening prices while learning the long-term pattern and performs better than ARIMA [9]
Stock price prediction using News sentiment analysis [14]	ARIMA, LSTM with stock price and textual information	Standard and Poor's 500 (S&P 500), Online newspaper articles	The models did fit the data with some considerable accuracy but did not perform well when the stock prices were less or excessively volatile [14]
Improved deep learning model for prediction using time series and sentiment [15]	Empirical wavelet transform (EWT), Outliner Robust Extreme Learning Machine (ORELM), LSTM, Particle Swarm Optimization (PSO)	S&P 500, Dow Jones Industrial Average (DJI)	The hybrid novel stock prediction model is introduced using EWT, ORELM, LSTM, and PSO. The proposed hybrid model has better accuracy than other deep learning models and any single model [15]
Ensemble of Deep Q-learning agents for stock market forecasting [10]	Q-learning, Double Q-learning,	S&P 500, German DAX	The proposed model doesn't learn from market ups and downs, hence is flexible against the chaotic behavior of stock price rather it learns to maximize the return over time [10]
An integrated framework of deep learning and knowledge graph for stock prediction [16]	Attention based bidirectional LSTM, Convolutional Neural Network (CNN), Deep Stock-trend Neural Network (DSPNN)	Chinese stock Exchange market data	The proposed model takes into consideration real transaction records data and utilizes CNN to extract daily group trading vectors. DSPNN outperforms LSTM because of the attention mechanism and bidirectional structure of DSPNN [16]

(continued)

Table 1 (continued)

Research work	Technique/Algorithm	Dataset	Observation
A Multiplicative self-attention Bidirectional LSTM for stock prediction [17]	Volume Weighted Average prediction using multiplicative self-attention layer	NIFTY 50	When lag and exogeneous features are considered, The proposed model has higher accuracy over LSTM and simple RNN [17]
Stock prediction using graph-based CNN-LSTM [18]	CNN, LSTM, new method named Stock Sequence Array Convolutional LSTM (SACLSTM)	Historical data of stocks in Taiwan and America	Convolutional layer used to extract financial features combined with LSTM achieves better performance than any of method individually [18]
Stock price pattern using Neural Network and Back Propagation (BP) algorithm [19]	BP algorithm, Fuzzy algorithm	Gree electric, Maotai of shanghai mainboard	With improved training speed of BP algorithm the accuracy is still better than fuzzy algorithm to predict the closing stock price

[a] NYSE = New York Stock Exchange, GOOGL—Google (historical stock price)

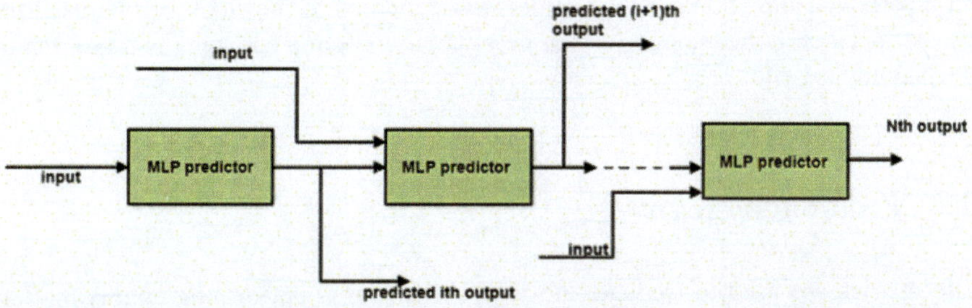

Fig. 1 Feed-forward MLP-regressor

after analyzing the sentiments of the financial news and the previous stock price for the next trading day and also predict the trend of the stock prices for upcoming n days. The problem has two tasks (1) what is the direction of the trend and (2) what is the amount by which the trend changes.

Direction of trend: If the stock price for the next day will be more than the stock price on the day preceding it, then the trend is positive, and if the stock price is less on the next day than the preceding day, then the trend is negative. If the stock price remains unchanged, then the trend is neutral.

$$T = \text{closing}(\text{day}i + 1) - \text{closing}(\text{day } i) \tag{1}$$

Date	Open	Low	High	Close
Jan 19-2010	544.84	536.24	547.83	537.41
Jan 18-2010	548.30	540.85	552.98	544.59
Jan 15-2010	557.61	547.34	564.10	550.16

Fig. 2 Historical reliance stock data

$$\begin{cases} T > 0, \text{trend} = \text{positive} \\ T < 0, \text{trend} = \text{negative} \\ T = 0, \text{trend} = \text{neutral} \end{cases}$$

3.2 Data Collection

The stock price data of reliance (Fig. 2) was collected from Jan 2010 to May 2020 from yahoo finance [20]. The closing price of each day along with the news sentiments applied on the financial news data of the same day using VADER sentiment analysis [21] from Natural Language Toolkit (NLTK) [22] and **label** was used to check the effect of the news sentiments on stock prices or future price of the stock.

Label defines the presence of the news of individual stock in the news collected on a particular day. Some important keywords related to the stock can be used to check the label. Label equals 1 when news about the same company is present and 0 when not present.

3.3 Evaluation Metrics

The metrics are used to evaluate the way our model is performing as this model classifies the data as per future predicted trend as well checks the amount of change. So, this paper employs metrics like MAPE to check the error in change detection which doesn't take the direction of change into consideration and accuracy which checks the classification correctness without considering the rate of change.

Accuracy In classification, accuracy is among the popular metric used. It represents the ratio of the number of correctly classified values to the total number of classified values [10].

$$\text{Accuracy}(X) = \frac{X^{(+)}}{|X|} \tag{2}$$

Precision tells how many predictions are really positive among all positively classified predictions.

Recall represents the model's ability to predict correct positives among all positives in the list.

F1-score model score which is the function of precision score and recall score. It is the harmonic mean of precision and recall score.

$$\text{F1-Score} = (2 * \text{precision} * \text{recall})/(\text{precision} + \text{recall}) \quad (3)$$

Mean Absolute Percentage Error (MAPE) It is average relative error of forecast predictions in percentage. It is one of the popular forecast prediction accuracy metrics [14, 23].

$$\text{MAPE} = \frac{100}{n}\sum_{j=1}^{n}\left|(y_j - y_j')/y_j\right| \quad (4)$$

where y_j is the original value and y'_j is the corresponding predicted value.

3.4 Model Used

This paper has used MLP-Regressor [24] which previously have not shown promising results in learning the long-term trend [25]. The MLP models without feed-forward tend to lose the historical information unlike LSTM models [26–28]. Our model (Fig. 3) uses feed-forward to remember the trend.

The output of one prediction is used as an input to the next prediction helping to preserve the previous trend and predict the next output in relation to the previous output instead of making the model work independently on sentiment score. Sentiment score affects the stock price but the effect is always shown to the previous price, not to a general or average price. Accuracy is measured by checking the real trend and predicted trend for each prediction. MAPE is measured to understand the percentage change with which the predictions vary to original output.

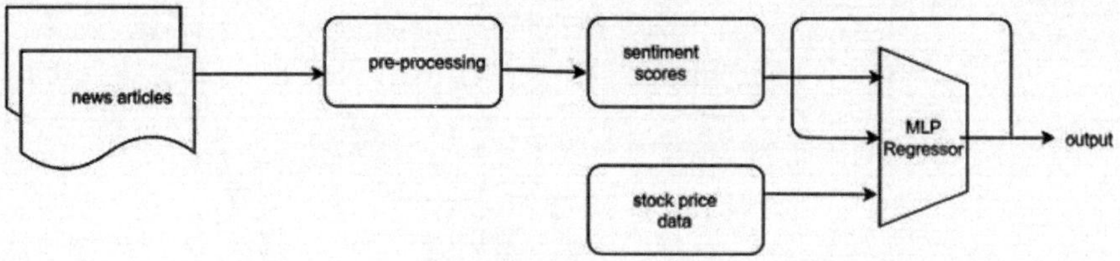

Fig. 3 Complete feed-forward MLP-Regressor with sentiment and stock input

3.5 Evaluation and Results

The proposed model was checked against 10-year stock and news data. The training and testing data were divided into an 8:2 ratio. The models were evaluated based on the input data keeping the model specifications same. The model was tested with label and without label to understand the effect of label parameter on the prediction which showed slight variation in accuracy. Hence label parameter does play role in improving the prediction accuracy of the model. Then, the model was tested against the sentiments available. The sentiments for same day (Fig. 4) were used in one case and sentiment score from the previous day (Fig. 5) was used in the other model to understand the time effect of the news on the prediction.

The model without label (Fig. 6) was unable to predict as accurately as done by with model with label. The models were tested for 7 days (Table 2), 30 days (Table 3), 300 days (Table 4), and 512 days (Table 5). Figures 4 and 5 show that both the models are able to learn the trend and follow it for a long time while previous day sentiment graph show more overlapping of predicted and original graph. The

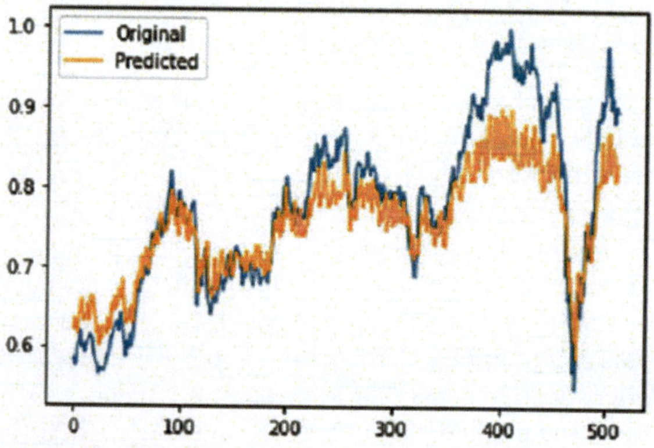

Fig. 4 Prediction using same day news sentiment

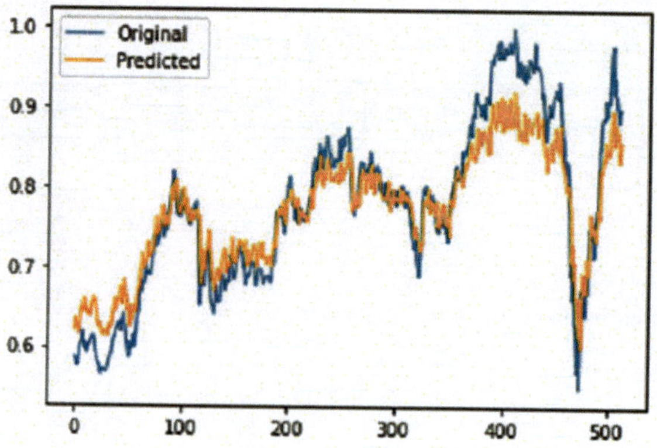

Fig. 5 Prediction using previous day news sentiment

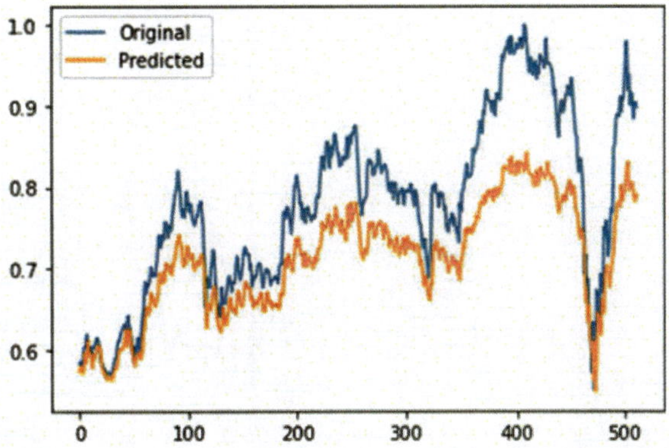

Fig. 6 Feed-forward MLP-Regressor without label on same day news data

Table 2 Accuracy metrics for different models for 7 days prediction

Models	Precision	Recall	Accuracy	F1-score	MAPE
MLP feed-forward with previous day data	0.80	0.80	**0.714**	0.714	**4.62**
MLP feed-forward with same day data	1.00	0.66	**0.714**	0.80	6.98
MLP without feed-forward	0.50	0.40	0.28	0.44	Infinite

Table 3 Accuracy metrics for different models for 30 days prediction

Models	Precision	Recall	Accuracy	F1-score	MAPE
MLP feed-forward with previous day data	0.56	0.64	**0.60**	0.60	**6.42**
MLP feed-forward with same day data	0.58	0.50	**0.60**	0.53	7.18
MLP without feed-forward	0.47	0.57	0.50	0.51	Infinite

Table 4 Accuracy metrics for different models for 300 days prediction

Models	Precision	Recall	Accuracy	F1-score	MAPE
MLP feed-forward with previous day data	0.55	0.52	0.52	0.53	**6.56**
MLP feed-forward with same day data	0.56	0.55	**0.54**	0.56	7.07
MLP without feed-forward	0.49	0.44	0.46	0.47	Infinite

Table 5 Accuracy metrics for different models for 512 days prediction

Models	Precision	Recall	Accuracy	F1-score	MAPE
MLP feed-forward with previous day data	0.53	0.53	**0.53**	0.53	**5.07**
MLP feed-forward with same day data	0.52	0.53	**0.53**	0.52	5.32
MLP feed-forwards without label for previous day data	0.50	0.51	0.50	0.50	9.16
MLP feed-forwards without label for same day data	0.50	0.50	0.50	0.50	8.09
MLP without feed-forward	0.51	0.49	0.51	0.50	Infinite

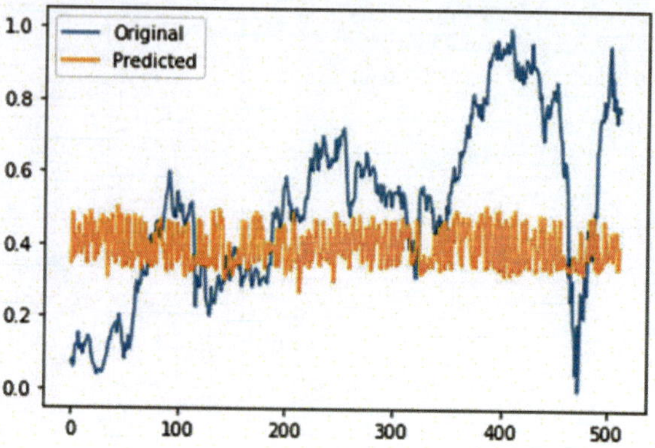

Fig. 7 MLP-Regressor without feed-forward on same day news data

model, MLP-Regressor without feed-forward (Fig. 7), was also tested and it can be concluded that these models are not able to follow the trend.

For predicting stock price of the upcoming 7 days, the model shows that using sentiments of same day (Fig. 4) does not increase the accuracy but decreases MAPE by 2.36%. But without feed-forward, the accuracy is as low as 0.28 and MAPE is infinite (Table 2). The previous day prediction was used to predict the next day which helped the models to remember and follow the trend. All models are plotted together show models are able to learn the trend (Fig. 8). The results suggest the MLP model with news sentiments from previous day shows better accuracy and lower MAPE (Fig. 9).

When predicting for 30 days, the accuracy in both feed-forward models the accuracy decreases by 0.11 while the MAPE doesn't show considerable change. And for the long-term prediction of 512 days, the MAPE decrease by 1–2% and accuracy by 0.07.

While comparing the results with label and without label (Table 5) the accuracy decrease by 0.03 and MAPE increase by around 4% showing that the label has a considerable effect on the amount of change in price for a particular stock. The

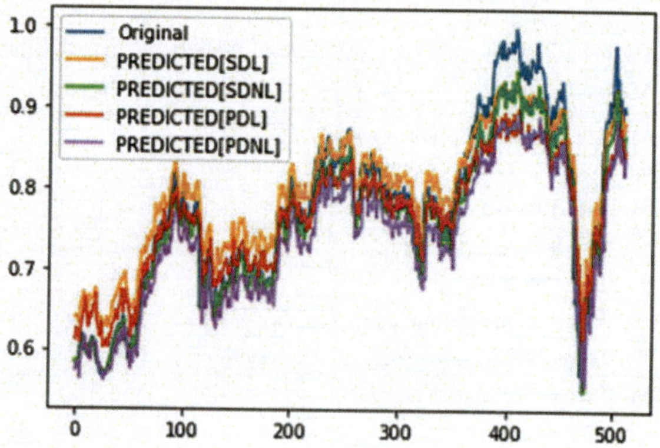

Fig. 8 Predicted models**

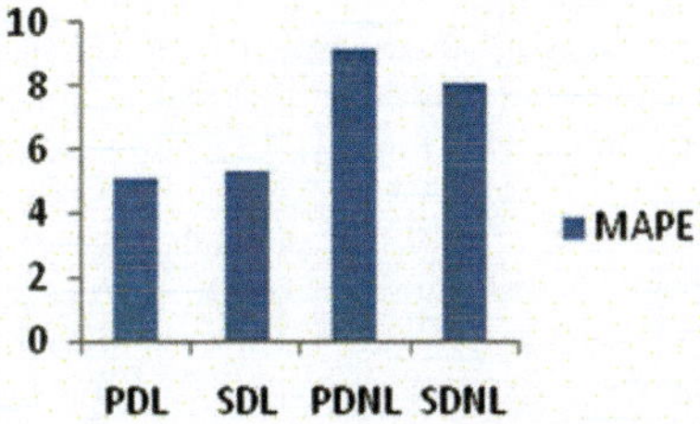

Fig. 9 MAPE comparison of models** (512 days) **Same Day news data with Label (SDL), Same Day news data without Label (SDNL), Previous Day news data with Label (PDL), Previous Day news data without Label (PDNL)

model without feed-forward is able to detect trend with an accuracy of 0.51. MAPE being infinite justifies Fig. 7 about the model without feedback not being able to follow or predict trend anywhere.

4 Conclusion and Future Scope

In this work, the individual stock price was predicted using sentiments of financial news and the time series data of the stock price. The graphs suggest that the label helps to fit the trend more accurately by decreasing the MAPE and show a high correlation between the stock price and news of a particular company than other news. The models were able to detect and follow a downward trend precisely while an upward trend was detected but the effect of the change was not followed when stocks were highly volatile. This paper shows that MLP-Regressor when employed with the feed-forward can provide promising results than MLP without feed-forward. The feed-forward can be used with more modifications in future work to predict the stock prices more accurately. There are various ways to predict stock price which can be explored in future work. Some of these include an ensemble of LSTM and other models where LSTM can be used to remember the trend and other models to predict the change rate. Different sentiment analyzing algorithms can be tested to check the compatibility of different sentiment scores against the ability to detect the trend of stock prices more accurately and tested against various models.

References

1. Yadav A, Jha CK, Sharan A (2020) Optimizing LSTM for time series prediction in Indian stock market. Proc Comput Sci 167(2019):2091–2100. https://doi.org/10.1016/j.procs.2020.03.257
2. Yudong Z, Lenan W (2009) Stock market prediction of S&P 500 via combination of improved BCO approach and BP neural network. Expert Syst Appl 36(5):8849–8854. https://doi.org/10.1016/j.eswa.2008.11.028

3. Khan W, Malik U, Ghazanfar MA, Azam MA, Alyoubi KH, Alfakeeh AS (2020) Predicting stock market trends using machine learning algorithms via public sentiment and political situation analysis. Soft Comput 24(15):11019–11043. https://doi.org/10.1007/s00500-019-04347-y
4. Zhang J, Cui S, Xu Y, Li Q, Li T (2018) A novel data-driven stock price trend prediction system. Expert Syst Appl 97:60–69. https://doi.org/10.1016/j.eswa.2017.12.026
5. Malagrino LS, Roman NT, Monteiro AM (2018) Forecasting stock market index daily direction: a Bayesian network approach. Expert Syst Appl 105:11–22. https://doi.org/10.1016/j.eswa.2018.03.039
6. Chong E, Han C, Park FC (2017) Deep learning networks for stock market analysis and prediction: methodology, data representations, and case studies. Expert Syst Appl 83:187–205. https://doi.org/10.1016/j.eswa.2017.04.030
7. Yeh CY, Huang CW, Lee SJ (2011) A multiple-kernel support vector regression approach for stock market price forecasting. Expert Syst Appl 38(3):2177–2186. https://doi.org/10.1016/j.eswa.2010.08.004
8. Turchenko V, Beraldi P, De Simone F, Grandinetti L (2011) Short-term stock price prediction using MLP in moving simulation mode. In: Proceedings of 6th IEEE international conference intelligence data acquisition advance computer system technology application IDAACS'2011, vol 2, no September, pp 666–671. https://doi.org/10.1109/IDAACS.2011.6072853
9. Moghar A, Hamiche M (2020) Stock market prediction using LSTM recurrent neural network. Proc Comput Sci 170:1168–1173. https://doi.org/10.1016/j.procs.2020.03.049
10. Carta S, Ferreira A, Podda AS, Reforgiato Recupero D, Sanna A (2021) Multi-DQN: an ensemble of deep Q-learning agents for stock market forecasting. Expert Syst Appl 164:113820. https://doi.org/10.1016/j.eswa.2020.113820
11. Nti IK, Adekoya AF, Weyori BA (2020) Predicting stock market price movement using sentiment analysis: evidence from Ghana. Appl Comput Syst 25(1):33–42. https://doi.org/10.2478/acss-2020-0004
12. Nti IK, Felix Adekoya A, Benjamin, Weyori A. A systematic review of fundamental and technical analysis of stock market predictions. Artif Intell Rev 53:3007–3057, 123AD. https://doi.org/10.1007/s10462-019-09754-z
13. Rapach DE, Strauss JK, Zhou G (2013) International stock return predictability: what is the role of the United States? J Finance 68(4):1633–1662. https://doi.org/10.1111/jofi.12041
14. Mohan S, Mullapudi S, Sammeta S, Vijayvergia P, Anastasiu DC (2019) Stock price prediction using news sentiment analysis. In: Proceedings—5th IEEE International Conference Big Data Serv Application BigDataService 2019, Work Big Data Water Resour Environ Hydraul Eng Work Medical, Heal Using Big Data Technol, pp 205–208. https://doi.org/10.1109/BigDataService.2019.00035
15. Liu H, Long Z (2020) An improved deep learning model for predicting stock market price time series. Digit Signal Process Rev J 102:102741. https://doi.org/10.1016/j.dsp.2020.102741
16. Long J, Chen Z, He W, Wu T, Ren J (2020) An integrated framework of deep learning and knowledge graph for prediction of stock price trend: an application in Chinese stock exchange market. Appl Soft Comput J. 91:106205. https://doi.org/10.1016/j.asoc.2020.106205
17. Hegde S (2021) Multiplicative self attention bi-directional long short-term memory network for volume weighted average, vol M, no 05, pp 2265–2272
18. Ming-Tai Wu J, Li Z, Herencsar N, Vo B, Chun-Wei Lin J. A graph-based CNN-LSTM stock price prediction algorithm with leading indicators, vol 1, p 3. https://doi.org/10.1007/s00530-021-00758-w
19. Zhang D, Lou S (2021) The application research of neural network and BP algorithm in stock price pattern classification and prediction. Futur Gener Comput Syst 115:872–879. https://doi.org/10.1016/j.future.2020.10.009
20. Yahoo Finance, Relaince. https://finance.yahoo.com/quote/RELI/history?p=RELI
21. Hutto CJ, Gilbert E (2014) VADER: a parsimonious rule-based model for sentiment analysis of social media text. In: Proceedings of the 8th international conference weblogs social media, ICWSM 2014, pp 216–225

22. Loper E, Bird S (2002) NLTK: the natural language toolkit. https://doi.org/10.3115/1225403.1225421
23. de Myttenaere A, Golden B, Le Grand B, Rossi F (2016) Mean absolute percentage error for regression models. Neurocomputing 192:38–48. https://doi.org/10.1016/j.neucom.2015.12.114
24. Pimprikar R, Ramachandra S, Senthilkuma K (2017) Use of machine learning algorithms and twitter sentiment analysis for stock market prediction. Int J Pure Appl Math 115(6):521–526
25. Ebrahimpour R, Nikoo H, Masoudnia S, Yousefi MR, Ghaemi MS (2011) Mixture of mlp-experts for trend forecasting of time series: a case study of the tehran stock exchange. Int J Forecast 27(3):804–816. https://doi.org/10.1016/j.ijforecast.2010.02.015
26. Selvin S, Vinayakumar R, Gopalakrishnan EA, Menon VK, Soman KP (2018) Stock price prediction using LSTM, RNN and CNN-sliding window model. In: 2017 International conference advance computer communication informatics, ICACCI 2017, vol 2017-Janua, no January 2018, pp 1643–1647. https://doi.org/10.1109/ICACCI.2017.8126078
27. Roondiwala M, Patel H, Varma S (2015) Predicting stock prices using LSTM. Int J Sci Res 6(4):2319–7064
28. Jiang Q, Tang C, Chen C, Wang X, Huang Q (2019) Stock price forecast based on LSTM neural network. Springer International Publishing

Framework for Context-Based Intelligent Search Engine for Structured and Unstructured Data

Ritendra R. Sawale, Lokesh Nagar, and Vaibhav K. Khatavkar

Abstract In today's time, determining the user's exact need through search queries is really a significant challenge as there is a tremendous amount of structure and unstructured data which is being produced daily. In structured data, it is easy to extract what we need as compared to unstructured data. There is a need to understand the semantics of the text given in such textual data. Natural language processing helps us to extract useful information from such unstructured textual data. Word embedding is one of the ways to overcome this issue. The implemented system aims to build a framework that will search based on the context hidden in the user query. As the context of keywords plays a vital role while extracting relevant search results from the database, the implemented system works on determining the context of the keyword in the query by using the GloVe word embedding technique. The embedded query is used to find out the most relevant documents from the database. This database consists of various text documents of different formats like pdf, word files, excel sheets, online crawled data, etc. All this data is stored in a database name as ElasticSearch. The proposed system can be used as an intranet searching system. The most relevant data is sent as an output quickly. The existing entity-based search engine is not having contextual capability which is provided by the implemented system. The result for search queries are based on the combination of entity and context-based search system.

Keywords Natural language processing · Artificial intelligence · Machine learning · ElasticSearch

---

R. R. Sawale (✉) · V. K. Khatavkar
Department of Computer Engineering, College of Engineering Pune, Pune, India
e-mail: sawalerr19.comp@coep.ac.in

V. K. Khatavkar
e-mail: vkk.comp@coep.ac.in

L. Nagar
SAS Research and Development, Pune, India
e-mail: Lokesh.Nagar@sas.com

P. K. Singh et al. (eds.), *Emerging Technologies for Computing, Communication and Smart Cities*, Lecture Notes in Electrical Engineering 875,
https://doi.org/10.1007/978-981-19-0284-0_6

1 Introduction

An unprecedented amount of data has been created daily. According to IDC, the digital universe will reach over 40ZB by 2020, which is approximately 1.7 MB of new information created for every human, every second, of every day. A large portion of this information is unstructured and in text format. This created a need for a system to analyze this unstructured data to get useful insights from it. Text mining (also called text data mining or text analytics) is a method for extracting useful information from unstructured data through the identification and exploration of large amounts of text. In other definition, text mining is a method for extracting structured data from unstructured text.

It becomes a need for various organizations to get useful information from such unstructured data. This unstructured data in any IT organization includes feedback from customers about specific products or services, comment on social media about the product, tweets related to the company, etc. All this data is in the form of unstructured textual data. It is a headache for data analysts to go through each feedback and comment manually to get the information about the updates or requirements of clients and users. Text mining helps us to determine the context of the unstructured data. Natural language processing plays a vital role in this scenario. The searching interface helps the user to get the most relevant data concerning the search query. Search query is the most used way for retrieving relevant data from the data repositories. As the size of data in just a repository is enormous, it becomes a challenge for a searching system to get only the relevant document as it also depends on the context of keywords used in the system. The proposed system's main aim is to determine the context of the keyword used in the search query by using artificial intelligence, NLP, and machine learning. There are various ways to determine the semantic of the text, like word embedding. Here, we will assign a vector value to the word according to its context, and words with similar context will be placed close to each other. So after getting the context of the text or keyword, the system will search for the most relevant documents from the data repository. As the keyword context plays a vital role while extracting relevant search results from the datasets or database, the proposed system would mainly work on determining the context of the keyword in the query by using different word embedding techniques.

The main motive of the implemented system is to get only the relevant documents using ElasticSearch. But we are using AI over the search query to understand many things like the context, spelling mistake, keywords, etc. Our system will also consider the user's role, depending upon which it will provide only authorized information, so such an intelligent search system for structured and unstructured data is proposed in this report. The system will also make the query more machine-readable so as to get relevant data from the documents. Further literature survey of related work is also done. The objectives and scope of this project are also defined along with system design. The data repository consists of data from different data sources. ElasticSearch is used to store all the documents as well as the sentences from the documents. GloVe embedding is used to find the vector of the sentence which focuses mainly on the context of the search query. So the sentence vectors are also stored in the ElasticSearch.

The existing system named as text exploration framework which is the product developed by SAS R&D is an entity-based search engine. The implemented system provides contextual capability to the existing search engine using word embedding, due to this relevancy of search result is improved efficiently. The combined effect of this two model result into a better search engine.

2 Related Work

There are various implementations of the searching system by many researchers, but AI-based searching systems are rarely done by anyone. Word embedding can be used for various applications in NLP. Many researchers have implemented various systems using different word embedding techniques. Word2vec, GloVe, Bert, fasText, etc., are the types of word embeddings.

Silva-Fuentes et al. [1] had implemented an information retrieval system using Word2vec to determine the semantic of information. The terms in the search query are used for searching along with the new terms added in the query, so the search query expansion is done by using word embedding. Salim et al. [2] have proposed a model which is a combination of recurrent neural network (RNN), long short-term memory (LSTM) and word embedding. Firstly, the words are represented into a vector by using word embedding. After this, the LSTM cells are initialized by using the attribute features. The work of neural networks is more significant than any other approach for detecting sarcasm. Within just 15 epochs, the model gives an accuracy of 88%. Yenala et al. [3] have proposed a novel deep learning technique that is used to detect the inappropriate language in the text. The inappropriate text includes various messages, feedbacks, or comments on social media posts. A deep learning model is proposed for detecting inappropriate suggestions in queries known as "Convolutional Bi-Directional LSTM" (BiLSTM). BiLSTM is the combination of two techniques, CNN and bi-directional LSTM. C-BiLSTM provides better results than LSTM, CNN, or any other deep-learning-based model if used individually. Jeffrey Pennington et al. [4] have developed the GloVe model. The model is trained on a massive corpus like Wikipedia with million of word tokens. The GloVe is used to assign vectors depending upon the global context of the word. It also outperforms related models on similarity tasks and named entity recognition. The model produces a vector space with a meaningful substructure as evidenced by its performance of 75% on a recent word analogy task. Srinivas and Jairaj [5] have developed a Siamese network; it learns the uncommon food dish names embeddings from scratch. Those names are out of vocabulary for traditional word embedding. So, for improving the performance of the food search system, this proposed method will help strongly. The dataset used for the proposed method is in-house food taxonomy data. Timo Spinde et al. [6] presented a system to analyze the context of the related (meaningly) words. They have used two training models of word embedding. One model is used to train on left-wings news articles, while another one is trained on right-wings news articles. The main idea is to prove that the context of the biased word varies in different news outlets than one of the non-biased words. This system is automated detection of media bias using the implicit context of bias words that will derive through fine-tuned word embedding. Two datasets are used—WordSim-353 and MEN and the Google analogy test set. Han and Zhu [7] used a framework known as Flask, which is a Python framework, to make a system for rapid retrieval and visualization of media data. The crawled network's raw data is initially stored in a non-relational database like MongoDB. A cluster is set up by ElasticSearch, where the data is imported. After this, map for data is created, and an inverted index is set up. Due to this, data retrieval becomes more accurate as well as fast. Kibana is used to display the data and to present it. Khatri et al. [8] have proposed a sarcasm detection system using machine learning and two-word embedding techniques. BERT and GloVe embeddings are used in this system. The GloVe is used to determine the response vector, while BERT is used to get the context vector. GloVe and BERT embeddings, when used together, produce a maximum F-measure score of 0.690 with logistic regression as a classifier.

3 Word Embedding: GloVe Embedding

The GloVe is one of the techniques used to find the word vector [10]. Word vector is the representation of words in vector space such that words with similar contexts cluster together while different words repel. The GloVe model productively influences statistical data by using only non-zero values of word–word cooccurrence matrix for training. Word–word cooccurrence is a sparse matrix. So only the non-zero elements are considered.The GloVe is used in this system as it assigns the vector depends upon the word's global context, unlike Word2vec which is dependent on the local context. The dimension of each vector can be of various dimensions, say 50, 100, 200, and 300. So GloVe provides a pre-trained model trained on big datasets like Wikipedia, Twitter, etc., which can be downloaded directly from the Web. It contains millions of words along with their vectors in various dimensions. But there can be some words from the corpus that are not in the pre-trained model, so the vector assigns them zero. There is a possibility that some words with vector zero are useful for users, and the result of the search becomes irrelevant. So it is important to train the GloVe model on the input corpus. So the implemented GloVe model is the combination of both pre-trained and trained GloVe models. While training the GloVe model on the user-defined corpus, various parameters like corpus length, vocabulary size, etc., need to be changed, and thus, it directly affects the quality of embedding. The prerequisite for getting good-quality embedding is the size of the input corpus. If the size of the corpus is huge like 10 million words, then the embedding would be better than the one with a corpus size of 10 thousand words. While training the model, one more

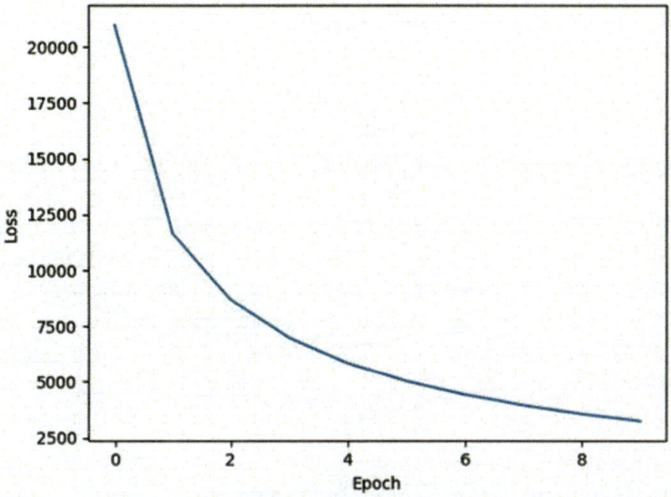

Fig. 1 GloVe training

parameter is important which is number of epoch. Figure 1 clearly depicts if more epochs are given, then the loss will further reduce and the embedding will be much better. This process is very time-consuming and requires huge computational power.

4 ElasticSearch

ElasticSearch is an open-source tool freely available on the Web; it is a full-text search and investigation engine. ElasticSearch provides the capability to store, search, and examine a vast amount of data in close to real-time. It is a highly efficient system for searching through unstructured data. It is utilized as the fundamental engine that powers applications with complex search requirements.

Elasticsearch is high-speed, scalable, and flexible, making the elastic stack a robust solution for many different use cases. Elastic search is useful for security purposes like finding threats and prevention, an enterprise search engine, and many other use cases. As there is a great need for a document-oriented database, ElasticSearch provides this capability. The indexing of the whole object graph needs to be done, but it is important to denormalize it first. This denormalization is important to improve the document retrieval performance as there is no need to join a query several times, but keeping all the instances consistent is important, and updating is quite difficult. Write-once-read-many-workload is suitable for it.

4.1 *Why Indexing in ElasticSearch?*

ElasticSearch searching for an index is much faster than searching for text into the documents, which is why ES is capable of performing fast searching. Searching through elastic search is like scanning the index of the documents related to the keywords and then retrieving the relevant data from the documents. Search for text on each and every page of a book is more difficult than searching for the index related to the keyword. This is known as an inverted index because it inverts the page-centric data structure, i.e., page -> words, to keyword-centric data structure, i.e., word to pages. ElasticSearch utilizes Apache Lucene for managing and creating the indexes. An index comprises at least one document, and the document is a collection of one or more fields.

4.2 *How Indexing Is Done in ElasticSearch?*

There are two types of indexing: 1. Forward Index and 2. Inverted Index.

Forward Index

It is the most basic way to build up an index. In forward index, a list of all terms from the documents are stored. It is very fast while indexing the documents. While appending a new document, there is no need to rebuild the index. But it is not so efficient when querying. The search engine needs to scan all the entries in the index searching for the specific term, so as to return all the possible documents containing that term. The example of the forward index is shown Fig. 2.

Inverted Index

It is a better option to sort the index by the term in the query so that the processing of the query becomes quicker. As this is the inversion of the forward index, it is named

Id	make	year
0	toyota	1996
1	mazda	1996
2	toyota	1996
3	ford	2002
4	toyota	2002
5	mazda	2002
6	toyota	2002
7	toyota	2009
8	ford	2009

Sorted value array.

Value	Index
ford	0
mazda	1
toyota	2

Forward index for "make" column.

id	value id
0	2
1	1
2	2
3	0
4	2
5	1
6	2
7	2
8	0

Fig. 2 Forward indexing example

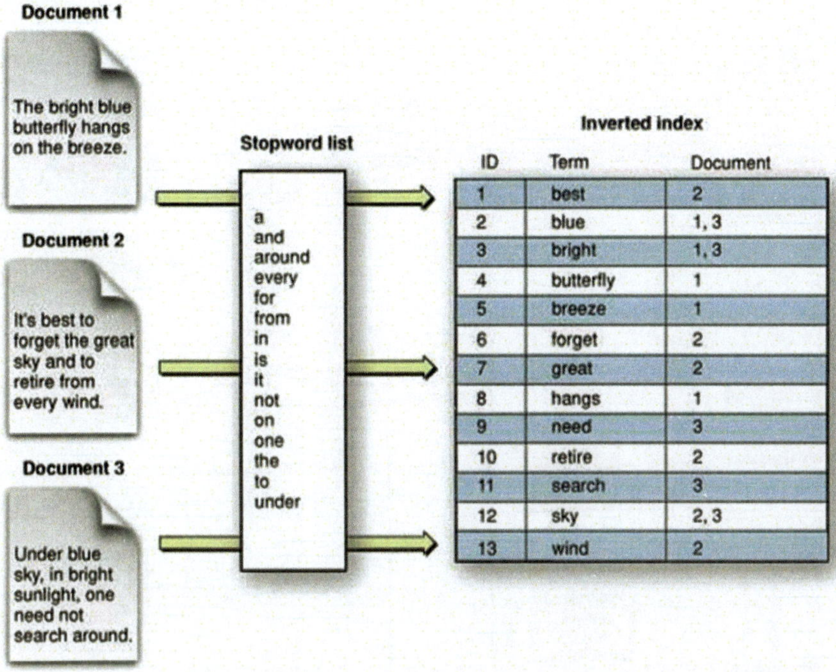

Fig. 3 Inverted indexing

as an inverted index. By using the inverted index, the machine will need to look for a term only once to successfully retrieve all the relevant documents containing that term. So ElasticSearch makes use of inverted index as as to provide faster retrieval of documents. An inverted index makes a list of all the unique words occurring in the documents, and it also identifies the documents where each word occurs in (Fig. 3).

The implemented system is divided into two parts Fig. 4. In the first part, all the modules run as a batch process that continuously works to learn new vocabulary from the fed documents. The user's input is in the form of a text document that includes PDF files, docs files, excel sheets, CSV files, etc. After accepting this input file, the first process is to convert the file into a regular text file. There is a need to extract only the text from the text file, and the images should be neglected as there is no need for an image; our system is not extracting text from image files. So the corpus is formed from the documents, but it contains a lot of unwanted text and symbols that are of no use. It is necessary to clean our corpus; only the required text should be there. So the next step is data preprocessing. In this step, we are removing unwanted symbols, punctuation marks, multiple spaces replaced with single space, converting whole text into lowercase, removing all the special characters, stopwords elimination, lemmatization, etc. After the data preprocessing step, the corpus contains only meaningful words. Now, the clean corpus is fed into

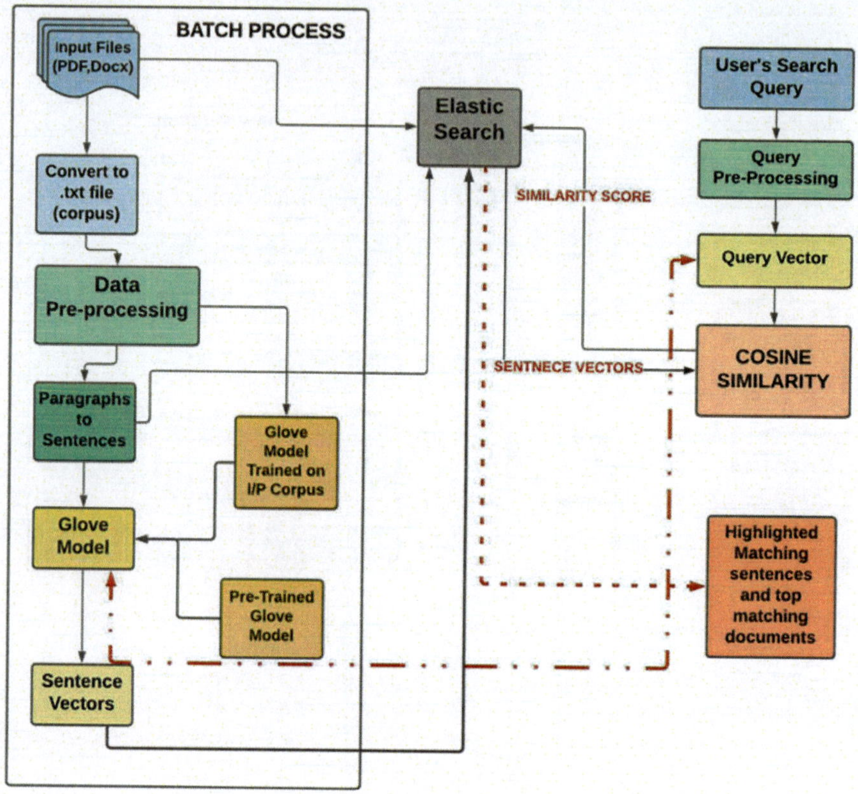

Fig. 4 Concrete system design

two modules simultaneously to paragraphs to sentences converter and GloVe model for training.The paragraph to sentences converter makes use of a sentence tokenizer from NLTK. The NLTK is a Python library and can be used for many purposes, mainly for NLP.

5 Implementation Details

Sent-tokenize [9] smartly split the paragraph into sentences with the help of punctuation marks. The sent-tokenize function uses an instance of PunktSentenceTokenizer from the nltk.tokenize.punkt module, which is already being trained and thus very well knows to mark the end and beginning of sentence at what characters and punctuation. So after converting every paragraph into sentences, we need the vector for the whole sentence, which can be done by using various word embedding techniques like Word2vec, GloVe, Fast text, Bert, etc. The GloVe is used in this system as it assigns the vector depends upon the word's global context unlike Word2vec which is dependent on the local context. The dimension of each vector can be of various dimensions, say 50, 100, 200, and 300. With the help of GloVe embedding, the sentence is embedded into a sentence vector and stored in a database for future use. The dimension of vectors are also important, so all the possiblitiese are checked while testing. ElasticSearch provides capabilities to search, store, and analyze vast amounts of data. This data can be structured as well as unstructured in nature. ElasticSearch is a high-speed, scalable, and flexible open-source tool, making the elastic stack a robust solution for many different use cases. Elastic search is useful for security purposes like finding threats and prevention, an enterprise search engine, and many other use cases. ElasticSearch is used to store all the documents. It exhibits useful properties like the ease in storing data, fast searching through the extensive database, and quick data retrieval. So ElasticSearch is the best option for storing all the user's documents to perform its searching task more effectively and add up to the search system's overall performance. So two separate indices are created within elasticsearch. The first index will store all the documents along with its names. Storing of data in ES is in json format, so document name and the extracted text from document is stored in this index. In the second index, various fields are stored like document names, sentences from the document, and their respective sentence vector. ElasticSearch helps to store the whole data efficiently without any loss of data as well as quickly. Retrieval of data is also very easy and fast as compared to traditional databases like MySql, Oracle, etc.

6 Results and Discussions

The result of the search system is dependent on the context of the search query which is the main motive of this project. With the help of word embedding, it was possible to determine the requirements of the user, and the relevant documents as well as the matching sentences from the document are displayed as output of the system. To test the searching system, a crawl dataset of COVID-19-related news article from Times Of India is created. The dataset consists of three columns, mainly headline, news article, and link of the article. So the articles are first preprocessed, and then, the GloVe model is trained on the corpus created by using the news data. After training, the model can embed out of vocabulary words like COVID-19, remdesiver etc., which are gathered from the news. Table 1 shows the query and the output documents names (news headline) from the dataset. To visualize the embedding of out of vocabulary words, a scatter plot of top-20 most similar words is created as shown in the figure. The figure shows the words which are most similar to the words like "Covid vaccine" shown in Fig. 5. It is clear from the figure that the embedding is based on context of the word used in the corpus.

Table 1 Query with output document

Query 1. Oxygen Concentrator news?
1. Companies shut ops to supply oxygen, provide bikes to medics fighting pandemic
2. PM Narendra Modi reviews oxygen status, production to be increased
Query 2. Covid vaccine related news
1. Serum Institute asks government for 403 million to boost AstraZeneca vaccine output
2. Double mutant of coronavirus fast replacing dominant N440K variant in south India Scientists
Query 3. Remdesivir drugs?
1. Overuse of steroids could be counter productive, says AIIMS chief
2. No Covid appropriate behaviour in Kumbh, Ramzan, says Amit Shah
Query 4. Daily death poll due to covid?
1. Covid At 685 deaths, India records highest daily toll in 5 months
2. With 89,129 new cases, India s daily coronavirus infections hit six month high

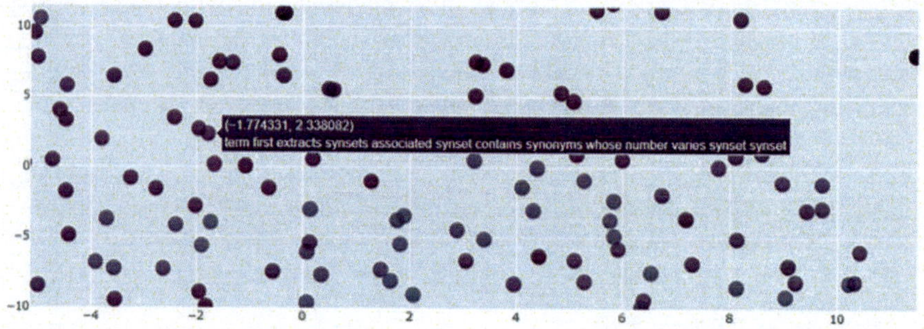

Fig. 5 Contextually similar sentences

The proposed system is also tested by providing different search queries as the input to the system. Later on, the result is checked manually to see whether relevant results are given as output or not. So some of the results are shown in Table 1. The table shows the queries and the output documents for the respective query.

From the result, it can be concluded that the system is considering the context of the query, and depending on that, the searching is done. In order to check the contextual similarity of sentences or words, a scatter plot shown in Fig. 6 is also built for some sentences to see if the vectors are calculated using the context words of the sentence. So the scatter plot easily shows that sentences with similar contextual meaning are forming a cluster and are at a distance from the one which is not contextually similar to them. To map the vectors which are of 50 dimensions into 2 dimensions, we make use of PCA and t-SNE. And these two-dimensional vectors are then used to plot the point on the scatter plot graph. Similarly, a scatter plot of contextually similar words is also built using the same technique. And both the scatter plot graph is showing appreciable output by clusteing similar words or sentences together to form a cluster (refer Fig. 6).

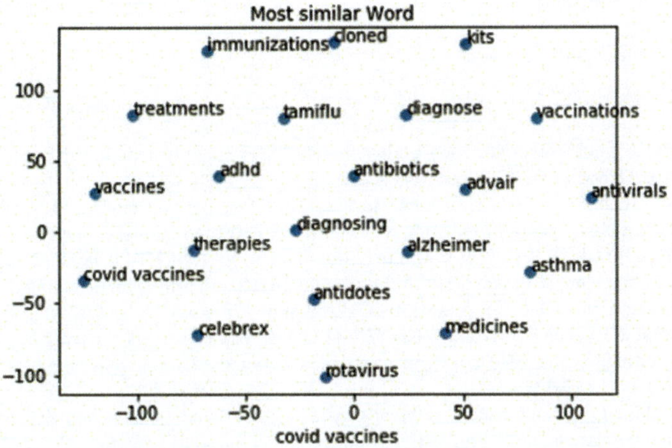

Fig. 6 Contextually similar words

7 Conclusion

With the help of word embedding, the main objective of determining the context of the search query is achieved. The trained GloVe model is able to embed out of vocabulary words from the corpus, which makes it more efficient. The search result gives relevant documents depending upon the context of the search query as well as by matching the term in the query. ElasticSearch makes the searching, storing, and retrieval of documents so faster and easier. The implemented system works efficiently on any kind of textual data like news articles, reviews datasets, etc. For testing purposes, the COVID-19 news article from TOI is used as a dataset by using a Web crawler, and the performance of the system appreciable. The overall performance of the implemented system is very appreciable; it is fetching relevant documents with lightning speed with minimal latency, which is very necessary for any search engine. The first future scope is integration with the existing system named as TEF. Exploring new word embedding techniques like the latest one Elmo, Bert, etc., but various other parameters are needed to be considered. Query expansion techniques is also a good addition for any search engine. Next word prediction is one of the important scopes for this project. The last one is to use other techniques like knowledge graph instead of word embedding.

References

1. Silva-Fuentes MA, Calderon-Vilca HD, Calderon-Vilca EF, Cárdenas-Mariño FC (2019) Semantic search system using word embeddings for query expansion. In: 2019 IEEE PES innovative smart grid technologies conference-Latin America (ISGT Latin America). IEEE, pp 1–6. https://doi.org/10.1109/ISGT-LA.2019.8894992
2. Salim SS, Ghanshyam AN, Ashok DM, Mazahir DB, Thakare BS (2020) Deep LSTM-RNN with word embedding for sarcasm detection on Twitter. In: 2020 international conference for emerging technology (INCET). IEEE, pp 1–4. https://doi.org/10.1109/INCET49848.2020.9154162
3. Yenala H, Jhanwar A, Chinnakotla MK, Goyal J (2018) Deep learning for detecting inappropriate content in text. Int J Data Sci Anal 6(4):273–286. https://doi.org/10.1007/s41060-017-0088-4
4. Pennington J, Socher R, Manning CD (2014) Glove: global vectors for word representation. In: Proceedings of the 2014 conference on empirical methods in natural language processing (EMNLP), pp 1532–1543
5. Vijjali R, Mishra A, Nagamalla S, Sathyanarayna J (2020) Semantic embeddings for food search using Siamese networks. In: Proceedings of the 4th international conference on natural language processing and information retrieval, pp 138–143. https://doi.org/10.1145/3443279.3443303
6. Spinde T, Rudnitckaia L, Hamborg F (2021) Identification of biased terms in news articles by comparison of outlet-specific word embeddings. In: Proceedings of the 16th international conference (iConference 2021). Springer Nature, Virtual Event China
7. Han L, Zhu L (2020) Design and implementation of elasticsearch for media data.. In: 2020 international conference on computer engineering and application (ICCEA), IEEE. https://doi.org/10.1109/ICCEA50009.2020.00036
8. Khatri A (2020) Sarcasm detection in tweets with BERT and GloVe embeddings. arXiv preprint arXiv:2006.11512. arXiv:2006.11512
9. AnalyticsVidhya (2018) https://www.analyticsvidhya.com/blog/2018/11/introduction-text-summarization-textrank-python/. Last accessed 1 Nov 2018
10. Pengyan510 (2021)https://github.com/pengyan510/nlp-paper-implementation. Last accessed 5 May 2021

Design and Development of Enhanced Secured Video Communication System for Medical Video Sensor Network

A. Sivasangari, D. Deepa, R. Vignesh, Suja C. Mana, and B. Keerthi Samhitha

Abstract Privacy and security of healthcare data are becoming important issue in healthcare field. The main issue to be considered during data transmission is the security of the data that can be transmitted through wireless medium to the remote server. The attacker can perform different types of security attacks on medical information due to broadcast nature of wireless communication. Thus, an effective security mechanism to safeguard the data against data alteration and deletion is the need of the hour. This sort of data tamper would result in wrong diagnosis by the physician, which is a life-threatening issue. WMSNs (Wireless Medical Video Sensor Networks) have the ability to store, process, and collect heterogeneous devices for multimedia data. Requirements such as high bandwidth, low packet loss ratio, and minimal delay have also been increased. To meet the security requirement, a secure key exchange method in WMSN is proposed in this paper. The main issue to be considered during data transmission is the security of the data, which can be transmitted through wireless medium to the cloud server. The attacker can perform different types of security attacks on medical information due to broadcast nature of wireless communication. Additionally, the collected multimedia data is stored in the cloud. Thus, an effective security mechanism to safeguard the data against data alteration and deletion is the need of the hour. In order to provide the security in three levels of data transmission in WMSN, the secure key exchange in WMSN (SK-WMSN) model is proposed in the present work. Also, we analyzed other proportional assessment issues in WMSN, for instance, incorporation and security issues. Finally, we inspected and masterminded the current off-the-rack contraptions, models, and testbeds executed for WMSNs.

Keywords Key exchange · WMSN · Queue scheduler

A. Sivasangari (✉) · D. Deepa · R. Vignesh · S. C. Mana · B. Keerthi Samhitha
Sathyabama Institute of Science and Technology, Chennai, India

P. K. Singh et al. (eds.), *Emerging Technologies for Computing, Communication and Smart Cities*, Lecture Notes in Electrical Engineering 875,
https://doi.org/10.1007/978-981-19-0284-0_7

1 Introduction

For a little while now, researchers and technology experts have been drawn to a popular use of WMSNs (wireless medical sensor networks), which consists of a large number of portable sensors with low memory space, processing capacity, and latency. Several study organizations and programmers have concentrated their efforts on the electronic-healthcare surveillance infrastructure, using CodeBlue, UbiMon, SPINE, LiveNet, MobiHealth, Alarm-Net, etc. For WSNs, the role of WMSNs in healthcare systems has resulted in significant developments in the healthcare sector in the twenty-first century. WSNs became a science fiction/movie-fantasy subject for the healthcare industry a few years earlier, but now they've become a fact and would provide much excellent performance. According to a survey taken place in 2008, the people worldwide ($\geq$65 years) were projected to be 502 million, and that number is expected to reach 1.4 billion by 2030. When more people reach that age, the risk of developing such serious and chronic conditions increases dramatically. For instance, Alzheimer's signs usually occur after the age of 65 years, cardiovascular disease and stroke symptoms rise just after age 60, and diabetes, such as those of several other conditions, increases after age of 60 (e.g., blood glucose, blood pressure levels, etc.). As a result, the elderly population wanted to bring independence and high-quality treatment without jeopardizing their convenience while lowering their medical expenses. In this sense, wireless sensor technologies could offer extremely useful resources for tracking the wellbeing of elderly and patients who require constant monitoring. As a result, healthcare using wireless technologies is an innovative and expanding area of research inquiry. In reality, the future of contemporary healthcare in an aging world would necessitate constant health screening with little actual contact between patients and doctors. Recent times, the concept WMSN (wireless medical sensor network) has been coined to pull together many experts from various backgrounds such as electronics, bioengineering, medicine, computer, etc. The architecture of a wireless healthcare framework presents several problematic situations, such as efficient data transfer, node transport protocols and fast incident tracking, accurate data distribution, power storage, node computing, and web services. However, introducing emerging technology in healthcare systems without understanding potential risks endangering patient safety. For example, the patients' health vital signs are especially susceptible (i.e., whether a patient does have a humiliating disease), so any leakage of specific access to healthcare can disgrace him/her. In reality, disclosing disease details will often lead to a person destroying his or her employment or making it difficult for her or him to seek health cover. Moreover, wireless medical sensor networks include a wide variety of healthcare uses, including medical data analysis, movement tracking in fitness clubs, player position recording, etc. As a result, WMSNs share the patient data with insurance providers, doctors, wellness coaches, and family. Such sharing of data can cause problems to and make patients vulnerable for external threats. In a clear case, a patient's sensors relay body information to a nursing staff; however, an intruder could be spying on the patient information as it is transmitted, compromising the patient confidentiality and the attacker can post the details of the patient to public using social media and it risks the privacy of the patient. After all, wireless healthcare can provide many benefits to monitoring patients; nevertheless, while an individual's physical and mental data is extremely sensitive, data protection becomes major issues for healthcare implementations, especially where new communication is used. More specifically, if HIPAA laws are not enforced correctly, a healthcare worker faces stringent criminal and civil sanctions (i.e., a penalty or prison sentences). As a result, medical confidentiality and safety are the primary concerns in smart healthcare.

Data security seems to be the method of securing servers, records, and identities on a system by implementing a range of controls, programs, and procedures that define the relative value of various databases, their vulnerability, and regulatory enforcement criteria, and afterward apply necessary safeguards to defend those properties. Data security, like other methods such as security controls, file safety, or user behavioral security, is not really the be-all and end-all of a security practice. It is one way of assessing and mitigating the risks associated with storing some kind of information. The data management technologies mentioned below are being used to deter attacks, minimize damage, and maintain safeguards.

The concern is not when a security violation will occur, nor when it might occur. Once forensic analysis is called on to investigate the underlying problem of a hack, getting a data auditing framework in place to monitor and document on access management modifications to data, who has access to confidential data, when this was obtained, file route, and so on is critical to the review. Alternatively, by implementing appropriate data auditing methods, IT managers can achieve the exposure required to avoid unwanted modifications and possible breaches.

Usually, this takes several days for a company to find a hack. Companies also learn about violations from their clients or third party rather than from their own IT teams. Through tracking data activity and unusual activities in full detail, you can detect security violations more easily, resulting in accidental damage, failure, modification, unintended exposure, or acceptance of data which is personal.

Data Risk Evaluation

Data vulnerability evaluations assist businesses in identifying their most vulnerable confidential data by providing dependable and repeatable measures to prioritize and address severe security threats. The procedure begins by defining confidential data obtained by global networks, expired data, and conflicting permissions. Risk analyses outline key conclusions, reveal data flaws, explain each flaw in depth, and provide prioritized remediation suggestions. Data Minimization.

The understanding of data has shifted over the last generation of IT management. Initially, having additional data always seemed to be better than having less. You will never know in advance what you would like to do about it. Data is now a liability. The possibility of a reputation-damaging data leak, damage in billions of dollars, or substantial regulatory penalties both perpetuate the notion that obtaining something other than the bare minimum of confidential data is highly risky.

At the end, follow best practices for data minimization and evaluate all data processing needs and processes from a market perspective.

Since the accessibility nodes send private data through wireless networks, any person may behave as an attacker and interrupt and tamper with the sensitive information being transmitted on the WSNs. Furthermore, the attacker can effectively decrypt data exchanges between clients and data centers. This might lead to a variety of threats, such as repeat, key imitation, stolen authenticator, and so on. Related types of threats are regarded, with an emphasis on the insecurity of wireless technology, in team to address the cloud-centric, multistage encryption as just a service scheme and also to investigate IoT-based post-disaster management mechanisms. Body sensor network (BSN) software has been one of the newest Internet of Things (IoT) innovations in healthcare. A patient could be tracked using this software by using a series of tiny controlled and portable wireless sensors. When such a method is seen in terms of protection, nevertheless, patient privacy is compromised. The main security standards in BSN-based medical systems is a stable IoT-based welfare system that introduces biometric-based approaches as the third encryption element, as well as a modern user-anonymous authentication method relying on WMSNs. To ensure safe and allowed communication in WMSNs, the symmetric key-based authentication mechanism is built. The following are the practical protection properties of WMSNs.

Mutual authentication: It enables the sensor, gateway, and medical practitioner to mutually verify one another.

Session-key agreement: It distributes the session key between patients and medical specialist in order to secure contact in WMSNs.

Known-key security: And if the attacker tampers with one user's session key, the attacker cannot breach the session key of yet another user.

Client anonymity: It masks the identity of people, such as a medical specialist and a patient. Stolen smart card attack unless the user wants to change his or her credential, he or she can include the login credentials and identity during the session-key-update process so that SC and GW access can determine if the inputs are correct or not.

2 Related Work

Iqtidar et al. [1] talk about a novel intrusion detection system to prevent attacks on personal medical devices. Authors talk about an efficient intrusion detection system which can identify the intrusion into personal medical devices. Authors reveal about some of the possibilities of vulnerabilities for these devices and how the proposed system effectively detects these intrusions in advance. In paper [2] Liu et al. discuss about regional medical data sharing using big data technologies. Quantin et al. [3] describe the implementation of an information system to facilitate the keep tracking of electronic medical records of people in European regions. Teodor Sumalan et al. [4] present a solution for the surveillance system which is based on emulating the data generated by some sign sensors. Paper [5] describes a video based monitoring based on wireless sensor network technology. This video monitoring station will act as a gateway between various processing nodes and the network. The software and hardware implementations are studied in detail in the paper by Yifan et al. Sodhro et al. [6]. They talk about energy efficient video transmission in wireless body sensor networks. Authors have proposed a sustainable algorithm to address this energy management issues. This algorithm proved to be effective in increasing the life of wireless body sensor networks.

Singh et al. [7] propose an energy efficient video streaming architecture applicable to wireless media sensor networks. This implementation tries to reduce the energy consumption in the network [7]. A novel energy consumption approach is proposed by the author [7]. In paper [8] by Abbas et al., authors describe about a video surveillance system suitable for wireless sensor networks. Authors also implemented a dynamic queue scheduler. An open source platform based on raspberry pi is the core component of this implementation [8]. Qian et al. [9] describe about a video based wi-fi sensor node. The core function of this node is the air target detection. This sensor is capable of detecting the air quality index which is a key measure to determine the level of air pollution [9]. A zigbee based WSN model is described in paper [10]. This work by A. Jasim et al. describe about centralizing the network management task at the coordinator of the network. It tries to remove loads from other nodes [10]. In paper [11], authors describe about traffic violation detection algorithm. In paper [12], authors talk about a wearable device applying the principle of wireless sensor networks. Yi et al. [13] propose a method to reduce the number of blackout nodes. This implementation will increase the data collected by selecting an appropriate video encoding methods. This proposed architecture is particularly applicable in solar powered wireless video sensor network. Raj et al. [14] talk about an iot-based electronic-healthcare system. In this work, authors have presented a cost-effective health sensor platform for rural health monitoring. Jamshed et al. [15] describe about a node scheduler which is capable of scheduling tasks depending on the priorities of tasks. This scheduler helps to attain better quality of service for the network.

3 Proposed Work

It involves three phases. They are (i) Cluster Formation (ii) data transmission between the SH and the BS and (iii) data transmission between the BS and the cloud storage.

3.1 Sensor Node Design

A sight and sound sensor hub comprises of an outsized number incorporating a detecting unit with a camera, a foreign correspondence unit, and a preparing unit; the whole sensor needs to be controlled by a unit of measurement as demonstrated in Fig. 1. The preparing unit is particularly vital for the sensor hub. The handling unit

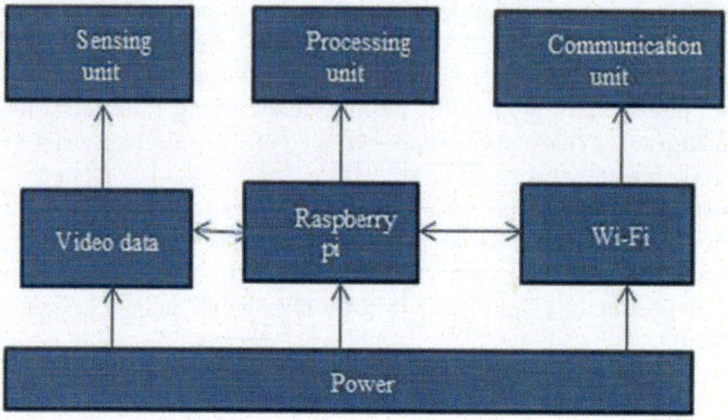

Fig. 1 Processing sequence sensor node

needs to have low force utilization, high velocity estimation, and furthermore need to be little. Thus, we use raspberry pi (RPi) (Raspberry Pi Establishment, Cambridge, Joined Realm) as a handling unit as demonstrated in Fig. 2a.The Raspberry pi 2 model B incorporates 1 GB of Smash and incredible 900 MHz quad-center ARM Cortex-A7 (ARM Holding, Cambridge, Joined Realm) central processing unit. It depends on Broadcom BCM2835 framework on a chip. Its HDMI attachment, a SD space card, a couple of USB connectors, and an Ethernet connector. With these associations, raspberry pi may be a good instructive work area PC. We use raspbian working framework,

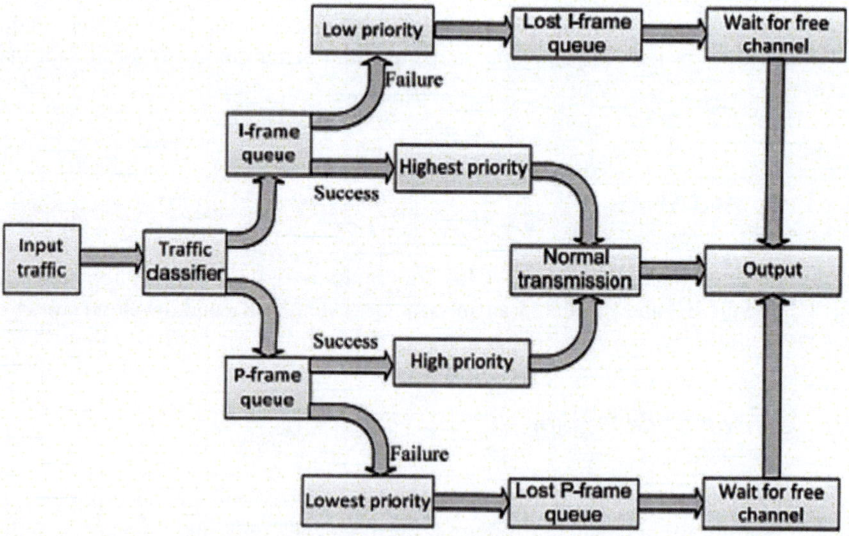

Fig. 2 Input to output queue steps

which depends on a debian conveyance. Raspbian working framework (raspbian-wheezy, Raspberry Pi Establishment, Cambridge, Joined Realm, 2015) may be a linux/GNU rendition (3.18) created for raspberry pi equipment. Our foundation comprises of seven raspberry pi model 2B sensor hubs.

The remote correspondence unit has the duty of sending video information in WMSNs. For remote correspondence, we utilize monetarily accessible EDIMAX wi-fi dongle as demonstrated in Fig. 2b. EDIMAX wi-fi dongle (EW-7811Un, Edimax, Taipei, Taiwan) is connected to USB port of each raspberry pi board. The dongle has information rates up to 150 Mbps, agrees with remote 802.11 b/g/n, and it upholds shrewd communicate power control and auto-inactive change. With low force utilizations, these dongles are especially appropriate for quick prototyping for remote sight and sound sensor network applications.

3.2 The Development and Transmission of Video Bundle

Our point is to send video transfers with high casing rate and great quality. Consequently, IVSP utilizes H.264 standard for video pressure. The worker application produces a video transfer from either live or put away source and converts video information in H.264 transfers. As indicated by H.264 standard, there are I, P, and B outlines. We do not utilize B outlines in tests, on the grounds that the deciphering and encoding of those edges are reliant upon the subsequent edge. In our analyses, we use I edges and P outlines. I outline need is in every case tons above P outline. The length of a bundle needs to be changed in order that it can satisfy the necessities of transmission. On the off chance that a parcel has extremely limited quantity of data, each hub must send rapidly. This may expand the likelihood of clog, energy waste, and more crashes in remote channel. On the off chance that a parcel has enormous measure of data, which can comprise of various edges, each bundle must stand by until all the video outlines are made. This may create more setbacks. Hence, length of parcel should be picked in suitable manner. We accept that the majority extreme parcel length is l. within the event that the length goes past this edge, the video information is going to be separated into numerous bundles. The parcel is going to be I outline bundle, if a bundle contains the knowledge of I outline. Something else, parcel is going to be P outline bundle. We realize that remote transmission is problematic, so parcel misfortune may be a conspicuous element. An I outline parcel misfortune will cause loss of all P outline bundles, since they will not be decoded needless to say and can be futile. Within the event that *P* outline parcels are lost, video deciphering will continue surely aggravations in video information. Within the event that I outline bundles are retransmitted rapidly, the results of misfortune parcels won't be conspicuous. Along these lines, retransmission of I outline bundles is significant for acceptable quality video information.

3.3 *Queue Scheduler*

As we talked about in Segment 4.1, various sorts of parcels have distinctive retransmission prerequisites. As per the extent of importance, parcels are often separated into four needs from high to low: The I outline bundles have most elevated need, P outline bundles have high need, lost I outline parcels that require retransmission have low need, and lost P outline bundles that needs retransmission have least need. Figure 4 shows the bundle stockpiling and transmission measure at a sensor hub. At the purpose when a hub produces or gets an information bundle, the parcel is moved to the I-Casing parcel line or P-Edge bundle line contingent upon its quality. These two lines are utilized to store the unverified parcels incidentally, sitting tight for the retransmission if necessary. I outline bundles are of critical significance in supporting the constant hour gridlock. Thusly, we give the best got to I outline bundles, on the grounds that our fundamental point is to make sure I outline parcels. Within the event that I outline parcel is lost, it goes into lost I outline line, and if a P outline bundle is lost it goes into lost P outline line. In our examinations, to manage traffic classes with various needs during a proficient manner, weighted cooperative scheduler is employed within the weighted cooperative scheduler, we dispensed a weight.

3.4 *Congestion Identification*

To quantify clog in remote sensor organizations, we present a mix of two diverse blockage tips that could precisely recognize blockage at every sensor hub. The most markers are support inhabitance T, and second pointer is cradle inhabitance change rate δ. The support inhabitance may be a vital marker of clog. It's received to assess the measure of cushion involved by the parcels at a hub over its start to end transmission. The upside of support inhabitance is that clog at every hub is often straightforwardly and immediately identified. Cradle inhabitance T are often communicated into three states: typical state, moderate state, and important state as demonstrated in above figure to acknowledge diverse cushion states, we utilize two unique limits T1 and T2, as demonstrated in figure Support inhabitance change rate δ is employed to differentiate the organization clog and may be characterized as:

$$\delta i = \mathrm{T\,c\,I} - \mathrm{T\,L\,I\,T}_{\max} - \mathrm{T\,c\,I},\ \mathrm{I\,e}\ \{\text{I frame, P frame}\} \tag{1}$$

in which T c I and T L I, I e{I Edge, P Frame} show the support inhabitance in current and last round, individually. From above condition, we will get the (Fig. 3)

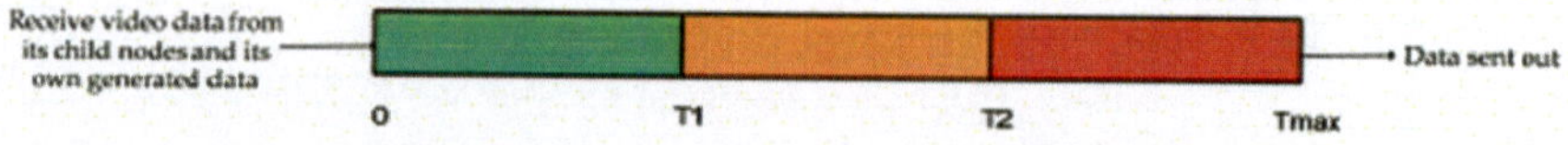

Fig. 3 Buffer states for congestion detection

$$\text{Cushion inhabitance change rate } \delta = \sum iwi \times \delta i \sum iwi, \text{ Ie \{I frame, P frame\}} \quad (2)$$

All need lines have an identical support size T_{max}. The cradle inhabitance change rate δ mirrors the inclination of cushion inhabitance. For a given support inhabitance, the larger the price of δ, the upper the likelihood of line flood is. Also, a negative δ shows that clog has been lightened and cushion inhabitance is diminished. As per these two blockage pointers, the sensor hubs have an aggregate of three states.

In the proposed approach, each SH is assumed to be a network connector that controls several sensor nodes utilizing a distributed clustering technique. The SH can proceed with an independent decision without the presence of any centralized management. The clustering approach can be manually exploited by taking the location and communication range between the SH and BS into account. The sensor nodes with minimum energy levels become members in a cluster while remaining nodes become CHs. The elected CHs are responsible for authentication of member nodes and receiving and forwarding of data to a BS (Base Station).

Secure Communication between SH and BS.

After electing the SH, the proposed scheme performs the operations are compression, encryption, and data transmission between the SH and BS. The multimedia data compression consists of four stages such as transform calculation, thresholding, quantization, and entropy encoding. The signal transformation is done with wavelet transform (WT). The Coefficients are quantized by using the uniform step size which depends on the maximum and the minimum values of signal matrix and the number of quantization levels. The Huffman encoding can be used for obviating this problem. Data can be encrypted by using RC7 encryption algorithm. Inverse transform and decompression are applied at the receiver side to reconstruct the original data.

Secure Communication between BS and Cloud storage.

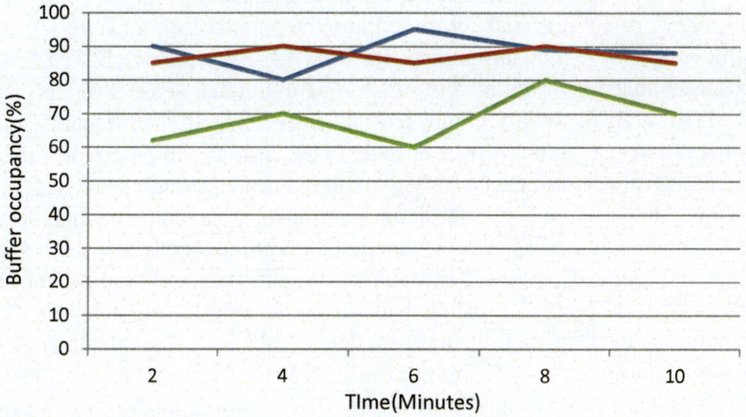

Fig. 4 Buffer occupancy with IVSP

In this phase, a novel lightweight Feistel Cipher structure is presented for encryption. The key management for the presented work is achieved through the use of Attribute based Key-Exchange (ABKE).

The BS nodes are preloaded with a set of attributes, which are ID, type of data and type of users. Each user is assigned an access structure and corresponding secret key SK. The session key is derived by applying hash value for the attributes, which are preloaded in BS nodes. The BS collects the information from SH and sends it to the cloud server. The users who satisfy the access policy are allowed to access to the patient information. The user derives the session key from their attributes and decrypts the master key, which is used for deriving the medical information. The key generation module provides a secure key exchange for ensuring authentication and is achieved by a trusted third party who is responsible for health care information monitoring and key distribution.

4 Performance Analysis

To assess entirely unexpected conventions and calculations relating for different systems administration layers (transport, network layer, MAC, or physical layer) of remote transmission locator organization or check various applications over the WMSN, specialists may perform scientific examination, direct tests, or use reenactments. Once in a while, Analytical examination neither offers a right model for such modern remote framework nor truly portrays the conduct of period remote organizations. Likewise, in a few cases, tests and examinations in remote identifier network for the most part and in remote transmission indicator network over all square method some way or another muddled and long, and grave to be re-conducted by various specialists. WMSN test beds square measure utilized for higher arrangement and fulfilling the reasonable and specialized difficulties of organizations sent in period frameworks. While testbeds turned into the popular technique for testing and assessing with remote transmission locator network applications, they furthermore give proposes that for bunch activity numerous individual sensors on an ordinary remote stage in an incredibly controlled and instrumented environmental factors. Subsequently, investigation on test testbeds with current equipment and PC code stages, licenses clients not exclusively to exhibit pertinence and assess application-level and organization level execution measurements (Ex. Jitter, streams, detection likelihood, end to end delay, etc.) in genuine conditions, anyway also to approve investigation models. Contrasted and leading period trials and field organizations, testbeds offer wide power in testing without doubt long-lasting analyses that are indispensable in troubleshooting, approval, and combination periods of dependable remote transmission indicator organizations. WMSN testbeds are ordered into 2 classes, PC code testbeds and Hardware testbeds. Table about six delineates the overall PC code and equipment testbeds found inside the writing and sums up their particulars and fundamental alternatives.

Hardware testbeds incorporate passing on of Hardware contraptions, like different sorts of cameras with absolutely different objectives and picture taking care of capacities, and remote correspondence equipment that will maintain various rules and absolutely special data rates. Other than that, hardware testbeds offer supporting code for data insight and program. Dependent upon the class wise organization maintained and supported by different network, Hardware testbeds are more segregated into single and multi-tier testbeds.

(a) Software Testbed:

Camera Resolution: WiSNAP, AER Emulator.

Wireless Mote: Added library like Agilent ADCM-1670,OV7649@30fps, 60fps.

Features: Matlab based testbeds, VisualC++ based testbeds, AE recognition.

(b) Hardware Testbed:

Camera Resolution:Meerkat, SenseEye, IrisNet.

Wireless Mote:Logitech QuickCam Pro, PTZ Sony SNC-RZ30N.

Features: Energy efficient, Multi level resolution, Surveillance applications.

Figures 4 and 5 shown about node packets and congestion probability ratio respectively,

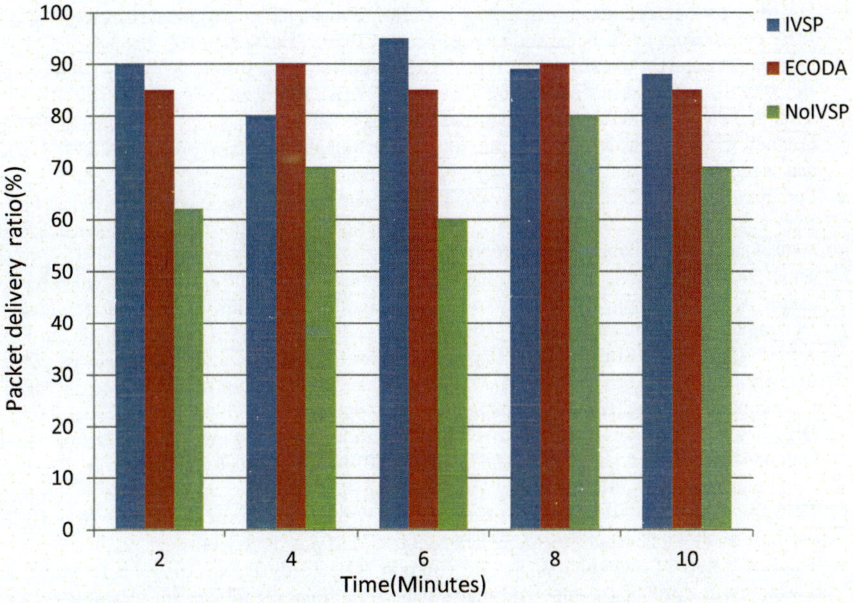

Fig. 5 Packet delivery ratio of frame packets

5 Conclusion

In this paper, here discussed about the issues and secure challenges in WMSNs. As well as examined overviewed and detailed design about WMSNs and also uses explained clearly. In this paper added security challenges based on criteria also analyzed detailed assessment carried on WSMS (Wireless Multimedia Sensor Networks) and investigation challenges and their issues in arranging estimations, shows, plans, and gear to wireless sensors. And here discussed about enormous segment of the current responses for WMSN at the different layers of the correspondence stack: physical, MAC, coordinating, transport, and application close by the possible cross layer execution. Feistal light weight cipher was used for encryption and decryption.

References

1. Newaz AI, Sikder AK, Babun L, Uluagac AS (2020) HEKA: a novel intrusion detection system for attacks to personal medical devices. In: 2020 IEEE conference on communications and network security (CNS), Avignon, France, pp 1–9
2. Liu Y, Cai H, Li G (2018) Research on sharing and application of medical big data. In: 2018 3rd international conference on smart city and systems engineering (ICSCSE), Xiamen, China, pp 771–774. https://doi.org/10.1109/ICSCSE.2018.00166
3. Quantin C, Allaert FA, Fassa M, Avillach P, Fieschi M, Cohen O (2007) Interoperability issues regarding patient identification in Europe. In: 2007 29th annual international conference of the IEEE engineering in medicine and biology society, Lyon, France, pp 6160–6160
4. Sumalan T, Lupu E, Arsinte R, Onaca E (2015) Low-cost wireless system for patient monitoring. In: 2015 E-health and bioengineering conference (EHB), Iasi, Romania, pp 1–4
5. Yifan K, Peng J (2008) Development of data video base station in water environment monitoring oriented wireless sensor networks. In: 2008 international conference on embedded software and systems symposia, Chengdu, China, pp 281–286
6. Sodhro AH, Fortino G (2017) Energy management during video transmission in wireless body sensor networks. In: 2017 IEEE 14th international conference on networking, sensing and control (ICNSC), Calabria, pp 655–660
7. Singh JP, Mishra MK, Khan MA (2017) Energy-efficient approach towards video-based sensor networks (wireless) beneath barrier coverage. In: 2017 8th international conference on computing, communication and networking technologies (ICCCNT), Delhi, India, pp 1–5
8. Abbas N, Yu F (2018) Design and implementation of a video surveillance system for linear wireless multimedia sensor networks. In: 2018 IEEE 3rd international conference on image, vision and computing (ICIVC), Chongqing, China, pp 524–527
9. Qian C, Cai W (2014) The design and implementation of video wireless WIFI sensor nodes with the function of air targets detection. In: 2014 international conference on wireless communication and sensor network, Wuhan, China, pp 222–225
10. Jasim A, Çeken C (2015) Video streaming over wireless sensor networks. In: 2015 IEEE conference on wireless sensors (ICWiSe), Melaka, Malaysia, pp 63–66
11. Mana SC, Keerthi Samhitha B, Jose J, Swaroop MV, Reddy PCR (2019) Traffic violation detection using principal component analysis and viola Jones algorithms. Int J Recent Technol Eng (IJRTE) 8(3). ISSN: 2277-3878
12. Ramamoorthy V, Divya S, Mana SC, Keerthi Samhitha B. Examining and sensing of artificial knee with multi sensors networks. J Adv Res Dyn Control Syst (JARDCS) 10(4):115–120

13. Yi JM, Yoon I, Noh DK (2017) Adaptive video coding selection scheme for solar-powered wireless video sensor networks. In: 2017 4th international conference on computer applications and information processing technology (CAIPT), Kuta Bali, Indonesia, pp 1–4
14. Raj C, Jain C, Arif W (2017) HEMAN: health monitoring and nous: an IoT based e-health care system for remote telemedicine. In: 2017 international conference on wireless communications, signal processing and networking (WiSPNET), Chennai, India, pp 2115–2119
15. Jamshed MA et al (2015) An energy efficient priority based wireless multimedia sensor node dynamic scheduler. In: 2015 12th international conference on high-capacity optical networks and enabling/emerging technologies (HONET), Islamabad, Pakistan, pp 1–4

Detecting Twitter Hate Speech Using Sentiment Analysis

Arpita Tripathy, Anshika Goyal, Urvashi Tyagi, and Poonam Tanwar

Abstract In current era due to the rapid increase in the consumption of the Internet by people of different cultural backgrounds, malicious content on the Internet has become an endless problem. In the automatic detection of malicious text content, distinguishing between hate speech and profanity will be a major issue. The police investigating strategies for positive or negative emotions in the text. Companies use it extensively to perceive emotions in social data, measure overall reputation, and understand customers. For example, if you use sentiment analysis to mechanically analyze more than 4000 reviews related to your products, you can more easily determine whether customers are satisfied with your customer service and pricing plans. The Twitter information set is usually used for comparative analysis of the model. Information science provides useful data based on large amounts of complex information or big data. Information science/engineering combines different fields of statistics and computer science to interpret information for decision-making purposes. The purpose of this article is to provide the method for detecting twitter the speech using Support vector machine and Machine Learning algorithm.

Keywords Sentiment analysis · Machine learning (ML) · Support vector machine (SVM)

1 Introduction

The global availability of the Internet has changed the way we greatly perceive the world. The social network (SM) is one of the youngest representatives of the global network, and it is also a gift in various forms: online gaming platforms, chemical analysis applications, forums, online news services, and social networks [1, 2]. Completely different social networks represent different goals: opinion delivery

---

A. Tripathy · A. Goyal · U. Tyagi · P. Tanwar (✉)
Department of CSE, MRIIRS, Faridabad, India
e-mail: poonamtanwar.fet@mriu.edu.in

Faculty, CSE, FET, MRIIRS, Faridabad, India

P. K. Singh et al. (eds.), *Emerging Technologies for Computing, Communication and Smart Cities*, Lecture Notes in Electrical Engineering 875,
https://doi.org/10.1007/978-981-19-0284-0_8

(Twitter or Facebook), business contact (LinkedIn), image exchange (Instagram), video delivery (YouTube), chemical analysis (Meetic), etc. [5].

Among the diverse existing social networks, Twitter presently ranks with inside the live performance of the leading structures and is one of the predominant critical information reasserts for researchers. Twitter can be a famous real-time public microblogging community where, frequently, the information seems earlier than then on legitimate fourth estate. Characterized via way of means of its brief message limit (now 280 characters) and unfiltered feed, its utilization has fast escalated, specifically amid events, with a median of 500 million tweets published in line with day.

In current era, social media (mainly twitter) can be used to spread hate posts/messages. Hate post/speech refers to a form of communication that makes one or more people who maintain their own sense of group belonging to the nose. It is mostly presented by race, sexual orientation, personality, disability, etc., religion, political affiliation, or opinion [14].

1.1 Twitter

Twitter can be a micro blog site can be used to share their posts in the form of opinions, experiences, and feelings using text messages (tweets). Tweet on behalf of users from various desktop and mobile clients and even other applications. Who posted the tweet [1–4]. Each user has unique identity their name and an identification number. Users can have followers; get the latest information about what users have posted. Other users can view, share, or in other words, repost the content of any posted tweet [1–4].

Data Retrieval

Tweets and related information (such as username, posting date and location etc., are usually retrieved from Twitter through several API. Access via authentication request [1, 2, 4, 10]. If you request access to one or more APIs, you will need to submit credentials issued by Twitter. To receive them, you need to register a request on Twitter [21].

- *Rate Limits*

Streaming API allows to send 5000 user IDs at the same time [11]. The duration of the API rate limit window is 15 min [12]. A user represented by an access token can make 180 requests/requests [11] per time window. Submit 450 enquiries/inquiries on your behalf every quarter.

1.2 Data Cleaning

Cleaning was performed with some iterations of regex syntax to induce obviate re-tweets, handles, and special characters. Duplicate tweets were also removed and lemmatization with a part of speech was also applied to any or all tweets. The last stage involved removing stop words and also words shorter than three characters as they are doing not usually carry very valuable meaning. However, stop words like 'but' and 'not' were kept for neural networks sequencing. We created two additional columns, 'tweet_length' and 'handle_count' to analyze whether these two factors have any impact on positive/negative language [5].

1.3 Text Mining

Intelligent text analysis is a method to retrieve the patterns and their relationships in a collection of unstructured data [9]. Text analysis is usually divided into three stages: pre-processing: delete, export, and mark stop words. Knowledge extraction-use machine learning tools to find hidden dimensions. And relationships that are difficult to detect manually.

Eliminating the ignored words will delete meaningless words, and they often appear in every text, such as B. articles and pronouns. Derivatives help to express features more densely by shortening the roots of curved words. Tokenization removes punctuation marks and breaks down sentences into simple combinations of words and letters. After the mandatory pre-processing is completed, statistical methods are used to analyze the content of the text to determine its attributes [2, 3]. Text data can be processed in different ways and at different levels.

1.4 Exploratory Data Analysis

The EDA section provided some useful insights into the very fabric of the words utilized in these tweets. Word clouds were created to showcase the foremost common 1, 2, and 3 g present within the text. Attaching below a comparison between positive and negative lexicon, larger words correspond to a higher frequency [1–4, 17, 19].

The duration distribution between positive and negative tweets is also analyzed, as shown in Figs. 2 and 3 [19], negative tweets are on average shorter than positive tweets. 1 represents negative tweets and 0 represents positive tweets as represented in Fig. 1. A simple t test confirms that when the *p*-value is less than 001, the difference in means is important.

In previous section, the association between the controls and aggressiveness is obtained by sketching the number of positive/negative results against the total number

Fig. 1 Negative tweets and positive tweets

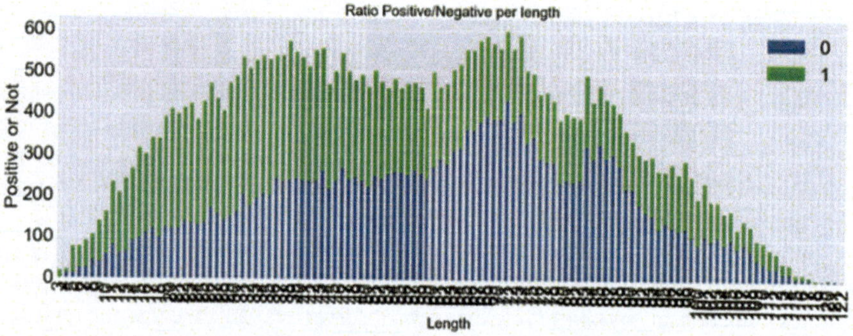

Fig. 2 Graph showing positive and negative tweets per length [19]

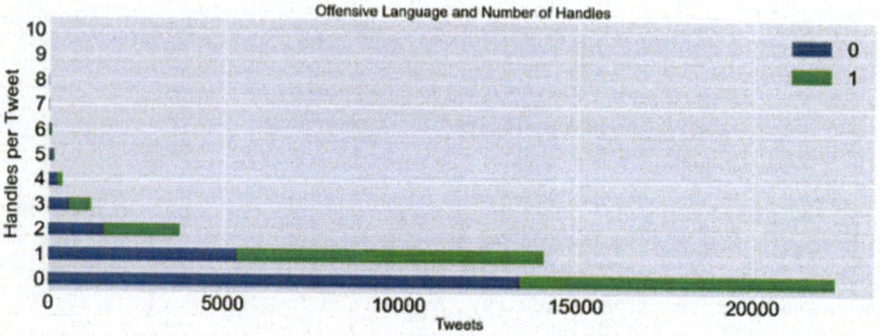

Fig. 3 Graph showing offensive language and number of handles [19]

of controls. Mainly, tweets contain 0–3 controls whereas 0 and 1 handler tweets. One can explain this by the fact that a curator can direct small talk to someone.

1.5 Modeling

The only predictor variable we use for modeling is the text itself, and we use the TFIDF method to vectorize a refined version of the text. TFIDF is more popular than word data packets, because in this case, the rarity of words has a certain meaning [7, 8]. Except for CNN (which only accepts token string input) and Naive Bayes, all models use the TFIDF array. This is where we tried to use the arched structure to see if power is needed. The performance of CNN is not added to the figure below, but the performance is very slow, only exceeding 50% accuracy. The neural network part requires a different job. You can try to implement RNN on an existing CNN at a price lower than the standard cost. As mentioned above, the most effective model is logistic regression with f-1 score. The test suite is 98.2%. Logistic regression is also very useful for describing the meaning of each word in the prediction result [15–17].

1.6 Twitter API and MySQL Storage

New tweets are periodically collected by making the request access to the Twitter portal in the Realm MySQL database, which is only created for the incoming tweet stream [5]. The Python Tweepy library was not used to create the Twitter API link. How to restore SQL database from unformatted JSON stream:

- Twitter ID, Time and Tweet Text.

1.7 Model Evaluation and Dashboard

The basic pipeline was developed by dragging and dropping recently uploaded tweet packets onto a separate laptop and placing them where the exact same pre-cleaning was applied (Fig. 4).

In addition, the final model and the objects selected by tf can be retrieved in fresh_tweets_df.ipynb, and the file can also be found in the 'notebook' section of the repository. 9000), and the dictionary of words is simply discarded, and the rest is left in the matrix. Then the probability prediction function can be used on vectorized matrix and greater than 0.9 are filtered to collect only the tweets that the model finds to be very offensive. As the last step, the new forecast data frame will be uploaded to another repository, which is located. All heroku implementations are sorted here, including creating configuration files and 'tweetresources.txt' files. Over time, tweets that mention the topic will appear in the final application, the most repeated words and the ratio of negative to neutral/positive tweets [21–23].

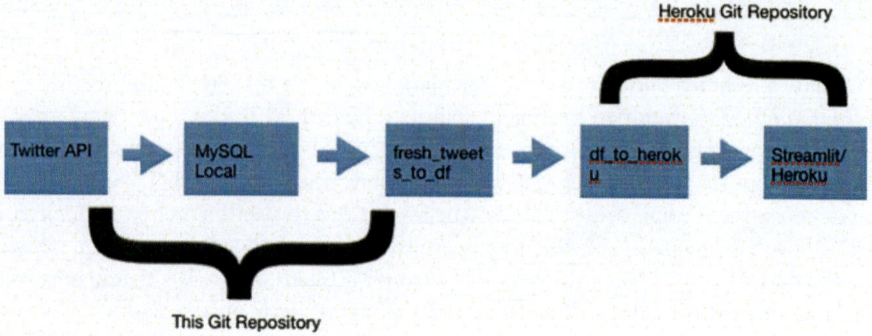

Fig. 4 Twit pipeline [19]

2 Existing Work and Problems Identified

The Internet allows you to easily access and share information at super fast speeds. This set of anonymous functions also makes it an effective means of distributing hateful or offensive content. In addition, many competitions are held to solve similar problems (e.g., publicly insulting, offensive or hateful/objectionable content in German, Spanish, or English).The basic method is to use a simple template. The most common method is a combination of feature extraction and the problem oriented machine learning algorithms [1–5]. The BoW (bag of words) method is used in combination with the naive Bayes classifier. Use BoW with Support Vector Machine (SVM). However, due to the many false positives of BoW where as other researchers have applied more complicate methods to provide the functionality of classic cars. Other deep learning techniques in this field include the use of convolution neural networks (CNN), recurrent neural networks (RNN), and deep learning. Another important step is the introduction of converters, especially the BERT converter, which provides seven. One of the ten most effective models in the subtask. In fact, this method has recently performed best in competitions between more than fifty participating teams (based on the average performance of all subtasks) [11–15].

The limitations of the method used are choices that are often difficult to understand, and it is difficult for people to explain the reasons for making these decisions. This is usually a practical problem, as systems that automatically review personnel voices may require manual complaint procedures. A new method of categorizing hate speech can better understand the selection content and show that it can even outperform existing methods for certain data sets. This may be effective, but these sources need to be maintained and updated so that you know that it may be a problem [23, 24].

3 Proposed Work and Research Method

The data pre-processing stage includes two methods: Bag of words (BoW) and Term Frequency Inverse Document Frequency (TFIDF) [1, 2].

The bag-of-words technique is often a simplified illustration employed in language process and data retrieval. In this method, due to its word pocket(s), text such as sentences or documents is displayed, regardless of grammar and uniform order, while maintaining the plural.

TFIDF can be a number of statistics whose purpose is to reflect the meaning of words in documents in an overwhelming collection. The use of weights considers knowledge search, text mining, and user-defined modeling [1, 2].

Before we archive computer files using various algorithms, we will clean them up because these tweets contain a lot of tenses, grammatical errors, unknown characters, # tags, and Greek characters [2].

Problem can be solved using abbreviations, stop word deletion, and skipping. Tonalization eliminates the inflection of the word ending and restores the word to its basic or lexical form. Keep promises [2, 20–24]

Following methods are used for data.

- Eliminate stop words: they can be articles or prepositions. These words can be deleted without any negative impact on the ending. A feasible model. The words 'es', 'is', etc., are usually used in West Germanic.
- Greek characters or special character deletion.
- Slang Language words deletion: example: 'luv', 'thnx', etc.
- Stemming: Through stemming, a word is abbreviated to the word root, so that the uppercase or beginning of the word is separated. Consider the list of common prefixes and suffixes contained in words with variable angles. Determine how the conditions should be reduced. Porter's algorithm is used in our data set.
- Lemmatization: Lemmatization is a method of categorizing different types of phrases in a word so that they can be analyzed together. Therefore, he connects words that are roughly similar to the words. The word 'run', so 'run' is the motto of these words.

Main classification algorithms like Decision Tree, Logistic Regression, Random Forest (RF) etc., can be used for bag of words and TFIDF.

In order to spot hateful tweets machine learning algorithms works well. 2 use cases were used to implement the system.

1. Positive tweet so as to properly classify it as a positive tweet.
2. Negative tweet so as to properly classify it as an annoying tweet.

After providing several labeled coaching SVMs for every category, we are able to classify the new text. Next, we will contend with the text classification problem. We will refine our training data and even try to fill it. Then, we continue to use SVM, which is a fast and powerful classification algorithm that can handle the limited amount of information that needs to be checked. The main goal of the project is to

Fig. 5 Flowchart showing proposed system

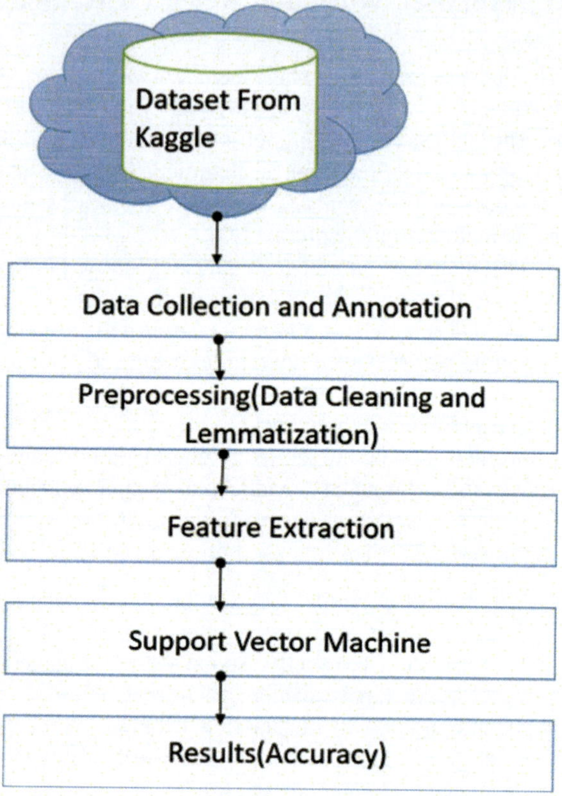

determine the hate speech, which will be a binary classification task. It is to classify that every sample into a non-hate speech or hate speech category. We estimate the likelihood of ranking each post based on the likelihood that each post contains hate speech. Therefore, we will test some of the most popular software and choose the one that works best. We will choose the naive Bayesian method because it can classify spam well and can similarly identify hate speech. You choose SVM and RF because they work well in the most difficult tasks (Fig. 5).

4 Results and Observation

The dataset consists of 20 K posts from Gab and Twitter. Each data point is annotated with one of the hate/offensives/normal labels, target communities mentioned, and snippets (rationales) of the text marked by the annotators who support the label. The proposed work has been analyzed using Kaggle record where a.csv file containing 31,962 tweets. The data set is severely skewed: 93% of Twitter data does not contain hate tags, while 7% contains Twitter data with hate tags.

The functions that cannot be called after data deletion will be called first. Those are:

1. Number of empty words
2. Pound signs used
3. Characters used
4. Number of uppercase letters.

After deleting the text, some features are extracted because they are more meaningful at this time. These are:

1. Word count: This feature indicates how many words are in the revision.
2. Number of characters: How many letters does the revision contain?
3. Average word length: The average number of letters in a word during over scan.

Although the performance of the final model is very good even when testing our data set, one of the biggest limitations of the project is that the performance of the model is measured in a new tweet. From a pragmatic point of view, people can simply view a few tweets that have been marked as negative, and subjectively regard them as unwelcome. In this context, this latest view raises another important question related to the inherent tendency of individuals who manually flag tweets.

The judgment also tried to incorrectly define the original basic facts, because offenses to some people might not be the case to others. The wordless dictionary may be one of the main shortcomings of this model. There are 9 million words in our original vocabulary. When it comes to word processing outside of the vocabulary, a recurrent neural network may be the easiest choice.

5 Conclusion

Cyber bullying and violent speeches on social platforms are some of the shortcomings of our time. Freedom of expression on the Internet can easily be reduced to insulting, unfounded and unconstructive criticism of sex, political, and religious beliefs. Both machine learning and the large amount of data available on social platforms are beneficial. An effective solution to alleviate this problem. We use SVM, which is a very simple but powerful supervised machine learning algorithm that can be used for classification and regression. It is suitable for small and medium data sets and is very easy to set up.

References

1. Pitsilis GK, Ramampiaro H, Langseth H (2018) Effective hate-speech detection in Twitter data using recurrent neural networks. Appl Intell 48:4730–4742. https://doi.org/10.1007/s10489-018-1242-y, https://www.researchgate.net/publication/326517890

2. Gaydhani A et al (2018) Detecting hate speech and offensive language on Twitter using machine learning: an N-gram and TFIDF based approach. In: IEEE international advance computing conference 2018. https://arxiv.org/pdf/1809.08651.pdf
3. Watanabe B, Ohtsuki (2018) Hate speech on Twitter: a pragmatic approach to collect hateful and offensive expressions and perform hate speech detection, 15 February 2018, pp 13825–13835. https://ieeexplore.ieee.org/stamp/stamp.jsp?arnumber=8292838
4. Zimmerman, Kruschwitz (2018) Improving hate speech detection with deep learning ensembles. In: Proceedings of the eleventh international conference on language resources and evaluation (LREC 2018), pp 2546–2553. https://www.aclweb.org/anthology/L18-1404.pdf
5. Pereira-Kohatsu JC (2019) Detecting and monitoring hate speech in Twitter. Published in Special issue sensors for affective computing and sentiment analysis. Sensors 19(21):4654. https://doi.org/10.3390/s19214654
6. Ziqi et al (2018) Hate speech detection using a convolution-LSTM based deep neural network. In: Proceedings of ACM the web conference (WWW'2018). ACM, New York, NY, USA, p 10. Article 4, https://doi.org/10.475/123
7. MacAvaney S et al (2019) Hate speech detection: challenges and solutions. PLOS One 14(8). https://doi.org/10.1371/journal.pone.0221152, https://journals.plos.org/plosone/article?id=10.1371/journal.pone.0221152#sec019
8. Salminen J et al (2018) Anatomy of online hate: developing a taxonomy and machine learning models for identifying and classifying hate in online news media. In: Proceedings of the twelfth international AAAI conference on web and social media (ICWSM 2018). https://ojs.aaai.org/index.php/ICWSM/article/view/15028/14878
9. Lee K, Palsetia D, Narayanan R, Patwary MMA, Agrawal A, Choudhary A (2011) Twitter trending topic classification. In: 2011 IEEE 11th international conference on data mining workshops (ICDMW), December 2011, pp 251–258. https://doi.org/10.1109/ICDMW.2011.171
10. Chih-Chung (2021) A library for support vector machines, 14 April 2021. https://www.csie.ntu.edu.tw/~cjlin/libsvm/
11. Kumar S, Morstatter F, Liu H (2013) Twitter data analytics. Springer, New York, NY, USA
12. Kamal O et al (2021) Hostility detection in Hindi leveraging pre-trained language models. In: CONSTRAINT 2021. https://arxiv.org/pdf/2101.05494v1.pdf
13. Leite JA et al (2020) Toxic language detection in social media for Brazilian Portuguese: new dataset and multilingual analysis. In: IJCNLP 2020. https://arxiv.org/pdf/2010.04543v1.pdf
14. Mollas I et al (2020) ETHOS: an online hate speech detection dataset. In: CIKM2020. https://arxiv.org/pdf/2006.08328v1.pdf
15. Raghad, Hend (2020) A deep learning approach for automatic hate speech detection in the Saudi Twittersphere. In: Proceedings of the fifth Arabic natural language processing workshop, December 2020, pp 12–23. https://www.aclweb.org/anthology/2020.wanlp-1.2.pdf
16. Mathew B et al (2021) HateXplain: a benchmark dataset for explainable hate speech detection. Assoc Adv Artif Intell. www.aaai.org, https://arxiv.org/pdf/2012.10289v1.pdf
17. Twitter Inc. Twitter developers. Accessed on 18 October 2019. 2019 Available online https://developer.twitter.com/en.html
18. Fan L et al (2020) Stigmatization in social media: documenting and analyzing hate speech foCOVID-19 on Twitter. Proc Assoc Inf Sci Technol 57(1). https://doi.org/10.1002/pra2.313. Epub 22 October 2020. https://asistdl.onlinelibrary.wiley.com/doi/10.1002/pra2.313
19. Tortella (2020) How to build a (quasi) real-time hate speech classifier for Twitter, June 2020
20. Vedanshu et al (2020) Live Twitter sentiment analysis. In: Proceedings of the international conference on innovative computing and communications (ICICC), May 2020. Available at SSRN https://ssrn.com/abstract=3609792 or https://doi.org/10.2139/ssrn.3609792
21. Tanwar et al (2016) A tour towards sentiments analysis using data mining. Int J Eng Res Technol 5(12)
22. Tanwar, Rai (2020) A proposed system for opinion mining using machine learning, NLP and classifiers. IAES Int J Artif Intell (IJ-AI) 9(4):726–733. ISSN 2252-8938. https://doi.org/10.11591/ijai.v9.i4.pp726-733

23. Kaustav et al (2020) A facial expression recognition system to predict emotions. In: 2020 international conference on intelligent engineering and management (ICIEM), June 2020
24. Tanwar P, Priyanka (2019) Spam diffusion in social networking media using latent Dirichlet allocation. Int J Innov Technol Explor Eng (IJITEE) 8(12). ISSN 2278-3075
25. Tanwar, Prasad, Dutta (2018) Natural language processing for hybrid knowledge representation. Int J Adv Paradigms 10(3)

Comparative Evaluation of Semantic Similarity Upon Sentential Text of Varied (Generic) Lengths

Richa Dhagat, Arpana Rawal, and Sunita Soni

Abstract Semantic similarity plays a vital role in natural language processing, information retrieval, text mining, question and answering system, text-related research, biomedical informatics, and plagiarism checking. Measuring similarity between sentences means estimating the degree of closeness in meaning between these sentences. Very recently, NLP research has observed much-inclined shift from word-lexicon and syntactic feature-based techniques to structural feature-based techniques of exploring NLP semantics. Rather, no stone is unturned by NLP community working on computation of text semantic similarity to formulate supporting feature spaces, say, word-to-word co-occurrences, lexical corpora and natural language grammar rule-based word order vectors as well as semantically parsed structures. The most popularly used WordNet lexical dictionary was exploited as hierarchical structures (semantic trees) to fetch all combinations of noun and verb families of word phrases, predominantly occupying free texts. In this paper, the computation of text semantic similarity is addressed by devising a novel set of generic similarity metrics based on both, word-sense of the phrases constituting the text as well as the grammatical layout and sequencing of these word phrases forming text with sensible meaning. The experiments performed on benchmark datasets created by Li and group explore semantically parsed dependency structures in feature extraction step and achieve very promising values of semantic closeness within sentence pairs. The model's performance is indicated achieving highest value of Pearson's correlation coefficient (0.89) with mean-human similarity scores against those obtained through closely competent structured approach models. However, these very promising results are contributed by word-sense similarity component of sentential similarity measure. A future scope of work is also incepted that shall also improve upon dependency (grammatical relations) component to raise the text similarity performance metric.

Keywords Natural language processing · Semantic similarity · Structured approach · Similarity matrix · Lexical dictionary

R. Dhagat (✉) · A. Rawal · S. Soni
Department of Computer Science and Engineering, Bhilai Institute of Technology, Durg, India

P. K. Singh et al. (eds.), *Emerging Technologies for Computing, Communication and Smart Cities*, Lecture Notes in Electrical Engineering 875,
https://doi.org/10.1007/978-981-19-0284-0_9

1 Introduction

Advances in NLP, preferably combining the field of computational linguistics, along with a large amount of text-based resources nowadays are publicly available in digital formats (e.g., online encyclopedias). These repositories allow newer NLP prototype or commercial tools to become more reliable and robust in accomplishing many natural language applications such as semantic search, summarization, question answering, document classification, sentiment analysis-cum-classification, plagiarism detection tasks and much more while performing machine learning tasks on Web and image datasets. All these are possible only if syntax-semantic-based intelligent techniques are employed to capture the meaning, concept and idea revealed within the text content. Diversified school of thoughts have been triggering various approaches and algorithms for over nearly two decades to find robust methods to arrive at precisely effective semantic similarity measures, and the work is been pioneered in the early decade of 2000 by Erkan and Radev [1], Li et al. [2], Mihalcea [3], Miranda [4]. Their works invariably differed in using feature spaces for learning the textual pattern similarities and machine learning approaches. In addition, they differed in using computational expressions for measuring text similarities that were used for varied learning (classification labeling or grouping) tasks with their own decided thresholds settings.

A major breakthrough in recently pursued research works on text similarity was unanimously agreeable opinion of representing the whole text document as sentences posing to reflect the actual meaning of the text in itself or augmented by adjacently lying sentences. By analyzing such works, we arrived at some of the major findings outlined as follows:

- Computations on word vectors, in any combinations, whether POS-tagged or not, were very computationally expensive;
- The conventional techniques were highly domain or corpus dependent, while a lot effort was being attempted to arrive at generic approach of semantic similarity computation;
- There was a need to adopt a robust method that maps the correctness of word order and validates the word-sense borne by word (phrases) in and around the sentences reflecting a theme or topic or idea or context narrated through.

All the above-mentioned observations were welcomed and drilled into by NLP research community as they could see sentential similarity approach of text mining, easily invading into sufficiently higher information retrieval effectiveness in application realms like web page retrieval, image caption retrieval from web, web page ranking and document summarization from the topic-relevant web pages at hand.

The paper is outlined into sections as follows: Sect. 2 presents the related research carried out using structural features of free text, albeit, various research groups explored these structural features with varied research objectives pertaining to different realms. Section 3 outlines the refinements in the methodology undertaken to compute word-sense similarity and grammatical similarity by taking a sample instance of sentence pair. Section 4 tabulates the experimental computations for a finite set of sentence pairs upon which results are already available due to approaches already dealt in the past. The concluding remarks on the result comparisons are presented in Sect. 5.

2 Semantic Similarity Computations: A Survey

2.1 Structured Approaches: State-of-the-Art

As rightly observed by Alzahrani and group [5, 6] that extraction of lexical, syntactic and semantic features from the experimental text is governed by some or the other pre-defined list of lexical dictionaries (databases) like some commonly available ones from the English language corpora are WordNet [7], Brown Corpus, Wikipedia Rewrite Corpus [8], METER corpus [9] and Webis Crowd Paraphrase Corpus [10]. In reality, these external knowledge resources may not account for capturing the thematic sense reflected in that text. Hence, they attempted to explore the structural features in two different perspectives, namely block-specific and content-specific textual features. There have been several works that have attempted to compute inter-document similarities by representing them into hierarchical levels of paragraphs, sentences and word phrases. It was observed by NLP research community that much more promising results were obtained when they performed text similarity experiments upon these structural feature space representation types.

The work done by Li et al. [2] claims to have devised hybrid approach to text semantic computation method based on semantic similarity (using cosine metric) over the information content associated with each participating sentence of the sentence pair. However, their work leaves an element of doubt unresolved, as initially they formulate the logic stating that their method does not have any domain-dependency issues, while, their experiments did make use of pre-compiled dictionaries, namely, WordNet for computing semantic similarity and Brown Corpus for calculating information content values.

Meanwhile, Islam and Inkpen [11] attempted to formulate a variant text semantic similarity model by incorporating string, word co-occurrence and optimal word order matching steps, however, augmented by enormous time consumption in constructing intermediary joint semantic word similarity matrices. Lee's [12] contribution differs from the above piece of work in the sense that the former emphasizes on corpus dependent word-to-word association weight (information content), while the latter supports the usage of distance-based (wup) similarity metric to arrive at overall semantic similarity of sentence pairs. Also, Croft et al. [13] replicated the work done by Li et al. by simply using cosine vector similarity over formulated weighted word vectors, where weights on these vectors were computed as an outcome of WordNet-based path similarity measure. The candidate texts were only caption and header titles and were not valid grammatically constructed sentences. Taieb et al. [14] proposed measures that aggregate noun and verb fragments together along with common word order parameter to obtain overall sentence similarity metric. The work extends the concept of other related works by also including matching of modality (in form of verb tenses) between candidate pair of sentences along with compound nouns. Although Pawar and Mago [15] proposed almost similar approach as Li et al. that considers both semantic and syntactic structures for text similarity using lexical knowledge base, their work distinguishes with others in the sense that they have used totally novel similarity metric. Vani and Deepa [16] have borrowed Li and Lee's contributions in hybridizing the text semantic similarity metrics with a variation in formulating logic for second contributive metric (sim^2) in computing over all sentence similarity. Here, the word-to-word association weight is replaced by proportionate count of common (matched) words (or allied synsets) found in the sentence pairs. Such related works motivated our research group to investigate further using multiple combinations of both syntactic and semantic similarity metrics and pursue comparative studies, if any.

2.2 *Structure-Based Approach Revisited*

Ozates et al. [17] claim to pioneer the exploring of a novel structure-based approach based on grammatical relations among all combinations of participating words in the sentence pairs. However, their work restricted to usage of only those grammatical (dependency) relations contributing to predominantly noun and verb family of word phrases and ignoring unimportant dependency relations, for instance, prep (prepositional), punct (punctuation), det (determiner) and possessive and many more, as stated by them. They emphasized on the merits of using dependency tree structures as they reflect the neighboring (may be adjacent or non-adjacent) word-senses within individual sentential units better than word vector-based or syntactic-based feature representation of texts. They used dependency grammatical structures in bigram structured formats as textual feature spaces in order to compute sentence semantic similarity. Each bigram structure is comprised of a dependent word, a head word and a typed-dependency tag expressing the type of relationship between them. The consequent semantic similarity measure was expressed in terms of Kernel expression formulations, namely simple approximate bigram kernel (SABK), TF-IDF-based approximate bigram kernel (TABK) and its variants. The sentence-similarity metric assigns weights to the grammatical relation tags using degrees of grammatical relation tag overlap. Also, weights to word phrases are assigned by computing *tf-idf* measures upon head as well as dependent arguments.

Wali et al. [18] also claims to develop a hybrid approach of computing sentence similarity upon dependency-parsed structural features extracted from sentence pairs. They name it as syntactico-semantic similarity based on thematic roles and semantic classes of distinct as well as common words and their synonymous equivalents, been matched using Jaccard coefficient metric.

Under this paradigm, Vakare et al. [19] have devised a novel metric to compute sentence similarity using dependency parsing. Their work attempts to learn grammatical tag relations by training the similarity scores on pre-defined datasets. This leaves a query as to what will be the learnt weights, if at all, the background corpus gets changed and indicates an element of uncertainty aroused due to domain-dependency nature of the undertaken problem objective.

The authors appreciate much closer work done very recently by Zhang et al. [20] who have proposed a sentence similarity computation model which uses a hybrid approach combining both syntactic-cum-semantic similarity between sentence pairs using grammatical dependency relations obtained from an appropriately selected semantic NL dependency parser. The sentential semantic similarity metric is named as kernel function that itself is expressed in terms of another (corpus-based) semantic 'wup' metric. However, the filtering out step of some of the dependency relations: {*'det', 'expl', 'goes with', 'possessive', 'preconj', 'prep', 'punct', 'ref'*} as unimportant ones need further justifications.

Farouk [21] distinguishes its work from Zhang's by formulating sentence graphs as structural features obtained from grammatical relations of parsed dependencies. Consequently, the semantic similarity component of overall text sentence similarity is computed through these semantic relation matrices.

After undertaking an exhaustive survey upon the above-mentioned similarity metrics used in both unstructured and structured approaches of semantic similarity computation, it was found that NLP researchers, who explored inter-sentential grammatical relations for feature space representations of sentential datasets, obtained competently better results in semantic similarity computations than those who did not use these structures.

3 Methodology

In this paper, we proposed a refined similarity score between two input sentential texts with a generic perspective to use this similarity for any machine learning task in NLP domain. In order to keep the configuration settings to the problem objective simple, we continue to borrow thematic (word) sense from the most popularly used WordNet (lexical database of semantic relations between words) dictionary. The methodology adopted in this paper can be considered as a refinement to the approach followed by Lee [12]. The current work chooses Li's benchmark datasets [22] following the same protocol of NLP scientists who have been working in this realm. It may be noted that the previous works had computed the noun vector by comparing each noun of the candidate sentence with the union of noun word-sets from both the sentences of sentence pair. The same applies for verb vector computation, making the steps more cumbersome and confusing. We have straight forwardly compared noun word-sets of candidate sentences with one another. Similar approach is repeated for verb word-sets of the candidate sentences.

3.1 Semantic Similarity Computation (Phase I)

We investigate the importance of the proposed similarity by a simple example. Let us consider a pair of sentences. Let S_A = '*A cemetery is a place where dead people's bodies or their ashes are buried.*' S_B ='*A graveyard is an area of land, sometimes near a church, where dead people are buried.*'

Step 1: The words in each sentence after part-of-speech tagging are categorized into noun and verb sets. (N_A, N_B), (V_A, V_B) are noun and verb sets of sentence *A* and sentence *B*, respectively. In this example, $N_A =$ ['cemetery', 'place', 'people', 'bodies', 'ashes'] and $N_B =$ ['graveyard', 'area', 'land', 'church', 'people']. $V_A =$ ['buried'] $V_B =$ ['buried']. '*m*' and '*n*' are count of noun sets in vector N_A and N_B. '*p*' and '*q*' are count of verb sets in vector V_A and V_B.

Step 2: For noun pair of each candidate sentence, (n_A, n_B) in the noun sets $\forall i, j : n_{iA} N_A, n_{jB} N_B$, we construct a semantic similarity matrix of dimensionality size, ($m \times n$) such that if $n_{iA} == n_{jB}$, path_similarity = 1 else extract synset lists for n_{iA} and n_{jB} from WordNet lexical corpus. Similar steps of computation ply for all verb pair (v_A, v_B) in the verb sets $\forall i, j : v_{iA} V_A, v_{jB} V_B$ belonging to sentences *A* and *B*. In this way, two synset lists are formed N_{A_syn} = synsets (n_{iA}, POS(n_{iA})), N_{B_syn} = synsets (n_{jB}, POS(n_{jB})); hence, there shall be four synset lists for each candidate word pair denoted by: N_{A_syn}, N_{B_syn}, V_{A_syn} and V_{B_syn}.

Step 3: Semantic similarity matrix of dimensionality size: ($p \times q$) between V_A and V_B is computed similar to semantic similarity matrix obtained between N_A and N_B of dimensionality size: ($m \times n$). For example, pair of sentences stated that the semantic similarity matrix '*S*' is denoted as follows (Fig. 1).

Step 4: This step details out the sequence of calculations that are needed to arrive at each of the synset-similarity values S_sim that form the semantic similarity matrix of the two sentences as a whole. This value is calculated as the maximum of the path similarity among all synset pairs of both, participating noun and verb word-sets. For instance, let synset lists of the word pair $\{n_{iA}$ = 'church', n_{iB} = 'ashes'$\}$ be *[Synset('church.n.01'), Synset('church.n.02'), Synset('church.n.04'), Synset('church_service.n.01')]and [Synset ('ash.n.01') Synset('ash.n.02'), Synset('ash.n.03')]*, respectively. The synset pair [*Synset ('church.n.04'), Synset ('ash.n.01'*)] provides the maximum path similarity (S_sim) out of all synset pairs as '0.1'. Similarly, the S_sim value for word pair ('area', 'place') is '0.5' (Fig. 2).

$$
S = \begin{matrix} \textit{graveyard} \\ \textit{area} \\ \textit{land} \\ \textit{church} \\ \textit{people} \end{matrix}
\begin{pmatrix}
\textit{cemetery} & \textit{place} & \textit{people} & \textit{bodies} & \textit{ashes} \\
1 & .2 & .083 & .1 & .09 \\
.166 & .5 & .125 & .33 & .125 \\
.166 & .33 & .5 & .2 & .166 \\
.076 & .142 & .2 & .5 & .1 \\
.083 & .166 & 1 & .25 & .125
\end{pmatrix}
$$

Fig. 1 5 × 5 Semantic similarity matrix (for noun vectors)

$$
S = \begin{array}{r|ccccc|c}
 & cemetery & place & people & bodies & ashes & SvecN_B \\
graveyard & 1 & .2 & .083 & .1 & .09 & 1 \\
area & .166 & .5 & .125 & .33 & .125 & .5 \\
land & .166 & .33 & .5 & .2 & .166 & .5 \\
church & .076 & .142 & .2 & .5 & .1 & .5 \\
people & .083 & .166 & 1 & .25 & .125 & 1 \\
SvecN_A & 1 & .5 & 1 & .5 & .166 &
\end{array}
$$

Fig. 2 Semantic noun vectors from noun vector semantic similarity matrix

Step 5: Finally, semantic vectors SvecN_A, SvecN_B are computed from semantic similarity matrices for both noun and verb sets of the two sentences. The vectors seek the maximum of the synset_similarity values in order of row and column dimensions to obtain the two vectors; in our example, SvecN_A = [1.0, 0.5, 1, 0.5, 0.166] and SvecN_B = [1.0, 0.5, 0.5, 0.5, 1], SvecV_A = [1] and SvecV_B = [1]; the calculations of example pair of sentences are shown in the above figure.

Step 6: As a result, semantic noun and verb portions of similarity scores can be expressed as

$$
\text{Sem\_}N_{A,B} = \frac{\sum_{i=1}^{m} \text{Svec}N_{iA}}{m+n} + \frac{\sum_{j=1}^{n} \text{Svec}N_{jB}}{m+n} \tag{1}
$$

where m and n are length of semantic vectors SvecN_A and SvecN_B, respectively.

$$
\text{Sem\_}V_{A,B} = \frac{\sum_{i=1}^{p} \text{Svec}V_{iA}}{p+q} + \frac{\sum_{j=1}^{q} \text{Svec}V_{jB}}{p+q} \tag{2}
$$

where p and q are length of semantic vectors SvecV_A and SvecV_B, respectively. In our case, semantic noun score, $\text{Sem\_}N_{A,B} = \frac{1+0.5+1+0.5+.0166}{5+5} + \frac{1+0.5+0.5+0.5+1}{5+5} = 0.66$ and semantic verb score, $\text{Sem\_}V_{A,B} = \frac{1}{2} + \frac{1}{2} = 1$.

Step 7: We retain the same expression (as related works discussed above) to compute overall word-sense semantic similarity between sentence pairs $(S_A, S_B) = \zeta \times \text{Sem\_}N_{A,B} + 1 - \zeta \times \text{Sem\_}V_{A,B}$; the reason behind is to compare the sentential similarity scores with the values obtained in the previous works with similar kind of experimental setup. The authors did not drill into the insights of 'ζ' parameter (also called exponential balance coefficient (EBC)) which is borrowed from the previous works and is usually set in the range [0.5, 0.1], and for our experiments, value of EBC is set to 0.65. In our case, the overall word-sense similarity $(S_A, S_B) = 0.65 \times 0.666 + 0.35 \times 1 = 0.78$.

3.2 Semantic Similarity Computation: Phase II

The concept of grammatical dependencies pioneered by Stanford research group led by manning describes grammatical relations augmented with word arguments lying in 2-tuple format. These binary arguments hold a governor argument (also known as a regent or a head) and a dependent argument (also known as tail). It may sometimes happen that two or more adjacently or non-adjacently lying words with in a sentence may jointly reflect a different meaning in thematic sense. Recently, lots of work is being done in carrying out text mining tasks using variant versions of semantic (dependency) parsers. One such novel method of obtaining this portion of semantic similarity is being discussed in the algorithm below.

Step 1: For all candidate sentence pair, (S_A, S_B) the sentences S_A, S_B are parsed into a set of grammatical (binary) relations denoted by $S_A = \{T_A^1, T_A^2, \ldots, T_A^m\}$, $S_B = \{T_B^1, T_B^2, \ldots, T_B^n\}$. T^i denotes the ith relation, where $1 \leq i \leq m$, $1 \leq j \leq n$, m and n denotes the number of participating relations in sentence A and B, respectively. $T_A^i = \{h_A^i, t_A^i, d_A^i\}$ and $T_B^i = \{h_B^i, t_B^i, d_B^i\}$, where h_A^i is the head, d_A^i refers to dependent, t_A^i denotes the participating grammatical tag.

Step 2: The grammatical portion of semantic similarity between two relations can be defined as

$$\text{sim}\left(T_A^i, T_B^j\right) = \left\{\alpha \times \text{arg\_sim}\left(h_{A,}^i h_B^j\right) + \beta \times \text{arg\_sim}\left(d_{A,}^i d_B^j\right)\right\} \times \text{tag\_sim}\left(t_A^i, t_B^j\right) \quad (3)$$

Here, 'arg_sim' denotes WordNet-based (path) similarity between corresponding head and dependent nodes, respectively. We have borrowed the expression (3) from the work by Zhang et al. [20] but with a disagreement toward assignment of parameters α and β. According to any language scholar, as none of the words can be ignored for reflecting semantics of a sentence, we provide equal importance to the sets of head and dependent arguments, thus providing equal weights (= 0.5) to both head and dependent arguments. Tag_sim is easily computed over corresponding grammatical relation tags such that $\text{tag\_sim}\left(t_A^i, t_B^j\right) = 1$ if $t_A^i == t_B^j$ else 0.

Step 3: The final semantic similarity computation is done at sentential level. Here, the greatest value of inter-sentential grammatical relation similarities is averaged over the count of participating grammatical relations as shown in Eq. (4).

$$\text{Grammatical similarity} = \frac{\sum_{i=1}^{m} \max\limits_{1 \leq j \leq n} \left\{\text{sim}\left(T_{A,}^i T_B^j\right)\right\}}{m+n} + \frac{\sum_{j=1}^{n} \max\limits_{1 \leq i \leq m} \left\{\text{sim}\left(T_{A,}^i T_B^j\right)\right\}}{m+n} \quad (4)$$

4 Experiments and Results

We have evaluated our methodology using preliminary baseline dataset comprising nine sentence pairs, borrowed by Lee [12]. The semantic similarity values obtained due to word-sense and grammatical-sense were tabulated in separate columns (3 and 4), some of which are shown in Table 1. The computed similarity measures now can be compared with the equivalent results obtained by Lee [12] and Alzahrani et al. [6]. As seen in the experimental tabulations of Table 1, for the sentence pairs that were found meaningfully highly similar, their high word-sense similarity scores justified their most similar semantics; however, their less grammatical-sense similarity scores indicated that while the first candidate of the pair was written in a clausal form, the second candidate was completely scribed in phrasal form. Such observations were made for sentence pairs: $\{1, 7\}$. Some other sentential pairs exhibited total dissimilarity and were assigned very low word-sense similarity scores, for instance, sentential pair numbers: $\{2, 4, 5, 8\}$. According to human judgments, these sentences narrated totally different domain of context. There was still another category of sentence pairs that were assigned equally high word-sense similarity scores and grammatical-sense similarity scores; such sentential pairs narrated on similar topic or context and also were found written in similar grammatical styles. For instance, in sentential pair ID: $\{3, 6, 9\}$, both the sentences were found scribed in clausal form of sentence construction.

The only way to authenticate the performance of the algorithm used in this paper was to experiment our methodology upon Li et al. [2] datasets, given that Islam and Inkpen [11], Croft et al. [13], Taieb et al. [14], Pawar and Mago [15], Zhang et al. [20] and Farouk [21] too had compared their methodologies upon the same dataset, setting it as benchmark standard for the aspiring semantic NLP researchers. It was observed that all these works followed WordNet-driven knowledge-based approach except by Islam and group. All of them used varied part-of-speech combinations of word vectors and/or word-embeddings as feature vector variants except that latter two work groups which used sentential parsed dependencies as structural features to explore sentential semantics.

Even between the last two closely competent approaches, Farouk [21] was able to measure the similarity of participating lexicons reflecting semantic inter-relations using parsed dependencies from C & C parser as against Stanford dependency parser used by Zhang et al. [20]. The overall similarity measures obtained from all the above-mentioned participating models were tabulated as shown in Table 2. These variations could only be reflected as minor deviations in the form of Pearson's correlation coefficient values as the only feasible performance metric to compare our model.

Their result comparisons were done by finding Pearson's correlation coefficient between both (word-sense similarity and mean-human similarity) as well as (grammatical-sense similarity and mean-human similarity) measures. It was found that our methodology yielded better semantic similarity scores than those obtained by

Table 1 Semantic similarity computations of sentence pairs (courtesy: Lee's, 2011 datasets)

Sentence A	Sentence B	Word-sense similarity	Grammatical-sense similarity	Lee (2011)	Alzahrani (2015)
If she can be more considerate to others, she will be more popular	She is not considerate enough to be more popular to others	0.65	0.46	0.9125	0.75
If she can be more considerate to others, she will be more popular	You are not supposed to touch any of the art works in this exhibition	0.08	0.05	0.0195	0
I will not give you a second chance unless you promise to be careful this time	If you could promise to be careful, I would consider to give you a second chance	0.68	0.69	0.938	0.933
I will not give you a second chance unless you promise to be careful this time	The obscurity of the language means that few people are able to understand the new legislation	0.22	0.18	0.419	0.357
About 100 officers in riot gear were needed to break up the fight	The army entered in the forest to stop the fight with weapon	0.47	0.37	0.695	0.877
Your digestive system is the organs in your body that digest the food you eat	Stomach is one of organs in human body to digest the food you eat	0.90	0.62	0.918	0.777
I do not think it is a clever idea to use an illegal means to get what you want	It is an illegal way to get what you want, you should stop and think carefully	0.72	0.544	0.918	0.777
The powerful authority is partial to the members in the same party with it	Political person sometimes abuses their authority that it is unfair to the citizen	0.38	0.17	0.591	0.718

(continued)

Table 1 (continued)

Sentence A	Sentence B	Word-sense similarity	Grammatical-sense similarity	Lee (2011)	Alzahrani (2015)
The fire department is an organization which has the job of putting out fires	An organization which has the job of putting out fires is the fire department	1	0.92	1	1

all six competent models. This was evident from the very promising value of correlation score as 0.891 (for word-sense similarity) as compared to similarity comparisons put forth in Zhang's [20] and Farouk's [21] work. These correlations can be seen tabulated in Table 3. The results obtained from the proposed methodology can be illustrated in scatter plots shown in Fig. 3.

5 Conclusion

The salient promising feature of the current work that can be drawn at the end is that the experiments are not performed by setting any kind of constraints on the input (free) text nor narrowing the semantic feature spaces by removal of function or stop words or filtering out certain specific grammatical relations representing the context of the topic narrated in the sentences.

In other words, there is a fair attempt to find a generic method to compare texts in the direction of NLP semantics. Unlike past pieces of works, the authors are still on the move to find a suitable expression to compute over all sentential semantic similarity contributed from the perspective of word-sense and grammatical-sense in totality. Moreover, the less correlation value (0.677) of grammatical-sense similarity measures for highly similar sentence pairs in experimental datasets needs further investigations that is being undertaken as the next scope of research in this direction.

Table 2 Varied sentential semantic similarities for performance comparisons

R&G number	Mean-human similarity	Li 2006	Islam 2008	LSS 2013	Pawar et al. 2018	Zhang et al 2020	Farouk 2020	Proposed similarity (Word-sense)	Proposed similarity (Grammatical)
1	0.01	0.33	0.06	0.18	0.02	0.04	0.11	0.12	0.13
5	0.01	0.29	0.11	0.20	0.07	0.07	0.13	0.11	0.27
9	0.01	0.21	0.07	0.28	0.01	0.03	0.07	0.08	0.13
13	0.11	0.53	0.16	0.17	0.29	0.07	0.20	0.25	0.28
17	0.13	0.36	0.26	0.32	0.36	0.14	0.28	0.33	0.37
21	0.04	0.51	0.16	0.32	0.23	0.10	0.21	0.33	0.31
25	0.07	0.55	0.33	0.22	0.28	0.13	0.27	0.39	0.33
29	0.01	0.33	0.12	0.22	0.13	0.07	0.21	0.13	0.33
33	0.15	0.59	0.29	0.32	0.76	0.08	0.33	0.35	0.42
37	0.13	0.44	0.20	0.28	0.1	0.09	0.24	0.39	0.30
41	0.28	0.43	0.09	0.32	0.05	0.11	0.23	0.30	0.21
47	0.35	0.72	0.30	0.20	0.16	0.46	0.30	0.25	0.23
48	0.36	0.65	0.34	1	0.54	0.42	0.35	0.28	0.44
49	0.29	0.74	0.15	1	0.30	0.39	0.34	0.30	0.36
50	0.47	0.68	0.49	0.8	0.25	0.49	0.28	0.36	0.24
51	0.14	0.65	0.28	0.8	0.3	0.1	0.21	0.30	0.26
52	0.49	0.49	0.32	1	0.84	0.31	0.40	0.46	0.29
53	0.48	0.39	0.44	0.47	0.89	0.40	0.40	0.52	0.47
54	0.36	0.52	0.41	0.8	0.78	0.05	0.26	0.30	0.09
55	0.41	0.55	0.19	0.8	0.31	0.07	0.32	0.54	0.29

(continued)

Table 2 (continued)

R&G number	Mean-human similarity	Li 2006	Islam 2008	LSS 2013	Pawar et al. 2018	Zhang et al 2020	Farouk 2020	Proposed similarity (Word-sense)	Proposed similarity (Grammatical)
56	0.59	0.76	0.47	0.8	0.98	0.38	0.49	0.61	0.37
57	0.63	0.70	0.26	1	0.48	0.37	0.32	0.49	0.35
58	0.59	0.75	0.51	0.8	0.89	0.56	0.44	0.48	0.50
59	0.86	1	0.94	1	0.86	0.86	0.87	1	0.87
60	0.58	0.66	0.60	0.8	0.90	0.43	0.51	0.52	0.40
61	0.52	0.66	0.29	0.8	0.93	0.37	0.30	0.48	0.34
62	0.77	0.73	0.51	1	1	0.52	0.51	0.78	0.55
63	0.56	0.64	0.52	1	0.7	0.45	0.44	0.50	0.38
64	0.96	1	0.93	1	0.87	0.93	0.87	1	1
65	0.65	0.83	0.65	1	0.85	0.36	0.55	0.71	0.33

Table 3 Pearson's correlation comparisons to human judgment scores

Algorithm	Pearson's correlation coefficient
Word-sense similarity (proposed work)	0.891
Grammatical similarity (proposed work)	0.677
Sentence similarity (Farouk M.)	0.881
Overall sentence similarity (Zhang et al.)	0.877
Sentence similarity (Pawar et al.)	0.781
LSS (Croft et al.)	0.807
STS (Islam et al.)	0.853
STASIS (Li et al.)	0.816

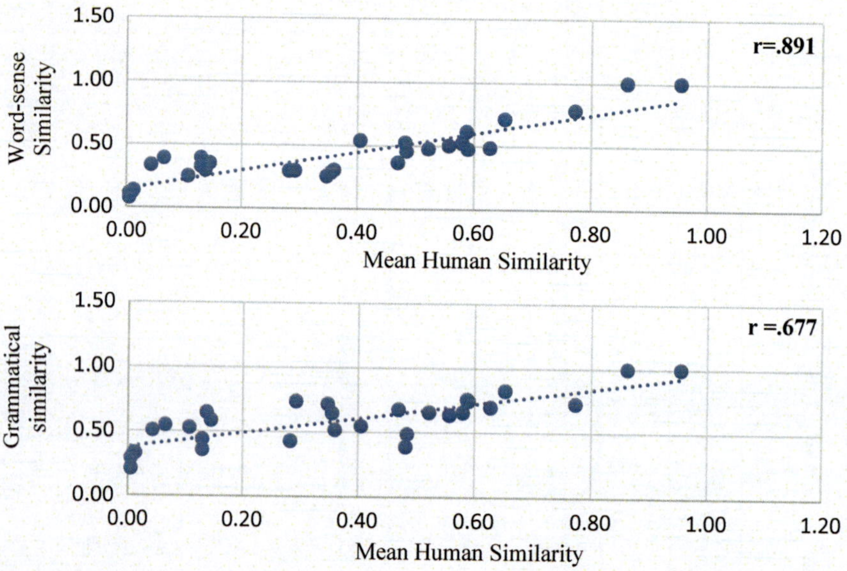

Fig. 3 Pearson's correlation (*r*-value) comparisons with mean-human similarity scores (Datasets: Li et al. 2006)

Acknowledgements The paper's findings are the part of an ongoing research being carried out at Research and Development cell of Bhilai Institute of Technology, Durg, Chhattisgarh. The authors express gratitude to the sponsoring organization for providing an opportunity and infrastructure for pursuing the experiments.

References

1. Erkan G, Radev R (2004) Lexrank: graph-based lexical centrality as salience in text summarization. Artif Intell Res 22:457–479
2. Li Y, McLean D, Bandar ZA, O'Shea JD, Crockett K (2006) Sentence similarity based on semantic nets and corpus statistics. IEEE Trans Knowl Data Eng 18(8):1138–1150
3. Mihalcea R, Corley C, Strapparava C (2006) Corpus-based and knowledge based measures of text semantic similarity. In: Proceedings of the 21st national conference on artificial intelligence. AAAI, Boston, Massachusetts, pp 775–780
4. Miranda Chong MY (2013) A study on plagiarism detection and plagiarism direction identification using natural language processing techniques. Doctoral dissertation, University of Wolverhampton, UK, pp 1–326. http://rgcl.wlv.ac.uk/papers/chong-thesis.pdf
5. Alzahrani S, Valade P, Salim N (2012) Using structural information and citation evidence to detect significant plagiarism cases in scientific publications. J Assoc Inform Sci Technol 63(2):286–312
6. Alzahrani S, Valade P, Salim N (2015) Uncovering highly obfuscated plagiarism cases using fuzzy semantic-based similarity model. J King Saud Univ Comput Inform Sci 27(3):248–268
7. Fellbaum C (ed) (1998) WordNet: an electronic lexical database. MIT Press, Cambridge, MA
8. Clough P, Stevenson M (2011) Developing a corpus of plagiarized short answers. Lang Resour Eval. Special Issue Plagiarism and Authorship Analysis 45(1):5–24
9. Gaizauskas, Foster J, Wilks Y (2001) The METER corpus: a corpus for analyzing journalistic text reuse. In: Proceedings of the corpus linguistics. Lancaster, UK, pp 214–223
10. Burrows S, Potthast M, Stein B, Eiselt A (2013) Webis crowd paraphrase corpus 2011 (Webis-CPC-11). Zenodo
11. Islam A, Inkpen D (2008) Semantic text similarity using corpus based word similarity and string similarity. ACM Trans Knowl Discovery Data (TKDD) 2(2):1–25
12. Lee MC (2011) A novel sentence similarity measure for semantic based expert systems. Exp Syst Appl 38(5):6392–6399. https://doi.org/10.1016/j.eswa.2010.10.043
13. Croft D, Coupland S, Shell J, Brown S (2013) A fast and efficient semantic short text similarity metric. In: 13th international proceedings on UK workshop on computational intelligence. IEEE, USA, pp 221–227
14. Hadj Taieb MA, Ben Aouicha M, Bourouis Y (2015) FM3S: features-based measure of sentences semantic similarity. In: 10th international proceedings on hybrid artificial intelligent systems. Springer International Publishing, Bilbao, Spain, pp 515–529
15. Pawar A, Mago V (2018) Calculating the Similarity between words and sentences using a lexical database and corpus statistics. IEEE Trans Knowl Data Eng 1–14
16. Vani K, Gupta D (2017) Detection of idea plagiarism using syntax—semantic concept extractions with genetic algorithm. Exp Syst Appl 73:11–26
17. Özates S, Ozgur A, Radev D (2016) Sentence similarity based on dependency tree kernels for multi-document summarization. In: 10th international proceedings on language resources and evaluation. European Language Resources Association (ELRA), Portorož, Slovenia, pp 2833–2838
18. Wali W, Gargouri B, Ben Hamadou A (2017) Sentence similarity computation based on WordNet and VerbNet Comput y Sistemas 21(4):627–635
19. Vakare T, Verma K (2019) Sentence semantic similarity using dependency parsing. In: 10th international proceedings on computing, communication, and networking technologies. IEEE, Kanpur, India, pp 2089–2092
20. Zhang P, Huang X, Zhang L (2020) Information mining and similarity computation for semi-/un-structured sentences from the social data. IEEE Internet Things J. 2352–8648.https://doi.org/10.1016/j.dcan.2020.08.001

21. Farouk M (2020) Measuring text similarity based on structure and word embedding. Cogn Syst Res 63(11):1–10
22. O'Shea J, Bandar Z, Crockett K, McLean D (2008) Pilot short text semantic similarity benchmark data set: full listing and description. http://www2.docm.mmu.ac.uk/STAFF/J.Oshea/TRMMUCCA20081_5.pdf

The Role of Artificial Intelligence and Data Science Against COVID-19

Saumya Tripathi and Dilip Kumar Sharma

Abstract Today, a massive outbreak of a deadly virus has challenged humanity in each aspect; we deemed ourselves powerful. Millions of people are at risk, and in some cases, even if the person has reported negative, it is relapsing. Globally, data scientists are also working to tackle this issue by diagnosing, predicting it with the help of machines. Artificial intelligence (AI) serves as a potential tool in fighting against the COVID-19 pandemic. Science the onset of a pandemic, the use of AI has exponentially increased to detect patients suffering or those who are critical. In this paper, the authors discussed the role of AI and the selective review of constraints associated with the functioning of AI. We will also discuss the shortcomings due to the excessive data available between the private and public organizations related to the health sector and their rigorous use without taking note of the data validity.

Keywords COVID-19 · Coronavirus · Data science · Artificial intelligence

1 Introduction

The virus named SARS-CoV-2 has been spreading terror in the hearts of people due to its deadly nature. It was first identified in Wuhan, China, as pneumonia of unknown cause on December 8, 2019 [1]. The spread of this disease has been so fast and so wide that the world is struggling to stabilize itself. As of April 25, 2020, there has been 2,802,118 cases of coronavirus and 195,443 deaths globally. The nations are fighting the coronavirus as best they can by expanding their healthcare facilities, stopping people from collecting at a particular place and in that they have also started utilizing the power of data science, artificial intelligence (AI) and machine learning. Since December 2019, a new type of coronavirus called novel coronavirus (2019-nCoV, or COVID-19) was identified in Wuhan, China. The COVID-19 has then

S. Tripathi (✉) · D. K. Sharma
GLA University, Mathura, India

D. K. Sharma
e-mail: dilip.sharma@gla.ac.in

P. K. Singh et al. (eds.), *Emerging Technologies for Computing, Communication and Smart Cities*, Lecture Notes in Electrical Engineering 875,
https://doi.org/10.1007/978-981-19-0284-0_10

rapidly spread all over China and the world. It can cause symptoms including fever, difficulty in breathing, cough and invasive lesions on both lungs of the patients [1]. It can spread to the lower respiratory tract and cause viral pneumonia. In severe cases, patients suffer from dyspnea and respiratory distress syndrome [2, 53]. There are total 175,622,44,582,692 cases, recovered case 159,171,963 and 3,788,723 death according to World O meter [16]. The World Health Organization (WHO) has declared that COVID-19 is responsible for the death of many individuals and termed a pandemic. Nations are responsible for the spread and one's own country wellness. The countries use AI and machine learning (ML) tools to avoid and secure their nations from this pandemic.

Many people who are suffering from COVID-19 infection are developing mild to moderate respiratory tract infection and uneasiness and inhaling oxygen. Older people suffering from another disease like blood pressure and diabetes other chronic illness are experiencing more threat to their lives. To avoid this disease, the best way that can be used is to prevent the spread of this disease as it is spreading from touch and smell. When an infected person comes in contact with a healthy person and if he sneezes or coughs that there are 99% chances that the healthy person might get infected from the virus. The droplets from the saliva of the infected person will get into the respiratory tract of a healthy person and may make him ill and suffered from COVID-19. Therefore, it is necessary to maintain hygiene and keep sanitizing hands as the guidelines provided by WHO indicates that either masking off and proper sanitization can keep you off from the virus.

The governments from various countries have a hard time-fighting coronavirus as they are putting all efforts into preventing the virus from spreading. The critical conditions are managed by applying emergencies and using technology from equipment to fight from it. The kits and their use contain artificial intelligence, data science, data analytics, machine learning, thermal sensors and other techniques. Furthermore, in Sect. 2, we have written about applications of data science, Sect. 3 is about the role of artificial intelligence in predicting COVID-19 patients, Sect. 4 discusses guidelines issued by WHO, Sect. 5 discusses the role of data science in managing critical patients, in Sect. 6, we listed some available datasets, Sect. 7 consists of the things that could be done to control COVID-19, Sect. 8 consists of a summary of techniques and challenges in applying them, and Sect. 9 deals with conclusion and future work.

2 Application of Data Science

In these times, many people feel helpless in the face of the pandemic. However, people could focus on how they can give back to society by helping people in need. Given below are some of the things that the data science community is doing in the face of the novel coronavirus.

2.1 Real-Time Visualization

Some of the people are making interactive visualizations by utilizing the data given by different governments and World Health Organization (WHO) with the help of various tools available. This is helping many people track accurate and up-to-date statistics.

2.2 Forecasting and Prediction

For making the database more accurate and suitable for finding results efficiently and provide proper suggestion and treatment regarding the avoidance of the spread of this virus.

2.3 Discovering Treatments

At Columbia University, two graduates are using machine learning to find treatment for this virus. They screen therapeutic antibodies, and the probability of success is quite high [3].

2.4 Calculating Risk Factors

The teams are trying to find various factors that can be responsible for the spread of COVID-19. They are also trying to predict how can we stop the spread and which measures are being helpful and which are not very helpful by using the data of different regions and precautions taken in those regions.

2.5 Identifying Who is Dying and Why

People are trying to find the pattern among the people who are dying because of COVID-19. This includes respiratory diseases, heart diseases, genetic component, environmental factors, etc. But this data is challenging to access because of HIPAA restrictions [4].

2.6 Drug Discovery

The overall research is going on in order to find the most vital drug that can fight the coronavirus. Though the scientist is still not successful but with the help of data science, the scientist can track what all discovery using some similar salts in creating a more appropriate drug is taking place.

2.7 Monitor Patients Health

Using artificial intelligence and data science, the health of patients can be monitored using a similar pattern of cell structure and changes in the patient's body and anatomy.

2.8 Track and Prevent Disease

By tracking the patients who are showing some signals regarding the flu and keep monitoring them till they are negative can help prevent the virus and stop its wide-spreading nature.

2.9 Robotics

In China, robots were used to help deliver medicines to patients so as to stop the spread of the coronavirus among doctors. They were also used to disinfect hospitals by using ultraviolet light.

2.10 Predictive Analytics

Making the patients aware of the symptoms and helping them to take certain kind of salt and another remedy to help cope up with the virus and prevent it from spreading.

2.11 Contact Tracing

Data science-based tools are used to track patients or people who are quarantined across the country. It is also used to trace contacts of the patients.

3 Role of Artificial Intelligence in Predicting COVID-19 Patients

The health industry needs a superpower to diagnose and make the decision to handle the virus efficiently and provide proper suggestion and treatment regarding the avoidance of the spread of this virus. AI serves as a boon to the health industry as it is capable of handling numerous amount of data and use its intelligence and diagnose the patient having certain issues common to categorize whether he/she is infectious of COVID-19 or not. It also helps in understanding the development of vaccine and their availability. This technology is used for prediction, analysis and tracking of patients currently suffering from the virus and helps the sector to be aware of the vast illness due to pandemic [56].

3.1 Early Detection

AI helps in monitoring and detecting disease using prior knowledge and through its experience. As the data is supervised to the machine, therefore it is speedy and more accurate rather than any medical doctor detection and diagnosis.

3.2 Decision Making

AI helps the patients to make the better decision during any critical situation as the patients are moped derived toward the access and dealing it using dataset monitoring.

3.3 Treating Disease

Healthcare workers seek help from the machine and the Internet to find accurate medicine by providing the symptoms of the patient. If the patient is well educated about the COVID-19 virus, the patient illness is slowed down from spread.

3.4 Associated Care

A small study in Wuhan [18] has shown that the risk of dying increases with age. Elderly people have more diseases such as heart disease, respiratory problems, diabetes and clotting of blood. The place visited by him should be sanitized and locked up, and people around him should be tested immediately.

3.5 *Checking Health Through Wearable*

This includes respiratory diseases, heart diseases, genetic component and environmental factors. It is made publicly available by AWS and is a repository that constitutes of datasets related to machines.

3.6 *End of Life Care*

If the patient is not treated well using in intelligence and machine learning tools gently and with care and all the medication is used well, then the infected person may lose his life. But if the tools are used efficiently, and the medication and care is taken properly, then the well-being of the patient can be maintained.

AI serves as a platform to train machines with huge dataset and analyze health sector data and categorize accordingly. It also helps doctors and health vectors to train machines and set algorithm to optimize the data related to the virus with speed and accuracy. These machines help in retrieving the more precise data and support the health workers by diagnosing the issue more efficiently.

4 Precautions to be Taken to Avoid Getting Infected

In order to avoid getting infected, whether from indoor or outdoor, certain measure and guidelines are published by World Health Organization (WHO). If a member is infected from one family, then avoid getting in touch with him, and keep necessary action when getting in touch with him. If one is moving out, then avoid getting in touch with the infected person and maintain proper distance. These are some of the limitations and guidelines to be followed to avoid virus outbreak in the country.

The guidelines issued by WHO indicates that a person should keep washing his hand whenever he is going outdoor or if there is an infected person indoor that wash hand while being indoor. Always have a mask on, and in the second wave, the double mask was changed in the guidelines. Always disinfect the area where the ill person is, and do not share your personal item with them. Social distancing is the key to preventing the virus from getting into one's respiratory tract. For children, special guidelines are indicated in which children above three years of age have to keep the mask on and keep sanitizing their hands. They are not allowed to go outdoors. All the schools and educational institutions are temporarily shut for children. For traveling in a four-vehicle like a car, try to keep proper ventilation, and if there are two-person traveling that tries to sit socially distant to avoid getting infected.

5 Role of Data Science in Managing Critical Patients

People affected by COVID-19 are requiring or critical respiratory care. Data science has come up to care for these patients. In [7], they have studied various researches and come up with a three-stage process that will categorize the search result in three states to help the patients in ICU. They have read studies from [7–11].

We can also determine who the people most vulnerable are so that they could be separated beforehand from the rest [38]. It will also help us in identifying the people who are more susceptible to death. A small study in Wuhan [37] has shown that the risk of dying increases with age. Elderly people have more diseases such as heart disease, respiratory problems, diabetes and clotting of blood.

6 Available Datasets

There are various datasets freely available on the Web to help combat COVID-19 (Table 1).

These are only some of the datasets that are available from the sea of them available online. Countries have shared their own datasets online for everyone to analyze and forecast. There are related to age groups, counts of confirmed, recovered and dead, hospital beds and testings, etc., depending on the country.

7 Controlling COVID-19

Worldwide, countries are quarantining people, imposing lockdown, sanitizing to stop community spread. Public health has been the top priority for countries currently, and they are willing to sacrifice their economy to save lives. Although there is no single measure that could be taken to stop the pandemic worldwide, there are still some steps that could be taken like.

7.1 Social Distancing

People should follow proper protocols to ensure a safe distance from each other. A minimum of 1 m distance is required.

Table 1 Available datasets for the COVID-19 data

S. No	Dataset name	Description
1	"Google Covid-19 Public Datasets" [39]	This dataset comprises various datasets that are provided by Google for free so that it is more accessible to researchers, data scientist and analysts. This contains JHU CSSE (Johns Hopkins Center for Systems Science and Engineering) dataset, OpenStreetMap data and global health data from the World Bank.
2	"Covid-19 Open Research Dataset" [40]	At the Allen Institute, the semantic scholar team is providing CORD-19 by partnering with groups that are leading in COVID-19 research. It is also available for free on the Web site
3	"The COVID tracking project collects and publishes the complete testing data available for US states and territories" [41]	As the name suggests, this has all the data published from US states and territories of testing constitutes of positive, negative and pending, deaths and total test results
4	"LitCovid" [42]	It is an organized hub of literature to track researches on COVID-19. It takes into account how many papers are getting published every week and from which country. It is the resource of NCBI
5	"ECDC data on Covid-19 cases Worldwide" [43]	ECDC is a European Centre for Disease Prevention and Control, and it delivers data on all geographical distributions of novel coronavirus
6	"COVID-19 Open Research Dataset Challenge (CORD-19)" [44]	It is a Kaggle dataset, constituting 57,000 scholarly articles, with over 45,000 including full text, about COVID-19, SARS-CoV-2, and related coronaviruses. To cope up with this, the medical research community needs different AI experts to find answers
7	"A public data lake for analysis of COVID-19" [45]	It is made publicly available by AWS and is a repository that constitutes of datasets related to characteristics and spread of COVID-19
8	"COVID19 Global Forecasting (Week 4)" [46]	It is a dataset launched by Kaggle and is one of the challenges to answering WHO and NASEM questions on the pandemic
9	"Research Resource Repository COVID-19" [50]	Related datasets for research and analytics. The data may be directly or indirectly related to coronavirus. It deals with some subsidiary issues also

(continued)

Table 1 (continued)

S. No	Dataset name	Description
10	"Complete our world in Data for COVID-19 dataset" [51]	It is a crucial metric to understand the pandemic from a positive perspective
11	"Activity, population and location-COVID-19" [52]	Deals with the location and population table in compliments and COVID-19 dataset
12	"Research initiatives in HRS for COVID-19" [48]	Data resource including the data for COVID-19 patients and questionnaire with the US State policyholder
13	"Covid-19 OPEN DATA–FIGSHARE" [47]	It deals with sharing the initiative of collaboration of universities and medical colleges
14	"ICPSR: Data Repository for COVID-19" [53]	It deals with the health industry and maintains a centralized repository for the medical record of the COVID-19 patients and store a secure enclave to study those records
15	"VIVLI-COVID Data" [54]	It deals with the clinical trials on COVID-19

7.2 Testing

Countries should start testing people in large volumes as the virus is spreading exponentially in the world. People should also ensure that if any symptoms occur, they should go and have themselves checked out.

7.3 Travel Measures

People should avoid traveling and getting in touch with people. Countries have sealed their borders, but also in-country travel should be avoided.

7.4 Containment

If a person has tested positive, he/she should be contained and quarantined so as to not spread the virus in other places. The place visited by him should be sanitized and locked up, and people around him should be tested immediately.

7.5 Personal Hygiene

Every person should wear a mask whenever going out and should wash hands regularly. Personal hygiene is necessary in this case as the virus enters our body from the eyes, nose, mouth, etc., and people should avoid touching their face. They should wash or sanitize their hands multiple times in a day.

8 Summary of Works Done by Researchers

See Table 2.

9 Conclusion and Future Work

AI serves as a potential tool in fighting against COVID-19 and many similar pandemic. Obviously data serves as the main center for such pandemic to give optimum results. The increase of documents from hospitals and patients pathological data has made us rely more on artificial techniques and machine learning to extract a more accurate data so that the diagnosis can be done rapidly. Extraordinary surveillance by the government to keep the track of records to get over the pandemic takes into lots of dataset and perfect coordination with the health workers. Due to the evolving pandemic, we are dependent on deep learning approaches and are developing a model that is more prognostic and empirical. These models help to retrieve datasets using AI models. These models help the health workers to guide and help patients to recover and establish a more reliable bond with their health workers as due to the crises patients and their relatives are more apprehensive. It is of great importance for the health workers to take care of their personal data and be justified with that. AI serves as a boon in the present crises to accelerate the digitization of economy and production activities.

In this paper, we have discussed about how data science, artificial intelligence and machine learning are used against the spread of COVID-19. It also helps doctors and health vectors to train machines and set algorithm to optimize the data related to the virus with speed and accuracy. These machines help in retrieving more of the various datasets that could be used to do the predictions and forecasting, real-time visualizations, finding contacts of people that are exposed to this virus and various other applications as discussed above in the paper. For future work, there can be various other applications that could be added to this paper. Every country has issued its own datasets. So, details about them can also be added to a paper. Furthermore, there are many things that could be done to stop the spread of them we have listed some of the main ones. All could be listed in detail.

Table 2 Authors and there related work regarding COVID-19

Authors	Data type	Technique/method	Challenges
Wouters et al. [32], Mathur et al. [9]	Population density, strategic data	Deep learning, clustering algorithm, DBSCAN	Low quality of data and issues with privacy factors
Pham et al. [55]	Public announcements and distribution data	Naïve Bayes model, machine learning and	Insufficient data and ambiguity, and there is no truthfulness of data being correct
Vaishya et al. [20], Wehbe et al. [26]	Radiology data (X-ray and CT scan)	Camera system and image recognition technique	Not able to explain the results and results taking a long time to train and Vulnerabilities need to involve medical experts to give correct results
Cresswell et al. [29], Abu-Raddad et al. [35]	Camera images (infrared images)	AI-based computer vision and FR (facial recognition system)	Need a high-quality system to capture images and noise present may make it difficult to train
Zhou et al. [11]	Satellite data (various location using GPS)	Non-parametric process using Gaussian technique	Noises in satellite images, phone data not synchronized
Soltan et al. [27]	Text data from social media	Deep learning, machine learning, speech recognition and synthesis of certain linguistic data Information extraction using machine learning	Insufficient data from time series, Privacy issues and invasion, deep learning models take a long time and are very heuristic Social media reports might not be of good quality and be multidimensional
Vijayashanthi et al. [12]	Healthcare reports, travel data	Dynamic programming heuristic data	To establish trust with autonomous system
Rahmatizadeh et al. [7]	Text data on various COVID-19 details	CNN-based model data, supervised learning, text extraction	Results to be verified, but it will become extremely tedious Efforts to train workers to work with autonomous systems

References

1. Who.int, 2020. [Online]. Available https://www.who.int/docs/default-source/coronaviruse/situation-reports/20200226-sitrep-37-covid-19.pdf?sfvrsn=2146841e_2. Accessed 17 April 2020
2. Song F, Shi N, Chan F, Zhang Z, Shen J, Lu H, Shi Y et al (2020) Emerging 2019 novel coronavirus (2019-nCoV) pneumonia. Radiology 295(1):210–217
3. Datascience.columbia.edu. 2020. DSI alumni use machine learning to discover coronavirus treatments. Data Science Institute. [online]. Available at https://datascience.columbia.edu/dsi-alumni-use-machine-learning-discover-coronavirus-treatments. Accessed 18 April 2020
4. HealthITAnalytics, 2020. Understanding the COVID-19 pandemic as a big data analytics issue. [online]. Available at https://healthitanalytics.com/news/understanding-the-covid-19-pandemic-as-a-big-data-analytics-issue. Accessed 18 April 2020
5. MarkTechPost, 2020. List of COVID-19 resources for machine learning and data science research Marktechpost. [online]. Available at https://www.marktechpost.com/2020/04/12/list-of-covid-19-resources-for-machine-learning-and-data-science-research/. Accessed 18 April 2020
6. Petropoulos F, Makridakis S (2020) Forecasting the novel coronavirus COVID-19. PloS One 15(3):e0231236
7. Rahmatizadeh S, Valizadeh-Haghi S, Dabbagh A (2020) The role of artificial intelligence in management of critical COVID-19 patients. J Cell Mol Anesth 5(1):16–22
8. Alimadadi A, Aryal S, Manandhar I, Munroe PB, Joe B, Cheng X (2020) Artificial intelligence and machine learning to fight COVID-19
9. Mathur P, Burns ML (2019) Artificial intelligence in critical care. Int Anesthesiol Clin 57(2):89–102
10. Chaplain X, Huet O (2019) Is artificial intelligence (AI) at the doorstep of intensive care units (ICU) and operating room (OR)? Anaesth Crit Care Pain Med 38(4):337–338
11. Zhou W, Li H, Tian Q (2017) Recent advance in content-based image retrieval: a literature survey. arXiv preprint arXiv:1706.06064
12. Vijayashanthi M, Krishna VV, Reddy G (2018) Survey on recent advances in content-based image retrieval techniques. J Innov Comput Sci Eng 7(2):41–48
13. Gutierrez G (2020) Artificial intelligence in the intensive care unit. Crit Care 24:1–9
14. Mupparapu M, Wu CW, Chen YC (2018) Artificial intelligence, machine learning, neural networks, and deep learning: futuristic concepts for new dental diagnosis. Quintessence Int (Berlin, Germany: 1985) 49(9):687–688
15. Bini SA (2018) Artificial intelligence, machine learning, deep learning, and cognitive computing: what do these terms mean and how will they impact health care? J Arthroplasty 33(8):2358–2361
16. Coronavirus Update (Live): 175,622,445 cases and 3,788,723 deaths from COVID-19 virus pandemic—Worldometer. Worldometers.info, 2021. [Online]. Available https://www.worldometers.info/coronavirus/. Accessed 11 June 2021
17. Lin E, Tsai SJ (2019) Machine learning in neural networks. Adv Exp Med Biol 1192:127–137
18. Hashimoto DA, Witkowski E, Gao L, Meireles O, Rosman G (2020) Artificial intelligence in anesthesiology: current techniques, clinical applications, and limitations. Anesthesiology 132(2):379–394
19. Ghassemi M, Celi LA, Stone DJ (2015) State of the art review: the data revolution in critical care. Crit Care 19(1):1–9
20. Vaishya R, Javaid M, Khan IH, Haleem A (2020) Artificial intelligence (AI) applications for COVID-19 pandemic. Diabetes Metab Syndr 14(4):337–339
21. Jamshidi M, Lalbakhsh A, Talla J, Peroutka Z, Hadjilooei F, Lalbakhsh P, Jamshidi M et al (2020) Artificial intelligence and COVID-19: deep learning approaches for diagnosis and treatment. IEEE Access 8:109581–109595
22. Zhou Y, Wang F, Tang J, Nussinov R, Cheng F (2020) Artificial intelligence in COVID-19 drug repurposing. Lancet Digit Health

23. Lalmuanawma S, Hussain J, Chhakchhuak L (2020) Applications of machine learning and artificial intelligence for COVID-19 (SARS-CoV-2) pandemic: a review. Chaos Solitons Fractals 25:110059
24. Summers RM (2021) Artificial intelligence of COVID-19 imaging: a hammer in search of a nail
25. Mushtaq J, Pennella R, Lavalle S, Colarieti A, Steidler S, Martinenghi CM, Palumbo D, Esposito A, Rovere-Querini P, Tresoldi M, Landoni G (2021) Initial chest radiographs and artificial intelligence (AI) predict clinical outcomes in COVID-19 patients: analysis of 697 Italian patients. Eur Radiol 31(3):1770–1779
26. Wehbe RM, Sheng J, Dutta S, Chai S, Dravid A, Barutcu S, Katsaggelos AK et al (2021) DeepCOVID-XR: an artificial intelligence algorithm to detect COVID-19 on chest radiographs trained and tested on a large US clinical data set. Radiology 299(1):E167–E176
27. Soltan AA, Kouchaki S, Zhu T, Kiyasseh D, Taylor T, Hussain ZB, Peto T, Brent AJ, Eyre DW, Clifton DA (2021) Rapid triage for COVID-19 using routine clinical data for patients attending hospital: development and prospective validation of an artificial intelligence screening test. Lancet Digit Health 3(2):e78–e87
28. Agarwal M, Saba L, Gupta SK, Carriero A, Falaschi Z, Paschè A, Danna P, El-Baz A, Naidu S, Suri JS (2021) A novel block imaging technique using nine artificial intelligence models for COVID-19 disease classification, characterization and severity measurement in lung computed tomography scans on an Italian cohort. J Med Syst 45(3):1–30
29. Cresswell K, Tahir A, Sheikh Z, Hussain Z, Hernández AD, Harrison E, Williams R, Sheikh A, Hussain A (2021) Understanding public perceptions of COVID-19 contact tracing apps: artificial intelligence–enabled social media analysis. J Med Internet Res 23(5):e26618
30. Vasileiou E, Simpson CR, Robertson C, Shi T, Kerr S, Agrawal U, Sheikh A et al (2021) Effectiveness of first dose of COVID-19 vaccines against hospital admissions in Scotland: national prospective cohort study of 5.4 million people
31. Perrotta D, Grow A, Rampazzo F, Cimentada J, Del Fava E, Gil-Clavel S, Zagheni E (2021) Behaviours and attitudes in response to the COVID-19 pandemic: insights from a cross-national Facebook survey. EPJ Data Sci 10(1):1–13
32. Wouters OJ, Shadlen KC, Salcher-Konrad M, Pollard AJ, Larson HJ, Teerawattananon Y, Jit M (2021) Challenges in ensuring global access to COVID-19 vaccines: production, affordability, allocation, and deployment. Lancet
33. Fan G, Yang Z, Lin Q, Zhao S, Yang L, He D (2021) Decreased case fatality rate of COVID-19 in the second wave: a study in 53 countries or regions. Transboundary Emerg Dis 68(2):213–215. Fallahgoul H (2021) Inside the mind of investors during the COVID-19 pandemic: evidence from the StockTwits data. J Financ Data Sci 3(2):134–148
34. Bendavid E, Oh C, Bhattacharya J, Ioannidis JP (2021) Assessing mandatory stay-at-home and business closure effects on the spread of COVID-19. Eur J Clin Investig 51(4):e13484
35. Abu-Raddad LJ, Chemaitelly H, Butt AA (2021) Effectiveness of the BNT162b2 COVID-19 vaccine against the B. 1.1. 7 and B. 1.351 variants. New England J Med
36. Cosgriff CV, Celi LA, Stone DJ (2019) Critical care, critical data. Biomed Eng Comput Biol 10.1179597219856564
37. Zhou F, Yu T, Du R, Fan G, Liu Y, Liu Z, Cao B (2020) Clinical course and risk factors for mortality of adult inpatients with COVID-19 in Wuhan, China: a retrospective cohort study. Lancet 395(10229):1054–1062
38. Arxiv.org, 2020. [Online]. Available https://arxiv.org/pdf/2003.07347.pdf. Accessed 25 April 2020
39. Google Cloud Platform. Console.cloud.google.com, 2020. [Online]. Available https://console.cloud.google.com/marketplace/browse?filter=solution-type:dataset&filter=category:covid19. Accessed 29 April 2020
40. CORD-19 | Semantic Scholar. Pages.semanticscholar.org, 2020. [Online]. Available https://pages.semanticscholar.org/coronavirus-research. Accessed 29 April 2020
41. Most recent data. The COVID Tracking Project, 2020. [Online]. Available https://covidtracking.com/data. Accessed 29 April 2020

42. LitCovid—NCBI—NLM—NIH. Ncbi.nlm.nih.gov, 2020. [Online]. Available https://www.ncbi.nlm.nih.gov/research/coronavirus/#data-download. Accessed 29 April 2020
43. Download today's data on the geographic distribution of COVID-19 cases worldwide. European Centre for Disease Prevention and Control, 2020. [Online]. Available https://www.ecdc.europa.eu/en/publications-data/download-todays-data-geographic-distribution-covid-19-cases-worldwide. Accessed 29 April 2020
44. COVID-19 Open Research Dataset Challenge (CORD-19). Kaggle.com, 2020. [Online]. Available https://www.kaggle.com/allen-institute-for-ai/CORD-19-research-challenge. Accessed 29 April 2020
45. A public data lake for analysis of COVID-19 data | Amazon Web Services. Amazon Web Services, 2020. [Online]. Available https://aws.amazon.com/blogs/big-data/a-public-data-lake-for-analysis-of-covid-19-data/. Accessed 29 April 2020
46. COVID19 Global Forecasting (Week 4) | Kaggle. Kaggle.com, 2020. [Online]. Available https://www.kaggle.com/c/covid19-global-forecasting-week-4/data. Accessed 29 April 2020
47. COVID-19.zbmed.de, 2020. [Online]. Available https://covid-19.zbmed.de/externalResources/literature/. Accessed 25 April 2020
48. Coronavirus outbreak is the world's worst crisis since Second World War. Dailytimes.com.pk, 2020. [Online]. Available https://dailytimes.com.pk/586812/coronavirus-outbreak-is-the-worlds-worst-crisis-since-second-world-war/. Accessed 29 April 2020
49. 2020. [Online]. Available http://www.unitedindiversity.org/information/425/coronavirus-or-covid-19/en. Accessed 29 April 2020
50. 2020. [Online]. Available https://datascience.nih.gov/covid-19-open-access-resources. Accessed 25 May 2020
51. Our World in Data, 2020. [Online]. Available https://ourworldindata.org/coronavirus-source-data. Accessed 25 May 2020
52. COVID-19 activity—location and population table, 2020. [Online]. Available https://data.world/covid-19-data-resource-hub/covid-19-activity-location-population-table. Accessed 25 May 2020
53. OPENICPSR, 2020. [Online]. Available https://www.openicpsr.org/openicpsr/covid19. Accessed 25 May 2020
54. VIVLI-Center for Global Clinical Research Data, 2020. [Online]. Available https://vivli.org/how-to-share-your-research-on-the-vivli-covid-19-portal/. Accessed 27 May 2020
55. Pham QV, Nguyen DC, Hwang WJ, Pathirana PN (2020) Artificial intelligence (AI) and big data for coronavirus (COVID-19) pandemic: a survey on the state-of-the-art
56. Agbehadji IE, Awuzie BO, Ngowi AB, Millham RC (2020) Review of big data analytics, artificial intelligence and nature-inspired computing models towards accurate detection of COVID-19 pandemic cases and contact tracing. Int J Environ Res Public Health 17(15):5330
57. Zhu H, Wei L, Niu P (2020) The novel coronavirus outbreak in Wuhan, China. Glob Health Res Pol 5(1):1–3

Spectral Analysis of DNA on 1-D Hydration Enthalpy-Based Numerical Mapping Using Optimal Filtering

Subhajit Kar, Madhabi Ganguly, and Aryabhatta Ganguly

Abstract The paper proposed a new numerical coding technique of DNA, based on hydration enthalpy, to find the coding regions of a gene. The genetic sequences when converted into the numerical sequence can be applied to a predesigned least square FIR filter to capture period three property of coding regions. Moving average filter is then used to eliminate high-frequency noise present in the spectrum due to long range correlation between nucleotide bases. The proposed algorithm has been applied successfully to various standard datasets to predict the exon boundaries. The performance of the proposed method is compared with other popular numeric encoding schemes, namely dipole moment and pseudo-EIIP, and found superior in terms of accuracy and other evaluation parameters. The proposed method achieved an accuracy of 92% on the benchmark gene F56F11.5 of C Elegans (Accession number AF099922).

Keywords Genomic signal processing · Exons · Digital filter · Hydration enthalpy · Coding regions

1 Introduction

DNA or deoxyribonucleic acid which consists of four nucleotide bases, namely adenine (A), cytosine (C), thymine (T), and guanine (G), is responsible for transferring heredity information from generation to generation. It is made up of genes and intergenic spaces (Fig. 1a). Each gene contains a particular set of instruction. Eukaryotic genes are further divided into coding and non-coding regions.

Discrimination of protein coding regions from non-coding regions is called splicing. Signal processing methods rely on period three property for that purpose [1]. The property states that a protein coding region of length N exhibits a relatively

S. Kar (✉) · M. Ganguly
Department of Electronics, West Bengal State University, Kolkata, India

A. Ganguly
Department of Multimedia, Brainware University, Kolkata, India

P. K. Singh et al. (eds.), *Emerging Technologies for Computing, Communication and Smart Cities*, Lecture Notes in Electrical Engineering 875,
https://doi.org/10.1007/978-981-19-0284-0_11

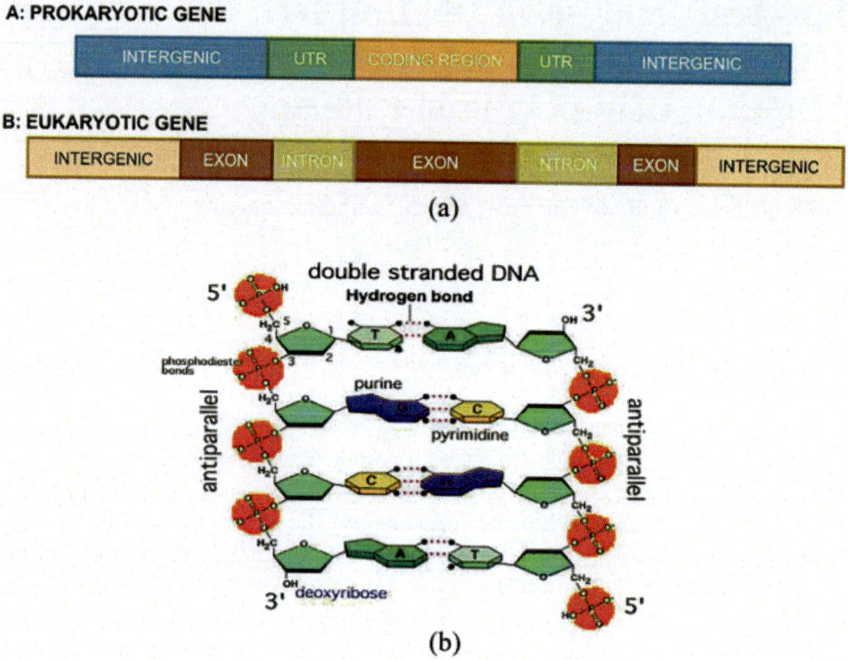

Fig. 1 **a** Eukaryotic and prokaryotic gene structures, **b** double helix structure of DNA

large value at discrete frequency $k = N/3$ but near zero value at other frequencies corresponding to magnitude spectrum.

Mapping of the DNA character sequence into its corresponding numeric sequence is eventual requirement of applying DSP mechanism. Numerical sequence mapping from DNA is classified into fixed mapping, DNA physicochemical property-based mapping and statistical property-based mapping. Each of these classifications comes with their own advantages and disadvantages [2–4]. The detection of correct exons vastly depends on these mapping techniques. Voss proposed a 4-D binary representation of DNA sequence which can be effectively used in exon finding problem [5]. But the method is computationally complex since it employed four unique binary sequences to solve the exon finding problem. To ease the time complexity, we have proposed the hydration enthalpy-based mapping method where the hydration enthalpies of the four nucleotide bases are derived using Monte Carlo simulation. The technique has yielded superior result in identifying these coding areas with great accuracy in eukaryotic genes. Since hydration enthalpy depends on various physicochemical property of nucleotide bases including size of the bases, their dipole moments and number of hydrophilic center in the particular base, so it can take account of multiple properties of DNA at a time.

In the DNA helix, the bases: adenine (A), cytosine (C), thymine (T) and guanine (G) are each linked with their complementary bases by hydrogen bonding (Fig. 1b) between themselves. All hydrogen bonds of nucleotide bases are strengthened and protected from solvent hydration by the hydrophobic stacking of the bases. At the time of hydration, the nucleotide bases would quickly form bridges with water molecules and hydrate with solvent water. Recent study proved that water molecules are responsible for the double helical structure of DNA, whereas hydrogen bonds are responsible for sorting the base pairs so that they can link together in correct order [6]. Thus, hydration plays a crucial role in stability of DNA. As the coding regions of DNA are more stable and conserved compared to the non-coding counterparts, hydration enthalpy can be a good discrimination property between these two regions.

2 Materials and Method

The proposed method to determine protein coding region in eukaryotic gene employs two stage digital filtering for spectral analysis. The purpose of the spectral analysis of nucleotide sequence is to compute its period three spectral components located at a frequency equal to $2\pi/3$ in the PSD spectrum. The nucleotide sequence is converted into numerical sequence using the hydration enthalpy of nucleotide bases. Generated sequence is now applied to least square FIR filter designed to recognize period three property. Removal of high-frequency noise present in the PSD is done by moving average filter. The output power spectra of designed filter will indicate the exons by certain peaks. The flowchart of the implemented algorithm comprises six steps which is depicted in Fig. 2.

2.1 DNA Sequence Database

In this work, several datasets such as HMR195, GENESCAN and Burset and Guigo are considered for analysis. The genes of NCBI accession numbers AF071552, AF061327, M62420, AF059734, AF099922 and D10914 were downloaded from NCBI Web site (https://www.ncbi.nlm.nih.gov/) in FASTA format. The nucleotide

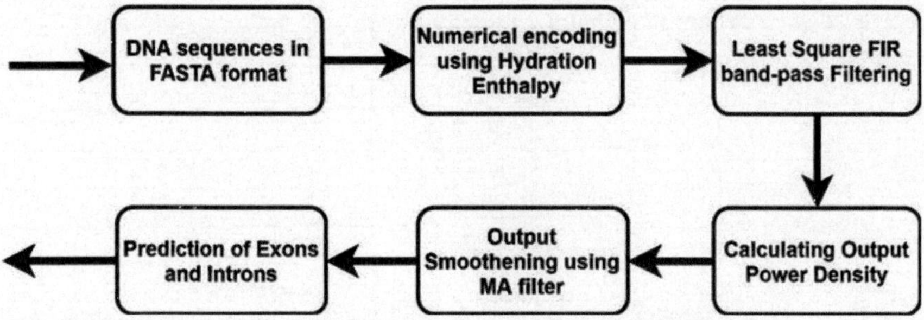

Fig. 2 Prediction of protein coding region using proposed numeric representation

sequences are chosen so that sequence length must be smaller than 10,000 bp and greater than 100 bp. Genes with different numbers of exons are chosen to detect the efficiency of the method. The encoded proteins by the coding regions can be observed from UNIPROT Web site (https://www.uniprot.org/uniprot/P18440).

2.2 *Numerical Mapping of DNA Sequence*

The mapping technique maps the DNA sequence into a one-dimensional numerical sequence. In this work, the enthalpy of hydration of the nucleotide bases is calculated using Monte Carlo simulation [7]. Enthalpy of hydration is defined as the quantity of energy produced upon dilution of one mole of gaseous nucleotide bases. The solvation of DNA bases into water constitutes couple of steps. In the first step, the hydrogen bonds between the water molecules break apart. It is an endothermic process requiring very small amount of heat energy. In the second step, the small water molecules bond to the various hydration sites of nucleotide bases. During the process, water molecules form hydrogen bonds with the hydrophilic centers of nucleic acid bases as well as between themselves. This step includes interaction between solute–solvent and solvent–solvent particles. New hydrogen bonds are created in the process releasing heat energy. Thus, the second step is an exothermic process. The total hydration energy of a nucleotide base can be found by the relation.

$$\Delta E_{\text{Total}} = \Delta E_{\text{Soln}} + \Delta E_{\text{Solv}} + \Delta E_{\text{int}} \tag{1}$$

where ΔE_{Soln}, ΔE_{Solv} and ΔE_{int} represent energy contributions from solute–solvent, solvent–solvent and intermolecular interactions, respectively.

Hydration energy of nucleotide bases is always negative due to the release of energy during this process. The solvation energy of DNA bases in water is in the following order depending upon the number of hydrophilic center present in the bases:

Etotal(Guanine) > Etotal(Thymine) > Etotal (Cytosine) > Etotal(Adenine).

The adapted hydration enthalpy of the four nucleotide bases is as follows.

Nucleotide base	Hydration enthalpy (Kcal/mol)
A	−39.28
C	−40.58
T	−40.63
G	−47.99

In this way, a sequence $x(n)$ = 'ACTAGAT' is transformed into its equivalent numerical representation as $x(k) = [-39.28\ -40.58\ \ -40.63\ \ -39.28\ \ -47.99\ -$ 39.28 −40.63].

2.3 Linear Phase Least Square FIR Filtering

The critical function of the algorithm is to design a precise band pass filter to capture the three base periodicities present in the coding regions. Designing linear phase FIR filters applying window methods is relatively simple yet lack of precise control of the critical frequencies such as w_p and w_s (pass-band and stop-band cut-off frequencies). Optimized filters are preferred to overcome the problem. One of them is least square FIR filter method which minimizes the weighted, integrated squared error between a desired and actual magnitude response of the filter over a set of desired frequency bands [8]. In our method, the desired signal is the ideal piecewise linear output of the designed filter. In order to predict the true exon locations in gene, a direct form FIR filter of order $N = 270$ has been designed. The goal of the filter is to detect three base periodicity, and hence, the center frequency of the filter should be $2\pi/3 = 0.667$ π radian.

The designed filter has the pass-band between 0.664 π radian and 0.672 π radian. Another very important criteria needed to be set is weight value of each band which controls the magnitude response of the filter.

A weighted moving average filter is implemented using Gaussian window of length 30 which gives output free of noise.

2.4 Calculation of Power Spectral Density

Power spectral density of filtered output with respect to different base position is obtained by squaring the filter response. For instance, if $Y(k)$ is the filtered output of a numerical sequence $x(n)$, then PSD of the sequence can be generated by [abs$(y)^2$] in MATLAB environment. Thus, the equation of power spectrum of output filter can be written as:

$$S(k) = [\text{abs}(Y(k))^2] \quad \text{where}\, k = 0, 1, 2, \ldots, N - 1 \tag{2}$$

2.5 Evaluation Criterion

Various evaluation parameters such as specificity (S_p), sensitivity (S_n), accuracy, Matthews correlation coefficient (MCC) and F1 score are measured to assess the performance of the coding scheme at nucleotide level. Similarly, various evaluation parameters at exon level like discrimination factor, miss rate and wrong rate are also measured and taken into consideration.

Finally, area under the curve of ROC is considered for evaluation purpose. ROC curve is a graph that measures performance and accuracy of a binary classifier system.

Here, the true positive rate (sensitivity) is plotted as a function of the false positive rate (100-Specificity) for different threshold points.

3 Results

The encoded genes from different dataset with the help of proposed numerical representation method are passed through specifically designed bandpass filter with center frequency $2\pi/3$ to get desired result. The output PSD plot of the experimented genes is given in Figs. 3 and 4. The dotted lines describe the boundaries of actual coding

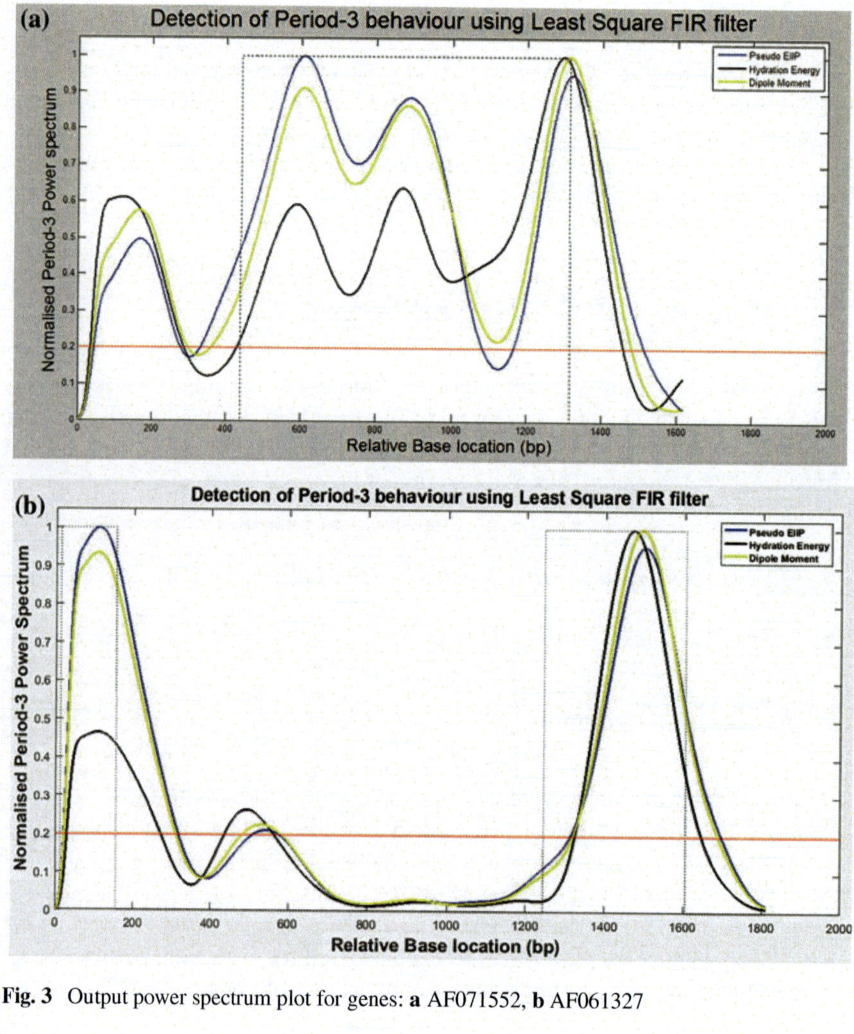

Fig. 3 Output power spectrum plot for genes: **a** AF071552, **b** AF061327

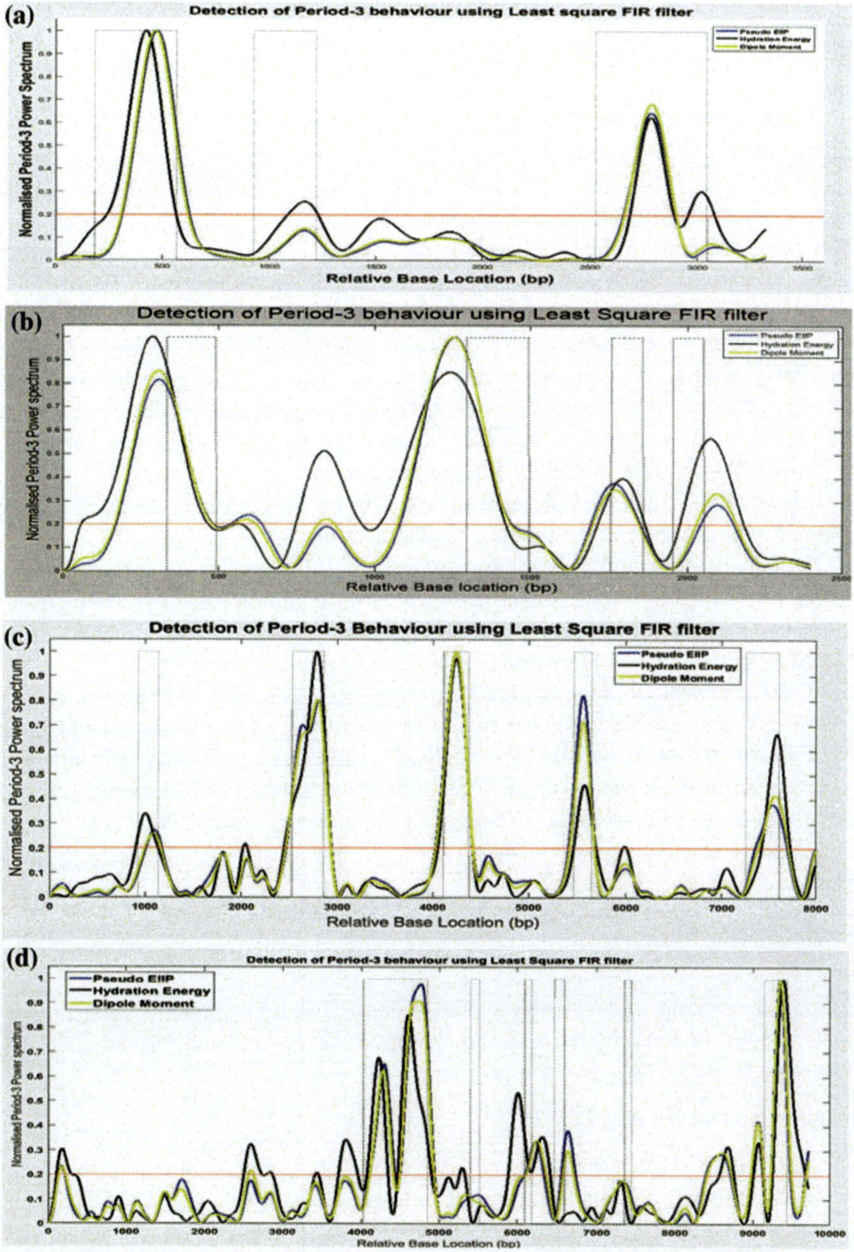

Fig. 4 Output power spectrum plot for different genes: **a** M62420, **b** AF059734, **c** AF099922, **d** HAMHG5

regions of genes as specified in NCBI Website. The blue, black and yellow plots are representing the PSD obtained from pseudo-EIIP, hydration enthalpy and dipole moments, respectively.

To assess the efficacy of the proposed numerical representation based on enthalpy of hydration, various evaluation parameters are computed from the output power spectral density plot and compared with two well-established coding measure, namely pseudo-EIIP and dipole moment [9, 10]. The detail results of predicted boundaries of gene F56F11.5 are provided in Table 1. Here, NA suggests there is no exon in the NCBI database. The detail evaluation results at nucleotide level are depicted in Fig. 5 with the help of bar graphs. The sensitivity and specificity for the benchmark gene having NCBI accession number AF099922 found to be 88% and 93%, respectively, which are best when compared to the other two encoding schemes. Receiver operating characteristic curve for gene AF099922 is plotted for all the three numerical representations, namely pseudo-EIIP, dipole moment and hydration enthalpy, for further evaluation purpose.

An exon level study of different evaluation parameter performed through Table 2 which reveals the efficacy of the proposed numerical model.

The experiment result showing great improvement in results as the miss rate and wrong rate for most of the exons is zero. The discriminating factors are also greater than one for most of the genes indicating that the exons and introns could be easily identified in the output PSD with respect to the threshold value.

We have measured the area under the curve (AUC) in case of hydration energy from Fig. 6, and it is found to be 0.8631 which is very good considering the fact that AUC close to 0.9 or above indicates excellent accuracy. A wide number of numerical representations have been evaluated in Akhtar et al. through AUC criterion, found best of the value of 0.81 using paired numeric representation of DNA [11].

4 Discussions

To show the superiority of the new hydration enthalpy-based mapping over the previously adopted encoding schemes, a comparative study is adopted for various genes according to the findings in previous literatures.

Gene Accession No AF071552:

For the gene AF071552, the output of the designed algorithm is comparable with the previous findings [12]. The proposed method produces MCC value of 0.55 which suggest good prediction accuracy of coding and non-coding regions. The sensitivity obtained is 100% means all the nucleotides in the coding regions predicted correctly.

Gene Accession No AF061327:

The specificity of 77% computed using the proposed algorithm is very good comparing to the others [13]. High specificity signifies that the algorithm successfully captured the non-coding regions i.e. introns.

Table 1 Predicted boundaries of exons for the selected genes. Figures in bracket enlist the original boundaries of exons as specified by NCBI

Sequence (NCBI ACCN. NO)	Gene description	Exon-1	Exon-2	Exon-3	Exon-4	Exon-5	Exon-6
AF071552	Single exon	428–1459 (441–1313)	NA	NA	NA	NA	NA
AF061327	Two exons	39–282 (13–153)	1333–1656 (1245–1604)	NA	NA	NA	NA
M62420	Three exons	278–591 (185–566)	1134–1220 (1112–1219)	2686–2908 (2608–3044)	NA	NA	NA
AF059734	Four exons	59–466 (335–491)	1016–1424 (1296–1495)	1712–1865 (1756–1857)	1975–2185 (1953–2051)	NA	NA
AF099922	Five exons	912–1147 (928–1039)	2455–2948 (2528–2857)	4062–4428 (4114–4377)	5470–5703 (5465–5644)	7429–7738 (7265–7605)	NA
HAMHG5	Six exons	4108–4919 (4028–4855)	5269–5348 (5399–5516)	5903–6161 (6098–6187)	6173–6444 (6473–6596)	– (7354–7440)	9252–9682 (9143–9437)

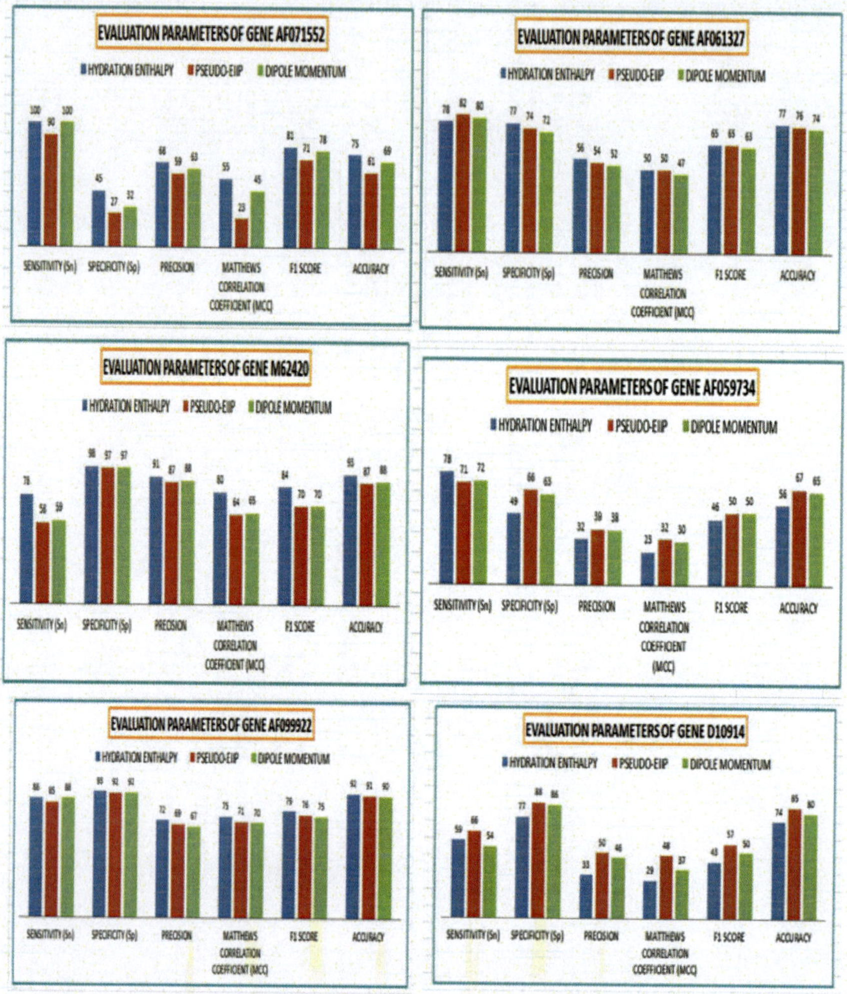

Fig. 5 Performance analysis of different coding rules using various matrices

Table 2 Different evaluation parameters at exon level

Gene Accn. No	Discrimination measure	Missing rate	Wrong rate
AF071552	1.66	0	1
AF061327	4	0	0.5
M62420	1.22	0	0
AF059734	0.72	0	0.2
AF099922	1.6	0	0
HAMHG5	0.65	0.17	0.4

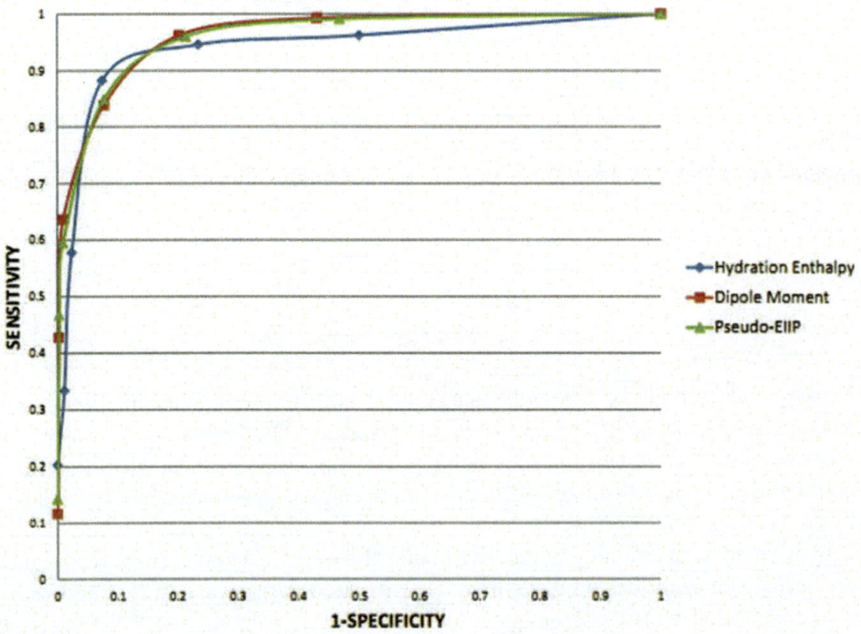

Fig. 6 ROC plot of exon prediction method based on three different mapping techniques

Gene Accession No M62420:

For the gene M62420, all the evaluation parameters at nucleotide and exon level are compared with other numerical representation method [10]. The miss rate and wrong rate for this gene are zero suggesting the algorithm have identified all the exons and introns correctly. Discrimination factor $D > 1$ signifies all exons identified without ambiguity. The sensitivity of the designed algorithm for this gene is 98% which is highest comparing to the previous findings [10].

Gene Accession No AF059734:

Proposed method derived significant improvement in exon prediction for the gene AF059734 which contain four exons. From Zhang et al. study, it is clear that the best method employing frequency of nucleotide occurrence generates sensitivity of 69% and specificity of 32%, while the proposed method resulted in 78% sensitivity and 48% specificity [12].

Gene Accession No AF099922:

The findings from the proposed hydration energy produce results of sensitivity 88% and specificity 93%. The proposed representation produces discrimination factor of unity. When comparing to the other numerical methods like tri-nucleotide, entropy, trigonometric mapping, our method derived excellent results [14–16].

Gene Accession No D10914 (HAMHG5):

The gene is chosen due to its diverse nature of exons. The gene contains very short exons (less than 100 bp) as well as long exons (close to 1000 bp). The gapping of the exons in the gene is not symmetric. Hence, effectiveness of the implemented method could be tested on this gene. The result obtained through this gene is very satisfactory as all the exons in the gene captured successfully.

5 Conclusions

Finding coding regions of various genes using DSP requires more in-depth study as there are still scope of improvement in every aspect. Numerical representation of DNA sequence along with genomic signal processing algorithms can identify hidden periodicities, nucleotide distribution and feature extraction which cannot be revealed by conventional methods. Very recently machine learning, deep learning and neural network models are gaining popularity in this domain of research. These techniques are very much accurate and developing rapidly. Hidden features of DNA and genes can be identified using these techniques when used simultaneously with digital signal processing methods and thus can be considered in future studies.

References

1. Anastassiou D (2001) Genomic signal processing. IEEE Sig Process Mag 18(4):8–20
2. Yu N, Li Z, Yu Z (2018) Survey on encoding schemes for genomic data representation and feature learning—from signal processing to machine learning. Big Data Mining Anal 1(3):191–210
3. Kar S, Ganguly M, Ghosal S(2021) Prediction of coding region and mutations in human DNA by effective numerical coding and DSP technique. In: 2021 international conference on computing, communication, and intelligent systems (ICCCIS), 19 February 2021. IEEE, pp 180–185
4. Vaegae NK (2020) Walsh code based numerical mapping method for the identification of protein coding regions in eukaryotes. Biomed Sig Process Control 58:101859
5. Voss RF (1992) Evolution of long-range fractal correlations and 1/f noise in DNA base sequences. Phys Rev Lett 68(25):3805
6. Feng B, Sosa RP, Mårtensson AK, Jiang K, Tong A, Dorfman KD, Takahashi M, Lincoln P, Bustamante CJ, Westerlund F, Nordén B (2019) Hydrophobic catalysis and a potential biological role of DNA unstacking induced by environment effects. Proc Natl Acad Sci 116(35):17169–17174
7. Monajjemi M, Ketabi S, Zadeh MH, Amiri A (2006) Simulation of DNA bases in water: comparison of the Monte Carlo algorithm with molecular mechanics force fields. Biochem Mosc 71(1):S1-8
8. Proakis JG (2001) Digital signal processing: principles algorithms and applications. Pearson Education India

9. El-Badawy IM, Gasser S, Aziz AM, Khedr ME (2015) On the use of pseudo-EIIP mapping scheme for identifying exons locations in DNA sequences. In: 2015 IEEE international conference on signal and image processing applications (ICSIPA), 19 October 2015. IEEE, pp 244–247
10. Meher JK, Panigrahi MR, Dash GN, Meher PK (2012) Wavelet based lossless DNA sequence compression for faster detection of eukaryotic protein coding regions. Int J Image Graph Sig Process 4(7)
11. Akhtar M, Epps J, Ambikairajah E (2007) On DNA numerical representations for period-3 based exon prediction. In: 2007 IEEE international workshop on genomic signal processing and statistics, 10 June 2007. IEEE, pp 1–4
12. Zhang R, Zhang CT (1994) Z curves, an intutive tool for visualizing and analyzing the DNA sequences. J Biomol Struct Dyn 11(4):767–782
13. Mabrouk M (2017) Advanced genomic signal processing methods in DNA mapping schemes for gene prediction using digital filters. Am J Sig Process 7(1):12–24
14. Singh AK, Srivastava VK (2020) A tri-nucleotide mapping scheme based on residual volume of amino acids for short length exon prediction using sliding window DFT method. Netw Model Anal Health Inform Bioinform 9(1):1–3
15. Das B, Turkoglu I (2018) A novel numerical mapping method based on entropy for digitizing DNA sequences. Neural Comput Appl 29(8):207–215
16. Das L, Nanda S, Das JK (2019) An integrated approach for identification of exon locations using recursive Gauss Newton tuned adaptive Kaiser window. Genomics 111(3):284–296

A Novel Ensemble Model for Breast Cancer Diagnosis

Aalif Hamid

Abstract Breast cancer is one of the world's second-leading causes of death. One out of every eight women experiences this illness at some point in their lives. Patients with early detection of the disease have a higher likelihood of survival and have a better chance of recovery. There is a critical requirement to classify tumors for malignancy. This study investigates automatic breast cancer prediction using a variety of machine learning techniques. The most prevalent type of classification is binary (benign cancer versus malignant cancer). Machine learning (ML) techniques are now being employed to detect breast cancer. They have a high degree of classification accuracy and a wide range of diagnostic capabilities. This research work presents a novel ensemble model for binary classification of breast mass tumors on the WDBC dataset. The outcomes of seven distinct machine learning individual models (logistic regression, KNN, SVC, etc.) are compared to the hybrid ensemble model of the above seven techniques. The evaluation of the model is done in terms of numerous performance indicators, such as accuracy, precision and recall. Compared to stand-alone models, the results demonstrate that the ensemble model performs remarkably well, with an accuracy of 0.98%.

Keywords Machine learning · Ensemble · Breast cancer · Support vector machine · Stacking classifier

1 Introduction

Cancer is a chronic condition caused by excessive growth and cell division that affects the cellular structure. It causes accelerated cell growth in some cases, depending on cancer type, and in others, cell growth is significantly decreased. This division abnormality affects the body in numerous states, resulting in cancers, undermining the immune system, and other significant complications. A record 15.5 million persons with a history of cancer were residents of the United States alone, according to the

A. Hamid (✉)
Department of Information Technology, Central University of Kashmir, Ganderbal, Kashmir, India

P. K. Singh et al. (eds.), *Emerging Technologies for Computing, Communication and Smart Cities*, Lecture Notes in Electrical Engineering 875,
https://doi.org/10.1007/978-981-19-0284-0_12

2018 report. According to the World Health Organization (WHO) [19], the illness will have a massive effect on the planet. About one in five individuals will face cancer diagnosis once in their lifetime by 2020. 18.1 million persons worldwide had cancer in 2018, and 9.6 million died of the disease. The figures will likely double by the year 2040. Accelerating global cancer containment by early diagnosis, detection, care and observation is the need of the hour. Pulmonary cancer is the most frequently diagnosed cancer (11.6%), followed by female breast cancer (11.6%) and bowel cancer (10.2%). The study's primary objective is to apply machine learning to diagnose the disease early, potentially saving millions of lives worldwide. The use of several machine learning techniques such as decision trees, artificial neural networks (ANN) and support vector machines (SVM) in cancer diagnosis yields positive outcomes. The adoption of machine learning in the medical field will transform disease diagnosis and help doctors get better insights into the disease.

Breast cancer is the most common type of cancer in women [2], affecting about 2.1 million women each year and resulting in the highest number of cancer-related fatalities. Breast cancer claimed the lives of around 15% of cancer-related deaths in 2018 or 627,000 women. Research conducted by [1] shows that one woman dies every 13 min due to breast cancer, while one case of breast cancer gets diagnosed every two minutes worldwide. Since the shift in clinical practice from cure-based evidence medicine to care-based evidence medicine, disease detection and identification have been given top importance while still in the early stages of development. Breast cancer death rates can be reduced if early identification of the disease is possible. An essential and often used diagnostic method is the triple-test; it consists of three medical examinations, i.e., self-examination, ultrasonography and fine needle biopsy (FNB) of the breast used to produce high accuracy in the disease diagnosis. FNB is an invasive technique that consists of obtaining material from the tumor directly. The examination of the content is performed under the microscope for the occurrence of cancer cells.

Data mining is a discipline in which computer science and analytics converge to detect correlations in the database. The data mining process's fundamental goal is to retrieve and shape valuable information from the data file into a coherent framework for potential use. It employs a combination of strong analytical abilities, domain knowledge and an explicit knowledge base to uncover hidden trends and patterns in heterogeneous information [7]. Many forensic organizations, pharmacies, hospitals and research institutions have many medical diagnosis data available. To make the system computerized and rapid to identify diseases, it is barely necessary to classify them [9]. Forecasting trends can aid in the development of new treatments by assisting in the identification of high-risk individuals based on known pathological and experimental risk factors. Accurate breast cancer risk assessments can help with lifestyle changes, immunotherapy, personalized screening and risk-stratified follow-up therapy [2]. The suggested research project is a study to see if a breast tumor is benign or malignant. In this regard, machine learning has been applied to the dataset. Machine learning is a series of instruments that are used to build and test algorithms that enable inference, identification of patterns and classification. Machine learning focuses on four steps: data collection, a compilation of models, model training and model verification [20]. The association between breast cancer and machine learning is not new. It has been used for decades to diagnose cancers and other autoimmune disorders, predict gene mutations that trigger cancer and evaluate the prognostic. Data mining and machine learning techniques are widely used in the medical industry since they are discovered to be quite useful in medical practitioners' decision-making processes. The implementation of multiple machine learning algorithms such as support vector machine (SVM), artificial neural network (ANN) and decision trees has shown promising results in the field of cancer detection. In addition to transforming illness identification, the adoption of artificial learning in medical imaging can enable doctors to obtain more in-depth insights into the disease.

The triple-test is an essential and frequently used diagnostic method; it consists of three medical tests, i.e., self-examination, ultrasonography and breast fine needle biopsy (FNB) used to diagnose the disease to achieve high precision. FNB is an invasive technique that consists of directly obtaining tumor material. The contents are checked for the presence of cancer cells under the microscope. The study work uses the WDBC dataset. The dataset consists of attributes that measure a digitized description of a fine needle aspirate from the image of mammary mass (FNA). The features were recorded by analyzing the characteristics of the cell nuclei under the microscope.

The following is a description of the paper's structure. The past work in breast cancer diagnosis using machine learning approaches is briefly described in Sect. 2. The materials and procedures employed in the research study are described in Sect. 3. The work's experimentation results are discussed in Sect. 4, while the paper's conclusion is discussed in Sect. 5.

2 Related Work

Large number of researches has been done in the past decade on the diagnosis of breast cancer. Some researchers have used mammograms, histopathological images, while others have worked on calculative data. The ability of the machine learning algorithms to find new hidden patterns in a large amount of data has shown quite great results in the healthcare sector. Cancer diagnosis at an early stage increases the chances of recovery in patients. For the diagnosis of breast cancer, researchers use a variety of machine learning methods such as artificial neural networks (ANN), decision trees, support vector machines (SVM), convolutional neural network and random forest. On the WDBC dataset, SVM and ANN were applied, and the results of the applied machine learning (ML) techniques were compared using performance measures.

In comparison with ANN, SVM demonstrated the best accuracy of 96% based on the performance criteria of the employed ML approaches [8]. The dataset WDBC was used to test a feed forward backpropagation neural network. The ANN in this study comprises nine input and one output neuron, with 1, 2, 3, 4 and 5 hidden layers. The results demonstrate that using the TANSIG transfer function, a 20-neuron feed forward backpropagation single hidden layer neural network using the TANSIG transfer function achieves the highest classification accuracy (98.9% on training set and 99% on test set) [10]. The methodology of adaptive feature selection was used, and classification was enabled using KNN, random forest and multilayer perceptron [22]. Results proved that before feature selection KNN, SVM and PNN obtained an accuracy of 100%, 100%, 96%, respectively, and after selection of best attributes, the accuracy got jumped to 97.19% for PNN. Breast Cancer Coimbra dataset was used on ANN, ELM, KNN and SVM algorithms by [23]; ELM is same as that of artificial neural network with one difference, and ELM consists of only one hidden layer instead of many hidden layers in case of ANN. Results showed that ELM outperformed the other two algorithms in terms of accuracy and training time and is preferable for more samples. The research [11] used six different algorithms, such as AdaBoost, random forest, extra tree classifier and the stacking classifier, on basic learners on the WDBC dataset and obtained an accuracy of 92.9% on voting classifier, with extra trees classifier receiving the highest score of 95.1739%, followed by XGBoost 95.1691% and AdaBoost 94.7343%. AlexNet architecture was used by [24] on the CBIS-DDSM dataset in which the CLAUDE algorithm for image enhancement was done with image segmentation to increase the sample size of the dataset. In the proposed work, linear SVM obtained the highest accuracy of 80.5%. Table 2 shows the brief summary of various methodologies adapted by researchers for breast cancer.

Table 1 Various breast cancer datasets

Database	Database size (GB)	Total patients
MIAS	2.3	161
DDSM	6	2620
CBIS-DDSm	70.5	237
ISPY1	76.2	237
Breast-MRI-NACT-Pilot	19.5	64
QIN-Breast	11.286	67
Mouse-Mammary	8.6	32
QIN Breast DCE-MRI	15.8	10
BREAST-DIAGNOSIS	60.8	88
RIDER Breast MRI	0.401	5
BCDR	1.4 GB	1734
TCGA-BRCA	53.92 (TB)	1098
BreakHis	4	82
Inbreast	40 MB	115
Coimbra Breast Cancer	7 Kb	116
WDBC	122 Kb	

There is a growing abundance of researchers drawing their interest attentions to ensembles. Moreover, it has solid proof that they can substantially change the classification performance. Six machine learning techniques were employed to analyze 8942 patients with breast cancer using the local dataset [12]. All methods, including decision trees, neural networks, extreme boost, logistic regression and support vector machines, produced very close accuracy of 79.80%, 82.7%, 81.7%, 81.7% and 81.7%, respectively, with random forest outperforming the others with an accuracy of 82.7%. Kadam et al. [25] suggested a feature ensemble model for categorizing breast cancer tumors into benign and malignant tumors based on sparse autoencoders and soft-max regression. In the analysis, ten-fold cross-validation was used that gave an impressive 98.60% result and showed that the proposed model performs other state-of-the-art techniques. The research presented by [26] is focused on the diagnosis of breast cancer using an ensemble learning algorithm based on SVM to decrease the diagnosis variance and increase the diagnosis accuracy. Twelve separate SVMs are hybridized using the weighted field under the receiver operating characteristic curve ensemble (WAUCE) approach. Wisconsin breast cancer, Wisconsin diagnostic breast cancer and the control, epidemiology and end outcomes (SEER) program breast cancer datasets have been studied to determine the efficacy of the suggested model. The results show that the WAUCE model achieves higher accuracy with a little reduced variance for breast cancer detection when compared to five additional ensemble mechanisms and two conventional ensemble models, namely adaptive boosting and bagging classification tree. The results show that the WAUCE model achieves higher accuracy with a little reduced variance for breast cancer detection when compared to five additional ensemble mechanisms and two conventional ensemble models, namely adaptive boosting and bagging classification tree. The datasets for breast cancer research are listed in Table 1.

3 Materials and Methods

The Wisconsin diagnostic dataset (WDBC) of breast cancer is used to measure the effectiveness of the proposed ensemble model for breast cancer diagnosis. This dataset has been collected by the University of Wisconsin Hospitals, Madison [27]. The dataset is available for download from the UCI machine learning repository. It is composed up of data derived from a digitized description of a fine needle aspirate of breast mass (FNA). The properties of the nuclei of the cells seen in Table 3 are identified by the characteristics. The dataset comprises a total of 32 attributes, of which 30 were used in the experiment as independent variables. There are a total of 569 instances of the patients recorded in the dataset.

Preprocessing is the method to remove the unwanted components from the dataset to fit the machine learning algorithms perfectly. WDBC dataset contains one independent categorical variable under the attribute name, "diagnosis". This column contains values in the form of benign or malignant (B/M). But the machine learning algorithm takes the input in the form of numbers (0/1); so, this column is preprocessed before

Table 2 Brief overview of diverse techniques for cancer diagnosis

Author	Technique	Dataset	Performance
[13]	Extracted characteristics using deep CNN architectures were proposed as a novel framework for the categorization of breast cancers	BreakHis	Accuracy = 97.5%
[14]	VGG-16 as a feature extractor combined with NN classifier for detection	MIAS-DDSM	Accuracy = 90.5%
[15]	Implemented CNN along with other machine learning algorithms	WDBC	Accuracy = 99.67%
[16]	The (SVM) model with radial basis function (RBF) kernel appeared to be the most successful classifier, according to the performance metrics of multicollinearity assessment on machine learning models	Coimbra Breast Cancer	Accuracy = 93.9%
[17]	Using a deep neural network with support value (DNNS), a novel technique to identify breast cancer was presented	Local Dataset	Accuracy = 97.21
[18]	For the diagnosis of breast cancer, an extreme learning machine model with a radial basis function (RBF) kernel was used	WDBC	AccuracyELM = 95.6 SVM = 88.9
[3]	Used convolutional neural networks, as well as several preprocessing strategies such as contrast scaling, dilation, cropping and decision fusion using an ensemble of networks, to tackle the challenge of breast microcalcification classification	DDSM	Recall = 97.3%

Table 3 Features of WDBC dataset [5]

Feature	Feature name
Feature 1	Patient ID
Feature 2	Diagnosis (B/M)
Feature 3	Nuclei radius
Feature 4	Nuclei texture
Feature 5	Nuclei perimeter
Feature 6	Nuclei area
Feature 7	Nuclei smoothness
Feature 8	Nuclei compactness
Feature 9	Nuclei concavity
Feature 10	Nuclei concave Points
Feature 11	Nuclei symmetry
Feature 12	Nuclei fractal dimension

fetching it into the algorithms. Label encoder which is available under Scikit-Learn package is used for the processing of the dependent variable. The dataset is divided using an 80:20 ratio, i.e., 80% of the data is used for algorithm training, while 20% of the data is the unseen data used to assess the performance assessment algorithms as suggested by [11]. Seven different machine learning algorithms like logistic regression, SVM extra trees classifier, Gaussian NB classifier, KNN, SGD classifier, etc., have been implemented in the research work. Combination of the said algorithms proved very beneficial and gave outstanding results. Implementation of work is done on the WDBC dataset which is preprocessed first and split into testing and training data.

3.1 Implementation of Work

The capability of computers to train from experience without being instructed personally is machine learning [21]. Both rules are coded in the standard programming style, and the computer can produce output depending on the logical argument, i.e., the rules. However, more rules need to be written as the structure gets complicated, so it becomes unfit to manage. However, it is assumed that machine learning can fix this dilemma. The computers here learn how the input and output data are associated and then write a law accordingly. Any time a new data or condition occurs, the programmer does not need to write new rules. The main focus of machine learning is learning and inference. Machine learning is graded into supervised, unsupervised and reinforcement learning [6]. In this research work, supervised machine learning algorithms such as logistic regression, random forest and K-nearest neighbor (KNN) are implemented.

Logistic Regression

By incorporating data onto a straight line, linear regression algorithms predict the values. It is said that these algorithms are unbounded since the value is not simply between 0 and 1. So, they are not ideal for problems with grouping, giving rise to "Logistic Regression". Logistic regression is a statistical model that is better adapted for problems of classification. "These algorithms use a mathematical function called "Sigmoid" to squeeze a number between 0 and 1.Two logistic regression algorithms were implemented in this research work, one with the solver equal to "saga" and the other with the solver parameter equal to "lbfgs".

Random Forest Classifier

The classifier averages the number of classifiers (decision-tree classifiers) on various data sub-samples. To boost prediction accuracy, it uses averaging of all classifiers. A parameter called "Criterion" is used in the random forest classifier. It is a feature that tests the efficiency of the division. This split parameter performs the task of splitting the data into clusters and should be so that a single class (0 or 1) outweighs each category. By applying variability, random forest models minimize the risk of overfitting by:

Constructing many trees (n-estimators)
Drawing substitute findings (i.e., a bootstrapped sample)

The best split nodes are split between a random subset of features chosen for each node.

A random forest classifier is implemented in the ensemble model with n_estimators set to 10 and criterion parameter set to "gini".

Extra Trees Classifier

This puts together a variety of randomized decision trees on various data sub-samples. The average estimate of individual classifiers is the approximation of this ensemble. We used two extra tree classifier models with distinct parameters in our ensemble model, as described in the previous classifier. Extra trees is like random forest. It creates several trees and breaks nodes using random subsets of characteristics, but with two key differences: It does not bootstrap observations (meaning it tests without replacement), and nodes are broken on random splits rather than best splits. An extra tree classifier is implemented in the ensemble model with the same n_estimators as the random forest classifier and criterion parameter set to "gini".

KNN

It is among the straightforward and non-parametric techniques for classification that stores and groups all data based on some similarity function (e.g., distance). The plurality vote of neighbors controls the classification/grouping of any data point. The number of neighbors eligible for voting is determined by the K. In this research work, K is set to 5.

Support Vector Classifier

It is a supervised learning technique which is most commonly used to solve classification problems. Data points reflect the dataset in the SVM. SVM constructs hyperplanes that have the highest margin in multi-dimensional space to categorize the results. A margin represents the longest interval between the closest data points. The linear kernel is proven to show promising results. SVC with the linear kernel is used in the model.

Gaussian NB Classifier

This model utilizes the Gaussian Naïve Bayes algorithm to predict/forecast the result. The constant values connected to each attribute are considered to have a Gaussian distribution here, i.e., Gaussian is said to be the likelihood of attributes. All default values are used for this algorithm in the model.

SGD Classifier

Stochastic gradient descent (SGD) is a simple but powerful optimization algorithm used to find the parameter/feature coefficient values that minimize a cost function.

In other words, it is used under convex loss functions such as SVM and logistic regression for discriminatory learning of linear classifiers. Since the update to the coefficients is done for each training instance, it has been successfully extended to large-scale datasets rather than at the end of cases. Each parameter is set to its default values for this algorithm in the proposed work.

Ensembling Approach

The research work combines the eight different sets of algorithms to boost the performance of the model. All the algorithms are merged and passed to the stacking classifier (voting classifier). A voting classifier is a machine learning model that trains on a range of diverse models and selects an output or class based on its most excellent chance of determining the class as the output. It essentially aggregates the outcomes of each classifier passed into the stacking classifier and, based on the most significant majority of votes, forecasts the output class. Rather than creating individual-specific models and finding each of them' performance, we build a specific single model that trains these models and forecasts output based on their cumulative plurality of votes for each type of output. It supports two types of voting: hard voting and soft voting. In hard voting, the predicted result class is a class with the most significant majority of votes in hard voting, i.e., the class that was most likely to be predicted by one of the classifiers. In soft voting, the output class is the forecast for soft voting, based on the average likelihood given to that class. Hard voting is implemented in the research work. The ensemble model consisting of eight algorithms surpasses the individual model performance and achieves an accuracy of 98.2%. The category of stand-alone models surpasses the individual models in cancer diagnosis, classifiers and gain competitive outcomes (Fig. 1).

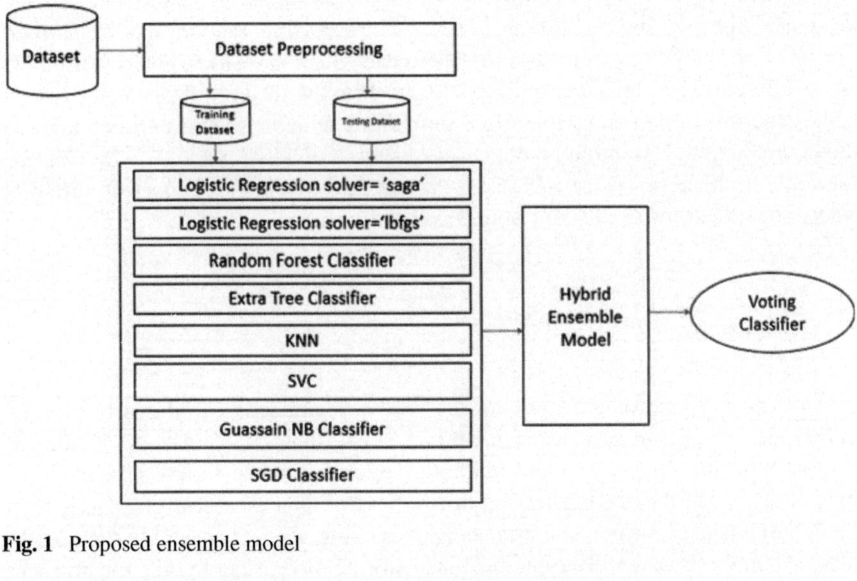

Fig. 1 Proposed ensemble model

4 Results and Discussion

4.1 Results Using Individual Classifiers

Each algorithm's confusion matrix is calculated. Following data preprocessing, the performance of the classifiers is shown using several performance parameters like accuracy, F1 score, recall, precision, etc.

The confusion matrix for the actual and forecast classes is expressed in true positive (TP), which indicates that the original and predicted classes were correctly classified [4]. True negative (TN) is incorrectly classified. False positive (FP) indicates that something has been misclassified. FN stands for false negative, which indicates that something has been misclassified. All classifiers' performance is evaluated using these formulas.

$$\text{Accuracy} = \frac{\text{TP} + \text{TN}}{(\text{TP} + \text{TN} + \text{FP} + \text{FN})} \tag{1}$$

$$\text{Precision} = \frac{\text{TP}}{(\text{TP} + \text{FP})} \tag{2}$$

$$\text{Recall} = \frac{\text{TP}}{(\text{TP} + \text{FN})} \tag{3}$$

$$\text{F1-score} = 2 * \left(\frac{\text{Precision} * \text{Recall}}{\text{Precision} + \text{Recall}} \right) \tag{4}$$

Before testing the model on the hybrid ensemble, the dataset is passed to each of the seven machine learning algorithms, i.e., logistic regression, random forest classifier, etc., to test the performance in terms of different evaluation metrics. Hybrid ensemble of the different algorithms is combined and passed to the stacking classifier the results of the experimentation on various algorithms along with the stacking classifier are as shown in Table 4. Ensemble model surpasses the individual models on the WDBC dataset with the accuracy of 0.98% for the prediction of breast cancer on different independent features of the dataset like radius, smoothness, concavity, etc.

5 Conclusion

Breast cancer is one of the most deadly diseases, impacting millions of women worldwide. As a result, any advancement in cancer illness diagnosis and prediction is vital for healthy living. Machine learning-based illness detection has proven to be quite valuable for the early detection of a variety of fatal diseases. This study uses the WDBC dataset for breast cancer diagnosis using a novel ensemble method. In the research work, seven different machine learning models are hybridized and sent

Table 4 Comparison of individual model with the ensemble model

Algorithm	Accuracy	F1-score	Recall	Precision
Logistic regression Solver = "lbfgs"	0.95	0.95	0.95	0.94
Logistic regression Solver = "saga"	0.89	0.88	0.86	0.91
Random forest classifier	0.96	0.95	0.95	0.96
Extra tree classifier	0.95	0.95	0.94	0.95
KNN	0.94	0.94	0.94	0.94
SVC	0.95	0.95	0.95	0.94
Gaussian NB classifier	0.93	0.93	0.93	0.93
SGD classifier	0.90	0.90	0.90	0.90
Stacking classifier	0.98	0.98	0.98	0.96

to a stacking classifier for malignancy diagnosis. The proposed hybrid ensemble was found to be more effective than the individual model like KNN, extra tree classifier, SVC, etc.

In future work, feature selection can be utilized to determine the most relevant features to provide to the algorithms and investigate their impact on the outcomes. The findings can also be tested against a range of breast cancer databases. In addition to breast cancer, the suggested ensemble model can be used to diagnose a variety of diseases. Moreover, in a future study, numerous optimization and parallel computing approaches can increase the model's performance.

References

1. The official NVIDIA blog. https://blogs.nvidia.com/blog/category/deep-learning/. Last accessed 2020/08/08
2. Ming C, Viassolo V, Probst-Hensch N, Chappuis PO, Dinov ID, Katapodi MC (2019) Machine learning techniques for personalized breast cancer risk prediction: comparison with the BCRAT and BOADICEA models. Breast Cancer Res 21(1):75
3. Sert E, Ertekin S, Halici U (2017) Ensemble of convolutional neural networks for classification of breast microcalcification from mammograms. In: 39th annual international conference of the IEEE engineering in medicine and biology society 2017, EMBC, IEEE, Korea
4. Mushtaq Z, Yaqub A, Hassan A, Su SF (2019) Performance analysis of supervised classifiers using PCA based techniques on breast cancer. In: International conference on engineering and emerging technologies 2019, ICEET, pp 1–6. IEEE, Pakistan
5. Osman A, Aljahdali H (2020) An effective of ensemble boosting learning method for breast cancer virtual screening using neural network model. IEEE Access 8(39):165–39174
6. Abdullah AJ, Hasan TM, Waleed J (2019) An expanded vision of breast cancer diagnosis approaches based on machine learning techniques. In: International engineering conference 2019, IEC. IEEE, Iraq, pp 177–181

7. Prabadevi B, Deepa N, Krithika LB, Vinod V (2020) Analysis of machine learning algorithms on cancer dataset. In: 2020 international conference on emerging trends in information technology and engineering 2020, IC-ETITE, India, pp 1–10
8. Bayrak EA, Kırcı P, Ensari T (2019) Comparison of machine learning methods for breast cancer diagnosis. In: 2019 scientific meeting on electrical-electronics and biomedical engineering and computer science 2019, EBBT, Turkey, pp 1–3
9. Islam MM, Haque MR, Iqbal H et al (2020) Breast cancer prediction: a comparative study using machine learning techniques. SN Comput Sci 1:290
10. Osmanović A, Halilović S, Ilah LA, Fojnica A, Gromilić Z (2019) Machine learning techniques for classification of breast cancer. In: Lhotska L, Sukupova L, Lacković I, Ibbott G (eds) World congress on medical physics and biomedical engineering 2018. IFMBE proceedings, vol 68/1. Springer, Singapore
11. Chaurasia V, Pal S (2020) Applications of machine learning techniques to predict diagnostic breast cancer. SN Comput Sci 1:270
12. Ganggayah MD, Taib NA, Har YC, Lio P, Dhillon SK (2019) Predicting factors for survival of breast cancer patients using machine learning techniques. BMC Med Inform Decis Mak 9(1):48
13. Khan S, Islam N, Jan Z, Din IU, Rodrigues JJPC (2019) A novel deep learning based framework for the detection and classification of breast cancer using transfer learning. Pattern Recogn Lett 125:1–6
14. Guan S, Loew M (2017) Breast cancer detection using transfer learning in convolutional neural networks. In: 2017 IEEE applied imagery pattern recognition workshop, AIPR, USA
15. Khuriwal N, Mishra N (2018) Breast cancer diagnosis using deep learning algorithm. In: International conference on advances in computing, communication control and networking, ICACCCN
16. Rahman MM, Ghasemi Y, Suley E, Zhou Y, Wang S, Rogers J (2020) Machine learning based computer aided diagnosis of breast cancer utilizing anthropometric and clinical features, IRBM
17. Vaka AR, Soni B, Sudheer Reddy K (2020) Breast cancer detection by leveraging machine learning. ICT Exp 6(4):320–324
18. Mojrian S, Pinter G, Joloudari JH, Felde I, Szabo-Gali A, Nadai L, Mosavi A (2020) Hybrid machine learning model of extreme learning machine radial basis function for breast cancer detection and diagnosis; a multilayer fuzzy expert system. In: RIVF international conference on computing and communication technologies 2020, RIVF. IEEE, Vietnam, pp 1–7
19. World Health Organization. https://www.who.int/cancer/prevention/diagnosisscreening/breast-cancer/en/#:~:text=In 2018, it is estimated, in nearly every region globally. Last accessed 2020/08/08
20. Amrane M, Oukid S, Gagaoua I et al (2018) Breast cancer classification using machine learning. In: Electric electronics, computer science, biomedical engineerings' meeting 2018, EBBT, pp 1–4
21. Bhardwaj R, Nambiar AR, Dutta D (2017) A study of machine learning in healthcare, vol 2. In: IEEE 41st annual computer software and applications conference 2017, COMPSAC, Italy, pp 236–241
22. Shargabi B, Shami F (2019) An experimental study for breast cancer prediction algorithms, vol 12. In: Proceedings of the Second international conference on data science, e-learning and information systems 2019. Association for Computing Machinery, pp 1–6
23. Aslan M, Celik Y, Sabanci K et al (2018) Breast cancer diagnosis by different machine learning methods using blood analysis data. Int J Intell Syst Appl Eng 6(4):289–293
24. Ragab DA, Sharkas M, Marshall S et al (2019) Breast cancer detection using deep convolutional neural networks and support vector machines. PeerJ 7
25. Kadam V, Jadhav S, Vijayakumar K (2019) Breast cancer diagnosis using feature ensemble learning based on stacked sparse autoencoders and softmax regression. J Med Syst 43(8)

26. Wang H, Zheng B, Yoon S et al (2018) A support vector machine-based ensemble algorithm for breast cancer diagnosis. Eur J Oper Res 267(2):687–699
27. Mangasarian O, Street W, Wolberg W (1995) Breast cancer diagnosis and prognosis via linear programming. Oper Res 43(4):570–577

A Novel Smart Aging Approach for Monitor the Lifestyle of Elderlies and Identifying Anomalies

Mohammed Shakil Malek, Pooja Gohil, Sharnil Pandya, Anand Shivam, and Kuldeep Limbachiya

Abstract For the past two decades, there is a paradigm shift in the field of automation (IoT), and it is affecting several aspects of life. "Activity Modeling" is a part of such automation, and it has several applications. One such application of "activity modeling" is in the field of distant well-being investigation and nursing. In proposed study, modeling and analysis of everyday activities of elderly persons are done by "recording of periodic behavior pattern of elderly people" on daily basis. Today, the challenge is to develop such a device that requires least human touch and provides automation in irregularity detection along with alarming arrangement in it. Presently, the investigation focuses on the diagnosis of elderly people's day-to-day behavioral patterns by observing their daily basis routine tasks from time, context, and location point of view. For the sake of precision, numeral sensing and actuator components are installed at retirement homely places. Through literature review, an exclusive sensing blend technique has been recommended in order to record physical, social, seasonal, and well-being-related notes about humdrum activities for elderly persons. As a part of research methodology, an innovative routine-based activity-learning system is recommended that can monitor relative data annotations of countless positions at an elderly home and can aware the custodians if any case of glitch recognition happens. Observations for 20 activities have been recorded for one month in two of elderly smart homes situated in the city having sample size of more than 5000. Analysis leads to the results that legalize the precision and the proficiency of the projected structure.

Keywords Health monitoring · Activity modeling · Behavior pattern · Sensor networks

M. S. Malek (✉)
F. D. (Mubin) Institute of Engineering and Technology, Gandhinagar, Gujarat, India

P. Gohil
Gujarat Technological University, Ahmedabad, Gujarat, India

S. Pandya · A. Shivam · K. Limbachiya
Symbiosis Institute of Technology, Symbiosis International (Deemed) University, Pune, India
e-mail: sharnil.pandya@sitpune.edu.in

P. K. Singh et al. (eds.), *Emerging Technologies for Computing, Communication and Smart Cities*, Lecture Notes in Electrical Engineering 875,
https://doi.org/10.1007/978-981-19-0284-0_13

1 Introduction

Recently, the trend of increasing number of nifty shelters across the world and computerization of everyday livelihood actions in people's life drags attention to the need for development of remote health monitoring systems. Amidst this rapidly rising era of Internet, where reciprocal communication has become customary, allowance for the involvement of users, far-flung well-being nursing is meant to aim a huge crowd of spectators at old-aged homes, which are now facing poor quality of health facilities [1]. In traditional far-flung well-being nursing activity, mainly, two tactics were been used: First is the employment of touch body beams to quantify physique constraints and second is the use of sensor systems in nifty homes for action modeling, which conveys information via report to therapeutic specialists for their opinions and verdicts [25–30]. Although the prime issue with initial tactic is that it subsidizes toward troublesomeness of elderly inhabitants, while the second one generally tallies with an outside healthiness professional, which in turn upsurges the repetitive well-being outlays [31–33]. As a result, there is demand in the field to develop such a device that requires least human touch and provides automation in glitch recognition along with watchful arrangement in it [2, 34–37]. The projected structure largely aims toward analysis of interactive patterns of aged individuals by observing their daily routine tasks being monitored at regular intervals of day, week, month, and year through an outline acknowledgment-based mechanism culture models and underpinning erudition methods. In present study, an innovative tactic is used to carry our daily well-being nursing of aging people and to distinguish, sense, and categorize irregularity situations by giving instant notice to the well-being specialists [3, 38–44]. The anticipated research work is further alienated into quatern segments: (i) proposal and experimental setup (ii) activity modeling procedure (iii) outcomes and debate (iv) impending augmentations.

2 Structure Planning

As portrayed below, entire well-being nursing structure planning comprises of five different layers, individually having its own practical position [45–49]. The first layer being physical layer consists of diverse hardware components, while next being the edge layer where events are classified into different categories. Moving on to processing layer through cloud layer, one can reach last and ultimate layer which is layer of application. Here, the final interactions take place between the elderly persons and the caretakers and medical specialists [50–56].

I. **Physical Layer/Level**—it is a lower most level in structure planning. It contains the individual hardware setups such as beams, ESP8266, conditioning circuits, and Raspberry Pi 3B+. Here, data are generated after detecting the situation and transferring it to the next layer [45–49, 57–61].

II. **Edge Layer/Level**—in this layer, concept of edge computing is utilized. The information is received from the beams using Wi-Fi, and system accomplishes the prior processing of information on micro-computer and then send it to the cloud layer [62–66]. The sub-components over here are numerous perception elements inside each room, and motion trials provided form base for motion identification by a classifier of an event. Now, information is shifted to cloud level using Internet [67–70] where MQTT is used rather than secure socket layer/level (SSL) for encoding safely data among MQTT clients and Raspberry Pi 3B+.

III. **Cloud Layer/Level (CL)**—the third level of health monitoring structure planning is cloud layer. This is made up of various segments for the function of data verification, instantaneous catalogue, file loading, and ML Kit for coating appliance culture on engendered C.S.V [71–74]. Here, firebase is used as BAAS: back end as a service. Facts are initially collected in catalog, wherein through cloud jobs, C.S.V files are generated and are put in a store unit. Later on, appliance erudition prototypical is functional for activity gratitude [74–77].

IV. **Processing Layer/Level**—The C.S.V files generated in CL are put as an input in this layer. Here, the established mechanism erudition prototypical is executed on C.S.V files, leading toward activity gratitude, activity framework peers, and also mapping results [78–80]. This activity pattern generated by this is used in glitch recognition for instantaneous scenarios, as that is prime objective of this entire arrangement. By the conclusion of daily work, report is ready based on the pattern created through the arrangement. This report is then utilized for far-flung well-being nursing by the caretakers.

V. **Application Layer/Level**—lastly, comes the practicality part. The last layer of the system is application layer, wherein the entire far-flung well-being nursing system relates thru caretakers, on basis of report engendered. Report creation at regular interval aids corresponding caretakers to deeply observe elder people's well-being without any help from remedial staff. This indirectly reduces the expense spent on elder person's health. System also alarms caretakers if there is detection of anomaly; thus, if required, doctor can be consulted well in time by sharing the reports [80–82]. Figure 1 represents a structure planning of the proposed system.

2.1 Action Health Nursing

The main motive of this study arrays from the delivery of low-altitude statistics to sensors followed by addition of superior quality information which in turn transfers information via combination of data-based and information-driven methodologies [4]. Latest periodicals have to recognize the effort on its own, as a revenue of digging high quality of data. However, general policy for all is as follows: They should be seen through non-professional activities like "cooking food," "bathing," or "watching television while sitting on a sofa" [5, 80–82]. The moment our tasks are appropriately

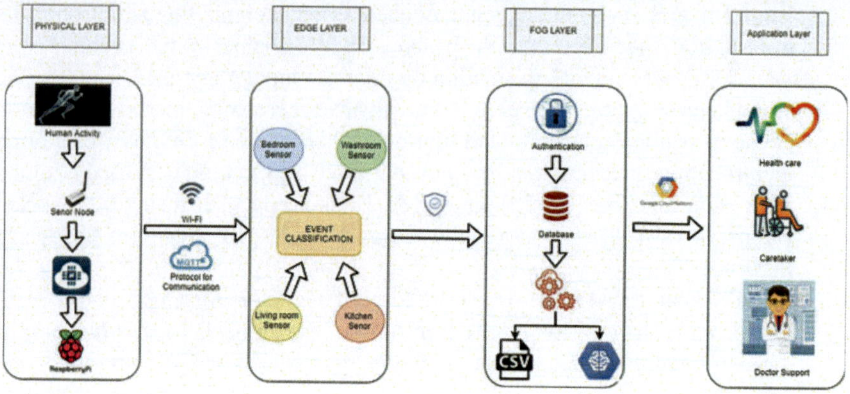

Fig. 1 Structure planning

acknowledged and automatically fixed, a variety of services are accessible just like medicinal emergencies, expert tips on overall lifestyle, pre-diagnosis, and physician assistance. Few of the hands-on examples are as follows: drop recognition system, mobile backup answer system and everyday well-being monitoring activities, and endorse lively existence resources [6, 80–83]. The novel facts so generated from beams, accommodates to a greater cause, serving mavens with treasured info to detect irregularities and ease affected ones.

2.2 *Irregularity Detection*

Apart from social modeling, looking for behavior change "anomaly detection" is likewise significant and challenging job too. Variations in regular living behavior are stalked through irregularity detection, and the intricacy of anomalous data is not considered to be the same as anomalous remoteness behavior [7, 84]. Recognition variations include the variations in several facets of a state like work order, strategy, fitness position, period overbearing, and so on [8, 85]. Fundamentally, communicative finding is contest of gaging humanoid behavior from beam figures and irregularity recognition to check the way to identify behavioral alterations that may contain best pilot of a model to arrest a regular routine. The dual practices to feel behavioral changes are as follows: (i) comment (ii) divider [9]. The comment practice emphasizes on overall behavior and reflects any new admissions that are erratic with the traditional average by means of an irregularity. Discernment reads discrepancy info on or afterward former figures and also examines meant at the similar strategy as new input result records to replicate differences. The comment plan is supplementary reachable by way of an anomaly facts are fewer observable trendy definite period, giving sample learning instances.

Though irregularity discovery is a novice among further approaches, it works in a crafty household and is vastly reputable front-line supplementary areas like interference gratitude, fake credit, negotiator recognition, commercial feature gratitude, picture dispensation, and manuscript data [10, 86]. A number of appliance knowledge methodologies are being used for irregularity detection like segmentation (resultant law, neuronal network, Bayesian, SVM), k-nearest neighbor, bunching, totaling, and evidence philosophy [11, 87]. Now-a-days, an audacious homely base atmosphere, variation recognition is been used broadly in safety awareness of senior citizens and assistive expertise for brain damage as lunatic patients. Statistics can be determined by given methods: (i) irregularity (ii) relative anomaly (iii) cohesive anomaly [12, 88].

2.3 Daily Living Activities

Daily living activities refer to belongings of our day-to-day life, say for an example, grooming, household chores, feeding, bathing, dressing, etc. These aerobics describe the capability near existing popular isolated families unconventionally.

ADL is crucial inhabited and vital humanoid workouts for people's everyday timetables including drinking, consumption, dressing, nodding, washing, and so on. They are attentive in two time primary assemblies, basic chores of everyday living. IADLs are diverse preps that are not substantial continually [13, 89]. Notwithstanding, IADLs ease the elderly and hampered people, for stuff like, household work, getting food ready, body toning, clearing, shopping, vocation, and pressing. The greeting of human workouts in enclosed situations is bothersome undertaking to achieve. ADL focuses on nursing the challenges and noticing reactions for thoughtful human crusade in intense environments [14, 89]. Persistent circumstances like nifty homes have stimulated everyday drills finding, facilitated associates with household ventures just to help incapacitated persons and elderly people for leaving peacefully and with liberty [15, 89].

2.3.1 Activity Beams for Daily Living

Basically, three kinds of beam are available along with the strategies for a person's act acknowledgment. They are as follows.

Physical Environment-Based Beams

Proximity, RFID, Zig-Bee, and Wi-Fi can be applied for spotting the overtone amid the individuals and flora surrounding them. Environment-based beams utilize raw data after detected objects to assume act-attempted past people [16, 90]. The carried plans detect act over constituents and their linking with substances. The information which is collected with the help of universal beams are then mentioned to a neighborhood server for added conduct.

Wearable Beams

Wearable beams like accelerometers, inertial beams, electro-mechanical switches, pedometers gyroscopes, and goniometers are body-appended beams and are viewed as broadly known devices aimed by individuals' deed acknowledgment. They are exploited to see actions of human being apart from the track of help for the people salutation expansions [17, 91]. These devices are designed to continually quantify physical and biomechanical information of the human body. Examining such available information can help in distinctive humanoid exercises in usual living besides making a construal of them in a central edifice employing enterprise acknowledgment.

Remaining Activity Recognition Beams

Camera stranded gadgets go on broadly secondhand for social act acknowledgment inside a restricted detection enclosure. These beams rely arranged the cameras footage as well as cinematic provisions toward perceive hominoid movement by the means of PC image calculations [18, 92]. Audiovisual instruments like RGB-D video, diffusion pictures measurement gadget, are basic kind of chromatic gadgets largely recognized in living being's movement acknowledged by decent nod magnitudes. Though these devices are expensive, they do have high vivacity consumption and furthermore need visit up keep and stand uncovered near shield linked alarms [19, 93].

Intellectual Computation

Intellectual computation idiom is often accustomed to express technological philosophies and podiums that are fundamentally built upon evolving arenas of artificial intelligence together with sign dispensation. These daises encompass all latest mechanism knowledge and in-depth learning zones varying from ordinary language dispensation, virtual reality, and cardinal signal processing for acoustic sorting to speech appreciation [18, 85, 88, 92].

In general, cognitive computing means all the modern hardware and software setups that are meant to thought-provoking an overall human mind functionality in order to support the policymaking by individuals and eliminate the logic of uncertainty in the structure [18, 59, 79, 85]. Cognitive computing systems integrate and blend the evidence from several technical causes, keeping into account the circumstantial hefts, contradictory sources, and information in order to propose and acclaim

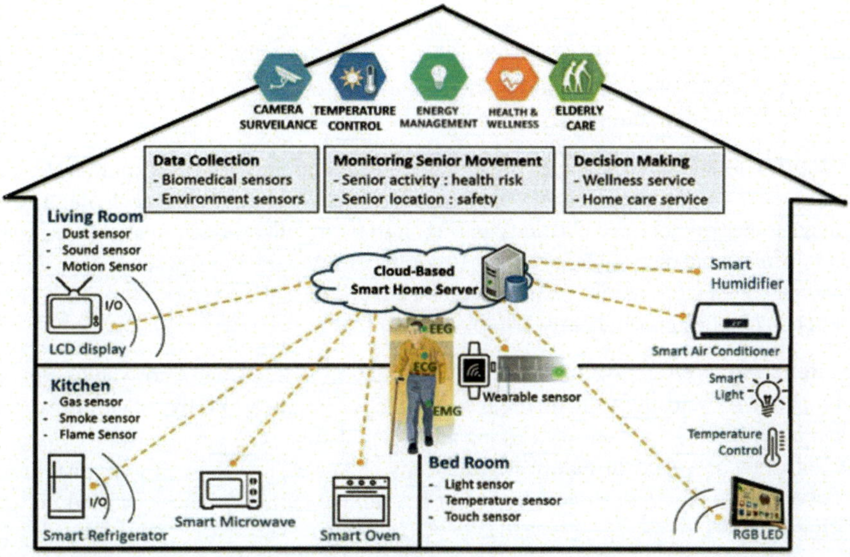

Fig. 2 Architecture of intellectual computation [84]

a superlative key at the end of process. For accomplishing efficiency and precision, intellectual computation integrates all possible kinds of self-education and evolving skills based on diverse categories like "Natural Language Processing" (NLP), statistics outline recognition, and fact mining with the purpose of successfully copying the functioning of human mind as well as behavior. [19, 50, 62, 76] As a matter of fact, the utilization of computers to resolve the issues, which are generally faced by humans, gives rise to inevitable need of hefty amount of structured and non-structured information which is fed to countless apparatus learning and in-depth learning set of rules. Over the period of time, intellectual computation has evolved and progressive methodology of categorizing forms so as to be proficient in anticipating and giving corresponding feasible solutions [20, 31, 41, 53, 66, 84]. Figure 2 represents the architecture of the intellectual computation. According to cognitive computing consortium, in order to gain growing competences, intellectual systems must have following five attributes:

Contextual: To have a thorough familiarity of background of a problem is necessary for problem-solving and is crucial in thought progression. One has to understand and categorize the contextual facts, then recognize and fetch elements such as processes, rules and regulations, handler contour, respective domain, and assigned aims and errands [21, 27, 35, 49]. These processes will extract and utilize various unstructured, structured, sensory, imagery, and audio–visual data [58, 69, 71, 80].

Interactive: Human–computer interactions (HCIs) are considered as main component of intellectual systems. Therefore, one of the vital necessities is to ensure that end handlers must bestowed link with systems with ease and effectively both for input as well as a feedback based on growing environmental circumstances under consideration [20, 47–55, 62–71, 81–85]. Diverse know-hows incorporated within the arrangement must bestowed link with other external devices, micro-processors, and far-off cloud podiums [21, 50, 60, 75].

Adaptive: Intellectual schemes are targeted to assist policymaking systems. Henceforth, schemes must be assembled in a manner that permits policymakers to iteratively gather material and have elasticity unified into them, allowing end users to learn as an informational data vagaries and goalmouths of schemes as a whole evolves. Thus, schemes must slog mainly on instantaneous data and make alterations to previously defined schemes consequently [21, 91–93].

Stateful and Reiterative: Intellectual computation systems and know-hows can be competent too and altered to distinguish and tactic complications by collecting excess data by enquiries and sensory instantaneous inputs incase accessible state evidence is deficient or vague in nature to reach any final verdict. In such circumstances, structures can address problem by use of any alike incident that has occurred in past [21, 84, 91–93].

3 Proposal and Experimental Setup

In an experimental setup, all the sensors are positioned at different places in the house, with an ESP8266 along with Wi-Fi connected. Here, basic data dispensation occurs, and then, the administered data are transported and charted into Raspberry Pi B+ server. To dodge doubling and flawed data, sensor information is pre-processed at ESP8266 level; thereby, safeguarding that Raspberry Pi is solitarily used as resident server, wherein fresh data are stored in overall setup. Charted data are then sent to cloud storage at regular intervals so as to ensure that backup of data is saved in case of any system shutdown, along with the logs of entire system's working, which can also be used at future stage for restoring the system.

PIR Sensor—it is utilized to detect human motion in a chamber. It chiefly senses the movement of a human being and sends data to cloud database (CDB) via local network.

IR-Sensor—it is alike individual's perceptive senses. It recognizes the restraints and send information to CDB where some hitches are classified like when an entrance is closed it identifies a hindrance.

Hall Magnetic Sensor—this sensor perceives a magnetic field in surroundings. When the sensor notices a magnetic field, it gets charged and sends data to CDB.

Pressure Mat Sensor—it is used to notice the pressure on any area particularly for the sitting places like chair, bed, and sofa. If it detects some pressure, it becomes lively and sends data to CDB.

Temperature–Humidity Sensor (dht-11)—it senses temperature and humidity level in a chamber. Generally, it takes average data of 1 min and send it to the CDB.

4 Outcomes and Debate

In this segment, we debate about the two datasets that represent the outcomes for various happenings being logged in a house through fitted sensory items for ADL of elder people, for cross-validation of an anticipated methodology.

4.1 DataSet for Activities of Different House Sensors

Data are collected for a house with single elderly occupant for a duration of about a week. Sensors installed in the architecture are encompassing sensors that are discreet in nature, with an added benefit of cheaply available in market. In the conducted experiments we have placed various temperature and humidity sensing units such as DHT11 in different corners of the smart home. Accordingly, doors and cupboard sensors are placed for corresponding data assortment.

Table 1 shows different types of activities mapped for a house with solo elderly occupant. All the given activities have their discrete sensors that map an individual's data for a duration of 1 week.

Figure 3 portrays data for humidity in atmosphere outside a house. It can be observed that due to rainfall at regular interval, by the end of July month, the humidity level reaches up to the level of 95%.

Figure 4 illustrates the usage of AC with respect to duration. As one can see, AC usage is maximum on 13th and 16th day of July month, while its minimum on 15th and 17th day of July month. This is due to effect of various factors like hall temperature, climate temperature, humidity in atmosphere, time of attendance in house and health state on corresponding week days.

Figure 5 represents frequency of opening a cupboard in a particular day. Actually, this data help in governing if any activity is ongoing inside bedroom, or person is simply sleeping or reading.

Table 1 Data of mapped activities

Type of activity	Start date	Finish date
Hall temperature	13/07/2019	19/07/2019
Climate temperature	13/07/2019	19/07/2019
Internal temperature of a house	26/07/2019	01/08/2019
Humidity in atmosphere outside a house	26/07/2019	01/08/2019
Bedroom	13/07/2019	19/07/2019

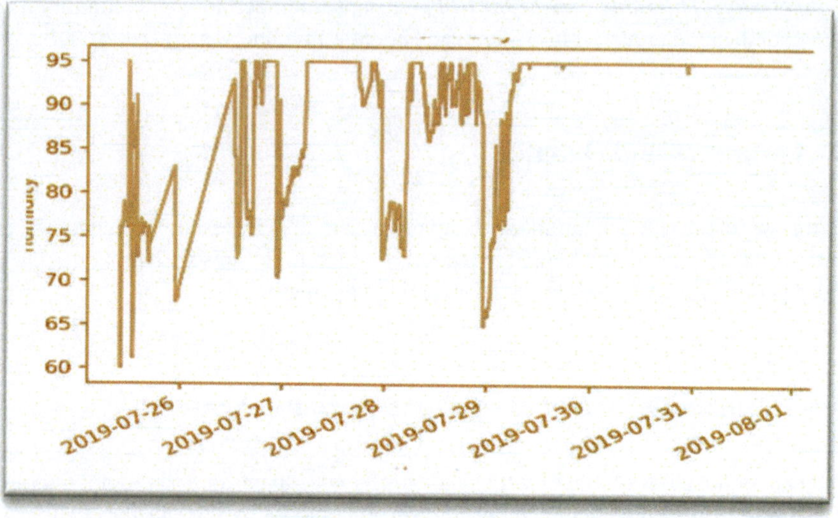

Fig. 3 Measure of humidity in atmosphere

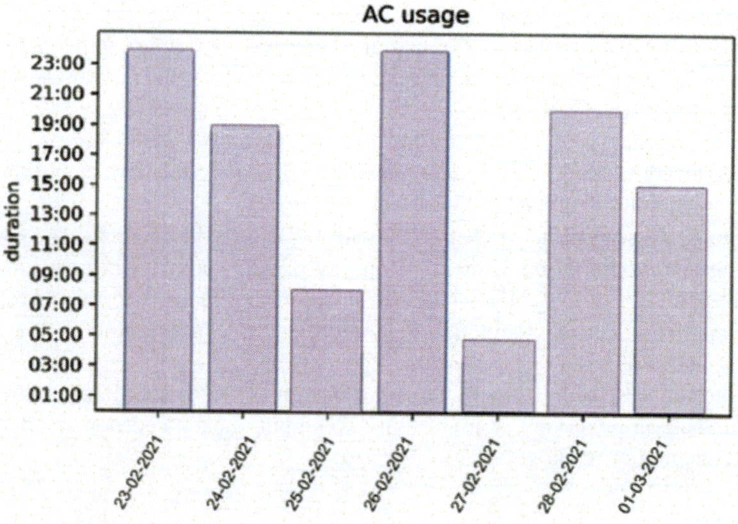

Fig. 4 Usage of AC

4.2 ADL Measurement

Data produced by the installed sensors are both cardinal as well as referend in nature. For instance, information created by hall magnetic sensor is cardinal and duple in nature and is in form of 0 or 1 only which indicates non-active and active states,

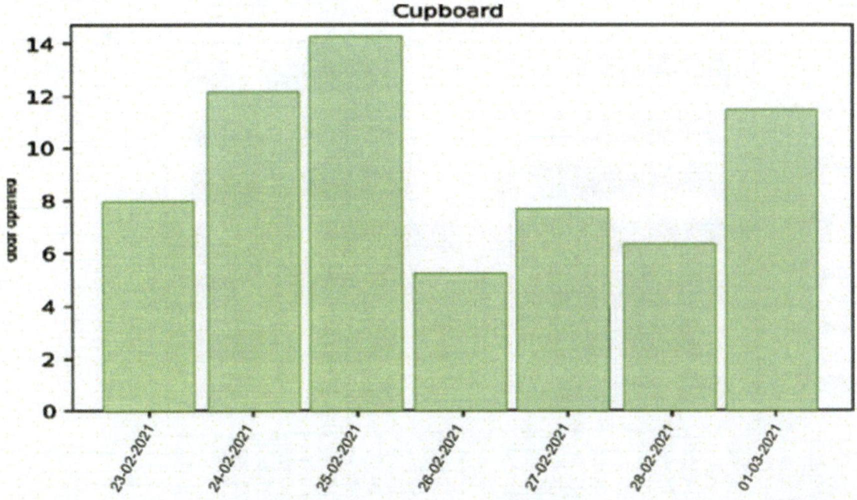

Fig. 5 Frequency of cupboard opening in a day

Table 2 ADL mapping

Type of activity	Start date	Finish date
Dining	13/07/2019	19/07/2019
Relaxing	13/07/2019	19/07/2019
Sleeping	13/07/2019	19/07/2019
Bathing	13/07/2019	19/07/2019

respectively. In the same way, the data produced by pressure mat is also cardinal and duple in nature. Activities done by aged persons in a house are read from a sensor reading only by applying apt logic which differs for each and every activity. In all, total 14 activities are recorded by the physical grid setup which includes relaxing, bathing, sleeping, dinning, reading, cooking, opening, and closing of doors and cupboards of a house, watching television, lavatory usage, and sitting on chairs and sofas. Each recorded activity has start date, finish date, duration, and location committed to it as depicted in Table 2. Figure 6 represents daily activities of elderly persons.

4.3 Comparison of Older and Younger People's Data

Equated to younger ones, elder person's health when gets deteriorate, their toilet visits reflect a significant rise from a usual pragmatic pattern. As recorded for about one week, during sickness in elders, i.e., on 17th and 18th July, there is noticeable

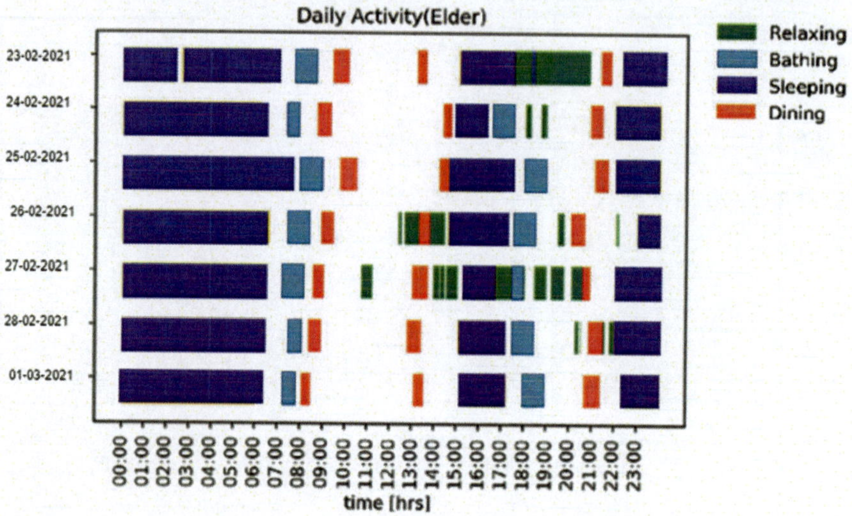

Fig. 6 Daily activities of elderly persons

change in their sum of toilet visits. Figure 7 represents a comparison of elder and younger person visits.

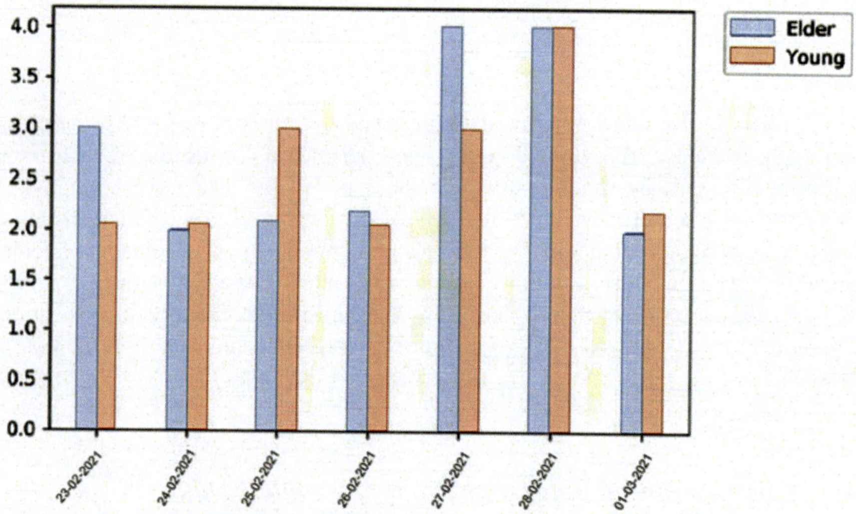

Fig. 7 Comparison of elder and younger person's lavatory visits

5 Impending Augmentations

The present work reflects the prevailing state of research for IOT by reviewing presently relevant and accessible masses of literature and highlighting recent drifts in the ongoing study field. The motive of the work is to intricate the current issues faced by the researchers in development of remote well-being nursing systems useful for prognostic well-being nursing, and how anxieties of confidentiality assault have disappointed systems based on pictorial systems grounded on concurrent computer visualization applications and thus motivate scholars to move forward to well-being nursing using non interfering sensor grids. The current study proposes a distant well-being nursing system reposed on a wireless sensor grid installed at an elderly resident's house aiming to monitor health of aging people staying inside house and predict in case of any irregularity based on analytical investigation of behavioral pattern of concerned individual which can otherwise be done by using several machine learning models such as advanced hidden Markov models and predictive time-series analysis.

Proposed arrangements can be successfully arrayed at various places such as old-age homes where there are a smaller number of caretakers compared to the elderly people residing over there; houses consisting of senior citizens staying alone, hospitals dealing with health monitoring, etc. Over here in all the cases, distant and prophetic health nursing can be a bonus as it is based on concept of edge computing and no touch body radars henceforth no requirement for a person under deliberation to put on sensor regularly. With the help of these sensors, medicinal personnel can also endlessly observe the behavioral reports generated by them on his/her mobiles or computers and can prescribe medicines based on past history of a patient under observation.

Some of the pitfalls of the present device are flawed data created by falsely rectified sensors, low prognostic power of current prevailing models, and prerequisite of huge quantity of data before model begins to bounce its analytical results. However, these all faults can be efficiently settled by using nearly advanced machine learning models and making copies skilled of developing it selves based on instantaneous situations and applying data collected at similar location for a group to be exploited for analysis for same group till substantial quantity of data is obtainable for current setup. Also, further exploration of the field of remote health monitoring is required for upgrading existing system and further progress the distant well-being nursing systems by integrating prevailing technologies in better ways.

References

1. Cadavid H, Garson W (2005) Towards a smart farming platform: from IoT-based crop sensing to data analytics. In: Radford T (ed) The guardian. Springer. https://doi.org/10.1007/978-3-319-98998-3_19. [Internet]. 2005 Available from https://www.theguardian.com/science/2005/mar/30/environment.research

2. Nandurkar S, Thool V, Thool R (2014) Design and development of precision agriculture system using wireless sensor network. In: International conference on automation, control, energy and systems (ACES), Hooghly
3. Andrew R, Malekian R, Bogatinoska D (2018) IoT solutions for precision agriculture. In: MIPRO, Opatija
4. Benyezza H, Bouhedda M (2018) Smart irrigation system based ThingSpeak and Arduino. In: International conference on applied smart systems, ICASS, Médéa
5. Mat I, Kassim M, Harun A (2015) Precision agriculture applications using wireless moisture sensor network. In: IEEE 12th Malaysia international conference on communications, Kuching
6. Fountas S, Aggelopoulou K, Gemtos T (2016) Precision agriculture: crop management for improved productivity and reduced environmental impact or improved sustainability. Supply Chain Manag Sustain Food Netw
7. Miles C (2019) The combine will tell the truth: on precision agriculture and algorithmic rationality. Big Data Soc 1–12
8. Malek S, Akalkotkar P (2014, December) Need of PPP in Indian road sector. Int J Emerg Technol Appl Eng Technol Sci. In: Proceedings of the international conference on information, knowledge and research in engineering management and sciences (IC-IKR-EMS), Kalol, Gujarat, India. KITRC, pp 300–305
9. Saiyed F, Pitroda J, Malek S (2016, April) Assessment of very critical risk factors in real estate sector by criticality index. In: Proceedings of the international conference on engineering: issues, opportunities and challenges for development, Umrakh, Gujarat, India, SNPITRC, pp 193–205
10. Ghayvat H, Mukhopadhyay S, Gui X, Suryadevara N (2015) WSN-and IoT-based smart homes and their extension to smart buildings. Sensors 15:10350–10379
11. Nautiyal V, Shah Y, Malek S (2019) Application of pavement management system in HDM-4: a critical review. Int J Manag Technol Eng 9(4):1850–1859. https://doi.org/16.10089.IJMTE.2019.V9I4.19.27237
12. Macwan J, Bhavsar A, Malek M (2019) Smart project scheduling for construction projects using AI. Int J Tech Innov Mod Eng Sci 5(5):14–19
13. Agrawal A, Malek M (2018) Principles of sustainability in construction project management. Int J Adv Innov Res 5(4):45–52
14. Gajera D, Malek M (2018) A Study on consumer behaviour in real estate for Vadodara city. Univ Rev 7(12):956–969
15. Kumbhani H, Malek M (2018) Cost management process in construction industry. Suraj Punj J Multi Res 8(12):151–158. https://doi.org/16.10089.SPJMR.2018.V8I11.16.3145
16. Upadhyay U, Malek M (2018) Identification of factors responsible for organizational transformation of GSRDC. Pramana Res J 8(12):189–201. https://doi.org/16.10089.PRJ.2018.V8I12.18.2063
17. Jariwala K, Malek M (2018) Total quality management in construction of textile market in Surat city. J Appl Sci Comput 5(12):699–706. https://doi.org/16.10089.JASC.2018.V5I12.453459.1500186
18. Bachwani D, Malek M (2018) Parameters indicating the significance of investment options in real estate sector. Int J Sci Res Rev 7(12):301–311. https://doi.org/16.10089.IJSRR.2018.V7I12.287391.2950
19. Rathod S, Malek S (2018) Six sigma in construction industry: literature review. Int J Manag Technol Eng 8(12):913–920. https://doi.org/16.10089.IJMTE.2018.V8I12.17.2102
20. Kochai Z, Malek M (2018) Causes of schedule delay and cost overrun in Afghanistan infrastructure projects. Int J Adv Innov Res 5(3):18–25
21. Araya SY, Malek S (2018) Application of building information modelling in Indian construction projects—a critical review. Int J Creative Res Thoughts 6(1):979–987
22. Tadewos SG, Malek S, Patel D (2018) Challenges of building construction in Mekelle city, Ethopia: from contractor's perspective. J Emerg Technol Innov Res 5(3):420–424. https://doi.org/10.6084/m9.jetir.JETIR1803080

23. Aynalem YF, Malek S (2018) Developing an effective project concept in 5D environment. J Emerg Technol Innov Res 5(5):372–380. https://doi.org/10.6084/m9.jetir.JETIR1805660
24. Pandya A, Malek S (2018) A review for the causes of delay in Indian construction projects. Int J Adv Eng Res Dev 5(3):1–6
25. Shah B, Malek S (2018) A recapitulation on exigency of smart villages in Indian ambience. Int J Adv Eng Res Dev 5(3):1–8
26. Patel V, Malek S (2018) A review on identification of safety factors in construction projects. Int J Adv Eng Res Dev 5(3):1–9
27. Parekh P, Malek S (2018) A critical literature review on implementing lean construction concepts in infrastructure projects. Int J Adv Eng Res Dev 5(3):1–7
28. Chaudhari K, Malek S (2018) Effectiveness of public private partnership in infrastructure development in India: a review. Int J Adv Eng Res Dev 5(3):1–12
29. Hada N, Malek S (2018) Identification of risk factors in government housing project: a review. Int J Adv Eng Res Dev 5(3):1–6
30. Jethwa B, Bhavsar A, Malek S (2017) An overview of factors affecting risk management in PPP based infrastructure projects. Int J Eng Dev Res 5(2):1737–1742
31. Jethwa B, Bhavsar A, Malek S (2017) Critical review on risk management in PPP based infrastructure projects. Int J Eng Dev Res 5(1):357–361
32. Parmar D, Pitroda J, Malek S (2017) Use of BIM (Building Information Modelling) as an integrated tool to plan, design and manage critical construction projects. Int J Adv Res Innov Ideas Edu 3(1):1145–1153
33. Malek S, Saiyed F (2017) A study on incipient trends: the real estate sector of Ahmedabad city (Gujarat). Int J Res Granthalayah 5(1):304–315
34. Patel D, Malek S (2016) Application of performance management system in infrastructure sector of construction industry. Int J Eng Dev Res 4(2):1998–2000
35. Kalkani K, Malek S (2016) Analyzing causes of delay in construction projects. Int J Innov Res Sci Technol 2(12):257–262
36. Upadhyay S, Malek S (2016) Adoption and implementation of total quality management in construction industry. Int J Eng Dev Res 4(2):1256–1260
37. Patel K, Malek S (2016) Planning, scheduling and resource management for residential real estate projects in Ahmedabad by using microsoft project. Int J Eng Dev Res 4(2):189–194
38. Malek S, Akalkotkar P (2016) Driving forces leading to the adoption of PPP: perspective from Gujarat (India), Hong Kong and Australian practitioners. Int J Innov Res Sci Technol 2(9):6–11
39. Malek S, Akalkotkar P (2016) Factors contributing to successful public private partnership projects for highway projects in Indian context. Int J Sci Res Dev 3(12):25–29
40. Malek S, Akalkotkar P (2016) Negative factors adopting public private partnership for highway projects: a case study of Gujarat. Int J Sci Technol Eng 2(8):30–35
41. Malek S, Akalkotkar P, Zala L (2016) Gujarat PPP: attractive factors for adopting public private partnership for highway projects. Int J Futuristic Trends Eng Technol 3(2):1–6
42. Saiyed F, Pitroda J, Malek S (2015) Reaping the returns: review on risk management in real estate. Int J Futuristic Trends Eng Technol 3(1):1–8. Malek M, Mohibali S, Bachwani D (2021) Identification, evaluation, and allotment of critical risk factors (CRFs) in real estate projects: India as a case study. J Project Manag 6(2):83–92. https://doi.org/10.5267/j.jpm.2021.1.002
43. Malek MS, Gundaliya PJ (2020) Negative factors in implementing public–private partnership in Indian road projects. Int J Constr Manag. https://doi.org/10.1080/15623599.2020.1857672
44. Malek M, Gundaliya P (2021) Value for money factors in Indian public-private partnership road projects: an exploratory approach. J Project Manag 6(1):23–32. https://doi.org/10.5267/j.jpm.2020.10.002
45. Malek M, Bhatt V, Patel A (2020) Global, local and national growth of road projects through PPP. Test Eng Manag 83:25837–25860
46. Memon A, Malek M, Patel A (2020) A review on risk and risk management technique in construction projects. Int J Disaster Recovery Bus Continuity 11(1):2695–2701
47. Patel R, Bachwani D, Malek M (2020) Human resource management, a vital tool for labour management in construction industry. Stud Indian Place Names 40(9):201–206

48. Parikh R, Bachwani D, Malek M (2020) An analysis of earn value management and sustainability in project management in construction industry. Stud Indian Place Names 40(9):195–200
49. Shah K, Malek M, Phugat S (2020) Alternate building materials used in construction industry. Stud Indian Place Names 40(9):184–189
50. Khan BR, Malek M, Tidke A (2020) Risk management in construction industry. Stud Indian Place Names 40(9):165–171
51. Khan S, Malek M, Phugat S (2019) Factors influencing time and cost overrun in construction of government buildings: Indore as a case study. CIKITUSI J Multi Res 6(4):288–295. https://doi.org/16.10089.CJMR.2019.V6I4.19.2857
52. Pandya S, Sur A, Kotecha K (2020) Smart epidemic tunnel—IoT based sensor-fusion assistive technology for COVID19 disinfection. Emerald
53. Patel NR, Kumar S (2017) Enhanced clear channel assessment for slotted CSMA/CA in IEEE 802.15.4. Wirel Pers Commun 95:4063–4081
54. Patel NR, Kumar S (2018) Wireless sensor networks' challenges and future prospects. In: 2018 international conference on system modeling and advancement in research trends (SMART), Moradabad, India, pp 60–65
55. Ghayvat H, Awais M, Pandya S, Ren H, Akbarzadeh S, Chandra Mukhopadhyay S, Chen C, Gope P, Chouhan A, Chen W (2019) Smart aging system: uncovering the hidden wellness parameter for well-being monitoring and anomaly detection. Sensors 19:766
56. Saket S, Pandya S (2016) An overview of partitioning algorithms in clustering techniques
57. Shah, JM, Kotecha K, Pandya S, Choksi DB, Joshi N (2017) Load balancing in cloud computing: methodological survey on different types of algorithm. In: 2017 international conference on trends in electronics and informatics (ICEI). https://doi.org/10.1109/ICOEI.2017.8300865
58. Ghayvat H, Pandya S, Shah S, Mukhopadhyay SC, Yap MH, Wandra KH (2016) Advanced AODV approach for efficient detection and mitigation of wormhole attack in MANET. In: 2016 10th international conference on sensing technology (ICST)
59. Pandya S, Shah J, Joshi N, GhayvatH, Mukhopadhyay SC, Yap MH (2016) A novel hybrid based recommendation system based on clustering and association mining. In: 2016 10th international conference on sensing technology (ICST)
60. Patel S, Singh N, Pandya S (2017) IoT based smart hospital for secure healthcare system. Int J Recent Innov Trends Comput Commun
61. Pandya SP, Prajapati MR, Thakar KP Assessment of training needs of farm women. Guj J Ext Edu 25(2):169–171
62. Pandya S, Ghayvat H, Sur A, Awais M, Kotecha K, Saxena S, Jassal N, Pingale G (2020) Pollution weather prediction system: smart outdoor pollution monitoring and prediction for healthy breathing and living. Sensors 20:5448
63. Pandya S, Ghayvat H, Kotecha K, Awais M, Akbarzadeh S, Gope P Smart home anti-theft system: a novel approach for near real-time monitoring and smart home security for wellness protocol. Appl Syst Innov 1(4):42
64. Patel RR, Pandya SP, Patel PK Characterization of farming system in north west Agro climatic zone of Gujarat state. Guj J Ext Edu 27(2):206–208
65. Pandya S, Ghayvat H, Kotecha K, Yep MH, Gope P (2018) Smart home anti-theft system: a novel approach for near real-time monitoring, smart home security and large video data handling for wellness protocol
66. Joshi N, Kotecha K, Choksi DB, Pandya S (2018) Implementation of novel load balancing technique in cloud computing environment. In: International conference on computer communication and informatics (ICCCI)
67. Patel W, Pandya S, Mistry V (2016) i-MsRTRM: developing an IoT based intelligent medicare system for real-time remote health monitoring. In: 2016 8th international conference on computational
68. Wandra KH, Pandya S (2012) A survey on various issues in wireless sensor networks. Int J Sci Eng

69. Saket JS, Pandya S Implementation of extended K-medoids algorithms to increase efficiency and scalability using large dataset. Int J Comput Appl
70. Bhola YO, Socha BN, Pandya SB, Dubey RP, Patel MK (2019) Molecular structure, DFT studies, Hirshfeld surface analysis, energy frameworks, and molecular docking studies of novel (E)-1-(4-chlorophenyl)-5-methyl-N′-((3-methyl-5-phenoxy-1-phenyl-1H-pyrazol-4-yl) methylene)-1H-1, 2, 3-triazole-4-carbohydrazide. Molecular Cryst Liq Cryst
71. Patel WD, Pandya S, Koyuncu B, Ramani B, Bhaskar S (2019) NXTGeUH: LoRaWAN based next generation ubiquitous healthcare system for vital signs monitoring and falls detection. In: 2018 IEEE Punecon
72. Dandvate HS, Pandya S (2016) New approach for frequent item set generation based on Mirabit hashing algorithm. In: 2016 international conference on inventive
73. Swarndeep SJ, Pandya S (2016) Implementation of extended K-medoids algorithm to increase efficiency and scalability using large datasets. Int J Comput Appl
74. Wandra K, Pandya S (2014) Centralized timestamp based approach for wireless sensor networks. Int J Comput Appl
75. Garg D, Goel P, Pandya S, Ganatra A, Kotecha K (2002) A deep learning approach for face detection using YOLO. In: 2018 IEEE Punecon
76. Sur A, Pandya A, Sah RP, Kotecha K, Narkhede S (2020) Influence of bed temperature on performance of silica gel/methanol adsorption refrigeration system at adsorption equilibrium. Part Sci Technol
77. Sur A, Sah RP, Pandya S (2020) Milk storage system for remote areas using solar thermal energy and adsorption cooling. Mater Today Proc
78. Cohen JM, Pandya S, Tangirala K, Krasenbaum LJ (2020) Treatment patterns and characteristics of patients prescribed AJOVY, Emgality, or Aimovig. In: HEADACHE
79. Cohen JM, Pandya S, Krasenbaum LJ, Thompson SF (2020) A real-world perspective of patients with episodic migraine or chronic migraine prescribed AJOVY in the United States. In: HEADACHE
80. Barot V, Kapadia V, Pandya S (2020) QoS enabled IoT based low cost air quality monitoring system with power consumption optimization. Cybern Inform Technol
81. Ghayvat H, Pandya S, Patel A (2019) Proposal and preliminary fall-related activities recognition in indoor environment. In: 2019 IEEE 19th international conference on
82. Akbarzadeh S, Ren H, Pandya S, Chouhan A, Awais M (2019) Smart aging system
83. Ghayvat H, Pandya S (2018) Wellness sensor network for modeling activity of daily livings—proposal and off-line preliminary analysis. In: 2018 4th international conference on computing
84. Awais M, Kotecha K, Akbarzadeh S, Pandya S (2018) Smart home anti-theft system
85. Patel M, Pandya S, Patel S (2017) Hand gesture based home control device using IoT. Int J Adv Rese
86. Pandya S, Yadav AK, Dalsaniya N, Mandir V Conceptual study of agile software development
87. Samani MD, Karamta M, Bhatia J, Potdar MB (2016) Intrusion detection system for DoS attack in cloud. Int J Appl Inform Syst
88. Ayanelam YF, Malek S (2018, 16–17 February) A critical review of effective project management using 5D BIM. In: Smart construction. 1st international conference on engineering and technology: smart engineering, Parul University, Baroda, India. McGraw Hill Education, pp 155–158
89. Kochai Z, Malek S (2018, 16–17 February) An assessment of time and cost overrun in Afghanistan's construction. In: Smart construction. 1st international conference on engineering and technology: smart engineering, Parul University, Baroda, India. McGraw Hill Education, pp 159–166
90. Chaudhari K, Malek S (2018, 16–17 February) Critical review on delay and cost overrun in construction projects. In: Smart construction. 1st international conference on engineering and technology: smart engineering, Parul University, Baroda, India. McGraw Hill Education, pp 167–172

91. Shah B, Malek S (2018, 28–30 June) Identification of significant factors affecting the design of smart village. In: Proceedings of the 2nd international conference on women in science and technology: creating sustainable career, Birla Vishvakarma Mahavidyalaya, Vallabh Vidyanagar, India
92. Saiyed F, Malek S (2017, 10–11 November) The realty growth trajectory of india and china-inexorable comparison. In: Proceedings of the 2nd international conference on construction, real estate, infrastructure and project management, NICMAR, Pune, India
93. Olatinwo S, Joubert T (2019) Enabling communication networks for water quality monitoring applications: a survey. IEEE 7(100332)

Spatial Analysis of Violent Crime Dataset Using Machine Learning

Falade Adesola, Ambrose Azeta, Sanjay Misra, Aderonke Oni, Ravin Ahuja, and Ademola Omolola

Abstract The monster called crime has been living with us from the beginning of human existence and impacts negatively on the general health of a nation. Different approaches were employed in the past studies for predicting occurrence of violent crime to aid predictive policing, which makes conventional policing more efficient and proactive. This paper investigates the accuracy of Machine Learning-based crime prediction approaches, which were used previously by other researchers. This study presents Machine Learning approaches to violent crime prediction. Five years' historical dataset between July 2014 and July 2019 were collected from Nigerian Police Lagos, analyzed and used for training the models built. Two different Machine Learning predictive models, Decision Tree and K-Nearest Neighbor, were implemented using IBM Watson Studio and violent crime prediction accuracy of 79.65%, and 81.45% were obtained, respectively, with the real-life dataset collected from Nigerian Police Obalende Lagos and online crime reported portal during violent crime prediction in Lagos. This could be used to enhance crime prevention and control strategies in curbing the worrisome crime rate in the country.

Keywords Machine learning · Crime · Prediction

F. Adesola · A. Azeta · A. Oni · A. Omolola
Covenant University, Ota, Nigeria
e-mail: adesola.falade@covenantuniversity.edu.ng

A. Azeta
e-mail: ambrose.azeta@covenantuniversity.edu.ng

A. Oni
e-mail: aderonke.oni@covenantuniversity.edu.ng

S. Misra (✉)
Ostfold University College, Halden, Norway
e-mail: sanjay.misra@covenantuniversity.edu.ng

R. Ahuja
Shri Viswakarma Skill University, Gurgaon, Hariyana, India

P. K. Singh et al. (eds.), *Emerging Technologies for Computing, Communication and Smart Cities*, Lecture Notes in Electrical Engineering 875,
https://doi.org/10.1007/978-981-19-0284-0_14

1 Introduction

Crime remains human security challenges across the globe. Different countries of the world have to grapple to contain the rising wave of violent crime such as banditry, kidnapping, rape, murder, and manslaughter which in some cases already assuming a different dimension. According to [1, 2], Nigeria is currently found in the web of crime manifesting in an upsurge of violent and non-violent crime. This worrisome trend has made Nigeria notable with high crime rate in the world [3].

Presently, the crime is now ravaging the country like tsunami and blowing a climate of anxiety and fear about our public safety. Negative effects of crime include and not limited to the following: increased poverty among citizenry, unemployment, and sudden death just to mention a few.

Following a report by National Bureau of Statistics, 2018, crime rate in Nigeria is increasing on a yearly of 3.4% over the last thirty years. This should be a thing of concern to all and sundry and demand urgent attention. Due to the increase in violent crime such as armed robbery, kidnapping, rape, murder, manslaughter, and aggravated assault across the length and breadth of the nation, anxiety of people in the community have become intensified. Predictive policing is the way to go as it is already being implemented in some countries of the world for example USA and China [3, 4].

In the year 2010 and 2011, there were more reports of armed robbery incidents in Nigeria compared to other years as shown in Fig. 1. Additionally, cult killings were also reported in various parts of Nigeria between year 2009 and 2018, and it is observed that year 2010, 2018, and 2019 had the deadliest cult killings in the country [4].

Notable authors in crime prediction domain [2–4] have done a lot of research work to improve on the techniques of crime prediction. Machine Learning and data

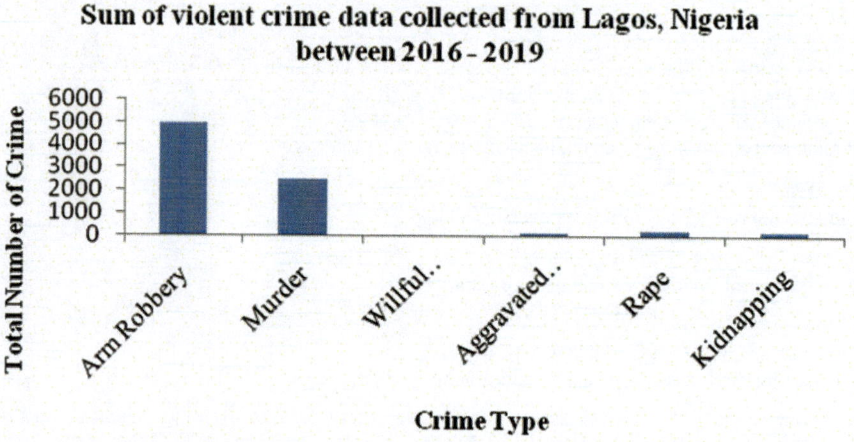

Fig. 1 Violent crime dataset summary for this study

mining approaches have been reportedly used by these authors. Their works have received diverse attention looking at the attendance benefits of crime prediction which is both beneficial to the police patrol team and the citizens at large. Occurrence of violent crime brings about death and very serious injuries in some cases. One useful application of crime prediction is that police patrols are undertaken by the police depending on the predicted crime hotspots in the vicinity.

The paper is organized as follows. Section 2 provides the literature review. Section 3 provides materials and methods. Section 4 provides the results and discussion and finally conclusion drawn in Sect. 5.

2 Related Work

In [5], an effort to make useful contribution to crime prediction, divided the entire region in consideration into grids and proposed a novel spatio-temporal crime network prediction model that was based on Convolutional Neural Network for automatic crime referenced features extraction. The model developed can forecast crime risk in each region of the urban areas for the next day using high volume of dimensional data. The model was evaluated and discovered to outperform other models in the previous studies. However, the gap observed is it requires high dimensionality of crime data to be able to make accurate prediction, which is the major shortcoming of deep neural network.

Authors in [6] developed a linear regression model use for forecasting crime trends in Bangladesh. Real crime dataset was collected from Bangladesh police website, and this data contained a collection of different crime types. The linear regression model developed was trained on the dataset, after which the model was used in forecasting future crime trends. In the empirical evaluation carried out, the linear regression model outperformed other three selected algorithms. The gap observed here is it takes time to make crime prediction.

Authors [7], in a research work titled "Crime Prediction Using Twitter Sentiment and Weather" developed a crime prediction model that was able to predict the location and time of occurrence of a particular type of crime by using sentiment analysis based on lexicon methods and understanding of categorized weather data. The obtained crime prediction by combining sentiment analysis based on lexicon method and Kernel Density Estimation using historical crime dataset via linear modeling approach. The result of the experiment showed that the accuracy of the model developed surpasses the benchmark model, which makes prediction using Kernel Density Estimation. The limitation of this approach is that it could not predict time slot and crime location for the occurrence of crime.

2.1 K-Nearest Neighbor Classifier

KNN usually stores the available objects during classification and then use similarity measure to classify new objects by seeking the nearest neighbor of the input values [8]. It is a classifier that makes a prediction based on the majority vote of the k-nearest samples on the feature vector space. KNN performs the following basic steps during training:

- The distance is calculated
- The closest neighbor is sought after
- The labels are voted for.

2.2 Decision Tree Classifier

In building the model for Decision Tree, the datasets are normally divided into smaller pieces to form the tree structure. During every singular stage of the process, the feature that best fit and splits the dataset is chosen with the assistance of two the important functions, namely information gain and Gini index. Gini index determines the likelihood of a random sample being classified incorrectly using the equation in (1)

$$I_G(p) = \sum_{i=1}^{k} p_i(1 - p_i) \tag{1}$$

Information gain assists in deciding which the features in the dataset will split next. Equation (2) is used in calculating the information gain using entropy.

$$H(T) = I_E = -\sum_{i=1}^{k} p_i \log(p_i) \tag{2}$$

3 Materials and Methods

Datasets used for this study were derived from the historical crime dataset of Nigerian Police Obalende Lagos and online crime reported portal. The violent crime datasets between July 2014 and July 2019 were pulled out from the general crime data made available for this research and this amounting to a total of 8234 turples with 14 attributes. After data cleaning and diligent application of principal component analysis, nine attributes which are relevant for violent crime prediction were made available and then applied. The various important attributes in the dataset are:

Table 1 Violent crime dataset summary used for the experiment

Crime type	Sum of data collected	Percentage (%)
Arm robbery	4984	60.5
Murder	2549	31.0
Willful manslaughter	82	1.0
Aggravated assault	146	1.8
Rape	273	3.3
Kidnapping	200	2.4
	Total	8234

crime description, crime id, time of crime, date crime was committed, number of deaths involved, type of crime, crime location, as well as crime hotspot. The first seven attributes are the predictor or dependent variables and the last attributes represent the target variable or the class label. Presented in Table 1 is the violent crime dataset summary pulled out between July 2014 and July 2019.

The violent crime dataset summary used for the study is also presented in Fig. 1.

Decision Tree and K-Nearest Neighbor algorithms were chosen for this study because of the outstanding performances of the duo in literature and quick adaptation to changes in dataset.

Watson Studio was used to realize the model implementation. Watson studio is an integrated and collaborative platform for data scientists and business analysts specifically built on open-source components and IBM added value tools [9]. It is available both in the cloud and on premise. It is built around open-source components such as Python, Scala, R, SQL, Apache Spark, and Notebooks (Jupyter and Zeppelin). IBM added value tools such as Watson Machine Learning, Watson Natural Language Processing, Watson Natural Language Understanding, SPSS Predictive analytics algorithms, Chatbots, and many more are provided in Watson Studio. It is endowed with set of tools to enable data scientists and business analysts to collaboratively and work easily.

Therefore, the Machine Learning models for both K-Nearest Neighbor and Decision Tree were built in IBM Watson Studio [10, 11]. Python programming language in Watson Studio was then used for violent crime prediction. Confusion matrix was used to evaluate the performances of the models developed with evaluation results presented in Table 2.

4 Results

During the empirical study, Decision Tree and K-Nearest Neighbor models or predicting the target column were built. This was done after the dataset was split into test and training set [12].

Table 2 Evaluation results of K-Nearest Neighbor classifier for different violent crime types

S/N	Evaluation results for K-Nearest Neighbor for different violent crime types							
	Violent crimes	TP rates	FP rates	Accuracy (%)	Precision (%)	Recall (%)	F1 score (%)	MSE
1	Armed robbery	0.64	0.425	83.72	81.62	80.53	79.21	0.0179
2	Kidnapping	0.69	0.389	81.95	78.75	81.14	78.78	0.0626
3	Rape	0.68	0.578	81.82	80.02	78.99	79.75	0.0709
4	Aggravated Assault	0.65	0.546	80.20	78.24	79.35	75.92	0.0478
5	Murder	0.67	0,255	79.45	77.75	76.55	81.89	0.0187
6	Ritual killing	0.67	0,555	81.55	71.15	80.15	78.39	0.0587

Evaluation Results of K-Nearest Neighbor Classifier for different violent crimes is presented in Table 2, and the accuracy was found to be 81.45%.

Figure 2 presents K-Nearest Neighbor prediction results comparison graph for different violent crime types.

Evaluation Results of Decision Tree Classifier for the violent crimes is presented in Table 3 and the accuracy was found to be 76.69%.

Decision Tree classifier delivers 76.69% accuracy. Figure 3 shows Decision Tree prediction results comparison for different violent crime types.

The Decision Tree classification has an accuracy of 76.69% in its prediction and 59% of the instances correctly classified, while 41% of the instances were incorrectly classified.

The summary of the evaluation results based on accuracy and training time as the standard evaluation techniques from literature [13, 14] is presented in Table 4.

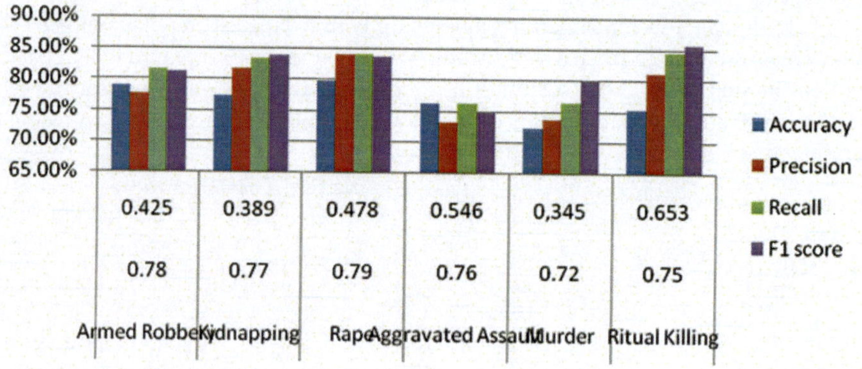

Fig. 2 Prediction performance comparison of KNN for violent crimes

Table 3 Evaluation results of Decision Tree classifier for different violent crime types

S/N	Evaluation results for Decision Tree							
	Violent crimes	TP rates	FP rates	Accuracy (%)	Precision (%)	Recall (%)	F1 score (%)	MSE
1	Armed robbery	0.78	0.425	78.98	77.62	81.53	81.22	0.0179
2	Kidnapping	0.77	0.389	77.25	81.75	83.24	83.78	0.0126
3	Rape	0.79	0.478	79.83	84.02	83.99	83.75	0.0109
4	Aggravated Assault	0.76	0.546	76.20	73.24	76.35	74.90	0.0178
5	Murder	0.72	0,345	72.45	73.75	76.50	79.89	0.0543
6	Ritual killings	0.75	0.653	75.40	81.25	84.46	85.82	0.0645

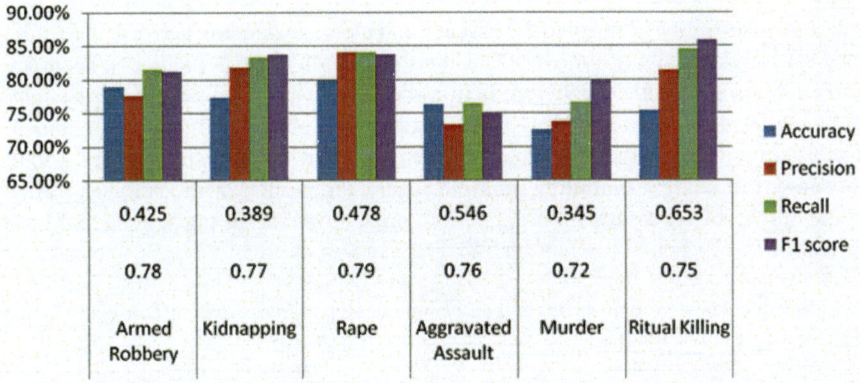

Fig. 3 DT prediction performance comparison between different violent crime types

Table 4 Evaluation results summary of the two models

S/N	Metrics	Decision Tree	KNN
1	Accuracy (%)	76.69	81.452
2	Training time (s)	217	209

Discussion

Several other Machine Learning algorithms were taken into consideration before chosen Decision Tree and K-Nearest neighbor models for this study. This decision was informed by the outstanding performances of both Decision Tree and K-Nearest Neighbor models from literature and their quick adaptation to changes in dataset. The models for both algorithms were developed using IBM Watson studio with Python as the programming language. The preprocessed dataset was trained on the models developed using the split of the training and the test sets. The results of the empirical study reveal that Decision Tree delivers 76.69% predictive accuracy of the unknown class labels. Also K-Nearest Neighbor model returns 81.45% predictive accuracy of the unknown class labels. These two results are good enough for any predictive system to rely on.

The dastard effects of violent crime include anxiety and sudden death among citizenry, bad economy, poor quality of life, and increase in poverty rates among others. During the experiment, over-fitting was avoided during training and testing by engaging in cross validation [8–10, 15–17] Testing of hypothesis is not included in the formulation of this model, rather Decision Tree and KNN Machine Learning techniques were engaged in the formulation of the prediction model.

5 Conclusion

The study has further confirmed the efficacy and more outstanding performance of K-Nearest Neighbor algorithm over Decision model with 81.45% predictive accuracy and Decision Tree with 76.69% prediction accuracy in a new context. By drastically reducing false positives and high true positives rates, a high accuracy was attained. The use of modern method for violent crime prediction has again being displayed in this empirical study. Combining multiple Machine Learning methods to see if they could deliver better results in violent crime prediction may be considered as future work.

Acknowledgements We appreciate the sponsorship and support from Covenant University for the overall success of this study.

References

1. NSW Criminal Law Handbook (2017). Available online https://www.booktopia.com.au/nsw-criminal-law-handbook-2017/book/9780455500058.html
2. Chen P, Yuan H, Shu X (2008) Forecasting crime using the ARIMA model. In: Proceedings of the 5th IEEE international conference on fuzzy systems and knowledge discovery, vol 5, pp 627–630
3. Liao R, Wang X, Li L, Qin Z (2010) A novel serial crime prediction model based on Bayesian learning theory. In: Proceedings of the 2010 IEEE international conference on machine learning and cybernetics, vol 4, pp 1757–1762
4. Wang P, Mathieu R, Ke J, Cai HJ (2010) Predicting criminal recidivism with support vector machine. In: Proceedings of the 2010 IEEE international conference on management and service science, pp 1–9
5. Lian D, Tao H, En C, Jianfeng Z, Chao G (2017) Deep convolutional neural networks for spatiotemporal crime prediction. In: International conference information and knowledge engineering, IKE'17

6. Mohler G, Short M, Brantingham P, Schoenberg F, Tita G (2011) Self-exciting point process modeling of crime. J Am Stat Assoc 106(493):100–108
7. Xinyu C, Youngwoon C, Suk Y (2015) Crime prediction using Twitter sentiment and weather. In: 2015 systems and information engineering design symposium, 24–24 April 2015. https://doi.org/10.1109/SIEDS.2015.7117012
8. Barnadas MV (2016) Machine learning applied to crime prediction. Thesis, Universitat Politècnica de Catalunya, Barcelona, Spain, September 2016
9. Cortes C, Vapnik V (1995) Support vector network. Mach Learn 20(3):273–297. https://doi.org/10.1023/A:1022627411411
10. Alves L, Ribeiro H, Lenzi E, Mendes R (2013) Distance to the scaling law: a useful approach for unveiling relationships between crime and urban metrics. PLoS One 8(8):1–8. https://doi.org/10.1371/journal.pone.0069580. PMID: 23940525
11. Ristea A, Leitner M (2018) Integration of social media in spatial crime analysis and prediction models for events. In: AGILE 2018, Lund, 12–15 June 2018
12. IBM Watson studio documentation and resources. Available online https://dataplatform.cloud.ibm.com/docs/content/wsj/getting-started/overview-ws.html
13. Adesola F, Azeta A, Oni A, Chidozie F (2019) Forecasting violent crime hotspots using a theory-driven algorithm. Int J Eng Res Technol (IJERT). ISSN 09743154
14. Adesola F, Misra S, Omoregbe N, Damasevicius R, Maskeliunas R (2019) An IOT-based architecture for crime management in Nigeria. In: Shukla R, Agrawal J, Sharma S, Singh TG (eds) Data, engineering and applications. Springer, Singapore, pp 245–254
15. Nicholas-Omoregbe OS, Azeta AA, Chiazor IA, Omoregbe N (2017) Predicting the adoption of e-learning management system: a case of selected private universities in Nigeria. Turk Online J Distance Edu-TOJDE 18(2):106–121
16. Azeta AA, Misra S, Azeta VI, Osamor VC (2019) Determining suitability of speech-enabled examination result management system. Wirel Netw 1–8
17. Isah AO, Alhassan JK, Misra S, Idris I, Crawford B, Soto R (2016, July) Network system design for combating cybercrime in Nigeria. In: International conference on computational science and its applications, pp 497–512

Analysis of Violent Crime Dataset Using Support Vector Machine Model

Falade Adesola, Ambrose Azeta, Sanjay Misra, Aderonke Oni, Ravin Ahuja, and Ademola Omolola

Abstract It is usually a challenging task predicting violent crime occurrences in space and time. Huge dataset are needed for accurate prediction of future violent crime occurrence, which in most cases were subjected to artificial intelligence or statistical methods. Most studies from literature adopted data mining techniques for violent crime prediction with some inherent limitation of accuracy as one of the gaps that needed to be filled. The study applied support vector machine model on the six different historical violent crime dataset gathered between July 2016 and July 2019 from Nigeria Police Lagos headquarter to predict spatio-temporal occurrences of violent crime in the state. The six different violent crime dataset used for the study are: armed robbery, rape, kidnapping, assault, murder and manslaughter. The dataset was preprocessed and fed into the support vector machine model built in Watson machine learning studio using python as a programming language. The model returned 82.12% prediction accuracy, which is assumed to be good enough for any prediction system. This result was evaluated using confusion matric, and tested against some results from literature, and was found to out-perform some machine learning models used in the previous studies. Based on this empirical study, the police authority could adopt this model to strengthen violent crime prevention strategies in order to mitigate violent crime occurrences in Lagos state, Nigeria.

F. Adesola · A. Azeta · A. Oni · A. Omolola
Covenant University, Ota, Nigeria
e-mail: adesola.falade@covenantuniversity.edu.ng

A. Azeta
e-mail: ambrose.azeta@covenantuniversity.edu.ng

A. Oni
e-mail: aderonke.oni@covenantuniversity.edu.ng

S. Misra (✉)
Ostfold University College, Halden, Norway
e-mail: sanjay.misra@covenantuniversity.edu.ng

R. Ahuja
Shri Viswakarma Skill University, Gurgaon, Hariyana, India

P. K. Singh et al. (eds.), *Emerging Technologies for Computing, Communication and Smart Cities*, Lecture Notes in Electrical Engineering 875,
https://doi.org/10.1007/978-981-19-0284-0_15

Keywords Machine learning · Support vector machine · Violent crime · Prediction

1 Introduction

Crime has been in existence from time immemorial and remains the number one enemy of humanity. Any act committed or omitted deliberately or otherwise that is punishable under the law of the land is regarded as crime [1]. Crime impacts dastardly on the economy of any nation and equally affect the quality of life of citizens.

Crime is more pronounced in the urban areas because of rapid population growth and social infrastructural development. In other words, rapid population growth in the urban cities has culminated into crime rate increase [2]. In addition, movement of inhabitants from villages to cities has also lead to the increase in crime rate. It is noteworthy that crime that is violent in nature has negative effect on the economic development of any nation. It is therefore pertinent at this juncture to create a safe and secured urban environment for everyone to live and benefit from. Any research or efforts toward reducing crime rate are always a welcome development. All of these are geared toward implementation of effective policing and police patrols to reduce crime rate. Following a report by National Bureau of Statistics, 2018, crime rate in Nigeria is increasing on a yearly average of 3.4% over the last thirty years. This should be a thing of concern to all and sundry and demand urgent attention. Due to the increase in violent crime such as armed robbery, kidnapping, rape, murder, manslaughter and aggravated assault across the length and breadth of the nation, anxiety of people in the community have become intensified. Predictive policing is the way to go as it is already being implemented in some countries of the world for example USA and China [3, 4].

In the year 2010 and 2011, there were more reports of armed robbery incidents in Nigeria compared to other years as shown in Fig. 1. Additionally, cult killings

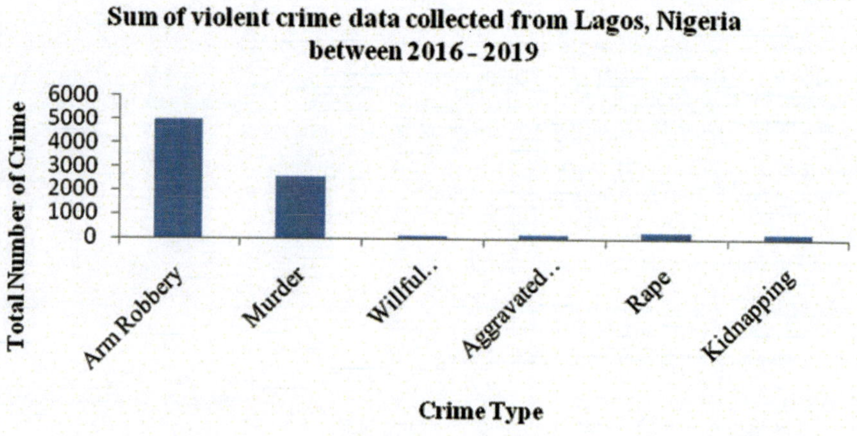

Fig. 1 Violent crime dataset summary for this study

were also reported in various parts of Nigeria between year 2009 and 2018. It can be observed that year 2010, 2018 and 2019 had the most deadly cult killings in the country [5].

The paper is organized as follows. The second section provide the literature review. The third section provide materials and methods. The forth section provides the results and discussion and finally conclusion drawn in Sect. 5.

2 Literature Review

Previous research works by Chen et al. [2], Liao et al. [3], Wang et al. [4], Alves et al. [6] on violent crime prediction attracted several attentions from literature by considering the potential benefits of crime free society to the country. Different machine learning algorithms were used by these authors to predict occurrence of violent crime with inherent limitation of accuracy to be improved upon. Data mining approached were equally engaged for dataset preparation, analysis and prediction.

In [7], authors developed a crime prediction model that was able to predict the location and time of occurrence of a particular type of crime by using sentiment analysis based on lexicon methods and understanding of categorized weather data. They obtained crime prediction by combining sentiment analysis based on lexicon method and Kernel Density estimation using historical crime dataset via linear modeling approach. The result of the experiment showed that the accuracy of the model developed surpasses the benchmark model which makes prediction using Kernel Density Estimation a good choice. The limitation of this approach is that it could not predict time slot and crime location jointly for violent crime occurrence.

In [8], an effort to make useful contribution to crime prediction, divided the entire region in consideration into grids and proposed a novel spatio-temporal crime network prediction model that was based on convolutional neural network for automatic crime referenced features extraction. The model developed can forecast crime risk in each region of the urban areas for the next day using high volume of dimensional data. The model was evaluated and discovered to outperform other models in the previous studies. However, the gap observed is it requires high dimensionality of crime data to be able to make accurate prediction, which is the major shortcoming of deep neural network.

Authors in [9] developed a linear regression model use for forecasting crime trends in Bangladesh. Real crime dataset was collected from Bangladesh police website, and this data contained a collection of different crime types. The linear regression model developed was trained on the dataset, after which the model was used in forecasting future crime trends. In the empirical evaluation carried out, the linear regression model out-performed other three selected algorithms. The gap observed here is it takes time to make crime prediction.

2.1 The Support Vector Machine (SVM)

Authors [10] discovered support vector machine model and was premised on structural risk management theory according to [11]. The author used decision boundaries to define decision planes. Group of objects with diverse class membership were separated. It works using constructed hyper-plane that uses linear models implementing class boundaries that has non-linear attributes which engages input vectors with non-linear mapping in high dimensional feature space [12]. The SVM has been engaged in different domains for prediction ranging from system intruder's detection, pattern recognition, estimation of age, face recognition, as well as telecommunications according to [13–16].

2.1.1 Methodology

Eight thousand, two hundred and thirty-four total dataset for violent crime were extracted from the huge collection of overall crime dataset made available by the statistics department of Nigerian Police, Obalende Lagos. The relevant attributes in the dataset are: crime description, crime_id, crimetimecommitted, crimedate_was committed, numberofdeaths_involved, crimelocation as well as typeofcrime. Presented in Table 1 is the violent crime dataset summary made available between July 2016 and July 2019 used for this study.

Summary of dataset used during the study is presented in Fig. 1.

Watson Studio was used to realize the model implementation. Watson studio is an integrated and collaborative platform for data scientists and business analysts specifically built on open source components and IBM added value tools [17]. It is available both in the cloud and on premise. It is built around open source components such as Python, Scala, R, SQL, Apache Spark and Notebooks (Jupyter and Zeppelin). IBM added value tools such as Watson machine learning, Watson natural language processing, Watson natural language understanding, SPSS predictive analytics algorithms, Chatbots and many more are provided in Watson studio. It is endowed with set of tools to enable data scientists and business analysts to collaboratively and easily work to produce data analysis reports.

Table 1 Total violent crime dataset summary

Crime type	Sum of data collected	Percentage (%)
Arm robbery	4984	60.5
Murder	2549	31.0
Wilful manslaughter	82	1.0
Aggravated assault	146	1.8
Rape	273	3.3
Kidnapping	200	2.4
Total	8234	

A total of eight thousand, two hundred and thirty-four (8234) violent crime dataset with seven important attributes were used for this study. These important attributes are: crime_description, crime_id, timeofcrime, dateofcrime, numberofdeaths_involved, crimelocation as well as typeofcrime. The dataset was preprocessed with independent variables and target columns identified. Then the SVM model was built in Watson studio [17] with python as a programming language for the purpose of training and testing with confusion matrix use for evaluation. The model returned an accuracy of 82.12% across all the violent crime dataset used. In addition, after engaging the stratified cross validation [4], 89.04% was classified rightly and 10.96% was classified wrongly.

3 Results and Discussion

During the empirical study, the dataset was split into 75% training and 25% test set according to [4], the target columns of violent crime hot spot location and time of violent crime occurrence were predicted using support vector machine.

The prediction results for the support vector machine model on the preprocessed dataset is presented in Table 2 and found to return an accuracy of 82.12%. This result out-performed the results of some machine learning approaches on crime dataset reported in literature.

Table 2 presents the empirical results of the support vector machine model trained using Watson studio with Python as a programming language. The attribute crime-location represents the violent crime hot spots area in the state, HotSpot and Predict-edtime columns are the support vector machine prediction results. Hot_Spot value 0 represents happening event not possible of violent crime, while HotSpot value 1 is to predict the occurrence of violent crime at a particular time reported that must warrant police intervention. The last column Predictedtime shows the likely time of violent crime occurring. The evaluation results using confusion matrix of different violent crime types is also presented in Table 3.

The support vector machine model used during the empirical study returned 82.12% accuracy. The results summary on different violent crime dataset is presented in Fig. 2.

4 Discussion

During the machine learning comparative analysis stage, support vector machine became a choice for this study because of its unique performances in different domains from literature and its quick adaptation to new dataset. Watson studio was engaged to develop the model using Python as a programming language. Both the split training and test dataset were applied on the SVM model. Consequently, the

Table 2 The model sample results

CrimeLocation	HotSpot	PredictedTime
10	0	0
11	0	0
12	0	0
13	1	8.101
14	1	16.101
30	1	11.101
31	1	8.101
32	1	14.101
51	0	0
52	1	6.101
53	0	0
54	0	0
32	0	0
34	0	0
35	1	19.122
60	1	9.122
61	1	10.122
62	1	6.251
55	1	9.122
63	1	11.201
64	1	9.201
81	1	15.201
82	1	9.201
83	0	0
90	1	9.201
91	1	9.201
57	0	0
92	0	0
93	0	0
84	1	10.001

experimented results showed the prediction of unseen class labels with 82.12% accuracy. This could represents a fairly good system that can premise on. This result out-performed the work of [17] with 79.2% accuracy for support vector machine using the same dataset.

The dastard effects of violent crime include anxiety and sudden death among citizenry, bad economy, poor quality of life and increase in poverty rates among others. During the experiment, over-fitting was avoided during training and testing by

Table 3 Evaluation results of SVM classifier

	Evaluation results for support vector machine							
S/N	Violent crimes	TP rates	FP rates	Accuracy (%)	Precision (%)	Recall (%)	F1 score (%)	MSE
1	Armed robbery	0.82	0.315	82.92	87.62	81.53	86.22	0.0227
2	Kidnapping	0.81	0.301	81.25	81.70	83.24	85.71	0.0326
3	Rape	0.79	0.478	82.43	83.12	82.92	85.75	0.0309
4	Aggravated Assault	0.76	0.546	84.26	83.24	86.05	84.09	0.0278
5	Murder	0.82	0.345	82.45	83.75	86.50	81.89	0.0245
6	Ritual killings	0.78	0.655	79.41	81.25	83.66	83.82	0.0445

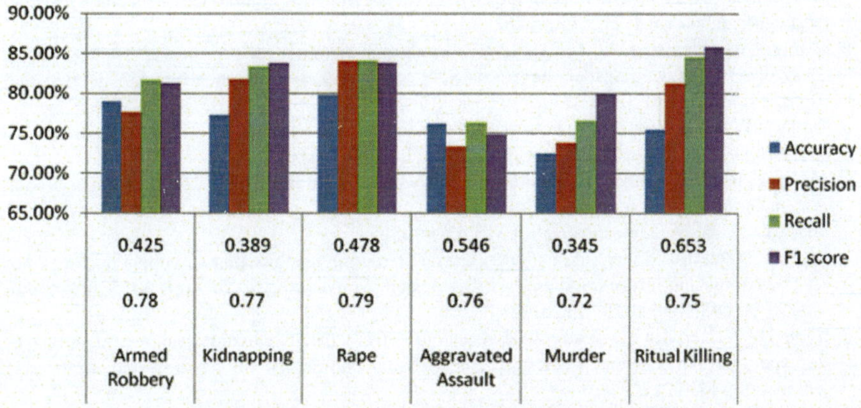

Fig. 2 SVM prediction performance comparison between violent crime types

engaging in cross validation [18–21]. Consequently, the training and testing produces an accuracy of 82.12%.

5 Conclusion

In the context of this study, support vector machine technique has demonstrated its efficacy as a prediction method among other machine learning techniques. In this study, support vector machine delivers 82.12% predictive accuracy across the six violent crime dataset used. The use of SVM as a modern method for violent crime prediction has again being proven. Combination of different artificial intelligence techniques may be considered to see if they could deliver a better predictive accuracy as a future work.

Acknowledgements We appreciate the sponsorship and support from Covenant University for the support received during this study.

References

1. NSW Criminal Law Handbook (2017) Available online: https://www.booktopia.com.au/nsw-criminal-law-handbook-2017/book/9780455500058.html
2. Chen P, Yuan H, Shu X (2008) Forecasting crime using the ARIMA Model. In: Proceedings of the 5th IEEE international conference on fuzzy systems and knowledge discovery. vol 5, pp 627–630
3. Liao R, Wang X, Li L, Qin Z (2010) A novel serial crime prediction model based on Bayesian learning theory. In: Proceedings of the 2010 IEEE international conference on machine learning and cybernetics, vol 4, pp 1757–1762
4. Wang P, Mathieu R, Ke J, Cai HJ (2010) Predicting criminal recidivism with support vector machine. In: Proceedings of the 2010 IEEE international conference on management and service science, pp 1–9
5. Nigeria Watch Project (2019) Fourth report on violence in Nigeria (2009–2018). Retrived rom www.nigeriawatch.org/media/html/NGA-Watch-Report14.pdf
6. Alves L, Ribeiro H, Lenzi E, Mendes R (2013) Distance to the scaling law: a useful approach for unveiling relationships between crime and urban metrics. PLoS One 8(8):1–8. https://doi.org/10.1371/journal.pone.0069580PMID:23940525
7. Xinyu C, Youngwoon C, Suk Y (2015) Crime prediction using twitter sentiment and weather, 2015 systems and information engineering design symposium, 24–24 April 2015. https://doi.org/10.1109/SIEDS.2015.7117012
8. Lian D, Tao H, En C, Jianfeng Z, Chao G (2017) Deep convolutional neural networks for spatiotemporal crime prediction. International conference information and knowledge engineering | IKE'17
9. Cortes C, Vapnik V (1995) Support vector network. Mach Learn 20(3):273–297. https://doi.org/10.1023/A:1022627411411
10. Chiu NH, Guao YY (2008) State classification of CBN grinding with support vector machine. J Mater Proc Technol 201:601–605
11. Awal A, Rabbi J, Rana I (2017) Using data mining technique to analyze crime of Bangladesh. IJCSN - Int J Comput Sci Netw 6(4):2277–5420
12. Elmi HE, Sallehuddin R, Ibrahim S, Zain AM (2014) Classification of sim box fraud detection using support vector machine and artificial neural network. Int J Innovat Comput 4(2):19–27
13. Guo G, Fu Y, Dyer CR, Huang TS (2008) Image-based human age estimation by manifold learning and locally adjusted robust regression. Trans Image Proc IEEE 17(7):1178–1188
14. Kumar M, Ghani R, Mei ZS (2010) Data mining to predict and prevent errors in health insurance claims processing. In: ACM 16th international conference on knowledge discovery and data mining, pp 65–74. Available from: https://doi.org/10.1145/1835804.1835816
15. Kirlidog M, Asuk C (2012) A fraud detection approach with data mining in health insurance. Proc Soc Behav Sci 62:989–994. https://doi.org/10.1016/j.sbspro.2012.09.168
16. Anwar S, Zain JM, Zolkipli MF, Inayat Z, Khan S, Anthony B, Chang V (2017) From intrusion detection to an intrusion response system: fundamentals, requirements, and future directions. MDPI Algor 10(2):1–24. https://doi.org/10.3390/a10020039
17. Keivan K, Reda A (2008) Effectiveness of support vector machine for crime hotspots prediction. Appl Artif Intell Int J 22(5)

18. Adesola F, Misra S, Omoregbe N, Damasevicius R, Maskeliunas R (2019) An IOT-based architecture for crime management in Nigeria. In: Shukla R, Agrawal J, Sharma S, Singh TG (eds) Data, engineering and applications. Springer, Singapore, pp 245–254
19. Nicholas-Omoregbe OS, Azeta AA, Chiazor IA, Omoregbe N (2017) Predicting the adoption of e-learning management system: a case of selected private universities in Nigeria. Turkish Online J Dist Educ-TOJDE 18(2):106–121
20. Azeta AA, Misra S, Azeta VI, Osamor VC (2019) Determining suitability of speech-enabled examination result management system. Wireless Networks, pp 1–8
21. Isah AO, Alhassan JK, Misra S, Idris I, Crawford B, Soto R (2016) Network system design for combating cybercrime in Nigeria. In: International conference on computational science and its applications, pp 497–512

Nigeria Human Population Management Using Genetic Algorithm Double Optimized Fuzzy Analytics Engine Approach

Abraham Ayegba Alfa, Sanjay Misra, Blessing Iganya Attah, Kharimah Bimbola Ahmed, Jonathan Oluranti, Ravin Ahuja, and Robertas Damasevicius

Abstract Governments and informal sectors require accurate perception concerning the impending size of various entities such as population, resources, demands, consumptions, and failures for purpose of planning activities. In order to acquire this information, the behavior of the associated variables is analyzed on the basis of past; thereafter utilize the outcomes to make imminent predictions of the targeted variable. Presently, statistical methods such as conventional and Bayesian has been deployed for population data analytic about regions and countries. More recently, heuristics and metaheuristic are being exploited for the purposes of forecasting population growth rate and optimization problems. The concept of staggered optimization of fuzzy analytic engine rules list have been undertaken in the benchmark studies to reduce redundancy in the rules lists and increase forecasting accuracy. There are still problems of redundancy and low accuracy, which were explored with evolutionary algorithm (that is, genetic algorithm (GA) and fuzzy analytics) approaches to manage Nigeria's human population changes. This proposed approach combines the outcomes of staggard GA optimizations of fuzzy analytics engines whose rules

A. A. Alfa · K. B. Ahmed
Kogi State College of Education, Ankpa, Nigeria
e-mail: abraham.alfa@kscoeankpa.edu.ng

S. Misra (✉)
Department of Computer Science and Communication, Ostfold University College, Halden, Norway

B. I. Attah
Federal University of Technology, Minna, Nigeria

J. Oluranti
Centre for ICT/ICE Research, Covenant University of Technology, Otta, Nigeria
e-mail: jonathan.oluranti@covenantuniversity.edu.ng

R. Ahuja
Shri Viswakarma Skill University, Gurgaon, Hariyana, India

R. Damasevicius
Kaunas University of Technology, Kaunas, Lithuania
e-mail: robertas.damasevicius@ktu.lt

P. K. Singh et al. (eds.), *Emerging Technologies for Computing, Communication and Smart Cities*, Lecture Notes in Electrical Engineering 875,
https://doi.org/10.1007/978-981-19-0284-0_16

lists were filtered to generate the finest fuzzy rules list. The outcomes showed that 12.92% error rate was observed against 17.82, 26.95%, and 42.32% errors observed in the benchmark works. This model developed offers useful insights to government agencies, development partners, and economic planners on the potent ways to manage population, birth, and death rates for improved resources allocation and well-being of populace throughout the country.

Keywords Population growth · Error rate · Heuristic · Rules-lists · Genetic algorithm · Fuzzy analytics · Birth rate · Death rate · Forecasts · SDG 2

1 Introduction

Several nations of the world are facing the problem of diminishing population and demographic dynamics, which have brought about economic and social interplays. The projections about population have implications on workface, economic activities, care-takers, aging and growth rates, and industrialization of nations. The need to comprehensively understudy population dynamics cannot be over-emphasized from management to finding answers to numerous problems of both under-and-over-population growths. Specially, the concept of graying crisis was utilized to explain rising life expectancy and falling birth rate over a period of time [1].

In the developing world (such as the African continent), the population changes are influenced greatly by factors such as birth rate, mortality rate, and migration rate. In general, these factors are capable of propelling or retarding economic growth. Aside that, the population figures of certain countries impact the global economy directly through disease spread or migration of people; or indirectly by geopolitics or environmental factors [2].

Geometric mean, logistic equations, and arithmetic mean are classical techniques for ascertaining future population trends using births, deaths, annexations, and migration for the purpose of investigating the pattern of growth [3]. Monte Carlo algorithm, Markov chain, and probability are stochastic-based techniques that have been deployed for estimating population growth [4]. Population forecasts are less-desirable for a diverse group of benefactors (including users, planners, policy-makers, and regulatory agencies) due to its inaccuracies, crude-approach, and uncertainties [5]. Population changes have diverse variations in the structure of age and spatial considerations, which impact the demography, environment, and economy [6, 7].

In the case of Nigeria, there is no reliable census data available caused by the failure to enumerate all citizens, incomplete documentation, and false estimates for people across age groups [8]. Again, the population figures are outdated or lopsided population data updates leading to inaccurate assumptions and inappropriate policies. Therefore, there is the need to deploy more accurate methodologies to ascertain the dynamics of population in future times in order to facilitate appropriate decision-making and national prosperity. This paper develops double optimization approach of the Genetic Algorithm on Fuzzy analytics to accurately and effectively forecast future population changes in Nigeria.

The remaining sections of the paper include: literature review in section two; research methodology in section three; discussion of results in section four; section five is the conclusion.

2 Literature Review

2.1 Theories of Human Population Changes

The Malthusian model noted that the growth in population is at a geometric proportion; production of food was rising at an arithmetic proportion. Also, Malthus in [9] opined that there is the need to put in place measures to regulate the sharp increases in population otherwise, the population tends to control itself naturally though in a positive perspective. Accordingly, the earth and its available resources are limited or scarce, while production of food cannot outlast population growth proportion, then, factors such as hunger, unrests, and natural occurrences were capable of impacting the growth in population [2]. Malthus proposed that since population growth is detrimental to economic prosperity of a state, serious and effective must be adapted to minimize unwarranted increases through quality education, moral suasion, marriage delays, birth controls, and family planning.

In the later part of the 1960s, Neo-Malthusians experts and proponents stretched the original idea of the Malthus by alluding that institutions within the framework of the society are incapable of tilting their economic situations to accommodate shocks triggered by uncontrollable growth in population and high concentration of humans, which are capable of causing havoc to the environment and social fabric. In contrast, these institutions make attempts to deal with the obvious changes whenever they are allowed to fester uncontrollably which is in agreement with two previously held views: the Distributionists, and the Neoliberals.

Simon in [10], a Neoliberal, pushed the fact that growth in population is never a serious challenge reason being that the institutions are capable of outlasting it, through technological innovations. These inventions could offer a new perspective to overcoming and reshaping the trends caused overpopulation situations. Again, Distributionists think that growth in population meant deliberate shifts from the pressing issues because institutions of the state have the ability to rollout poverty alleviation and equity programmes for the purpose of adjusting the resource distribution gaps.

In the late 1980s, the Revisionist opined that to achieve population growth adjustments, countries having high populations are expected to establish and operate efficient contemporary institutions in order to withstand debilitating negative impact of supposed population explosion. Nevertheless, the school of thought warned that ineffectiveness of these institutions could inevitably result in catastrophic economic conditions. Majority of the population policy arguments favored the revisionist ideas that the prospects in population explosion on the nation's economy were inconsequential according to Birdsall, Kelley, and Sinding in [11].

During the study of developing economies of 1986, National Academy of Sciences reported that on-balance and sluggish population growth were considered to be advantageous to the economic progress at large according to Landau and Rosenberg [12]. Significantly, this idea contributed to a weakening in the policy and political priority aimed at controlling growth in population but concentrated efforts towards revamping economies in these nations.

2.2 *Population Forecasting Techniques*

Heuristic Based Algorithms. These algorithms are built on the concept of several natural processes, which offer sub-optimal solutions from obtainable huge piles of data within a satisfactory frame of time. In particular, Evolutionary Algorithms (EAs) are a class of metaheuristic algorithms that imitate biological evolution and steered by the Darwinian principle of *the fittest survives*. The Genetic Algorithm (GA) is in the class of EAs introduced by John Henry Holland in 1975 [13]. It makes use of binary vectors for the purpose of encoding the solution. It is an artificial intelligence system reproducing the evolutionary principle of survival-of-the-fittest process in order to produce progressively better solutions to a problem by Khmeleva [14]. Generally, EAs is an optimizing system because it finds input arrangement that produce the best outputs. They are most suitable for decision-making situations where thousands or even millions of solutions are probable. More importantly, it is easy to find and evaluate possibilities in a faster and more thoughtfully style rather than humans [15]. However, the algorithm's parameters relied on outcomes from independent experiments which may fall short in optimal for a particular problem. In future, the parameters might be retested in order to verify their suitability for the given problems [14].

Fuzzy Logic Analytics. This is utilized for assessing, decision-making, and mechanical control systems such as automobile controls, air conditioning, smart houses, and industrial process controllers. The notable real-world adoption of fuzzy logic was in the numerous applications, in Japan as process controllers, even though the foremost fuzzy control developments were carried out in Europe. Though, the British engineer Ebrahim Mamdani was the foremost person to practicalize fuzzy sets in a control system unpremeditatedly. In the early 1970s, Mamdani was working on an automated control system for stream engine by means of human operator's expertise with Bayesian decision theory (that is, techniques of eliciting possibilities in uncertain situations) that focused on events after the fact to modify prediction concerning imminent outcomes [16].

Consequent upon the poor performance of the work that artificial intelligence method of enhancing the learning controller known as rule-based expert system through the combination of human proficiency with sequence of logical rules for utilizing the knowledge. Mamdani put forward five kinds of systems capable of profiting from fuzziness concepts. These include [17]:

1. Complex system that is impossible or difficult to model;
2. Human experts-controlled systems;
3. Complex and continuous input and output systems;
4. Systems making use of human observation for inputs or basis for rules creation; and
5. Naturally vague systems similar to behavioral and social sciences.

The input and output fuzzy variables are specified during fuzzy rule-based modeling. For each fuzzy variable, its units are defined; the scope of discourse and a set of membership functions are specified in order to describe the specific fuzzy concepts linked with the fuzzy variable. The challenge of the model is the selection of technical indicators required to form the inputs to the decision support system [18].

2.3 Related Studies

ARIMA model was applied to 145 nations of the world to ascertain the population of COVID-19 infections across 6 continents using ARIMA parameters such as population (per million), cases of infection, and polynomial functions. Other sub-models were developed to forecast behavior of COVID-19 such as climate, culture, and humidity [19].

The geographically weighted regression (GWR) mechanism was utilized in determining the interplay between change in population and various driving factors for the purpose of understanding the spatial variations in the relationships with small-area population forecasting in Wisconsin, USA. The outcomes could be improved in terms of accuracy of climate change and disasters [20].

The role of correct population estimation with stochastic methods in the design and planning of public sector and critical infrastructure was highlighted in the study by Patel and Gundalia [5]. It is established that incorrect population forecasts tamper future services developments such as transportation, water resources, sewage, and water treatment plant. Modeling of prospective extremes by means of deterministic technique for measuring population growth in order to account for total equalization and current differentiation. The main idea is to prepare for inequalities of the future geographical decision-making process across European countries according to Bleha [21].

Fuzzy controller has been deployed to control movement of robots in farm environment during the application of herbicide by Bala et al. [22]. The outcomes revealed that the superiority of the model over existing approaches, and could improve the production of food crops and agribusiness. An SIR model for forecasting COVID-19 infection was built using parameters such as infection rate, death rate, and recovery rate by Abdy et al. [23]. The member functions were constructed and used to train the model with high accuracy in revealing the possibilities of spread of COVID-19 in Indonesia.

A hybrid correction model composed of Fuzzy logic and ARIMA was proposed by Xie et al. [24] for forecasting transportation traffic volume. The results indicated the suitability of the model for long-term forecasts of waterway volume of traffic in more accurate and reliable manner. A logic-related system for assessing the learning achievements of students, detection of faulty teaching methods timely by their teachers was developed by Rojas et al. [25]. The results offered the possibility of early remediation to teaching practice and educational support space for schools in Colombia.

3 Methodology

3.1 The Proposed Model

This paper proposed the double optimization approach composed of GA-based optimizations of the antecedents [26] and the consequents [27] for fuzzy analytics engine. These best solutions are further combined to produce the finest rules list for forecasting population changes as discussed in the next subsection. The events and processes engaged by the proposed model are illustrated in Fig. 1.

From Fig. 1, the complete description of the proposed model for population management is as follows:

Event 1. The raw dataset is collected and formatted properly by removing redundancy and unnecessary information to fit it for analytics.

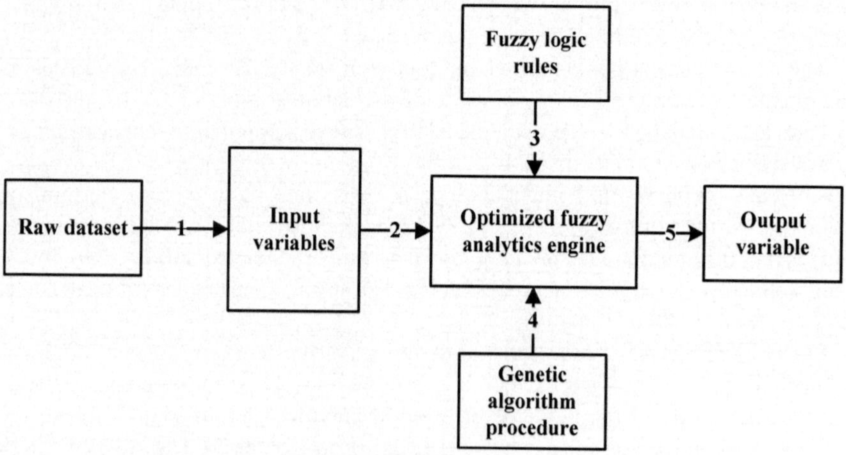

Fig. 1 A double GA-optimized fuzzy analytic engine model

Event 2. The input variables and corresponding values are used for model training.

Event 3. The original fuzzy rules list generated are passed to genetic algorithm procedure block for optimization processes.

Event 4. The rules lists are obtained for optimized antecedents, consequents, and combined outcomes known as double optimized rules list.

Event 5. The test dataset is used to evaluate the performance and effectiveness of the distinct models built with optimized rules lists using standard metrics in Sect. 3.3.

3.2 Fuzzy Rule List Construction

The inputs variables are Birth rate and Death rate (crude per 1000 people) and the population rate served as output variable, which was obtained for the period (1960–2021) from United Nations—World Population Prospects dataset. The structural relationships between the factors for constructing the fuzzy analytics engine rules-list are illustrated in Fig. 2.

The degrees of input conditions, combination, and their outcomes using structural relationships represented in Fig. 2 are shown in Table 1.

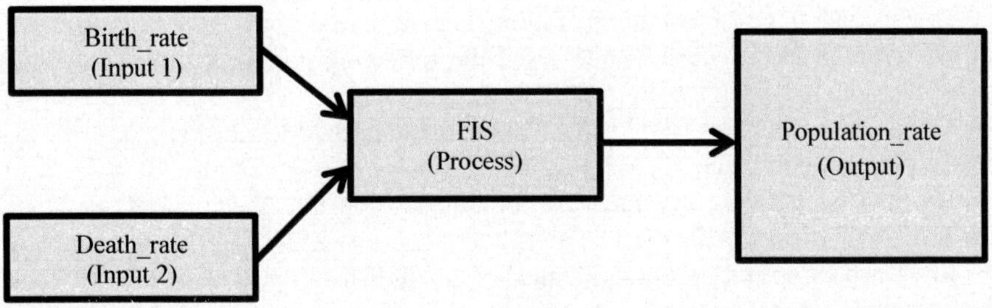

Fig. 2 The structure of the fuzzy analytics rules-list

Table 1 Fuzzy decision table

Antecedents	Condition indices	Maximum-minimum values
What is today's birth rate per thousand people?	High = 3 Medium = 2 Low = 1	[36.86–47.24]
And, what is today's death rate per thousand people?	High = 3 Medium = 2 Low =	[11.38–31.15]
Consequents		
Therefore, population rate per billion for today will be?	Over = 3 Stable = 2 Under = 1	[37.86–211.40]

Table 2 Unoptimized rules-lists indices

Rule number	Birth_rate	Death_rate	Population_growth
1	3	3	3
2	3	2	2
3	3	1	3
4	2	3	3
5	2	2	2
6	2	1	2
7	1	3	3
8	1	2	2
9	1	1	1

From Table 1, the first or unoptimized rules-list for the fuzzy analytics engine is presented as shown in Table 2 [26, 27].

From Table 2, the double GA-optimization procedures are performed on the antecedents, and consequents separately as represented in Table 3.

From Table 3, the values of antecedents and consequents, which are [333231232221131211] and [323322321] form the chromosomes for the first stage of the GA-optimizations whose outcomes are shown accordingly.

The second stage of GA-optimization (known as the double GA-optimization of rules-list) proposed in this paper used the following optimized fuzzy rules-list derivable with mean square error (MSE) fitness function:

Antecedents optimized = [233 132 133 313 112 332 233 312 221].

Consequent optimized = [331 322 313 232 222 213 133 122 113].

The antecedents and consequent are combined to construct the required rules-lists for fuzzy analytics, which is [231 132 133 312 112 333 233 312 223].

After inspection and further refinement of the rules-lists, these rules: 1, 2, 3, 7, and 9 are excluded due to redundancy and inconsistency in order to realize 5 rules-lists

Table 3 GA-optimizations of rules-list indices

S/N	Antecedents (Before)	Consequent (Before)	Antecedents (After) [26]	Consequent (After) [27]
1	33	3	23	1
2	32	2	13	2
3	31	3	13	3
4	23	3	31	2
5	22	2	11	2
6	21	2	33	3
7	13	3	23	3
8	12	2	31	2
9	11	1	22	3

as final double-GA-optimization outcomes. The new rules-lists, that is [312 112 333 312 223], is used to build the fuzzy analytics for determining the population growth based on the birth rate and death rates.

The datasets depict the raw values of variables or factors influencing population changes, which are used to create the membership functions and subsequently trained with the double-optimized rules lists of antecedents and consequents in Tables 1 and 3.

3.3 Experimental Setting

The simulation is carried out using MATLAB (R2019b) with system specifications for hardware and software as follows: Hardware (×64-based processor, AMD E1-1200 APU with Radeom™, HD Graphics, 1.40 GHz, 4 GB RAM, and 160 HDD). Software (Windows 8, and 64-bit Operating System).

3.4 Performance Evaluation Metrics

This paper adopted a number of performance evaluation parameters including the Mean Square Error (MSE), Root Mean Square Error (RMSE), Mean Absolute Percentage Error (MAPE), and Relative Absolute Error (RAE) (Alfa et al. in [26, 27]). The mathematical representations are provided as follows:

The mean square error is given by Eq. 1.

$$\text{MSE} = \frac{1}{m}\sum_{k=1}^{m}(A - B)^2. \tag{1}$$

The root mean square error is depicted by Eq. 2.

$$\text{RMSE} = \sqrt{\frac{1}{m}\sum_{k=1}^{m}(A - B)^2}. \tag{2}$$

The mean absolute percentage error is represented by Eq. 3.

$$\text{MAPE} = \frac{1}{m}\sum_{k=1}^{m}\left|\frac{A - B}{A}\right| \times 100 \tag{3}$$

The relative absolute error is represented by Eq. 4.

$$\text{RAE} = \left|\frac{A - B}{A}\right| \times 100 \tag{4}$$

where,
A is real values of training dataset.
B is observed values of training dataset.
m, k are the first and last items of the training dataset.

4 Discussion of Results

The outcomes of the double-optimization procedures of genetic algorithm on the fuzzy analytics engine are presented in Table 4.

From Table 4, the performance of proposed double GA-optimization of antecedents and consequents carried out together outperformed benchmark studies by 12.92% to 42.32%, 17.82% respectively. This implies that there is a minimal error rate for the proposed model when used to forecast population changes.

Similarly, the proposed approach outperformed the remaining metrics such as RAE, MAPE, MSE, and RMSE. The distribution of forecasts from initial forecasts, double GA-fuzzy optimization, final GA-fuzzy, and the target datasets are plotted in Fig. 3.

From Fig. 3, the proposed approach provided the nearest-best solution to the problem of population management in the future time due to smallest errors. It is caused by minimal rules lists in the fuzzy analytics engine offered by the optimization procedures of GA. Therefore, the birth_rate and death_rate are valuable factors in determining changes in population of countries and regions of interest. It follows that a growing population must keep birth rate on the increase while making efforts to keep citizens safe from untimely deaths such as diseases, wars, and low-living standards.

Table 4 The outcomes of GA-fuzzy analytics engine optimizations

Metrics	Double GA-fuzzy (Unrefined)	Final double GA-fuzzy (Proposed)	GA-fuzzy [26]	GA-fuzzy [27]
RAE	0.3451	0.2389	3.768072451	2.365752510
MAPE	0.0173	0.0119	0.091904206	0.057701281
MSE	0.00066309	0.00031780	0.001041135	0.000438377
RMSE	257.50	178.27	0.032266628	0.020937461
No. of fuzzy rules	8	5	7	5
GA operation	Both	Both	Consequent	Antecedents
Error rate (%)	26.95	12.92	42.32	17.82

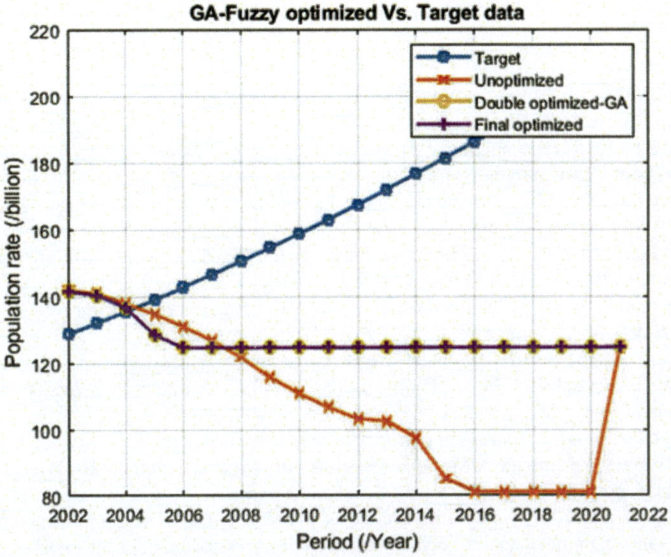

Fig. 3 The performance of GA-fuzzy analytics compared

5 Conclusion

The population size management impacts numerous factors such as possible supply of inexpensive unskilled and skilled labor, capability to consume and raise aggregate demand, market demand, and the Gross Domestic Product (GDP) growth, geopolitical status, tax and revenue, and infrastructural provisions. There is a need to control the rate of population growth in order to match the pace of economic growth. When this is achieved, many crisis situations can be avoided including balance of payment, and high youth unemployment caused by surplus net imports, inflation, corruption, food insecurity, pollution, deforestation, social vices, and environmental degradations (such as dwindling natural resources and global warming).

This study adopted GA optimized fuzzy rules lists for both antecedents and consequents, which serve as basis for generating new double optimized fuzzy rules-list. The outcomes showed superior performance of the proposed model against comparable approaches [26, 27] in case of Nigeria's population forecasts. However, fuzzy logic analytics-based models cannot be easily modeled, aside high error rates. Several simulations and finetunings are needed to make it more suitable for or crisp system, or mathematical precision, and linear modeling. There is the need to consider other optimization techniques for the purpose of enhancing the effectiveness of fuzzy analytics especially in the construction of its rules-lists for forecasting tasks.

Acknowledgements The authors appreciate the Covenant University through its Centre for Research, Innovation and Discovery for Financial assistance.

References

1. Sigurðsson EÖ (2017) Impacts of population aging in modern Japan and possible solutions for the future. Universitatis Islandiae, Sigillum
2. Anudjo EY (2015) The population growth-economic growth nexus: new evidence from Ghana. University of Ghana, Legon
3. Kumar V (2020) Application of mathematical modeling in population forecasting. Uttaranchal Univ J Appl Life Sci 3(1):50–57
4. Zhu D, Jiang K (2018) Population forecasting model based on monte carlo algorithm. ACM ICCDE 2018:67–72
5. Patel N, Gundalia M (2020) Comparative study of population forecasting methods for Surat City (Gujarat-India). J Interdisc Cycle Res 12(10):633–636
6. Uddin GA (2016) Population changes and implications for economic growth and the environment in Australia. A Ph.D. Thesis, Department of Commerce, University of Southern Queensland, Australia, pp 1–201
7. Pettersen AM (2017) Youth employment in Uganda: challenges and survival strategies for young women in Kampala. A Master's Thesis, Centre for Peace Studies, The Arctic University of Norway, pp 1–66
8. Kusumasindra F (2017) Comparing sources of population data in health information systems: a case study of Zambian health sector. Master's Thesis, Department of Informatics, University of Oslo, Norway, pp 1–113
9. Malthus TR (1798) An essay on the principle of population (Printed for J. Johnson)
10. Simon JL (1977) The economics of population growth
11. Birdsall N, Kelley AC, Sinding S (2001) Population matters: demographic change, economic growth, and poverty in the developing world: demographic change, Economic Growth, and Poverty in the Developing World, Oxford University Press
12. Landau R, Rosenberg N (Eds) (1986) The positive sum strategy: harnessing technology for economic growth. National Academies Press
13. Holland JH (1975) Adaptation in natural and artificial systems. MIT Press
14. Khmeleva E (2016) Evolutionary algorithms for scheduling operations, A Ph.D. Thesis, Sheffield Hallam University, pp 1–337
15. Baltzan P (2015) Business driven information systems, 5th edn. McGraw-Hill, New York
16. Kayacan E (2011) Interval type-2 fuzzy logic systems: theory and design. A Ph.D. Thesis, Department of Electrical and Electronic Engineering, Bogazici University, Instanbul, Turkey, pp 1–149
17. Kotta A (2015) Condition monitoring using computational intelligence methods. unpublished M.Sc. Thesis, Department of Computer Engineering, University of Technology Chemnitz, Germany, pp 1–92
18. Delnavaz B (2014) Forecasting of the stock price index by using fuzzy-neural network and genetic algorithms. J Appl Sci Agric Am-Eurasian Netw Sci Inf J 9(9):109–117
19. Hernandez-matamoros A, Fujita H, Hayashi T, Perez-Meana H (2020) Forecasting of COVID19 per regions using ARIMA models and polynomial functions. Appl Soft Comput J 96:106610. https://doi.org/10.1016/j.asoc.2020.106610
20. Chi G, Wang D (2017) Small-area population forecasting : a geographically weighted regression approach. In: Swanson DA (ed) The frontiers of applied demography, applied demography serie, pp 449–471
21. Bleha B (2020) Future population developments in Europe. Is the concept of convergence indisputable? Towards the role of geographical thinking in population forecasting. Appl Spat Anal Pol. https://doi.org/10.1007/s12061-019-09330-6
22. Bala JA, Folorunso TA, Olaniyi OM, Daniya E (2021) Design of an agribusiness innovative and autonomous robot system for chemical weed control for staple food crops production in sub-saharan Africa. In: Che FN, Strang KD, Vajjhala NR (eds) Opportunities and strategic use of agribusiness information systems, IGI Global, pp 234–262

23. Abdy M, Side S, Annas S, Nur W, Sanusi W (23) An SIR epidemic model for COVID-19 Spread with fuzzy parameter: the case of Indonesia. Adv Diff Equ 105, https://doi.org/10.1186/s13662-021-03263-6. BioMed Central Ltd Part of Springer Nature
24. Xie Y, Zhang P, Chen Y (2021) A Fuzzy ARIMA correction model for transport volume forecast. Math Prob Eng Article ID 6655102. Hindawi. https://doi.org/10.1155/2021/6655102
25. Rojas JA, Espitia HE, Bejarano LA (2021) Design and optimization of a fuzzy logic system for academic performance prediction. Symmetry 13(133):1–20. https://doi.org/10.3390/sym13010133.MDPI
26. Alfa AA, Yusuf IO, Misra S, Ahuja R (2020) Enhancing stock prices forecasting system outputs through genetic algorithms refinement of rules-lists. In: 1st IC4S 2019, Lecture Notes in Networks and Systems, vol 12. Springer Nature Singapore Pte Ltd, pp 669–680
27. Alfa AA, Misra S, Bumojo A, Ahmed KB, Oluranti J, Ahuja R (2020) Comparative analysis of optimisations of antecedents and consequents of fuzzy inference system rules lists using genetic algorithm operations. In: Chillarige RR (ed) ICACII 2019, Lecture Notes in Networks and Systems, Advances in Computational Intelligence and Informatics, vol 119. Springer Nature Singapore Pte Ltd, pp 373–379

Flower Species Detection System Using Deep Convolutional Neural Networks

Arun Solanki and Tarana Singh

Abstract A system that correctly identifies the name of a flower species may be beneficial for botanists, camping enthusiasts, and researchers. Previously, classification was only done based on a flower's shape, geometry, and texture, which is not enough for an efficient system. Some significant challenges in this classification task include inter-species similarity, intra-class variation, and the same objects such as leaves or grass around a flower, making this task a research topic. This research has developed an efficient and robust deep learning flower classifier to overcome these problems and limitations based on the current state of the art convolutional neural networks and transfer learning. This research has utilized the Oxford-102 flower dataset having 8189 images of 102 flower species. The proposed method is divided into two different steps. Firstly, the flower images are segmented, and secondly, these segmented images are fed as an input to a convolutional neural network for classifying the species of the flowers. This work has used the PyTorch library for recognition purposes. The flower's dataset uses various pre-trained models on the ImageNet dataset such as AlexNet, VGG, DenseNet, Inception v3, and GoogLeNet. Out of these, DenseNet achieved the highest classification accuracy of 97.92% when trained on GPU provided by Google Collaboratory. This classifier can be integrated with a mobile application to provide an accurate real-time flower species prediction.

Keywords Convolution neural network (CNN) · Segmentation · Cropping · Augmentation · Transfer learning · ReLU activation function

1 Introduction

Plant species recognition based on flower recognition remains a challenge in the field of image processing and computer vision, primarily due to their widespread presence, complex structures, and unpredictable species in nature. Due to this natural complexity, it is highly undesirable to segment or extract regular features or combine

A. Solanki (✉) · T. Singh
Department of Computer Science and Engineering, Gautam Buddha University, Greater Noida, U.P., India

P. K. Singh et al. (eds.), *Emerging Technologies for Computing, Communication and Smart Cities*, Lecture Notes in Electrical Engineering 875,
https://doi.org/10.1007/978-981-19-0284-0_17

shape, texture, and color features, resulting in moderate accuracy in benchmark datasets increase. Several feature extraction techniques that combine global and local feature descriptors have achieved the highest accuracy in flower classification, but automatically identify and recognize large flower species in complex environments [1]. Still needs a powerful and efficient system. This paper also uses mechanism of transfer learning to save our time and resources. For this, we have utilized Oxford-102 dataset of images having 8189 flower images belonging to 102 kinds of different flower species. The proposed method is divided into two major steps [2]. Firstly, the flower images are segmented, and secondly, these segmented images which act as input afterward go into a convolutional neural network for classifying the species belonging to different flower categories. We have also pre-processed the flower images which we will discuss in the later section of this paper. To facilitate our proposed work, we have employed PyTorch library for the recognition purposes and various pre-trained models on ImageNet dataset such as AlexNet, VGG, DenseNet, Inception v3, and GoogLeNet were used on the flower's dataset. Out of these, DenseNet achieved the highest classification accuracy when trained on GPU provided by Google Collaboratory [3]. This classifier can be built into a mobile application so that it can provide an accurate flower species prediction in real-time environment. For the implementation of the flower species recognition system, we have used python 3.6.6 version. We have employed PyTorch library for the development of the code based on transfer learning. The whole training of the model is done on Google Collaboratory which is free GPU provided by Google. We have obtained the dataset which is used in our proposed methodology is from Science and Engineering department of University of Oxford. This dataset is known as 102 flower category datasets [4]. This flower's dataset is having 102 types of different flower species. Each category of flower contains the images ranging from 40 to 258 images.

CNN comes under the picture of an artificial neural network that has wide recognition in image classification. Several layers in a CNN includes convolutional layers, pooling layers, and fully connected layers [5]. Operation in the series format is applied to the data, which acts as input to the CNN network to find a particular pattern in the image. This network does the processing of the data with a grid-like topology [6]. A CNN model uses the image pixel in the form of an array as input. The input data is processed through a hidden layer, and the final output is shown or given by the output layer. The primary function of the hidden layer is feature extraction that deals with the calculation and manipulation of the data [7]. This work is based on the convolution layer, which filters the matrix and performs the convolution operations to help pattern finding in the image. Hidden layers can be many depending upon the architecture of the network like we can have a ReLU layer or a pooling layer or convolution layer [8]. At the end of each process, we get feature maps convolved, rectified, and pooled. The modified pixel values have to get passed through the fully connected layer where the real object detection occurs [9]. Figure 1 illustrates the overall architecture of a convolutional neural network. This also has the depiction of all the layers contained in the network [10].

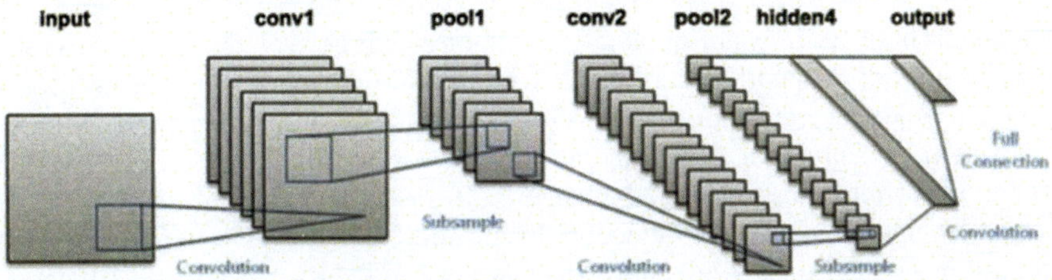

Fig. 1 Architecture of convolutional neural network [11]

When reviewing previous studies, several flower identification methods have been suggested [1, 6, 12]. These methods usually consist of four steps: pre-processing, segmentation, manual design feature extraction, and classification [13]. Due to the complex background of flower images, this task can be very time consuming, and for many types, in particular, the accuracy obtained is still low. Recently, learning feature representations using convolutional neural networks (CNNs) has been very successful in various areas of computer vision, including object detection, segmentation, and visual image classification [9]. Feature learning methods provide a natural way to capture clues using many codewords (sparse coding) or neurons (deep networks) [14, 15]. All of these are useful clues because you can capture the natural features of the object. Therefore, this article examines and presents the efficiency of deep convolutional neural networks, which may more effectively identify plant species based on flowers [16, 17].

This whole paper is organized into eight sections. In section one, introduction and the motivation of the work are presented. In the second section of this paper, a literature survey of the related domain is presented. In section three, the proposed architecture of the system is discussed. The process flow chart of the proposed system is discussed in section four. Then, in section five of this chapter, the pseudo-code of the proposed system is presented, followed by the step-by-step discussion of the implementation of the proposed algorithm in section six. Section seven discusses the results of the proposed system, followed by the comparison of the proposed work with the existing systems. At the end of the paper, a conclusion is given, followed by future work.

2 Literature Survey

Krizhevsky [18] brings out the phenomenal results on the ILSVRC2012 through developing a deep convolutional neural network. The top-1 error rate was 37.5% and the top-5 error rate was 17%. This method was certainly better than other methods in the past for the same domain. A system based on a convolutional neural network was a build-up of deep layers ensemble with the structural network containing eight layers. To avoid overfitting problem [19], there is the incorporation of the essential features. These features are pooling layers along with normalizing layers with the functionality of dropout. According to Sermanet [20], using CNN for object location and object detection in images will boost classification accuracy. It will also increase the accuracy of detection and location tasks. This method is the winner of the localization task on the challenge of ILSVRC2013 through the developed integrated approach used for detection, localization, and recognition [21, 22]. This algorithm gave brilliant results through classification accuracy. Szegedy [10] developed and designed the architecture of a deep convolutional neural network which is called inception and there is seen great classification and detection results for the challenge ILSVRC2014 [23]. The author in [20] states that for the benefit of classification depth representation is essential. With the substantial increase in the intensity, good results can be achieved on the ImageNet dataset using a conventional CNN.

We can use a convolutional neural network for the segmentation of the images and can be employed to detect the objects in the images. Segmentation through CNN has been achieved through the paper's fully convolutional networks (FCN) concept [20]. Several methods extend the concept of CNN to allow object detection tasks with good accuracy on benchmark datasets. These methods are R-CNN [24] which is region proposals with CNN. Another advancement is fast R-CNN explained in [25]. Later on, there is the development of the architecture of Faster R-CNN [26] and YOLO [27]. The results are similar if we compare FCN with these methods when using CNN's architectures, including AlexNet [28] and VGG-16 [10].

3 The Architecture of the Proposed System

Figure 2 shows the framework, which is the designed architecture for our proposed method employed to deal with flower species recognition.

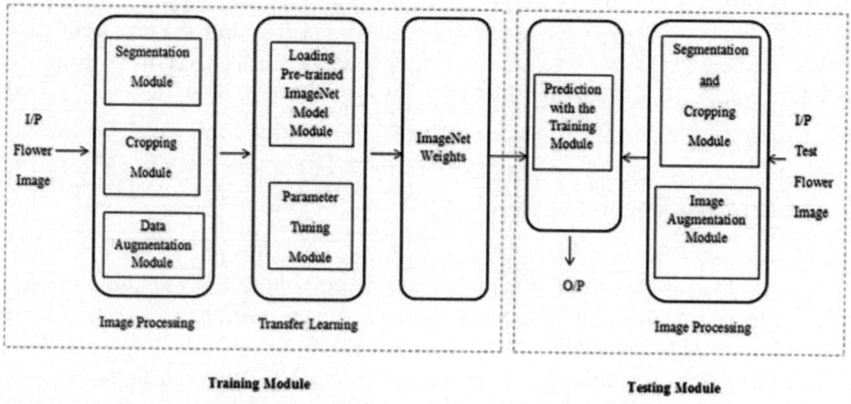

Fig. 2 Architecture of the proposed work

The efficient and robust system we have developed to classify different types of flower species is depicted in Fig. 2. This figure shows the overall framework, which is also the architecture of the proposed method. Architecture is composed of modules, blocks, and sub-modules [29]. Architecture describes the flow of the code right from data collection to the prediction of an unknown set of flower examples by the trained classifier. Here, we have utilized the architecture type of convolutional neural network, DenseNet, a pre-trained model on the ImageNet dataset [30]. This is called a transfer learning mechanism. There are two significant modules in the architecture are:

- **Training Module**: The training module of the architecture proposed contains three blocks and five modules which are described below. The input to this module is the raw flower images one by one. This module has three blocks: Image Processing Block, which is mainly responsible for the data preparation for the training and contains three modules: segmentation module, cropping module, and data augmentation module. The second is the Transfer Learning Block, which focuses on the transfer learning mechanism and comprises two major modules: loading pre-trained model module and parameter tuning module. And the third one is the **ImageNet Weight Block, which has the weights of the ImageNet dataset** used in our flower classification problem.
- **Testing Module**: The testing module of the architecture proposed contains two blocks and three modules which are described below. The output from this module is the class label, the predicted species of an unknown flower image, which is the input to this testing module. This module has three further modules: Predictions with the Training Module, Segmentation and Cropping Module, and Image Augmentation Module.

4 Process Flow Chart of the Proposed System

The process flow carried out in the proposed work is systematically explained by the below flowchart, which contains all the steps of execution to accomplish the research (Fig. 3).

4.1 Oxford 102 flower's Dataset

We have the oxford 102 flower's dataset at our disposal, which has to go into our classification model for flower species prediction [31]. Figures 4 and 5 are the depiction of variability between flower species and variability within flower types.

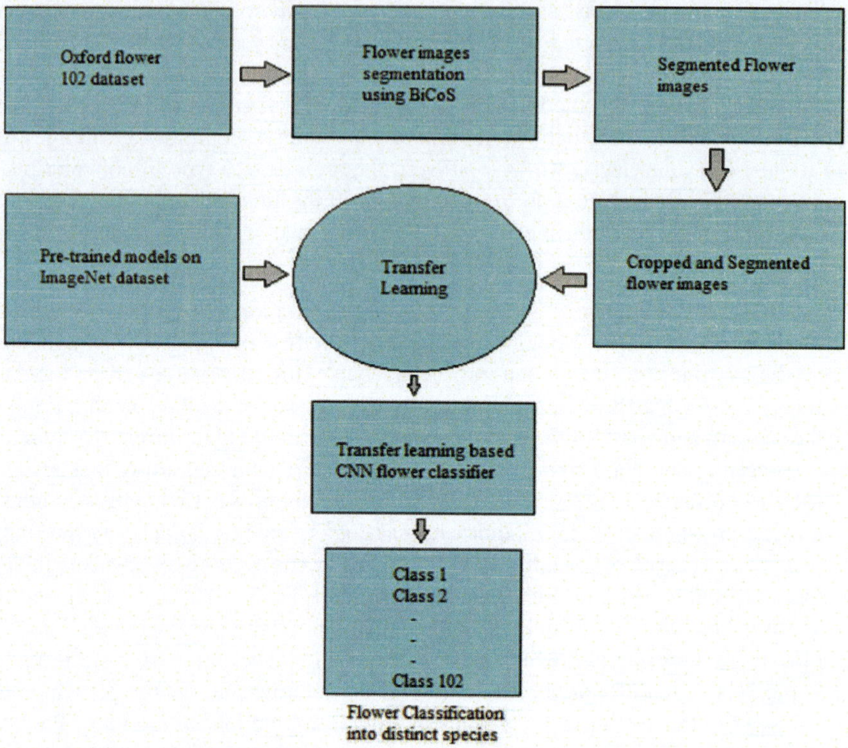

Fig. 3 Process flow of the proposed method

Fig. 4 Variability between classes

Fig. 5 Variability within classes

4.2 *Flower Images Segmentation Using BiCoS*

To improve classification accuracy, there is a need to separate foreground (the flower structure) from the complex background due to leaves and grass, which is useless for the classification task. So, we do the flower images segmentation to obtain the desired segmented images of the flowers.

4.3 *Cropped and Segmented Flower Images*

After segmentation, the background becomes black. We need to remove that, so we crop the segmented images using a python script to obtain the cropped and segmented images of flowers.

4.4 *Pre-trained Models on ImageNet Dataset*

There are many pre-trained models like DenseNet, VGG-16, or AlexNet that can be loaded from the Torchvision module of PyTorch.

4.5 *Transfer Learning*

This mechanism of transfer learning is gaining huge popularity in deep learning. We load the pre-trained models on the ImageNet dataset into our code. This mechanism is called transfer learning which is reusing things that have very high standards. Pre-trained networks generally contain two things. One is feature detectors, and the other is a classifier. Feature detectors extract from each image the information [32, 33]. Our classifier will learn the input given by feature layers, and therefore we will freeze the feature layers to avoid any modification. If we talk about the most commonly used pre-trained models on the ImageNet dataset, these are AlexNet, DenseNet, VGG-16 and many more that have gained popularity over recent years.

5 Pseudo-Code of the Proposed System

Step 1: Loading Oxford 102 Dataset.
Step 2: Essential Libraries Loading.
Step 3: Data Segmentation and Cropping of Flower Images.
Step 4: Data Augmentation.
Step 5: Loading pre-trained model.
Step 6: Classifier Building.
Step 7: Model Training.
Step 8: Model Testing.
Step 9: Save Model Checkpoint.
Step 10: Load Model Checkpoint.
Step 11: Processing Images.
Step 12: Class Prediction.
Step 13: Sanity Check.

6 Implementation of the Algorithm

Step 1: Loading Oxford 102 Dataset—In this step, we load the Oxford 102 dataset of flowers images into our code to apply our model for the prediction of flower species. This dataset contains 102 species of flower images and is divided into training and test sets.

Step 2: Essential Libraries Loading—This process involves the loading of essential libraries and packages to make use of the functions in the modules of these packages.

Step 3: Data Segmentation and Cropping—We have to remove the complex background, which contains leaves and grass and these things create significant confusion for the flower classification task. Therefore, we have segmented the flower images using a technique called BiCoS segmentation for image classification. Then, this segmented image is cropped with the help of a python script to improve the accuracy of the network. Figure 6 shows the conversion of the original image of the flower picked up from the Oxford 102 species to the segmented image and the conversion from segmented image to cropped image.

Step 4: Data Augmentation—As our dataset is not very large, we need to augment the dataset of flower images. This is because we want our program to learn as much as it can. So, we must apply some random transformations to build a robust and efficient flower classifier. For this, we have to train our model on various variations of the

Fig. 6 Cropped image obtained from the segmented image

original image. These variations include rotation, translation, scaling, cropping, or flipping of the original image.

The transformations are done on each image and these images are passed through a neural network in each epoch. This will allow the network to train on a more significant number of images. Therefore, we have increased the variety of our training data which will further reduce overfitting. This also improves the capability of generalization by the classifier, so there is a sure shot increase in the model's accuracy. To normalize the image values before entering them into the network, we have to provide the network with mean and standard deviation. If we look at the dimensions of image tensors, we can have the values of mean and standard deviation. PyTorch library allows us to do the data transformation through its torchvision package. Inside torchvision package, we have the module named transform, which has several functions helpful in transforming the images.

Step 5: Loading pre-trained model—We load the pre-trained models on the ImageNet dataset into our code. This mechanism is called transfer learning which is reusing things that have very high standards. This mechanism of transfer learning is gaining huge popularity in the field of deep learning. Pre-trained networks generally contain two things. One is a feature detector and the other is a classifier. Feature detectors extract from each image the information. Our classifier will learn the input given by feature layers and therefore, we will freeze the feature layers to avoid any modification. If we talk about the most commonly used pre-trained models on the ImageNet dataset, these are AlexNet, DenseNet, VGG-16, and many more that have gained popularity over recent years. All we save through the power of these trained feature models on huge datasets is our precious time and the computer's resources. These models provide cutting edge results on the smaller datasets when reused.

Step 6: Classifier Building—We will build our classifier and this classifier has to be replaced by the model pre-trained on ImageNet. We will freeze the feature layers of the network. This is because we will provide the input according to our dataset. The feature layer will learn these inputs and we don't want these layers to learn the same inputs as of ImageNet dataset. We will set the output size of the classifier as 102 as we have these many species in our dataset. We will achieve this task with the help of defining a precise function.

Step 7: Model Training—In this step, it is time to train our classifier on our flower's 102 category dataset. We will train the final layers of the network. We only train the classifier parameter while the feature parameters are kept frozen. We can change our optimizer as well as a scheduler in the piece of our code.

Step 8: Model Testing—In this step, our trained model is evaluated to measure the performance of the test images of our dataset. At the end of this step, we obtain the percentage accuracy, which means how many flower test images are correctly classified.

Step 9: Save Model Checkpoint—We will save our model in the directory created. This is done to ensure the backup of our created and trained model. This will come in handy when we have to use this trained model on some unknown images of flowers.

Step 10: Load Model Checkpoint—We load our trained model to use this on the unknown flower images to predict their species name.

Step 11: Processing Images—We will carry out the processing of the images because we will take this image as the unknown image for which we need to predict the class label. So, there is a need of related data transformations.

Step 12: Class Prediction—We will predict the class of flower species of the given image unknown to the model. This whole process is carried out in the probability that a particular flower type belongs to that class.

Step 13: Sanity Check—All the earlier parts of this code are combined in a function. This function performs the plotting, or we can say just graphing where the models predict with uncertainty.

7 Discussion of Result of the Proposed System and Comparison of the Results with the Existing Work

7.1 Number of Epochs Versus Classification Accuracy

In Graph 1a, we plotted 60 values of classification accuracies measured at epochs ranging from 1 to 60. This graph depicts the dependency of several epochs on the classification accuracy of our deep learning trained classifier. As the number of epochs increases, the accuracy also increases and reaches the maximum of 97.92 at epoch 25. From then onwards, accuracy declined and then increased again. Then, the graph follows a pattern of increase and decreases until epochs 60. But we observed that the accuracy could not cross the value of the maximum, which is 97.92% at any epochs after 25. So, we conclude that increasing the epochs after 25 do not affect or increase the classification value of the model. Graph 4 shows the results of the Proposed System.

In Graph 1, we have plotted the number of epochs on the x-axis and training accuracy on the y-axis. We plotted 100 values of training accuracies measured at the epochs, ranging from 1 to 100 in the training phase. This graph illustrates that training accuracy shows a sudden increase till epoch 15. Still, from there onwards, the accuracy doesn't show a desirable increase and there is also not drastic decrease in the training accuracy. This means that after epochs 15 the accuracy curve seems to

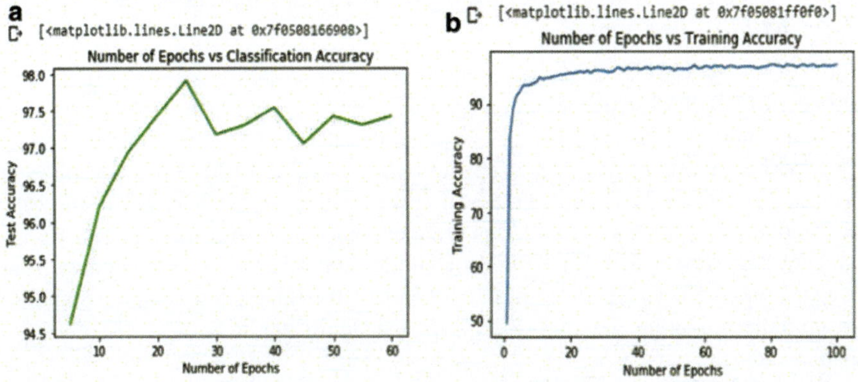

Graph 1 **a** and **b** Number of epochs versus classification accuracy and training accuracy

be constant as it keeps on increasing and decreasing only by a very small or negligible value.

7.2 *Number of Epoch Versus Training Loss*

In Graph 2, we have plotted the number of epochs on the *x*-axis and training loss on the *y*-axis. We have plotted 100 values of training loss measured at the epochs, which range from 1 to 100 in the training phase. This graph illustrates that training loss shows a sudden decrease till epochs 10, but the loss doesn't show a desirable decrease and there is also no drastic increase in the training loss. This means that after epochs 10 the loss curve seems to be constant as it keeps on decreasing and increasing but only by a very or negligible small value.

Graph 2 Number of epochs versus training loss

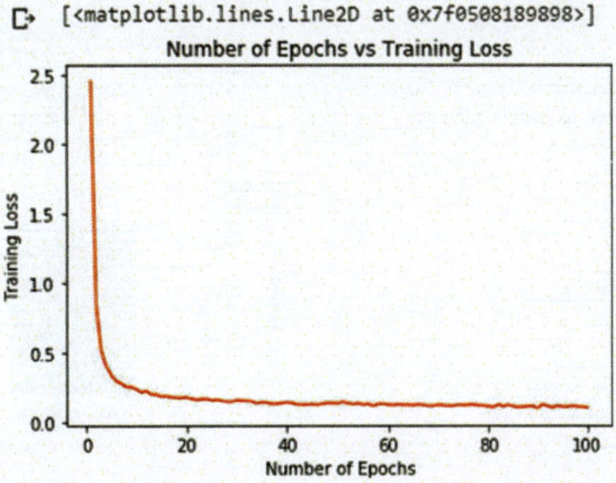

7.3 Validation Curve

Number of Epochs versus validation accuracy.

Graph 3a shows the dependency of the number of epochs on the validation accuracy. We have plotted 100 values of epochs on the x-axis and 100 values of validation accuracy on the y-axis. As the number of epochs increases, the accuracy of the validation points also increases. Still, this accuracy value increases only until the epoch value of 8 at a fast rate. Then, from epoch 8 the validation accuracy fluctuates. That is it keeps on increasing and decreasing till epochs 100. Overall, we can say that after epoch 8 the validation accuracy doesn't show a significant increase or also does not display a drastic decrease.

Graph 3b shows the dependency of the number of epochs on the validation loss. We have plotted 100 values of epochs on the x-axis and 100 values of validation loss on the y-axis. As the number of epochs increases, the loss of the validation points decreases, but this value of loss decreases only till the epoch value of 8 at a very fast rate. Then from epoch 8 the validation loss fluctuates. That is, it keeps on decreasing and increasing till epochs 100. Overall, we can say that after epoch 8 the validation loss doesn't show a significant decrease or also does not display a drastic increase.

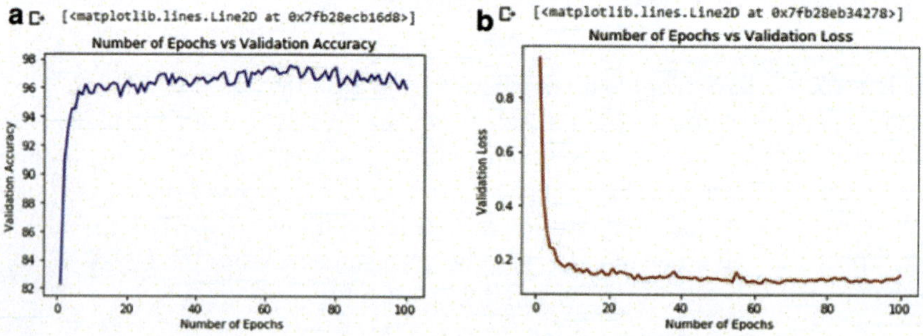

Graph 3 **a** and **b** Number of epochs versus validation accuracy and validation loss

Graph 4 Existing work versus proposed work

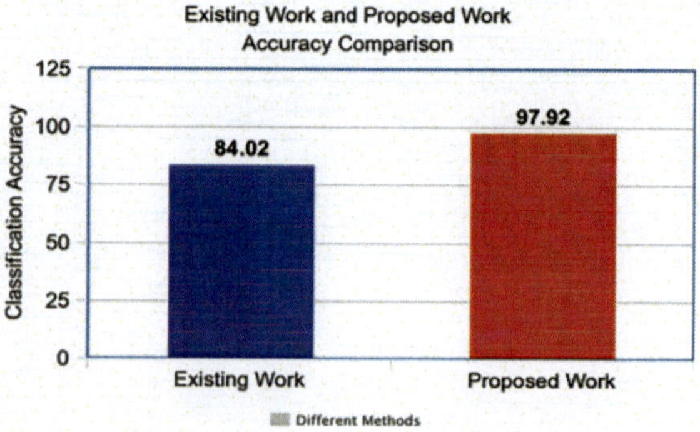

7.4 Comparison of the Results of the Proposed System with the Exiting System

Graph 4 compares the existing work [34] in flower classification with the proposed method, which is robust and efficient. Both of the work is performed on the Oxford-102 dataset. The existing work achieved a classification accuracy of 84.02% with the use of five convolutional layers. Our developed deep learning flower classifier system has set the really high standards in this domain by achieving a very high recognition rate of 97.92%.

8 Conclusion and Future Work

This work has developed an efficient and robust flower species recognition classifier based on deep learning. We have used the dataset from the University of Oxford, which is the Oxford-102 flower dataset with a total of 8189 images of different categories of flower species [35, 36]. We have divided our dataset into training sets and validation set for evaluation purposes. We have employed PyTorch library by Facebook to code our research work. DenseNet161, a pre-trained model of the ImageNet dataset, was loaded to use its weights and later applied to our flower's dataset [11, 37]. This all result was achieved through transfer learning mechanism, which is gaining popularity in deep learning. We have developed a four-step novel approach for the classification of the 102 categories of flower species which is below;

1. The data Augmentation for better training of flower classifier.
2. The Flower Image Segmentation using the BiCoS method for removing the complex background.
3. The cropping of segmented flower images using python script.
4. The model training using the pre-trained model—DenseNet.
5. For training purposes, we have used Jupyter Colab Notebook, a free graphics processing unit (GPU) provided by Google Collaboratory. Our proposed method achieved very high accuracy on the flower's dataset, which is 97.92% classification accuracy. This is one of the best results obtained in the domain of flower species classification.

A deep learning-based CNN classifier is being developed in this work, one of the most robust and efficient with 97.92% classification accuracy on the benchmark dataset. But still, there exists some more work in this domain that can be done in future to use the system in the real world with high accuracy. Some future work points are an extension in the dataset having more categories, integration with mobile applications, and increase in the training data.

References

1. Singh SP, Solanki A, Singh T, Tayal A (2021) Internet of intelligent things: injection of intelligence into IoT devices. In: Artificial intelligence to solve pervasive internet of things issues. Academic Press, pp 85–102
2. Issa MB, Daraghmeh M, Jararweh Y, Al-Ayyoub M, Alsmirat M, Benkhelifa E (2017) Using logistic regression to improve virtual machines management in cloud computing systems. In: 2017 IEEE 14th international conference on Mobile Ad Hoc and Sensor Systems (MASS), 22–25 Oct. 2017
3. Zeiler MD, Fergus R (2014) Visualizing and understanding convolutional networks. In: Fleet DJ, Pajdla T, Schiele B, Tuytelaars T (eds) ECCV, volume 8689 of Lecture Notes in Computer Science. Spring, pp 818–833
4. Redmon J, Divvala S, Girshick R et al (2016) You only look once: unified, real-time object detection. In: Proceeding IEEE conference computer vision and pattern recognition, Las Vegas, NV, June 2016, pp 779–788
5. Singh T, Nayyar A, Solanki A (2020) Multilingual opinion mining movie recommendation system using RNN. In: Proceedings of first international conference on computing, communications, and cyber-security (IC4S 2019). Springer, Singapore, pp 589–605
6. Luong DTA, Chandola V (2017) A K-means approach to clustering disease progressions. In: IEEE Conference on 14 September 2017
7. Solanki A, Singh T (2021) COVID-19 epidemic analysis and prediction using machine learning algorithms. Emerging technologies for battling Covid-19: applications and innovations, pp 57–78
8. Rong F (2017) Audio classification method based on machine learning. In: IEEE Conference on 21 September 2017
9. Singh T, Mishra J (2021) Learning with artificial intelligence systems: application, challenges, and opportunities. In: Impact of AI technologies on teaching, learning, and research in higher education. IGI Global, pp 236–253
10. Simonyan K, Zisserman A (2015) Very deep convolutional networks for large-scale image recognition. In: Proceeding of international conference learning representations, San Diego, CA, May 2015, arXiv preprint arXiv:1409.1556
11. Hu W, Hu R, Xie N et al (2014) Image classification using multi-scale information fusion based on saliency driven nonlinear diffusion filtering. IEEE Trans Image Process 23(4):1513–1526
12. Nitta T (1993) A back-propagation algorithm for complex numbered neuralnetworks. In: Proceedings of 1993 International Joint Conference on Neural networks, IJCNN' 93-Nagoya. 25–29 Oct. 1993
13. Nilsback M, Zisserman A (2008) Automated flower classification over a large number of classes. In: Proceeding sixth indian conference computer vision, graphics and image processing, Bhubaneswar, India, December 2008, pp 722–729
14. Nilsback M, Zisserman A (2006) A visual vocabulary for flower classification. In: Proceeding IEEE conference computer vision and pattern recognition, New York, NY, June 2006, vol 2, pp 1447–1454
15. Pandey S, Solanki A (2019) Music instrument recognition using deep convolutional neural networks. Int J Inf Technolhttps://doi.org/10.1007/s41870-019-00285-y (SpringerPublication)
16. Singh T, Solanki A, Sharma SK (2021) Role of smart buildings in smart city—components, technology, indicators, challenges, future research opportunities. Digital Cities Roadmap: IoT-Based Architecture and Sustainable Buildings, pp 449–476
17. Bhardwaj N, Solanki A (2016) An efficient algorithm for color image segmentation. Selforganizology 3(3):87–99
18. Krizhevsky A, Sutskever I, Hinton G (2012) ImageNet classification with deep convolutional neural networks. In: Pereira F, Burges C, Bottou L et al (eds) Advances in neural information processing systems (Curran Associates, Inc., Red Hook, NY, USA, 2012), pp 1097–1105
19. Hinton GE, Srivastava N, Krizhevsky A, Sutskever I, Salakhutdinov R (2012) Improving neural networks by preventing co-adaptation of feature detectors, CoRR, abs/1207.0580

20. Szegedy C, Liu W, Jia Y et al (2014) Going deeper with convolutions. arXiv preprint arXiv: 1409.4842
21. Yang M, Zhang L, Feng X et al (2014) Sparse representation based Fisher discrimination dictionary learning for image classification. Int J Comput Vis 109(3):209–232
22. Priyadarshni V, Nayyar A, Solanki A, Anuragi A (2019) Human age classification system using K-NN classifier. In: Luhach A, Jat D, Hawari K, Goa XZ, Lingras P (eds) Advanced informatics for computing research. ICAICR 2019. Communications in computer and information science, vol 1075. Springer, Singapore
23. Khan F, van de Weijer J, Vanrell M (2012) Modulating shape features by color attention for object recognition. Int J Comput Vis 98(1):49–64
24. Shelhamer E, Long J, Darrell T (2017) Fully convolutional networks for semantic segmentation. IEEE Trans Pattern Anal Mach Intell 39(4):640–651
25. Girshick R, Donahue J, Darrell T et al (2014) Rich feature hierarchies for accurate object detection and semantic segmentation. In: Proceeding IEEE conference computer vision and pattern recognition, Columbus, OH, June 2014, pp 580–587
26. Ren S, He K, Girshick R et al (2017) Faster R-CNN: towards real-time object detection with region proposal networks. IEEE Trans Pattern Anal Mach Intell 39(6):1137–1149
27. Girshick R (2015) Fast R-CNN. In: Proceeding of IEEE international conference computer vision, Santiago, Chile, December 2015, pp 1440–1448
28. Zhou H, Zheng J, Wei L (2013) Texture aware image segmentation using graph cuts and active contours. Pattern Recog-nition 46(6):1719–1733. https://doi.org/10.1016/j.patcog.2012.12.005
29. Xie L, Wang J, Lin W et al (2017) Towards reversal-invariant image representation. Int J Comput Vis 123(2):226–250
30. Hsu T, Lee C, Chen L (2011) An interactive flower image recognition system. Multimedia Tools Appl. 53(1):53–73
31. Mottos A, Feris R (2014) Fusing well-crafted feature descriptors for efficient fine-grained classification. In: Proceeding IEEE international conference image processing, Paris, France, October 2014, pp 5197–5201
32. Chai Y, Rahtu E, Lempitsky V et al (2012) TriCoS: a tri-level class discriminative co-segmentation method for image classification. In: Proceeding of European conference computer vision, Florence, Italy, October 2012, vol I, pp 794– 807
33. Chen Q, Song Z, Hua Y et al (2012) Hierarchical matching with side information for image classification. In: Proceeding IEEE Conference. Computer Vision and Pattern Recognition, Providence, RI, June 2012, pp 3426–3433
34. Liu Y, Tang F, Zhou D et al (2016) Flower classification via convolutional neural network. In: Proceeding of IEEE international conference functional-structural plant growth modeling, Simulation, Visualization and Applications, Qingdao, China, November 2016, pp 110–116
35. Chai Y, Lempitsky V, Zisserman A (2011) BiCoS: a Bi-level co-segmentation method for image classification. In: Proceeding of international conference computer vision, Barcelona, Spain, November 2011, pp 2579–2586
36. Qi X, Xiao R, Li C et al (2014) Pairwise rotation invariant co-occurrence local binary pattern. IEEE Trans Pattern Anal Mach Intell 36(11):2199–2213
37. Ciresan DC, Meier U, Schmidhuber J (2012) Multi-column deep neural networks for image classification. In: CVPR, pp 3642–3649

Data Localization and Privacy-Preserving Healthcare for Big Data Applications: Architecture and Future Directions

Ashwin Verma, Pronaya Bhattacharya, Yogesh Patel, Komal Shah, Sudeep Tanwar, and Baseem Khan

Abstract With the exponential rise of generated big data in Healthcare 4.0 ecosystems, the data is shared among multiple stakeholders for analysis and data collection process. During the data collection process, a public data set is released from privately collected data through anonymizing critical and sensitive attribute fields. However, existing privacy preservation techniques are often susceptible to linkage and statistical-based attacks that releases sensitive attributes of the patient. Thus, researchers worldwide have shifted attention towards privacy preservation of healthcare data. Owing to the risk of release of critical attributes, well-defined privacy-preservation techniques and ethical data collection process are much-sought issues. Thus, in this paper, we present a systematic survey of data localization issues coupled with privacy preservation mechanisms, specifically in the context of the healthcare industry. In privacy preservation, we present key techniques like K-anonymity, l-diversity, t-closeness, aggregation, removal of linkage-based records, homomorphic encryption, and noise-based schemes. The survey intends to help existing healthcare practitioners, stakeholders, and researchers design effective data localization and privacy preservation mechanisms to safeguard critical and sensitive patient data.

A. Verma · P. Bhattacharya (✉) · Y. Patel · K. Shah · S. Tanwar
Institute of Technology, Nirma University, Ahmedabad, Gujarat, India
e-mail: pronoya.bhattacharya@nirmauni.ac.in

A. Verma
e-mail: ashwin.verma@nirmauni.ac.in

Y. Patel
e-mail: 20mcec13@nirmauni.ac.in

K. Shah
e-mail: 20mcei18@nirmauni.ac.in

S. Tanwar
e-mail: sudeep.tanwar@nirmauni.ac.in

B. Khan
Hawassa University, Institute of Technology, Awasa, Ethiopia
e-mail: khanbaseem@hu.edu.et

P. K. Singh et al. (eds.), *Emerging Technologies for Computing, Communication and Smart Cities*, Lecture Notes in Electrical Engineering 875,
https://doi.org/10.1007/978-981-19-0284-0_18

Keywords Anonymization · Big-data · Data localization · Healthcare · Privacy preservation

1 Introduction

Privacy preservation in healthcare and big data has received much attention from industry as well as from academic researchers. The problem has become more prevalent due to the widespread storage of data on the network. Electronic medical care has improved a lot from a manual paper-based framework, where every detail of patient is stored in files towards digital electronic health record (EHRs) stored in databases [1]. This provides remarkable benefits such as high performance, better precision, and extensive accessibility. The hospital chain or set of hospitals have their data set, and sharing medical records among the chain of hospitals exploits the privacy of patients personal data [2]. Thus, data sharing among all those entities should be done securely without record linkages and violation of security concerns. For the same, the data is required to be localized properly, and at the same time, privacy needs to be addressed [3]. There are multiple entities such as doctors, patient family members, nurses, data providers, patients, pharmaceuticals, and many others. Each one of them has access privileges to shared data, and only a few entities are authorized to view, update and delete the EHR, as EHR information is sensitive, and access should be in consensus to the patient only.

EHR contains patients' medical records as well as private information such as name, gender, age, guardian detail, and address, and loss and modification of such data might create a severe impact on patient life and hospital management system. In today's technological era, everything is stored on the cloud or Web server, or centralized database. The attacker can exploit the loopholes of the system and manipulate the data if proper security and privacy preservation techniques are not adopted. Thus, EHR needs a system that ensures the confidentiality, integrity, and availability of information. In recent times, blockchain-based ledgers are considered a potential solution to managing distributed healthcare records [4]. Thus, blockchain-based healthcare solutions are increasingly applied to monitor even the recent COVID-19 pandemic situations, where chronological contact-tracing ledgers are proposed [5]. So a chain of the hospital need to provide authorized access to their system to ensure security, as well as no other entity in the system, can view the identity of the patient from EHR. The existing system mostly uses a machine learning model in clouds to preserve privacy. The privacy of EHR is very crucial in the current novel coronavirus (COVID-19) pandemic, where different state governments require statistical data to estimate the population of the region and uses the national population register (NPR) [6]. It requires data sharing among research centres, hospitals, pharmacy, and government entities, and localizing the data require privacy, confidentiality, and security on patients data [7]. Thus, EHR must not be disclosed with other entities while sharing its network.

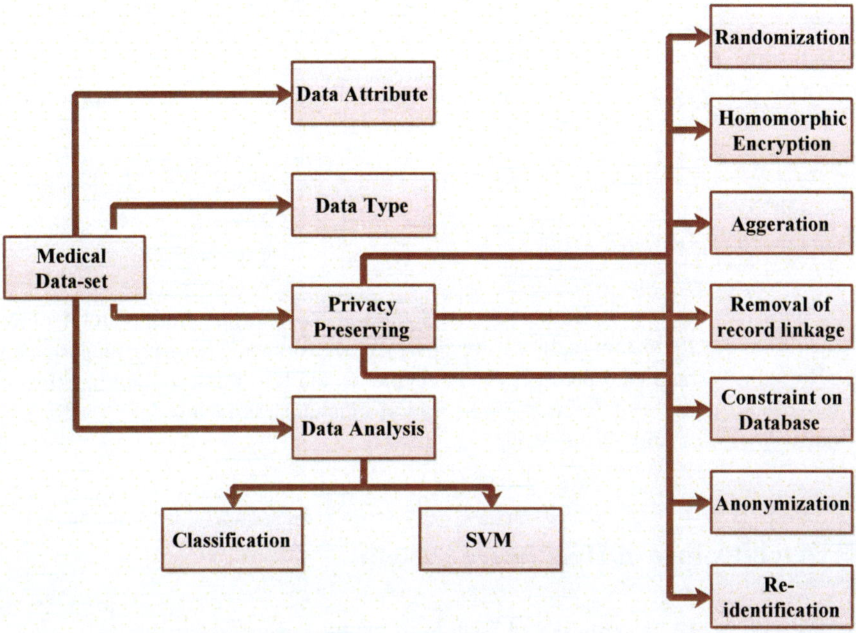

Fig. 1 A systematic classification of privacy-preserving techniques in healthcare

There is a huge amount of data generated every second from different sensors to monitor a patient's health, several medical test reports generated after diagnosis, and we store that information along with EHR and HIS. The data may be structured or unstructured, and currently, the traditional database cannot handle this voluminous data, so we need a sophisticated system to store this data; such data are referred to as big data. Healthcare data are one of the driving forces of big data, as one single genome of human deoxyribonucleic acid (DNA), that is, double helix genetic structure, takes 100–160 gigabyte (GB) of disk space. The expected size of big data in healthcare in 2022 is around 100 zettabytes (ZB). Figure 1 presents a systematic classification of privacy preservation techniques in healthcare and highlights the underlying issues and challenges that healthcare stakeholders face while ensuring accurate analytics (more requirement of data) and privacy preservation.

1.1 Research Contributions

Following are the research contribution of the survey

- We present the overview of data localization in generic healthcare architecture and highlight the importance of privacy preservation for healthcare users.

- We present the overview and solution taxonomy of privacy preservation techniques in healthcare.
- An overview of data anonymization (DA) techniques in privacy preservation and supporting reference architecture are presented.

1.2 Article Layout

The survey is divided into four sections. Section 2 presents the generic architecture of healthcare systems that includes sensor-based networks and decision analytics on EHR data sets. Section 3 presents the overview of the big-data-oriented healthcare scheme and presents the privacy preservation mechanisms in big data healthcare. Finally, Sect. 4 concludes the paper.

2 Architecture of Healthcare System

In this section, we present the architecture of the healthcare ecosystem. Depending on the way healthcare stakeholders process the EHRs, many reference architectures are followed. In some enterprise solutions, EHRs are shared and managed at the discretion of the patient's end, and other stakeholders monitor and manage it in a central fashion [8]. Figure 2 depicts the major components of any healthcare enterprise and can be considered as a generic solution to handle EHR at different levels. The major components of the system are as follows.

- *Communication Network*: The main network that connects all other parts in the system is considered as the communication network.
- *Sensor units*: Patients' body sensors that monitor different body part movements and their day-to-day activities, and information about the different parameters to be considered as wireless body area networks.
- *Healthcare entities*: Different entities of the healthcare system (patients, doctor, insurance provider, pharmacist), who generally access the EHR and interact with other entities to provide life-saving services to the patient.
- *Underlying protocols*: The main communication channel and underlying protocols that connect all the entities to create one single and unified ecosystem, through which information is shared to remote entities or centralized systems, where predictive analysis is done.

The main concern of any healthcare system is the privacy and security of hospital information systems (HIS) where data are categorized into three different views, namely view-1, view-2, and view-3. In view-1, we consider the patient with the sensor units strategically placed at different body locations, which also consists of the source nodes in the sensor environment and the sink nodes. Overall, view-1 presents the body area network view. In view-2, we consider the communication

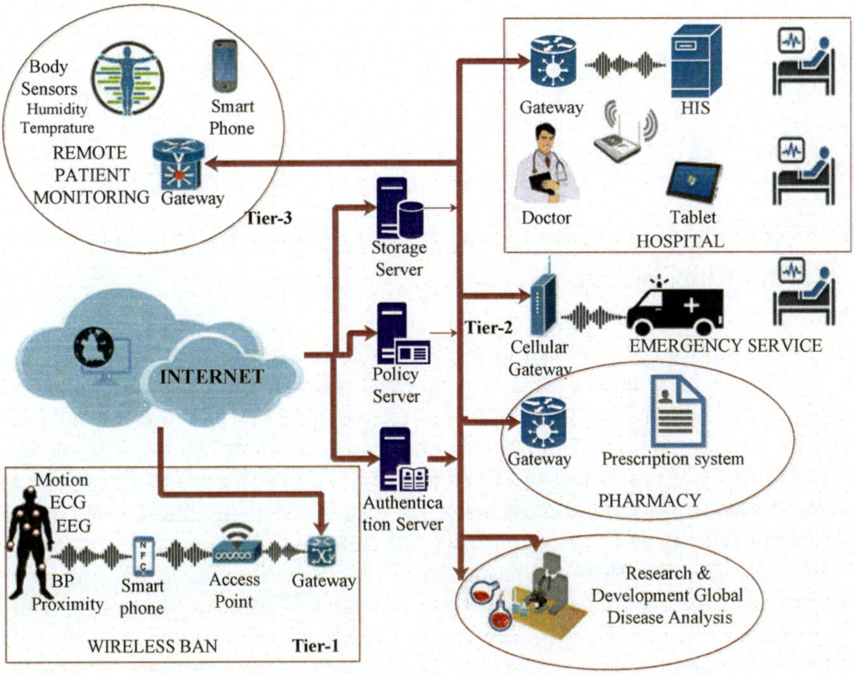

Fig. 2 Basic architecture of the healthcare system [8]

channel and the cloud service providers (CSP), or edge service providers (ESP), that communicate through low-powered communication networks like Zigbee, z-wave, and Bluetooth in a personal range. To address the issues of network management, researchers have presented a solution that combines fifth-generation (5G) services for low-latency communication like ultra-reliable low-latency service (uRLLC), and in the case of real-time response, services like tactile Internet (TI) is considered with the near-real-time response and high availability of 99.99999% [9]. In view-3, we consider the healthcare stakeholders like doctors, nurses, drug research labs, and pharmaceutical companies. The collected data is shared among the different entities in the system with different access rights, so we cannot impose a single security mechanism on the entire enterprise. In such cases, the data view remains dynamic, and every view is materialized with an association [10].

Each component view is different from others and requires different handling of EHR records. The complete healthcare system and EHRs are required to be protected from different threats at different attack points in the communication channels, protocols, and the core network to safeguard the IoT analytics over the ingested big data. The body area network and communication channel to mobile devices have a threat environment. The mobile devices collect the data and prepossess it before sending it to centralized or edge systems. Thus, strong and efficient encryption algorithms are required to enable secure data transfer. Moreover, authorization of data requires signing from different stakeholders in the healthcare chain [11]. Once the data arrives securely to the main network, privacy preservation, users anonymity, and access control are required to be established so that exclusive access control rules are designed.

3 Privacy Preservation in Big-Data-Oriented Healthcare: A Solution Taxonomy

Big data analytics has provided promising aspects in recent Healthcare 4.0 solutions, where a high amount of data is ingested, cost-effective confidentiality and privacy-based big data healthcare schemes are required, that can improve and simplify the machine learning and deep learning-based fusion model designed for predictive analysis of healthcare big data and at the same time manage the privacy of users. Moreover, in big data ecosystems, the data is constantly transformed owing to the high velocity and veracity of data [12]. Thus, solution providers are working towards services that fetch and process data for inside learning to the entities so that they can take important decisions with a more integrated and preventive model [13]. Intelligent use of data can speed up decision-making capabilities and provide better engagement to patients. The data can provide important drivers of healthcare in different segments of the population with solid proof. Gender, ethnic, and cultural prejudices, whether inherent or deliberate, can be eliminated via the proper use of healthcare data. Federal law regulates the information and is not available for all. Lack of digitization and trained employees to handle the big data creates a barrier to systematic solutions. The different issues related to this are discussed below.

3.1 *Privacy Preservation Techniques*

In case EHRs are communicated among different components and entities, then it is important to ensure security and privacy among the different entities; only a defined set of stack-holders are authorized to read, write, and manipulate the data. There are different ways we can preserve privacy in big data such as K-anonymity, L-diversity, T-closeness, Randomization, Data distribution, and cryptographic techniques like homomorphic encryption and multidimensional sensitivity-based anonymization (MDSBA). Each technique has its pros and cons as per system requirement and usage. The data set can be centralized or decentralized, and based on that environment, we identify the technology which is most suitable to the requirement. Figure 3 presents a solution taxonomy of different privacy preservation techniques in healthcare-oriented big data ecosystems. The details of the techniques are discussed in the following subsections in detail.

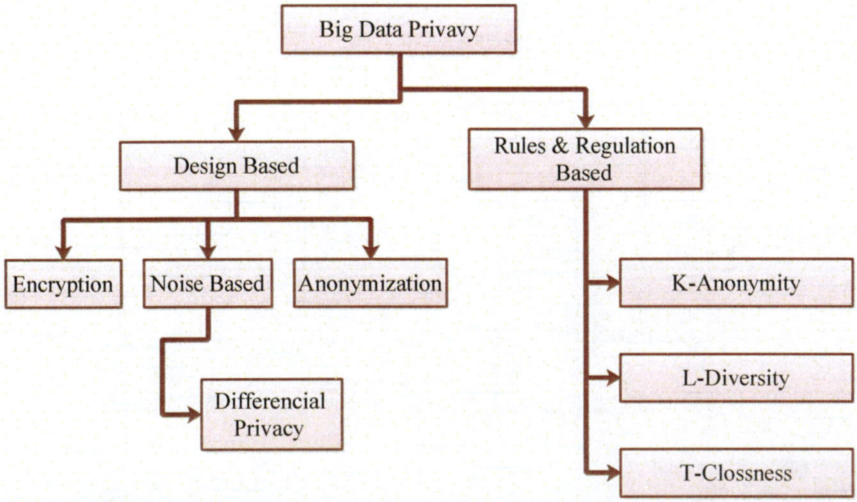

Fig. 3 A solution taxonomy of privacy preservation in big-data-oriented healthcare

Data Anonymization: Data anonymization (DA) refers to the process of hiding sensitive data either by deleting or encrypting the information that connects to the explicit identification of the patient through his public viewable EHR. Through DA, the linkage and cross information of patient identity are not revealed, and thus, other stakeholders can effectively share the EHRs among themselves. Moreover, the publicly available EHRs collected as healthcare data sets can be used for research even by third parties. Thus, DA is a critical step that is required before the data is shared with some untrusted third party [14]. Figure 4 presents the overview of DA in healthcare ecosystems.

K-**anonymity**: K-anonymity is a security model that normally applies to protect the information in data-sharing scenarios where data is collected in large groups by masking the identity of the user [15]. In various privacy preservation frameworks, the ultimate objective is anonymity for the unlabelled data subjects [16]. Elimination of explicit user identifiers is not sufficient to assure DA. Anonymity information can be re-distinguished by connecting information with another publicly available data set. The information might incorporate snippets of data that are not themselves exceptional identifiers but rather can become distinguishable when joined with other data sets; these are known as semi-identifiers. Thus, K-anonymity ensures the following conditions:

- The data of interest should not uncover data that was redacted in the unspecified partitions or columns. There is a scenario of medical data set where similar kind of diseases has been caught to men or women, then we can redact the gender parameter from there [16].
- The sensitive data represented in columns are not different either way for a specific classification of K. When sensitive parameters are not different in either way for

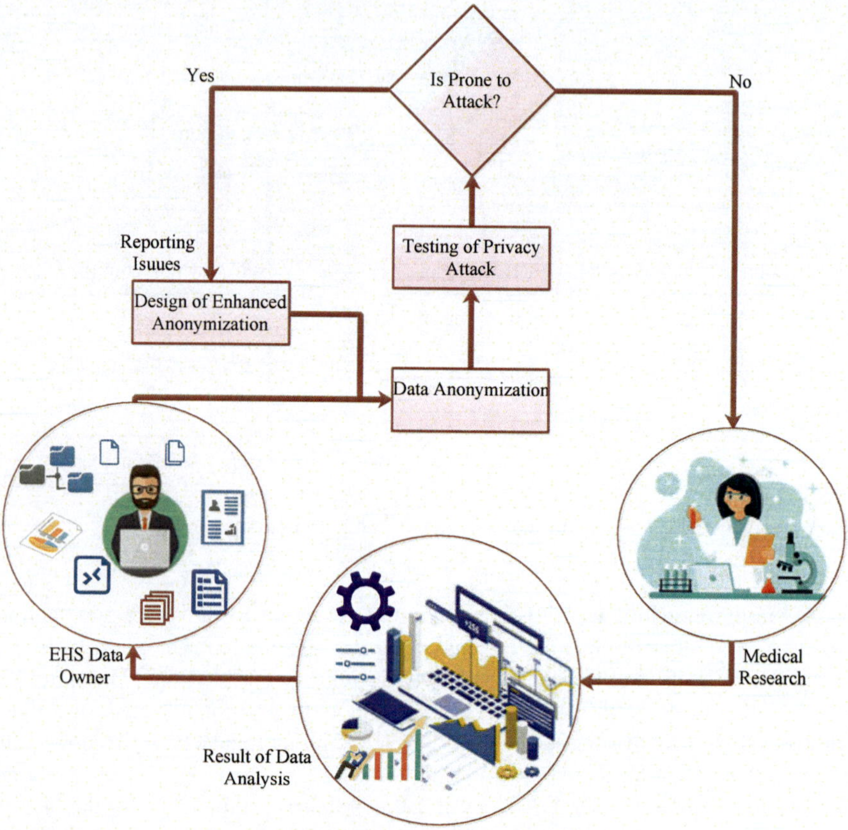

Fig. 4 Overview of DA in healthcare ecosystems

a bunch of K records that share the semi distinguishing credits, at that point, the data set is defenceless against a supposed homogeneity attack. In a homogeneity assault, the assailant utilizes how it is sufficient to discover the gathering of records that the individual has placed with if every one of them has a similar worth. For example, if we consider that all men more than 60 in our data set have malignancy, and if some person X age, denoted as $\text{Age}(X) > 60$, in the data set, then a user might infer that X has a particular disease D explicitly.

- The dimensionality of the information should be adequately low. There are some high-dimensional attributes like time-series information; it turns out to be very difficult to give similar protection like low-dimensional information. It is feasible to recognize an individual by focussing on different sensitive parameters. Additionally, the dimensionality of information increases frequently along with the focussed information are scantily circulated. This makes it hard to bunch records without intensely contorting the information to accomplish K-anonymity.

l-**diversity**: l-diversity, or l-diversity, refers to the security conservation technique where the homogeneity assaults are covered. l-diversity is preposterous to carry out on each type of different data set [3]. The l-diversity model is an augmentation over the K-anonymity model that diminishes the granularity of information portrayal and utilizes strategies that include speculation and concealment. The end goal is that a given record should map onto at least $K - 1$ other records in the information [15]. The l-diversity model handles the shortcomings of the K-anonymity model where ensured personalities to the level of K-people is not comparable [17]. A K-anonymity result is supposed to be l-diverse with the assumption that each of the identity classes in the table has event 'l' that represents values for every quality [18].

T-**closeness**: An improvement of the l-diversity model is the T-closeness strategy that diminishes the unprocessed deciphered information [19]. The spectator's degree of information on explicit information is restricted whereas the information is not restricted to the general table containing the data sets. This lessens the connection between the semi-identifier parameter and the sensitive properties [15]. The distance between the propagation is estimated through the earth mover's distance (EMD). For a categorical parameter, EMD is utilized to scale the distance between the qualities in it as per the base degree of speculation of qualities in the area of importance.

Randomization: Randomization presents the user random reaction models, termed as k-mix, instead of utilization of the standard activities i.e. speculation, concealment, or added substance clamour. In k-mix, conceivable deniability for utilizing a combination of attributes by the data reconstruct system is achieved [20]. k-mix component reports the value that draws from the diffusion of the genuine perceptions with a likelihood p, and it also reports value inspected from other pre-decided (mix) allocations, which all cover the whole range of the genuine appropriation with a likelihood of $1 - p$ altogether.

Homomrophic Encryption—The cryptographic solution to data privacy: Homomorphic encryption (HE) can perform significant computations on encoded information without decoding the information, and the HE framework can be utilized to encode the information to address the data protection issue [21]. HE provides an answer of extraordinary encryption that can settle these issues permitting any outsider to work on scrambled information without earlier decoding [7].

HE assures proper security for sharing of clinical EHRs. It scrambles clinical pictures and performs valuable procedures on them without breaking the classification picture. There is a protected and high-vision clinical picture system to ensure the protection and security. In HE, we perform reversible data hiding (RDH) technique that implants private information into clinical pictures.

We assume that an encryption scheme G is represented as (P, C, K_s, E, D), where P is the plaintext space, C is the ciphertext space; K_s is the key space, E is the encryption algorithm, and D is the decryption algorithm. $\oplus$ represents the $C - P$ operator. Based on the notations, different HE types are presented as follows.

Definition 1 Normal HE: Assuming P and L are operations on plain text of data set $P = \{p_1, p_2, p_3, \ldots, p_n\}, k_s \in K_s$, it satisfies $P(E_{ks}(p_1), E_{ks}(p_2), \ldots, E_{ks}(p_n)) = E_{ks} = (L(p_1, p_2, p_3, \ldots, p_n))$.

Definition 2 Addictive HE: For plaintext $p_i, p_j \in P$, the corresponding ciphertext is $c_i = E(p_i), c_j$ and $c_i, c_j \in C$. If $E(p_i + p_j) = E(pe_i) \oplus E(p_j)$ or $D(E(p_i) \oplus E(p_j)) = p_i + p_j$, then the condition is termed as additive HE in G.

Definition 3 Multiplicative HE: For any plaintext $p_i, p_j \in P$,*the corresponding ciphertext is* $c_i = E(p_i), c_j$ and $c_i, c_j \in C$. If $E(p_i.p_j) = E(p_i) \oplus E(p_j)$ or $D(E(p_i) \oplus E(p_j)) = p_i.p_j$ is satisfied then, it is known to be an multiplicative HE in G.

Multidimensional Sensitivity-Based Anonymization: In the multidimensional sensitivity-based anonymization (MDSBA) approach, we ensemble the K-anonymity model with l-diversity and (X, Y) anonymity. With basic MDSBA, one-dimensional sets are trained and validated to represent the multi-dimensional plots. It combines the approach of top-down anonymization and presents considerable authority in information based on collected user scores. It applies to the big data MapReduce scheme. In a two-stage multi-dimensional top-down approach, it separates the enormous information into small-sized lumps and provides anonymized data. Every piece of information requires n seasons of emphasis to reflect the best score.

In MDSBA, it provides base anonymization conveyed in equal cycles, and it diminishes the MapReduce to reduce the overall complexity. The algorithm of MDSBA includes a data proprietor, that decides on the identity of item-sets, denoted as $Q(IDs)$, the item delicate characteristics S, and provides the anonymized $Q(IDs)$.

Noise-based techniques: The name itself suggests that privacy is preserved by adding some amount of noise to the data sets. The noise added must be so much that it hides the sensitive information of the patient. One of the popular techniques is differential privacy. In terms of privacy preservation in healthcare, differential privacy works very well. It improves the systems' speed, and it does not allow any person to interfere even if he knows the data of other members [22]. It proposes the method that makes sure that the output of the two different data sets should be approximately the same. Thus, the attacker is not able to intercept the differences between two information sets, and thus, explicit user identifiers are not linked to a particular user.

4 Conclusion

In this paper, we present a systematic overview of DA and privacy preservation techniques in big-data-oriented healthcare ecosystems. The generic reference architecture of the healthcare system is highlighted, and major components are discussed. Then, big-data-oriented healthcare and privacy preservation are discussed, and a solution taxonomy is presented. Next, we present the overview of DA and presented a reference to DA architecture in healthcare. Finally, issues and challenges of ensuring privacy and security are discussed. Thus, the survey assures that effective healthcare ecosystems are designed with an assurance of secured privacy matrices for patient EHRs as well as assure correct and responsive analytics.

References

1. Bhattacharya P, Tanwar S, Bodkhe U, Tyagi S, Kumar N (2021) Bindaas: blockchain-based deep-learning as-a-service in healthcare 4.0 applications. IEEE Trans Netw Sci Eng 8(2):1242–1255. https://doi.org/10.1109/TNSE.2019.2961932
2. Chen L, Yang JJ, Wang Q, Niu Y (2012) A framework for privacy-preserving healthcare data sharing. In: 2012 IEEE 14th international conference on e-health networking, applications and services (Healthcom). pp 341–346. https://doi.org/10.1109/HealthCom.2012.6379433
3. Jin H, Luo Y, Li P, Mathew J (2019) A review of secure and privacy-preserving medical data sharing. IEEE Access 7:61656–61669. https://doi.org/10.1109/ACCESS.2019.2916503
4. Bodkhe U, Tanwar S, Bhattacharya P, Verma A (2021) Blockchain adoption for trusted medical records in healthcare 4.0 applications: a survey. In: Singh, PK, Wierzchoń ST, Tanwar S, Ganzha M, Rodrigues JJPC (eds) Proceedings of second international conference on computing, communications, and cyber-security. Springer Singapore, Singapore, pp 759–774
5. Zuhair M, Patel F, Navapara D, Bhattacharya P, Saraswat D (2021) Blocov6: a blockchain-based 6g-assisted uav contact tracing scheme for covid-19 pandemic. In: 2021 2nd International Conference on Intelligent Engineering and Management (ICIEM), London, United Kingdom, pp 271–276. https://doi.org/10.1109/ICIEM51511.2021.9445332
6. Verma A, Bhattacharya P, Zuhair M, Tanwar S, Kumar N (2021) Vacochain: blockchain-based 5g-assisted uav vaccine distribution scheme for future pandemics. IEEE J Biomed Health Inf 1–1. https://doi.org/10.1109/JBHI.2021.3103404
7. Sinha K, Majumder P, Ghosh SK (2020) Fully homomorphic encryption based privacy-preserving data acquisition and computation for contact tracing. In: 2020 IEEE international conference on advanced networks and telecommunications systems (ANTS). pp 1–6. https://doi.org/10.1109/ANTS50601.2020.9342834
8. Sahi MA, Abbas H, Saleem K, Yang X, Derhab A, Orgun MA, Iqbal W, Rashid I, Yaseen A (2018) Privacy preservation in e-healthcare environments: state of the art and future directions. IEEE Access 6:464–478. https://doi.org/10.1109/ACCESS.2017.2767561
9. Bhattacharya P, Tanwar S, Shah R, Ladha A (2020) Mobile edge computing-enabled blockchain framework—a survey. In: Singh PK, Kar AK, Singh Y, Kolekar MH, Tanwar S (eds) Proceedings of ICRIC 2019. Springer International Publishing, Cham, pp 797–809
10. Verma A, Bhattacharya P, Bodkhe U, Ladha A, Tanwar S (2021) Dams: Dynamic association for view materialization based on rule mining scheme. In: Singh PK, Singh Y, Kolekar MH, Kar AK, Chhabra JK, Sen A (eds) Recent innovations in computing. Springer Singapore, Singapore, pp 529–544
11. Bhattacharya P, Mehta P, Tanwar S, Obaidat MS, Hsiao KF (2020) Heal: a blockchain-envisioned signcryption scheme for healthcare iot ecosystems. In: 2020 International Conference on Communications, Computing, Cybersecurity, and Informatics (CCCI), Sharjah, United Arab Emirates. pp 1–6. https://doi.org/10.1109/CCCI49893.2020.9256705
12. Ghayvat H, Pandya SN, Bhattacharya P, Zuhair M, Rashid M, Hakak S, Dev K (2021) Cp-bdhca: blockchain-based confidentiality-privacy preserving big data scheme for healthcare clouds and applications. IEEE J Biomed Health Informat 1–1 (2021). https://doi.org/10.1109/JBHI.2021.3097237

13. Chauhan R, Jangade R (2016) A robust model for big healthcare data analytics. In: 2016 6th international conference—cloud system and big data engineering (Confluence), pp 221–225. https://doi.org/10.1109/CONFLUENCE.2016.7508117
14. Mahesh R, Meyyappan T (2013) Anonymization technique through record elimination to preserve privacy of published data. In: 2013 international conference on pattern recognition, informatics and mobile engineering, pp 328–332 https://doi.org/10.1109/ICPRIME.2013.6496495
15. Afifi MH, Zhou K, Ren J (2018) Privacy characterization and quantification in data publishing. IEEE Trans Knowl Data Eng 30(9):1756–1769. https://doi.org/10.1109/TKDE.2018.2797092
16. Srinivasan NSC, GN (2015) A robust privacy preserving model for data publishing. In: 2015 International Conference on Computer Communication and Informatics (ICCCI), pp 1–6. https://doi.org/10.1109/ICCCI.2015.7218095
17. Zeng P, Zhang Z, Lu R, Choo KKR (2021) Efficient policy-hiding and large universe attribute-based encryption with public traceability for internet of medical things. IEEE Int Things J 1–1. https://doi.org/10.1109/JIOT.2021.3051362
18. Rajendran K, Jayabalan M, Rana ME (2017) A study on k-anonymity, l-diversity, and t-closeness techniques focusing medical data. Int J Comput Sci Netw Secur 17
19. Binjubeir M, Ahmed AA, Ismail MAB, Sadiq AS, Khurram Khan M (2020) Comprehensive survey on big data privacy protection. IEEE Access 8:20067–20079. https://doi.org/10.1109/ACCESS.2019.2962368
20. Abouelmehdi K, Beni-Hssane A, Khaloufi H, Saadi M (2017) Big data security and privacy in healthcare: a review. Proc Comput Sci 113:73–80 (2017) https://doi.org/10.1016/j.procs.2017.08.292, https://www.sciencedirect.com/science/article/pii/S1877050917317015, the 8th international conference on emerging ubiquitous systems and pervasive networks (EUSPN 2017)/The 7th international conference on current and future trends of information and communication technologies in healthcare (ICTH-2017)/affiliated workshops
21. Wang Q, Zhou D, Yang S, Li P, Wang C, Guan Q (2019) Privacy preserving computations over healthcare data. In: 2019 international conference on internet of things (iThings) and IEEE green computing and communications (GreenCom) and IEEE cyber, physical and social computing (CPSCom) and IEEE smart data (SmartData). pp 635–640. https://doi.org/10.1109/iThings/GreenCom/CPSCom/SmartData.2019.00123
22. Begum SH, Nausheen F (2018) A comparative analysis of differential privacy vs other privacy mechanisms for big data. In: 2018 2nd International Conference on Inventive Systems and Control (ICISC), pp 512–516 (2018). https://doi.org/10.1109/ICISC.2018.8399125

Network and Computing Technologies

Design of Facial Recognition Based Touchless Biometric System

Priya Mishra, Shashwat Singh Mann, Mohit Sharma, and Madhulika Bhatia

Abstract In today's era, usage of biometric systems for the entrance in a particular premise is getting common. Fingerprint biometrics is one of the tools which is usually practiced a lot. Face identification with ID card recognition is a method which is still not introduced in various schools or universities. The paper presents an approach for real time face detection and scanning ID cards to identify a person's identity. Algorithm like Principal Component Analysis (PCA) is used for reduction of face space dimension and later used to obtain the image characteristics using Fisher Liner Discriminant (FLD), generally known as Linear Discriminant Analysis (LDA). LBP (Local Binary Pattern) is yet another technique used for face detection. The matched temporary data with the already existing data set is used for system training. The system testing is basically done by the process of feature vector and pattern matching. Haar feature-based Cascade Classifier is a machine learning based approach where a cascade function is trained from a lot of positive and negative images and detects the objects in other images. Optical Character Recognition (OCR) is an effective technique for image to text detection with Pytesseract being one of the main tools used in OCR. Just like the techniques used for the face recognition, the results found out through image to text techniques are compared with the existing dataset. Combining both the techniques (face recognition as well as image to text recognition) will result out in the design of a touchless biometric system.

Keywords Face recognition · Text recognition · Biometric system · Machine learning

P. Mishra (✉) · S. S. Mann · M. Sharma · M. Bhatia
Amity University, Amity Road, Sector-125, Noida, Uttar Pradesh 201301, India

S. S. Mann
e-mail: shashwat.mann@student.amity.edu

M. Sharma
e-mail: mohit.sharma31@student.amity.edu

M. Bhatia
e-mail: mbhadauria@amity.edu

P. K. Singh et al. (eds.), *Emerging Technologies for Computing, Communication and Smart Cities*, Lecture Notes in Electrical Engineering 875,
https://doi.org/10.1007/978-981-19-0284-0_19

1 Introduction

Due to the increased use and improved proficiency of face recognition systems, a lot of study work has been proposed to make it more accurate and advanced in how the detection algorithm works. The latest revolution being the Real Time Face Detection, that is the capability to detect faces. Applications of face detection and recognition aren't just limited to features of entertainment and sharing photograph. The recent advances in this work have shown implications of face recognition in many areas such as security, biometrics, personal safety, law- enforcement, etc.

With the rise of a new pandemic on the whole world in 2020, people have become anxious to touch the objects which do not belong to them personally. Corona virus has affected human minds mentally as well as physically. People have been scared of even touching small objects. A touched biometric system is a type of biometric which uses human interaction physically to access a particular object. For example, fingerprint system is used in various places for identification of a person. From Aadhar Card centers to various school and colleges, people have been performing this activity from a long time. What if we want to improvise and change the method of biometric system to touchless i.e., without physical human interaction? What will be the need and tools for it? To answers all these questions which arise in our minds, we came up with a motivation of designing a touchless biometric which is a lot easier to use and is user friendly where a person will need to show his/her face to a camera in order to get entrance to a respective institute/school. Further, the method can be used for marking attendance in a premise. Text recognition is yet another effective method of using a touchless biometric system for various purposes where the text written in the ID card of a person is compared with the dataset of that person already stored.

The basic idea is to design a touchless biometric system that is capable of identifying or verifying a person using face and text scanning techniques. A database is created which contains all the details of an individual, and the model is trained with the existing dataset. If the biometric recognizes the face of a person with the matching dataset that contains his/her information in it, the system will allow the person to enter the premise or mark their attendance depending upon where the system will be used for. The key strength of using a touchless biometric system is to reduce human effort physically and use this system for security purposes like giving permission to entry to a particular location only to the people an organization wants.

The challenges which a touchless biometric can come across is the accuracy of detection of a face as well as taking a lot of time to detect the face. To overcome these problems, the number of images to be trained must be more than 300 for a single person in different angles and expressions. For the time taking problem, several algorithms have been discussed and compared so that the least time taking techniques are implemented for a better working of the touchless biometric system.

Basic steps involved in the working of the model are as follows:

1. Capturing: Image of the person/text is captured and further stored in the dataset for comparison. A person needs to get near the camera and show his face so that the camera captures the image and trains the image for further use.
2. Feature extraction: Feature extraction describes the relevant shape information contained in a pattern so that the task of classifying the pattern is made easy by a formal procedure. Facial feature extraction is the process of extracting face component features like eyes, nose, mouth, etc. from a human face image. Facial feature extraction is a very important method for the initialization of face recognition.
3. Comparison: It is very important to check the image captured and the ID card information is matching with the existing dataset. This step involves the comparison of the person who is in front of the touchless biometric system with the pre-existing dataset which is stored in the system containing his/her information.
4. Match/non-match: After the comparison, the permit/non-permit is implemented according to the output of the detection.

Later, the trained dataset and the image are matched and compared with the processed dataset available for the face recognition. Once the image is captured, the person will have to scan his/her ID too. The person will only be given attendance or be permitted if the image and ID card details matche with the dataset else the system will reject the person's entry.

The applications of touchless biometric system can be used in:

1. Various schools, colleges and institutes for the entry and exit purposes inside the premise
2. A private company for keeping record of employee's daily routine
3. For marking attendance

Figure 1 shows the architecture of how the touchless biometric system will work. The first step includes the face image as well as ID card image input to our system. The

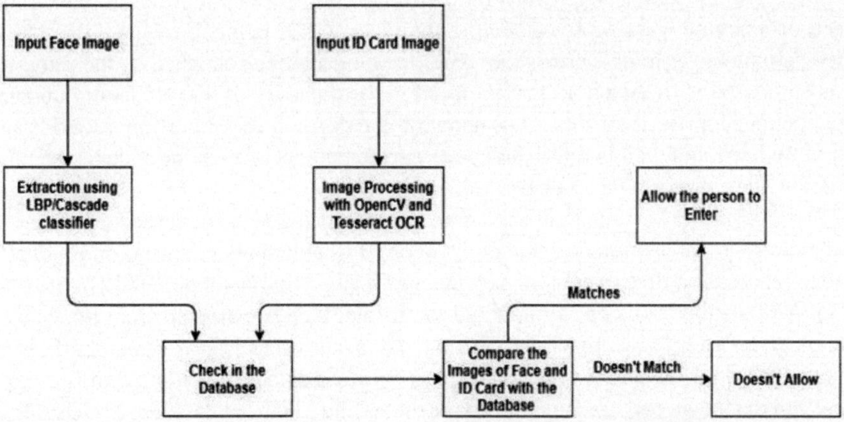

Fig. 1 Overview of the steps involved for implementing a touchless biometric system

extraction of face image is done either through Local Binary Pattern or Haar Cascade technique. On the other hand, the ID card scanned is processed using OpenCV with Optical Character Recognition technique. Both the inputs are then compared with the existing data stored in the database and if the inputs get matched with the pre-existing data of that particular person, the system will allow that person to enter a premise or vice versa.

2 Related Work

The touchless biometric system works on the different concepts and techniques used in field of Machine Learning and Deep Learning. Machine Learning is technique where we train a machine/system to perform a particular work by itself without the need of human interaction. Deep Learning uses the help of machine learning in the working of algorithms. Various methodologies are being used to get better result accuracy for face as well as for text recognition using Machine Learning. Some research shows the use of MATLAB (Matrix Laboratory) being used for the working of face recognition. A subfield of Machine Learning called NLP (Natural Language Processing) is a technique of processing texts and words which is widely used for many purposes but is yet another effective technique that can be performed for text extraction. NLP uses Tokenization which is the splitting of a particular sentence and then assigning each word into a particular category by applying the Parts Of Speech Tagging (POS) to each split text.

There is a lot of scope for face recognition technique and can be used almost anywhere including hospitals, a home, schools, etc. In the upcoming digital world, a touchless system will be used almost everyday in a person's life. There is a huge demand of cost-effective machines capable of identifying person's identity and that can be done using face recognition and text recognition of an individual's ID card.

The technique of CNN (Convolution Neural Network) which is an algorithm used in Machine learning for detecting the faces. CNN is much fast and increases the efficiency to eliminate errors too. After the face has been detected by the camera, it is compared with the photo that is stored in the database of that student to update the attendance. An excel sheet will update the record of attendance on a weekly or monthly basis and then that file can be sent to the parents of each individual students to know their performance in class [1].

A proposal showed an attendance management system based on face recognition technology in the institute/organizations in order to obtain attendance of individuals and to record their time of entry and exit using OpenCV library and FLTK (Light Tool Kit). An image is taken as an input and the frontal face is extracted from the image which is in RBG (Red Blue Green) format and is converted to gray scale of 50×50 pixels. The algorithm is used for finding patterns in the data for calculating similarities between the image captured and the image that is already in the database [2]. Usually, MATLAB deals with small images for processing the image faster [3]. The average of all face images from converted trained image in matrix form and then calculating the eigen faces value is done by Fisher Face method by applying PCA (Principal Component Analysis) on the image and then applying LDA algorithm to calculate the average value of each person [4]. Eye detection algorithm is used to reduce false face detection rate. The facial recognition performances are also greatly proved by using facial components alignment, contrast enhancement and image smoothing [5]. A module consisted of text area extraction by using segmentation process and OCR (Optical Character Recognition) process which will compare the character recognition model with the trained data using tesseract tools [6]. Normalization is performed where a printed document is converted into a gray scale image and then finally converted into a binary image for image to text detection [7].

2.1 Literature Review

See Table 1.

3 Methodology

The methodology is divided into 3 parts:

(a) Methodology for Face Recognition:

Three different tools are used for face recognition namely Haar Cascade Classifier, OpenCV library and LBP (Local Binary Pattern) algorithm (Fig. 2).

The first step is to acquire the image of an individual and then convert it into grayscale by using OpenCV.

The second step is to detect the eyes, mouth and face of the person using Haar cascade classifier/LBP and then normalize face images and train the data and store information in a doc or XML file.

(b) Methodology for image to text recognition:

For text recognition, two tools are used namely OpenCV library, OCR (Optical Character Recognition) (Fig. 3).

Firstly, the ID card image is scanned, and the method of noise removal and normalization is implemented. Later, the process of segmentation feature extraction and classification is adopted using optical character recognition. The extracted text data is stored in a word or excel file.

After the image is scanned the most important step would be pre-processing the module which includes Noise removal and Normalization of image, which will increase the quality of image and also check for any rotational or uniform size in the image text for better result in text recognition. After pre-processing, the filtered image is later converted which can be understood by the computer. This is done by segmentation, feature extraction and classification, where first individual character

Table 1 A brief review of related literature papers

S. No.	Aim of the work	Methods used in the work	Limitations	Type of recognition
[1]	Focus on making an automated attendance system proposal using face recognition by capturing the face of each student and storing it in the database for the attendance	CNN (Convolutional Neural Network)	Lack of accuracy and only a single machine learning module used	Face recognition
[2]	An attendance management system based on face recognition technology in the institute/organizations in order to obtain attendance of individuals and to record their time of entry and exit	PCA (Principal Component Analysis), OpenCV, FLTK (Light Tool Kit)	Limited usage of the methodology in different areas of professional work	Face recognition
[5]	Focus on real time face recognition using many techniques at once as it deals with more complex background	ADA boost and cascade classifier	Lack of result analysis and accuracy prediction	Face recognition
[7]	Focus on how we can do the recognition and extraction of words written in images	Artificial neural networks, support vector machine, etc. as classifiers and linear discriminant analysis, principal component analysis, etc. for feature extraction	Hard to interpret the model and does not perform well on small data sets	Text recognition
[6]	ID card detection by using image recognition techniques for mainly Indonesian citizens by collecting their data from the citizen ID card an individual has	Optical character recognition, morphological transformation technique	Difficult to collect the data from citizens and less accuracy for text size less than 12 Points	Text recognition
[3]	A proposed solution for face recognition of an individual person	Dilation, Sobel operation, MATLAB	Lengthy process which can be time consuming for detection of an image	Face recognition

(continued)

Table 1 (continued)

S. No.	Aim of the work	Methods used in the work	Limitations	Type of recognition
[4]	Focus on fisher face method for recognition and detection of face	Fisher face method, principal component analysis, linear discriminant analysis	Complex process and the paper only focused on how to improvise the method using fisher face method	Face recognition

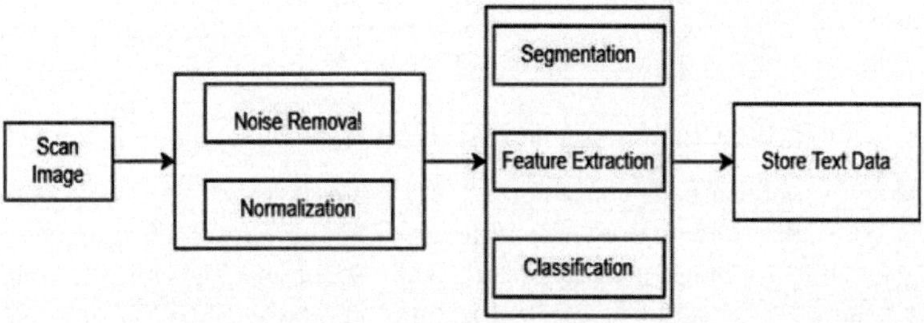

Fig. 2 Detecting and recognizing a face processing

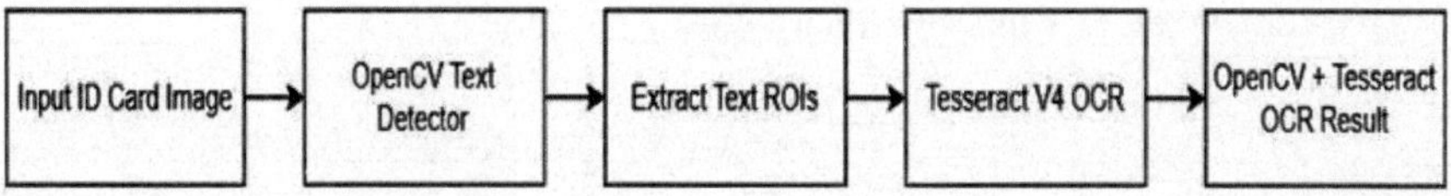

Fig. 3 Text recognition process

is separated from image and then feature extraction is performed on that to extract important data from the raw data.

(c) Comparing with the dataset:

The final step is comparing the extracted data of a person collected from the face recognition and image to text recognition with the existing database. If the data is matched, it will permit the person to access inside a premise or mark the attendance according to where the system is used, else the access is denied.

3.1 Techniques and Tools

(a) OpenCV:

OpenCV is a python library used for object detection which includes face as well as image to text detection. OpenCV is used for implementing the LBP algorithm as well as the used for the Haar Cascade Classifier. Also, an OCR tool known as Pytesseract is used for text recognition which uses OpenCV library for implementation. OpenCV is basically used for real time face vision.

(b) LBP (Local Binary Pattern) Algorithm:

LBP algorithm is an effective algorithm used for detection of face by using the result in form of binary number by calculating the pixels around the object we want to detect. Due to less complexity, this algorithm turns out to be one of the least time taking algorithm for detecting face.

LBP histogram can be defined as:

$$H_i = \sum x, yI\{fl(x, y) = i\}, i = 0, \ldots, n - 1 \tag{1}$$

where x, y are the neighborhoods and H represents the histogram.

(c) ADA boost and Cascade Classifier:

ADA boost algorithm is used in face detection for gaining faster face recognition output as well as for better accuracy result. The ADA boost algorithm is used to get the best features to check the faces. Best features are first selected as weak classifiers and then calculated together as a weighted combination of these features to make a strong classifier. The strong classifier can be used to detect faces. In the equation down below, h_f is the threshold by the strong classifier to detect a face.

Given down below is the formula used for ADA Boost:

$$h_f(x) = \left\{ \begin{array}{ll} 1 & \text{if} \sum_{t=1}^{T} \alpha_t h_t(x) \geq \frac{1}{2} \sum_{t=1}^{T} \alpha_t \\ 0 & \text{otherwise} \end{array} \right\} \tag{2}$$

In Eq. (2),

t initial time,
T final time,
h_f final hypothesis,
x training data value.

Cascade classifiers are basically trained on ADA Boost here. The series of classifier consists in a series of tests on the input features. The selected features are divided into many stages and each stage is trained to become a strong classifier with the help of the weak classifiers. At every stage, the Haar cascade classifier checks if there is any image present or not and if not, the whole window will be discarded. Here, the process of execution will be faster and the number of stages will give a predetermined accuracy. Haar Cascade Classifier is used for face recognition just like in the local binary pattern technique yet not as effective as LBP, it produces a great accuracy for the time taken to recognize the face.

(d) Optical Character Recognition with Pytesseract:

OCR is a technique used for text recognition and extraction. It is used to identify and recognize the text written inside an image by visualizing it and converting it into an electronic form. Pytesseract is a tool used with OCR to scan and then recognize the text written in the ID card and extract it in a text file format.

OCR cannot be used for face recognition as it is limited for extraction of words written on an object. OCR is preferred over deep learning techniques just because of the simplicity and better accuracy of extracted text image.

Different libraries are used in Python for the working of Face recognition and text recognition code. The effective libraries used for implementing the touchless biometric system are:

1. numpy: This python library is used for storing multi-dimensional arrays and matrices. The text as well as face recognition method will require a large amount of data to be stored which can be done easily using numpy library.
2. Tensor Flow: Tensor Flow is an effective python library which is used for object detection. This library is used for text image detection.
3. Python Imaging Library (PIL): Python Imaging Library is used to enhance a text written on an image. It increases the sharpness and contrast of the captured text image so that the extraction of the words written on it can be done easily without producing any errors which can happen if the text is blur or if the text written on it is not sharp enough.

4 Results and Discussion

After running the setup of face detection using Haar Cascade and LBP (Local Binary Pattern), it was found that the model runs successfully with somewhat no instantaneous changes between the faces in the image. With several alterations and researching through the web, even though the face detection for still image is very accurate, for moving frame, the speed of changing the frames multiplies the difficulty exponentially. The LBP algorithm was found to be the least time taking algorithm to detect a face as compared to Haar Cascade classifier.

After successfully comparing the sample images from Fig. 4 and taking the mean of the time taken, the results were as follows (Fig. 5).

The above graph represents the time graph of three different face samples. After comparing the time taken with both the algorithms, it can be seen that LBP algorithm is taking comparatively less time to detect the face as compared to Haar cascade classifier resulting in better accuracy (Table 2).

5 Conclusion

The combination of face recognition with text recognition is a unique touchless biometric technique which can be used in various places according to its need,

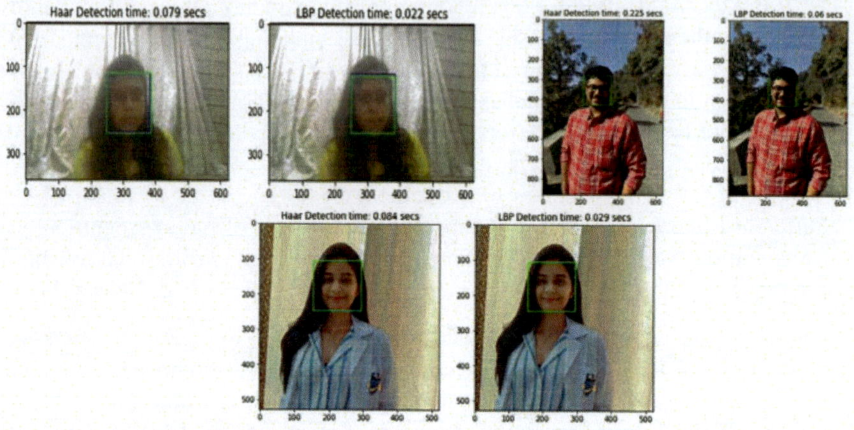

Fig. 4 Comparing Haar Cascade and LBP detection time

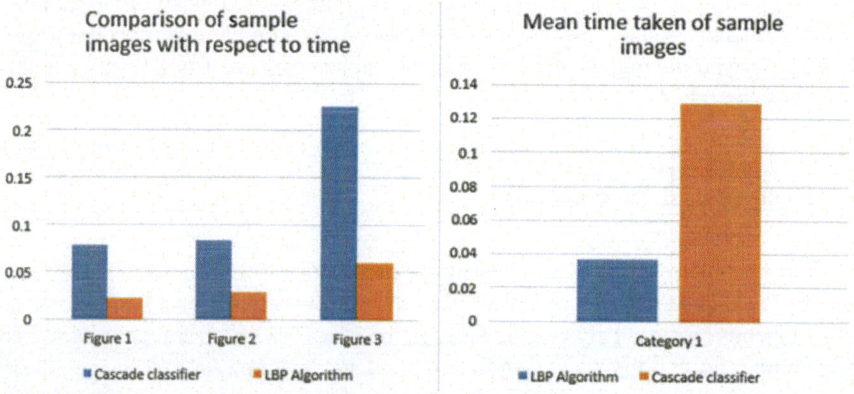

Fig. 5 Graphs comparing the time taken to detect the images

Table 2 Comparison of the techniques used with respect to time

Sr. No	Technique used	Mean accuracy of the time taken (From sample images taken from the above)
1	Haar cascade classifier	0.129 s
2	LBP algorithm	0.037 s

whether it's for personal use or for entry in a college/school. For any security purposes, this system can be an advantage as it will not only check the facial detection but will also scan the ID cards if necessary. It turns out to be user friendly because if a person forgets to bring his ID card along, he/she can just use the facial recognition tool to enter the premise. This system can also be used for marking attendance in schools or in a company too. Combining face recognition with image to text recognition can be beneficial as they give a positive output and combining both of them is what makes the touchless biometric recognition system different than other systems out there.

References

1. Nandhini R, Duraimurugan N, Chokkalingam SP (2019) Face recognition based attendance system. Int J Eng Adv Technol (IJEAT) 8(3S). ISSN: 2249-8958
2. Saha A, Kar N, Karma M (2012) Study of implementing automated attendance system using face recognition technique. Int J Comput Commun Eng 1(2)
3. Ronanki S, Gundu S, Baratam R, Mounika P, Kumar JR (2017) Face detection and identification using svm
4. Anggo M, Arapu L (2018) Face recognition using fisher face method. J Phys Conf Ser 1028(1):012119. https://doi.org/10.1088/1742-6596/1028/1/012119
5. Zhang X, Gonnot T, Saniie J (2017) Real-time face detection and recognition in complex background. https://doi.org/10.4236/jsip.2017.8200
6. Satyawan W (2019) Citizen ID card detection using image processing and optical character recognition. In: The 3rd international conference on computing and applied informatics 2018 IOP Conf Ser: J Phys Conf Ser 1235:012049. IOP Publishing https://doi.org/10.1088/1742-6596/1235/1/012049
7. Manwatkar PM, Singh KR (2015) Technical review on text recognition from images. In: IEEE sponsored 9th international conference on intelligent systems and control (ISCO), Coimbutur, https://doi.org/10.1109/ISCO.2015.7282362

Seeker Optimization for Linear Phase Fir Filter Design

Harmandeep Kaur, Satish Saini, and Amit Sehgal

Abstract In this paper, filter design procedure is analyzed using seeker optimization technique to design an optimal finite impulse response (FIR) filter with desired parameter specifications. Using these algorithms, the optimum impulse response coefficients of the different finite impulse response filter are determined to fulfill their exact output. A fitness function based on the error is generated in which we can specify the desired levels of δ_p and δ_s individually. Simulative results of these techniques are presented and compared in conformity with their select band and reject band ripples. The techniques developed are mainly useful for digital signal processing applications as clearly indicated by the results that it forms a good platform for newer applications.

Keywords Seeker · Bat · FIR filter

1 Introduction

With the various advancements in the transformation of digital signal, filters based on the digital techniques are gaining much more importance instead of analog filters. Digital filtrate is a system that uses digital signal as input and provides digital signal as output after filtering. These are categorized into two: FIR and IIR filters reliant to size of its impulse. Each filter possesses its own advantages and disadvantages. Designing parameters for FIR filter, which are to be considered during the designing of filters, are mentioned as: select band frequency, reject band frequency, ripples, fading, filter coefficients, filter order, etc. [1–4]. FIR filter has numerous advantages over IIR filter which leads to its popularity among the researchers. The implementation of FIR filter is much simpler as it requires no feedback. By making the coefficients of the filter to be

H. Kaur (✉) · S. Saini
School of Engineering (Electronics and Communication Engineering), RIMT University, Mandi Gobindgarh, Punjab, India

A. Sehgal
School of Engineering and Technology, Sharda University, Greater Noida, UP, India

P. K. Singh et al. (eds.), *Emerging Technologies for Computing, Communication and Smart Cities*, Lecture Notes in Electrical Engineering 875,
https://doi.org/10.1007/978-981-19-0284-0_20

symmetrical, linear phase response can be achieved; thus, it allows consistent group delay [5]. Filter designing involves determining the filter coefficients and filter order which satisfy the required filter stipulations, for example, the frequencies of select band and reject band, select band and reject band width, and select band and reject band ripples as well as attenuation. The FIR filters feasibly be projected using various techniques as an illustration window design technique [6–9], frequency sampling [6, 10] weighted least squares design [6], and Parks–McClellan method (equiripple or optimal method) [2, 11]. Traditionally, windows method is the most common of all these. But, in case of windows method, a fabricator will have to be negotiable between one or two parameters.

So, optimization techniques gained a lot of interest of the candidate researchers to design the filter with appropriate design conditions such as evolutionary optimization techniques. Previously, the swarm-based optimization techniques and genetic algorithms (GA) were used to implement FIR filter design such as particle-based optimization (PSO). Various techniques in particular Parks–McClellan method, genetic algorithm (GA) [12, 13], particle swarm optimization (PSO) [12, 14, 15], differential evolution (DE) optimization [16–18], teaching–learning-based optimization (TLBO) [19–21] seemed to be exerted for the projection procedure of FIR filters. Yang proposed an innovative evolutionary optimization technique identified to be Bat algorithm [22]. Severino et al. designed a FIR filter by the usage of PSO and Bat techniques [23]. It is being used in almost every field such as optimization [22, 24], filter designing [23], image processing [25, 26], data mining, feature selection, fuzzy logic [27, 28], artificial neural networks [29, 30], and many more. SOA [31] mimes the doings of individual hunt populace to resolve actual optimization issues. It works on the basis of human searching to reach at the desired optimum solution. A group of humans known as population are taken. Each individual in this are known as seekers. Each seeker has its own center position, search radius, trust degree, and search direction according to which it updates its position. Seeker optimization finds its use in almost every field of study especially in engineering and technology and are successfully implemented in various problems such as in [32] SOA method has also been used for optimal reactive power dispatch problem. In [33], it has been used for designing IIR filter, and as per the literature review, it is not been used much for optimizing FIR filter parameters.

In the current paper, comparison is done between the seeker, Bat and PSO optimization algorithms to design the FIR filter. The simulation results based on this design are presented here. The article is organized as per the aforementioned manner. Module 1 presents the introduction. Module 2 describes the FIR filter model. In module 3, filter design using optimization technique is discussed. In module 4, outcome of the simulative environment is presented. Finally, module 5 concludes the paper.

2 Optimization Algorithms for FIR Filter

The major aim in the present study is to project an optimal digital FIR filter with better design considerations with the use of seeker optimization technique. The symmetrical nature of the coefficients requires only half of the coefficients to be optimized which are then combined to form $N + 1$ count of coefficients. Various parameters that are considered during the optimum filter designing are pass band and stop band frequencies (w_p and w_s), flat pass band, highest reject band attenuation, pass band and stop band ripples (δ_p and δ_s), and short transition width.

Error function/fitness function/objective function is the major parameter to be considered in all optimization algorithms. While designing the filter, coefficients are changing the values in a manner with the purpose of the minimization of error function. The Parks–McClellan algorithm uses the weighted approximate error for FIR filter design as presented in (1).

$$E(w) = M(w)[K(w_k) - K_i(w_k)] \tag{1}$$

$K(w_k)$ illustrates the frequency output of desired filter, and $(K_i(w_k))$ symbolizes the frequency output of actual filter. M (w) represents the weighting function. Fixed proportion of δ_p/δ_s is the key disadvantage of PM technique. An umpteen number of error equations have been and are being used in variant researches as given in [34–36]. The Lp standard approximation error is indicated as:

$$\text{Error} = \left\{ \sum_{i=0}^{k} [||K_d(w_k)| - |K_i(w_k)||]^p \right\}^{\frac{1}{p}} \tag{2}$$

Minimum value of the error equation will be interpreted as the optimum one. A new set of coefficients will be generated using the optimal value of fitness function. By individually specifying the required values for δ_p and δ_s, flexibility in the error equation may further be improved by specifying the error equation as presented in [3]

$$J_1 = \max_{w \le w_p} \left(|E(w)| - \delta_p\right) + \max_{w \ge w_s} (|E(w)| - \delta_s) \tag{3}$$

2.1 PSO Algorithm

PSO algorithm is a swarm-dependent technique based upon the swarm of birds or school of fish that follow the path by learning from their fellow mates. Kennedy and Eberhart [14] established it in 1995. It has the advantage of not getting trapped in the local optimal solution but solves the problem at a global approach [15]. Every individual varies their parameters in accordance with the group of individuals (known as swarm), especially position which has two values—personal or local best (pbest) and group (global) best (gbest). Each particle modifies the location and velocity of its own in consonance with the following equations.

$$\mathrm{Ve}_j^{(g+1)} = w * \mathrm{Ve}_j^g + C_1 * r_1 * \left(\mathrm{pbest}_j^g - S_j^g\right) + C_2 * r_2 * \left(\mathrm{gbest}^g - S_j^g\right) \quad (4)$$

Ve_j^g symbolizes the *j*th particle vectors velocity at *g*th iteration; w states weighting factor; C_1 and C_2 are the cognitive functions known as positive weighting functions; r_1 and r_2 represent the random values in within 0 and 1; S_j^g is the occurant position of *j*th particle vector $h(n)$ at *g*th iteration; pbest_j^g represents the local best of the *j*th particle at the gth iteration; gbest^g represents the global best of the entire group at the *g*th iteration.

The position is updated in correspondence to the undermentioned equations:

$$X_j^{(g+1)} = X_j^g + V_j^{(g+1)} \quad (5)$$

2.2 *Bat Optimization Algorithm*

Bat optimization is a search algorithm depending upon the behavior of bats and their echolocation capability. It is proposed by Yang [22] and works according to the search behavior of bats for their food. It senses the distance using echolocation property and takes advantage of frequency equation. To utilize this algorithm for any optimization problem, initialized values for velocity, position, minimum frequency, wavelength, and loudness (speech intensity) values are set to find out the target.

Bat Motion and Variations of Loudness and Pulse Rates

Each bat has its initialized velocity v_i, position p_i, and pulse frequency q_i in a solution space.

$$q_i = q_{\min} + (q_{\max} - q_{\min})\beta \quad (6)$$

where $q_{\max}$ and $q_{\min}$ are maximal and minimal emission frequencies which are assigned a value uniformly. Initially, value of a frequency is haphazardly assigned for each bat and is drawn consistently from $[q_{\max}, q_{\min}]$. The values for the vector $\beta \in [0, 1]$ are considered to be the random value taken from a uniform distribution.

The new velocity position v_i^t and position z_i^t of the bats are updated at every time step t in accordance with the velocity and position equations as follows:

$$v_i^t = v_i^{t-1} + \left(z_i^{t-1} -- z*\right) f \quad (7)$$

where z^* states the present best position (result) globally that is taken after the comparison of the entire solutions of n count of bats.

$$z_i^t = z_i^{t-1} + z_i^t \tag{8}$$

A fresh solution is bring forth locally for each bat utilizing arbitrary walk after the selection of a solution among the present best solutions.

$$z_{\text{new}} = z_{\text{old}} + \varepsilon L^t \tag{9}$$

where ε symbolizes random count selected within -1 and 1, i.e., $\varepsilon \in [-1, 1]$

$$L_i^{t+1} = \alpha L_i^t \tag{10}$$

$$p_i^{t+1} = p_i^0\left[1 - \exp(-\gamma t)\right] \tag{11}$$

assuming L_i and p_i as the loudness value and pulse rate and are needs to be updated in accordance with the proximity of the prey with an increase in the number of iterations. On reaching near the prey, the decrease in loudness value is seen, but on the contrary, the speed of emitted pulse rises. The values of loudness are specified as per the convenience of the problem to be solved; usually, $L_o = 1$ and $L_{\text{min}} = 0$ are chosen for any problem where $L_{\text{min}} = 0$ indicates that the search of a bat for its prey is successful and is currently not sending any pulses for further processing. α and γ are constant values, and generally, for almost all simulative analysis, these are taken to be equal, i.e., $\alpha = \gamma$.

2.3 *Seeker Optimization Algorithm*

Seeker optimization (SOA) mimes the doings of individual hunt populace to resolve actual optimization problems. It works on the basis of human searching to reach at the desired optimum solution. A group of humans known as population are taken. Each individual in this are known as seekers. Each seeker has its own center position, search radius, trust degree, and search direction, and in accordance with these parameters, each seeker updates its position [31–33]. The final decision-making is done while considering these four parameters. Each seeker is initialized with a random value, and afterward, their positions are changed as per the following equation:

$$Z_{\text{id}}(t+1) = Z_{\text{id}}(t) + \alpha_{\text{id}}(t)\beta_{\text{id}}(t) \tag{12}$$

where $\alpha_{\text{id}}(t)$ and $\beta_{\text{id}}(t)$ are the step length and search direction of the ith seeker and dth dimension or variable. $\beta_{\text{id}} = 1$ indicates that the ith seeker moves in the positive direction on the dimension, whereas $\beta_{\text{id}} = -1$ indicates its movement in negative direction and $\beta_{\text{id}} = 0$ shows that no movement of the ith seeker. Step length and search direction are updated at each iteration depending upon the following factors.

2.3.1 Filter Design Using Seeker Algorithm

For designing FIR filters, undermentioned steps are implemented:

Step 1: Determine the controlling parameters for FIR filter such as select band and reject band frequencies, order of filter, and select band and reject band ripples.
Step 2: Specify the parameters for seeker optimization such as maximum population number (n_p) of seeker strings, where each string has $(N/2 + 1)$ count of filter coefficients, maximal count of loops, and initialization for the values for s number of seekers. Also, initialize the values of highest and lowest frequency and minimum and maximum value of membership degree value (μ_{min} and μ_{max}).
Step 3: Initialize the arrays for the position values.
Step 4: Calculate the fittingness value for any and all individuals.
Step 5: Upgrade the best suited location of all individuals and select adjacent individual of all seekers.
Step 6: Compute the hunt trajectory as well as step size for all seekers, after that amend the location of each seekers.
Step 7: Upgrade the location for all seekers and measure the fitness from these updated values. Update the new solutions otherwise repeat the process from step 4.
Step 8: Generate the filter coefficients using these updated solutions by selecting the best seeker among the entire group.

3 Results

In the current module, the outcome of the codes simulated on MATLAB environment for designing an optimal FIR filter of the all four kinds of filters is presented. Filter order is kept to be 40 after comparing the different orders for the same problem. The sampling frequency is equal to $f_s = 1$ Hz, and count of frequency samples is 512. The required guidelines of the filter to be projected using optimization techniques are taken as: $\delta p = 0.1$, $\delta s = 0.01$. The codes of all the optimization techniques are run for 100 iterations to procure the optimal output, i.e., number of iterations considered for the optimization problem = 100. For low pass and high pass filter, cut off frequency (normalized) is taken as 0.7 and 0.3, respectively. For band pass and band stop filters, lowest and highest edge frequencies (normalized) are 0.3 and 0.7. Population size for this design procedure is taken as 20. For PSO optimization, personal learning and cognitive acceleration coefficient is considered to be equal to 2. Bat parameters are taken as: Loudness is 0.5; pulse rate is 0.5; minimum and maximum frequency is 0 and 2, respectively. In case of seeker, optimization parameters are as follows: minimum and maximum membership degree is given as 0.1 and 0.4; minimum and maximum frequency is 0.2 and 0.6, respectively.

Figures 1, 2, 3, and 4 compare the magnitude responses of actual filter and the filters projected with PSO, Bat, and seeker optimization algorithms for FIR—low pass, high pass, band pass and band select filters successively.

Figures 5, 6, 7, and 8 depict a comparison of convergence profile, i.e., the error function for FIR-LP, FIR-HP, FIR-BP, FIR-BS filters consecutively.

Tables 1, 2, 3, and 4 successively present the various other comparative outcomes of performance parameters of all techniques for 40th order FIR—low pass, high pass, band select, and band reject filters. Almost lowest reject band obstacle and select band ripple is achieved for seeker optimization algorithm while designing all the kinds of filters. For low pass, high pass, select band, reject band filters seeker

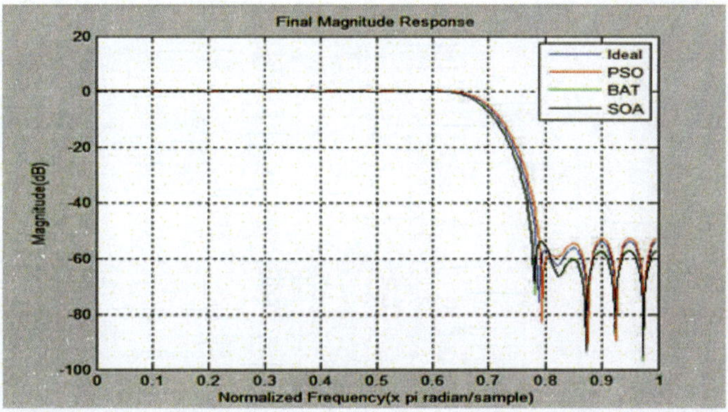

Fig. 1 Comparison of actual filter and filter projected with PSO, Bat, and seeker algorithm for low pass filter

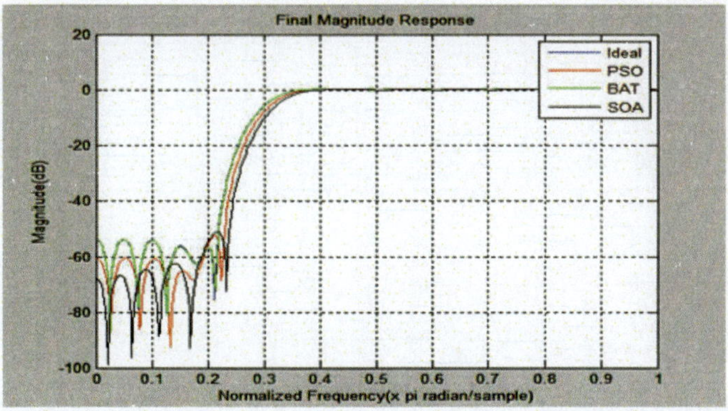

Fig. 2 Comparison of actual filter and filter projected with PSO, Bat, and seeker algorithm for high pass filter

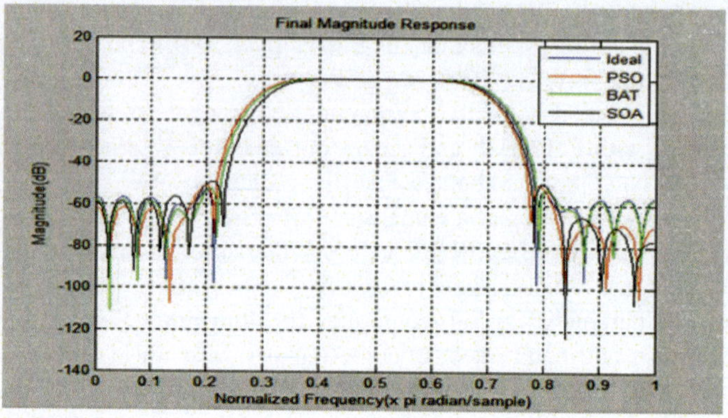

Fig. 3 Comparison of actual filter and filter projected with PSO, Bat, and seeker algorithm for band pass filter

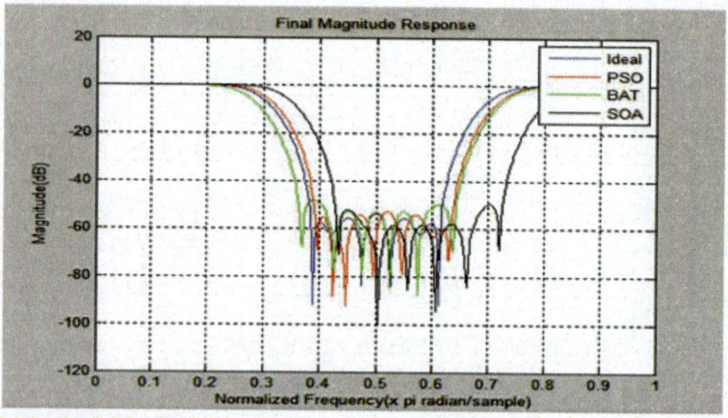

Fig. 4 Comparison of actual filter and filter projected with PSO, Bat, and seeker algorithm for band stop filter

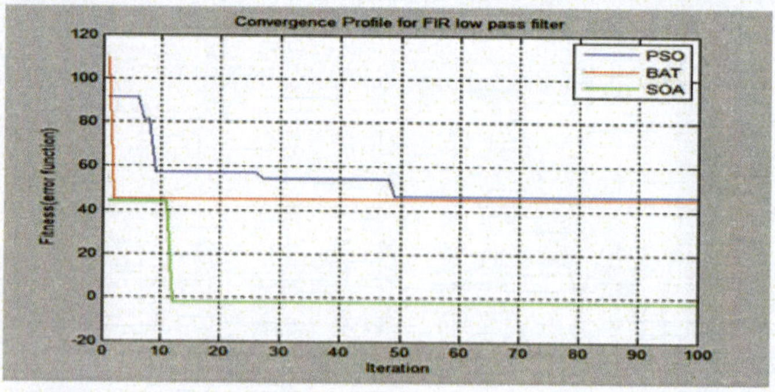

Fig. 5 Evaluation of convergence profile for low pass filter using PSO, Bat, and seeker

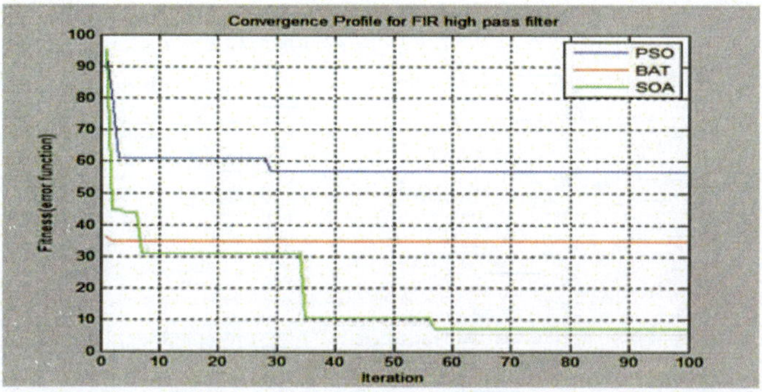

Fig. 6 Evaluation of convergence profile for high pass filter with PSO, Bat, and seeker

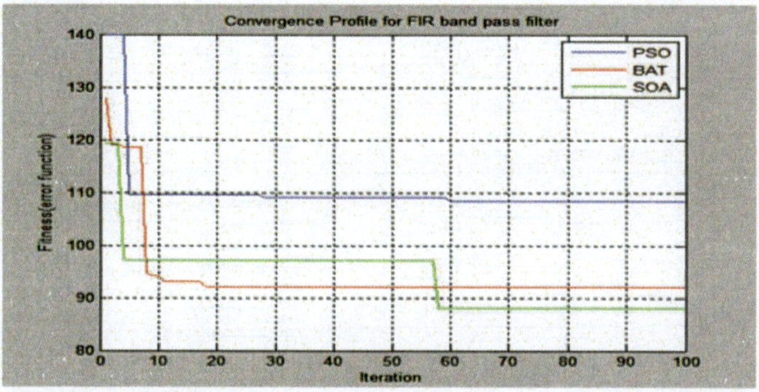

Fig. 7 Evaluation of convergence profile for band select filter using PSO, Bat, and seeker

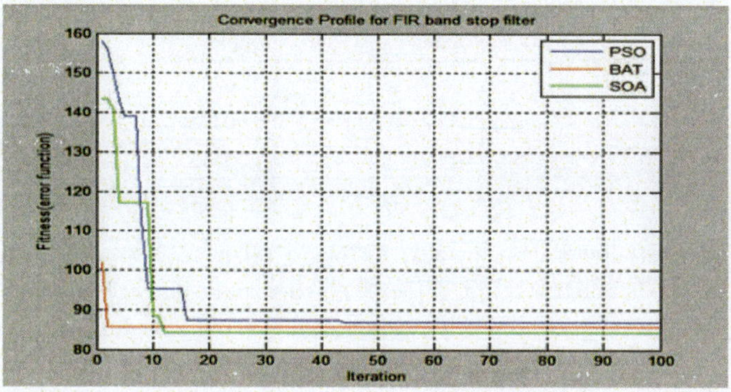

Fig. 8 Evaluation of convergence profile for band reject filter using PSO, Bat, and seeker

Table 1 Analysis of comparative variables for FIR-LP filter projected with various techniques

Technique	Max reject band attenuation(dB)	Max select band ripple	Max reject band ripple	Error fitness
PSO	−52.5	0.0027	0.002322	46.26
BAT	−53.76	0.0029	0.002053	45.17
SEEKER	−53.76	0.0029	0.002051	−2.006

Table 2 Analysis of comparative variables for FIR-HP filter projected with various techniques

Technique	Max reject band attenuation(dB)	Max select band ripple	Max reject band ripple	Error fitness
PSO	−53.12	0.0028	0.002208	56.96
BAT	−53.73	0.0029	0.002053	34.96
SEEKER	−53.73	0.0029	0.00201	7.18

Table 3 Analysis of comparative variables for FIR-BP filter projected with various techniques

Technique	Max reject band attenuation(dB)	Max select band ripple	Max reject band ripple	Error fitness
PSO	−49.71	0.0056	0.003268	108.4
BAT	−9.87	0.0044	0.002461	92.02
SEEKER	−52.18	0.0061	0.003426	87.9

Table 4 Analysis of comparative variables for FIR-BS filter projected with various techniques

Technique	Max reject band attenuation (dB)	Max select band ripple	Max reject band ripple	Error fitness
PSO	−52.97	0.0019	0.002241	86.96
BAT	−53.73	0.005	0.002059	85.71
SEEKER	−53.73	0.9997	0.003353	84.4

technique presents lowermost, or near to lowest reject band ripples of 0.002051, 0.00201, 0.003426, and 0.003353, respectively. Seeker design converges to the least value of error fitness value among all techniques. For low pass, high pass, band pass and band stop filter, it converges to min error count of −2.006 in 11, 7.18 in 57, 87.9 in 58 and 84.4 in 12 iterations. Thus, it performs best among all the design approaches discussed in this works

Statistical parameters of FIR filters obtained by the projection procedure for all the variant optimization algorithms are given in Tables 5, 6, 7, and 8 successively.

Table 5 Statistical data for FIR low pass filter with variant algorithms

Algorithm	Select band ripple				Reject band attenuation (decibels)			
	Max	Average	Variance	S D	Max	Average	Variance	SD
PSO	0.0027	0.001	0.000000776	0.000881	−52.5	−54.086	6.40958	2.531715
BAT	0.0029	0.001614	0.000003481	0.00059	− 53.76	−54.954	2.41288	1.553345
SOA	0.0029	0.002414	0.0000000881	0.000297	− 53.76	−57.2325	7.588291667	2.754685

Table 6 Statistical data for FIR high pass filter with variant algorithms

Algorithm	Select band ripple				Reject band attenuation (decibels)			
	Max	Average	Variance	SD	Max	Average	Variance	SD
PSO	0.0028	0.002486	0.000000048095	0.000219	−53.12	−58.176	10.48643	3.238276
BAT	0.0029	0.001867	0.0000002667	0.000516	−53.73	−54.938	2.46017	1.568493
SOA	0.0029	0.001643	0.0000003329	0.000577	−53.73	−54.938	2.46017	1.568493

Table 7 Statistical data for FIR band pass filter with variant algorithms

Algorithm	Select band ripple				Reject band attenuation (decibels)			
	Max	Average	Variance	SD	Max	Average	Variance	SD
PSO	0.0056	0.005	0.00000063	0.000794	−49.71	−58.1544	26.0031278	5.099326
BAT	0.0044	0.0039	0.0000003333	0.000577	−9.87	−53.1988	332.4233268	18.23248
SOA	0.0061	0.005433	0.0000009733	0.000987	−52.18	−59.4488	25.94118393	5.093249

Table 8 Statistical data for FIR band reject filter with variant algorithms

Algorithm	Select band ripple				Reject band attenuation (decibels)			
	Max	Average	Variance	SD	Max	Average	Variance	SD
PSO	0.0019	0.0014	0.0000001533	0.000392	−52.97	−55.7717	11.43669	3.381819
BAT	0.005	0.003422	0.000001701920	0.001305	−53.73	−57.6767	35.29782667	5.941197
SOA	0.9997	0.749125	0.248754	0.498753	−53.73	−57.6767	35.29783	5.941197

4 Conclusion

In the present article, the projection procedure of FIR digital filters is presented with the aim to diminish the error function by optimizing the variant filter parameters with the use of PSO, Bat, and seeker optimization techniques. The procedure runs for 100 iterations to minimize the error function by optimizing the filter coefficients. Different types of analysis are performed on the outcomes of the codes run on MATLAB environment such as statistical analysis, analysis of convergence profile and magnitude response analysis, and comparison of the variant parameters of filters, namely ripples, attenuation, and as forth. Comparison is performed between the PSO, Bat, and seeker optimization, and seeker optimization presents the best outcomes in terms of all the criterions.

References

1. Ababneh JI, Bataineh MH (2008) Linear phase FIR filter design using particle swarm optimization and genetic algorithms. Dig Sig Proc 18(4):657–668
2. Abdel-Rahman EM, Ahmad AR, Akhtar S (2012) A metaheurisic bat- inspired algorithm for full body human pose estimation. In: Ninth Conference on Computer and Robot Vision, pp 369–375
3. Aggarwal A, Rawat TK, Upadhyay DK (2015) Design of optimal digital FIR filters using evolutionary and swarm optimization techniques. Int J Electron Commun 70(4):373–385
4. Alam SM, Hasan MT (2010) Performance analysis of FIR filter design by using optimal, blackman window and frequency sampling methods. Int J Electr Comput Sci 10(1):13–18
5. Antoniou A (n.d.) Digital filters: analysis and design. McGraw Hill, New York
6. Shenoi BA (2006) Introduction to digital signal processing and filter design. Wiley, Canada
7. Coulombe S, Dubois E (1996) Multidimensional windows over arbitrary lattices and their application to FIR filter design. In: Proceeding IEEE international conference on acoustics, speech and signal processing. Atlanta, GA, pp 2383–2386
8. Dai C, Chen W, Zhu Y (2006) Seeker optimization algorithm. In: International conference on computational intelligence and security. Guangzhou, pp. 225–229
9. Dai C, Chen W, Zhu Y (2010) Seeker Optimization algorithm for digital IIR filter design. IEEE Trans Industr Electron 57(5):1710–1718
10. Dai C, Chen W, Zhu Y, Zhang X (2009) Seeker optimization algorithm for optimal reactive power dispatch. IEEE Trans Power Syst 24(3):1218–1231
11. Du ZY, Liu B (2012) Image matching using a bat algorithm with mutation. Appl Mech Mater 203(1):88–93
12. Gholami A, Ansari HR (2017) Estimation of porosity from seismic attributes using a committee model with bat-inspired optimization algorithm. J Petrol Sci Eng 152(2):238–249
13. Harris FJ (1978) On the use of windows for harmonic analysis with the discrete fourier transform. Proc IEEE 66(1)
14. Ifeachor E, Jervis B (2002) Digitalm signal processing: a practical approach. Pearson, UK
15. Jaddi NS, Abdullah S, Hamdan AR (2015) Optimization of neural network model using modified bat-inspired algorithm. Appl Soft Comput 37(C):71–86
16. Karaboga N (2005) Digital IIR filter design using differential evolution algorithm. J Adv Sig Proc 8(4):1269–1276
17. Karaboga N, Cetinkaya B (2006) Design of digital FIR filters using differential evolution algorithm. Circ Syst Sig Proc 25(5):649–660

18. Kennedy J, Eberhart R (1995) A new optimizer using particle swarm theory. In: Proceeding IEEE sixth international symposium on micro machine and human science. Nagoa, Japan, pp 39–43
19. Kennedy J, Eberhart R (1995) Particle swarm optimization. Proceeding IEEE international conference on neural networks. Perth, pp 1942–1948
20. Lemma TA, Bin Mohd Hashim F (2011) Use of fuzzy systems and bat algorithm for energy modelling in a gas turbine generator. In: IEEE colloquium on humanities, science and engineering, pp 305–310
21. Litwin L (2000) FIR and IIR digital filters. IEEE Potentials, pp 28–31.
22. Liu G, Li YX, He G (2010) Design of digital FIR filters using differential evolution algorithm based on reserved gene. In: Proceedings of the IEEE conference on evolutionary computation. Barcelona, Spain, pp 1–7
23. Liu G, Li YX, He G (2010) Design of digital FIR filters using differential evolution algorithm based on reserved gene. In: IEEE Conference on Evolutionary Computation. Barcelona, Spain, pp 1–7
24. Najjarzadeh M, Ayatollahi A (2008) FIR digital filters design: particle swarm optimization utilizing LMS and minimax strategies. In: IEEE international symposium on signal processing and information technology. Sarajevo, Bosnia and Herzegovina, pp 129–132
25. Nayak MR, Nayak CK, Rout PK (2012) Application of multi-objective teaching learning based optimization algorithm to optimal power flow problem. In: 2nd International conference on communication, computing and security, pp 255–264
26. Parks TW, Burrus CS (1987) Digital filter design. Wiley, New York
27. Parks TW, McClellan JH (1972) Chebyshev approximation for non-recursive digital filters with linear phase. IEEE Trans Circ The 19(2):189–194
28. Proakis JG, Manolakis DG (2000) Digital signal processing-principles, algorithms and applications. Prentice-Hall, New Delhi, India
29. Proakis J, Ingle V (2011) Digital signal processing using MATLAB, 3rd edn. Cengage, USA
30. Rao RV, Savsani VJ, Vakharia DP (2011) Teaching–learning-based optimization: a novel method for constrained mechanical design optimization problems. Comput Aided Des 43(3):303–315
31. Severino AG, Linhares LL, DeAraujo FM (2015) Optimal design of digital low pass finite impulse response filter using particle swarm optimization and bat algorithm. In: 12th international conference on informatics in control, automation and robotics, pp 207–214
32. Singh R, Verma HK (2014) Teaching-learning-based optimization algorithm for parameter identification in the design of IIR filters. J Inst Eng 94(4):285–294
33. Slowik A, Bialko M (2007) Design of IIR digital filters with non-standard characteristics using differential evolution algorithm. Bullet Pol Acad Sci Tech Sci 55(4):359–363
34. Tamiru AL, Hashim FM (2013) Application of bat algorithm and fuzzy systems to model energy changes in a gas turbine. Stud Comput Intell 427:685–719
35. Yang XS (2010) A new metaheuristic bat-inspired algorithm. Nat Insp Cooperat Strateg Optim 284(2):65–74
36. Yang XS, Deb S (2010) Eagle strategy using L´evy walk and firefly algorithms for stochastic optimization. Stud Comput Intell 284:101–111

Cloud Supported Secure e-Card College Management System

Garima Jain and Rashmi Mishra

Abstract As the term, smart campus attracts professionals and academics from multiple disciplines, and the technology keeps intervening in every aspect of life; it becomes inevitable for the smart campus to take place and deploy the future vision of smart cities. As a first step to achieving this vision, it is very important to develop a clear understanding of a smart campus. The student e-card management system (SEMS) is a barcode-based cutting-edge technology and innovative software deployed in the AWS cloud environment. It uses the core infrastructure of the AWS cloud, which includes AWS DynamoDB, AWS EC2 cluster, AWS S3, and the REST API. There are many manual activities in the existing college management system and lead to problems as they are time-consuming and expensive. The proposed idea aims to design a student smart card that configures a bar code with each user's ID card, and that bar code can be read through the web, mobile app, and card reader. With that said, the user's ID card will be multi-functioned now and will serve many purposes. The users of this application can be anyone who is associated with the college. The ID card is used for financial transactions within the college's boundary, which ultimately serves the biggest commitment of having cashless transactions across the country. The same ID card can be used to have books from the library and many more activities. As this paper is a full-fledged cloud computing-based system, we are also proposing an OTP-based transaction to provide the end-to-end reliability of the financial transaction. This feature shows the perspective and influence of SEMS and its versatility, practicality, and usability. This research opens the doors for future studies to gain a deeper insight into the type of decisions needed to transform a traditional campus into a smart campus.

Keywords Bar code · REST API · Cloud computing · DynamoDB · Payments · SEMS

G. Jain (✉)
Noida Institute of Technology and Engineering, Gr. Noida, UP 201306, India
e-mail: garimajain@niet.co.in

R. Mishra
GL Bajaj Institute of Technology and Management, Gr. Noida, UP 201306, India

P. K. Singh et al. (eds.), *Emerging Technologies for Computing, Communication and Smart Cities*, Lecture Notes in Electrical Engineering 875,
https://doi.org/10.1007/978-981-19-0284-0_21

1 Introduction

In the current scenario, the smart card can play the role of a minicomputer that can process the data and store it efficiently with the help of a chip called a microprocessor. Student Management System is a framework for tracking all sorts of tasks done by students on the premise of presence in class, bus tracking, hostellers can use for food choice, accessibility of books online in library, college notifications, and events. As the university has a clear mission of education, it is certainly a huge business for enterprise, involving immense management activity. Technology changes the utilization of information and the education system of fellows, which assists with making their data and reports centralized.

It has to provide many different jobs for many people who are engaged in an enormous variety of activities. The administrations need to work smoothly as per the general learning of their clients; they should help, not disturb, the legal client exercises, and they ought to be cost-effective.

Different university administrations incorporate payments, for example, for parking, printing, and canteens. Cashless buy is definitely liked for money installment, as it reduces the expenses of dealing with and moving money. Shrewd card utilized as an electronic satchel gives a method for eliminating out money flow from university exchanges. The possibility of scholars and staff flexibility between colleges has increased a wide acknowledgment as communicated by the Bologna Declaration.

SEMS is a barcode-based multi-platform application software fully deployed on cloud clusters and supporting all the available user interfaces like android, IOS, web, etc. A bar code will be configured with each user's ID card. After the successful read of the bar code, users are provided endpoint URLs performing specific tasks through REST. In the era of cutting-edge technologies and increasing demand for saving physical storage and maintaining the application's complexity, SEMS uses full-fledge clusters of AWS cloud to deploy the application. A light-weighted no SQL database DynamoDB maintains the user's record, AWS S3 cloud storage to keep the documents, classroom notes, placement records, etc. Amazon MQ is operational to send emails to the users if any notification email is needed to trigger users [1–3].

Users: Each entity of the college can be the user of this application. There is a registration page for every new user. On successful completion of registration, the user will be given a user ID which normally is the user's card ID.

Barcode: This is the SEMS application's main key. A bar code will be created after successful registration, and that bar code will be configured on the user's ID card. For all the transactions, mainly financial transaction, this bar code will serve as the backbone and makes this application unique from all the existing software of the college management system.

REST API: It provides endpoint URLs to access the application and its database and features.

- GET: Retrieves record from the table.
- POST: Inserts record into the table.

- PUT: Modifies an existing record into a table.
- DELETE: Deletes a record from the table.

Encryption: To ensure the security of data and documents, the RSA encryption/decryption technique will be applied to document sets before uploading or updating them to cloud storage, ensuring that attackers will not be able to steal the papers.

AWS Cloud: AWS offers reliable, scalable, and inexpensive cloud computing services.

DynamoDB: This application uses the Amazon DynamoDB database to store and process the data. DynamoDB is a no SQL and light-weighted database that provides fast and predictable performance with seamless scalability and saves you from the burdens of operating and scaling a distributed database.

Elastic load balancing: To scale the application and balance the load, elastic load balancing has been used to distribute incoming application traffic across multiple targets automatically. It can handle the varying load of SEMS application traffic in a single availability zone or across multiple availability zones.

AWS Lambda: AWS Lambda is a serverless compute service that runs SEMS code in response to the events and automatically manages the underlying compute resources for you.

Clusters: A cluster is a defined action for the combination of tasks or services or related activities. SEMS uses an EC2 cluster that groups different container instances.

Amazon MQ: To manage the active queue from the incoming messages for one of the modules "chat with experts," this application uses Amazon MQ services.

Amazon s3: SEMS application uses Amazon s3 features to create a bucket for each user to personalize the application based upon their requirements.

The working methodologies of Fig. 1 are described as below:

1. Every user of the college has an ID card associated with them. In this step, we will configure a bar code with that ID card.
2. On the successful read of bar code by the system, the user requests go to the cloud environment through a REST API call. These calls could be of any type depends upon the environment
   (a) If this is a GET call, then the requested data is fetched from the DynamoDB database.
   (b) For the POST rest call, the data sent from the user is first encrypted and stored in the DynamoDB database in the corresponding table.
   (c) PUT call to rest API updates the existing record of the tables in the DynamoDB database.
   (d) DELETE removes the data from a database.
3. The entire SEMS application is deployed in the AWS EC2 cluster, and elastic load balancing is used to distribute the load and traffic across several targets.
4. As this application needs huge resources as the number of students increases rapidly, AWS lambda is used to manage the reserve, which computes the resources at run time.

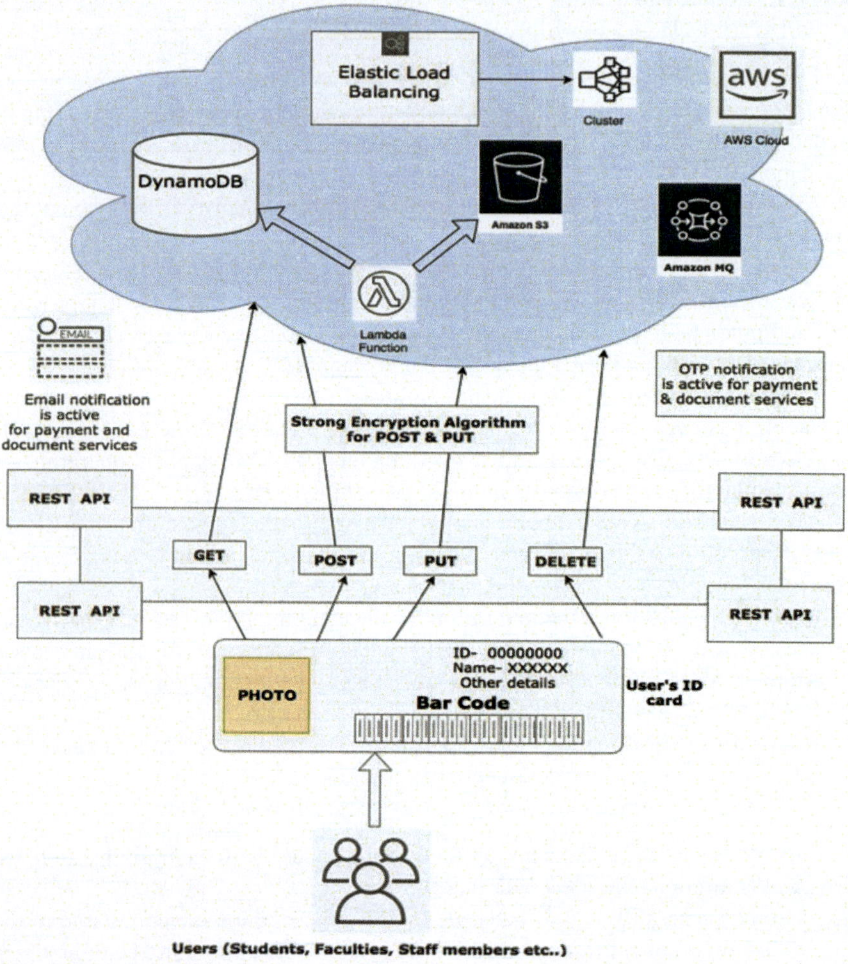

Fig. 1 High-level flow of SEMS

5. SEMS software provides the feature of the automatic triggering of emails and messages with the help of Amazon MQ.
6. To keep the user documents in the cloud environment, each user is provided a bucket titled "Users ID" with the help of Amazon S3. This bucket will be password-protected and will keep sensitive documents in the encrypted format only. The user will have to tell the system which document s sensitive and which is not.
7. The proposed software supports OTP-based communication for all the financial transactions and access to all the critical information.

2 Literature Survey

The current 'university management system' [4] is either not automated or bulky. They proposed a system that gives an idea to overcome multiple pitfalls and uses the user's ID as an important entity for financial transactions throughout the college boundary. The problem may be face problems in the crashing system as they work on a single point of contact. In [5] "Multi-Purpose Student Card System Using Smart Card Technology Education," the author shown in the implementation is just one sector where can adopt smart cards. Others can also take on the adoption to improve their functionality and usability. It may increase security issues. In [6], author explained that comparison of planned and implemented schedules, systematic schedule adjustments, and the survival models applied to ridership. In [7], author said that the card is useful for the students in places like the library, canteen, stationery shops, and online storage of important documents. From there, the potential and power of smart cards their versatility and usability. In [8], the author works on the design concept, architecture, and functionality of the "Students' Electronic Card" computer system. In [9], the author develops a linear Internet addiction (LIA) model and a neural network Internet addiction (NIA) model to calculate students' Internet addiction levels, respectively. In paper [10], Zigbee technology is made use of, which requires one of the processors to get connected to the Zigbee and therefore connect to the server for marking the attendance. It can increase the hardware cost and maintenance cost. In paper [11], the authors used Wsn's for marking the attendance, which is ideally very costly for colleges to implement in real-time just for attendance. Most colleges do not invest a lot in the attendance marking process as they rely on professors to do the same. In [12], the author performs a survey about the previously proposed systems and thereby analyzes the drawbacks of those systems for proposing the advanced and efficient solution to automation of attendance. In [13], the study is a research in progress using Smart PLS-SEM about the adoption mechanism of SPS utilizing the data mentioned above. None of the research work on data that is accessible from virtually any Internet-connected device also requires higher security. In [14], the author had given a significant study on smart campus initiative; they took data from official university sources as a lack of resources. They give a comparison result on a green campus, smart research, smart learning and smart management in which the highest number is associated with smart management that is 58%. In [15], the author explains smart campus in which he considers University Tun Hussein Onn Malaysia data. According to its conceptual framework, some pillars like the student pillar, research pillar, academic pillar etc., are considered as a parameter. The result from the pilot study shows that Cronbach's Alpha coefficient is 0.993. In [16], author mainly focusing on factors choosing e-learning were: ease of use, cost, flexibility, functionality, and range of features. In [17], the author designs into a student information system that can use to input student scores online at SMPN 1 Curug. Authors mention that the advantage of designing a system is to help teachers to enter grades when system design is created; in [18], author proposed Integrated Cloud Education System through a single-window interface called, namely multi-purpose electronic card which will help the student to avail online education as per their schedule. In [19], author explains that how efficiently IoT and cloud infrastructure restructure the traditional education and learning methods. IoT and cloud computing technologies can provide solutions for a smart and sustainable campus to improve student's learning methods and improve the efficiency of everyday activities in the Institution. In [20], their article focuses on introduces technologies like cloud computing and then analyzes the needs of the smart campus service platform. On this basis, they proposed a smart campus service platform based on cloud computing technology to promote the development of smart campuses in China. But their study fails to work with security issues. In [21], the author studies reviews of research in different smart card authentication methods and proposes an improved authentication method to cover existing security weaknesses, including the security of verification. Finally, they compared and analyzed their related work.

3 Modules

Figure 2 explains the detailed view of the modules. All the modules are using distributed database so that the communication among those modules is not time-consuming and will be decoupled from each other. "Home," "faculty," "student," "administrator," "alumni cell," "library," "payment," "others" are the main modules of the proposed software where can access the first five modules through a web or an app. In contrast, the barcode is made mandatory for the last three modules. We can add more modules in the future on the demand of the organization.

Activities performed by the first five modules are mentioned below: -

- **"Home"**: This module is the home page of the proposed idea.
- "**Faculty**": This module is particularly used by the faculties of the college and facilitates the essential features for the faculty.
- "**Students**": This module is particularly used by the students of the college and facilitates the essential features for them.
- "**Administrator**": All the administrator and management will use these modules to perform there day to day activities.
- "**Alumni Cell**": This feature is self-explanatory and can also be accessed by the college's ex-students and administrator and faculties members. They can help the current students of the college to get placed in the companies. With this module, the features like job referral, events provide communication between the ex-students and the current students of the college. The module "events" is a chart server between the ex and current college students where the current students ask help from their seniors for a job, guide, aptitude, interview questions, etc.

The sub-activities performed by these five modules are described as follows:-

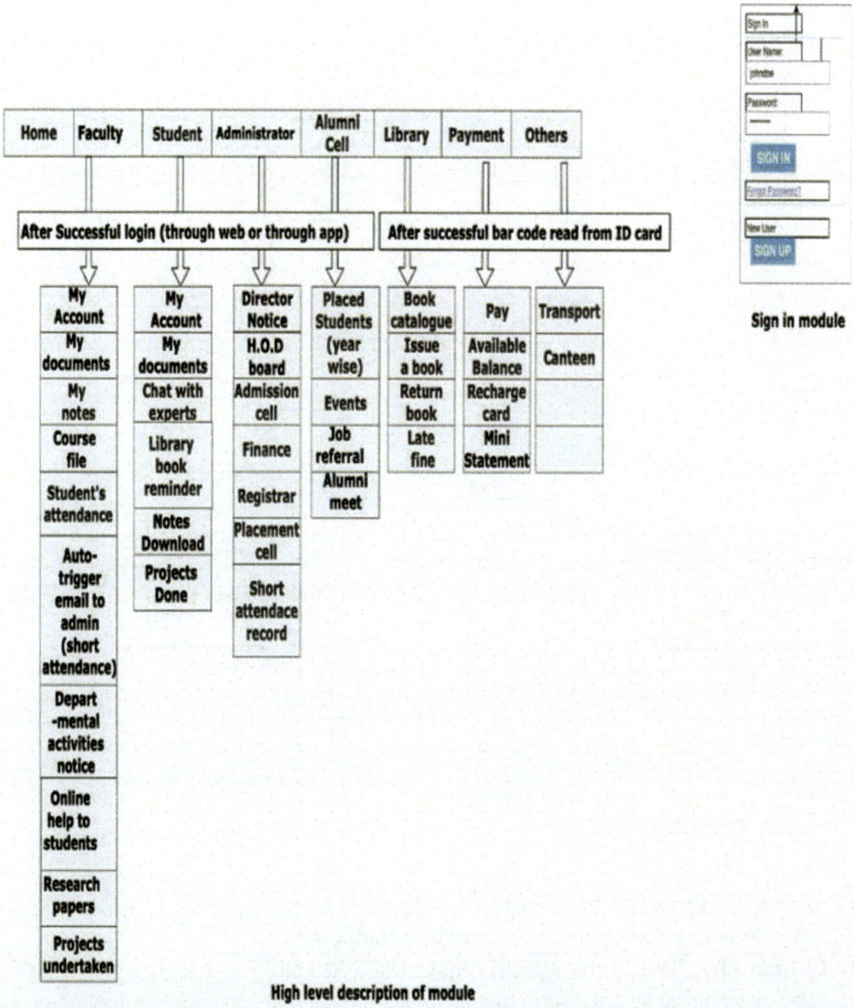

Fig. 2 Modules of SEMS

- **"My Account"**: This tab gives the features like account nickname, Aadhar number, email, mobile number, password, profile photo, university rank, and any other personal details related to students, teachers, and staff members.
- **"My Documents"**: It has two features: Document upload and document download.

  The document upload: Process is shown in Fig. 3.
  Document download: To download any document, through REST API DynamoDB is hit. To download it successfully, the user will have to provide OTP sent on a registered mobile number.

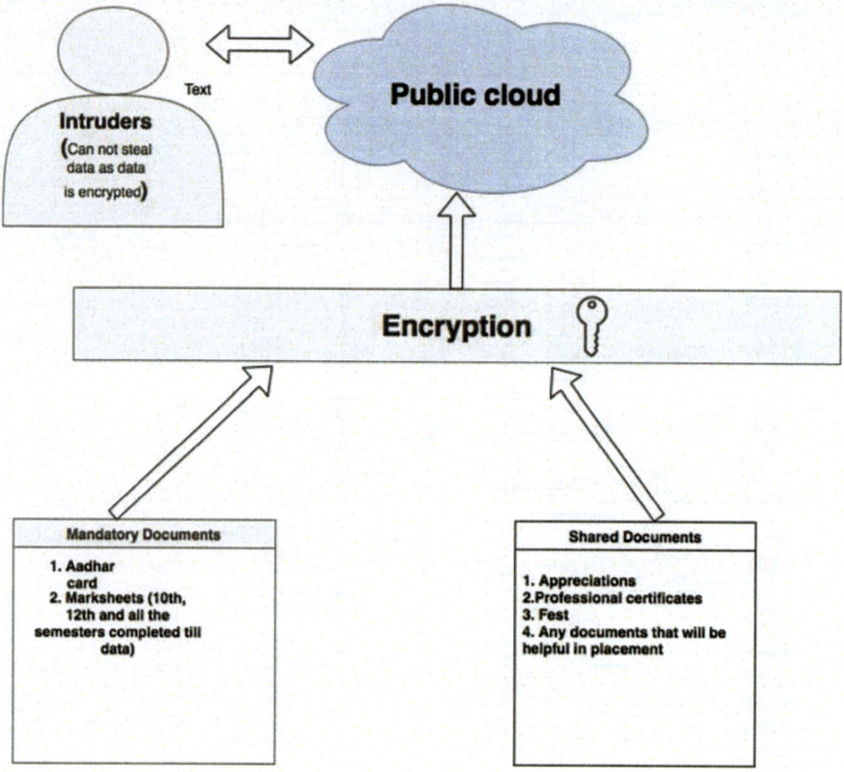

Fig. 3 Document upload of SEMS

- **My notes**: These are the classroom notes uploaded by faculty members which can be downloaded by the corresponding students.
- **Course files**: This is the teacher's day-to-day teaching activity for a particular subject teaching in the particular semester. Once the course is announced as completed, the H.O.D can track the status.
- **Attendance**: The emails, text messages, or WhatsApp messages will be automatically triggered to the student's guardian if their attendance is less than a particular percentage.
- **Departmental activities**: This feature enables a particular teacher to know the departmental activities like FDP, teacher's meeting, etc.
- **Online help to students/chart with experts**: If any student seeks online help from a faculty member or subject matter experts, the student can go for the "chart with experts," and the proficient can help the student through "online help to students." This feature will be very helpful for the students doing a PhD or doing research work in any field. If the students want remote help, they will highly appreciate this feature.

- **Research papers/projects undertaken**: This module is for the research papers/patent work/projects done or projects get completed by the students under that teacher's guidance. All these documents can be kept in the amazon s3 bucket with password protected after proper encryption.
- **Library book reminder**: It reminds the last date of the issued book which must be returned before the last date.
- **Director's notice**: Any notice issued by the director's office will be available for the concerned departments, students, etc.
- **Admission cell/finance/registrar/placement cell**: All these terms are self-explanatory and will be used by the corresponding departments.
- **Short attendance record**: This feature is similar to the "Attendance" module, and they will be sharing the same database so, no manual communication will be done.
- **Alumni cell**: This feature is self-explanatory and can also be accessed by the college's ex-students and administrator and faculties members. They can help the current students of the college to get placed in the companies. With this module, the features like job referral and events provide communication between the ex-students and the current students of the college. The module "events" is a chart server between the ex and current college students where the current students ask help from their seniors for the job, guide, aptitude, interview questions, etc.
- Library, payment, and others will be read through the barcode.

The activities performed by these three modules are defined as below:-

- **Library**: All the library-related activities like issue a book, return a book, late fine, etc., will be handled by this module. Now, there is no need to do the manual entry or no need to read the database tables. Just one bar read to ID card, and the proposed software will show all the history of that card.
- **Payment**: This is the most magical part of the software and the heart of this application.

  Pay: Fig. 4 explains the working of the pay module. There are the following steps to make a payment with the help of SEMS:-

Step 1: The user will go to the cash counter. The cash counter could be anything in college like the canteen, financial department, library, etc.

Step 2: The user's ID card will be scanned on the cash counter. Step 3:- On the successful scan of the user ID card, its details and the available balance on the card will be displayed on the screen. Else "Please add valid ID card" message will be displayed.

Step 3: On the successful scan of the user ID card, its details and the available balance on the card will be displayed on the screen. Else "Please add valid ID card" message will be displayed.

Step 4: If the appropriate balance to make a payment is available on the card, then a successful transaction is done. Else the "Insufficient balance, please recharge your card" message will be displayed on the screen.

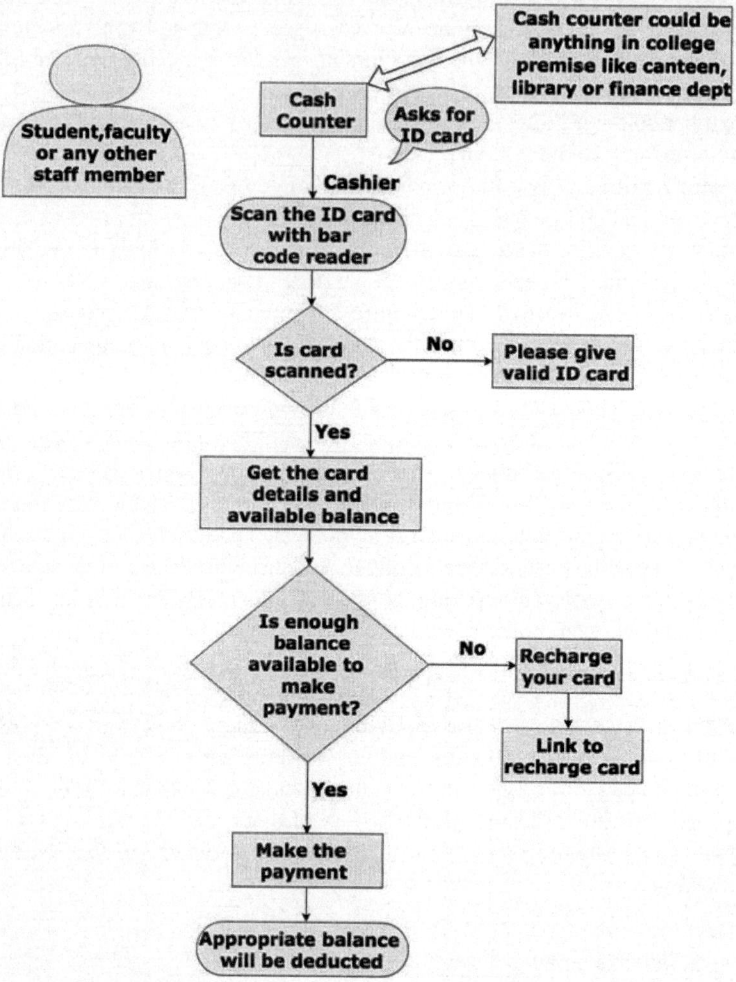

Fig. 4 Pay to a party of SEMS

- **Recharge a card**: Figure 5 explains the working of the recharge a card module. There are the following steps to recharge a card with the help of SEMS:-

  Step 1: SEMS application will read the barcode from the user's ID card.

  Step 2: After the successful verification of ID card, registered email ID and registered mobile are needed to proceed further. For an invalid ID card, a message is displayed "Please first make a Registration."

  Step 3: Valid user will get the OTP on both:—Registered email ID and on a registered phone number.

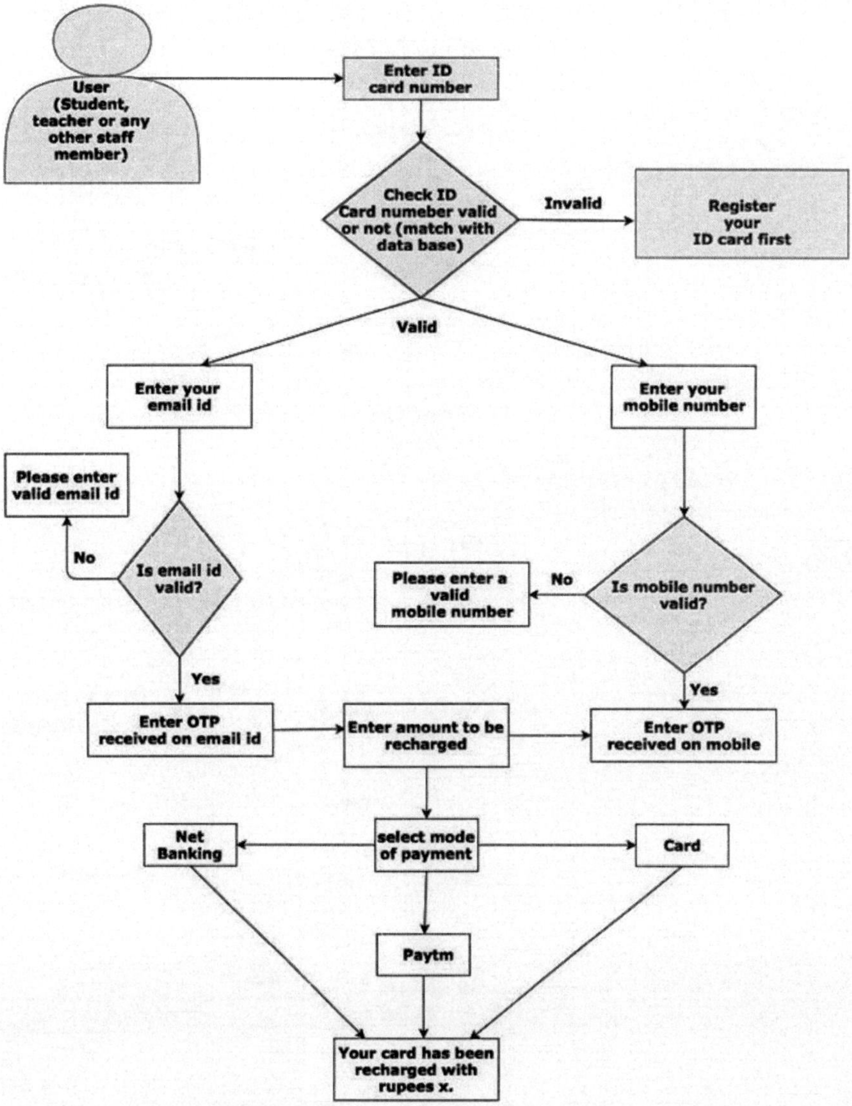

Fig. 5 Recharge a card of SEMS

Step 4: On successful OTP verification, will choose the mode of payment, e.g., net banking, Paytm, debit /credit card, etc., and the message displayed that "Your card has been recharged with rupees X."

- **Available balance**: This module shows the balance available on a card.
- **Mini statement**: Figure 6 explains the working of the mini statement module. There are the following steps to Mini statement with the help of SEMS:-

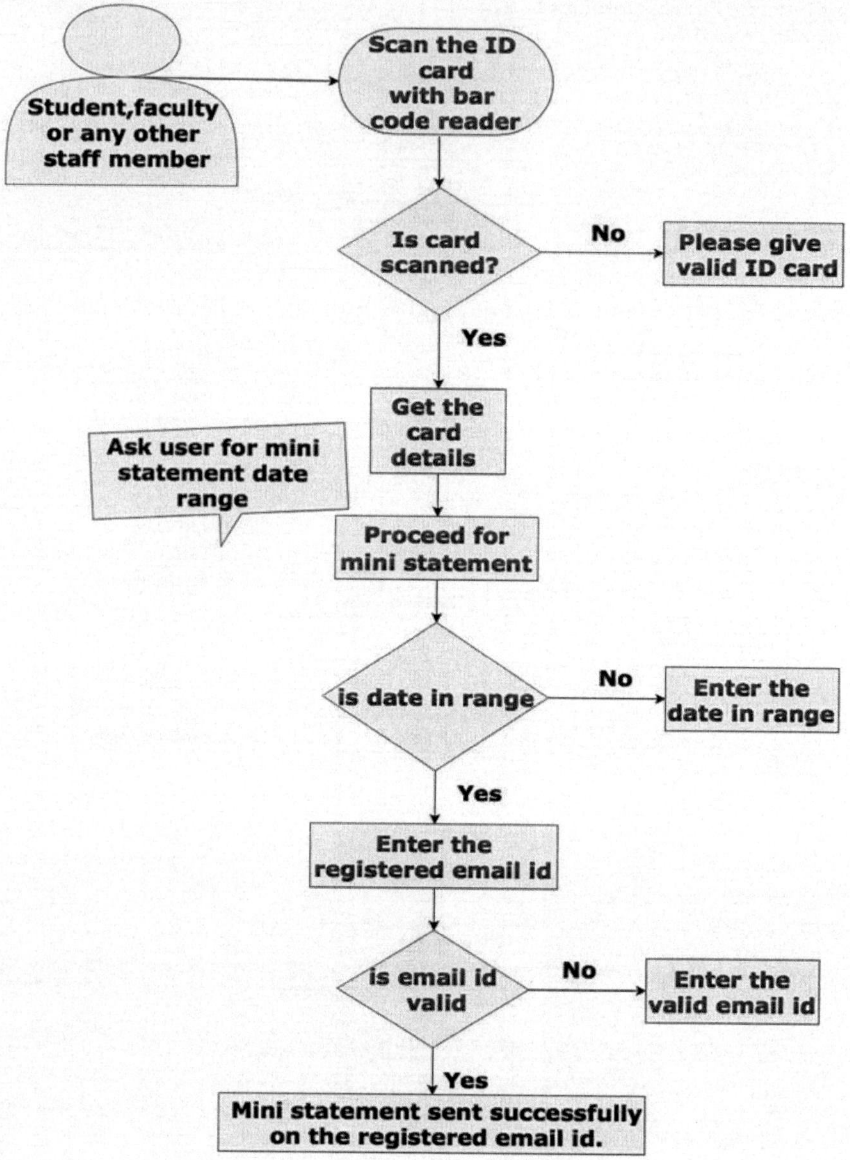

Fig. 6 Flow of mini statement

Step 1: User's ID card is scanned from the barcode reader.

Step 2: On the successful read of the barcode, the date range to get a mini statement of the associated card is entered.

Step 3: For the dates which are in range, the mini statement of the associated card is sent on the registered email ID and a message will be prompted to the user as "mini statement sent successfully on registered email ID."

- **Canteen**: The ID card with bar code will be used to make any payment at the canteen too and have the same flow as the "pay" sub-modules of the payment module.
- **Transport**: The ID card with bar code will be used to make any payment at the transport desk and have the same flow as of "pay" sub-modules of the payment module.

4 Analysis the Gathered Data

Therefore, we had seen that IoT, cloud computing, and the latest technology like big data analytics had proven their efficiency in participating smart applications into a campus design to transform it into a smart campus. Although, despite these technologies' existence, little has been reported in the literature to define the complete set of criteria that underpin the development of a smart campus. The smart card enables secure, fast, flexible, and easy accessibility facilities in a smart campus.

As a case study for presenting the smart card in a campus facility that can store a photo ID and e-wallet and enable library lending privileges, recreation, medical center services, e-learning, alumni data, student data, and controlled access to university facilities. Furthermore, some of the other applications addressed in the literature are summarized in Fig. 7. Therefore, we can conclude that a smart card is an essential

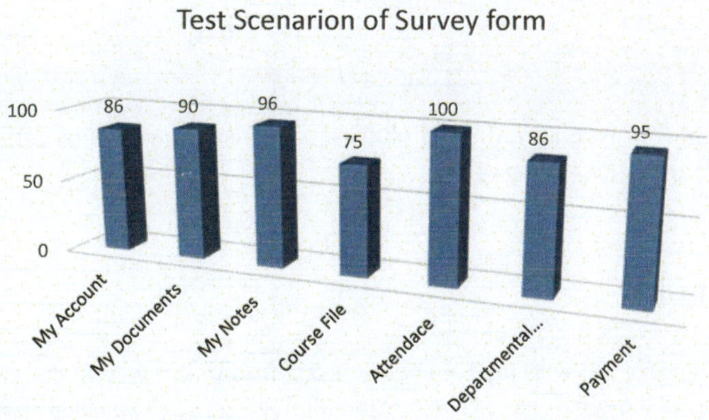

Fig. 7 Test scenarios from survey result using questionnaire

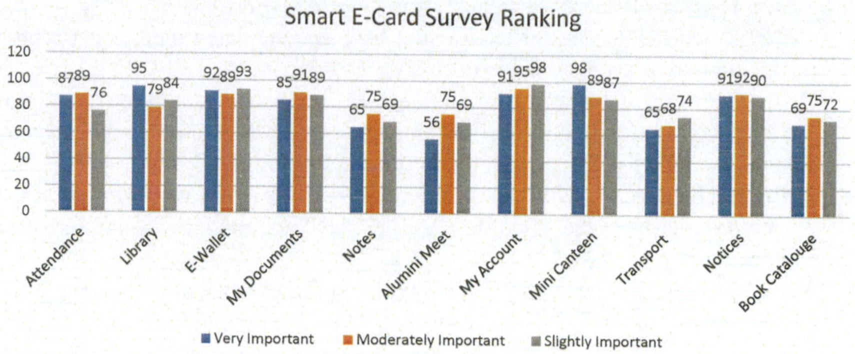

Fig. 8 Smart e-card survey ranking using questionnaire

criterion for the smart campus framework, allowing for quick identification of all transactions and a personal database possible to log in via the cloud.

These results, therefore, indicate that students are more interested in using electronic cards to facilitate and streamline services and processes than in physical mobility and accessibility. So, Fig. 8 shows smart E-carding ranking.

It shows the average score attributed to the e-card criterion and its importance within the different facilities of the university campus, such as its importance for monitoring student attendance in a classroom, student residential activities, library activities, loan, as a wallet to process payments and recode student data (e.g., student information, admission, transcript, degree information, student, and activity records). The results showed that the use of electronic cards to personal record data scored the highest among respondents, followed by their use for library activities and payments. For analyzing the consistency, the concluded result is analyzed by applying Cronbach's alpha (coefficient of reliability) using SPSS by using a formula.

$$\alpha = \frac{M\tilde{x}}{\tilde{y} + (M - 1)\tilde{x}} \tag{1}$$

The coefficient is analyzed to be 0.995, as we know the resulted reliability is efficient when it is more than 0.70.

5 Conclusion

An ID card configured with the bar code-based software deployed in the AWS EC2 cluster using all the core features of cloud computing has been presented. The system offers the advanced version of the recently proposed applications. It introduces the least complex, time, and space-saving software that will be useful for all the users of the college/university. AWS Dynamo DB no SQL database has been used to facilitate incremental scalability and high performance with the ease of cloud administration, reliability, and table data model. REST API has been used to support the client–server framework and facilitate communication between the user and the cloud. From the point of view of management, this innovation will make their day-to-day activity easier. At the same time, the students will enjoy the software's core features to boost up their career, and it will be helpful for students' knowledge enrichment. The user ID card serves many purposes: as can be treated as a firearm for all transactions, including financial transactions inside the college premises. A few modules like chat with experts and payment through ID card have been introduced with any college management software. In the COVID scenario, this system provides. Thus, the proposed education system model provides smart, economic, and environmentally sustainable campus.

6 Future Scope

In the future, we willing to use blockchain-based digital technology and IoT in student smart cards, which is an immense revolution as it helps educational institutions to keep a clear digital transcript, a ledger of records, and payments for each student. Furthermore, the technology RFID, which many researchers use and very costly, can be replaced by Bluetooth with IoT to estimate the overall cost for optimal design. In IoT, we will use sensors and actuators to flow data between sender and receiver.

References

1. Narula S, Jain A, Prachi (2015) Cloud computing security: amazon web service. In: 2015 Fifth international conference on advanced computing and communication technologies, Haryana, pp 501–505. https://doi.org/10.1109/ACCT.2015.20
2. Varia J, Mathew S (2014) Overview of amazon web services. Amazon Web Services, pp 1–22
3. Alqahtani A, Gull H (2018) Cloud computing and security issues—a review of amazon web services. Int J Appl Eng Res 13(22):16077–16084
4. Jain S, Garg R, Bhosle V, Sah L (2017) Smart university-student information management system. In: International conference on smart technologies for smart nation (SmartTechCon), pp 1183–1188. https://doi.org/10.1109/SmartTechCon.2017.8358555
5. Omar S, Djuhari H (2014) Multi-purpose student card system using smart card technology. In: Information technology based proceedings of the fifth international conference on higher education and training, pp 527–532. https://doi.org/10.1109/ITHET.2004.1358229
6. Pelletier M-P, Trépanier M, Morency C (2011) Smart card data use in public transit: a literature review. Transp. Res. Part C: Emerg. Technol. 19(4):557–568. https://doi.org/10.1016/j.trc.2010.12.003
7. Singh A, Chopra A, Imran M, Singh R (2017) Designing and implementation of smart card technology for the students of higher education. Int J Comput Sci Mob Comput 6(1):144–149
8. Materka A, Strzelecki M, Dębiec P (2009) Student's electronic card: a secure Internet database system for university management support. In: On internet–technical development and applications. Springer, Berlin, Heidelberg. pp 59–72. https://doi.org/10.1007/978-3-642-05019-0_8

9. Peng W, Zhang X, Li X Using behavior data to predict the internet addiction of college students. In: International conference on web information systems and applications. Springer, Cham, pp 151–162. https://doi.org/10.1007/978-3-030-30952-7_17
10. Badhe S, Chaudhari K, Kale S, Mane T (2016) Smart attendance management system. Int J Comput Appl (0975–8887) Nat Conf Adv Comput Inf Technol 4(5):10–14
11. Alassery F (2019) A smart classroom of wireless sensor networks for students time attendance system. In: IEEE Integrated STEM Education Conference (ISEC), Princeton, NJ, USA, pp 324–331. https://doi.org/10.1109/ISECon.2019.8881998
12. Upadhye PV, Nishigandha SP, Patmase BA, Vyawahare AR, Bharate A (2019) A survey paper on various smart attendance systems. Int J Res Appl Sci Eng Technol 7(11):755–758
13. Morita M, Naruse K, Uesugi S, Okada H (2018) A study on adopting smart payment system. In: 5th international conference on behavioral, economic, and socio-cultural computing (BESC). IEEE, pp 144–147. https://doi.org/10.1109/BESC.2018.00037
14. Musa M, Ismail MN, Md Fudzee MF (2021) A survey on smart campus implementation in Malaysia. JOIV: Int J Inf Vis 5(1):51–56
15. Musa M, Ismail MN, Md Fudzee MF (2021) Smart campus implementation in Universiti Tun Hussein Onn Malaysia: towards a conceptual framework. J Phys Conf Ser 1860(1). IOP Publishing
16. Puška A et al (2021) Students' satisfaction with E-learning platforms in Bosnia and Herzegovina. Technol Knowl Learn 26(1): 173–191
17. Fauziah Z, Latifah H, Rahardja U, Lutfiani N, Mardiansyah A (2021) Designing student attendance information systems web-based. Aptisi Trans Technopreneurship (ATT) 3(1):23–31
18. Paul, S, Bandyopadhyay K, Roy A (2021) A study on integrated cloud education system. In: Advances in electrical and computer technologies: select proceedings of ICAECT 2020. Springer Singapore
19. Revathi R, Suganya M, Gladiss Merlin NR (2020) IoT based cloud integrated smart classroom for smart and a sustainable campus. Proced Comput Sci 172:77–81
20. Li Y (2020) Research on building smart campus based on cloud computing technology. In: 2020 5th International Conference on Mechanical, Control and Computer Engineering (ICMCCE). IEEE
21. Dowlatshah K et al (2020) A secure and robust smart card-based remote user authentication scheme. Int J Internet Technol Sec Trans 10(3):255–267

An Optimization of VM Allocation and Migration in Cloud Computing Systems for High Throughput and Low Tolerance

Iram Warsi, Hiresh Gupta, and Monika

Abstract Virtualization is among the most popular procedures in the Cloud Computing scenario since it involves creating an effective virtual server operating system and storage devices. This will help the user by providing numerous machines at a similar time and allowing different users to be assigned to different physical requests of the programme. The efficient scheduling and optimization scenario for resource allocations and virtualization will be discussed in this research paper. Cloud Virtualizations also help with task burdens by replacing out-of-date computing procedures with more accessible, cost-effective, and efficient alternatives. The efficiency of the system is measured in terms of throughput, execution time, downtime, and Service Level Agreements (SLA) violations. The proposed method is the ability to accomplish high performance and is well suited to achieving the desired results.

Keywords Service level agreements (SLA) violations · VM migrations · Virtualizations · Cloud infrastructure

1 Introduction

Cloud computing is the transfer of on-demand services, such as applications, storage, and processing power, through the web and on a pay-per-use basis. Cloud computing services today include everything from Natural language processing (NLP) to storage, connectivity, and computing power, artificial intelligence (AI), and ordinary desktop apps. Cloud systems can now distribute any service that does not require

I. Warsi (✉) · H. Gupta
Shri Ram Murti Smarak College of Engineering and Technology, Bareilly, Uttar Pradesh 243202, India

H. Gupta
e-mail: hiresh.gupta@srmscet.edu

Monika
Deen Bandhu Chottu Ram University of Science and Technology, Murthal, Sonipat, Haryana 131001, India

P. K. Singh et al. (eds.), *Emerging Technologies for Computing, Communication and Smart Cities*, Lecture Notes in Electrical Engineering 875,
https://doi.org/10.1007/978-981-19-0284-0_22

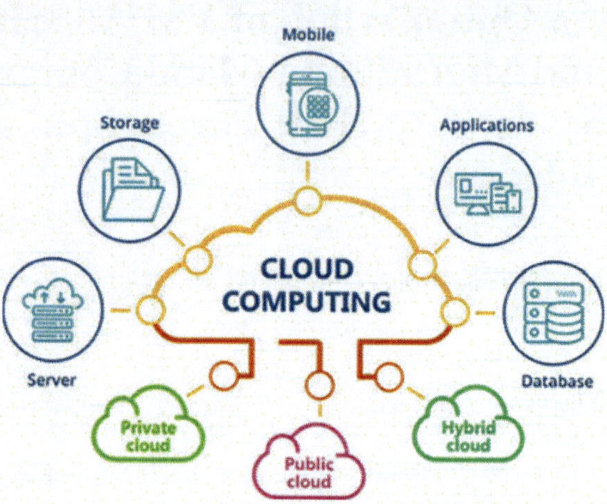

Fig. 1 Cloud computing connectivity

being physically adjacent to the processor hardware. Cloud computing provides a wide range of services, from consumer features, for instance, aim of data analysis on multimedia devices to enterprise benefits such as hosting big data and running all of their workloads in the cloud. Cloud computing is becoming increasingly appealing as a default option for various software programme manufacturers, who are gradually giving their presentations as online services rather than discrete goods or models. However, the drawback to cloud computing processes in the sense that additional costs and risks for concerns, perhaps increasing some real-time process failures [1, 2] (Fig. 1).

2 Cloud Computing Diversities

There are mainly three types of diversities in the cloud system and are discussed below:

2.1 Infrastructure-as-a-Service

There are several types of cloud computing models to choose from. This model describes the fundamental framework of a cloud scenario, which includes virtual servers, data storage backup, and networking. This is fantastic for businesses which are dealing with modest applications from the beginning and wish to control almost each but every component on their own, but it does necessitate that businesses have the practical capabilities to control services at that level [3].

2.2 *Platform-as-a-Service*

Platform-as-a-Service is the layer up version which deals with the storage, virtual servers and comprises of the tools and software that cloud developers need to build software-based applications that includes middleware, management of the databases and tools used for the development [4].

2.3 *Software-as-a-Service*

It is concerned with the delivery of services based on applications that the majority of people use on a regular basis. The final customer will interact with the service via online applications or browsers; hence the basic hardware and operating system are inadequate [5, 6].

3 Related Work

The VM Placement problem is a multi-objective optimization problem that tries to optimize the conflicting objectives while also making competent use of multidimensional resources; this was formulated by Xu and Fortes [7]. Users may avoid using hotspots in virtualized data center environments using this technology, which also helps to reduce energy usage. Users are able to deal effectively with the large solution space by the proposed genetic algorithm modified. The innovative methodology is based on the Fuzzy multi-objective evaluation, which assisted the user in combining goals that were determined to be contradictory. In comparison with single objective methods and bin-packing algorithms, this method shows better performance. Jamali and Malektaji [8] recommended a grouping of genetic algorithms that is improved. A technique was introduced which is considered competent for encoding and generating solutions. This strategy reduces the difficulty of managing a large number of VMs, as well as resource waste and power consumption. The crossover method has provided a positive result as there is a significant reduction in the cost and operation efficiency. Here, the VM Placement is modelled is considered as a vector packing problem which finds solutions to alleviate the utilization of power, turn off the vacant servers and increase the competence of resources. A sub-optimal SLA aware resources allocation optimization system was designed by Huang et al. [9]. To estimate the amount of resources that are utilized considering the SLA specification more accurately with the help of prediction mechanism that uses support vector regression (SVRs). Then, using a genetic algorithm-based technique, VM needs can be changed dynamically. By using a system that uses SVR as a prediction mechanism and then applies a genetic algorithm, we are able to lower the total waiting time of the apps that are running. By providing the promised Quality of Service (QoS) to the client, cloud providers can attain more profits by the proposed method. An Optimal VM placement algorithm in which expense is reduced for hosting the VMs in the environment which has several cloud providers taking into consideration as per the demands of future and uncertainty of price; this was introduced by Chaisiri et al. [10]. It arrived at its decision after solving stochastic integer programming with two stages of resources. The effectiveness and performance of this method can be evaluated with the help of Numerical analysis and simulations. As these tests show, this method may be utilized for effective provisioning, it fits the requirements of both customers and suppliers.

A novel approach was proposed by Tordsson et al. [11] which try to optimize the placement of VMs across numerous cloud providers bearing in mind the criteria specified by the clients an interoperability layer is provided by the cloud broker. It aids in the placement of price–performance tradeoffs. Constraints with respect to load balance, maximum budget and minimum performance. etc., are some client specifications by which the allocation of VM is steered. When we look at the evaluation findings, we can see that when there is consumption in more than one cloud, efficiency increases.

4 Proposed Work

This section discusses a potential approach for efficient virtualization for balancing the load and reduced execution time for high-performance assessments. The proposed solution is providing efficient solutions for the virtualization process in cloud systems. The cloud system mainly facing the problem of overloading and under loading hosts which are producing failures of the executions of the tasks and processes. The load is also a major problem, as it is one of the most important parameters for controlling the operation of cloud systems. Firstly the job specifications are initialized and then the machines are deployed with the VM. The machines here are termed as the hosts on which the VM are deployed. Then the execution of the services as requests starts through which the CPU utilizations are achieved. Through CPU utilization the under load and overload of the hosts are obtained. After evaluating the overload of the hosts then min–max scheduling is processed for the completion of the tasks or the processes through which the minimum execution time can be achieved but it can increase the overhead consumption in the migration process through which the system's effectiveness can be degraded. To overcome this situation the overhead consumption is reduced using moth flame optimization process which increases the throughput of the virtualization process and also decreases the downtime which reduces the energy consumption of the hosts and balances the load. This improves the overall performance of the suggested solution and reduces the likelihood of failure.

5 Methodology Flow Diagram

See Fig. 2.

5.1 Step by Step Execution of the Research Methodology

Step 1: User requests for the service by initializing the job specifications.
Step 2: Initializing the VM and VM's specifications for real-time scenarios, users and their requests for different services can be in any number.

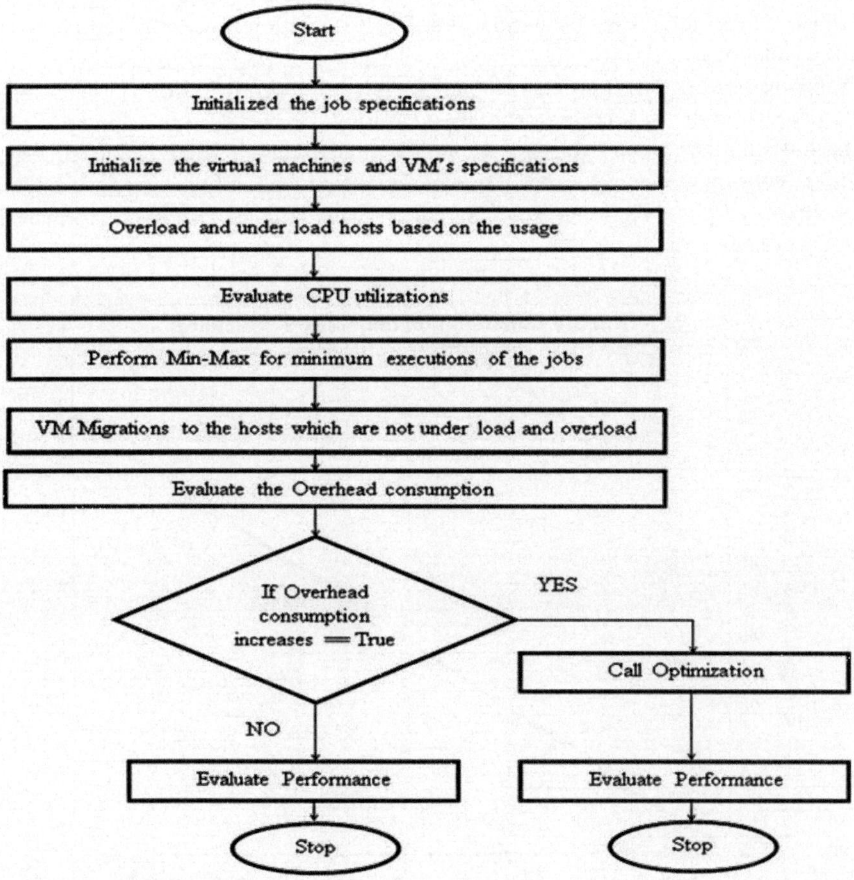

Fig. 2 Proposed methodology flow chart

Step 3: Analyzing the overload and under load hosts/consumptions based on the usage by evaluating CPU utilizations. Perform min–max for minimum execution of the jobs.
Step 4: If overload consumption increases.
Then call optimization.
Step 5: Else evaluate performance. Stop.

6 Result and Discussions

This section covers the simulations and results obtained for the proposed research. The implementations are done in MATLAB tool and the cloud scenario is generated which will be helpful for the virtualization and migration process for appropriate SLA violations.

Figure 3 shows the proposed methodology is able to achieve low energy consumption, which is the desired outcome, by measuring the energy consumption of the machines for the executed jobs per second. For high executions of dynamic requests, energy consumption must be low. If the energy consumption is high then the machine gets overloaded and there can be a lot of failures which can degrade our performance of the system.

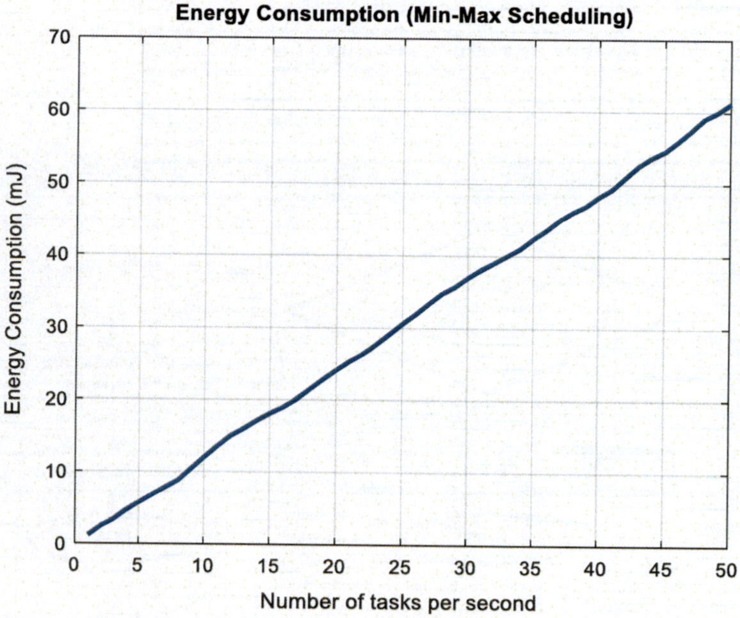

Fig. 3 Energy consumption

Figure 4 shows these machines Ids which are getting overloaded and under loaded. The machines which are consuming more than 85% CPU utilizations then will raise the flags and extract the ids of the machines which are getting overloaded.

Figure 5 shows the migration time of the machines which are getting overloaded to the host which are ideal and is having high capacity and bandwidth to execute

Fig. 4 Machines overloaded and under loaded

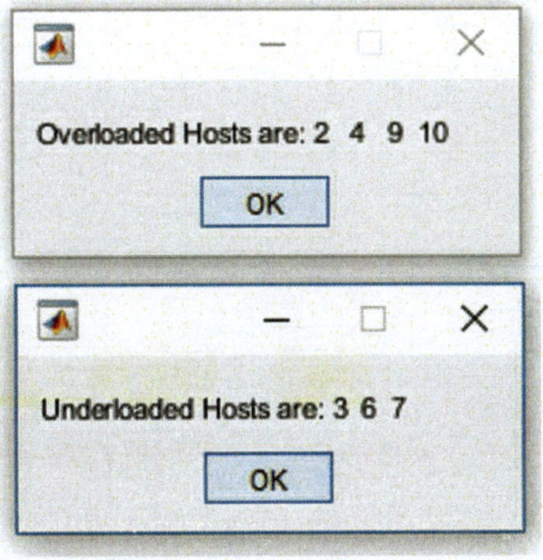

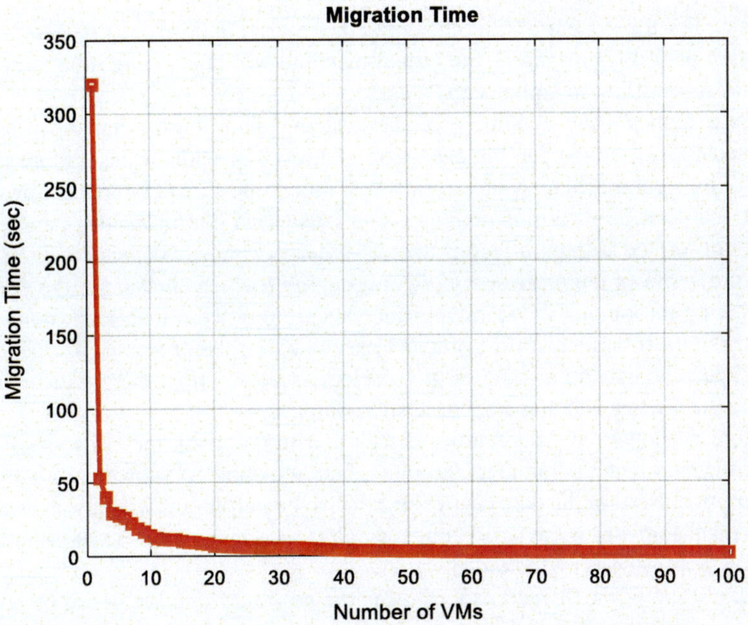

Fig. 5 Migration time

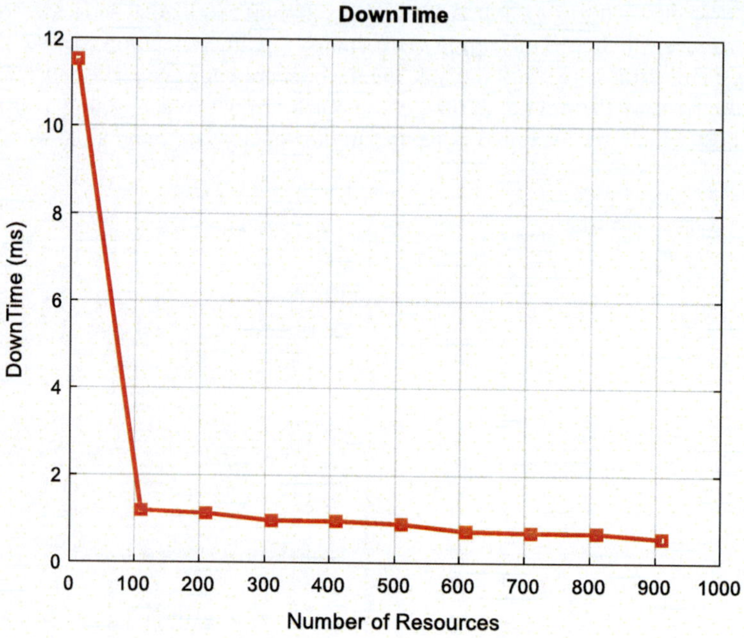

Fig. 6 Down time

the tasks. For the system to run efficiently, the migration time must be short. If the migration period is prolonged, the virtual machine will take a long time to migrate to the host, resulting in high energy usage.

Figure 6 shows the downtime of the system which shows that the proposed approach is well suited for the execution of the tasks without any failure of the hosts and virtual machines which means that downtime should be low If downtime grows, real-time applications running on various cloud platforms may have a high number of service failures, affecting the real-time environment dependent on servers.

Figure 7 shows the proposed service throughput which shows that the solution is achieving throughput in migrations and execution process with high uptime and throughput must be high for the efficient executions of the processes. If due to low throughput, the execution time of the services increases then there will be heavy loads in the queues which is not our desired output.

Figure 8 shows the data centres energy consumption on the daily basis. It is one of the significant parameters through which we come to know the total energy consumption used for the data centres. It can also be helpful in the analysis based on power management for the data centres which is the key component of running the real-time applications on the cloud.

Figure 9 shows the SLA violations using the proposed solution. It can be noticed from Fig. 9 that the SLA violations are less which shows that there are less chances of software failures as per the agreements. Service level violation occurs when the

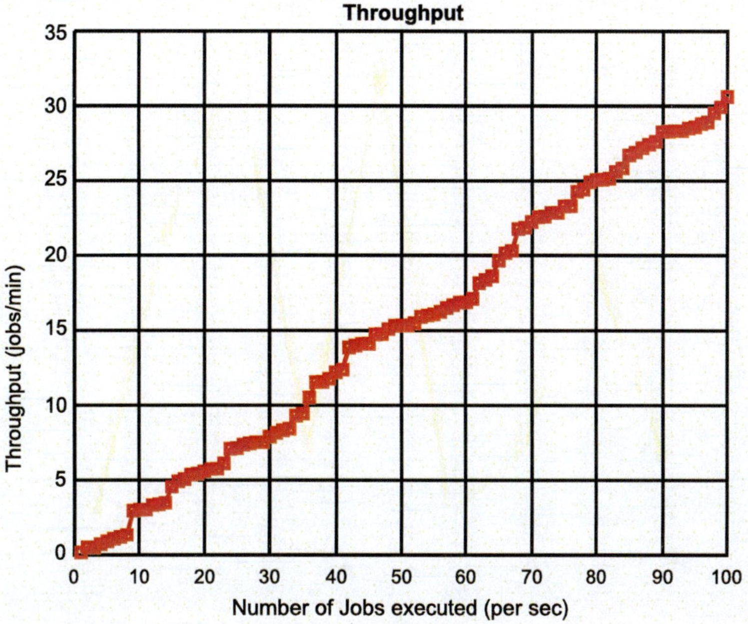

Fig. 7 Throughput

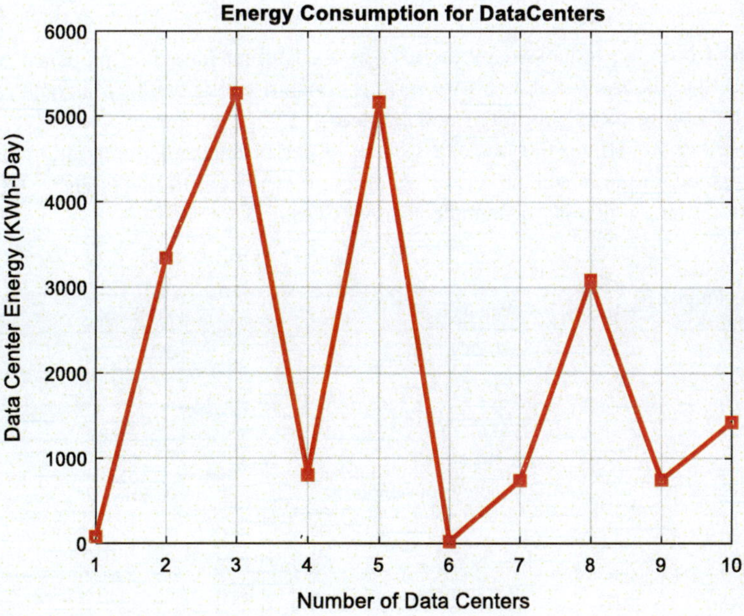

Fig. 8 Data centres energy consumption

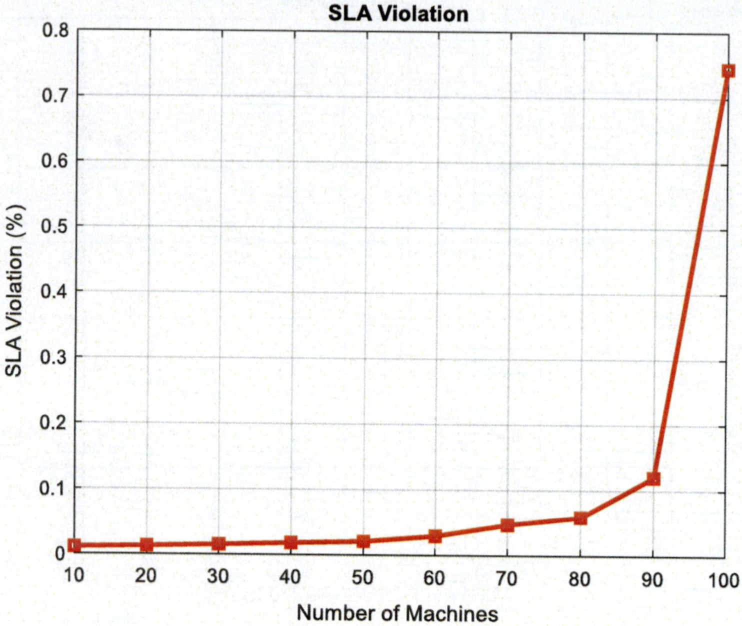

Fig. 9 SLA Violations

services which are requested by the user or a customer have not happened on the decided time interval. SLA is a very serious concern which can be managed for the smooth workings of the contracts and services.

Table 1 demonstrates the suggested system's performance, demonstrating that the proposed technique is well suited for an efficient system with low latency and little migration time, which is our intended output (Table 2).

Table 1 Proposed performance

Parameters	Proposed
Throughput	30
Downtime	1.5 ms
Migration time	0.01 s
SLA violation	0.008
Energy consumption	60 mJ

Table 2 Performance comparison

Parameters	Base	Proposed
Throughput [11]	3.5	30
SLA violation [12]	0.015	0.008

7 Conclusion

In a cloud computing context, virtualization is a strategy for connecting existing resources in a network by dividing existing bandwidth into multiple channels, each distinct and independent.

Virtual machine migration is one of the most important VM management tasks, as its success has a direct impact on cloud data centre (DCs) energy efficiency [13]. Large-scale virtualized data centres were employed for VM's power management, with resource management groups divided into local and global managers [14]. Virtualization refers to the creation of a virtual platform for the server operating system and storage devices in Cloud Computing, and it provides a virtual stage of server operating arrangements and storage solutions. The evaluation of a high-performance cloud computing scenario is carried out in the proposed methodology. The new framework can provide a stable solution for efficient virtualization and high throughput in real-time cloud scenarios.

References

1. Jouini M, Rabai LBA (2019) A security framework for secure cloud computing environments. In: Cloud security: concepts, methodologies, tools, and applications. IGI Global, pp 249–263
2. Jonas E, Smith JS, Sreekanti V, Tsai CC, Khandelwal A, Pu Q, Shankar V et al (2019) Cloud programming simplified: a Berkeley view on serverless computing. arXiv preprint arXiv:1902.03383
3. Manvi SS, Shyam GK (2014) Resource management for Infrastructure as a Service (IaaS) in cloud computing: a survey. J Network Comput Appl 41:424–440
4. McGrath PM, Hicks M, Wiest T, McPherson DC (2014) Controlling utilization in a multi-tenant platform-as-a-service (PaaS) environment in a cloud computing system. U.S. Patent 8,850,432. Last accessed 30 Sept 2014
5. Ilager S, Wankar R, Kune R, Buyya R, GPU PaaS computation model in Aneka cloud computing environments. In: Smart data: state-of-the-art perspectives in computing and applications
6. Valls MG, Cucinotta T, Lu C (2014) Challenges in real-time virtualization and predictable cloud computing. J Syst Archit 60(9):726–740
7. Xu J, Fortes JAB (2010) Multi-objective virtual machine placement in virtualized data centre environments: green computing and communications (GreenCom). In: IEEE/ACM International conference on & International conference on cyber, physical and social computing (CPSCom). IEEE
8. Jamali S, Malektaji S (2014) Improving grouping genetic algorithm for virtual machine placement in cloud data centers. In: 4th International e-conference on computer and knowledge engineering (ICCKE). IEE
9. Huang JC et al (2013) An adaptive resource management scheme in cloud computing. Eng Appl Artif Intell 26(1):382–389
10. Chaisiri S, Lee B-S, Niyato D (2009) Optimal virtual machine placement across multiple cloud providers. In: Services computing conference, APSCC 2009, IEEE Asia-Pacific. IEEE
11. Tordsson J et al (2012) Cloud brokering mechanisms for optimized placement of virtual machines across multiple providers. Future Gener Comput Syst 28(2):358–367
12. Patel N, Patel H (2017) Energy efficient strategy for placement of virtual machines selected from under loaded servers in compute cloud. J King Saud Univ Comput Inf Sci

13. Mohammad M, Hemn K (2020) Efficient VM migrations using forecasting techniques in cloud computing: a comprehensive review. Cluster Comput 23:2629–2658
14. Kim MH, Lee JY, Shah Raza SA et al (2021) Min-max exclusive virtual machine placement in cloud computing for scientific data environment. J Cloud Comp 10:2

GraDex—Graph-Based Data Analytics for Extractive Text Summarization

Maher Thakkar, Siddhant Patel, and Jai Prakash Verma

Abstract This paper aims to brief the reader about different Automatic Text Summarization methods and their efficiency when it comes to providing meaningful summaries. In this paper, we have conducted comparative research between the BERT model for text embeddings along with K-Means clustering to identify sentences closest to the centroid for summary selection, and a Word Frequency algorithm that computes the frequency of appearing word, assigns appropriate weights and selects sentences based on a threshold score. The purpose of this is to compare the two different approaches, for Reviews and Feedback Analysis of different texts and their summaries. Through our research, we were able to find that BERT outperforms the Word Frequency model according to all the evaluation metrics and this is clearly demonstrated in the following sections of the paper.

Keywords Extractive summarization · BERT · Word frequency · Natural language processing · Machine learning · Big data

1 Introduction

In recent years, there has been an exponential increase in the amount of textual data over the internet. Different sources of information often result in redundancy, repetition, and difficult comprehension for the average user. In such a scenario, manual summarization can be supplanted by Automatic Text Summarization (ATS) to make up for the time consumption and related costs. An ATS system provides a computer-generated summary of the given input, covering important information and discarding redundant and unnecessary data. The last few years have witnessed tremendous research in this domain through a wide range of approaches [1]. Text Summarization can be done in two ways: Extractive Summarization (ES) and Abstractive Summarization (AS). In the former, one identifies important sections in the text, and the summary generated is a subset of the original in the input. In the latter, the latter

M. Thakkar · S. Patel · J. P. Verma (✉)
Institute of Technology, Nirma University, Ahmedabad, India
e-mail: jaiprakash.verma@nirmauni.ac.in

P. K. Singh et al. (eds.), *Emerging Technologies for Computing, Communication and Smart Cities*, Lecture Notes in Electrical Engineering 875,
https://doi.org/10.1007/978-981-19-0284-0_23

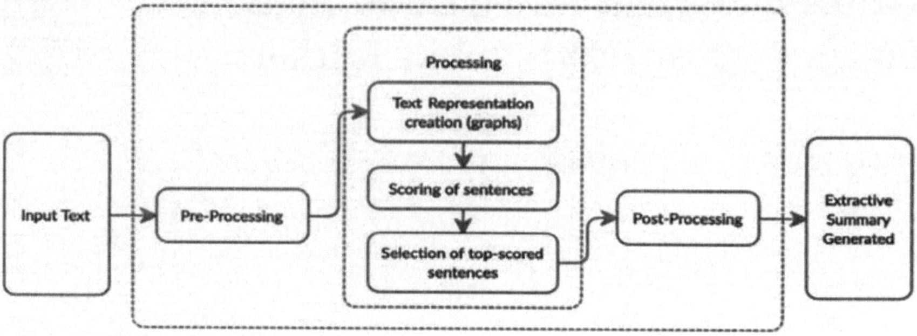

Fig. 1 General overview of the ATS system

uses advanced Natural Language Processing (NLP) methods to reproduce important parts of the text after interpretation, examination, and analysis. It offers more cohesion than the other approach and resembles the human approach, but the ease in the process fetches the attention of a majority towards ES (Fig. 1).

We can observe many steps that constitute parts of the ATS system:

- Pre-Processing: Using various NLP techniques like stemming, lemmatization, tokenization, etc. to filter the data for processing.
- Processing: Once unnecessary data has been filtered out, we perform the following steps to identify and rank important sentences from the given text.
  - Text Representation: Simplification of text analysis is preceded by an appropriate representation of sentences through graph, Bag of Words (BOW) model, etc.
  - Scoring of Sentences: Through chosen parameters, sentences are scored in importance from the highest to the lowest, to reflect their importance in the data.
  - Selection of top-scored sentences: With things like user need in terms of number of words and number of sentences, the top-ranked sentences are chosen to match the desired length.
- Post-Processing: Once the important sentences are obtained, re-ordering and concatenation can be applied as refinement techniques to obtain the final computer-generated summary.
- Compare scores for generated summaries: The Recall-Oriented Understudy for Gisting Evaluation (ROUGE) scoring algorithm is used to calculate the similarity between generated summaries and the golden handwritten summaries provided in the dataset. Use the ROUGE score to understand the quality of document translation and summarization model [2].

In this paper, we present a BERT model-based Extractive Summarizer and compare it with a Word Frequency Model that utilizes a different approach for generation of summary and compare the results to derive meaningful inference that could help identify the best approach for the Extractive Summarization. The two models are compared on the basis of a Review and Feedback Analysis on the Opinosis Dataset which will be discussed in the subsequent sections of this paper. We discuss the entire pipeline, wherein the input is taken, pre-processed, processed, and lastly post-processed for the final resulting summary. The organization of this paper is as follows: Related Work discusses the literature, related work, and recent advancements made in relation to text summarization techniques; Methodology and Concepts describe the methods used for review and feedback analysis; and the Execution section involves the necessary graphs and tables supporting the analysis and inferences done by us.

2 Related Work

Several media platforms have contributed to the huge amount of textual data and researchers have since been simultaneously researching the proposal and improvements in several techniques of text summarization for concise and accurate summaries [3].

There are various methods that are used for ATS; these include graph-based, statistical-based, DL-based, semantic-based, Machine Learning (ML)-based, and sentence centrality. The commonly used traditional approach for ATS is as follows:

1. Find the required features to be kept as weights, and calculate the value for each.
2. Assign scores to every sentence using the feature and corresponding weights, as shown in Eq. 1.

$$\text{Score}(S) = \sum_{i=1}^{n} W_i * F_i(S) \tag{1}$$

As a part of the earlier approaches, Krishnaven proposed a local scoring and ranking approach using heading and subheadings in given data [4]. Sentences are selected based on heading; upon compression ratio, we get an equal number of sentences from each heading.

El-Kassas et al. [5] adopts a combination of various ATS methods (graph, statistical, semantic, and centrality based) for the extraction of summary. The unsupervised nature it follows does not call for the requirement of any training data. Here, Recall-Oriented Understudy for Gisting Evaluation (ROUGE) is used as an evaluation metric, and successful implementation obtained the best results among state-of-the-art ATS models.

Fang uses an unsupervised graph-based model to compute the relationship between words and corresponding sentences [6]. They have performed the experiment of real-life datasets; the use of complex matrix operations and a redundancy elimination technique provide theoretical convergence and an improvement in quality.

Liu used a Deep Learning (DL) approach using Generative Adversarial Networks (GANs) with reinforcement learning, undergoes prediction of the abstractive summary and discrimination from the ground truth summary for the evaluation results [7]. In comparison, this model turned out to generate more abstractive and diverse summaries from the given text.

A similar approach was adopted by Song, where they combined Long short-term memory (LSTM) and Convolutional Neural Network (CNN) for an abstractive summarization of the given text [8]. The focus here is shifted to semantic phrases for a fine-grained approach, and the DL execution paves way for a more syntactic and semantic approach in the end (Table 1).

3 Methodology and Concepts

The following subsections in this section will brief the reader regarding the framework of the ES model used as well as the dataset and application considerations for the same. It will also include necessary figures and flowcharts to make the understanding of the framework much better and easier.

3.1 Dataset Considerations

"GraDex—Graph-based Data Analytics for Extractive Text Summarization": The application of this type of summarization has been done using a dataset that contains multiple different, long reviews with multiple human written feedbacks each called the Opinosis Dataset. In all, there are 51 topics with every topic having approximately 100 sentences (on average). The reviews were taken from various sources—Tripadvisor (hotels), Edmunds.com (cars), and Amazon.com (various electronics). The dataset file also comes with gold standard summaries used as a reference for the summarization scores which will be discussed in the subsequent sections. The graph-based BERT approach is seen to have a higher score when utilized over various iterations of different reviews yielding successful analysis results which will be discussed in the following sections (Fig. 2).

3.2 Text Pre-processing

Figure 3 gives an overview of the methodologies we have adopted as part of our proposed approach. For better performance of our algorithm, it is essential for us to convert the text into a more understandable form. The proposed tasks go as follows:

Table 1 Comparison of previous implementations of text summarization

Author	Year	Objective	Cons	1	2	3	4	5	6
Fang et al. [9]	2016	Extractive text summarization for word sentence co-ranking	Currently limited to single document implementation	Yes	Yes	No	No	Yes	Yes
Krishnaveni et al. [10]	2017	Coherent ATS using current scoring and ranking	Restricted to single document, heading wise summary generation	No	Yes	No	No	Yes	Yes
Chen et al. [11]	2018	Text image summarization using multi-model RNN	Absence of pre- and post-processing steps of implementation	No	Yes	No	Yes	No	No
Liu et al. [12]	2018	Use of GANs for abstractive text summarization	Absence of pre- and post-processing steps of implementation	No	No	Yes	Yes	No	No
Mackey et al. [13]	2018	Using big data frameworks for ATS	Lack of complete automation for the summarization process	No	Yes	No	No	Yes	Yes
Gunawan et al. [15]	2019	Using maximum marginal relevance and text rank for multidocument summarization	Low *F*-scores in certain categories of document	No	No	Yes	No	Yes	Yes
El-Kassas et al. [5]	2020	Using a graph-based framework for ATS	Complex approach due to incorporation for multiple approaches	Yes	Yes	No	No	Yes	Yes

1. Graph-based approach; 2. Extractive summarization; 3. Abstractive summarization; 4. DL approach; 5. Post-processing setup; 6. Pre-processing set up

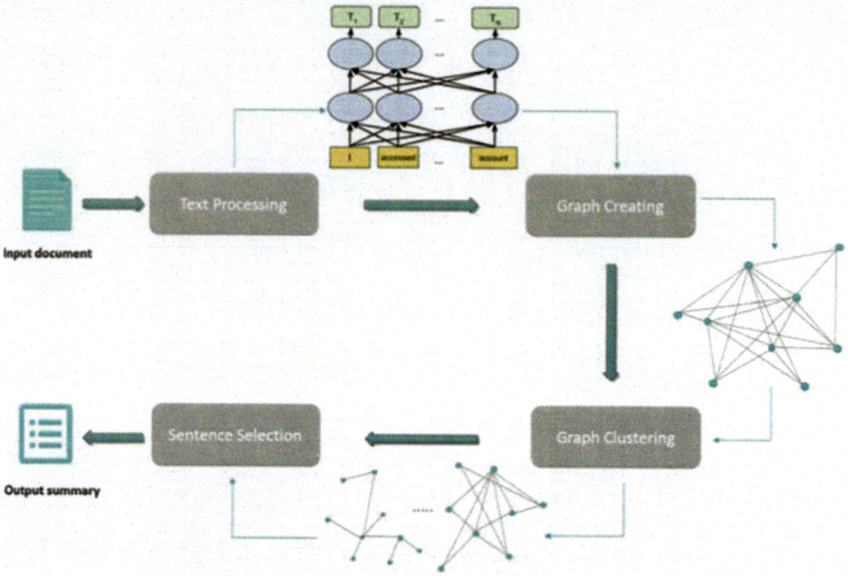

Fig. 2 Framework for graph-based extractive summarization

Fig. 3 Pre-processing methodologies adopted

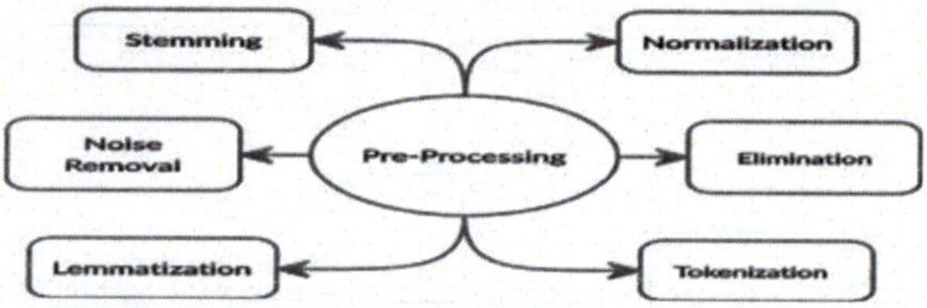

- Stemming and Lemmatization help reduce words to their bases in order to prevent duplication and redundancy. There is a lot of unnecessary noise in the form of tags, stop words, connectors, conjunctions, etc. which must be removed and eliminated for better efficiency of our model.
- Tokenization is essential for the splitting of text into smaller constituents called "tokens".
- Following this, normalization (where necessary) helps in converting all text to the same lower case, to have everyone on the same field.

3.3 Processing

This section contains all the necessary information regarding the methodologies explained in the previous section along with the libraries and methods utilized for the successful analysis of the Feedback and Reviews [14].

3.3.1 Word Frequency Model

It gives us the frequency of the word in each document in the corpus. It is the ratio of the number of times the word is seen in a document analyzed to the total number of words in that document. It gets increased as the number of instances of that word within the document is increased.

For pre-processing, NLTK (stands for Natural Language Toolkit) is used for building python programs that endeavor with human language data for putting in statistical natural language processing (NLP). It accommodates text processing libraries for tokenization, parsing, classification, stemming, tagging, and semantic reasoning [12].

Libraries Used:

- We imported stopwords from the nltk.corpus package to remove the stopwords defined in the English dictionary. The stopwords are English words which does not add much meaning to a sentence. They can be ignored without abandoning the meaning of the sentence.
- We also imported word_tokenize and sent_tokenize from the nltk.tokenize package. Tokenization in NLP is the process through which a big quantity of text is split into shorter parts called tokens. In word_tokenize() method, we extract the tokens from a string of characters by utilizing tokenize. It gives back the syllables from one word. One word can contain one or two syllables. In sent_tokenize we will be splitting words from sentences.
- We imported PorterStemmer from the nltk.stem package. Stemming is the process of producing lexical alternatives of a root/base word.

Functions used:

- **create_frequency_table(text_string)**: In this, we will take the input which will be a paragraph as a text_string. We will tokenize text into words, stem words and then remove the stopwords. We will get a bag of word dictionaries (word: frequency) as the output.
- **score_sentances(sentences, freqTable)**: In this, we will take the input as a list of sentences and word frequency table. We will be computing the score for each sentence which would be equal to total word value upon the word count. As the output, we will get a sentence score dictionary (sentence: score).
- **find_average_score(sentenceValue)**: In this, we will be taking a sentence score dictionary as the input. Then we compute the average sentence score equal to the total sentence score upon the sentence number. We should get the average sentence score threshold as the output.
- **generate_summary(sentences, sentenceValue, threshold)**: In this, we will take the input as the list of sentences and the sentence value dictionary. We should get the output as the sentence whose score is greater than the threshold as the summary.
- **summarize_text_wf(text)**: In this, we will use a paragraph of text. The output will be a summary of text according to the word frequency algorithm (Fig. 4).

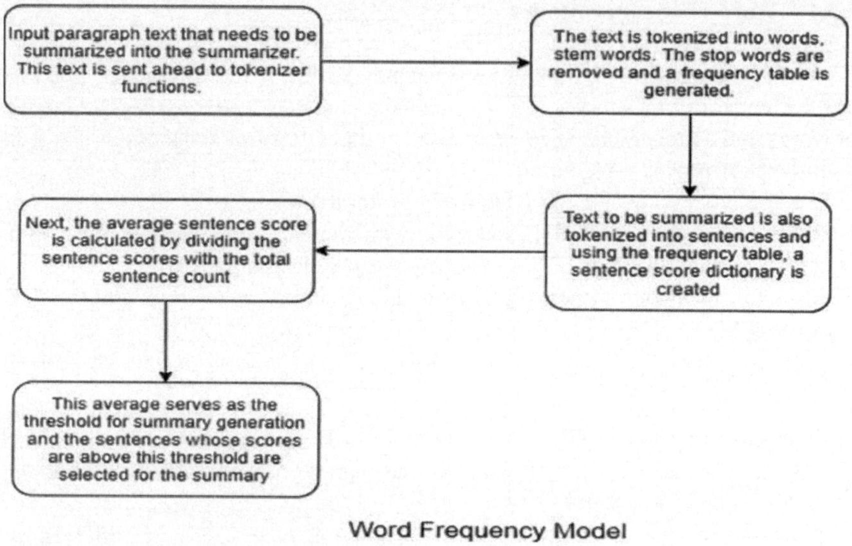

Fig. 4 Flow of word frequency model

3.3.2 BERT Model with *K*-Means Clustering

It is easier to cluster text documents with the help of BERT and *K*-means. We can apply the *K*-means algorithm on the BERT embeddings to cluster documents. Similar sentences are clustered based on their sentence embedding similarity [25].

In this approach preprocessing will not change the output predictions. There are different cases:

1. **Stemming or Lemmatization**: BERT uses BPE (Byte-Pair Encoding) to shrink its vocab size, so words like run and running will eventually be decoded to run + ##ing without the need of any separate function for stemming.
2. **De-Capitalization**: BERT provides two models (lowercase and uppercase) where one changes your sentence into lowercase, and others will not transpose related to the capitalization of your sentence. So, you do not require to do any changes here, just choose the model for your use case.
3. **Removing high-frequency words**: BERT takes the Transformer model, which does work on the attention principle. So, when you fine tune it on any problem, it will look only at those words which will impact the output and not on words which are common in all data.

Like the previous approach, we will be importing modules from the NLTK library as well as the Scikit-learn library for the *K*-means clustering algorithm.

Libraries Used:

- Here we also imported Torch which is the open-source machine learning library, a scientific computing framework, and a script language.

- Just like before, we have made use of the word_tokenize and sent_tokenize modules from the nltk.tokenize package.
- We will be importing BertTokenizer and BertModel from pytorch_pretrained_bert to instantiate a BERT pretrained model with the help of the BertModel class.
- We will be importing K-Means and pairwise_distances_argmin_min from sklearn.metrics. The k-means clustering method is an unsupervised machine learning technique used to get to know clusters of data objects in a dataset. pairwise_distances_argmin_min is used to calculate minimum distances between one point and a set of the points. This function calculates for each row in X, the index of the row of Y which is closest (according to the specified distance). The minimal distances are also returned.
- We will be importing numpy. NumPy is used for matrix manipulations and data handling which is one of the main purposes of this library.

Functions used:

- **bertSent_embedding(sentences)**: We will take a list of sentence tokens as the input. A list of latent vectors, each vector is a sentence representation should be the output. Bert model produces 12 layers of latent vector, the 'last layer' method is used here, other choices include average last 4 layers, average all layers, etc.
- **kmeans_sumIndex(sentence_embedding_list)**: We will take a list of embedded sentence vectors as the input. The output will be a list of indices of sentences in the paragraph, representing the clustering of the key sentences. K Means is used here for clustering.
- **bertSummarize(text)**: We will take a paragraph as string as the input. The output will be the summary including a few key sentences using BERT sentence embedding and clustering.

4 Results and Evaluation

This section contains the results of the Feedback and Review Analysis done on the Opinosis Dataset using different models and their evaluation metrics. The necessary plots and data tables are included in the following subsections along with the inferences (Fig. 5).

4.1 Principle and Evaluation Metric

For researching the two approaches for generation of an extractive summary, we first consider a single review, out of the 51 different types of reviews each having approximately 100 lines of text, from the dataset as the reference input to be given to the two models. In the dataset, we also have the gold standard human-written

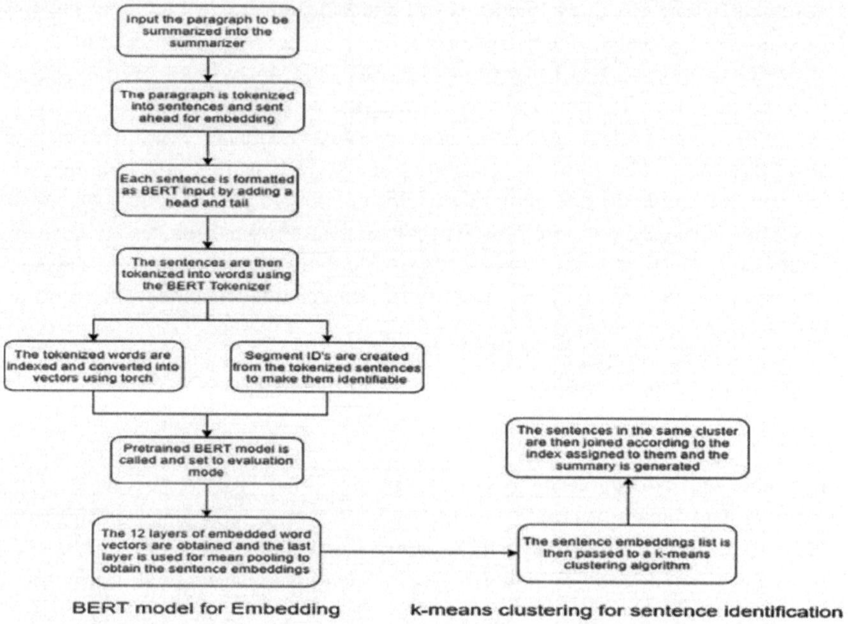

Fig. 5 Flow of BERT model with *k*-means clustering

summaries for each review to compare our model-generated summaries with. Now, for the evaluation of the generated summaries, we choose the evaluation metric known as ROUGE, i.e., Recall Oriented Understudy for Gisting Evaluation. ROUGE is a popularly used set of metrics used for evaluating automatic text summarization and machine translation software in the field of natural language processing. The metrics compare a system-generated summary against a reference.

ROUGE has five different evaluation metrics viz.:

- **ROUGE-N:** Overlap of *N*-grams between the generated and reference summaries, e.g.,
  - ROUGE-1 mentions the overlap of *unigram(each word)* in between the system and reference summaries.
  - ROUGE-2 mentions the overlap of *bigrams* between the system and reference summaries.
- **ROUGE-L**: Longest Common Subsequence (LCS) based statistics. Longest common subsequence problem takes into account sentence level structure similarity naturally and identifies longest co-occurring in sequence *n*-grams automatically.
- **ROUGE-W**: Weighted LCS-based statistics that favor side-by-side LCS(s).
- **ROUGE-S**: Skip-bigram-based co-occurrence statistics. Skip-bigram is any set of words in their sentence order.

- **ROUGE-SU**: Skip-bigram plus unigram-based co-occurrence statistics.

In our implementation, the ROUGE-N and ROUGE-L metrics were evaluated, and the results were compared. ROUGE score is composed of:

- **Precision**: # of overlapping words/total words in the reference summary
- **Recall**: # of overlapping words/total words in the model generated summary
- **F_1-score**: # measure of model's accuracy on dataset.

4.2 Evaluation Results

For the comparison of evaluation results, the subsequent parts of this section will include tabular data and waveform analysis for the same.

Tables 2, 3 and 4 depict the precision, recall and F_1-score of ROUGE-1 gram, ROUGE-2 gram, and ROUGE-L evaluation metrics for both the models. We can observe that the Word Frequency Model performs slightly better in terms of ROUGE scoring than the BERT and k-means model. However, on further analysis done by calculating the mean ROUGE score for the entire Opinosis dataset which contains 51 different reviews with each of their handwritten summaries for reference, we find that the BERT model performs considerably better than the Word Frequency Model. The tabular data for the mean ROUGE-1 score, for the entire dataset, of the two models are as shown in Table 5.

In Fig. 6, we can see a bar graph which depicts the ROUGE-1 mean F_1-score's comparison for the BERT model and the Word Frequency Model as well as a distribution plot that gives us an idea about the ROUGE-1 F_1 score density for the same. We can clearly see that the BERT model yields a higher mean F_1 score as well as a better density curve with wider distribution. Since the F_1 score is a weighted average of the precision and recall values, where an F_1 score reaches its best value at 1 and worst

Table 2 Comparison of performance parameters of ROUGE-1 gram evaluation metric

Parameter	BERT and K-means cluster model	Word frequency model
F_1-score	0.1111	0.1176
Precision	0.0639	0.0666
Recall	0.4230	0.5

Table 3 Comparison of performance parameters of ROUGE-2 gram evaluation metric

Parameter	BERT and K-means cluster model	Word frequency model
F_1-score	0.0102	0.0182
Precision	0.0058	0.0103
Recall	0.04	0.08

Table 4 Comparison of performance parameters of ROUGE-L evaluation metric

Parameter	BERT and K-means cluster model	Word frequency model
F_1-score	0.1406	0.1560
Precision	0.0849	0.0924
Recall	0.4090	0.5

Table 5 Comparison of mean performance parameters of ROUGE-1 evaluation metric for the entire dataset of reviews

Parameter	BERT and K-means cluster model	Word frequency model
F_1-score	0.0867	0.05336
Precision	0.0483	0.0282
Recall	0.5839	0.7329

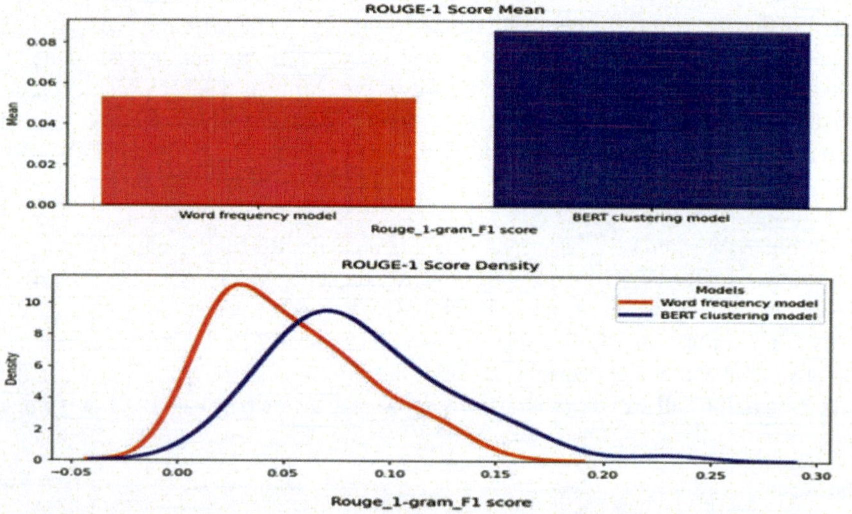

Fig. 6 ROUGE-1 mean F_1 score and score density for the two models

value at 0, we can say that the BERT model even without hyperparameter tuning performs better than a conventional frequency model for summary generation.

Similar to the analysis done in Figs. 6 and 7 shows similar results that indicate that the BERT model has a higher mean precision score as well as a better distribution meaning that the BERT model was able to generate summaries that were more precise as per logic and understanding rather than the literal meaning of correct prediction.

Figures 6 and 7 both indicate that when a larger amount of data is to be analyzed and summary needs to be generated, the F_1 score and precision of the BERT are much better when the entire data is considered.

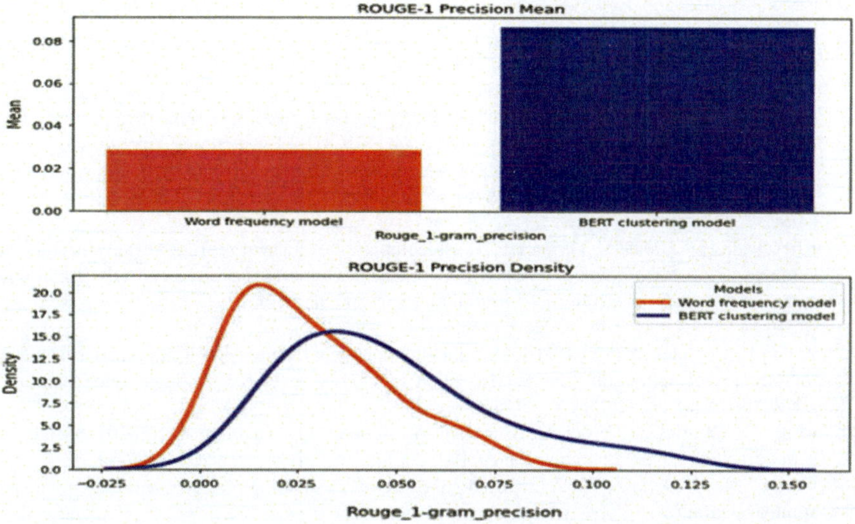

Fig. 7 ROUGE-1 mean precision and precision density for the two models

5 Conclusion

The purpose of this research paper was to study the performance of the Word Frequency Model and the BERT and *K*-means model in the generation of an extractive summary when a single text as well as multiple texts of large sizes are applied to each of them. The data gathered was tabulated and plotted on waveforms to derive useful inferences that may help in further research or the usage/application of the two models in their respective areas of requirement. The study that was conducted utilized a BERT model that was not fine-tuned to provide ideal results, however, the generated summaries and their respective performance parameters still proved to be better than the Word Frequency Model in the case of larger data size. With that, we can conclude that in areas of higher precision and quality is required, the BERT model is the ideal choice of ATS system at the cost of more resource utilization. However, in scenarios of limited hardware/software capabilities, the Word Frequency Model proves to be more than sufficient in the classification and generation of summaries.[15]

References

1. Virapat K, Bowen T, Yiming N (2020) Automatic text summarization of COVID-19 medical research articles using BERT and GPT-2. https://arxiv.org/abs/2006.01997
2. Verma JP, Atul P (2017) Evaluation of unsupervised learning based extractive text summarization technique for large scale review and feedback data. Indian J Sci Technol 10(17):1–6. ISSN: 0974-6846

3. Verma JP, Atul P (2017) An extractive text summarization approach for analyzing educational institution's review and feedback data. Int J Comput Appl, June 2016
4. Wolf T, Sanh V, Rault T (2019) PyTorch pretrained BERT: the big & extending repository of pre-trained transformers. https://github.com/huggingface/pytorch-pretrainedBERT
5. El-Kassas WS, Salama CR, Rafea AA, Mohamed HK (2020) Edge-summ: graph-based framework for automatic text summarization. Inf Process Manag 57(6):102264. https://doi.org/10.1016/j.ipm.2020.102264; http://www.sciencedirect.com/science/article/pii/S0306457320301047
6. Kheruwala HA, Shah JV, Verma JP (2020) Comparative study of sentiment analysis and text summarization for commercial social networks. In: Gupta S, Sarvaiya JN (eds) Emerging technology trends in electronics, communication and networking. Springer, Singapore, pp 213–224
7. Zhang JJ, Chan HY, Fung P (2007, December) Improving lecture speech summarization using rhetorical information. In: IEEE Workshop on automatic speech recognition & understanding, 2007 ASRU, pp 195–200
8. Ming Z, Pengfei L, Yiran C, Danqing W, Xipeng Q, Xuanjing H (2020) Extractive summarization as text matching. In: ACL 2020. https://arxiv.org/abs/2004.08795
9. Fang C, Mu D, Deng Z, Wu Z (2016) Word-sentence co-ranking for automatic extractive text summarization
10. Krishnaveni P, Balasundaram SR (2017) International conference on computing methodologies and communication (ICCMC) - automatic text summarization by local scoring and ranking for improving coherence. [IEEE 2017 international conference on computing methodologies and communication (ICCMC) - Erode, India (2017.7.18–2017.7.19)]
11. Chen J, Zhuge H (2018) Abstractive Text-Image Summarization Using Multi-Modal Attentional Hierarchical RNN
12. Liu L, Lu Y, Yang M, Qu Q, Zhu J, Li H (2017) Generative adversarial network for abstractive text summarization
13. Mackey A, Cuevas I (2018) Automatic text summarization within big data frameworks. J Comput Sci Coll 33(5):26–32
14. Wei L, Pei-Ran S, Rui-Li J (2020) A BERT based single document extractive summarization model. J Comput 31(2):241–249
15. Gunawan D, Harahap SH, Fadillah Rahmat R (2019) Multi-document summarization by using textrank and maximal marginal relevance for text in Bahasa Indonesia. In: 2019 International conference on ICT for smart society (ICISS), vol 7, pp 1–5. https://doi.org/10.1109/ICISS48059.2019.8969785

Smart Traffic Monitoring with Fog and Cloud Computing

Varsha Choudhary, Yashwant Singh, and Pooja Anand

Abstract The connectivity among billions of real and virtual things to the Internet with unique identities, brought the term Internet of Things (IoT) which is constantly changing the world. IoT penetration in major services is making our lives easier by managing things from anywhere at any time. Consequently, this ubiquitousness has vigorously led to exponential increase of data, questioning its efficient storage and analysis. However, Cloud Computing has been playing a substantial role to manage the colossal army of data but becomes inefficient for time-sensitive applications like Smart Traffic Monitoring. Perhaps, this would result in cloud overloading and brings bandwidth of network at saturation point. This necessitates the use of fog computing with Smart systems to aid efficient utilization of cloud computing paradigm. Fog computing is a decentralized computing paradigm that extends cloud services to the edge of the network. In this article, Fog and Cloud computing based integrated solutions for smart traffic monitoring are studied, to overcome the downsides of real-time analytics.

Keywords IoT · Fog computing · Cloud computing · Smart traffic · Monitoring

1 Introduction

Internet of things (IoT) has merged the physical and the digital world together by changing the physical world unit into the smart world via the global connectivity to control anything from anywhere around the world. Taking the former, the Internet is

V. Choudhary (✉) · Y. Singh · P. Anand
Department of Computer Science and Information Technology, Central University of Jammu, Samba, J&K 181143, India

Y. Singh
e-mail: yashwant.csit@cujammu.ac.in

P. K. Singh et al. (eds.), *Emerging Technologies for Computing, Communication and Smart Cities*, Lecture Notes in Electrical Engineering 875,
https://doi.org/10.1007/978-981-19-0284-0_24

a global network of networks made up of millions of personal, institutional, educational, commercial, and government networks ranging in size from local to worldwide, all connected via a wide range of electrical, wireless, and optical communication technology. With the latter, the things which may be considered any material or person in the physical universe. Thus, we can define IoT as an interconnected worldwide network that brings interactions between people, things, and things-to-things, which may be anything else in the world, by assigning them a distinct identity or name [1].

IoT is more than just linking devices to the Internet. Data analytics methods are used to process data generated by IoT devices in order to extract useful information about the system and its users, the operations, and surroundings. Cloud servers store and analyze the vast majority of IoT data. Cloud computing systems are highly scalable and may be adjusted on-demand in a "pay-per-use" approach, lowering the expenditure required to construct the necessary analytics application. Existing data analytics techniques can handle massive amounts of data stored in centralized cloud storage. However, for a variety of reasons, cloud data processing fails to match the current requirements of smart systems [2]. Generally, the cloud servers being located at a larger physical distance causes a delay, proving to be unreliable for real-time applications. Moreover, these billions of smart nodes must be generating data, overburdening the data center, and saturating the network bandwidth. To exclusively meet the IoT dynamics, fog computing must assist the cloud to provide better response time and higher transfer rate. Mainly the fog nodes are just like in between placed intermediators sharing the load of cloud nodes for efficient systems.

In this article, several cloud and fog computing based integrated approaches are studied in context to Smart Traffic Monitoring system. The sensor nodes gather multiple sources of traffic information for data analysis, manage traffic processes, and save the proper choice for future information presenting by tracking traffic flow, managing traffic processes, and storing the proper choice for future information presentation. These computing-based systems aid in optimizing the road networking systems, which allowed quick and better communication systems with better response time in terms of data analysis [3]. The rest of this paper is organized as follows. Section 2 summarizes the computing technologies for Smart Traffic Monitoring, followed by Sect. 3 covering the computing technologies-based solutions for smart traffic monitoring and their relative tabular comparison. Section 4 presents the emerging challenges in smart traffic monitoring techniques, followed by conclusion.

2 Computing Technologies-Based Solutions for Smart Traffic Monitoring

The authors have studied several research papers published from 2016 to date on IoT, fog, and cloud computing in the traffic monitoring system for critical analysis. They have been taken from prominent publications and databases such as Springer,

Elsevier, IEEE, ResearchGate, Google Scholar, etc. They are detailed as under this section. Ruimin Ke et al. suggested a technique for identifying and estimating traffic flow metrics (such as velocity, intensity, and volume) using UAV footage that involves four phases. They are Ego-motion irregularity, poor estimate precision in busy traffic, and powerful computational complexity being addressed by the suggested structure. The work showed positive results [4]. Diao et al. proposed a unique hybrid model for accurately forecasting passenger flow with many steps ahead of the view of time, origin-destination spatial, recurrence and self-similarity, and historical probability distributions. The findings revealed that the hybrid model can enhance accuracy on average of 20–50% [5].

In another noticeable work, Alsrehin et al. concentrated on transportation management techniques that relied entirely on machine learning and data mining technology for identifying and forecasting traffic. The steps for establishing intelligent transportation and control systems were given by the authors. The model also specifies the relationships between road intersections, such as how traffic lights near to one another contribute to one another [6]. Tunio et al. presented their work by employing various image processing methods to govern real-time traffic in the MATLAB tool. The purposed model uses LEDs to display the red and green light signals, as well as seven segments to display the signal green's decrementing timer. Furthermore, the device outperforms current conventional systems in terms of precision, performance, and ease of use [7].

In [8], Bui et al. proposed the RSDC, inspired by recent breakthroughs in city sound classification with CNN-based approaches, a novel methodology for congestion classification utilizing road sound datasets has been developed. The implementations, in particular, show promising results, with accuracies ranging from 92 to 95% for classifying traffic congestion over various time slots. Huang et al. [9] described a high-level architecture for vehicular Fog Computing. Also, the security and forensic concerns, as well as mitigating measures were focused. The application examples of vehicular Fog Computing and their potential advantages were also discussed [9]. Sookhak et al. had given a definition for Fog Vehicle Computing (FVC) as to supplement the fog computing's compute and storage capabilities, as well as a complete framework for the FVC and a variety of important technologies. Effective results have been found. Some open problems and potential directions in the sense of FVC were also highlighted for future study [10].

Additionally, Nahri et al. proposed IoV networked infrastructure for real-time traffic data analytics based on three main layers. The proposed framework consisted of real-time data collection and analytics modules. Advantages, disadvantages, results, security issues were also discussed [11]. In [12], Naha et al. addressed Fog and Cloud Computing patterns along with their technological differences. In this paper, Fog Computing Taxonomy was also suggested, with debating on the current research papers and their limitations. The authors have explored open problems and weaknesses considering Fault Tolerance, Resource Scheduling and Allocation, Tool Simulation, and Fog-based Microservices. Mohammad et al. introduced the importance of fog computing in relation to cloud computing and IoT with a model for mitigation traffic congestion to lower the latency. The authors presented a highly dispersed infrastructure and applications that were backed up by the use of location identification and mobility. Specifically, focusing on the drawbacks of cloud computing in context to smart environments [13].

Darwish et al. presented a three-dimensional structure (intelligent computing, real-time big data analytics, and Internet of Things) to provide real-time ITS big data analytics in an IoV context. Furthermore, the advantages and disadvantages of IoV and intelligent computing systems have also been mentioned. In addition, the different edge computing technologies were compared [14]. Qiong and others demonstrated an intelligent traffic light management system based on fog computing. It analyzed and reported the existing traffic conditions at the crossroads and nearby intersections using a Fog Computing platform with promising results [15]. In another work, Sangmek et al. used a microcontroller board to measure and manipulate data in a wireless environment using the architecture design of the Fog-Cloud Computing traffic model. It entailed surveying relevant studies for a Fog-Cloud Computing traffic model research, establishing architecture for USN, describing the Fog-Cloud Computing traffic model design documentation, and evaluating progress [16]. Erskine et al. proposed the Cuckoo search algorithm (CSA). Several hybrid optimization algorithms were used comprising firefly algorithm (FA), firefly neural network, and main distribution establishment (KDE) to authenticate both network and node tier against all attacks for truthfulness in VANET. Furthermore, the results were also evaluated [17].

Ning et al. developed a three-layer VFC model in which vehicles serve as Fog nodes, allowing for real-time traffic management, congestion detection, and other functions. Followed by some research issues and problems related to VFC-enabled traffic management were described and highlighted [18]. In [19], Hussain et al. proposed an Integer Linear Programming (ILP) design for determining the best location, power, and quantity of Fog Computing Nodes (FCN) in order to reduce total network latency and energy usage, with illustrating the operation of the VFNP model and its related constraints. In recent work, Bhardwaj et al. presented a unique smart Fog Computing based taxonomy to control vehicular traffic. Mobile edge computing and mobile Cloud Computing, cloudlets, microdata centres were used in their work. The role of smart Fog Computing and important features of its architecture were also discussed with significant results and comparative analysis [20]. Following this work, Vergis et al. used fog computing technologies for low-cost vehicular traffic monitoring and proposed a three-tier architecture-based system compromising of mobile tracking system (MTS), Information gathering system (GPS), and Fog devices (FD) [21]. Thakur et al. discovered vehicle congestion detection strategies relied on connected vehicles technology [22]. Tang et al. developed a VFC framework that was studied from a variety of perspectives, including latency optimization, utility maximization, and so on. The research shows positive results [23].

In Table 1 few basic traffic monitoring techniques are compared. Some of them address the use of Machine Learning (ML) and other algorithms like Data Mining (DM), and Artificial Intelligent (AI). It is found that unmanned aerial vehicles (UAVs) show high accuracy in terms of performance. Additionally, using these algorithms shows positive results in traffic monitoring. Table 2 presents the relative contrast of

Table 1 A relative comparison of the basic traffic monitoring techniques

Authors	Algorithms used	Software/Tools/Model used	Year	Result
Ke et al. [4]	Optical flow, *K*-means clustering and connected graph algorithms	Unmanned aerial vehicles (UAVs)	2019	96 and 87% accuracy in estimating average speed and vehicle count
Diao et al. [5]	Predictive analytics algorithm	Novel hybrid DTMGP model, Gaussian process	2019	20–50% accuracy improvement
Alsrehin et al. [6]	ML, data mining, and AI algorithms	Traffic intelligence tool, OpenTraffic, MOTUS, Waze, CARTO-Waze connector, SUMO	2019	–
Tunio et al. [7]	Image processing algorithm	MATLAB, OTSU's method	2020	Accurate, time-saving, and ease of use
Bui et al. [8]	ML algorithms (KNN, SVM, RF, XGboots)	CNN architecture using SoftMax	2020	92–95% accurate

Table 2 A comparison of the Fog and Cloud computing based smart traffic monitoring techniques

Authors	Algorithms used	Software/Tools/Model used	Year	Result
Huang et al. [9]	Traffic data fusion algorithm, Intelligent traffic light control, and Dynamic routing algorithm	Fog-assisted traffic control system	2017	–
Darwish and Bakar [14]	Image processing algorithms, Vehicle detection, and length estimation algorithm	OpenStreetMap, MongoDB, NodeMCU, Thinger. Io, Magnetic Sensors, LCD Unit	2018	Energy efficient
Qiong et al. [15]	Intelligent control algorithm, Optimization algorithm	JAVA compiler tool My Eclipse	2018	Reduced traffic congestion
Erskine and Elleithy [17]	Cuckoo search algorithm (CSA), firefly algorithm (FA)	MATLAB	2019	Convenient for reducing delay
Hussain et al. [19]	Optimization algorithm	GAMS CPLEX solver	2019	Limit energy usage
Vergis et al. [21]	Data reduction algorithms	GPS technologies	2020	Provide accurate estimation
Tang et al. [23]	Greedy offloading decision algorithm,	Heuristics based scheduling strategies	2020	Beneficial results

the different works in the literature. The readers might quickly recognize already accessible contributions in IoT-based Fog and Cloud Computing smart traffic monitoring systems in this table. They primarily focused their efforts on the Internet of Things (IoT) enabled fog and cloud computing architectures for smart traffic system, algorithms used in the process, and the latest technologies and tools used. From the studied literature it is noticed that there are few works, which precisely emphasize Image processing algorithms with higher accuracy. Furthermore, these works in the literature provide insight into IoT, Fog Computing, and Cloud Computing and the latest techniques augmented for more satisfactory results.

3 Computing Technologies for Smart Traffic Monitoring

Smart Traffic Monitoring is all about controlling the traffic in a smart way with the use of IoT and AI. A smart traffic management system includes smart bridges, smart highways, smart street lights, sophisticated 3D vision cameras, and automated traffic signaling systems. In most countries, in cities and urban regions, transportation congestion is now becoming a huge problem. Traffic congestion puts a strain on municipal infrastructure and has a detrimental influence on people's socioeconomic life as a result of time spent stuck in traffic. The most serious issue with traffic congestion is delaying of HPVs, in which the driver will have difficulty reaching his or her destination in the shortest possible time, particularly in urban areas. Priority Vehicles (HPV) have been significantly impacted, resulting in vehicles in massive number to reach at their destinations, in the circumstances in which human lives are at stake. Since the HPV has no special priority, a set of traffic lights with a predetermined schedule creates more difficulties in showing up on time [2].

A centralized traffic monitoring and management system is needed to solve this issue which involves infrastructure that allows for the recording and updation of congestion data [2]. Smart traffic monitoring prioritizes moderate traffic conditions by analyzing real-time traffic situations, providing congestion-free traffic, and enhancing conventional ticketing with an electronic E-bill payment system with speed sensors to notify passengers about speeding. Additionally, provides safe and punctual public transportation, reduces emissions, and aids in moving in the right traffic directions via GPS and GIS at intersections and narrow road ends. Figure 1 depicts the smart traffic monitoring system.

3.1 Cloud Computing for IoT

Cloud Computing implies, in the simplest words, storing and accessing data and programmes over the internet from a remote location or device instead of the hard drive of our computer. This so-called remote location has many features that are

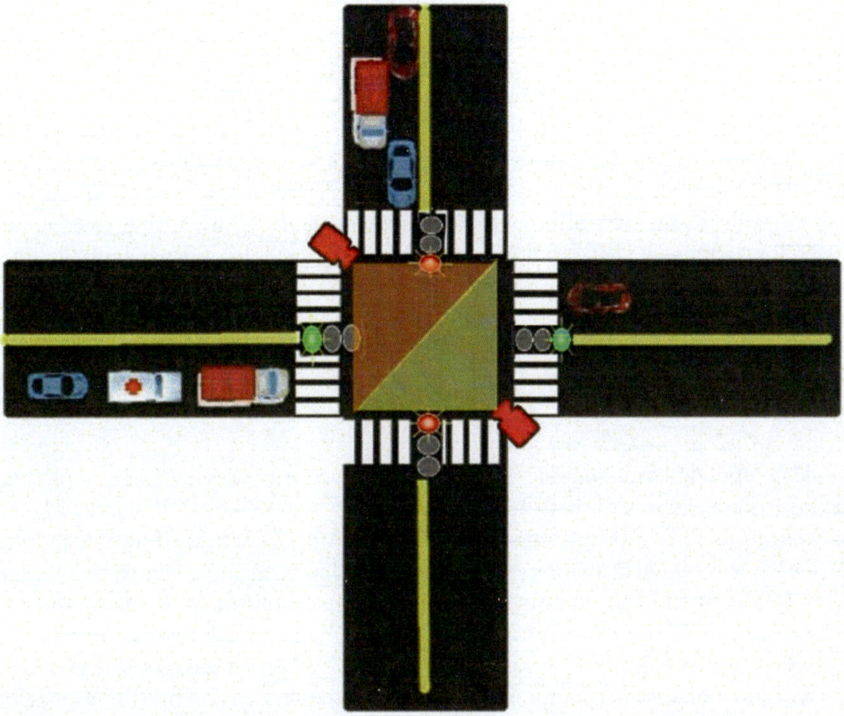

Fig. 1 Smart traffic monitoring system

substantially different from a traditional remote computer, such as scalability, elasticity, etc. The Cloud is just an Internet metaphor. That is called Local Storage and Computing when we store data on or run a programme from the local computer's hard drive. We need to access our data or programmes over the Internet for it to be called Cloud Computing. The end result is the same, but Cloud Computing can be done anywhere, and via any computer with an online connection [24]. Facebook, Instagram, Twitter, Dropbox, Google Images, Google Drive, iCloud, Gmail, etc., are the simple examples of Cloud Computing used by common people in everyday life. It provides scalability, versatility, agility, and simplicity, which is why its use in companies is growing rapidly [25].

Vehicular Cloud Computing is designed to aid vehicles with traffic protection as well as performance. There is tremendous potential for vehicular Cloud Computing to shift the structure of contemporary vehicular communication. Explicitly, vehicles underused services may be shared with other vehicles in order to control traffic during congestion. Vehicle Clouds are divided into three large groups, that is, the V2V Cloud, the V2I Cloud which is vehicle to infrastructure, and the combination of vehicle Clouds with other commercial clouds, i.e., integrated Clouds. V2V clouds are the one in which on the road or in parking lots, vehicles form a cloud to share data for common services. They are further divided into dynamic vehicular clouds and static clouds. Dynamic clouds include moving cars and vehicular sensor clouds. A cloud of automobiles parked in a parking lot is an example of a static vehicular cloud. These clouds are often utilized for processing and storage. V2I Clouds can be used for participatory sensing and cooperative sensing. Integrated clouds are formed when certain clouds, including such mobile computing and internet clouds, are linked to vehicular clouds [26]. Via Cloud Computing, network service providers can transfer tens of millions or even billions of messages instantly. Thus, Cloud Computing technology will be the IoT promoter with so many features like cost-saving, flexible, and secure [25].

3.2 Fog Computing for IoT

Cloud computing has numerous intrinsic abilities and is now extensively employed, having a huge impact on both businesses and people. However, it also has some drawbacks, including its restriction of communication between the Cloud and the end devices. A new digital technology, Fog Computing, has been implemented to solve the problems facing Cloud Computing. In Fog Computing applications, the data collected is not sent to the Cloud server but to nearby processing devices thereby improving the service quality and also reducing the latency [13]. Fog Computing is characterized as a paradigm of distributed computing that expands the services offered to the edge of the network by the Cloud [27, 28]. Cisco suggests the idea of Fog Computing where fog nodes are composed to form a Fog network. Any computer having processing, storage, and network access can be these nodes. The Fog nodes, with main road or maybe an oil plant, can be deployed anywhere, anywhere on the lane. Within milliseconds, they function on IoT data. These nodes transport processing information to the cloud allowing historical study and relatively long-time preservation. This method is mostly conducted for productivity purposes (latency reduction and network bandwidth conservation), as well as for safety and enforcement causes [29]. Figure 2 depicts IoT-based integrated Fog and Cloud Computing structure in smart traffic monitoring.

4 Emerging Challenges

- *Downtime*: Downtime is often cited as one of the Cloud server's main drawbacks. The traffic monitoring system is unable to cope with the effects of an outage or slowdown. In Cloud Computing, every component is available online, exposing potential vulnerabilities. Even the most powerful teams are subjected to tremendous breaches and privacy violations from period to period. This whole system is expensive too as smart vehicles are costly [9].
- *Technological faults*: Geofencing, like any technology, has its faults. Most geofencing systems have found that it has worked most, if not all, of the time.

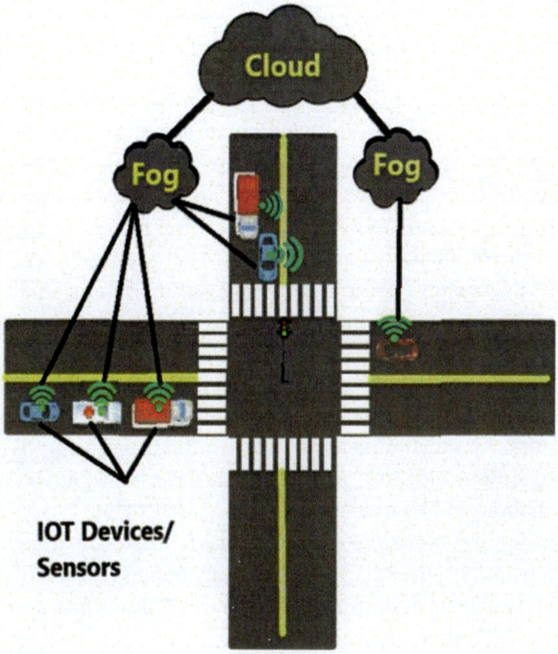

Fig. 2 Fog and cloud computing structure in smart traffic monitoring

Sometimes geofence technology has consistent problems. Geofence can also cause privacy concerns. It can drain cell phone data and battery life too [13].

- *Cost*: In order to become more sensitive to the traffic entering the intersection, there could be additional costs associated with improving traffic light control. However, the Gaussian process extensively used is computationally expensive [20].
- *Network*: In addition to mobility and pollution metrics, one of the challenges of smart traffic monitoring is to ensure that network metrics perform well. Standard vehicular communications produce negative effects on the output of all metrics in the congestion situation. The highest rate is anchored by these conventional vehicle communications, while emissions are impacted by vehicle rises [30].
- *Interoperability*: It refers to the capacity of two devices to communicate with one another, and it is a major issue in traffic management systems. It takes time to read information during Vehicle to Vehicle (V2V) contact because a link can be lost, resulting in data loss [31].
- *Security*: In VANETs and IoV, there are many kinds of threats that can pose a major issue to vehicular networks. The use of sensor nodes have some drawbacks like sensor node devices' fairly simple operation makes them easy to deploy, and it also exposes them to fraudulent security assaults, as there are often no comprehensive security systems for these sensors. UAVs are also vulnerable to hacking [11, 32].

5 Conclusion

The Internet of Things (IoT) has now become a popular topic of discussion and marketing strategy in today's world with its exponential rise and growing sustainability issues. By witnessing its applications everywhere, we have explored its usability in traffic systems. One of the biggest challenges of the twenty-first century is traffic jam and therefore there is an urgent need to combat the traffic congestion issue with automated smart solutions. The intelligent traffic control system isn't limited to decreasing traffic congestion, air pollution, and regulating road traffic; it also needs continuous monitoring and preserving the public's safety and welfare. This paper covered the introduction to Smart Traffic Monitoring, brief idea of Fog and cloud computing in context to their usage in this environment. Then the basic smart traffic monitoring techniques based on fog and cloud computing were studied. Some important techniques for smart traffic monitoring like VANET, GPRS, UAVs, Geo fencing, smart fog computing, etc., were also analyzed. Additionally, the emerging challenges like privacy, downtime, attacks, etc., in computing technologies for smart traffic monitoring were highlighted. Via a few improvements and technical implementations, the smart traffic management system will provide intelligent information and analysis to current problems. As a prominent tool in every field, IoT is growing and an intelligent traffic management system will be the future traffic management scheme for the smart city.

References

1. Madakam S, Ramaswamy R, Tripathi S (2015) Internet of things (IoT): a literature review. J Comput Commun 3(5):164–173
2. Lah AAA, Latiff LA, Dziyauddin RA, Kaidi HM, Ahmad N (2017) Smart traffic monitoring and control architecture and design. In: 2017 IEEE 15th Student conference on research and development (SCOReD)
3. Rabby MdKM, Islam MM, Imon SM (2019) A review of IoT application in a smart traffic management system. In: Proceeding of the 2019 5th international conference on advance in electrical engineering (ICAEE), Sept 2019
4. Ke R, Li Z, Tang J, Pan Z, Wang Y (2019) Real-time traffic flow parameter estimation from UAV video based on ensemble classifier and optical flow. IEEE Trans Intell Transp Syst 20(1):54–64
5. Diao Z, Zhang D, Wang X (2019) A hybrid model for short-term traffic volume prediction in massive transportation systems. IEEE Trans Intell Transp Syst 20(3):935–946
6. Alsrehin NO, Klaib AF, Magableh AWS (2019) Intelligent transportation and control systems using data mining and machine learning techniques. IEEE Access 07:49830–49857
7. Tunio MH, Memon I, Mallah GA, Shaikh NA, Shaikh RA, Magsi Y (2020) Automation of traffic control system using image morphological operations. In: IEEE Xplore, 2020 International conference on information science and communication technology
8. Bui KNAM, Oh H, Yi H (2020) Traffic density classification using sound datasets: an empirical study on traffic flow at asymmetric roads. IEEE Access 08:125671–125679
9. Huang C, Lu R, Choo K-KR (2017) Vehicular fog computing: architecture, use case, and security and forensic challenges. IEEE Commun Mag 105–111

10. Sookhak M, Yu FR, He Y, Talebian H, Sohrabi Safa N, Zhao N, Khan MK, Kumar N, Fog vehicular computing: augmentation of fog computing using vehicular cloud computing. IEEE Veh Technol Mag 12(3):55–64, 7994678
11. Nahri M, Boulmakoula A, Karima L, Bath AL (2018) IoV distributed architecture for real-time traffic data analytics. Procedia Comput Sci 130:480–487
12. Naha R, Garg S, Georgakopoulos D, Jayaraman P, Gao L (2018) Fog computing: survey of trends, architectures, requirements, and research directions. IEEE Early Access 06:47980–48009
13. Mohammed TS, Khan OF, Ibrahim AS, Mamlook R (2019) Fog computing-based model for mitigation of traffic congestion. Int J Simul Syst Sci Technol. ISSN: 1473-804x online, 1473-8031 print
14. Darwish TSJ, Bakar KA (2018) Fog based intelligent transportation big data analytics in the IoV environment. IEEE Trans Cybern 06:2570–2580
15. Qiong W, Fanfan H, Xiumei F (2018) The intelligent control system of traffic light based on fog computing. Chin J Electron 27(6)
16. Sangmek C, Boonnam N (2019) Fog–cloud computing traffic model and performance evaluation for ubiquitous sensor network infrastructure, Oct 2019. IEEE. https://doi.org/10.1109/ICSEC47112.2019.8974817
17. Erskine SK, Elleithy KM (2019) Secure intelligent vehicular network using fog computing. MDPI Electronics 8:455. https://doi.org/10.3390/electronics8040455
18. Ning Z, Huang J, Wang X (2019) Vehicular fog computing: enabling real-time traffic management for smart cities. IEEE Wirel Commun 26(1):87–93
19. Hussain M, Alam M, Bega M (2020) Vehicular fog computing-planning and design. Procedia Comput Sci 167:2570–2580
20. Bhardwaj A, Goundar S (2019) IoT enabled smart fog computing for vehicular traffic control. EAI Endorsed Trans Internet Things 5(17):162221
21. Vergis S, Komianos V, Tsoumanis G, Tsipis A, Oikonomou K (2020) A low-cost vehicular traffic monitoring system using fog computing. Smart Cities 3(1):138–156. https://doi.org/10.3390/smartcities30100008
22. Thakur A, Malekian R (2019) Fog computing for detecting vehicular congestion, an internet of vehicles-based approach: a review. IEEE Intell Transp Syst Mag 11(2):8–16
23. Tang C, Wei X, Zhu C, Wang Y, Jia W (2020) Mobile vehicles as Fog nodes for latency optimization in smart cities. IEEE Trans Veh Technol 69(9):9364–9375
24. Chandrasekaran K, Gujarati D, Porter D (2010) Essentials of cloud computing, p 577. ISBN: 978-0-07-337584-7
25. Srivastava P, Khan R (2018) A review paper on cloud computing. Int J Adv Res Comput Sci Softw Eng 8(6)
26. Ahmad I, Noor RMd, Ali I, Imran Md, Vasilakos A (2017) Characterizing the role of vehicular cloud computing in road traffic management. IJDSN 13(5)
27. Dastjerdi AV, Gupta H, Calheiros RN, Ghosh SK, Buyya R (2016) Fog computing: principle's architectures and applications. arXiv:1601.02752
28. Yi S, Li C, Li Q (2015) A survey of fog computing: concepts, applications and issues. In: Proceedings of the 2015 workshop on mobile big data. ACM, pp 37–42
29. Dhingra S, Madda RB, Patan R, Jiao P, Barri K, Alavi AH (2020) Internet of things-based fog and cloud computing technology for smart traffic monitoring. ScienceDirect JID: IOT 09(11)
30. Butun I, Österberg P, Song H (2020) Security of the internet of things: vulnerabilities, attacks and counter measures. IEEE Commun Surv Tutorials 22(1):616–644
31. Mehmood Y, Ahmad F, Yaqoob I, Adnane A, Imran M, Guizani S (2017) Internet-of-things-based smart cities: recent advances and challenges. IEEE Commun Mag 55(9):16–24
32. Hussein ARH (2019) Internet of things (IOT): research challenges and future applications. Int J Adv Comput Sci Appl (IJACSA) 10(6):77–82

A Review of Precision Agriculture Methodologies, Challenges, and Applications

Mohammedshakil Malek, Bachwani Dhiraj, Deep Upadhyaya, and Dixit Patel

Abstract Internet of things (IoT) furnished recent cultivated procedures can alter the prevailing agriculture mechanisms by computerizing the complete technique of farming since crop-management, aquatic irrigation to creating superior judgments in accordance with real-time monitoring of conservational circumstances, situations of soil and surroundings of landscape during the prevailing periods, technology spurred preciseness agriculture explanations have spurred a framework towards stagnant and instruction physical agriculture procedures to automatic preciseness positioning agricultural procedures by the current innovations like agricultural IoT and AI, WSN assisted harvest checking along with its control and cloud computing. During the projected assessment, a demanding and elaborated calculation has been performed to recognize the scope of the study and study the modern knowledge sparked applications and policies. Furthermore, in this paper, five strands IoAT-based PA model is presented. The first strands represent the physical level maneuvers, second strands define safety rules, third strands explain effective data-management performs, fourth strands deliver operative irrigation prototypes, and the last strands deliberate technology-aided facilities for aquatic management. Lastly, alongside upcoming guidelines, definite investigation of the performed literature review in the visual form accompanied by acquired reviews and academic training were delineated.

Keywords Internet of agricultural things (IoAT) · Precision agriculture (PA) · Crops irrigation

M. Malek (✉)
F. D. (Mubin) Institute of Engineering and Technology, Gandhinagar, Gujarat, India

B. Dhiraj
Civil Engineering Department, Parul University, Vadodara, Gujarat, India

D. Upadhyaya
Civil Engineering Department, GEC, Godhara, Gujarat, India
e-mail: deep.upadhyaya@gtu.edu.in

D. Patel
Civil Engineering Department, GEC, Surat, Gujarat, India

P. K. Singh et al. (eds.), *Emerging Technologies for Computing, Communication and Smart Cities*, Lecture Notes in Electrical Engineering 875,
https://doi.org/10.1007/978-981-19-0284-0_25

1 Introduction

By 2050, the number of inhabitants on Earth is assessed to ascend to a huge amount of 9100 million occupants. The existing population of 7300 million has plundered high right around two-thirds of Earth's assets [1]. Over the most recent 60 years, because of popularity for wares, like fiber, petroleum, food, wood, and new water, extra-terrestrial has been allocated to horticulture as compared to eighties and nineties time period. The realities disclose that there is presently an unmistakable short-fall in regular assets and if the continuous consumption of assets isn't managed in the correct manner, the people in the future won't have anything to encourage [2]. Rainwater system is an essential worry for the improvement of agribusiness in agricultural nations. To determine the water system problem, individual specialists have put forth a few attempts by proposing imaginative strategies, for example, drip water system, mechanized sprinkler innovation, and smart. "Smart Farming technique (otherwise called accuracy agribusiness (PA) utilize most recent innovations", for example, IoAT, cyber physical system, big data analytics, intellectual sensing and frameworks, and AI procedures to resolve the concern towards asset management using demonstrating effective administration of accessible assets.

Besides, most recent advanced methodologies have additionally contributed colossally in taking care of the issue of water deficiency and in the improvement of irrigation system approaches, for example, improvement of a computerized framework for agriculture, which advises farmer of specific vulnerabilities ahead of time and supports, they control such bothersome circumstances [3, 4]. A central segment of the PA framework is field planning. The idea of accuracy farming isn't old; it was induced before 25 years when inventive innovations, for example, remote sensing networks and GPS were introduced [5].

PA is a creative strategy for gathering, scrutinizing, and highlighting information, which assumes a vital part in settling on crop the executives' choices [6]. Present-day agriculture is as of now confronting another period with expanding creation needs and lessening accessibility. The remarkable expansion in population carries with it the need for a more prominent measure of creation; the capacity to control assets precisely permits more noteworthy control of costs/benefits and assists with checking the exhaustion of characteristic assets, for example, streams and soil supplements [7]. PA permits observing of environment entities and to have an exact reaction on visualization of plantation behavior to tackle explicit issues for each situation [8]. As demonstrated in Fig. 1, four primary parts of accurate agriculture cycle which involve every one of the various procedures utilized are Sense-Analyze-Predict-Act. Figure 1 represents the transformation of a cloud stage administration, is incorporated by sensors, which thus forecasts the outcome using the assistance of various models and ultimately advises the farmer which assists them with making vital moves and further their innovation advanced farming. As portrayed in Table 1, four parts of the accurate agribusiness cycle, characterize how the cycles are completed and beneath we see what each parameter accomplishes.

Fig. 1 Precision agriculture life cycle

Table 1 Statistical display of articles according to their research fields

Major category	Sub-category	Sub-sub category	No. of articles
Technology	Hardware	Type of sensors	31
		NFC	7
		Sensor networks	28
	Software	Middle-ware	24
		Database	23
	Architecture	Hardware/Network	12
		Software	9
		Process	10
		System	12
		General	12
Application	Smart infrastructure		46
Challenges	Security challenges		8
	Privacy challenges		6
	Legal challenges		6
	General challenges		18
Business models			7
Future directions			30
Overview/Survey			14

1.1 Precision Agriculture Policies and Performs

1.1.1 **Irrigation System**

Taking everything into account, innovative water system frameworks are vital in the whole harvest managing process, which incorporates planning of seedbeds, usage of supplements, pesticides and insect sprays, growth, harvest and development, and re-development of ranch. A profoundly exact and flexible irrigation system framework is highlighting a significant part of the PA procedure.

1.1.2 Quality Land Management System

It is the utilization of tasks, performs, and medicines to ensure soil and improve its exhibition. It is significant, straightforwardly and by implication both, to efficiency and natural manageability of crops [9] and the board assists us with controlling significant agricultural parameters, i.e., vegetal well-being, its shelter, and soil dampness that assists with giving a greater depiction to mix of advancements and farming related sensors to reduce expenses and increment harvests, for example, sensors for pH [10]. This sensor is able to gauge the focus on intensities of oxygen, particle and hydrogen that will help ranchers during anticipating land squabble circumstances [11].

1.1.3 Weather Anticipating System

The exact ground-level climate data support for surveying the usefulness of the rancher's grounds and becoming further proficient in pest and infectious prevention, by staying away from obliteration of harvests and expanding the functionality and productivity in their activities [12, 13]. Expanding necessity towards IoT to interrelate every one of the parts of both the virtual and real sphere has brought about organizations and buyers to vigorously depend on IoT-related advances to create headways in their individual arenas.

1.1.4 Data Insight

Expansion in proficiency and usefulness is the fundamental target of IoT-initiated advances. Gathering adequate measures of information to utilize it as an instrument to increase buyer involvement is the Holy Grail that most organizations depend on. The assortment of sensor investigation to build the information extension guarantees numerous advantages in cultivating, chiefly decrease in delays because of outside correspondence, which at last aides ranches arranged in country regions with helpless availability [14, 15].

1.1.5 Sustainable Computing

With the developing requirement aimed at sustainability of assets, the objective of bigger rated associations is to make the entire world a superior spot. In the wake of investigating numerous utilizations of IoT advances, the Forum of World Economic tracked down that 84% were tending to, or might actually discourse, the Sustainable Development Goals of UN. Sustainable processing assists with vanquishing a portion of the world's developing obstacles, furthermore, handles those issues by giving arrangements, like Smart Grid. A savvy network comprises of all the most recent creating/created IoT manageability innovations. It identifies and responds to nearby alterations in use using an electric inventory system which utilizes computerized correspondences innovation.

1.1.6 Data Security

Data Security has a prime job in such turns of events. The extent of information and abilities to incites, overall aides cultivate the fundamental correspondence between each one of those worried about food creation and the climate based on study by the ranchers and ground laborers in the overall logic, wherever [16]. Playing out a portion of the information investigation straightforwardly inside the ranch additionally has number of advantages from the data security perspective. Just a synopsis of the homestead checking information will be shared by outsider. This will complicate getting explicit touchy data from the information shared.

1.1.7 **Data Analytics**

Data scrutinization helps with creating significant, preemptive, and beneficial choices which can expand openings and effectiveness in the whole cycle of smart cultivating. In spite information examination require relevant statistics to be effective, and information which is inadequate or inaccurate will give experiences that are not completely dissected. Data from sensors fixed at filed, assortment of information at every stage, and financial elements of choices will keep on being basic for achievement of information examination.

1.1.8 **Cyber Physical Systems**

Digital actual frameworks will shape an incorporated climate of digital and actual items which helps ranchers progressively observing of homesteads, soil management and crop-management by demonstrating assortment of data.

2 Legacy Systems

2.1 Crop Monitoring System Using IoAT

The 3D organizes hub associated with the yield distinguishing proof component addresses different IoAT frameworks. This framework manages the claim normal for IoT and highlights the prospective it needs to increase the accuracy agribusiness strategies. The mishap the anticipated model has the consideration of water, environment and soil sensor in a similar class as yield recognizable proof, supplement regulator and energy regulator.

2.2 Camera-Chessboard Array Preparation

The principle obstacle in such sort of plan is assisting to manage the overlaying of IoT data using each single item on the perspectives on the camera. This design course of action goes through a specific strategy to see if some forecast blunders occur in the arrangement of cameras. The product, utilized to adjust such cameras, is a prior MATLAB tool compartment program that supports in discovering the camera forecast blunders. Subsequent to handling its programs and making virtual square outlines of their showcase and associations, a realistic of the vegetal using camera perception is made, that assists the ranchers with for all intents and purposes looking to their crops daily.

2.3 *Rural Architectural Network*

It is a WSN empowered organization intended to expand crop-management doings to consider justified harvesting choices [17, 18]. The proposed architecture gives versatility and covers huge topographical areas. It utilizes *6LoWPAN* centered WSN grid for detecting and inciting associated tasks.

Progressing upon past surveys for a multi-camera course of action on a smaller ranch, in this research the measure increment and notice various sensors hubs dissipated through an enormous farmland region. These hubs play out the detecting and impelling tasks all the while, expanding the adaptability and the inclusion range. The utilization of IPv6 for this arrangement has been fundamental. In addition to the fact that it solves the location space and auto-setup entanglement, but on the other hand is joined by the amendment level to present the *6LoWPAN* convention heap. Little parcel dimension, little data transfer capacity (amongst 250/40/20 kbps), and other lattice geography downsides are just the beginning of the detriments of the *6LoWPAN* organization. Despite the fact that it has its own merits, for example, energy redeemable overutilization of little fueled attending hubs, the key drawback of *6LoWPAN* is the absence of use that uses *6LoWPAN* due to the broad preparation which it needs to acquire acclimated with the innovation and the information on *IPv6* convention. Information on heap and the usefulness of *IPv6* is an absolute necessity aimed at the finishing clients dealing with such design. Directing arrangements, execution investigation, the structural intricacy of the proposed CLAD convention, impacts of climate over the *6LoWPAN* related organization and the confirmation of nature of administration are for the most part completely gave in the led study to really carry out haze and distributed computing innovations in their anticipated system. In the new occasions, the utilization of applications related to IoAT has expanded dramatically because of its low-energy prerequisites and inclusion huge topographical locales. Somewhat recently, LoRaPAN and LoRaWAN advances have supplanted customary correspondences conventions, for example, Wi-Fi and Classic Bluetooth [19–22].

For the most part talking, in the haze processing models, IoAT based gadgets are associated with heterogeneous gadgets in closeness to IoAT clients. The significant difficulties in current applications of IoAT in mist registering climate are the executives of accessible assets and booking of routine assignments [18, 23].

2.4 Irrigation System Through Aerial Sight and Central Pivot

This system, despite the fact that is the nearest thing we need to amplify IoT execution, ordinarily the last uses of these advances are not easy to understand. The comprehension of the huge potential outcomes of these innovations is a vital factor in finding and applying them. Moderation of the information hole is vital to utilize the tested innovation proficiently and suitably. With this, security issues, information misfortune, and control above lengthier diffusions are additionally extraordinary concerns that must be managed. This sort of system is exceptional in the constant dissemination of the amount of water in the whole farmland region [24, 25]. Each proposed framework enjoys its benefits and burdens, a couple of significant benefits and drawbacks that accompany the execution of this framework are for watering bigger geological fields. The proposed design requests just 60% of water when contrasted with the regular water system frameworks with minimal expense and higher productivity.

It is basic that customary overhauling and upkeep ought to be given to such frameworks to stay away from the chance of glitches, which requires colossal starting capital venture [24, 26]. Once more, if the extraordinary issues of this proposed model are relieved and vital advances, for example, consideration of sponsorships to ranchers are taken to spur them to roll out progressive improvements to their homestead and, that is the point at which the agriculture area will begin to develop the affirmative manner. Further, we take a gander at keen smart water management, which consolidates different advancements, with the objective to add the effort of IoT innovations and reason any continuous issues with the current designs. Brilliant water the board framework can give constant notice of the water system and yield conditions to ranchers, which may help ranchers in taking in the nick of time choices.

2.5 Anticipated Methodology

This part examined an IoAT framework that has been anticipated in this study to give an outline of an IoT initiated PA framework, which stages using an order of levels, adjustable administrations, and administrations that can be reproduced for normal IoT frameworks alike. As of small sensor establishment over the farm to definite overall benefit of end clients, such design gives an outline or a plan that can be used for comparable IoT frameworks.

I. **Stage I: Physical Stage**

   An IoT actual stage gadget, for example, actuators, sensors, and regulators communicate and robotizes the interaction of treatment, aquatic system and the usage of insecticides and insect poisons for superior harvest development the executives [27]. The IoAT framework consolidates a few customary advances, for example, NFC, distributed computing, RFID, WSNs to computerize the harvest the executives measures [28]. WMSN and WSN models transfer data using remote networks comprising of numerous interconnected hubs that are utilized to detect, measure, and speak through IoT incorporated innovations, which assumes a vital part in any IoT incited tests [29].

These designs are extremely exact as they follow a particular diagram of the relative multitude of required segments, which must be incorporated for effective signal transmission [30]. WMSN or WSN enjoy their benefits and impediments and though these segments are the innovators of node processing, expansion in proficiency, usability and consistency is consistently a worry [31]. In addition, to comprehend the administration of anticipated IoT Data prototypes, it is fundamental to dive further into rules independent entity-management, which happens at a reasonable stage. WQM is a significant obstacle in WSN related systems. Such frameworks request dependability, manageability, and ideal nature of administration to convey WQM information over longer distances [32–34].

When all is said in done, IoAT empowered WSN frameworks comprises of various correspondence frameworks that are needed for directing, power source reinforcement, actuators, and sensors for communicating data associated to crop, and easy to understand GUI-related plans [35]. The preparing chunk comprises of a miniature regulator, which benefits, in legitimate and number-crunching activities, whose outcomes are put away in the memory. In addition, the result of the preparing chunk is utilized with the correspondence chunk through an outsider application so the rancher can settle on better choices relying on their space of attention. The sensor chunk encloses various kinds of sensors, for example, pH sensors, moistness sensors, hotness sensors; which guarantees that the information collected is solid and can be examined deliberately utilizing numerical calculations [36]. As a rule, the regular IoAT framework incorporates three significant advancements: (a) actuators and sensors (b) mist registering and edge figuring (c) distributed calculating stockpiling. These advancements are able to give applicable data to GUI related gadgets and it additionally intersects the diverse IoAT engineering [37].

II. **Stage 2: Network Stage**

This stage manages information assortment, the board, information protection and security, and IoAT device management [38, 39]. Information obtaining, security, and the executives: Protocol and programming segments for information procurement are the vital attributes of this stage furthermore to safety and device-management purposes. This stage uses communication agents like Mosquitoes. Mosquitoes are proficient to execute IoT based conventions like *MQTT*. *MQTT* is a lightweight community IoT convention, which helps IoAT frameworks in intersecting the regulators, center products, switches, doors, workers, and detecting gadgets. MQTT is a generally utilized convention for asset and energy requirement IoAT applications [40, 41]. In the existing times, technologies, for example, LoRaWAN has empowered long-range remote correspondences around the scope of 25–30 km. It needs low force and lesser transmission capacity for composite remote correspondences. Irrespective, it is a troublesome errand for *LoRaWAN* to pledge present multimedia coursing above inferior transmission capacity for uses, like climate observing [42, 43]. LoRaWAN utilizes IPV6 convention to speak with IoAT based frameworks. Moreover, somewhat recently, with the appearance of various correspondence conventions IFTTT, Z-Wave, NFC, Thread, RFID, SigFox, ZigBee, and interlinking of digital and actual articles have gotten simpler. Thread and Z-Wave are brilliant correspondence convention which is being utilized in shrewd houses that needs to interlink house machines, for example, ACs, fans, freeze, lights, dishwasher, etc. [44, 45]. NFC is a label-based innovation which is utilized for distinguishing different gadgets and short-range interchanges. NFC is able to interrelate and regulate detecting gadgets straightforwardly through advanced mobile phones [46]. ZigBee is a two-way correspondence convention that is generally utilized for uses, for example, tree directing, vector steering, and low-rate information interchanges [47].

III. **Stage 3: Data Management Stage**

This stage manages information conveyance, handling, and capacity. It uses software management firmware stage-like *FIWARE*. *FIWARE* is a stage for the turn of events and worldwide sending of Internet utilization of the upcoming time. It gives an absolutely free, public, and open architecture just as a bunch of details which permit designers, specialist co-ops, organizations, and different associations to foster items [48, 49]. *FIWARE COSMOS* is the orientation execution of the *FIWARE*'s big data study general enabler. It is expected to convey implies for investigating stream and batch information [50, 51].

FIWARE ORION and *FIWARE CYGNUS* is an apache flume related center product that is liable for overseeing constant information stockpiling, which is generally utilized for overseeing ancient data. It primarily manages information assortment, information handling, and setting mindful data trade [48]. *FIWARE ORION* is a C++ execution of *FIWARE* stage that is competent to deal with the whole lifetime pattern of information the board. It incorporates capacities, for example, overseeing questions, routine updates and memberships. *FIWARE QUANTUM LEAP* is a component that can accumulate information as time-arrangement, for example, ngsi-tsdb [48].

IV. **Stage 4: Irrigation and Water Distribution Methodologies**

Ordinary farming plans are required to utilize robots and soil sensors to get logical data identified with soil conditions. The traditional models likewise use calculation insight methods for quality of water administration and cloud computing investigation for wellbeing or execution of output based uses [19]. These methodologies use various logical techniques and instruments to give constant relevant data to farming specialists [52, 53].

(a) **Edge Computing**: Edge figuring is a productive information securing method, which is generally utilized in mechanization related IoAT applications to give ongoing scientific calculations and notices identified with soil conditions, natural conditions, and landscape-related data [43, 45, 54]. It is a portable detecting system which gives constant checking of distant areas which will help rural specialists' in settling on farmland related choices [55]. Somewhat recently, individual analysts have proposed different approaches to carry out edge and fog registering created IoAT frameworks [56].

(b) **Irrigation Models**: A water system mechanism (drip irrigation model) comprises various water systems (drip irrigation) techniques which are utilized by agricultural experts in the existing eras. The five strategies to the water system methodology are (i) exterior water system (ii) sprinkling water system (iii) drop water system (iv) sub-surface water system and (v) water system using drone. On account of surface water system, the water is consistently appropriated above the dirt apparent because of the impact of gravitational stream. Sprinkling water system looks like a rain style activity where water is splashed around the ranch field. Trickle irrigation is a methodology wherein water is conveyed as drops or little streams. Smart trickle water system frameworks are boon to execute for ranchers and cause exceptionally less initial expense. In such frameworks, because of mechanization, no human mediation is required which likewise brings about moderation of water wastage [57–59]. Ranchers can oversee and control such frameworks utilizing smartphones. On account of subsurface water system method, the water is provided beneath the layer of soil. The water is generally provided inside the zone of plant root. Drone technology helped water utilizing IoAT helped innovations, for example, WSNs [60–62]. Agribusiness explicit drone in keeping an elevated perspective on crop creation and assists screen with trimming development. Improving the general ranch proficiency, the robots likewise give an arrangement where ranchers can screen their yields occasionally, as indicated by their preferring. The pesticides are the principle obstacle drones can intercept [63, 64]. Bugs are identified, and various spaces of farm are gotten in practically no time. The lone concern drone helped water system makes is that of protection. There is no authority required for drones entrance, passing over somebody's possessions using a camera and mic appended to it, is said as simple as possible, and could be a possible protection infringement.

V. **Stage 5: Water Use Facilities**

Easy-to-use GUI-related plans are given to ranchers and merchants to improved agreement and quicker admittance to water use administrations. Next, the pecking order of IoT in farming is trailed, the last year carries the appropriation of such developed administrations, just as how nicely the anticipated framework encounters client prerequisite and satisfies designer and client assumptions [65]. The engineers utilize the updated information assembled through this trial to additionally progress their IoT environment segments, and the clients (ranchers) who are actually occupied by such innovation on an everyday premise, advantage from the use of these administrations and help give significant client experience information towards the executed test. Because of energy imperatives of water quality administration detecting gadgets a solid energy reaping philosophy is needed to deal with the force reinforcement prerequisites. Be that as it may, the productive execution of energy gathering technique is as yet an open issue for individual analysts [32]. A productive steering convention is needed to oversee smooth directing of data between detecting gadgets, regulators, switches, entryways, and distributed computing based information stockpiling workers [66]. On the new occasions, it has been seen that the land is contracting slowly which presses the regular assets. In such situations, most recent advances can help ranchers in getting context-oriented data of assortment of yields for in the nick of time crop the board choices [66].

Besides, the factual portrayal of all explored information has appeared in the additional areas, which provides understanding of the definite directed overview of most recent IoAT related reviews [67, 68]. In the wake of assessing altogether the handling stages and determining how an IoT framework should work for explicit assignments, investigation of the inborn association among IoAT-related frameworks and information is addressed [69]. In the decision-making process, investigation is the utmost significant advance. All the information created through IoT gadgets is just of worth on the off chance that it is exposed to examination. Unlawful perceptions and significant ends are the things which IoAT frameworks plan to remove while carrying out an information examination (DA) measure. As demonstrated in Table 1, this paper has done the conveyance and characterization of the IoAT articles as for various classifications, which are explains subsequently. It additionally helps with recognizing the difficulties which undermine IoT scattering and open exploration issues which steers innovation the correct way for coming eras [70, 71].

(a) **Technology**: IoT is a network, which is implanted with innovation and can gather and trade information [71]. Like an environment, it connects different frameworks with the shared factor of IoT installed innovations and assists us with broadening IoT advancements farther than our PCs and cell phones.

(b) **Applications**: Uses of this layer in such engineering have a tremendous window for development. IoT gadgets are probably going to be valuable in association amongst individuals and the gadgets nearby them [72]. Based on following everyday exercises of a person, to observing their rest for more proficient and sound living, the utilizations of IoT are boundless and consistently have an extent of additional development [73, 74].

(c) **Challenges**: The difficulties in the field of IoT are largely scattered. To effectively carry out IoT implementations, these difficulties must be managed and insignificant or totally removed [75, 76]. From security, protection, and legitimate difficulties to challenges, which overall influence each of these viewpoints, are to be searched from this engineering.

Problem related to encryption, communication compatibility, identity management, universal liability problems, and data possession are the significant challenges which are to be engaged in.

(d) **Business Models**: Through modifications in IoT advances to expand usefulness and effectiveness in everyday life, plans of action likewise must be adjusted to match IoT assumptions [77]. For ecological development and upcoming improvements in the arena of accuracy farming, rural-centered enterprises are needed to distinguish feasible innovation-related agricultural plans that can give ongoing checking of harvests and give most recent soil conditions. The directed review distinguishes the deficiencies between the traditional horticulture based frameworks and developments received by farming based ventures [78].

(e) **Future Directions**: Despite the fact that anticipating the eventual fate of farming-related architecture is disputable, the idea of this headway is the main thrust for the development of the Web of Things [79, 80]. For coordination of clever items with this present reality, the simplicity of the cycle of advancement, expansion in combination among gadgets and defeating difficulties to build the attainability of thoughts are the great components, which must be encountered through.

(f) **Survey/Overview**: Here, comprehensive definite study of IoAT technologies are discussed and presented in Table 1.

Table 1 depicts statistical data distribution about different categories and subcategories with respect to the surveyed articles and discussions. Figure 2 signifies a graphical presentation of the quantity of papers considered for journal and conference papers. Similarly, Fig. 3 illustrates the dispersal of all the various sources based on which the study and analysis have been incorporated.

This statistics demonstration contributes various insights addicted to the prevailing investigation which headways and its records were used for this study and all

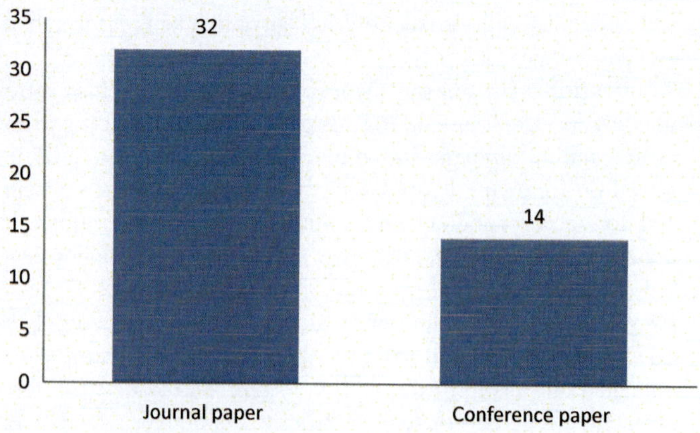

Fig. 2 Type of paper reviewed for the study

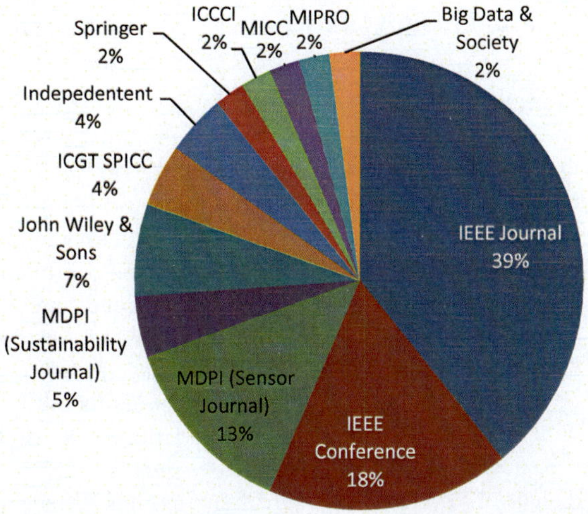

Fig. 3 Pie chart representation of different publications

the synchronized reference replicas based on which this study has been extracted. Figure 4 elaborates a global depiction of the previous studies of the similar research works.

Fig. 4 Global scenario for similar research work

3 Conclusions and Future Enhancement

Imperative agribusiness initiatives affect the advanced existing practices, methods, and applications. The most recent exploration discoveries additionally recommend that suitable utilization of accuracy based practices can make an incredible effect on the efficiency grade and sustainability factors [81–83]. This research paper expects to examine the various innovation-based water system frameworks (irrigation systems), ongoing modern practices and developments, different accuracy agribusiness-based stages alongside the conversation of a recently presented methodology. This paper discussed various contextual investigations that were likewise talked about to evaluate the adequacy of the inheritance frameworks[84–92].

In this domain, maximum research were focused on regulator of several agricultural crops and the remote monitoring, necessities of its resource, like required level of water, pesticides and insecticides quantity, and fertilizers usage to ascertain the indices of vegetation. Notwithstanding, this paper prepared a certified endeavor to dissect the current techniques as well as have proposed a viable model to accomplish better outcomes.

Though, the paper has finished a sincere effort for analyzing the prevailing approaches and also suggested an efficient mechanism to attain superior outputs. In future, a secure precision agriculture framework can be implemented to address IoT-related security challenges.

Conflicts of Interest The authors declare no conflicts of interest.

References

1. Shah B, Malek S (2018) Identification of significant factors affecting the design of smart village. In: Proceedings of the 2nd international conference on women in science and technology: creating sustainable career, Birla Vishvakarma Mahavidyalaya, Vallabh Vidyanagar, India
2. Nandurkar S, Thool V, Thool R (2014) Design and development of precision agriculture system using wireless sensor network. In: International conference on automation, control, energy and systems (ACES), Hooghly
3. Ayanelam YF, Malek S (2018) A critical review of effective project management using 5D BIM. In: Smart construction. 1st International conference on engineering and technology: smart engineering, Parul University, Baroda, India. McGraw Hill Education, pp 155–158
4. Kochai Z, Malek S (2018) An assessment of time and cost overrun in Afghanistan's construction. In: Smart construction. 1st International conference on engineering and technology: smart engineering, Parul University, Baroda, India. McGraw Hill Education, pp 159–166
5. Chaudhari K, Malek S (2018) Critical review on delay and cost overrun in construction projects. In: Smart construction. 1st International conference on engineering and technology: smart engineering, Parul University, Baroda, India. McGraw Hill Education, pp 167–172
6. Fountas S, Aggelopoulou K, Gemtos T (2016) Precision agriculture: crop management for improved productivity and reduced environmental impact or improved sustainability. In: Supply chain management for sustainable food networks
7. Zhang L, Dabipi I, Brown W (2018) Internet of things applications for agriculture. Wiley

8. Ghayvat H, Mukhopadhyay S, Gui X, Suryadevara N (2015) WSN-and IoT-based smart homes and their extension to smart buildings. Sensors 15:10350–10379
9. Ananthi N, Divya J, Divya M, Janani V (2017) IoT based smart soil monitoring system for agricultural production. In: IEEE
10. Malek M, Mohibali S, Bachwani D (2021) Identification, evaluation, and allotment of critical risk factors (CRFs) in real estate projects: India as a case study. J Proj Manag 6(2):83–92. https://doi.org/10.5267/j.jpm.2021.1.002
11. Malek MS, Gundaliya PJ (2020) Negative factors in implementing public–private partnership in Indian road projects. Int J Constr Manag. https://doi.org/10.1080/15623599.2020.1857672
12. Malek M, Gundaliya P (2021) Value for money factors in Indian public-private partnership road projects: an exploratory approach. J Proj Manag 6(1):23–32. https://doi.org/10.5267/j.jpm.2020.10.002
13. Malek M, Bhatt V, Patel A (2020) Global, local and national growth of road projects through PPP. Test Eng Manag 83(March–April 2020):25837–25860
14. Memon A, Malek M, Patel A (2020) A review on risk and risk management technique in construction projects. Int J Disaster Recovery Bus Continuity 11(1):2695–2701
15. Patel R, Bachwani D, Malek M (2020) Human resource management, a vital tool for labour management in construction industry. Stud Indian Place Names 40(9):201–206
16. Parikh R, Bachwani D, Malek M (2020) An analysis of earn value management and sustainability in project management in construction industry. Stud Indian Place Names 40(9):195–200
17. Khan S, Malek M, Phugat S (2019) Factors influencing time and cost overrun in construction of government buildings: Indore as a case study. CIKITUSI J Multi Res 6(4):288–295. 16.10089.CJMR.2019.V6I4.19.2857
18. Nautiyal V, Shah Y, Malek S (2019) Application of pavement management system in HDM-4: a critical review. Int J Manag Technol Eng 9(4):1850–1859. 16.10089.IJMTE.2019.V9I4.19.27237
19. Macwan J, Bhavsar A, Malek M (2019) Smart project scheduling for construction projects using AI. Int J Tech Innov Mod Eng Sci 5(5):14–19
20. Agrawal A, Malek M (2018) Principles of sustainability in construction project management. Int J Adv Innov Res 5(4):45–52
21. Gajera D, Malek M (2018) A study on consumer behaviour in real estate for Vadodara city. Univers Rev 7(12):956–969
22. Kumbhani H, Malek M (2018) Cost management process in construction industry. Suraj Punj J Multi Res 8(12):151–158. 16.10089.SPJMR.2018.V8I11.16.3145
23. Saiyed F, Malek S (2017) The realty growth trajectory of India and China—inexorable comparison. In: Proceedings of the 2nd international conference on construction, real estate, infrastructure and project management, NICMAR, Pune, India
24. Upadhyay U, Malek M (2018) Identification of factors responsible for organizational transformation of GSRDC. Pramana Res J 8(12):189–201. 16.10089.PRJ.2018.V8I12.18.2063
25. Malek S, Akalkotkar P (2014) Need of PPP in Indian road sector. Int J Emerg Technol Appl Eng Technol Sci. In: Proceedings of the international conference on information, knowledge & research in engineering management and sciences (IC-IKR-EMS), Kalol, Gujarat, India. KITRC, pp 300–305
26. Saiyed F, Pitroda J, Malek S (2016) Assessment of very critical risk factors in real estate sector by criticality index. In: Proceedings of the international conference on engineering: issues, opportunities and challenges for development, Umrakh, Gujarat, India. SNPITRC, pp 193–205
27. Jariwala K, Malek M (2018) Total quality management in construction of textile market in Surat city. J Appl Sci Comput 5(12):699–706. 16.10089.JASC.2018.V5I12.453459.1500186
28. Bachwani D, Malek M (2018) Parameters indicating the significance of investment options in real estate sector. Int J Sci Res Rev 7(12):301–311. 16.10089.IJSRR.2018.V7I12.287391.2950
29. Rathod S, Malek S (2018) Six sigma in construction industry: literature review. Int J Manag Technol Eng 8(12):913–920. 16.10089.IJMTE.2018.V8I12.17.2102

30. Pandya SP, Prajapati MR, Thakar KP, Assessment of training needs of farm women. Gujarat J Ext Educ 25(2):169–171
31. Pandya S, Ghayvat H, Sur A, Awais M, Kotecha K, Saxena S, Jassal N, Pingale G (2020) Pollution weather prediction system: smart outdoor pollution monitoring and prediction for healthy breathing and living. Sensors 20:5448
32. Kochai Z, Malek M (2018) Causes of schedule delay and cost overrun in Afghanistan infrastructure projects. Int J Adv Innov Res 5(3):18–25
33. Malek S, Akalkotkar P (2016) Factors contributing to successful public private partnership projects for highway projects in Indian context. Int J Sci Res Dev 3(12):25–29
34. Pandya S, Ghayvat H, Kotecha K, Awais M, Akbarzadeh S, Gope P, Smart home anti-theft system: a novel approach for near real-time monitoring and smart home security for wellness protocol. Appl Syst Innov 1(4):42
35. Araya SY, Malek S (2018) Application of building information modelling in Indian construction projects—a critical review. Int J Creative Res Thoughts 6(1):979–987
36. Tadewos SG, Malek S, Patel D (2018) Challenges of building construction in Mekelle city, Ethiopia: from contractor's perspective. J Emerg Technol Innov Res 5(3):420–424. https://doi.org/10.6084/m9.jetir.JETIR1803080
37. Aynalem YF, Malek S (2018) Developing an effective project concept in 5D environment. J Emerg Technol Innov Res 5(5):372–380. https://doi.org/10.6084/m9.jetir.JETIR1805660
38. Barot V, Kapadia V, Pandya S (2020) QoS enabled IoT based low cost air quality monitoring system with power consumption optimization. Cybern Inf Technol
39. Ghayvat H, Pandya S, Patel A (2019) Proposal and preliminary fall-related activities recognition in indoor environment. In: 2019 IEEE 19th International conference on
40. Pandya A, Malek S (2018) A review for the causes of delay in Indian construction projects. Int J Adv Eng Res Dev 5(3):1–6
41. Akbarzadeh S, Ren, Pandya S, Chouhan A, Awais M (2019) Smart aging system
42. Shah B, Malek S (2018) A recapitulation on exigency of smart villages in Indian ambience. Int J Adv Eng Res Dev 5(3):1–8
43. Saiyed F, Pitroda J, Malek S (2015) Reaping the returns: review on risk management in real estate. Int J Futuristic Trends Eng Technol 3(1):1–8
44. Patel V, Malek S (2018) A review on identification of safety factors in construction projects. Int J Adv Eng Res Dev 5(3):1–9
45. Malek S, Pathan N, Mal H (2013) Risk management in construction industry. Indian J Appl Res 3(6):377–379
46. Parekh P, Malek S (2018) A critical literature review on implementing lean construction concepts in infrastructure projects. Int J Adv Eng Res Dev 5(3):1–7
47. Chaudhari K, Malek S (2018) Effectiveness of public private partnership in infrastructure development in India: a review. Int J Adv Eng Res Dev 5(3):1–12
48. Hada N, Malek S (2018) Identification of risk factors in government housing project: a review. Int J Adv Eng Res Dev 5(3):1–6
49. Ghayvat H, Pandya S (2018) Wellness sensor network for modeling activity of daily livings—proposal and off-line preliminary analysis. In: 2018 4th International conference on computing
50. Jethwa B, Bhavsar A, Malek S (2017) An overview of factors affecting risk management in PPP based infrastructure projects. Int J Eng Dev Res 5(2):1737–1742
51. Awais M, Kotecha K, Akbarzadeh S, Pandya S (2018) Smart home anti-theft system
52. Jethwa B, Bhavsar A, Malek S (2017) Critical review on risk management in PPP based infrastructure projects. Int J Eng Dev Res 5(1):357–361
53. Patel M, Pandya S, Patel S (2017) Hand gesture based home control device using IoT. Int J Adv Res
54. Parmar D, Pitroda J, Malek S (2017) Use of BIM (building information modelling) as an integrated tool to plan, design and manage critical construction projects. Int J Adv Res Innov Ideas Educ 3(1):1145–1153
55. Malek S, Saiyed F (2017) A study on incipient trends: the real estate sector of Ahmedabad city (Gujarat). Int J Res Granthaalayah 5(1):304–315

56. Patel D, Malek S (2016) Application of performance management system in infrastructure sector of construction industry. Int J Eng Dev Res 4(2):1998–2000
57. Kalkani K, Malek S (2016) Analyzing causes of delay in construction projects. Int J Innov Res Sci Technol 2(12):257–262
58. Pandya S, Yadav AK, Dalsaniya N, Mandir V, Conceptual study of agile software development
59. Malek S, Akalkotkar P, Zala L (2016) Gujarat PPP: attractive factors for adopting public private partnership for highway projects. Int J Futuristic Trends Eng Technol 3(2):1–6
60. Garg D, Goel P, Pandya S, Ganatra A, Kotecha K (2002) A deep learning approach for face detection using YOLO. In: 2018 IEEE PuneCon
61. Sur A, Sah RP, Pandya S (2020) Milk storage system for remote areas using solar thermal energy and adsorption cooling. Mater Today Proc
62. Cohen JM, Pandya S, Krasenbaum LJ, Thompson SF (2020) A real-world perspective of patients with episodic migraine or chronic migraine prescribed AJOVY in the United States. HEADACHE
63. Sur A, Pandya S, Sah RP, Kotecha K, Narkhede S (2020) Influence of bed temperature on performance of silica gel/methanol adsorption refrigeration system at adsorption equilibrium. Part Sci Technol
64. Cohen JM, Pandya S, Tangirala K, Krasenbaum LJ (2020) Treatment patterns and characteristics of patients prescribed AJOVY, Emgality, or Aimovig. HEADACHE
65. Wandra K, Pandya S (2014) Centralized timestamp based approach for wireless sensor networks. Int J Comput Appl
66. Upadhyay S, Malek S (2016) Adoption and implementation of total quality management in construction industry. Int J Eng Dev Res 4(2):1256–1260
67. Patel WD, Pandya S, Koyuncu B, Ramani B, Bhaskar S (2019) NXTGeUH: LoRaWAN based NEXT generation ubiquitous healthcare system for vital signs monitoring & falls detection. In: 2018 IEEE PuneCon
68. Dandvate HS, Pandya S (2016) New approach for frequent item set generation based on Mirabit hashing algorithm. In: 2016 International conference on inventive
69. Swarndeep SJ, Pandya S (2016) Implementation of extended K-Medoids algorithm to increase efficiency and scalability using large datasets. Int J Comput Appl
70. Pandya S, Shah S, Joshi N, Ghayvat H, Mukhopadhyay SC, Yap MH (2016) A novel hybrid based recommendation system based on clustering and association mining. In: 2016 10th International conference on sensing technology (ICST)
71. Patel S, Singh N, Pandya S (2017) IoT based smart hospital for secure healthcare system. Int J Recent Innov Trends Comput Commun 5
72. Ghayvat H, Pandya S, Shah S. Mukhopadhyay SC, Yap MH, Wandra KH (2016) Advanced AODV approach for efficient detection and mitigation of wormhole attack in MANET. In: 2016 10th International conference on sensing technology (ICST)
73. Swarndeep Saket J, Pandya S, Implementation of extended K-Medoids algorithms to increase efficiency and scalability using large dataset. Int J Comput Appl
74. Bhola YO, Socha BN, Pandya SB, Dubey RP, Patel MK (2019) Molecular structure, DFT studies, Hirshfeld surface analysis, energy frameworks, and molecular docking studies of novel (E)-1-(4-chlorophenyl)-5-methyl-N′-((3-methyl-5-phenoxy-1-phenyl-1H-pyrazol-4-yl) methylene)-1H-1, 2, 3-triazole-4-carbohydrazide. Mol Cryst Liq Cryst
75. Patel W, Pandya S, Mistry V (2016) i-MsRTRM: developing an IoT based intelligent Medicare system for real-time remote health monitoring. In: 2016 8th International conference on computational
76. Wandra KH, Pandya S (2012) A survey on various issues in wireless sensor networks. Int J Sci Eng
77. Shah JM, Kotecha K, Pandya S, Choksi DB, Joshi N (2017) Load balancing in cloud computing: Methodological survey on different types of algorithm. In: 2017 International conference on trends in electronics and informatics (ICEI). https://doi.org/10.1109/ICOEI.2017.8300865
78. Malek S, Akalkotkar P (2016) Driving forces leading to the adoption of PPP: perspective from Gujarat (India), Hong Kong and Australian practitioners. Int J Innov Res Sci Technol 2(9):6–11

79. Patel RR, Pandya SP, Patel PK, Characterization of farming system in North West agro climatic zone of Gujarat state. Gujarat J Ext Educ 27(2):206–208
80. Pandya S, Ghayvat H, Kotecha K, Yep MH, Gope P (2018) Smart home anti-theft system: a novel approach for near real-time monitoring. smart home security and large video data handling for wellness protocol
81. Malek S, Akalkotkar P (2016) Negative factors adopting public private partnership for highway projects: a case study of Gujarat. Int J Sci Technol Eng 2(8):30–35
82. Ghayvat H, Awais M, Pandya S, Ren H, Akbarzadeh S, Chandra Mukhopadhyay S, Chen C, Gope P, Chouhan A, Chen W (2019) Smart aging system: uncovering the hidden wellness parameter for well-being monitoring and anomaly detection, smart aging system: uncovering the hidden wellness parameter for well-being monitoring and anomaly detection. Sensors 19:766
83. Saket S, Pandya S (2016) An overview of partitioning algorithms in clustering techniques
84. Andrew R, Malekian R, Bogatinoska D (2018) IoT solutions for precision agriculture. In: MIPRO, Opatija
85. Benyezza H, Bouhedda M (2018) Smart irrigation system based ThingSpeak and Arduino. In: International conference on applied smart systems. ICASS, Médéa
86. Patil K, Kale N (2016) A model for smart agriculture using IoT. In: International conference on global trends in signal processing, information computing and communication. IEEE
87. Ashwini BV (2018) A study on smart irrigation system using IoT for surveillance of crop-field. Int J Eng Technol 7:370–373
88. Shah K, Malek M, Phugat S (2020) Alternate building materials used in construction industry. Stud Indian Place Names 40(9):184–189
89. Khan BR, Malek M, Tidke A (2020) Risk management in construction industry. Stud Indian Place Names 40(9):165–171
90. Patel K, Malek S (2016) Planning, scheduling and resource management for residential real estate projects in Ahmedabad by using Microsoft Project. Int J Eng Dev Res 4(2):189–194
91. Pandya S, Sur A, Kotecha K (2020) Smart epidemic tunnel—IoT based sensor-fusion assistive technology for COVID19 disinfection. Emerald
92. Joshi N, Kotecha K, Choksi DB, Pandya S (2018) Implementation of novel load balancing technique in cloud computing environment. In: Computer communication and informatics (ICCCI)

Power Quality Enhancement and Low Voltage Ride Through Capability in Hybrid Grid Interconnected System by Using D-Fact Devices

T. Joshiram and S. V. R. Laskhmi Kumari

Abstract One of the major difficulties detected in the distribution system in present days has been power quality. These days, most individuals are utilizing the urbane electrical devices which are dependent on the semiconductor devices, these devices humiliate the power quality. Hence, there is a need to recover the voltage profile. In this paper, the photovoltaic (PV) plant and the wind turbine generator (WTG) are connected to the same point of common coupling (PCC) with a nonlinear load. The unified power quality conditioner (UPQC) [1] is familiar as the best solution for moderation of voltage sag associated problems in the highly taped distribution system. This effort grants the simulation modeling and analysis of innovative UPQC system for solving these problems. UPQC to increase the power quality and recover the low voltage ride through (LVRT) capability of a three-phase medium voltage network connected to a hybrid distribution generation (DG) system. UPQC associated to same PCC. Unlike fault condition setups are tested for improving the efficiency and the quality of the power supply and compliance with the requirements of the LVRT grid code. It inserts voltage in the distribution line to reserve the voltage profile and guarantees constant load voltage. The simulations were led in MATLAB/Simulink to show the UPQC-based future approach's usefulness to smooth the distorted voltage due to harmonics [2].

Keywords Active power · DC link voltage DFIG · Unified power quality conditioner · LVRT · Power factor · Photovoltaic voltage stability · Reactive power · Reactive components · Total harmonic distortion · Sag · Swell · D-FACTS

1 Introduction

Power Quality—The Problem

If possible, the goal of power industry is to supply a purely sinusoidal voltage at fixed amplitude and fixed frequency [1, 2]. Whereas it is the responsibility of the dealer to deliver a nearly sinusoidal voltage with smaller amount variation in amplitude and

T. Joshiram · S. V. R. Laskhmi Kumari (✉)
VR Siddhartha Engineering College, Kanuru, Vijayawada, Andhra Pradesh 520007, India

P. K. Singh et al. (eds.), *Emerging Technologies for Computing, Communication and Smart Cities*, Lecture Notes in Electrical Engineering 875,
https://doi.org/10.1007/978-981-19-0284-0_26

frequency, the customer additionally has a section to play in making such a situation [3]. At the PCC, both the service organization and the client have a few components to conform to. To overcome these constraints and assurance the stability of the electric power system related with a great deal of variable energy resources, the executives of energy system have relied upon examining distinctive particular courses of action [4]. Quite possibly the main grid code necessities are the low voltage ride through (LVRT) capacity, which means that the renewable energy power change system should remain associated during grid faults and supply reactive power to help the grid [5].

The prospects of power generation from hybrid energy systems are winding up being incredibly promising and dependable [2]. A DFIG and flywheel energy storage system was premeditated in [3] and the future control system was planned to ensure that the grid power is remote from wind power output fluctuations. To boost the LVRT capability of a grid-integrated DFIG-based wind farm [3]. Additional devices usually used in distribution networks to defend grave loads counter to voltage disturbances are known as D-FACTS and comprise: DSTATCOM (Static Dispensing Compensator), dynamic voltage restorer (DVR) and unified power quality conditioner (UPQC) [1]. Unified power quality conditioner (UPQC) is one of the most prevalent solutions cast-offs currently [5]. The UPQC as isolated device in as quick control device to control the active power and reactive power and it keeps the sensitive load from preeminent disturbances [5]. The UPQC incorporates of joined activity of (DSTATCOM) and dynamic voltage restorer (DVR) [1]. The DSTATCOM is compensating the reactive power and harmonics in the load side and DVR mitigates the voltage sags/swell in the source side [5]. From this future method the foremost disturbances are condensed and too the control voltage sag instigating from the supply side [5]. To reimburse for the harmonics in the load current by injecting the required harmonic currents. To normalize the power factor by injecting the required reactive current [6]. This paper dowries a simulation study to regulate the worth of the UPQC to restrained voltage disturbances, reduce their effect on the total stability of the transmission and distribution network, and recover the LVRT capability when the network is connected to a hybrid PV-wind system [7].

2 Future Simulated Methods and Explanation

The ideal of photovoltaic energy or wind turbine or together relies upon the availability of the sustainable asset after some time and furthermore is bounty at the spot of establishment [4]. The future topography is displayed in Fig. 1. It contains of a 500 kW PV farm interconnected to a distribution system through a 3 phase PWM inverter with a 3 phase AC choke filter and a move forward transformer. The DFIG has a negligible produce power of 500 kW and is associated with the matrix at the PCC through a move step-up transformer and providing the heap. Therefore, the appraised total power delivered by the hybrid system is 1 MW [2] (Fig. 2).

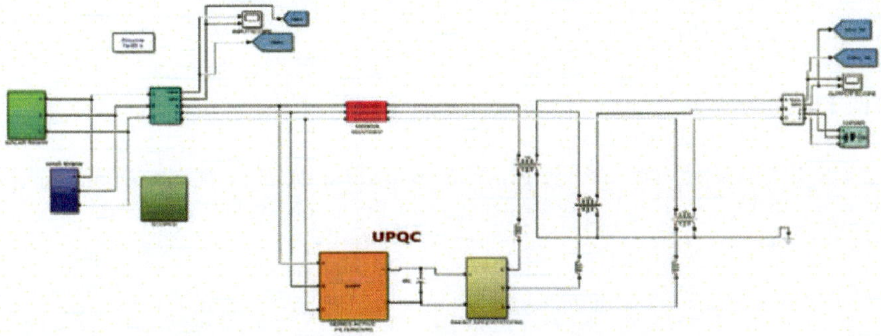

Fig. 1 MATLAB simulation with UPQC and a load connected to grid

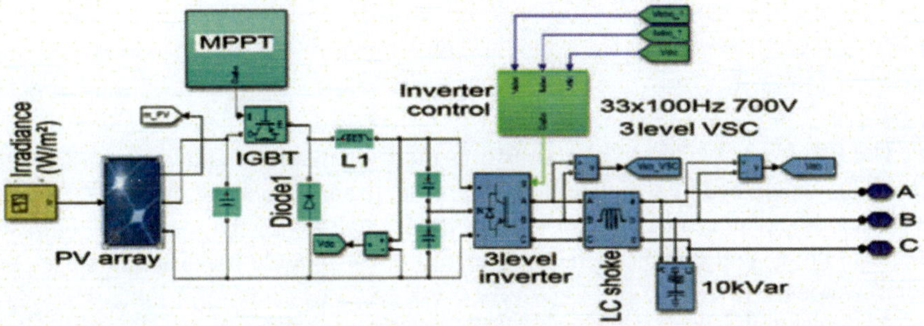

Fig. 2 PV farm of 500 kW connected to grid via an inverter associated to 500 kVA – 400 V/30 kV transformer [2]

The PV farm is exposed to the solar irradiance shown in Fig. 3 and the WTG is consecutively with a wind speed of 12 m/s throughout the simulation time of 5 s additional information required by the volume editor [1].

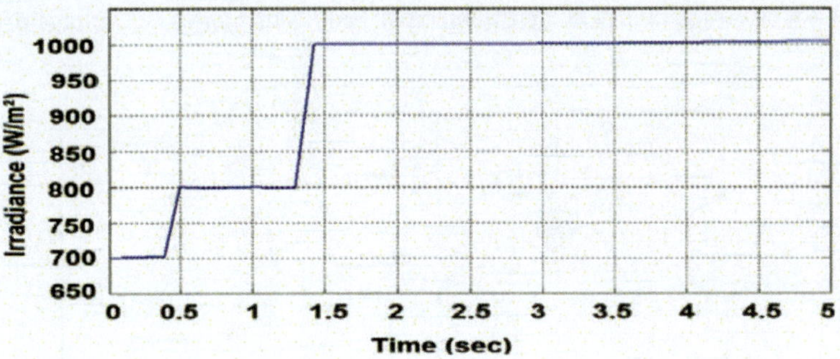

Fig. 3 Solar irradiance at 25 °C [5]

2.1 Photovoltaic Installations

The PV plant array contains of 16 series modules and 102 parallel strings (model: SunPower SPR-305-WHT) [2].

The PV model castoff in the paper is created on the 2-diode equivalent circuit revealed in Fig. 4. The PV cell total current in the equivalent circuit revealed in Fig. 4 is stated by [2].

$$I = I_{\text{Ph}} - I_{\text{Ph}}\left(e^{q\frac{z+IR}{nkt}} - 1\right) - I_{s1}\left(e^{q\frac{y+IR}{nkt}} - 1\right)\frac{v_{+IR}}{R_{\text{P}}} \tag{1}$$

where n is the idealist factor. Presumptuous that all the cells are equal and working under the similar operating circumstances [8].

$$R_{\text{p.field}} = \frac{G_{\text{s}}}{G_{\text{p}}} R_{\text{p.cell}} \tag{2}$$

The PV power transformation control is made on maximum power point tracker (MPPT) which endures high efficiency power transfer and be dependent upon together the solar irradiance and the electrical qualities of the load [9]. Figure 5 displays the I–V and P–V characteristics of the PV module for different stages of solar irradiance [5].

As depicted in Fig. 5, the most extreme force is positive by the significant square shape region, PMP = VMPIMP, reachable from the I–V trademark. The synchronizes of the VMP are start [8].

$$\left.\frac{\mathrm{d}P}{\mathrm{d}E}\right|_{E=\text{Emp}} = \left.\frac{\mathrm{d}(IE)}{\mathrm{d}E}\right|_{E=\text{Emp}} = \left.\left(1 + \frac{\mathrm{d}I}{\mathrm{d}E}\right)\right|_{E=\text{Emp}} = 0 \tag{3}$$

Then I_{MP} is resolute by evaluating Eq. 1 at $E = E_{\text{MP}}$ [10].

The PV array model is mounted to contain 12 series modules and 102 equal strings sought after to ship 500 kW at an irradiance of 1000 W/m^2 and a DC voltage of VDC

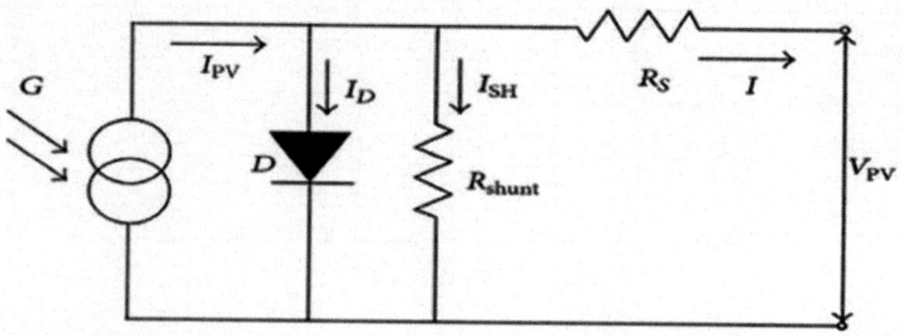

Fig. 4 PV cell circuit model [2]

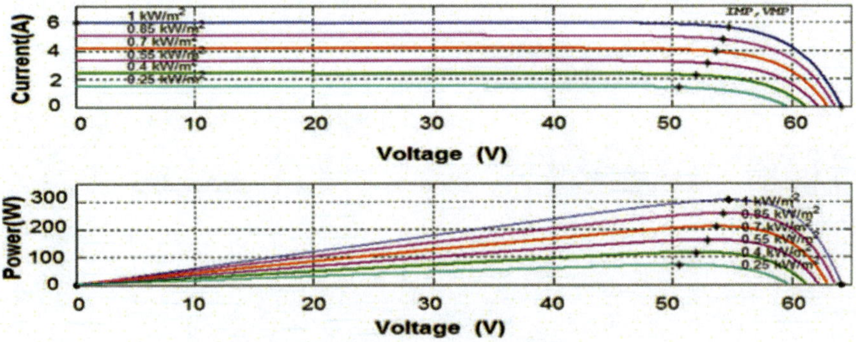

Fig. 5 *V–I* and *P–V* characteristic of PV module under dissimilar solar irradiance and at 25 °C with MPPT points in strings [5]

= 896 V applied to the boost converter as uncovered in Fig. 7. There are various MPPT control techniques future underway, some of them are a lot of the same in relations of their functioning rule [9] (Fig. 6).

The INC strategy can be seen as a superior kind of the Perturb and Observe (P and O) [11]. The point of the power bend is gotten from:

$$\frac{\mathrm{d}P_{\mathrm{PV}}}{\mathrm{d}E_{\mathrm{PV}}} = 0 \Rightarrow \frac{\mathrm{d}I_{\mathrm{PV}}}{\mathrm{d}E_{\mathrm{PV}}} = -\frac{I_{\mathrm{PV}}}{E_{\mathrm{PV}}} \tag{4}$$

$$\frac{\mathrm{d}P_{\mathrm{PV}}}{\mathrm{d}E_{\mathrm{PV}}} = \frac{\mathrm{d}(I_{\mathrm{PV}}E_{\mathrm{PV}})}{\mathrm{d}E_{\mathrm{PV}}} = I_{\mathrm{PV}} + E_{\mathrm{PV}}\frac{\mathrm{d}I_{\mathrm{PV}}}{\mathrm{d}E_{\mathrm{PV}}} \tag{5}$$

Multiplying 2 sides of Eq. (5) by $1/E_{\mathrm{PV}}$ gives:

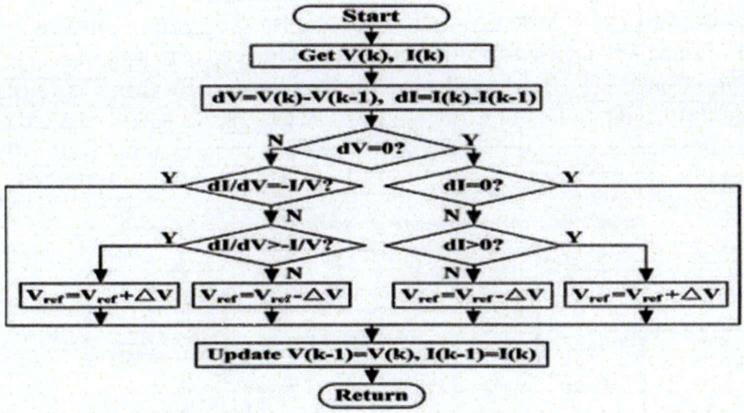

Fig. 6 Flowchart of the incremental conductance algorithm [2]

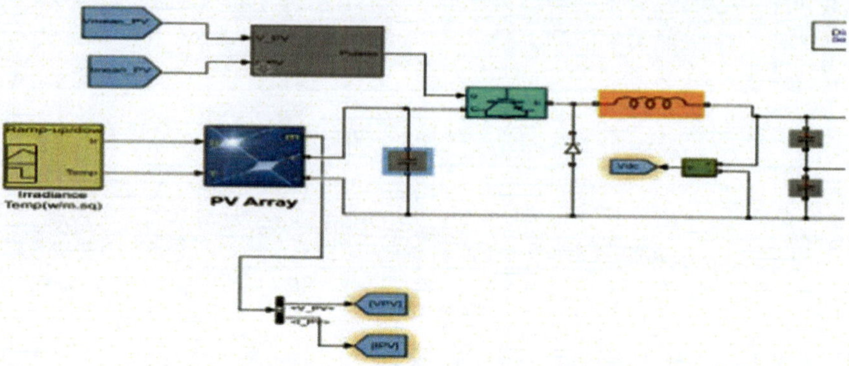

Fig. 7 Displays the circuit model of the DC–DC boost converter used in this effort [7]

$$\frac{1}{E_{\mathrm{PV}}}\frac{\mathrm{d}P_{\mathrm{PV}}}{\mathrm{d}E_{\mathrm{PV}}} = \frac{I_{\mathrm{PV}}}{E_{\mathrm{PV}}} + \frac{\mathrm{d}I_{\mathrm{PV}}}{\mathrm{d}E_{\mathrm{PV}}} = G_{\mathrm{PV}} + \mathrm{d}G_{\mathrm{PV}} \tag{6}$$

where G and dG mean the conductance and gradual conductance correspondingly. A flowchart identifying with the INC calculation is uncovered in Fig. 6. The calculation can follow the MPP and remains there till an adjustment of [dI]_(PV) or [dV]_PV occurs because of a modification in climatic conditions [11].

The grave upsides of the inductor and capacitances of the expected lift converter are: $L_1 = 30$ mH, $C = 100$ µF, $C_1 = C_2 = 120$ mF. To affirm greatest force reflection from the PV source, the converter interfacing the PV framework to the network should be cultivated of self-changing its own limitations progressively. The 3-level voltage source inverter topology revealed in Fig. 8 is simulated in this work [9].

The Vdc boost converter orientation output voltage is customary at 714 V and the IGBT 3-level inverter uses PWM technique, (3.3 kHz carrier frequency) converting DC power from 714 Vdc source to 400 Vac, 50 Hz. The grid is connected to the inverter through an inductive grid filter and a low frequency transformer to step-up the voltage from 0.4 to 3 0 kV in order to reduce losses when PV energy is transmitted to the grid [9] and to filter out harmonic frequencies. The 12 pulses required by the inverter are generated by the discrete three phase PWM generator [14] (Fig. 9).

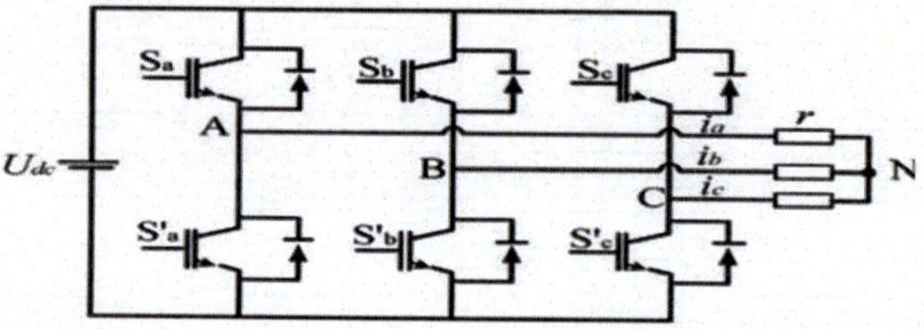

Fig. 8 3-level inverter topology [25]

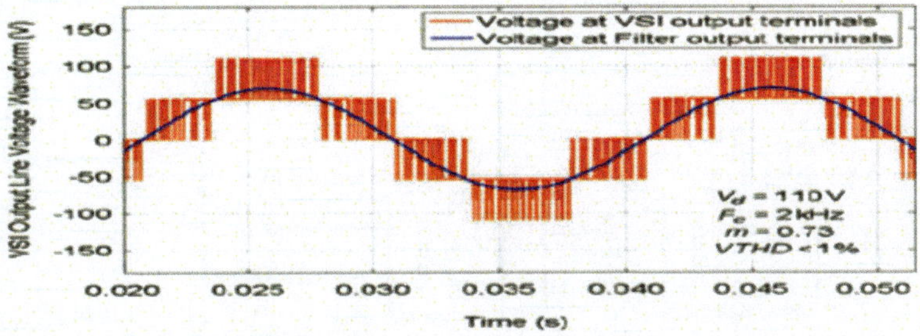

Fig. 9 Output three-level inverter unfiltered and filtered voltage waveforms [5]

2.2 Modeling of Wind Plant

1. Aerodynamic modeling of the wind turbine

The unique energy from the wind is caught by the wind turbine and changed over to mechanical force P_{m} [9, 12]. The wind power plant involves of a solitary DFIG-based wind turbine delivering 500 kW with 400 Vac produce voltage. The energy or power of a wind turbine might be indefatigable by various means [27]. The force P_{m} captured by the wind turbine is a control of the sharp edge sweep, the pitch point, and the rotor speed [7] (Fig. 10).

The power or torque of a wind turbine may be resolute by numerous income. The power P_{m} taken by the wind turbine is a purpose of the blade radius, the pitch angle, and the rotor speed [13]. Figure 13 displays the simulated power curves for dissimilar wind speeds [5] (Figs. 11 and 12).

$$P_{\mathrm{m}} = \frac{1}{2}\pi\rho C_{\mathrm{P}}(\lambda, \beta)R^2\vartheta^3 \tag{7}$$

$$C_{\mathrm{p}}(\lambda, \beta) = 0.22\left(\frac{116}{\lambda_i} - 0.4\beta - 5\right)e^{\frac{21}{\lambda_i}} \tag{8}$$

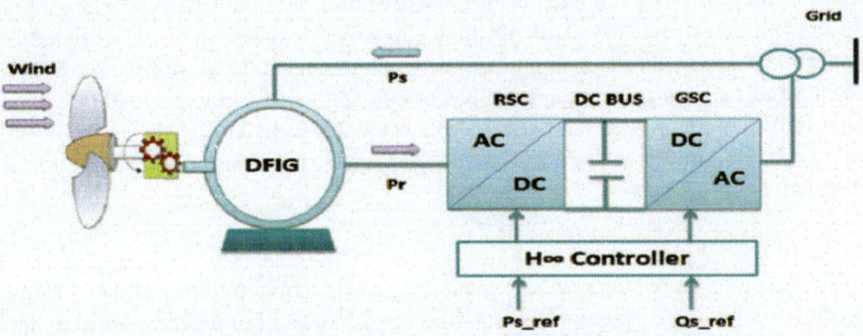

Fig. 10 Doubly fed induction generator [3]

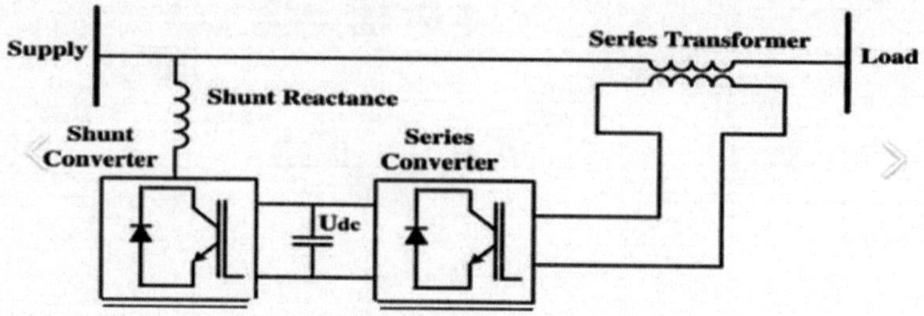

Fig. 11 Basic schematic representation of the UPQC

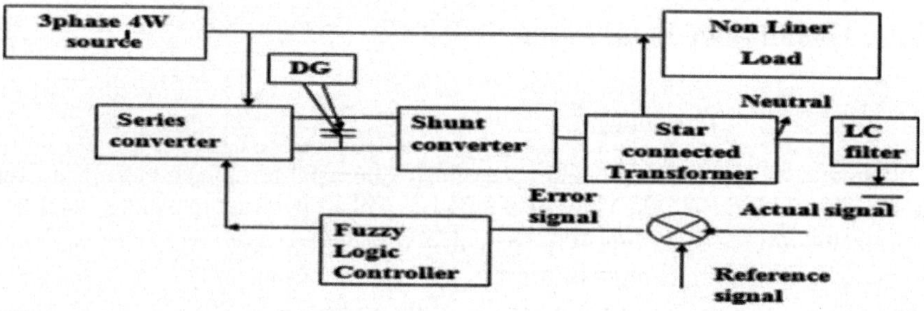

Fig. 12 Proposed UPQC model [8]

$$\frac{1}{\lambda_i} = \frac{1}{\lambda + 0.08\beta} - \frac{0.035}{\beta^2 + 1} \quad (9)$$

2.3 *Double Fed Induction Generator (DFIG) Modeling*

The Doubly Fed Induction Generator (DFIG)-based wind turbine with variable-speed variable-pitch control game plan is the most extreme common wind power generator in the wind power industry. This machine can be worked also in grid associated or independent mode. In this task an exhaustive electromechanical model of a DFIG-based wind turbine associated with power grid or just as independently worked wind turbine system with incorporated battery energy stockpiling is set up in the MATLAB/Simulink area and its adjusting generator and turbine control erection is executed. Natural [8].

Model of DFIG

The DFIG contains of stator winding and the rotor twisting outfitted with slip rings. The stator is giving 3-stage protected windings produce up a picked post plan and is associated with the matrix through a 3-stage transformer. The same to the stator, the rotor is additionally worked of 3-stage protected windings. The rotor windings

Fig. 13 FLC-based series converter controller

are associated with an outside fixed circuit through a bunch of slip rings and brushes [11].

3 UPQC Topology

The arrangement of the both DSTATCOM and DVR can handle the power quality of the source current and the load bus voltage. Furthermore, if the DVR and DSTATCOM are associated on the DC side, the DC bus voltage can be managed by shunt associated on the DSTATCOM while the DVR supplies the necessary energy to the load if there should arise an occurrence of the transient's disturbance in source.

The setup of such a device is displayed in Fig. 11. The DG is related in among the dc connection of the UPQC. The recreation of the arranged method has been endorsed out by MATLAB/SIMULATION [11].

4 Formation of Hybrid UPQC

The arranged technique embraces of the consolidated rating star connected transformer-based UPQC appended with DG. The 3 phase source is taken from the 1 MW system. The series converter is associated through the series reactor from the line correspondingly through the shunt reactor is shunt converter. The series and shunt converter is connected with common DC link connection and capacitor. The arranged UPQC model is displayed in Fig. 12 [13].

4.1 Series Converter Control

The various disturbances like switching operation and different faults occur in the distribution system causes voltage sags and swell. It influences the customer equipment cruelly. The series converter compensates the voltage sags and swells in the distribution system. The fuzzy logic controller-based series converter controller is displayed in Fig. 13. The DC link limited voltage is contrasted and the reference voltage by comparator [7]. The blunder signal acquired from the comparator is handled with FLC 1. The real worth of voltage in phase a, b, c is prepared with the magnitude of the injected voltage in series converters. The real value of voltage in phase a, b, c is handled with the magnitude of the injected voltage in series converters. This output value is contrasted and output of FLC 1 by comparator. The amplitude of voltage is utilized for reference current estimation [7].

4.2 Shunt Converter Control

Because of expanding in nonlinear load and power electronic equipment in distribution system causes harmonics. This harmonic and is compensate by the shunt converter. The dc link voltage is detected and contrasted with reference voltage [11]. The error signal is handled and it considered as the magnitude of the 3 phase supply current references. The reference current is determined by utilizing the unit vector in phase, with the real supply voltage the 3 phase unit vector in phase is inferred as in [12].

$$u_{sa} = \frac{v_{sa}}{v_{sm}}, u_{sb} = \frac{v_{sb}}{v_{sm}}, u_{sc} = \frac{v_{sc}}{v_{sm}} \tag{10}$$

where v_{sm} is the amplitude of supply voltage. v_{sa}, v_{sb}, v_{sc} are the three phase supply voltage. u_{sa}, u_{sb}, u_{sc} are the multiplication of three phase unit current vectors. The 3 phase shunt current for compensation of harmonics as shown in 10 [12]. The design of UPQC depends on the parameter specification of the distribution system. The 1 MW grid is considered in the system. The fifth, seventh and eleventh order harmonics are made in this plan. The decreased rating star connected transformer is connected with the UPQC, whereas the industrial and domestics loads are associated in close to the shunt converter side [7] (Fig. 14).

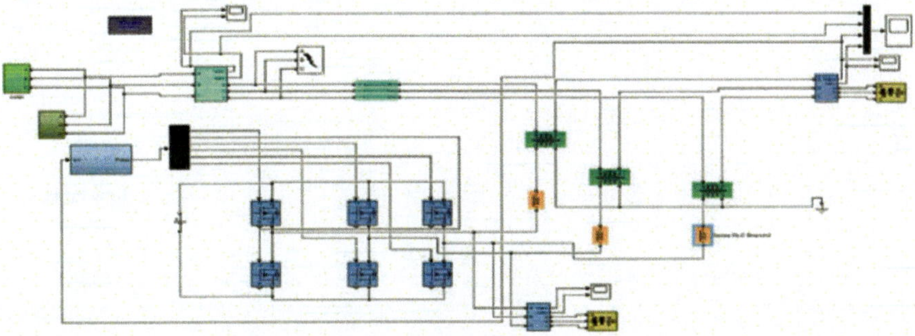

Fig. 14 MATLAB Simulink model of the DVR [2]

5 Results and Discussion

The UPQC has simulated using the proposed hybrid UPQC with DG. The source voltage waveform before and after connecting the UPQC are analyzed. It noticed that the source voltage is distorted before connecting the UPQC and it becomes sinusoidal after connecting the UPQC. The voltage waveform on source side without UPQC is shown in Fig. 15 and with UPQC is shown in Fig. 16. It has clearly shown that the voltage sag and swell present in the waveform is compensated after connecting the UPQC. The voltage sags and swell present in the load side are also reduced, due to source side compensation [9]. Hence, the power quality of the system can be improved (Figs. 17, 18 and 19; Table 1).

DVR is proved to compensate voltage levels under faulty conditions. Voltage harmonics has been reduced considerably. Harmonics generated at source side has THD of 30.5% which has been compensated to 3.6% at load end. Even the voltage sag during fault duration has also been compensated to a desired level [1]. UPQC is proved to compensate current and voltage levels under faulty conditions. Voltage and current harmonics have been reduced considerably. Current harmonics generated

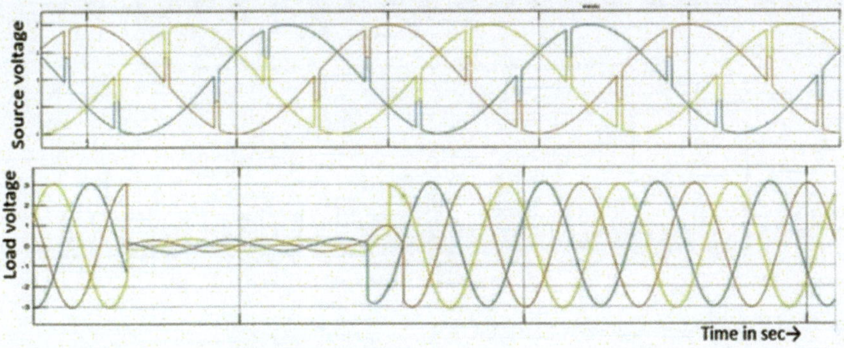

Fig. 15 Load voltage waveform without UPQC

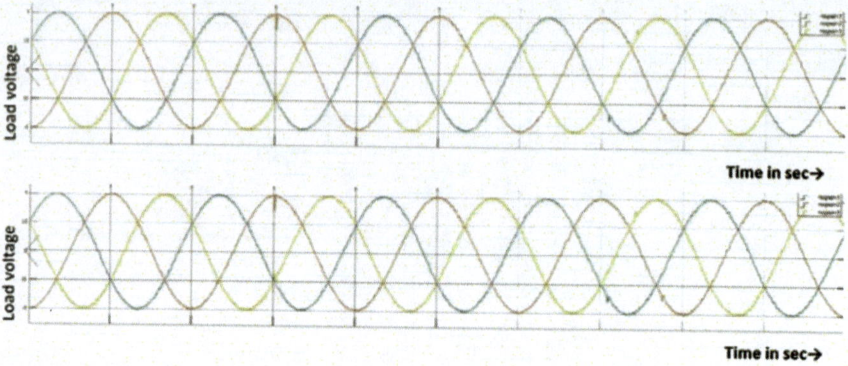

Fig. 16 Load voltage waveform with UPQC and DVR [2]

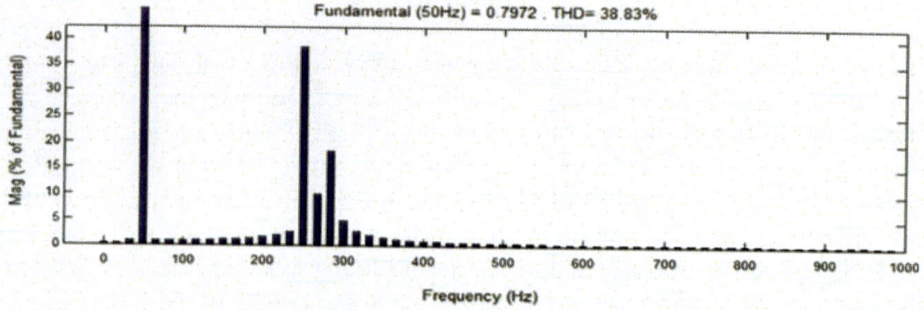

Fig. 17 THD without UPQC in load side [6]

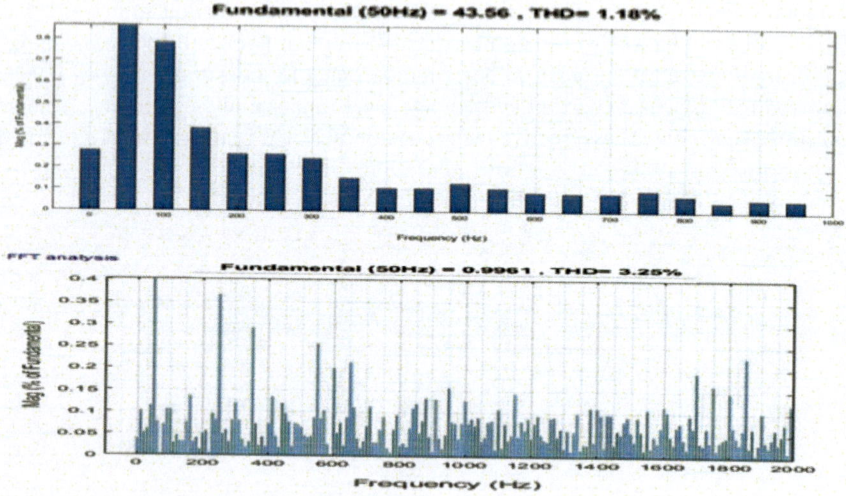

Fig. 18 THD with UPQC and DVR on load side

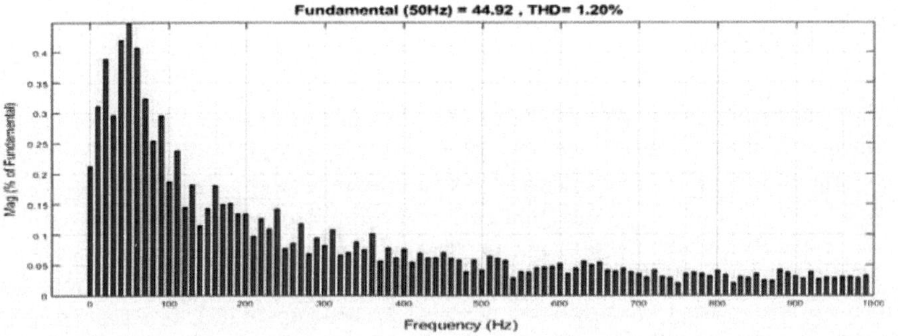

Fig. 19 THD with UPQC in source side

Table 1 THD % comparison with and without D-FACTS

Location	Without UPQC (%)	With DVR (%)	With DSTATCOM (%)	With UPQC (%)
THD in source side	30.5	30.5	30.5	1.1
THD in load side	38.83	3.69	4.12	1.21

at load side has THD of 30.24% which has been compensated to 1.21% at PCC. Voltage harmonics generated at source side has THD of 1.45% which has been compensated to 1.06% at load end [2]. The power quality is improved and the power oscillation overshoot reduction control of rotor speed and preventing the system from having a DC link overvoltage and thus increasing the stability of the power system in accordance with LVRT requirements [2].

Table 2 shows the system parameters.

Table 2 System parameter [2]

	Parameters	Values
MV grid	Grid L–L voltage	30 kV
	Frequency [2]	50 Hz
DFIG	Power	500 kW
	Frequency	50 Hz
	Nominal L–L voltage	400 V
Photovoltaic	Plant capacity	500 kW
	Vdc	896 V
	Parallel strings	102
	Series modules	12
UPQC	Capacitances	1100 uF
	Filter inductance	30 mH
	Filter capacitances	28 uF
	Star connected transformer	20 kV

6 Conclusion

This paper presents a hybrid UPQC and DVR in distribution systems for simultaneous compensation of load current harmonics, voltage sag/swell and source neutral current. The performance of proposed UPQC and DVR has been investigated through extensive simulation studies. From these studies it is observed that the proposed scheme completely compensated the source current harmonics, load current harmonics, voltage sag/swell and neutral current [2]. Even the current and voltage level during fault duration has also been compensated to a desired level [3]. Future scope. The more advanced controllers such as fuzzy controller, artificial neutral network, AUPF, ISCT, AGCT, IGCT theories can also be used with UPQC to make the system more effective [9].

References

1. Abas N, Dilshad S, Khalid A, Power quality improvement using dynamic voltage restorer. IEEE Access. https://doi.org/10.1109/ACCESS.2020.3022477
2. Benali A, Khiat M, Allaoui T, Denaï M, Power quality improvement and low voltage ride through capability in hybrid wind-PV farms grid-connected using dynamic voltage restorer. https://doi.org/10.1109/ACCESS.2019
3. Karthikeya P, Gonsalves R, Senthil M (2019) Comparison of UPQC and DVR in wind turbine fed FSIG using asymmetric faults. Int J ELELIJ 3(3)
4. Pota HR, Hossain J (2019) Robust control for grid voltage stability high penetration of renewable energy, 1st edn. Springer, Berlin, pp 1–11
5. Swain SD, Ray PK (2019) Improvement of power quality using a robust hybrid series active power filter. IEEE Trans Power Electron 32(5)
6. Improvement of power quality using a hybrid UPQC with distributed generator. In: 2016 International conference on circuit, power and computing technologies (ICCPCT). IEEE. 978-1-5090-1277-0/16/$31.00©2021
7. Dosela MK, Arson AB, Gülen U (2019) Application of STATCOM-supercapacitor for low-voltage ride-through capability in DFIG-based wind farm. Neural Comput Appl 28(9):2665–2674
8. Kosala M, Arson AB (2021) Transient modelling and analysis of a DFIG based wind farm with supercapacitor energy storage 78:414–421
9. Noureddine O, Ibrahim AMA (2021) Modelling, implementation and performance analysis of a grid-connected photovoltaic/wind hybrid power system. IEEE Trans Energy Convers 32(1):284–295
10. Rashid G, Ali MH (2021) Nonlinear control-based modified BFCL for LVRT capacity enhancement of DFIG-based wind farm. IEEE Trans Energy Convers 32(1):284–295
11. Dosela MK (2021) Enhancement of SDRU and RCC for low voltage ride through capability in DFIG based wind farm 99(2):673–683
12. Dosela MK (2021)Nonlinear dynamic modelling for fault ride-through capability of DFIG-based wind farm 89(4):2683–2694
13. Ghosh S, Malasada S (2021) An energy function-based optimal control strategy for output stabilization of integrated DFIG flywheel energy storage system. IEEE Trans Smart Grid 8(4)

Optimal LFC Regulator for Frequency Regulation in Multi Area Power System

V. Santoshi Raju and P. Venkatesh

Abstract In a power system, frequency control is more sensitive to the variations of load. So, it needs proper balancing of generation and load demand. Linear controllers such as proportional integral (PI) do not account the non-linearity of the system. Their performance with non-linear power system may be ineffective. This project will present Sliding Mode Control (SMC) which has distinguished properties of accuracy, robustness and easy tuning performance. Due to these advantages SMC technique is effectual for LFC and most vigorous to the base parameter and uncertainties. The anticipated controller is set side by side with Proportional-Integral Derivative (PID), Tilt-Integral Controller (TID) and Fractional Order PID (FOPID) controllers for performance analysis and Grey Wolf Optimization (GWO) will be used to tune the parameters of controllers in Load Frequency Control (LFC). The proposed controller is applied to two area single unit LFC power system consist of thermal unit. To validate the usefulness of controllers, Integral of the Time Weighted Absolute Error (ITEA) performance index will be considered and 1% step load change will be applied. The resulting power system will be modelled and simulated in MATLAB/SIMULINK environment. The proposed work is effective in maintaining the frequency deviation, undershoots and overshoots of the power system to zero in a considerably less time with SMC compared to other controllers.

Keywords Proportional-integral-derivative (PID) · Tilt-integral-derivative (TID) · Fractional order-proportional-integral-derivative (FOPID) · Sliding mode controller (SMC) · Grey wolf optimization (GWO)

1 Introduction

Frequency deviation control in the power system is the most general strategy for the efficient operation of power systems [1–5]; hence the power demand should be coordinated to the power generation whenever source and load varies [1]. The Load

V. Santoshi Raju (✉) · P. Venkatesh
VR Siddhartha Engineering College, Kanuru, Vijayawada, Andhra Pradesh 520007, India

P. K. Singh et al. (eds.), *Emerging Technologies for Computing, Communication and Smart Cities*, Lecture Notes in Electrical Engineering 875,
https://doi.org/10.1007/978-981-19-0284-0_27

Frequency Control (LFC) will make sure that the steady state errors in the power system will be maintained zero throughout the operation for a two area power system, where the two areas are coupled all the way through a tie-line [6]. The operation of LFC also includes to curtail the unexpected tie-line power flows between interrelated adjacent areas and also to curb the transient variations of multi area frequencies. The most intricate problem in LFC is when power transaction takes place between interconnected multi areas, when uncertainties rises in distributed generator and model parameters [7]. In a conventional LFC, Area Control Error (ACE), also termed as control input signal compromises of multi area tie-line power exchanges and local area frequency [8]. In order to meet up with the grid performance levels, the widely used controller to control the ACE is Proportional-Integral (PI) controllers. In order to tune the parameters of PI controller many intelligent optimization techniques are used which will enhance dynamics of the controller and also used to improve the robustness of the controller in conditions power system function state variations [9] and [10].

In the contemporary days, bunch of recent direct strategy came into existence like optimal control techniques [2], distributed control techniques [3], robust control technique [11] and hybrid control algorithms, are significantly used for LFC design [12]. On comparing all the controllers, Sliding Mode Control (SMC) dynamics has a unique feature which can be modelled regardless of disturbances and system parameters, which will enhance the robustness and response speed of LFC [4]. For a two area single unit LFC, second-order SMC and extended disturbance observer is proposed in [13].

This paper proposes a second-order SMC algorithm with an additional extended disturbance observer for a two area LFC scheme. In order to trim down the intricacy of the power system, the load change and the tie-line power are considered as one single parameter, i.e. lumped disturbance parameter so that the order of the power system will be reduced then the extended disturbance observer is used to estimate the lumped disturbance parameter. SMC requires a sliding surface calculated through state variable transformation which compels the frequency deviation to zero without an integral unit. Here, sliding surface matrix is used to tune the system dynamics and desirable sliding surface matrix can be calculated through optimal sliding manifold design or Eigen value assignment method. During load variations, if the scheduled power requires any changes, the modelled scheme will work more efficiently. In order to eliminate the chattering effect, the second-order sliding mode control technique along with super-twisting algorithm is engaged which will compel the sliding surface to reach the sliding surface, respectively. Therefore, the modelled robust LFC can effectively utilize the benefits of SMC and thus it permits very low real-time computational burden.

2 Grey Wolf Optimization (GWO)

Grey wolf optimization (GWO) technique is the latest meta-heuristic methods proposed by Mirjalili et al. in 2014 [5]. Grey wolfs lives in packs and is considered to be at the peak of the food chain. In Grey wolfs, they are further divided into 4 sub-groups, i.e. alpha (α), beta (β), delta (δ), and omega (ω) and alpha (α) category is considered to be the top of all the wolf's in ranking and hence alpha (α) is considered to be the healthy solution and the second, third and fourth best are beta (β), delta (δ) and omega (ω) [14].

Generally, Grey wolf's follow a particular and rare strategy to attack the prey. Initially, they hunt, chase and attack the prey, and then they blockade the prey and finally attack the prey.

The encircling behaviour of Grey wolf can be represented in Eqs. (1) and (2):

$$\vec{D} = \left|\vec{C}.\vec{X}_{\mathrm{p}}(\mathrm{t}) - \vec{X}_{\mathrm{p}}(\mathrm{t})\right| \tag{1}$$

$$\vec{X}(t+1) = \vec{X}_{\mathrm{p}}(t) - \vec{A}.\vec{D} \tag{2}$$

where T is present iteration, C, D and A is vector coefficient, X_{p} is prey's tracking vector and X is Grey wolf's tracking vector.

The vector A and C can be estimated as shown in Eqs. (3) and (4):

$$\vec{A} = 2\vec{a}.\vec{r}_1 - \vec{a} \tag{3}$$

$$\vec{C} = 2.\vec{r}_2 \tag{4}$$

r_1 and r_2 are the arbitrary vectors and 'a' decreases from 2 to 0 during the repetition. The chase is leaded by alpha (α), beta (β), delta (δ) and omega (ω). One of the advantages of GWO algorithm is it is simple to apply due to its simplex composition, low required memory and computational necessity [14].

3 Fractional Order Controllers

PI controller is the most conventional controller and is widely used to tune the parameters of LFC in power system. Day by day, as the order of LFC increases and multiple areas are inter-connected through the tie-line which enhances intricacy of the power system and degrading the performances of the orthodox controllers. In order to increase the effectiveness of the power system, a non-integer order control or a fractional order control (FOC) came into existence which is purely based on fractional calculus. There are different kinds of FOC's like Tilt Integral Controller (TID), fractional order PID (FOPID).

3.1 Tilt-Integral-Derivative (TID) Controller

TID controller is the FOC which is used to finely tune the LFC parameters. Since it is a non-linear controller, TID generally works on three controllers (*T*, *I*, *D*) and an additional parameter (*n*) is used here for tuning purpose. TID controller is almost same like a PID controller, but the proportional characteristics are substituted by tilt a proportional characteristic (represented by the transfer function $\left(s^{-\frac{1}{n}}\right)$) which provides a frequency function called feedback gain and it is tilted with reference to the gain of the traditional controller. On comparison to PID, TID controller offers high level of flexibility to control variables.

$$g(s, \beta) = \frac{T}{S^{\frac{1}{n}}} + \frac{T}{S} + Ds \tag{5}$$

where

$$\beta = \begin{bmatrix} T \\ I \\ D \\ n \end{bmatrix} \in R^4 \tag{6}$$

and $n \in R$ and $n \neq 0$ and $2 < n > 3$, $u(s)$ is control signal, $r(s)$ is reference signal $e(s)$ is error signal, $y(s)$ is output signal and $g(s, \beta)$ is transfer function of TID controller $[s \in Z, \beta \in R]$.

3.2 Fractional Order PID (FOPID) Controller

FOPID is the extension of PID controller which is purely based on fractional order differential calculus. FOPID gives better response than conventional controllers due to presence of five parameters that gives good design flexibility to design the derivative and integral components. λ and μ are additional parameters to the conventional PID in FOPID and can be expressed as $PI^{\lambda}D^{\mu}$. The two additional parameters λ of integration and μ of derivative also made the tuning of the new FOPID controller more flexible.

The transfer function of FOPID controller can be represented as Eq. (7):

$$G(s) = \frac{U(s)}{E(s)} = K_{\rm p} + K_{\rm I}\frac{1}{s^{\lambda}} + K_{\rm D}s^{\mu} \tag{7}$$

where $K_{\rm p}$ is proportional gain, $K_{\rm D}$ is differential gain, $K_{\rm I}$ is integral gain, λ is degree of integration and μ is degree of differentiation.

The $PI^{\lambda}D^{\mu}$ controller is more accurate and gives a prospect to further regulate the variations in control system.

4 Sliding Mode Controller (SMC)

The ith area system dynamics of a multi area inter-connected power system can be represented as shown below (Eqs. 8–11) [4]. The frequency and the power exchange between the LFC's of inter-connected areas in a multi area power system should be maintained constant throughout the operation [4].

$$\Delta\dot{f}_i = \frac{1}{2H_i}\Delta P_{mi} - \frac{1}{2H_i}\Delta P_{Li} - \frac{D_i}{2H_i}\Delta f_i - \frac{1}{2H_i}\Delta P_{\text{tie},i} \tag{8}$$

$$\Delta\dot{P}_{mi} = \frac{1}{T_{ti}}\Delta P_{gi} - \frac{1}{T_{ti}}\Delta P_{mi} \tag{9}$$

$$\Delta\dot{P}_{gi} = \frac{1}{T_{gi}}\Delta P_{ci} - \frac{1}{R_i T_{gi}}\Delta f_i - \frac{1}{T_{ti}}\Delta P_{gi} \tag{10}$$

$$\Delta\dot{P}_{\text{tie},i} = 2\pi \sum_{j=1, j\neq 1}^{N} T_{ij}\left(\Delta f_i - \Delta f_j\right) \tag{11}$$

This LFC model is as it is taken from [4] because the research studies reveal that this LFC model is more practical and reliable and will give results without disturbing the accuracy of the system.

Where i is the area of the system, Δf_i is fluctuations in system frequency, ΔP_{mi} is output of synchronous machine, ΔP_{gi} is position of valve, ΔP_{ci} is output of the controller, ΔP_{Li} is load variations, T_{ij} is tie-line coefficient, H_i is synchronous machine inertia, D_i is damping coefficient of machine, T_{gi} is the governor time constant, T_{ti} is the turbine time constant, R_i is the speed drop, $\Delta P_{\text{tie},i}$ is the deviation between the actual and the scheduled power flows.

$\Delta P_{\text{tie},i}$ Can be evaluated as:

$$\Delta P_{\text{tie},i} = \sum_{j=1, j\neq 1}^{N} \left(\Delta P_{\text{tie,act}j} - \Delta P_{\text{tie,sched}j}\right) \tag{12}$$

In the classical matrix form, the system changes can be written as:

$$\dot{x}_i(t) = \overline{A}_i x_i(t) + \overline{B}_i u_i(t) + \overline{F}_i \Delta P_{Li} \tag{13}$$

where

State variable matrix: $x_i(t) = \left[\Delta f_i \;\; \Delta P_{mi} \;\; \Delta P_{gi} \;\; \Delta P_{\text{tie},i}\right]^{\mathrm{T}}$,

Control input: $u_i = P_{ci}$,

$$\overline{A}_i = \begin{bmatrix} -\frac{D}{2H_i} & \frac{1}{2H_i} & 0 & -\frac{1}{2H_i} \\ 0 & -\frac{1}{T_{ti}} & \frac{1}{T_{ti}} & 0 \\ -\frac{1}{R_i T_{gi}} & 0 & -\frac{1}{T_{gi}} & 0 \\ 2\pi \sum_{j=1, j\neq 1}^{N} T_{ij} & 0 & 0 & 0 \end{bmatrix}, \overline{B}_i = \begin{bmatrix} 0 \\ 0 \\ \frac{1}{T_{gi}} \\ 0 \end{bmatrix}, \text{ and } \overline{F}_i = \begin{bmatrix} -\frac{1}{2H_i} \\ 0 \\ 0 \\ 0 \end{bmatrix}$$

The transfer function for hydro turbine is

$$G_{\text{hydro}} = \frac{1 - T_{ti}s}{1 + 0.5T_{ti}s}$$

Now the LFC model of a multi area power system can be inscribed as:

$$\dot{x}_i(t) = \widehat{A}_i x_i(t) + \widehat{B}_i u_i(t) + \widehat{F}_i \Delta P_{Li} \tag{14}$$

where

The state variable matrix: $x_i(t) = \left[\Delta f_i \; \Delta P_{mi} \; \Delta P_{gi} \; \Delta P_{\text{tie},i}\right]^{\text{T}}$,
Control input: $u_i = P_{ci}$,

$$\widehat{A}_i = \begin{bmatrix} -\frac{D}{2H_i} & \frac{1}{2H_i} & 0 & -\frac{1}{2H_i} \\ \frac{2}{R_i T_{gi}} & -\frac{2}{T_{ti}} & \frac{2}{T_{ti}} + \frac{2}{T_{gi}} & 0 \\ -\frac{1}{R_i T_{gi}} & 0 & -\frac{1}{T_{gi}} & 0 \\ 2\pi \sum_{j=1, j\neq 1}^{N} T_{ij} & 0 & 0 & 0 \end{bmatrix}, \widehat{B}_i = \begin{bmatrix} 0 \\ -\frac{2}{T_{gi}} \\ \frac{1}{T_{gi}} \\ 0 \end{bmatrix} \text{ and } \widehat{F}_i = \begin{bmatrix} -\frac{1}{2H_i} \\ 0 \\ 0 \\ 0 \end{bmatrix}$$

The ultimate purpose of the LFC is to uphold the system frequency invariable, i.e. the change in frequency $\Delta f_i = 0$. Equations (13) and (14) shows when there are alteration in the system load, ΔP_{Li}, and scheduled tie-line power, $\Delta P_{\text{tie},i}$, Δf_i should be driven to zero by regulating the generator control output $\Delta P_{ci} = \Delta P_{Li} + \Delta P_{\text{tie},i}$ and therefore the alteration in the system, load frequency and the scheduled tie-line power together dumped into the single parameter, i.e. extended disturbance. The power system variations can be expressed as (Fig. 1):

$$\dot{x}_i(t) = A_i' x_i(t) + B_i' u_i(t) + F_i' \Delta P_{di} \tag{15}$$

where

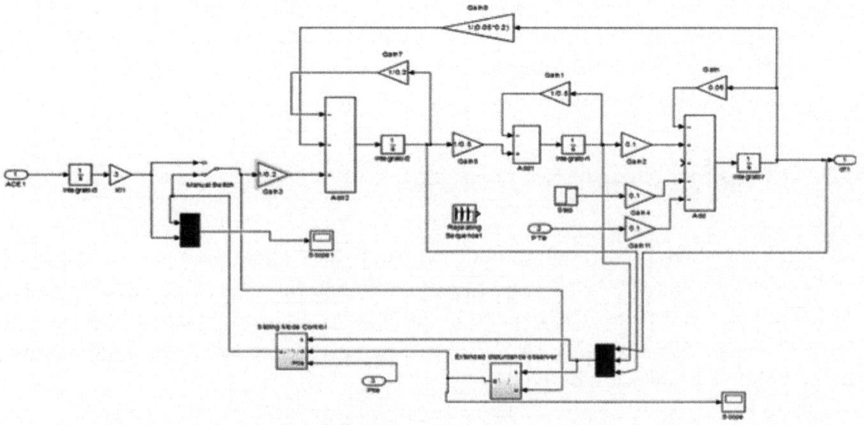

Fig. 1 Thermal unit with extended disturbance observer and SMC

$$A_i' = \begin{bmatrix} -\frac{D}{2H_i} & \frac{1}{2H_i} & 0 \\ T_{g1i} & -T_{t1i} & -T_{t2i} \\ -\frac{1}{R_i T_{gi}} & 0 & -\frac{1}{T_{gi}} \end{bmatrix}, B_i' = \begin{bmatrix} 0 \\ T_{g2i} \\ \frac{1}{T_{gi}} \end{bmatrix}, B_i' = \begin{bmatrix} 0 \\ T_{g2i} \\ \frac{1}{T_{gi}} \end{bmatrix} \text{ and } \Delta P_{di} = \Delta P_{Li} + \Delta P_{\text{tie},i}$$

On the basis of the new LFC system proposed in Eq. (15) and the estimated disturbance observer obtained from disturbance observer, a novel sliding mode is proposed using system state variable transformation.

4.1 Disturbance Observer to Estimate Extended Disturbance

From Eq. (8), the state variables can be modelled as Eq. (16):

$$\dot{x}_i(t) = A_i' x_i(t) + B_i' u_i(t) + F_i' \widehat{d}_i + L(y_i - \widehat{y}_i) \quad (16)$$

$$\widehat{y}_i = C\widehat{x}_i \quad (17)$$

where L = feedback matrix and C = output matrix [4].

A disturbance observer vector can predict the disturbance using the estimated states as in Eq. (18):

$$\left.\begin{aligned} \widehat{d}_i &= \beta_i + M\widehat{x}_i \\ \dot{\beta}_i &= -M\left(A_i' x_i + B_i' u + F_i' \widehat{d}_i\right) + \widehat{d}_i \end{aligned}\right\} \quad (18)$$

where $\dot{\beta}_i$ is a supplementary variable and M is a gain matrix constant [4].

Here, the first order derivative of disturbance is negligible, i.e. zero due to the slow system load changes during the LFC operation.

4.2 Design of Sliding Surface

The main objective of the sliding mode control is to regulate the system and further arrive at the sliding surface. Designing of sliding surface is completely dependent on the control objective. The desired conditions of control objective is $\Delta f_i = 0$, $\Delta P_{mi} = \Delta P_{di}$, $\Delta P_{gi} = \Delta P_{di}$ and $\Delta P_{ci} = \Delta P_{di}$. In order to satisfy the requirements, the new state variables are derived:

$$\left.\begin{aligned} \Delta f_i &= \Delta f_i - 0 \\ \Delta \eta_i &= \Delta P_{mi} - \Delta P_{di} \\ \Delta \xi_i &= \Delta P_{gi} - \Delta P_{di} \end{aligned}\right\} \tag{19}$$

By substituting (19) in (15), the power system dynamic equation can be rewritten as:

$$\begin{gathered} \begin{bmatrix} \Delta \dot{f}_i \\ \Delta \dot{\eta}_i \end{bmatrix} = \begin{bmatrix} -\frac{D_i}{2H_i} & \frac{1}{2H_i} \\ T_{g1i} & -T_{t1i} \end{bmatrix} \begin{bmatrix} \Delta f_i \\ \Delta \eta_i \end{bmatrix} + \begin{bmatrix} 0 \\ T_{t2i} \end{bmatrix} \Delta \xi_i \\ \Delta \dot{\xi}_i = \begin{bmatrix} -\frac{1}{R_i T_{gi}} & 0 \end{bmatrix} \begin{bmatrix} \Delta f_i \\ \Delta \eta_i \end{bmatrix} - \frac{1}{T_{gi}} \Delta \xi_i + \frac{1}{T_{gi}} u_i - \frac{1}{T_{gi}} \Delta P_{di} \end{gathered} \tag{20}$$

The state variable $\Delta \xi_i$ is directly proportional to the input variable u_i. In order to derive the sliding surface, the sliding variable can be selected as:

$$s_i = \Delta \xi_i - k \begin{bmatrix} \Delta f_i \\ \Delta \eta_i \end{bmatrix}, K \in R^{1\times 2} \tag{21}$$

When the system changes are only limited to the sliding surface $s_i = 0$, the modified reduced model can be given as Eq. (22):

$$\begin{bmatrix} \Delta \dot{f}_i \\ \Delta \dot{\eta}_i \end{bmatrix} = \left(\begin{bmatrix} -\frac{D_i}{2H_i} & \frac{1}{2H_i} \\ T_{g1i} & -T_{t1i} \end{bmatrix} + \begin{bmatrix} 0 \\ T_{t2i} \end{bmatrix} K \right) \begin{bmatrix} \Delta f_i \\ \Delta \eta_i \end{bmatrix} \tag{22}$$

The parameters of K can be calculated with propose methods i.e. Eigen value assignment method [4].

4.3 The Control Law Based on Super-Twisting Algorithm

The super-twisting algorithm based LFC have 2 objectives: an Equivalent controller to compensate the power system dynamics and the sliding mode control to compensate the unmodelled dynamics and disturbance arising due to errors present in the modelled parameter.

The equivalent controller can be given in Eq. (23):

$$u_i = \left(\begin{bmatrix} -\frac{1}{R_i} & 0 \end{bmatrix} + K + T_{gi}K(A_{11} + A_{12}K)\right)\begin{bmatrix} \Delta f_i \\ \Delta \eta_i \end{bmatrix} + T_{gi}\left(-\frac{1}{T_{gi}} - KA_{12}\right)s_i - T_{gi}v_i \quad (23)$$

Equation (20) is transformed to Eq. (24) as:

$$\begin{bmatrix} \Delta \dot{f}_i \\ \Delta \dot{\eta}_i \end{bmatrix} = (A_{11} + A_{12}K)\begin{bmatrix} \Delta f_i \\ \Delta \eta_i \end{bmatrix} + A_{12}s_i \quad (24)$$

$$\dot{s}_i = v_i + \tilde{f}(\Delta f_i, \Delta \eta_i, t) \quad (25)$$

where $\tilde{f}(\Delta f_i, \Delta \eta_i, t)$ = the system disturbance, unmodelled changes and system uncertainties. The sliding mode controller based on the super-twisting algorithm is given by Eq. (26):

$$v = -k_1 \emptyset_1(s) - \int_0^t k_2 \emptyset_2(s)\mathrm{d}t \quad (26)$$

$$\emptyset_1(s) = |s|^{\frac{1}{2}}\mathrm{sign}(s) \quad (27)$$

$$\emptyset_2(s) = \emptyset_1{}'(s)\emptyset_1(s) = \frac{1}{2}\mathrm{sign}(s) \quad (28)$$

where

$$\mathrm{sign}(s) = \begin{cases} 1, & s > 0 \\ 0, & s = 0 \\ -1, & s < 0 \end{cases}$$

If $\left|\tilde{f}(\Delta f_i, \Delta \eta_i, t)\right|$ has a boundary condition $\rho|s|^{\frac{1}{2}}$, where $\rho > 0$, the condition for the globally asymptotic at origin $s_i = 0$ is given in Eqs. (29) and (30):

$$k_1 > 2\rho \quad (29)$$

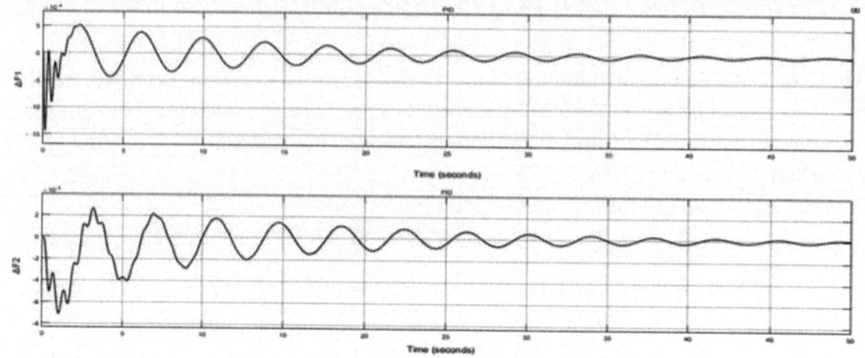

Fig. 2 Frequency deviations of two areas using of PID controller

$$k_2 > k_1 \frac{5\rho + 4\rho^2}{2k_1} \tag{30}$$

5 Results

Parameters of non-reheat turbine of area-1: $T_{t1} = 0.5$ (s), $T_{g1} = 0.2$ (s), $H_1 = 5$ (s), $D_1 = 0.6$ (pu/Hz), $R_1 = 0.05$ (pu/Hz) and $K_{Ii} = 0.3$. Parameters of non-reheat turbine of area-2: $T_{t1} = 0.6$ (s), $T_{g1} = 0.3$ (s), $H_1 = 4$ (s), $D_1 = 0.3$ (pu/Hz), $R_1 = 0.0625$ (pu/Hz) and $K_{Ii} = 0.3$.

Figure 2 shows the obtained graph when PID controller is used to tune the proposed two area power system and Fig. 3 shows the obtained graph when FOPID controller is used to tune the power system and it is showing best results in comparison with PID.

Figure 4 shows the obtained graph when the TID controller is used to tune the two area power system and Fig. 5 shows the obtained graph when SMC controller is used to tune the proposed two area power system.

6 Conclusion and Future Scope

A second-order SMC algorithm with an additional extended disturbance observer for a two area LFC scheme is proposed in this paper. For the proposed two area single unit power system, the overshoot and undershoot of the SMC is less than the PID, FOPID and TID controllers (shown in Table 1) which enables the more efficient operation of LFC. Though the settling time of TID controller is nearby SMC but on comparing the overall overshoots, undershoots and settling times of all the controllers, SMC yields

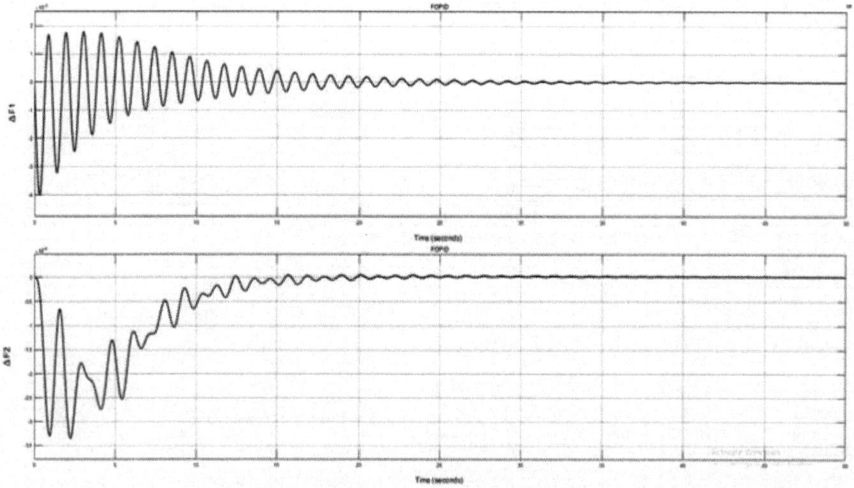

Fig. 3 Frequency deviations of two areas using FOPID controller

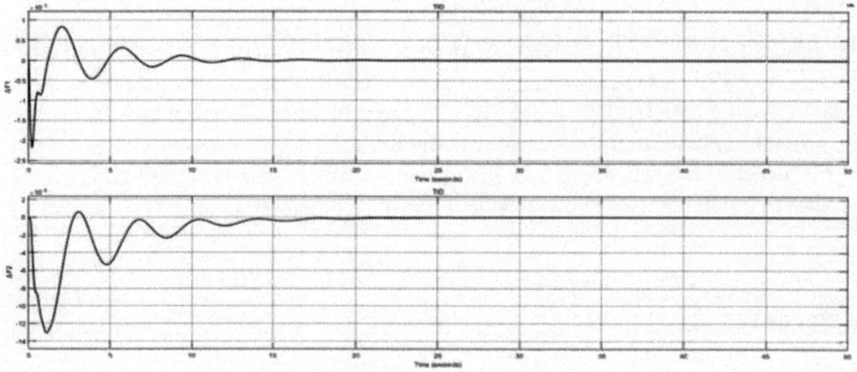

Fig. 4 Frequency deviations of two areas using TID controller

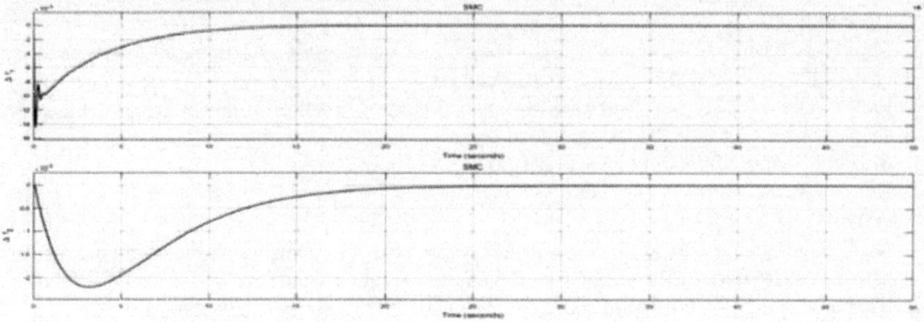

Fig. 5 Frequency deviations of two areas using SMC

Table 1 Controller parameters tuned using GWO algorithm

Controllers	K_P		K_T		K_I		K_D	
	ΔF_1	ΔF_2	ΔF_1	ΔF_2	ΔF_1	ΔF_2	ΔF_1	ΔF_2
PID	5.4062	25.5691	–	–	41.2698	12.864	8.69693	5.06956
FOPID	2.0842	0.53467	–	–	2.6327	6.1092	1.0024	1.4542
TID	–	–	5.4062	25.5691	41.2698	12.864	8.69693	5.06956

Table 2 Comparison of settling times, overshoots and undershoots for various controllers

Controllers	Settling time (in s)		Overshoot		Undershoot	
	ΔF_1	ΔF_2	ΔF_1	ΔF_2	ΔF_1	ΔF_2
PID	48	50	0.0005	– 0.0014	0.00003	– 0.00007
FOPID	40	30	0.0019	– 0.0004	0	– 0.0034
TID	18	24	0.0003	– 0.0022	0.00001	– 0.00013
SMC	16	26	0	– 0.0014	0	– 0.000022

the efficient results. The proposed work is effective for maintaining the frequency deviation of the power system to zero in a considerably less time and also reduces the overshoots and undershoots with SMC compared to other controllers which enables the efficient operation of the power system (Table 2).

Further, FACTS devices can be incorporated and also the proposed work can be applied to the deregulated power system for more efficient operation.

References

1. Siti MW, Tungadio DH, Sun Y, Mbungu NT, Tiako R (2019) Optimal frequency deviations control in microgrid interconnected systems
2. Daneshfar F, Bevrani H (2012) Multi objective of load frequency control using genetic algorithm
3. Singh VP, Kishor N, Samuel P (2016) Distributed multi-agent system based load frequency control for multi-area power system in smart grid
4. Liao K, Xu Y (2017) A robust load frequency control scheme for power systems based on second-order sliding mode and extended disturbance observer
5. Mirjalili S, Mirjalili SM, Lewis A (2014) Grey wolf optimizer
6. Hossain MdM, Peng C (2021) Observer-based event triggering H1 LFC for multi-area power systems under DoS attacks
7. Ali H, Madby G, Xu D (2021) A new robust controller for frequency stability of interconnected hybrid microgrids considering non-inertia sources and uncertainties
8. Khodabakhshian A, Edrisi M (2007) A new robust PID load frequency control
9. Liu F, Li Y, Cao Y, Jinhua S, Wu M (2015) A two-layer active disturbance rejection controller design for load frequency control of inter connected power system
10. Tan W (2009) Unified tuning of PID load frequency controller for power systems via IMC
11. Yousef HA, Al-Kharusi K, Albadi MH, Hosseinzadeh N (2013) Load frequency control of a multi-area power system: an adaptive fuzzy logic approach

12. Bevrani H, Daneshmand PR, Babahajyani P, Mitani Y, Hiyama T (2013) Intelligent LFC concerning high penetration of wind power: synthesis and real-time application
13. Chen C, Zhang K, Yuan K, Wang W (2017) Extended partial states observer based load frequency control scheme design for multi-area power system considering wind energy integration
14. Mohanty S, Subudhi B, Ray PK (2015) A new MPPT design using grey wolf optimization technique for photovoltaic system under partial shading conditions

Optimal Power Flow Using Firefly Algorithm with Solar Power

Thangella Aravind and Balusu Srinivasa Rao

Abstract One of the most intractable problems in power system networks is the optimal power flow problem (OPF). The firefly algorithm (FA), among the most popular meta-heuristic nature-inspired algorithms, is used to solve the OPF problem. This research uses FA to solve the optimal power flow problem with the addition of a solar energy system. The goal of this study is to reduce total fuel cost, minimize L-index (voltage stability index) and minimizing real power loss. The effect of incorporation of renewable energy system into OPF problem is studied on 30-bus IEEE test system. The proposed method has been implemented in MATLAB program, and these results are compared with various algorithms available in the existing literature.

Keywords Firefly algorithm (FA) · Optimal powerflow (OPF) · Solar energy system · Voltage stability index (L-index) · Transmission losses

1 Introduction

One of the very hard problems in power system networks is the optimal power flow (OPF) problem. During the span of time, many researches came into existence in OPF to reduce the optimization problems using different methods. In recent years, the OPF is a major task in renewable energy sources [1]. OPF problem is the main intention on three major conflicting objectives, i.e. minimization of generation cost, transmission losses, L-index [2]. In 1962, the OPF is first discussed in Carpentier. The power system network has to satisfy several constraints while maintaining generation costs as low as in an electrical network. There are two types of system constraints in a network: inequality and equality constraints [3]. An equality constraint is defined as to maintain the power balance equations, and the various inequality constraints

T. Aravind (✉) · B. S. Rao
V R Siddhartha Engineering College, Kanuru, Vijayawada, A.P. 520007, India

B. S. Rao
e-mail: balususrinu@vrsiddhartha.ac.in

P. K. Singh et al. (eds.), *Emerging Technologies for Computing, Communication and Smart Cities*, Lecture Notes in Electrical Engineering 875,
https://doi.org/10.1007/978-981-19-0284-0_28

of a power system network are required to maintain the system operating limits and security limits.

Predictable and artificial intelligence (AI) these is the solution of OPF problem methods. OPF is made up of a variety of universal techniques and has some drawbacks [4], i.e. continuous-time, slow convergence, and qualitative features are very weak in handling and operation is slow. Many authors are most preferred in artificial intelligence method since to get the optimal solution in global or approximate global. These approaches have a number of advantages, including the ability to deal with a variety of qualitative constraints, a single execution to obtain a large number of optimal solutions, the ability to solve multi-objective optimization problems, and the ability to find a global optimum solution [5]. The firefly algorithm is employed in this study to solve the multi-model optimization problem discovered by Xinshe Yang's [6]. It stands on the flashing behaviour of the bugs, including light emission, light absorption and the mutual attraction. There are various types of meta-heuristic algorithms that are differential evolution (DE) algorithm, artificial bee colony (ABC) algorithm, particle swarm optimization (PSO), clonal selection (CS) algorithm which are also similar to the proposed firefly algorithm [7]. FA is more useful for controlling parameters and also local searching ability, robustness, fast convergence [8]. The latest crowd intelligence gathering that utilizes firefly algorithm (FA) is proffered to determine the solution of OPF problem.

Wang Yi-BO [9] this paper presents the under structure of analysing steady-state characteristics of photovoltaic (PV) system connected to the power grid. Basically, the PV system consists of power converters. A PV system is separated into three basic modules: alternative current (AC) module, direct current (DC) module and inverter module.

This chapter is structured into seven sections as follows. The mathematical modelling of OPF problem formulation is presented in second section. Modelling of solar energy system is discussed in Sect. 3. The concept of FA is explained in fourth section. Section 5 discusses how to include the FA into OPF. In Sect. 6, FFA results obtained with MATLAB program are discussed. In Sect. 7, valid conclusions are discussed, and the last one is references.

2 Mathematical Problem Formulation of OPF

In any power system network, some objectives are reduced, and they met inequality and equality constraints. The OPF is a disordered optimization problem. Below equation represents the basic form of OPF problem.

$$\text{Minimize} : f(l, m)$$
$$\text{Subject to} : \begin{Bmatrix} g(l, m) \leq 0 \\ h(l, m) = 0 \end{Bmatrix} \tag{1}$$

where m—Independent control variables; l—Dependent state variables; $f(l, m)$—OPF objective function; $g(l, m)$—Inequality constraint; $h(l, m)$—Equality constraints.

2.1 Dependent State Variables

These types of variables in a power network can be expressed by vector l as:

$$l = \left[P_{G_1}, E_{L_1 \ldots} E_{L_{\mathrm{NL}}}, Q_{G_1 \ldots} Q_{G_{\mathrm{NG}}}, D_{l_1 \ldots} D_{l_{\mathrm{nl}}}\right] \quad (2)$$

where P_{G_1}—Slack bus generator real power; Q_{G_i}—Generator reactive power at ith bus; E_{L_p}—Magnitude of bus voltage at pth bus (load bus); D_{l_q}—Line loading of qth line; NL—Total transmission lines; Nl—Total load buses.

2.2 Independent system Control Variables

In a network controlling, the power flow depends on the variables presented in the below equation.

$$m = \left[P_{G_2 \ldots} P_{G_{\mathrm{NG}}}, E_{G_1 \ldots} E_{G_{\mathrm{NG}}}, Q_{c_1 \ldots} Q_{c_{\mathrm{NC}}}, T_{1 \ldots} T_{\mathrm{NT}}\right] \quad (3)$$

where P_{G_i}—Generator real power at ith bus; E_{G_m}—Voltage magnitude of PV bus; Q_{c_j}—shunt compensation at jth bus; T_i- ith branch transformer taps setting.

2.3 Constraints

Various types of constraints which are to be satisfied by this OPF problem are discussed in the following section.

2.3.1 Equality Constraints

These constraints are intertwined with both active and reactive power, as seen by the equations below.

$$P_{G_i} - P_{D_i} - E_i \sum_{j=1}^{\mathrm{NB}} E_j\left[K_{\mathrm{ij}} \cos\left(\delta_{\mathrm{ij}} \right. + B_{\mathrm{ij}} \sin\left(\delta_{\mathrm{ij}}\right)\right)\right] = 0 \forall i \in \mathrm{NB} \quad (4)$$

$$Q_{G_i} - Q_{D_i} - E_i \sum_{j=1}^{NB} E_j \left[K_{ij} \sin(\delta_{ij} - B_{ij} \cos(\delta_{ij}) \right] = 0 \forall i \in NB \quad (5)$$

where E_i – Voltage magnitude of bus i;δ_{ij}—are the voltage angles between the buses j and i; NB—Total no. of buses; P_D—Load demand of active power; Q_D—Load demand of reactive power; K_{ij}—Transfer conductance which is connected to ith bus; B_{ij}—Susceptance which is connected to jth bus.

2.3.2 Inequality Constraints

It is represented to maintain sustainable limits in a power system as shown in below equations.

(a) Generator Constraints: These constraints apply to both real and reactive power outputs, with the following upper and lower bounds limiting generator voltages:

$$E_{G_l}^{\min} \leq E_{G_l} \leq E_{G_l}^{\max} \forall l \in NG \quad (6)$$

$$P_{G_n}^{\min} \leq P_{G_n} \leq P_{G_n}^{\max} \forall n \in NG \quad (7)$$

$$Q_{G_n}^{\min} \leq Q_{G_n} \leq Q_{G_n}^{\max} \forall n \in NG \quad (8)$$

(b) Transformer constraints: Minimum and maximum limits limited these constraints in a transformer setting, expressed as follows.

$$T_i^{\min} \leq T_i \leq T_i^{\max} \forall i \in NG \quad (9)$$

(c) Shunt compensator constraints: These constraints are illustrated in reactive power injected at different buses and maintain upper and lower limits.

$$Q_{c_j}^{\min} \leq Q_{c_j} \leq Q_{c_j}^{\max} \forall j \in NC \quad (10)$$

(d) Security constraints:

$$E_{L_p}^{\min} \leq E_{L_p} \leq E_{L_p}^{\max} \forall p \in NL \quad (11)$$

$$S_{l_q} \leq S_{l_q}^{\max} \forall q \in nl \quad (12)$$

Equation (11) represents the voltage magnitudes at pth bus, and Eq. (12) represents the transmission line loading at qth bus.

2.3.3 Objective Functions

The following are the three major objective functions that were considered in this study in order to find the solution of the OPF problem:

a. Minimize cost of generation: This aims to decrease the generation cost of interrelated generation units. The stranded quadratic expression is given as follows.

$$f\left(P_{G_k}\right) = \sum_{k=1}^{\text{Ng}} \alpha_k + \beta_k P_{G_k} + \gamma_k P_{G_k}^2 [\$/\text{hr}] \tag{13}$$

where $\alpha_k, \beta_k, \gamma_k$—Cost coefficients of the kth generator.
$f\left(P_{G_k}\right)$—Fuel cost function; P_{Gk}—Generator power output at kth bus.
Ng—Total generators.

b. Voltage Stability index (L-index): A power system to maintain voltage of load buses L-index is used to avoid the voltage fall down point. This can be attained by minimization of L-index [10], expressed as shown in below equation.

$$L = \min\{E_j \quad j = 1, \ldots F_{\text{PQ}}\} \tag{14}$$

where F_{pQ}—total load buses.

c. Minimization of transmission losses: In this objective, to decrease the real power losses and it is denoted by P_{Loss}.

$$P_{\text{Loss}} = \sum_{i=1}^{N_L} \frac{r_k}{r_k^2 + x_k^2}\left[E_i^2 + E_j^2 - 2E_i E_j \cos\left(\delta_i - \delta_j\right)\right] \tag{15}$$

where N_L—Number of transmission lines;
r_k—Resistance of kth transmission line;
E_i,E_j—Voltage at ith and jth bus;
δ_i,δ_j—Angles at ith and jth bus.

3 Modelling of Solar Energy System

In the contemporary years, photovoltaic power generation is more developed in the power system network which reduces the pollution and PV generation has more vital social and economic advantages. One of the main boons of photovoltaic (PV) system is that it directly converts the solar irradiance into electricity. The PV system gradually improves the technology and reduces the cost and many countries adopted the PV generation system in order to reduce the harmful emissions which are dangerous

Fig. 1 Equivalent transformation of AC circuit

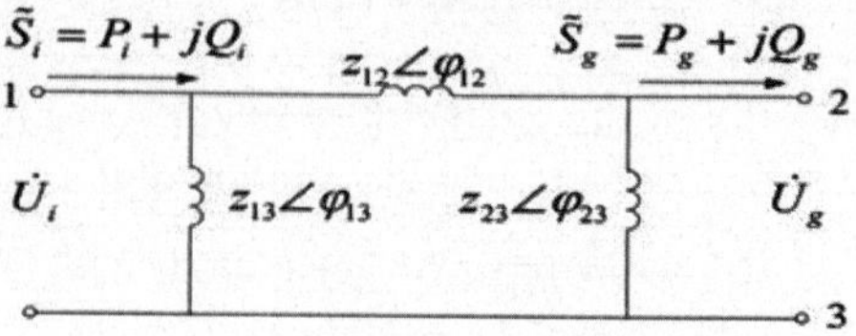

for the environment. Commonly, PV system is integrated by power electronics converters.

One form of renewable energy source is solar energy, when sunlight energy is directly converted into the electricity using PV panels. When PV panels are made up on mostly semiconductor materials since this is more gain of sunlight energy comparison of insulator materials. For the calculation of AC circuit, output power by using the parallel–series and star/delta transformation is shown in Fig. 1. The PV panel power output is transformed from inverter, and output of the inverter can be further transformed into grid as shown in below equations.

$$P_i = 3E_i\left[\frac{E_i}{Z_{12}}\cos\varphi_{12} + \frac{E_i}{Z_{13}}\cos\varphi_{13} - \frac{E_g}{Z_{12}}\cos\left(\theta_g - \alpha - \varphi_{12}\right)\right] \tag{16}$$

$$Q_i = 3E_i\left[\frac{E_i}{Z_{12}}\sin\varphi_{12} + \frac{E_i}{Z_{13}}\sin\varphi_{13} + \frac{E_g}{Z_{12}}\sin\left(\theta_g - \alpha - \varphi_{12}\right)\right]. \tag{17}$$

$$P_g = 3E_g\left[\frac{E_i}{Z_{12}}\cos\left(\alpha - \theta_g - \varphi_{12}\right) - \frac{E_g}{Z_{12}}\cos\varphi_{12} - \frac{E_g}{Z_{23}}\cos(\varphi_{23})\right] \tag{18}$$

$$Q_g = 3E_g\left[\frac{E_i}{Z_{12}}\sin\left(\alpha - \theta_g - \varphi_{12}\right) + \frac{E_g}{Z_{12}}\sin\varphi_{12} + \frac{E_g}{Z_{23}}\sin(\varphi_{23})\right] \tag{19}$$

4 Firefly Algorithm

Several innovative algorithms for solving engineering optimization problems have been introduced in the last couple of decades. Among all these new algorithms, it has been expressed that firefly algorithm (FA) is the most appropriately planned in dealing with global optimization problem [6]. FA, which is based on the shining pattern and social interactions of fireflies, was created in 2007 and 2008 by XinShe Yang at Cambridge University, including light absorption, light emission and mutual attractiveness [8, 11]. For the flexibility of new meta-heuristic FA, three major idealized rules are indicated [12–14].

(1) Generally, fireflies are unisexual; i.e. each firefly will be attracted to the other firefly in the group despite the sex.

(2) Attractiveness α brightness, i.e. any two shinning fireflies, the firefly that is less luminous will approach the firefly that is brighter. As the distance between them grows, the brightness's appeal reduces, and vice versa. If there isn't a brighter firefly nearby, it will migrate at random.
(3) The brightness of a firefly algorithm will be resoluted from the landscape of the objective function.

These three idealized principles are based on, and FA may be clarified in a step-by-step approach that can be presented as the pseudo-code [15].

5 Firefly Algorithm for Solving OPF Problem

This algorithm is mainly considered two major issues: The first one is a divergence in light intensity I, while the second is an expression of attraction β. Any brilliant firefly in a specific point z can be chosen at random as:

$$I(z) \propto \frac{1}{f(z)} \tag{20}$$

The firefly light intensity I is proportional to distance r. That is,

$$I(z) = I_o e^{-\gamma r} \tag{21}$$

whereI_o—Starting luminous intensity.

γ—Absorption ratio.

The light intensity observed by surrounding fireflies is related to the attraction of fireflies; i.e. a firefly's attractiveness can be calculated as:

$$\beta(r) = \beta_o e^{-\gamma^m} m \geq 1 \tag{22}$$

where

β_o—Attractiveness at distance $r = 0$; M—Total fireflies.

The firefly i that is less brilliant goes towards the firefly j that is less luminous. The updated position of firefly i can be represented as in Eq. (23):

$$z_i = z_i + \beta_0 e^{-\gamma r_{\mathrm{ij}}^2}\left(z_j - z_i\right) + \alpha(\mathrm{rand} - 0.5) \tag{23}$$

with

$$r_{\mathrm{ij}} = \left|z_i - z_j\right| \sqrt{\sum_{k=1}^{d} (z_{i,k} - z_{j,k})^2} \tag{24}$$

where r_{ij}—Parting between the two fireflies j and i at locations z_j and z_i.

α—Randomness parameter.

6 Results and Discussions

The propounded FA method has been practised on a standard 30-bus IEEE system with a solar energy system for single-objective optimization problem. This test system included 41 branches, 6 generator buses and twenty-four load buses, 4 transformers, and 9 shunt compensations on various buses. The test system consists of six thermal generators (TG) which are placed on the 1st (Slack), 2nd, 3rd, 4th, 5th and 6th buses. Table 1 shows the minimum and maximum real power generating limits, and cost coefficients of total generators. Table 2 lists the minimum and maximum voltage magnitudes, transformer tap settings, and reactive power injections. The overall load demand is 283.4 MW and 126.2MVAR. This manuscript includes three conflicting objectives such as total cost, L-index and power loss for optimization. The proposed FA is applied to find a solution to single-objective optimization with and without solar energy system.

Case 1-Without solar energy system: Initially, without considering the solar energy system, each objective function was considered separately for single-objective optimization using the FA technique. Table 2 shows that the FA is successful in decreasing total fuel cost, L-index and real power loss. Table 2 shows the optimal settings for all control variables for 30-bus IEEE system without solar energy. Fig. 2 depicts the convergence plots of these objectives in the absence of a solar energy system.

Case 2-With Solar energy system: In this part, the proposed FA is used to solve a single-objective OPF problem with the three objectives mentioned above and the incorporation of a solar energy system. At 7th bus of 30-bus IEEE system solar generator is placed. The optimal values of all the control variables obtained using FA when optimized separately with solar energy system are shown in Table 3. Figure 3 depicts the convergence curves of these objectives with a solar energy system.

Table 1 Cost coefficients

Generator	α_i	β_i	γ_i	$P_{G_i}^{\min}$	$P_{G_i}^{\max}$
TG1	37.5	200	0	0.5	2.0
TG2	175	175	0	0.2	0.8
TG3	83.4	325	0	0.1	0.35
TG4	250	300	0	0.1	0.3
TG5	625	100	0	0.15	0.5
TG6	250	300	0	0.12	0.4
Solar	0	160	0	0.1	0.2

Table 2 Best solution with FA when optimized separately without solar system (Case-1)

Variables names	Max	Min	Fuel cost	L-index	Loss
P_{G_1}(MW)	200	50	177.03	51.46	51.24
P_{G_2}(MW)	80	20	48.71	79.98	80.00
P_{G_3}(MW)	50	15	21.08	35.00	35.00
P_4(MW)	35	10	11.88	29.99	30.00
P_{G_5}(MW)	30	10	21.30	49.98	50.00
P_{G_6}(MW)	40	12	12.00	40.00	40.00
V_1(p.u.)	1.10	0.95	1.1000	1.1000	1.1000
V_2(p.u.)	1.10	0.95	1.0878	1.1000	1.0975
V_3(p.u.)	1.10	0.95	1.0693	1.1000	1.0867
V_4(p.u.)	1.10	0.95	1.0906	1.1000	1.0945
V_5(p.u.)	1.10	0.95	1.0618	1.0937	1.0798
V_6(p.u.)	1.10	0.95	1.1000	1.1000	1.1000
T_1	1.10	0.90	1.0376	1.0485	0.9838
T_2	1.10	0.90	0.9411	1.0289	1.0411
T_3	1.10	0.90	0.9643	0.9982	0.9728
T_4	1.10	0.90	0.9589	0.9548	0.9706
Q_{c10}(p.u.)	0.05	0.0	0.0023	0.0500	0.0006
Q_{c12}(p.u.)	0.05	0.0	0.0500	0.0500	0.0500
Q_{c15}(p.u.)	0.05	0.0	0.0500	0.0500	0.0500
Q_{c17}(p.u.)	0.05	0.0	0.0500	0.0500	0.0499
Q_{c20}(p.u.)	0.05	0.0	0.0500	0.0500	0.0500
Q_{c21}(p.u.)	0.05	0.0	0.0495	0.0500	0.0485
Q_{c23}(p.u.)	0.05	0.0	0.0258	0.0500	0.0216
Q_{c24}(p.u.)	0.05	0.0	0.0268	0.0500	0.0270
Q_{c29}(p.u.)	0.05	0.0	0.0244	0.0432	0.0232
Cost($/hr)	–	–	**799.0345**	967.2860	966.7782
L-index	–	–	0.1163	**0.11012**	0.1160
Loss(MW)	–	–	8.61	3.01	**2.8467**

The comparison of results both (without solar and with solar) by using the FA method is shown in Tables 2 and 3. The overall cost is lowered from 799.0345$/hr to 759.4226$/hr when a solar energy system. The L-index is slightly increased from 0.11012 to 0.11148 with solar energy system. Finally, with a solar energy system the total power loss is reduced from 2.8467 to 2.4 MW.

Table 4 shows that the proposed FA results for case 1 best among all other techniques currently available in the literature. However, the results obtained with incorporation of solar energy systems are not compared with the literature as there is no similar work found for case 2.

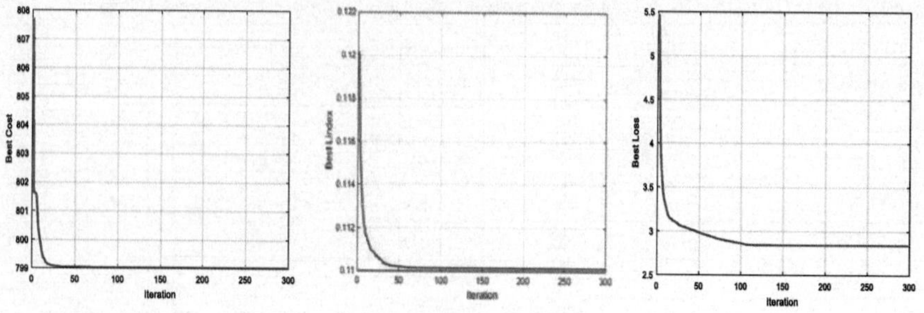

a) Cost b) L-Index c) Loss

Fig. 2 Convergence curves without solar system (Case-1)

7 Conclusion

In this paper, a current robust crowd intelligence built on FA with a solar energy system to work out the OPF problem. The FA was effectively implemented to solve the OPF problem to optimize the generation cost, L-index and active power loss. The proposed method is tested on standard 30-bus IEEE system. The FA results compared with and without solar energy system. The result analysis of the given test system shows that the proposed FA method is well suitable for handling single-objective OPF problems using solar power. The future scope of this research will be a multi-objective OPF problem combining solar and wind power.

Table 3 Best solution with FA when optimized separately with solar energy system (Case-2)

Variable names	Max	Min	Fuel cost	L-index	Loss
P_{G_1}(MW)	200	50	167.05	107.78	31.03
P_{G_2}(MW)	80	20	46.12	42.70	79.84
P_{G_3}(MW)	50	15	15.26	34.70	35.00
P_4(MW)	35	10	10.11	29.75	30.00
P_{G_5}(MW)	30	10	20.49	17.81	50.00
P_{G_6}(MW)	40	12	12.00	38.79	39.94
P_{G_7}(MW)	50	0	20.00	16.96	19.98
V_1(p.u.)	1.1	0.95	1.100	1.0612	1.0373
V_2(p.u.)	1.1	0.95	1.0868	1.0504	1.0358
V_3(p.u.)	1.1	0.95	1.0660	1.0600	1.0239
V_4(p.u.)	1.1	0.95	1.0827	1.1000	0.9991
V_5(p.u.)	1.1	0.95	1.0592	1.0224	1.0187
V_6(p.u.)	1.1	0.90	1.0999	1.0808	1.0491
V_7(p.u.)	1.1	0.90	1.0655	1.0449	1.0228
T_1	1.1	0.90	0.9493	1.0207	1.0226
T_2	1.1	0.90	1.0573	0.9617	0.9311
T_3	1.1	0.90	0.9661	0.9291	0.9695
T_4	1.1	0.90	0.9506	0.9558	0.9601
Q_{c10}(p.u.)	0.05	0.0	0.0202	0.0499	0.0004
Q_{c12}(p.u.)	0.05	0.0	0.0472	0.0499	0.0540
Q_{c15}(p.u.)	0.05	0.0	0.0485	0.0499	0.0570
Q_{c17}(p.u.)	0.05	0.0	0.0498	0.0498	0.0611
Q_{c20}(p.u.)	0.05	0.0	0.0497	0.0500	0.0873
Q_{c21}(p.u.)	0.05	0.0	0.0495	0.0498	0.0040
Q_{c23}(p.u.)	0.05	0.0	0.0283	0.0495	0.0155
Q_{c24}(p.u.)	0.05	0.0	0.0102	0.0496	0.0273
Q_{c29}(p.u.)	0.05	0.0	0.0121	0.0500	0.0185
Fuel cost($/hr)	–	–	**759.4226**	818.7200	951.3025
L-index			0.1147	**0.11148**	0.1254
Loss(MW)	–	–	7.67	5.2	**2.4**

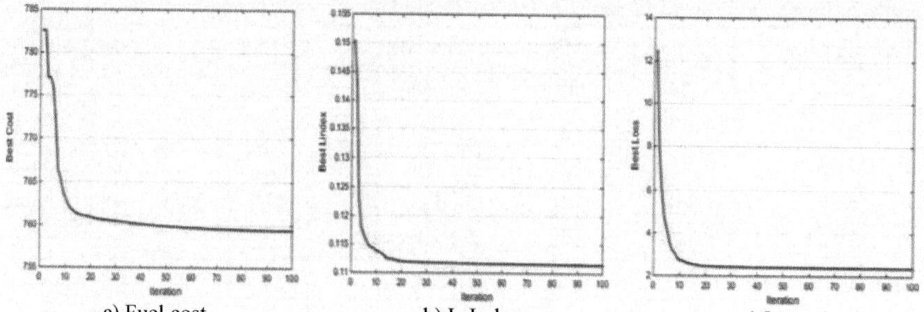

a) Fuel cost b) L-Index c) Loss

Fig. 3 Convergence curves with solar system (Case-2)

Table 4 FA results compared to prior research for the 30-bus IEEE system

Objectives	Algorithm	Fuel cost	L-index	Loss
Cost objective	FA	**799.0345**	0.1163	8.61
	MSA [16]	800.5099	0.13833	9.0345
	BSA [17]	799.0760	0.1273	8.6543
	ARCCBO[18]	800.5159	0.1385	9.0255
	SKH [19]	800.5141	0.1382	9.0282
	DE [20]	799.0827	0.1277	8.63
	GEM [21]	799.0463	0.1264	8.6257
L-index objective	FA	967.2860	**0.11012**	3.01
	SKH [19]	814.0100	0.1366	9.9056
	GEM [21]	816.9095	0.1257	6.2313
	DE [20]	915.2172	0.1243	3.626
Loss objective	FA	966.7782	0.1160	**2.8467**
	MSA [16]	967.6636	0.13832	3.0844
	DSA [22]	967.6493	0.12604	3.0954
	ARCBBO[18]	967.6605	0.1386	3.1009
	APFPA [23]	965.6590	–	2.8463
	GEM [21]	966.7473	0.1265	2.8863

References

1. Muller SC, Hager U, Rehtanz C (2014) A multiagent system for adaptive power flow control in electrical transmission systems. IEEE Trans Ind Inf 10(4):2290–2299
2. Rao BS (2017) Application of adaptive clonal selection algorithm to solve multi-objective optimal power flow with wind energy conversion systems. Int J Power Energy Conver 8(3):322–342
3. Biswas PP et al (2018) Optimal power flow solutions using differential evolution algorithm integrated with effective constraint handling techniques. Eng Appl Artif Intell 68:81–100

4. Bouchekara HREH (2014) Optimal power flow using black-hole-based optimization approach. Appl Soft Comput 24:879–888
5. Ponsich A, Jaimes AL, Coello Coello CA (2012) A survey on multiobjective evolutionary algorithms for the solution of the portfolio optimization problem and other finance and economics applications. IEEE Trans Evol Comput 17(3):321–344
6. Yang XS (2013) Firefly algorithm: recent advancements and application. Int J Swarm Intell 1:36–50
7. Mishra N, Pandit M (2013) Environmental/economic power dispatch problem using particle swarm optimization. Int J Electron Comput Sci Eng (IJECSE) 2(2):512–519
8. Yang X-S (2009) Firefly algorithms for multimodal optimization. Stochastic Algor: Found Appl SAGA 5792:169–178
9. Wang Y-B et al (2008) Steady-state model and power flow analysis of grid-connected photovoltaic power system. 2008 IEEE international conference on industrial technology. IEEE
10. Tuan TQ, Fandino J, Hadjsaid N, Sabonnadiere JC, Vu H (1994) Emergency load shedding to avoid risks of voltage instability using indicators. IEEE Trans Power Syst 9(1):341–351. https://doi.org/10.1109/59.317592
11. Sarbazfard S, Jafarian A (2016) A hybrid algorithm based on firefly algorithm and differential evolution for global optimization. Int J Adv Com Sci Appl 7(6):95–106
12. Lukasik S, Zak S (2009) Firefly algorithm for continuous constrained optimization tasks. In: Nguyen NT, Kowalczyk R, Chen SM, eds. Proceedings of the international conference on computer and computational intelligence (ICCCI '09), vol 5796. Springer, Wroclaw, Poland, pp 97–106
13. Yang XS (2010) Firefly algorithm, Levy flights and global optimization. Research and development in intelligent systems XXVI. Springer, London, UK, pp 209–218
14. Yang XS (2010) Firefly algorithm, stochastic test functions and design optimization. Int J Bio-Inspired Comput 2(2):78–84
15. Subramanian R, Thanushkodi K (2013) An efficient firefly algorithm to solve economic dispatch problems. Int J Soft Comp Eng (IJSCE) 2(1):52–55
16. Mohamed AAA, Mohamed YS, El-Gaafary AA, Hemeida AM (2017) Optimal power flow using mothswarm algorithm. Electric Power Syst Res 142:190–206
17. Chaib AE et al (2016) Optimal power flow with emission and non-smooth cost functions using backtracking search optimization algorithm. Int J Electr Power Energy Syst 81:64–77
18. Kumar AR, Premalatha L (2015) Optimal power flow for a deregulated power system using adaptive realcoded biogeography-based optimization. Int J Electric Power Energy Syst 73:393–399
19. Pulluri H, Naresh R, Sharma V (2018) A solution network based on stud krill herd algorithm for optimal power flow problems. Soft Comput 22(1):159–176
20. Shaheen AM, El-Sehiemy RA, Farrag SM (2016) Solving multi-objective optimal power flow problem via forced initialized differential evolution algorithm. IET Gen Trans Distrib 10(7):1634–1647
21. Bouchekara HREH, Chaib AE, Abido MA (2016) Multi-objective optimal power flow using a fuzzy based grenade explosion method. Energy Syst 7(4):699–721
22. Abaci K, Yamacli V (2016) Differential search algorithm for solving multi-objective optimal power flow problem. Int J Electr Power Energy Syst 79:1–10
23. Mahdad B, Srairi K (2016) Security constrained optimal power flow solution using new adaptive partitioning flower pollination algorithm. Appl Soft Comput 46:501–522

Designing of Learning Environment of an Intelligent Tutoring System for Academic Learning Needs of Learning-Disabled Learners Based on Survey Report of Region-Specific Target Group

Neelu Jyothi Ahuja, Monika Thapliyal, Anand Nayyar, and Adarsh Kumar

Abstract The twenty-first century is known for exponential growth in the technological as well as the education domain. However, conventional educational tools are still reliable to understand the actual scenario of performance and efficiency of the brain of young minds. If combined with technology this could play a vital role in getting a clear perception about what they feel and how they improve the educational methodology. Studies on dyslexia, dysgraphia and dyscalculia have shown that it is very tough and sometimes impossible to identify these learners without the help of professionals. Unlike physical challenges, the challenges from these disabilities are not measurable in quantified terms. For this, perception-based studies play a vital role. There are various studies, which suggest questionnaire-based survey, or direct interviews with the target group are a more reliable source of information. Also, regional-level data sometimes plays an important role to understand geographical and environmental impacts on the target group. Various reports were studied to understand the similarities. Authors collected information as a pilot project through a survey on 292 learners (learning-disabled and non-learning-disabled) in different institutes. 142 of them were dealing with single or multiple learning disabilities. Study helped in identifying the most affected learning domains and related multiple-criteria affecting the learners. This eventually is implemented in the domain model

N. J. Ahuja · M. Thapliyal · A. Kumar
Department of Systemics, School of Computer Science, University of Petroleum and Energy Studies, Dehradun, India
e-mail: neelu@ddn.upes.ac.in

A. Kumar
e-mail: adarsh.kumar@ddn.upes.ac.in

A. Nayyar (✉)
Graduate School, Duy Tan University, Da Nang 550000, Vietnam
e-mail: anandnayyar@duytan.edu.vn

Faculty of Information Technology, Duy Tan University, Da Nang 550000, Vietnam

P. K. Singh et al. (eds.), *Emerging Technologies for Computing, Communication and Smart Cities*, Lecture Notes in Electrical Engineering 875,
https://doi.org/10.1007/978-981-19-0284-0_29

of an intelligent tutoring system to develop the four learner-centric learning environments. Results show that practice-based learning environment was the most relatable learning environment followed by visual-based learning environment.

Keywords Dyslexia · Dysgraphia · Dyscalculia · Intelligent tutoring system · Domain model · Learning disability · Survey · Learning environments

1 Introduction

Learning disabilities (LDs) refer to the condition in which learner repeats similar errors and faces difficulty in academic learning skills related to reading, writing, speaking and solving mathematics. The learner may have one or multiple problems. Although there is no problem with the intelligence of these learners, repetition in mistakes make them look like 'problematic' or 'uninterested' person. Some characteristics features in these learners are slow learning speed, limited interest, speech defect, lack of muscular coordination, adjustment problem and social anxiety. Dyslexia, dysgraphia and dyscalculia are the most common of all these learning disabilities. 5–15% of children deal with learning disability of some form. Dyslexia is most common LD [1]. More than being learning-disabled, these learners are 'differently abled'.

The cases of learning disability first get the attention of the world when a psychologist Samuel Kirk coined the term 'Learning Disability' in Chicago conference in 1963. Until then, these learners were misunderstood as mentally challenged. In the literature, learning disability (LD) is also mentioned as learning difficulty and specific learning disability. The condition of intellectual disabilities like autism, down syndrome and attention deficit hyperactivity disorder are very much different from LD. The learning needs are very different in these two learning conditions. Intellectual disabilities are characterized based on intelligent quotient (IQ), whereas learning disabilities are characterized on problems in academic learning skills. LD learners have normal to high IQ. Until recent times, it is hard to diagnose due to lack of awareness among parents and teachers and lack of proper assessment tools [2].

1.1 Learning Disabilities: Dyslexia, Dysgraphia and Dyscalculia

Dyslexia concerns to reading-related problems. Despite having normal vision and normal IQ learner find it difficult to read alphabets, words, sentences or paragraphs. Learner adds, omits, substitutes or reverse letters or words with similar structure. Frequent mistakes make learner hesitate of reading in front of class or in stage. Since learner reads incorrectly, he learns incorrectly, which leads to frequent mistakes during writing. Learners lack in literacy skills and phonological awareness [3].

Dysgraphia concerns to writing-related problems. Learner has weak grip over pencils or has lack of muscular coordination. Handwriting is illegible therefore despite knowing correct answer learner fails to express in written form. These learners show positive results in verbal examinations. Some dysgraphia learner has problem in expression of thoughts and is weak in sentence and word expressions. Some find it hard to organize information of pictures, graphs, maps have visual confusion of left–right align and directions and also observed having weak visual–spatial response [4].

Dyscalculia concerns to math-related problems. Learners have difficulty in understanding the concept of basic math. He may find difficulty in some of the following problems: counting, backward counting, basic calculation like addition, subtraction, multiply, division, telling time, geometry and reasoning. Learner gets confused with mixed sums and finds more difficulty in understanding word problems [5].

Sometimes these learning disabilities are a result of some other intellectual disability or learning condition. For example, a learner with hyperactive disorder may tend to show the dysgraphia symptoms due to the effect on sensory–motor coordination. Also, learning-disabled learners have word-finding difficulty (WFD). It means child knows the word but is not able to use it in conversation or find it difficult to identify when shown in pictures. A learner with stuttering or social anxiety may show similar symptoms. Assessment through professionals or special educator is required to understand the exact condition of learner. Research has shown that special educator can diagnose very much accurately the condition of their student during early age.

1.2 Domain Model of an Intelligent Tutoring System

An intelligent tutoring system is a software designed with the ability to assess the learning needs of user. It reflects the developed rules and designed curriculum in preferential learning styles of users while adapting as per their response and activity during learning [6].

Out of four models of an ITS: student model, domain model, pedagogy model and user-interface model, the domain model is responsible for the resources, assessments, inventories, instruction plan and designed environment. A learner-centric approach based on individual skill and competency improves the learning [7, 8].

2 Objectives of Paper

The objectives of the paper are the following:

i. Study of learners to get regional quantified data on dyslexia, dysgraphia and dyscalculia to understand academic learning needs of learners.

ii. To perform a survey to identify target group in the target agency/school, understand their clinical condition and nature of learning disability, gauge their academic performance in a traditional learning environment followed by a thorough need assessment and analysis phase through written and verbal examination.
iii. Developing learner-centric learning environments for domain model of an intelligent tutoring system corresponding to affect learning domains and academic learning needs of target learners.

Here in this paper, the next section will describe the literature review section. In the following section, a brief study was conducted to identify regional-specific target group and to study their academic learning needs. Then, in research methodology section, a design of four learning environment for domain model of an intelligent tutoring system is presented. The later part of paper consists of result, discussion and conclusion.

3 Review of Literature

Learning is a dynamic process. It depends on various factors like academic level, prerequisite knowledge, level of interest about topic and learning style of learner. A child having learning disability has tendency to lose focus and interest in study. In order to teach such learner effectively, proper assessment is required. A detailed test in the form of written and verbal examination is necessary to understand the requirements and problems of learners. Following are some tests taken by professional to get detailed report of learner: Group intelligence test, individual intelligence test, achievement test, attainment test, teacher's/special educator's opinion.

Singh et al. [1] did a 5-year study in Indian students mainly from English medium schools and concluded most of the children with LD have dyslexia, dysgraphia and dyscalculia. Most children were reported to have more than one LD. 38.56% had ADHD along with LD [1].

Karande et al. [2] India stated SpLD as invisible handicap and suggests early intervention for all children dealing with LD. However, for proper diagnosis of LD a team of specialists should include a paediatrician, counsellor, clinical psychologist, special educator and child psychiatrist [2].

Gregely. K et al. [3] in their study concluded that a paediatrician can play a crucial role in diagnosis and therapy of LD learners. They can help parents by providing a detailed report of problems and solutions. Parents are confused about the learning disabilities and their impact on studies and daily life of their child. A paediatrician along with a special educator can be helpful in the most effective14 manner [3].

Nagyova [6] emphasizes the monitoring of student's learning activities in various learning environments to categorize students into studious and practical types. These learners were then provided learning material with different learning approaches [6].

Brandenburger et al. [9]in their study on the preference of data visualization during online teaching found out that learners find it more attractive if the classic form of data is presented. It helps in reducing distraction. Colour schemes were found not influencing the learners perception [9].

Karampa and Paraskeva [10] gave the concept for smart learning environment by blending information and communications technology and smart pedagogy. Thus, created SLE would be portable, affordable [10].

Nair et al. [11] emphasized on the individual and interdisciplinary early intervention with focus on skills of phoneme awareness, phonics, reading comprehension, spelling, organization of information [11].

Taur et al. [12] performed six-psychomotor computerized test on 25 learners with LD. Finger test, reaction test, sorting tests were performed. Weak attention and motor coordinations were reported [13]. LD learners also found it difficult to adjust to social norms. Social anxiety, weak adjustment, low confidence are common in these learners [12, 14].

González-Valenzuela and Martín-Ruiz [15] discussed the effect of written and verbal interventions in LD learners. They conclude that the early, systematic and planned interventions improve the writing skills. Learners with dysgraphia can get benefit when remedies are provided at an early age [15].

Beckmann and Minnaert [16] reviews the 23 publications on gifted children with learning disabilities of some form. Among these literatures, most common characteristic in this learner was frustration. In addition, a need for counselling was found for these learner due to the dual nature in cognitive abilities [16].

Gorker [2014] in their work concluded that problem in one learning domain can affect another learning domain. A learner with comorbid learning disorder will tend to perform poor in word problems of math due to poor reading skills. Such learner needs different guidance than others performing weak in math [17].

Asthana and Hazela [18] explored the role of machine learning in ITS to improve the learning environment. They propose the idea of using supervised and unsupervised learning to customize learning, automate assessments and provide feedbacks [18].

Rose et.al [19] in their work for children with learning disabilities in rural and urban areas of southern states of India emphasize on the benefits of inclusive education. The learning space in rural areas requires a great attention, guidance, counselling and professional support to fulfil the learning needs [19].

Kumar et al. [20] identified types of learners as acoustic, imagistic, intuitive and active. This is later used in their ITS designed for university students about the study of seismic waves to adapting as per learner's preferential learning style [20].

Russell et al. [21] of UK found that learning-disabled people tend to have weaker physical and mental health and found it difficult to explain their condition. Learning-disabled people find it difficult to analyse information, making decisions, making plans, logical sequencing, processing information at normal speed, solving problems, reasoning and visuospatial reasoning. Russell & team also found that mild and moderately learning-disabled were even not registered. Hence, recognition of such people is necessary to improve their health [21].

Chavan and Rozatkar [22] of India pointed out the drastic difference in data regarding disabled in India. Census 2001 has data of 2.13% people (21 million people) with disabilities, and National Sample Survey Organization (NSSO) has 1.8% people with disabilities. Also, this sample included visual, hearing, locomotor and mental disabilities. IQ tests are used to test these learners but IQ tests have limitations and are not helpful in finding the exact learning needs of learners [22].

A similar study by Polat et al. [23] in Turkey stated the unavailability of diagnosis tool and direct help to such learners. Their research was focused on learners with dyslexia, dysgraphia and dyscalculia. Their research involves children, teachers and parents. They concluded that there is a lack of information about LD in society and learners needs help which can be provided through a web-assisted system that can be used inside and outside classrooms [23]. Taymans and Kosaraju [24] in a study in USA pointed out the need of assessment and help for college students and adults dealing with learning disabilities for better job opportunities [24].

Mehta [25] points out the tremendous stress a learner with learning disability goes through. Parents and teachers pressurize the child for better score all the time. This results in low self-esteem and lack of self-confidence for such learners. Also, private treatment session costs high. Central Board of Secondary Education (CBSE) has given provisions of extra time in examinations, quota in higher studies and exemption of third language to these learners. This is beneficial for the needy children and is a good step but some highly ambitious parents are reported in misuse of certification by getting it with unethical way for their intellectually disabled children [25].

Ariffin et al. [26] studied the impact of mobile application for learners with dyscalculia. Seven learners with dyscalculia participated in experiment. The application designed for Malaysian kids was found helpful and effective [26].

4 Research Methodology

Learners were observed individually through questionnaire and direct interview in the presence of their special educators. Participants were from various cities and villages from in-and-around Dehradun District of Uttarakhand. Phone calls and mails were sent to schools and education centres. Survey was conducted in online and offline mode. Learners from LKG to class 7th in various schools of state board and Central Board of School Education (CBSE) participated in the survey. The schools were chosen out of 21 existing schools dealing with various forms of disabilities. A series of descriptive interviews, questionnaires and observations were taken with these learners and their educators or guardians. Target group identified was consisted of 292 learners during initial assessment.

In detailed assessment, it was found that 98 learners out of 292 were non-LD learners. These non-LD learners had weak academic performance but they were not among LD. 52 learners out of 292 learners were physically challenged or had some other intellectual disability like autism and attention deficit hyperactivity disorder

(ADHD). These learners are taken into group of others and not counted as LD learners.

Questionnaires were created based on NCERT syllabus and books after discussion with professionals. Five questionnaires were made for learners to measure academic learning skills related to reading, writing and mathematics: (KG), (I–II), (III–IV), (V–VI), (VII). No IQ test was taken. To determine the dyslexia-related problem, verbal tests and interview were done. For dysgraphia and dyscalculia written tests were given. Questionnaire involves the following types of questions:

1. Verbal questions (to check speech-related problems like pronunciations, phonological awareness, rhyming, expression of thoughts and visual confusions).
2. Written questions (to check handwriting, writing skills, calculations, series and patterns).
3. Inventories for guardian/special educator (to check social and personal skill of learner).

Questionnaire were given according to the educational level and prerequisite knowledge after discussion with special educator of learner. LD-specific cognitive strengths are taken into consideration [27]. The questions are mapped corresponding to the nine domains of cognitive ability:

Long-term memory, short-term memory, attention, concentration, orientation, abstraction/rational thinking, language ability, visual perception and fluency [18].

Academic learning skills and cognitive strengths are taken by us cover these nine domains of cognitive abilities.

Language ability and fluency are overlapping with problems of dyslexia as academic learning skills (phonological awareness and literacy skills of dyslexia) [28] and hence not considered as cognitive strength. Performance is checked on a scale from 1 to 5, 1 being bad, 2 being below average, 3 being average 4 being good and 5 being very good. Question unanswered is marked as 0 (zero). X is marked for questions not applicable according to age limit of learner. Observation sheets were made for learners (to be filled by project team based on verbal response of learners).

Dyslexia: To find out academic learning skills related to dyslexia, i.e. literacy skills and phonological awareness following tests were taken—identification of letters/words, phonological awareness, reading letters/words/non-sense words/sentences, rapid naming of colour/objects and rhyming words.

Dysgraphia: To find out academic learning skills related to dysgraphia, i.e. handwriting, sentence and word expression, visuospatial response following tests were taken—dictation, rewrite the sentence, jumbled words/sentence, directions and storytelling by picture.

Dyscalculia: To find out academic learning skills related to dyscalculia, i.e. counting, basic calculation, reasoning following tests were taken—identification of numbers, basic sums, pattern, forward/backward series and word problems.

All questions were as per educational level of learner and were suggested by their special educators. If learners find it very difficult to perform test of their grade level, a lower grade-level questionnaire was provided to them.

5 Results

Result section presents the result for the research objectives. It comprises of two sections:

5.1 Research Objective 1: Identification of Specific Target Group Through Various Assessments

Table 2 shows the list of most visible academic learning problems in written test and verbal test of learners. Figure 1 shows the associated problems observed by survey team while the learner was performing the test (Table 1).

Academic learning skills related to dyslexia (DL) are shown in Fig. 1 as literacy skills (LS), phonological awareness (PA) and reading (Rdng). LS consists of the identification and reading of letters/words/sentences and spellings PA consists of reading non-sense words, rhyming and Rdng involves fluency and Rapid Naming.

Academic learning skills related to dysgraphia (DG) are shown in Table 3 as handwriting (HW), visuospatial response (VSR) and sentence and word expression (SWE). HW consists of writing ability of learner to form words/sentences during dictation and rewriting/copying the sentences. VSR consists of organizing and outlining information from pictures, expressing their thoughts and storytelling. SWE includes arranging jumbled words, jumbled sentences, punctuations, etc.

Academic learning skills related to dyscalculia (DC) are shown in Table 1.3 as counting numbers (CN), basic calculation (BC) and reasoning (Rsng). CN consists of identification of numbers and ability to count. BC consists of solving basic problems of addition/subtraction/multiplication/division as per academic level of learner. Rsng involves series and pattern recognition.

In Table 3, problems from the literature were identified and observed during survey in learners. The classification of Yes and No is based on the mod of observations. If more than 50% learners with the subtype identified with the problem, then they are marked as Yes.

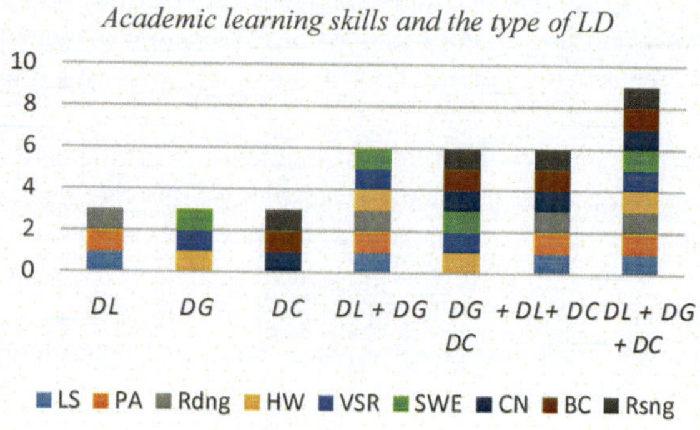

Fig. 1 Academic learning skills of LD learners

Table 1 Distribution of identified learner with learning disability

Learner	DL	DG	DC	DG + DC	DL + DC	DL + DG + DC	Non-LD	Other disability	Total learners
No. of learners	36	30	21	14	16	07	98	52	292
Girls	20	12	08	04	09	02	42	33	137
Boys	16	18	13	10	07	05	56	19	155
Rural	13	19	07	09	10	05	54	38	164
Urban	23	11	14	05	06	02	44	14	128

Table 2 Mapping of academic learning problems

Learning disability	Type of questions
Dyslexia related question (verbal and written)	• Literacy skill (letter identification—word reading—letter writing—listening comprehension)
	• Phonological awareness (rhyme—phoneme replacement), rapid naming (picture naming)
	• Fluency (semantic fluency-verbal fluency)
	• Language processing
	• Storytelling
Dysgraphia related questions	• Motor ability (grip- control- muscle coordination), dictation (alphabets/words)
	• Handwriting (font size, legibility, font shape), rewrite the sentence
	• Expression of thoughts in written form
	• Visuospatial response (left–right confusion, direction problem)
	• Word or sentence formation
Dyscalculia related questions	• Number sense, Counting (number system, place value), number operations
	• Basic calculation (addition, subtraction, multiplication, division),
	• Word problems and formal written numeracy
	• Organizing information
Affective related questions	• School adjustment
	• Social anxiety
	• Frustration
	• Self-confidence

5.2 *Research Objective 2: Developing Learner-Centric Learning Environments for Domain Model of an Intelligent Tutoring System*

Learner's preferred learning style differ time to time and learning in more than one way retain the knowledge. Same content in different learning environment in domain model can help learners in learning gain and retaining the knowledge for long time. This impact is proposed to be measure in the future work. The learning model used in various platforms is based on: personality, information processing, social interaction and multidimensional factors. The most renowned learning model are Kolb model (active/reflective/observation/experience), Felder–Silverman Model (active/reflective, sensory/intuitive, visual/verbal, sequential/global) and VARK model (visual, aural, read, kinesthetic) [20].

Table 3 Observation grade sheet (by survey team)

S no	Problems	DL	DG	DC
i	Omits a word/letter	Yes	No	Yes
ii	Substitutes a word/letter	Yes	No	Yes
iii	Adds a word/letter	Yes	Yes	No
iv	Mispronounces a word/letter	Yes	Yes	Yes
v	Spelling aloud & blending known words	Yes	Yes	Yes
vi	Spelling aloud & blending unknown words	No	No	No
vii	Ignores punctuation	No	No	Yes
viii	Weak storytelling	Yes	Yes	No
ix	Too much time to respond	No	No	Yes
x	Stutter when attempts hard	No	No	Yes
xi	Extra loudness/softness in voice	Yes	No	Yes
xii	Reading too slow/too fast	No	No	Yes
xiii	Distance between books and eyes too near	No	Yes	Yes
xiv	Ask for help repeatedly	No	Yes	No
xv	Answers only when asked each question	No	No	No
xvi	Illegible handwriting	Yes	Yes	No
xvii	Weak grip on pen/pencil	No	Yes	No
xviii	Rapid naming	Yes	Yes	Yes
xix	Colour identification	Yes	Yes	Yes

Based on cognitive ability and psychomotor skills, four types of learning environment are considered in this project. These intelligent learning environments are supposed to provide content according to most suited style.

1. Case-Based Learning Environment [CBLE]: With similar ability of a Reflective learner of Kolbe's learning model, CBLE focuses on good memory and attention as learner's strength. Cognitive weakness for these learners is that they give little time in thinking as they are quick to respond. Taking time to review and revise will help them in retaining knowledge later. This environment will help the learners to be part of the topic and be involved in problem. This will improve their imagination.
2. Video-Based Learning Environment [VBLE]: With similar ability with aural and visual learner of Felder–Silverman model or visual/auditory learner of VARK model, VBLE focus on the audio–verbal and visual perception of learner. These learners learn best through pictures, diagrams, videos. Here, learner is mapped with video-based learning environment to get content in video format using demonstration method.
3. Game-Based Learning Environment [GBLE]: With similar ability of active learner of Kolbe's learning model and Felder–Silverman GBLE focuses on rational and logical thinking ability of learner. Discussion, problem-solving

Table 4 Distribution of learning environments among target group

Learning environment	DL	DG	DC	DL + DG	DG + DC	DL + DC	DL + DG + DC	Total
GBLE	2	12	8	1	3	4	1	31
CBLE	9	0	5	4	0	1	1	20
PBLE	8	17	6	8	11	4	3	57
VBLE	17	1	2	5	0	7	2	34
Total	36	30	21	18	14	16	7	142

and working with other help these learners to learn. Here, learner is supposed to get content as a game. This helps learner to be a part of problem and learn in fun environment.

4. Practice-Based Learning Environment [PBLE]: With similar ability of kinaesthetic learner of various model PBLEs focuses on practice and repetition. The concept is similar to learning by doing. Learner is provided practice sets. For example, learner with dysgraphia has weak psychomotor skills [29]. Their problems need to be addressed by practice and repetition. Same practice is required for the learner having mild LD (learner with accuracy but less than average speed).

These above learning environments are designed as a part of study of ITS developed for LD. The above-stated four LEs were developed and during the pilot study with 142 learners. Following results were observed. Table 4 shows the learners and their corresponding learning environments.

(a) Out of 142 LD learners, 62% had single learning disability.
(b) Maximum 57 learners were identified with PBLE.
(c) Second most common LE is VBLE.
(d) GBLE and CBLE had very low difference of three learners only.
(e) Maximum learners with DL and DL + DC find VBLE more relatable.
(f) Maximum learners with DG, DL + DG, DG + DC and DL + DC identify with PBLE.
(g) Maximum learner with DC and DL + DC identified with GBLE.
(h) CBLE did not identify with maximum learners but still it has second-highest number of learners in DL and DC groups.

6 Discussion

A human tutor utilizes his/her intelligence, in order to investigate the learners, in terms of their competency levels, learning needs, and delivers instruction in accordance with this. Human tutor engages with the learner, to tutor as per the comfort level of the learner and has capability to seamlessly, adjust the tutoring style, to suit the learners [30]. In order to make an ITS adaptive, a content style on the basis of learner's academic profile and cognitive abilities is required. A domain of ITS consisting of learning material in different styles can address user problems in best-suited style. Researches proved that learning in more than one style maximizes the learning.

In this study, the most visible following cognitive ability groups were taken into consideration for the developing learning environment:

i. Memory and attention,
ii. Visual perception and
iii. Rational and logical thinking (processing)

And the following academic learning skills have been considered in the proposed work:

i. Dyslexia – Literacy Skill, Phonological Awareness, Rhyming
ii. Dysgraphia – Handwriting, Sentence and Word formation, Visuospatial Response
iii. Dyscalculia – Counting Numbers, Basic Calculation, Reasoning

Later the study, observations and series of tests and responses of target group help us identify with the following four learning environments for strengthening of learner-centric domain model. Table 5 shows the learning environment mapped with their type of learners, cognitive strengths and learning models.

Figure 2 shows the overall distribution of learning environments. PBLE was provided to 40% of learners as first choice. VBLE scored 24% and GBLE scored 22%. CBLE had least 14% of learners as first choice. Maximum learners identified with PBLE. The various reasons for this are illegible handwriting, slow speed and inaccuracy in similar questions. The time taken by LD learners was more than average learners. This makes them suitable performer for practice-based learning environment. PBLE works on accuracy and speed of learners along with writing ability [29]. Minimum learners identified with CBLE. The reason could be that CBLE is provided

Table 5 Learning environment model

Learning environment	Types of learner	Learning model (based on)	Cognitive strength
Case-based learning environment (CBLE)	Reflective learner	Reflective observant (Kolbe's learning model)	Good memory and attention
Video-based learning environment (VBLE)	Visual learner	Input (aural /visual) (Felder–Silverman model)	Good audio–verbal and visual perception
Game-based learning environment (GBLE)	Active learner	Processing (Felder–Silverman model)	Good rational and logical thinking
Practice-based learning environment (PBLE)	Kinaesthetic learner	VARK model	To improve: 1. Psychomotor skill 2 Speed and accuracy

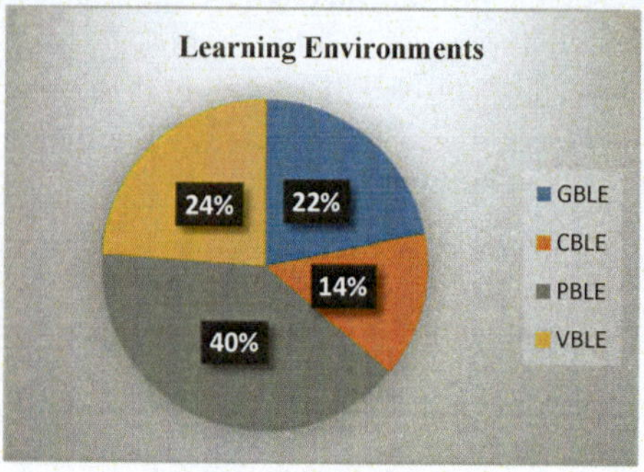

Fig. 2 Distribution of learning environments with LD learners

to learners with average memory and attention span. Most of LD learners had short attention span and weak retention of knowledge.

7 Conclusion

The study explored various combinations of learning disability, intellectual disability and their most relatable learning environments. We conclude that learners were found with different academic learning problems, and the effect of same learning problem has different intensities. Some important conclusions were:

- 62% of learner had single LD while rest LD learners have multiple LD.
- Most of LD learners could not perform questionnaire of their own grade levels and were given questionnaire of one or two less grade level on recommendation of their special educators.
- Mispronunciations were common problems in all learners with LD.
- Most LD learners had lack of phonological awareness.
- Some LD learners were able to identify the letters/words but were unable to read.
- During reading paragraph or sentences, LD learners became stressed and started to stutter.
- Learners with symptom of dyslexia were noticed to speak in very low voice and were not audible.
- Many of them show disinterest for solving a word problem.
- The created learning environment based on their performance of offline and online mode match their problems and traditional solutions and preferences.

In further studies, we are interested in in-depth evaluation of these remedial environments. The study on improvement on various learning disabilities due to these environments can be measured and compared with other form of learning disability. Also, the impact on different learning skills can be measured using modified tests.

Acknowledgements This work is carried out at the University of Petroleum and Energy Studies (UPES), Dehradun, under grant bearing reference number SEED/TIDE/133/2016. The authors thankfully acknowledge the funding support received from Science for Equity Empowerment and Development (SEED) Division, Department of Science and Technology (DST) for the project. The authors thank the management of the University of Petroleum and Energy Studies for supporting the work and granting permission to publish it.

References

1. Singh S, Sawani V, Deokate M, Panchal S, Subramanyam AA, Shah HR, Kamath RM (2017). Specific learning disability: a 5 year study from India. Int J Contemp Pediatr 4(3):863. https://doi.org/10.18203/2349-3291.ijcp20171687
2. Karande S, Sholapurwala R, Kulkarni M (2011) Managing specific learning disability in schools in India. Indian Pediatr 48(7):515–520. https://doi.org/10.1007/s13312-011-0090-1
3. Gergely K, Lakos R (2013) Role of pediatricians in the diagnosis and therapy of dyslexia, dysgraphia and dyscalculia. Orv Hetil 154(6):209–218. https://doi.org/10.1556/oh.2013.29526
4. Dutt S, Ahuja NJ (2020) A novel approach of handwriting analysis for dysgraphia type diagnosis. Int J Adv Sci Technol 29(3):11812. http://sersc.org/journals/index.php/IJAST/article/view/29852
5. Reisman F, Severino L (2020) Dyscalculia assessment. In: Using creativity to address dyslexia, dysgraphia, and dyscalculia. Routledge, pp 89–120. https://doi.org/10.4324/9781003038313-10
6. Nagyová I (2017) E-learning environment as intelligent tutoring system. AIP Conf Proc 1863(1). https://doi.org/10.1063/1.4992244
7. Bisht A, Ahuja NJ (2020) Design and development of competency-based instructional model for instruction delivery for learning disabled using case based reasoning. Int J Rec Technol Eng 8(6):1847–1858. https://doi.org/10.35940/ijrte.f7998.038620
8. Zydney JM, Warner Z, Angelone L (2020) Learning through experience: using design based research to redesign protocols for blended synchronous learning environments. Comput Edu 143. https://doi.org/10.1016/j.compedu.2019.103678
9. Brandenburger J, Constapel M, Hellbrück H, Janneck M (2020) Analysis of types, positioning and appearance of visualizations in online teaching environments to improve learning experiences. In: Advances in Intelligent Systems and Computing, vol 963. Springer Verlag, pp 355–366. https://doi.org/10.1007/978-3-030-20135-7_35
10. Karampa V, Paraskeva F (2020) Smart learning environments: a blend of ICT achievements and smart pedagogy for the world sustainable development. In Advances in intelligent systems and computing, vol 1018. Springer Verlag, pp 482–488. https://doi.org/10.1007/978-3-030-25629-6_75
11. Nair MKC, Prasad C, Unni J, Bhattacharya A, Kamath SS, Dalwai S (2017) Consensus statement of the Indian Academy of Pediatrics on evaluation and management of learning disability. Indian Pediatr 54(7):574–580. https://doi.org/10.1007/s13312-017-1071-9
12. Bandla S, Mandadi GD, Bhogaraju A (2017) Specific learning disabilities and psychiatric comorbidities in school children in South India. Indian J Psychol Med 39(1):76–82. https://doi.org/10.4103/0253-7176.198950
13. Taur S, Karande S, Saxena AA, Gogtay NJ, Thatte UM (2014) Use of computerized tests to evaluate psychomotor performance in children with specific learning disabilities in comparison to normal children. Indian J Med Res 140:644–648
14. Padhy SK, Goel S, Das SS, Sarkar S, Sharma V, Panigrahi M (2016) Prevalence and patterns of learning disabilities in school children. Indian J Pediatr 83(4):300–306. https://doi.org/10.1007/s12098-015-1862-8
15. González-Valenzuela MJ, Martín-Ruiz I (2020) Effects of early intervention on the writing skills of pupils who present a risk of learning disabilities within spain's early years and

primary education system. Revista de Educacion 2020(388):85–106. https://doi.org/10.4438/1988-592X-RE-2020-388-448
16. Beckmann E, Minnaert A (2018) Non-cognitive characteristics of gifted students with learning disabilities: an in-depth systematic review. In: Frontiers in psychology. Frontiers Media S.A. https://doi.org/10.3389/fpsyg.2018.00504
17. Görker I (2020) The prevalence and gender differences in specific learning disorder. In: Learning disabilities—neurological bases, clinical features and strategies of intervention. IntechOpen. https://doi.org/10.5772/intechopen.90214
18. Asthana P, Hazela B (2020) Applications of machine learning in improving learning environment. In: Intelligent systems reference library, vol 163. Springer Science and Business Media Deutschland GmbH, pp 417–433. https://doi.org/10.1007/978-981-13-8759-3_16
19. Rose R, Narayan J, Matam S, Reddy Sambram P (2021) A comparison of provision and access to inclusive education for children with disabilities in a metropolitan city and a rural district in Telangana State, India. Edu Sci 11(3). https://doi.org/10.3390/educsci11030111
20. Kumar A, Singh N, Ahuja NJ (2017) Learning styles based adaptive intelligent tutoring systems: document analysis of articles published between 2001 and 2016. Int J Cogn Res Sci Eng Edu 5(2):83–98. https://doi.org/10.5937/IJCRSEE1702083K
21. Russell AM, Bryant L, House A (2017) Identifying people with a learning disability: an advanced search for general practice. Br J Gen Pract 67(665):e842–e850. https://doi.org/10.3399/bjgp17X693461
22. Chavan B, Rozatkar A (2014) Intellectual disability in India: charity to right based. Indian J Psychiatry. Medknow Publications. https://doi.org/10.4103/0019-5545.130477
23. Polat E, Adiguzel T, Akgun OE (2012) Adaptive web-assisted learning system for students with specific learning disabilities: a needs analysis study. Kuram ve Uygulamada Egitim Bilimleri 12(SUPPL. 4):3243–3258
24. Taymans JM, Kosaraju S (2012) Introduction to the journal of learning disabilities special issue: adults with learning disabilities in adult education. J Learn Disabil https://doi.org/10.1177/0022219411426860
25. Mehta M (2011) Use and abuse of specific learning disability certificates. Ind Psychiatry J 20(2):77. https://doi.org/10.4103/0972-6748.102471
26. Ariffin MM, Halim FAA, Arshad NI, Mehat M, Hashim AS (2019) Calculic kids© mobile app: The impact on educational effectiveness of dyscalculia children. Int J Innov Technol Explor Eng 8(8):701–705
27. Fanning E (2005) Formatting a paper-based survey questionnaire: best practices. Prac Assess Res Eval 10(12). Available online: http://pareonline.net/getvn.asp?v=10&n=12
28. Rao C, Sumathi TA, Midha R, Oberoi G, Kar B, Khan M, Singh NC (2021) Development and standardization of the DALI-DAB (dyslexia assessment for languages of India—dyslexia assessment battery). Ann Dyslexia. https://doi.org/10.1007/s11881-021-00227-z
29. O'Connor ES, Mahvi DM, Foley EF, Lund D, McDonald R (2010) Developing a practice-based learning and improvement curriculum for an academic general surgery residency. J Am Coll Surg 210(4):411–417. https://doi.org/10.1016/j.jamcollsurg.2010.01.017
30. Felder RM, Soloman BA (2017) Felder & Soloman: learning styles and strategies. Retrieved from http://www4.ncsu.edu/unity/lockers/users/f/felder/public/ILSdir/styles.htm

Trusted 6G-Envisioned Dynamic Spectrum Allocation: A Reference Model, Layered Architecture, and Use-Case

Pronaya Bhattacharya, Farnazbanu Patel, Vishaka Raleganakar, Bhaumik Thakkar, Sudeep Tanwar, and Mohammad Abouhawwash

Abstract Spectrum allocation among multiple telecom providers is challenged with a fair spectrum allocation process and collusion among multiple telecom parties involved in spectrum bids. As spectrum licensing have shifted toward decentralization, blockchain-based spectrum allocation can address the limitations through a fair and trusted bidding process, and spectrum allocation which is transparent to all bidders. Moreover, the spectrum allocation process has shifted to dynamic spectrum allocations, and thus in the coming future, owing to the inherent advantages of sixth-generation (6G) networks, with high network reliability, user bandwidth, high-precision, ultra-high reliability, and extreme flexibility, all presented as networked-in-a-box service, dynamic spectrum allocation in 6G-envisioned networks is a reality in near future. In light of these facts, we present in this paper, a comprehensive examination of the integration of a blockchain-based scheme for dynamic spectrum access in 6G-envisioned communications. We present the layered reference architecture and highlight a case study of future blockchain-based spectrum access for 6G-serviced network access. The paper serves as starting foundation toward the build of effective blockchain-based spectrum allocation schemes with effective networked serviced applications.

P. Bhattacharya (✉) · F. Patel · V. Raleganakar · B. Thakkar · S. Tanwar
Institute of Technology, Nirma University, Ahmedabad, Gujarat, India
e-mail: pronoya.bhattacharya@nirmauni.ac.in

F. Patel
e-mail: 19mcei04@nirmauni.ac.in

V. Raleganakar
e-mail: 20mced18@nirmauni.ac.in

B. Thakkar
e-mail: 20mcec17@nirmauni.ac.in

S. Tanwar
e-mail: sudeep.tanwar@nirmauni.ac.in

M. Abouhawwash
Durban University of Technology, Durban, South Africa
e-mail: abouhaww@msu.edu

P. K. Singh et al. (eds.), *Emerging Technologies for Computing, Communication and Smart Cities*, Lecture Notes in Electrical Engineering 875,
https://doi.org/10.1007/978-981-19-0284-0_30

Keywords 6G · Blockchain · Spectrum sharing · Decentralization · Telecommunication industry

1 Introduction

Over the past decade, there has been a surge in telecommunications networks that have surged the requirements of spectrum allocation bands among telecom service providers. Currently, with a large number of users, fourth-generation (4G) long-term evolution (LTE) networks are facing bottlenecks to service the growing demands. By 2021, 4.5 billion mobile subscribers are registered globally. Figure 1 presents the scenario. Owing to the shift, the telecom industries have shifted toward spectrum licensing in the fifth generation (5G) bands. 5G offers effective service orchestration through a combination of different band frequencies to increase the coverage range. 5G commercial networks are expected to operate in the 3.3–3.8 gigahertz (GHz) range, with support of lower bands that include 1500 megahertz (MHz), 2.1 GHz, and 2.3 GHz for poor connection areas [2]. Thus, 5G is expected to provide faster and reliable network services that would support different verticals in smart cities, like smart factories, smart and autonomous vehicles, and healthcare industries. However, it also requires telecom providers to access higher-frequency bands to make the vision a reality.

However, spectrum frequencies are limited resources, and thus, an effective sharing mechanism is required. With the advent of a shift of network services at the

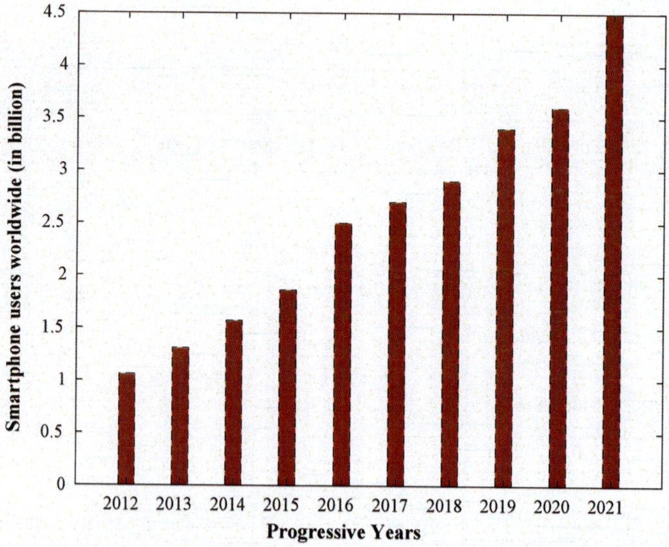

Fig. 1 Global increase of mobile users [1]

edge [3], latency in networked applications has also become a prime requirement. 5G services like ultra-reliable low-latency communications (eMBB) and massive machine-type communications (mMTC) offer an end-to-end latency of 5 ms and high connection density of 1 million devices/km^2. However, with the rise of automation, and an increase in massive device-to-device (D2D) connectivity in Internet-of-anything (IoE) ecosystems, networks would require extreme dense connections, edge intelligence support, and high reliability.

Thus, researchers have shifted toward sixth-generation (6G) networks, that is envisioned to support ultra-high data rates in the range of terahertz (THz) range, radio latency of 100 μs, and connection density of 10^7 devices/km^2 [5]. 6G services can be easily stacked to support the spectrum access. Moreover, 6G fine-tunes the edge computing requirements through artificial intelligence (AI)-enabled radio access, and thus, industries have shifted toward investment in 6G projects [6]. 6G supports effective features like omnipresent global coverage in space–air–ground–water communication, at ultra-high reliability of 99.9999999 %. 6G is expected to support verticals like holographic and 3D integrations as well [7]. Table 1 presents an overview of the mobile communication shift from second generation (2G) communication to 6G.

In terms of application viewpoint, 6G would support low-rate and long-distance Internet-of-things (IoT) applications, process automation of cyber-physical systems in the industry, digital twins, holography, AI support with the complex machine and deep learning models, extended, virtual, and mixed reality applications, and automatic vehicular networks. Owing to the AI-enabled radio, it resolves the issues of fixed spectrum allocation in decentralized environments and covers for underutilized spectrum limitations. However, static spectrum allocation is mostly centralized, and thus, blockchain is a viable choice of fair spectrum allocations owing to the inherent benefits of fairness, immutability, and chronological access [8]. Moreover, in spectrum allocation, we consider a permissioned blockchain, where government, telecom providers, and spectrum licensing stakeholders are allowed to participate in the bidding process.

1.1 Research Contributions

Following are the research contributions of the paper.

- A reference model of 6G-envisioned blockchain-based spectrum allocation is presented, and a layered stack reference is proposed.
- A case application based on the proposed reference scheme is presented that discusses a reputation-based scorecard registration of new user in the proposed ecosystem.

Table 1 An overview of mobile communications shift from 2G to 6G [4]

Generation	2G	3G	4G	5G	6G
Year	1990	2000	2010	2018	2030
Technology	DBN-CD	DBB PD (IPv4)	LAN/WAN, Unified IPv4/IPv6, WLAN	LAN/WAN/PAN, Unified IPv4, WLAN, advanced OFDM	Blockchain-based spectrum, artificial intelligence, quantum communication, laser and VLC, THz communication
Technique	SISO	SISO	MIMO	m-MIMO	SM-MIMO, UM-MIMO
Frequency	850–1900 MHz	1.6–2.5 GHz	2–8 GHz	3–300 GHz	95 GHz–3 THz
Spectral efficiency	0.17 bps/Hz	1.3 bps/Hz	30 bps/Hz	120 bps/Hz	5–10 times that of 5G
Latency	300–1000 ms	100–500 ms	< 15 ms	< 100 ms	$< 10-100\,\mu s$
Connection density	Limited	100/km^2	1000/km^2	1,000,000/km^2	10,000,000/km^2
Mobility	–	–	350 kmph	500 kmph	1000 kmph

1.2 Article Structure

This paper is divided into five sections. Section 2 presents the state-of-the-art schemes that are proposed related to 6G and blockchain-based schemes. Section 3 presents the layered reference model of 6G-envisioned blockchain-assisted dynamic spectrum allocation, which is supported by a layered reference stack architecture in Sect. 4. Section 5 presents the case-study of the proposed scheme, and finally Sect. 6 concludes the paper.

2 State of the Art

In this section, we present the recent state-of-the-art schemes that integrate blockchain and 6G in telecommunications. Saravanan et al. [9] proposed the integration of blockchain for telecom providers to simplify their phone usage charging and billing operations. Via blockchain, the third-party intermediaries are removed, and inconsistencies in the management of large customer databases are simplified. The paper proposes that blockchain ledger can manage user call records in an immutable manner, and through smart contracts, roaming agreements between inter-telecom providers are also managed, and balance transfers are automated. This reduces the overall transactional fees of third-party payment gateways and improves the complexity of the overall billing ecosystem. Xu et al. [10] proposed a resource management scheme for spectrum allocation for mobile operators and presented a reference framework that manages resources and sharing in 6G-IoE ecosystems. The authors proposed a network slicing-based approach in 6G, and a slice-broker-based scheme to manage the 6G resource orchestration. The resource transfer is managed as transactional ledgers in the blockchain. Zhou et al. [11] presented a privacy-preserved 5G human-to-human (H2H), and machine-to-machine (M2M) scheme, where a cost-effective solution is presented to optimally utilize the spectrum resources. The paper introduces a two-phased scheme. In the first phase, H2H users and 5G-enabled base stations execute a smart contract for transactional payments, and spectrum is released. The spectrum is allocated to M2M devices, with an incentive-based design.

Zhang et al. [12] proposed a distributed citizens broadband radio access (CBRS) spectrum sharing scheme to address the limitations of administrative costs, and privacy-based attack scenarios by an adversary. The authors include a low-powered consensus mechanism known as proof-of-strategy that finalizes the spectrum allocation, even in case of node failures. Patel et al. [14] proposed a 6G-based blockchain-based spectrum allocation scheme between dynamic service operations in a cell-free spectrum. The paper proposes a dynamic auction and bidding process of spectrum allocation. Hewa et al. [13] proposed a survey that introduces blockchain potential in 6G verticals such as health care, Internet-of-vehicles, infotainment, augmented and virtual reality, and M2M communication. The challenges of 6G and potential pitfalls are identified, and blockchain-based solutions are proposed to allow distributed 6G protocols and standards.

Table 2 State-of-the-art approaches of integration of blockchain and 6G in telecommunications

References	Year	Major contributions	Application domain
[9]	2017	Blockchain-based consumer balance transfers, carrier billing, roaming settlements through decentralized application deployments	5G-based telecom service
[10]	2020	The complexities of adapting blockchain for spectrum sharing in 6G are discussed	6G-envisioned spectrum sharing and access and blockchain for license transfers
[11]	2020	H2H and M2M enabled spectrum sharing ecosystem, with blockchain-based transactional ledgers	Fair spectrum sharing through blockchain-based 5G dynamic spectrum access
[12]	2020	Optimal CBRS design service model through blockchain and 6G-radio access model	6G-CBRS allocation scheme with trusted blockchain ledgers
[13]	2020	6G technical key performance indicators (KPIs) compared against 5G KPIs in terms of specification, regulation, standardization, and design in different driving verticals	6G-benefits in IoT, industry 4.0, virtual reality, and autonomous driving
[14]	2021	In 6G ecosystems, blockchain-enabled trusted dynamic frequency spectrum allocation and auction scheme among telecom providers	6G-based dynamic spectrum allocation and bidding process via blockchain-based auctions
[5]	2021	6G-benefits and reference architecture in telecom servicing, IoE, and inherent limitations, with architectures of blockchain to mitigate the limitations	6G-IoE standards, interfaces, and protocol regularization via blockchain

Jiang et al. [5] proposed different 6G frontiers in different verticals of smart cities and discussed the requirement of 6G to handle a high volume of data traffic. Potential use cases and scenarios are discussed, and a tentative roadmap of 6G standardization is presented. The details of the recent schemes, their contributions, and application domains are discussed in Table 2.

3 A Reference Model of Blockchain-Based 6G-Spectrum Allocation

In this section, we present the founding concepts of blockchain and 6G services. The section presents a reference architecture that discusses the potential benefits of blockchain in 6G applications that handle the issues of trust, privacy, and secure transfer of resources among communicating entities. We start initially with the discussion of 6G emergence and then move toward the usage of blockchain to support secured 6G services. The details are presented as follows.

3.1 The Emergence of 6G and Blockchain in Communications

With the increase of communication networks, and stringent requirements of bandwidth, latency, and availability of resources, to support applications like extended reality, autonomous driving, Internet-of-drones, real-time sensing and control, an eight-year program, termed as *6Genesis* Flagship started with an estimated fund of 290 million dollars. The project started in 2018 by Finland, and soon researchers worldwide started with the design of protocols and standards for 6G communication networks. The key visions of 6G communication highlighted in Table 3.

Initially, started as cryptographic ledgers [15], blockchain gained prominence owing to its inherent benefits of trust, immutability in block creation, and verification and thus has become a driving force in different verticals like smart grids, autonomous vehicles, and Internet-of-things [16]. To secure the 6G connectivity perimeter, blockchain can mitigate security attacks like distributed denial-of-service, impersonation, replay, and certificate-based attacks [13]. Thus, blockchains empower decentralized cooperative applications and also ensure that data is exchanged by all parties involved.

Table 3 Visions of 6G communication [4]

Key parameters	6G vision
Mobility (km/h)	1000
Peak spectral efficiency (b/s/Hz)	60
End-to-end latency (ms)	0.1
Reliability	10^{-9}
Connection density (device/km^2)	10^7
Area traffic capacity (Mbps/m^2)	1000
Channel bandwidth (GHz)	100
Spectral efficiency (b/s/Hz)	3
Energy efficiency (Tb/J)	1
User data rate (Gbps)	>10 Gb/s
Peak data rate	>100 Gb/s
Receiver sensitivity	< -130 dBm
Coverage	>99%
End-to-end delay	<1 ms

3.2 A Proposed Reference Model

In this subsection, we present the reference model of blockchain-based 6G-spectrum access. Figure 2 presents the details of the proposed model. In the proposed model, we consider entities $E = \{E_{BS}, E_{SMS}, E_{AC}, E_{SL}, E_{TP}, E_{SB}, E_{SV}, E_{SR}\}$, where E_{BS} denotes the base stations (BSs) that are integrated with 6G services to support

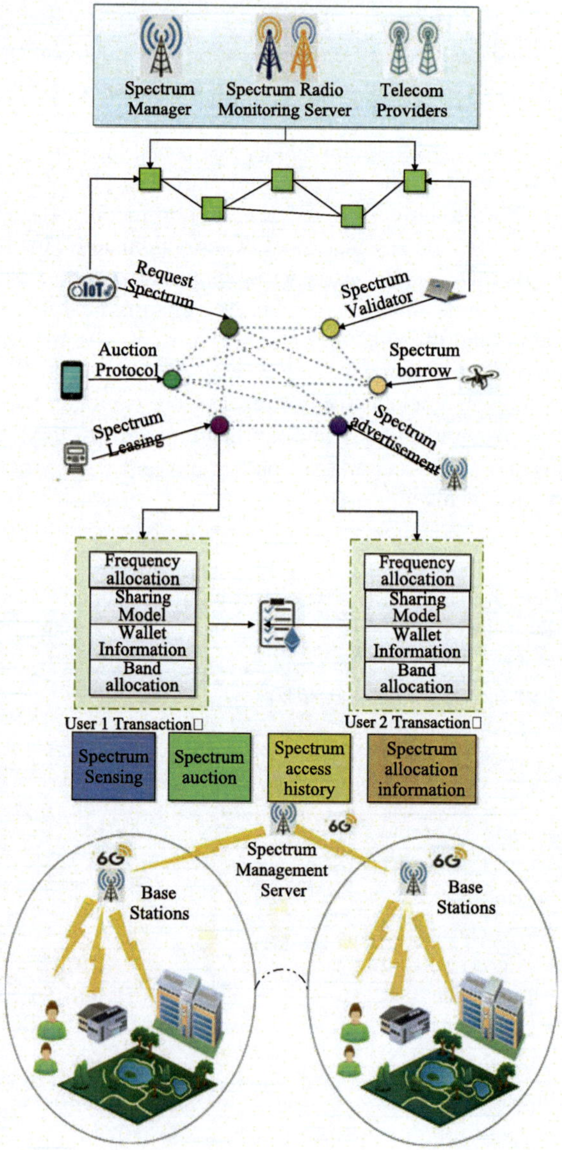

Fig. 2 A reference model of blockchain-based 6G-spectrum access

dynamic spectrum access. E_{SMS} denotes the spectrum management server, E_{AC} denotes the spectrum auctioneer, E_{SL} denotes the spectrum leaser, E_{TP} denotes the telecom provider, E_{SB} denotes the spectrum borrower, E_{SV} denotes the spectrum validators, and E_{SR} denotes the spectrum requester, respectively. Depending on the specifics of spectrum resource allocations, the utilization of spectrum is only done post the auction process. To maintain the regulations and control in spectrum allocation, we consider E_{SRMS} that denotes the spectrum radio monitoring server, which is a government regulating body to manage and distribute the spectrum to $E_{\text{TP}}, E_{\text{RS}}, E_{\text{BS}}$. As the considered ecosystem is a multi-party decentralized system with different stakeholders like industrial applications, spectrum auctioneers, and buyers, borrowers, spectrum leases, and spectrum advertisers, we require trust in the ecosystem. For the same, we consider a consortium-based permissioned approach, where the transaction updates are shared only by registered stakeholders in the chain.

For the spectrum ledger, we consider the ledger L with the required fields, namely $\{\text{AO, AI, SMod, FA}, W_{\text{U}}\}$, where AO denotes the asset (spectrum resource) ownership, AI denotes the asset meta-information, SMod is defined as the sharing model (competitive or collaborative), FA denotes the frequency allocated, and W_{U} denotes the wallet information of the user. The transaction ledgers are maintained through distributed offline storage (interplanetary file systems), where the ledger records are accessible by the IPFS key and the private key of the user only. We hash the stored record, $H(R)$, and store the record R indexed with its hash-pair $H(R)$. In the main chain, we store $H(R)$ only as a transaction, so that the record R may be retrieved by a search of the hash in the chain. Moreover, effective consensus schemes are required to be designed for E_{SV}, so that their incentives are maximized. Validators E_{SV} are chosen based on a reputation score R so that they add the transactions in a fair manner in the blockchain [17].

In the reference architecture, we consider servicing E_{BS} that provides network service to user sets $U = \{U_1, U_2, \ldots, U_n\}$. We consider a cell-based 6G grant spectrum access scheme, and two regions, R_1, and R_2, respectively. Any nth user in region R_1 is mapped to E_{BS} through a mapping $M_1 : U_n \rightarrow \text{BS}_n^1$, and similarly, any user U_n in R_2 is mapped to BS_n^2, through mapping $M_2 : U_n \rightarrow \text{BS}_n^2$. The user requests $R = \{R_1, R_2, \ldots, R_n\}$ are collected region-wise and send to E_{SMS} through directed 6G uplink frequency f_{u}. At E_{SMS}, the collected requests R are serviced as digital assets, and spectrum allocation requirements are advertised in the network, termed as spectrum advertisements. E_{SMS} handles the function of intelligent spectrum sensing through AI models and maintains spectrum access historical ledger entries by corresponding E_{BS}. The spectrum auction $A(S)$ is initiated at E_{SMS} depending on the base network requirements sent by E_{BS}. For the same, a list of freely available frequency $F(f_{\text{r}})$ is maintained, which is collected through network entities like IoE networks, satellites, free users, vehicular networks, and others. The finalization and broadcast of available spectrum bands are termed spectrum leasing. For the leasing process, an auction strategy is set up between spectrum bidders and spectrum borrowers, with the peer-profit optimization strategy. The auction can be modeled through a cooperative game-theoretic approach, to maximize the incentives of both bidders and borrowers, through the designated set of auctioneers. Once the spectrum auction process is over, the spectrum grant is maintained as transactional ledgers in IPFS and meta-information are chronologically recorded in the blockchain. To spectrum transaction information is maintained in consortium blockchain, and the available usage and regulations are reflected all authorized nodes in the chain by E_{SRMS}. This ensures the transparency of spectrum allocation to all E_{TP} and mitigates the possible collusion among malicious bidder nodes.

4 The Layered Reference Stack of Blockchain-Based 6G-Spectrum Allocation Scheme

In this section, we propose the layered stack model of the proposed reference architecture that handles the issues of static spectrum allocation. Figure 3 presents the details. We consider a four-layered scheme, and the details are presented as follows.

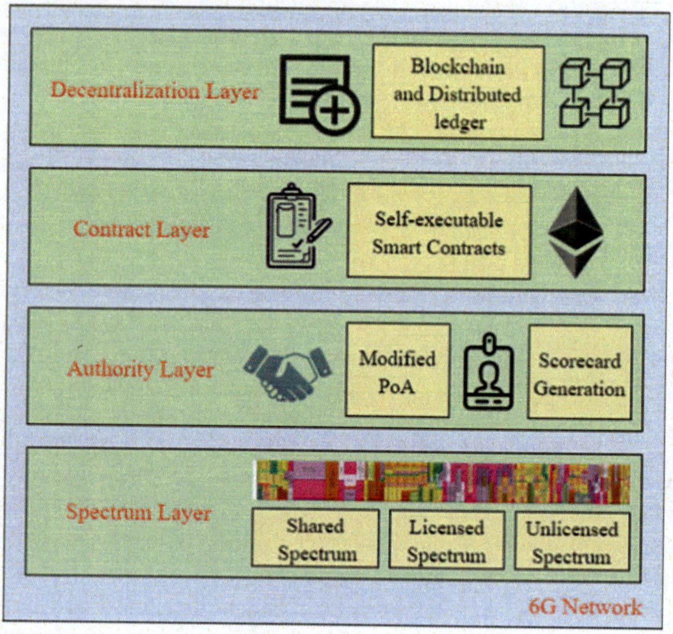

Fig. 3 Spectrum allocation using blockchain

4.1 Layer 0: The Spectrum Layer

At *Layer 0*, we assume the spectrum details are present, which is a cluster of frequency ranges $R(f)$, and consist of electromagnetic waves. Through $R(f)$, different communication devices such as TV, radio, mobile to send wireless messages across a certain distance d. The details of available spectrum bands are managed by E_{SMS}, and the allocation of bands to different servicing E_{BS}, of different E_{TP} is leveraged through a spectrum validator E_{SV}. The spectrum band is mainly divided into three regions as follows.

- **Licensed**: In the licensed band, a chunk of the radio spectrum is assigned to E_{SMS}, or E_{SRMS}, and is licensed as asset ownership by AO. Any user has to send a spectrum access request to AO, and the spectrum grant is defined for a definite time period T. Here, the access request is placed by E_{TP}, so they buy the licensed frequency ranges from $F(f_{\mathrm{r}})$, for a given price, and allocate frequencies to E_{BS} through the servicing downlink F_{d}.
- **Unlicensed**: In this band, the available frequencies can be used by any user, and normal users also have access to the unlicensed spectrum. This type of spectrum does not involve a specific type of permission from either E_{SMS}, or E_{SRMS}. The applications of the unlicensed spectrum are IEEE 802.11 access, TV white spaces, and wireless personal area networks, like IEEE 802.15.x.
- **Shared**: In this band, the frequencies are shared among different users, and each user utilizes a chunk of the frequency band. This type of paradigms helps the users and devices to completely utilize the spectrum band.

4.2 Layer 1: The Authority Layer

The shared spectrum suffers from a lot of obstacles. Generally, the practice involves the centralization of shared spectrum management (by CBRS) [18]. Here, an intermediary is needed to manage the complete flow of control in the shared environment. The centralized systems suffer from various issues such as lack of adaptability, overburdening on the central authority, one-sided communication, and biases in decision making [14]. These issues lead to poor utilization of resources and a less secure system. In the proposed scheme, the authority layer validates the authority of the users in the ecosystem. For consensus, we consider a modified version of the Proof-of-Authority (PoA) consensus mechanism. The primary PoA works by allowing nodes to create initial blocks that have demonstrated their authority. Any new user in PoA has to prove the identity to get access to the spectrum. Once the identity authentication gets done, a scorecard is generated for the user. This process is iterated for each user in the network to prove the genuineness of the identity of the users.

4.3 Layer 2: The Contract Layer

To break the tie created by a centralized environment, the user needs a mechanism that can automate the flow of taking decisions in a very honest manner. For that, we use smart contacts. Smart contracts are self-executing code without any third-party (such as humans) interactions. In the proposed model, smart contracts ensure the storage of authorized user data published on IPFS, and meta-information stored in distributed ledgers. The access of IPFS is restricted through identity authorization and IPFS key.

4.4 Layer 3: The Decentralization Layer

At *Layer 3*, we consider the distributed blockchain ledger. New blocks are added only after E_{SV} validates the transaction entries. Every authorized user has a copy of ledger L, and L gets updated once the state of IPFS changes, to reflect new contracts executed in the network. Through 6G, ease of access and scalability of node communication are improved.

5 Case Study: A Scorecard and Reputation-Based Spectrum Allocation

In this section, we propose a case study that presents the usage of the shared spectrum. Figure 4 presents the details. The shared spectrum can be allocated to the user using the integrated technology discussed in Sect. 4.

In the use-case, we consider entity A_1 that wishes to access the joint spectrum for communication purposes. A_1 first registers himself in the network and has to undergo the PoA consensus where E_{SRMS}, or E_{SMS} validates A_1 identity to all users. Then, A_1 is granted access to spectrum resources. This whole registration process is automated via a *DApp* that executes a smart contract at the back-end between E_U and E_{SMS} and publishes the transactional state to IPFS. Also, other E_U ledgers are updated with the new entry in their ledgers. Here, A_1 is presented with a scorecard, and based on future transactions performed by A_1, the reputation score increases, and the access-grant time of shared spectrum is reserved for A_1 also increases. This reward-based technique ensures the authenticity of the users is managed in real-time through 6G service sets.

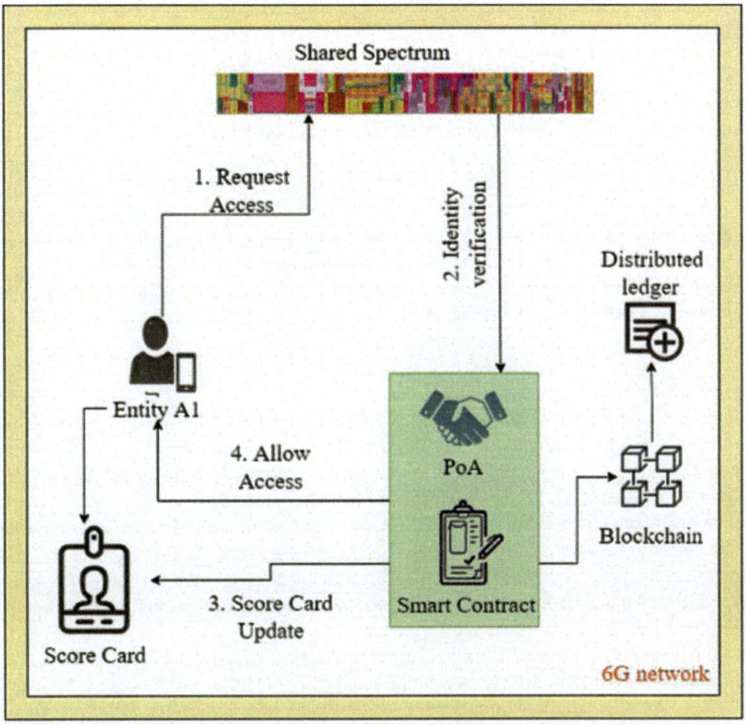

Fig. 4 Spectrum allocation using blockchain

6 Conclusion

The spectrum allocation process among competitive telecom providers and users is a complex problem. The problem is further intensified in decentralized environments owing to the issues of trust, alterations, and collusion-based attacks. Thus, in this paper, we have presented a reference model for blockchain-assisted dynamic spectrum access at the backdrop of 6G-envisioned communications. Through blockchain, a trusted chronology is maintained among distributed telecom stakeholders, and provenance is established. Owing to the high influx of network traffic, and users, 5G services would face bottlenecks in the near future. Due to this, we considered a 6G service set that provides intelligent and real-time network orchestration to users in the proposed ecosystem. A reference model is presented, and a supportive layered stack model is also proposed. Then, we present a reputation-based scorecard for registration of new users in the ecosystem that ensures the genuineness and transparency via PoA consensus in the spectrum allocation ecosystem.

As part of the future scope, the authors would investigate a deep reinforcement learning framework that manages the reputation of a user in the ecosystem and also would propose a cooperative game-theoretic approach to model and maximize incentives of the auction process.

References

1. Statista (2021) Number of smartphone users worldwide from 2012 to 2023. https://www.statista.com/statistics/330695/number-of-smartphone-users-worldwide/, May 2021
2. Gorla P, Chamola V, Hassija V, Ansari N (2021) Blockchain based framework for modeling and evaluating 5g spectrum sharing. IEEE Network 35(2):229–235
3. Bhattacharya P, Tanwar S, Shah R, Ladha A (2020) Mobile edge computing-enabled blockchain framework—a survey. In: Singh PK, Kar AK, Singh Y, Kolekar MH, Tanwar S (eds) Proceedings of ICRIC 2019. Springer International Publishing, Cham, pp 797–809
4. Alsharif MH, Kelechi AH, Albreem MA, Chaudhry SA, Zia MS, Kim S (2020) Sixth generation (6g) wireless networks: vision, research activities, challenges and potential solutions. Symmetry 12(4)
5. Jiang W, Han B, Habibi MA, Schotten HD (2021) The road towards 6g: a comprehensive survey. IEEE Open J Commun Soc
6. Huang T, Yang W, Wu J, Ma J, Zhang X, Zhang D (2019) A survey on green 6g network: architecture and technologies. IEEE Access 7:175758–175768
7. David K, Elmirghani J, Haas H, You X-H (2019) Defining 6g: challenges and opportunities [from the guest editors]. IEEE Veh Technol Mag 14(3):14–16
8. Patel SB, Bhattacharya P, Tanwar S, Kumar N (2021) Kirti: a blockchain-based credit recommender system for financial institutions. IEEE Trans Network Sci Eng 8(2):1044–1054
9. Saravanan M, Behera S, Iyer V (2017) Smart contracts in mobile telecom networks. In: 2017 23RD Annual international conference in advanced computing and communications (ADCOM). IEEE, pp 27–33
10. Xu H, Klaine PV, Onireti O, Cao B, Imran M, Zhang L (2020) Blockchain-enabled resource management and sharing for 6g communications. Digit Commun Networks 6(3):261–269
11. Zhou Z, Chen X, Zhang Y, Mumtaz S (2020) Blockchain-empowered secure spectrum sharing for 5g heterogeneous networks. IEEE Network 34(1):24–31
12. Zhang H, Leng S, Chai H (2020) A blockchain enhanced dynamic spectrum sharing model based on proof-of-strategy. In: ICC 2020-2020 IEEE International conference on communications (ICC). IEEE, pp 1–6
13. Hewa T, Gür G, Kalla A, Ylianttila M, Bracken A, Liyanage M (2020) The role of blockchain in 6g: challenges, opportunities and research directions. In: 2020 2nd 6G Wireless summit (6G SUMMIT). IEEE, pp 1–5
14. Patel F, Bhattacharya P, Tanwar S, Gupta R, Kumar N, Guizani M (2021) Block6tel: blockchain-based spectrum allocation scheme in 6g-envisioned communications. In: 2021 International wireless communications and mobile computing (IWCMC), pp 1823–1828
15. Srivastava A, Bhattacharya P, Singh A, Mathur A, Pradesh U, Pradesh U (2018) A systematic review on evolution of blockchain generations. Int J Inf Technol Electr Eng 7(6):1–8
16. Bhattacharya P, Tanwar S, Bodkhe U, Kumar A, Kumar N (2021) Evblocks: a blockchain-based secure energy trading scheme for electric vehicles underlying 5g-v2x ecosystems. Wirel Pers Commun, pp 1–41
17. Shyamsukha S, Bhattacharya P, Patel F, Tanwar S, Gupta R, Pricop E (2021) PoRF: proof-of-reputation-based consensus scheme for fair transaction ordering. In: 2021 13th International conference on electronics, computers and artificial intelligence (ECAI), pp 1–6
18. Zafaruddin SM, Bistritz I, Leshem A, Niyato D (2019) Distributed learning for channel allocation over a shared spectrum. IEEE J Sel Areas Commun 37:2337–2349

Experimental Evaluation of QFT Adders on IBM QX Hardware

Simran Jakhodia, Divyanshu Singh, and Babita Jajodia

Abstract Quantum Fourier transform (QFT) offers a versatile way to perform modular and non-modular quantum arithmetics and provide compact quantum circuits. QFT is the quantum implementation of the classical discrete Fourier transform (DFT) over the vector of amplitudes of a quantum state (wavefunction). In this paper, the authors experimentally evaluate the performance of QFT adders on various IBM Quantum Experience (QX) hardware against Quantum Information Science Kit (Qiskit) Simulator. The authors have constructed quantum circuits to represent one-qubit and two-qubit QFT adders using Qiskit and run on IBM QASM Simulator and eight IBM real quantum devices: YorkTown, Santiago, Athens, Quito, Belem, Lima, Manila and Melbourne. Experimental analysis of one-qubit QFT adder shows that IBM Lima and IBM Athens provide better accuracy results of 80.2% and 80.125%, respectively, in comparison to other five-qubit and fifteen-qubit quantum hardware. Experimental results of two-qubit QFT adder demonstrate that IBM Athens and IBM Manila provide better accuracy results of 25.575% and 17.52%, respectively, in comparison to other five-qubit and fifteen-qubit quantum hardware.

Keywords IBM · Quantum computing · Quantum adder · Quantum gates · Quantum information science kit (Qiskit) · Quantum Fourier transform (QFT) · Inverse quantum Fourier transform (IQFT)

S. Jakhodia · B. Jajodia (✉)
Indian Institute of Information Technology Guwahati, Bongora, Guwahati, Assam, India
e-mail: babita@iiitg.ac.in

S. Jakhodia
e-mail: simran.jakhodia@iiitg.ac.in

D. Singh
Gautam Buddha University, Greater Noida, Uttar Pradesh, India
e-mail: 19bhp011@gbu.ac.in

P. K. Singh et al. (eds.), *Emerging Technologies for Computing, Communication and Smart Cities*, Lecture Notes in Electrical Engineering 875,
https://doi.org/10.1007/978-981-19-0284-0_31

1 Introduction

Quantum computing along with its inherent quantum principles of superposition and entanglement are believed to bring revolutionary changes overcoming the limitations of classical computing systems [1]. Constant advancements are being made, focussing on developments of higher-qubit real quantum systems.

Researchers from academia and research companies long predicted that future developments in scientific computing would be enabled by the advent of software-programmable quantum computers and solve real-world problems that are NP-hardness (non-deterministic polynomial-time hardness) and not so easily solvable by classical computers.

Quantum addition is the fundamental of quantum arithmetics, and its use cases are in quantum cryptography, quantum modular arithmetics and multiplications. It forms the basis of a majority of high-computational arithmetic computations. Shaik and Rangaswamy [2] explore implementation of quantum addition circuits using conventional gate-based approach [1, 3], but at the cost of higher number of qubits. Beauregard [4] proposed a generalized circuit of Quantum Shor Algorithm [5] using quantum addition circuit depicted by Draper [6]. Here, in [6], quantum Fourier transform (QFT) [7], with exponential speed-up played an important part in implementation of quantum addition circuits [8]. In brief, QFT transforms the state of qubit from computational basis to Fourier basis [9]. Therefore, in this work, the authors have put an attempt to propose n-qubit QFT addition circuits with a minimum of $(2n + 1)$ qubits only, taking an example of one-qubit and two-qubit quantum addition for easy illustration. An experimental evaluation study of one-qubit and two-qubit QFT adders [7] will be discussed by its execution on real IBM quantum devices. Later, this work infers and derives accuracy rates of quantum circuits on various IBM real quantum systems and draw conclusions based on the performance. These contributions are made so that this work help other researchers to improvise upon this using real quantum computers to obtain better results.

The paper is organized as follows: Sect. 2 provides a brief background to QFT and inverse QFT (IQFT), arbitrary single-qubit and basic quantum gates that will be used in the development of QFT adders [10]. Section 3 discusses the proposed work on QFT adder taking an example of one-qubit and two-qubit quantum addition for better illustration. Section 4 discusses the experimental evaluation study of executing an illustrative one-qubit and two-qubit QFT adder on various IBM quantum experience (IBM QX) [11]. This is followed by conclusions and future works in Sect. 5.

2 Background

2.1 *QFT and IQFT [1]*

Quantum Fourier Transform (QFT) [7, 9] is the fundamental building block of various quantum algorithms such as for computing discrete logarithms, Shor's algorithm for factoring which is used in quantum cryptography, the quantum phase estimation algorithm which is used for estimating eigen values of a unitary operator. QFT is the quantum implementation of the discrete Fourier transform (DFT) over the amplitudes of a wave function [1]. DFT acts on vectors $(x_0, x_1, x_2, \ldots, x_{N-1})$ and maps it to the vector $(y_0, y_1, y_2, \ldots, y_{N-1})$ as

$$y_k = \frac{1}{\sqrt{N}} \sum_{n=0}^{N-1} x_n \omega_N^{-kn}, \quad (1)$$

where $k = 0, 1, 2, \ldots, N-1$ and $\omega = e^{2\pi i/N}$. Similarly, the QFT operator can be defined as

$$\text{QFT}\,|x\rangle = \frac{1}{\sqrt{N}} \sum_{k=0}^{N-1} e^{i.2\pi \frac{xk}{N}} \,|k\rangle \quad (2)$$

considering a $N-1$ dimensional system with states $|x\rangle$ from the computational basis $|0\rangle, |1\rangle, \ldots, |N-1\rangle$. The QFT helps to encode a number x in the relative phases of the states of a uniform superposition consisting in the sum of all the states $|k\rangle$ in the computational basis, with the amplitude of $\frac{1}{\sqrt{N}}$. It can also be stated that QFT transforms states in the computational (Z) basis to the Fourier basis. Applying QFT on $|0\rangle$ and $|1\rangle$ in Z basis, the obtained states are $|+\rangle$ and $|-\rangle$, respectively, in X basis. The IQFT can also be defined as

$$\text{IQFT}\,|k\rangle = \frac{1}{\sqrt{N}} \sum_{x=0}^{N-1} e^{-i.2\pi \frac{xk}{N}} \,|x\rangle\,. \quad (3)$$

The quantum circuit implementation of QFT and IQFT consists of Hadamard (H) gates and controlled U_1 gates [12]. The Hadamard (H) gate when applied to a qubit in a circuit it puts it in to a superposition of states such that when it is measured, it could be 0 or 1 with equal probability. The controlled U_1 (CU_1) gate implements rotation angle around the Z-axis (phase) of the target qubit if the control qubit is 1. The matrix representations of basic quantum gates (H, X and CU_1) are given in Sect. 2.3.

2.2 Arbitrary Single-Qubit Quantum Gates

An arbitrary single-qubit quantum gate, $U_3(\theta, \phi, \lambda)$ can be implemented by setting three parameters, θ, ϕ and λ, and the matrix representation is

$$U_3(\theta, \phi, \lambda) = \begin{pmatrix} \cos(\frac{\theta}{2}) & e^{-i\lambda}\sin(\frac{\theta}{2}) \\ e^{i\phi}\sin(\frac{\theta}{2}) & e^{i(\lambda+\phi)}\cos(\frac{\theta}{2}) \end{pmatrix} \tag{4}$$

Here, the three constraints for θ, ϕ and λ are $0 \le \theta \le \pi$, $0 \le \phi < 2\pi$ and $0 \le \lambda < 2\pi$, respectively. This is the most general form of a single-qubit unitary.

U_2 and U_1 are two additional single-qubit gates and are special cases of U_3 gates. The matrix representations of U_2 and U_1 are

$$U_2(\phi, \lambda) = U_3\left(\frac{\pi}{2}, \phi, \lambda\right) = \frac{1}{\sqrt{2}}\begin{pmatrix} 1 & e^{i\lambda} \\ e^{i\phi} & e^{i(\lambda+\phi)} \end{pmatrix} \tag{5}$$

$$U_1(\lambda) = U_3(0, 0, \lambda) = \begin{pmatrix} 1 & 0 \\ 0 & e^{i\lambda} \end{pmatrix} \tag{6}$$

2.3 Quantum Gates

The matrix representations of various quantum gates using the computational basis of $|0\rangle$ and $|1\rangle$ used in the development of QFT adders are:

H Gate (which puts qubits in superposition state)

$$H = U_2(0, \pi) = \frac{1}{\sqrt{2}}\begin{pmatrix} 1 & 1 \\ 1 & -1 \end{pmatrix} \; -\boxed{H}- \tag{7}$$

X Gate (which flips the state of qubits)

$$X = U_3(\pi, 0, \pi) = \begin{pmatrix} 0 & 1 \\ 1 & 0 \end{pmatrix} \; -\boxed{X}- \tag{8}$$

Controlled U_1 (CU_1) gate

$$\mathrm{CU}_1 = \begin{pmatrix} 1 & 0 & 0 & 0 \\ 0 & 1 & 0 & 0 \\ 0 & 0 & 1 & 0 \\ 0 & 0 & 0 & e^{i\phi} \end{pmatrix} \tag{9}$$

3 Proposed Work on QFT Adder Circuits

This section illustrates the design and implementation of n-qubit QFT adders using $(2n + 1)$ qubits only taking two examples of one-qubit and two-qubit additions.

3.1 One-Qubit QFT Adder

Figure 1 illustrates the quantum implementation of one-qubit QFT adder. Here, a_0 and b_0 are the two inputs; the output qubits are a_1 and a_0 that can be considered as sum qubits, respectively. Table 1 shows the truth table of one-qubit QFT adder for every possible combinations of inputs.

3.2 Two-Qubit QFT Adder

Figure 2 shows the quantum implementation of two-qubit QFT adder and Table 2 shows the truth table for every possible combinations of inputs. Here, a_0, a_1 and b_0, b_1 are the two inputs to adder and a_2, a_1, a_0 is the output of the circuit (Fig. 2).

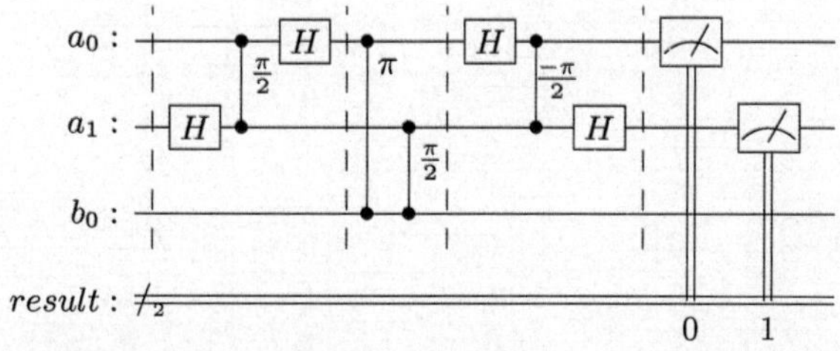

Fig. 1 Circuit diagram of one-qubit QFT adder

Table 1 Truth table of one-qubit QFT adder

Inputs		Outputs	
$A(a_0)$	$B(b_0)$	$A(a_1a_0)$	
a_0	b_0	a_1	a_0
0	0	0	0
0	1	0	1
1	0	0	1
1	1	1	0

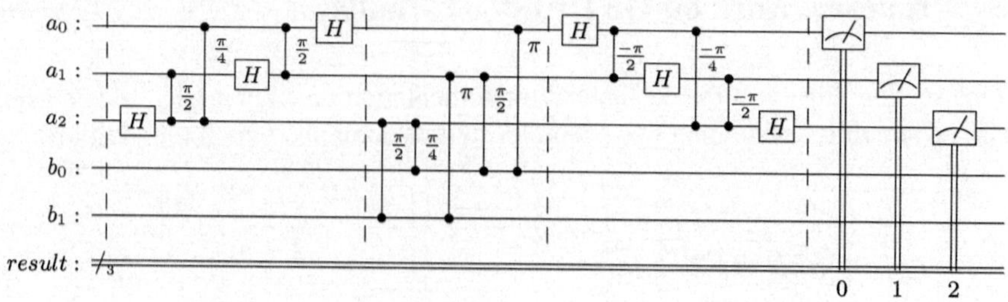

Fig. 2 Circuit diagram of two-qubit QFT adder

Table 2 Truth table of two-qubit QFT adder

Inputs				Outputs		
$A(a_0a_1)$		$B(b_0b_1)$		$A(a_2a_1a_0)$		
a_1	a_0	b_1	b_0	a_2	a_1	a_0
0	0	0	0	0	0	0
0	0	0	1	0	0	1
0	0	1	0	0	1	0
0	0	1	1	0	1	1
0	1	0	0	0	0	1
0	1	0	1	0	1	0
0	1	1	0	0	1	1
0	1	1	1	1	0	0
1	0	0	0	0	1	0
1	0	0	1	0	1	1
1	0	1	0	1	0	0
1	0	1	1	1	0	1
1	1	0	0	0	1	1
1	1	0	1	1	0	0
1	1	1	0	1	0	1
1	1	1	1	1	1	0

4 Experiments on Quantum Additions Using One-Qubit and Two-Qubit QFT Adders

Experiments were conducted on one-qubit and two-qubit QFT adders (Figs. 1 and 2) on IBM Quantum Assembly Language (QASM) Simulator [11] and eight IBM real quantum devices: YorkTown, Santiago, Athens, Quito, Belem, Lima, Manila and Melbourne. The IBM Melbourne system supports fifteen qubits, whereas the rest of the seven quantum hardware supports only five qubits. The authors have considered 1024 shots in each run and a total of ten runs on IBM devices. Results from each run are used to compute the probabilities of all the possible states and determine the accuracy of quantum circuits on eight IBM Quantum Systems.

The accuracy rate can be calculated by taking the summation of all probabilities of all expected output states divided by the number of all possible input states [13]. *Mathematically, the accuracy rate can be calculated as*

$$\text{accuracy\_rate} = \frac{1}{2^{2N}} \sum_{j=1}^{2^{N+1}} \sum_{i=1}^{2^{2N}} p_{\text{ideal}}^{(i,j)} \times p_{\text{real\_hw}}^{(i,j)} \quad (10)$$

where p_{ideal} *and* $p_{\text{real\_hw}}$ *are the probabilities of the ideal simulator and that of the real quantum hardware, respectively, for N-qubit adders*; i *and* j *represents the possible input states and output states respectively.*

The experimental analysis of accuracy rate on one-qubit and two-qubit QFT adder circuits over IBM real quantum systems are as follows:

4.1 Experimental Analysis of One-Qubit QFT Adder

The authors have created all possible inputs states using X gates, tested the output and then analysed the results by plotting histogram. Figures 3, 4, 5, 6, 7, 8, 9, 10 and 11 show the histograms of the results obtained by experimental evaluation of one-qubit QFT adder (Fig. 1) for all nine IBM quantum systems. The accuracy rate is calculated by comparing the results of histogram with the truth table (Table 1) according to expected outputs. Table 3 gives the accuracy rate of one-qubit QFT adder on different IBM quantum systems. The QASM simulator [14] is the ideal simulator and provides 100% accuracy (Table 3). From Table 3, it can be concluded that IBM Lima and IBM Athens provide better accuracy results of 80.2% and 80.125%, respectively, amongst available five-qubit and fifteen-qubit quantum hardware systems not taking into consideration the QASM simulator (ideal simulator). Note: The text colour in Table 3 are referred the same as the colour of histograms.

4.2 Experimental Analysis of Two-Qubit QFT Adder

Figures 12, 13, 14, 15, 16, 17, 18, 19 and 20 show the results obtained by experimental evaluation of two-qubit QFT Adder (Fig. 2) for available IBM quantum systems. The probabilities are calculated for the possible output states with reference to input states as per the truth table (Table 2). Table 4 provides accuracy results of two-qubit QFT adders. From Table 4, it can be clearly stated that IBM Athens performs the best with accuracy rate of 25.575% amongst the presently available systems not taking into consideration the QASM simulator [14] (ideal simulator) that provides 100% accuracy.

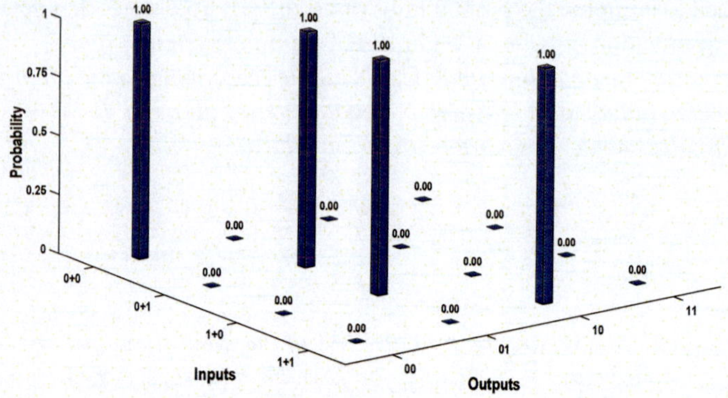

Fig. 3 Experimental results of one-qubit QFT adder on IBM quantum device: IBM QASM (32 qubits) [15]

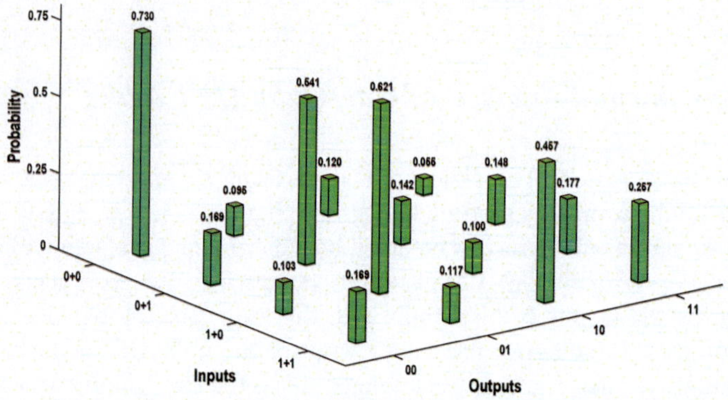

Fig. 4 Experimental results of one-qubit QFT adder on IBM quantum device: IBM YorkTown (5 qubits) [15]

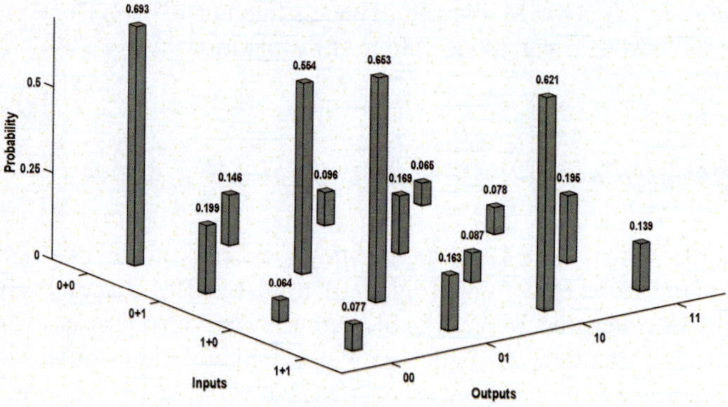

Fig. 5 Experimental results of one-qubit QFT adder on IBM quantum device: IBM Santiago (5 qubits) [15]

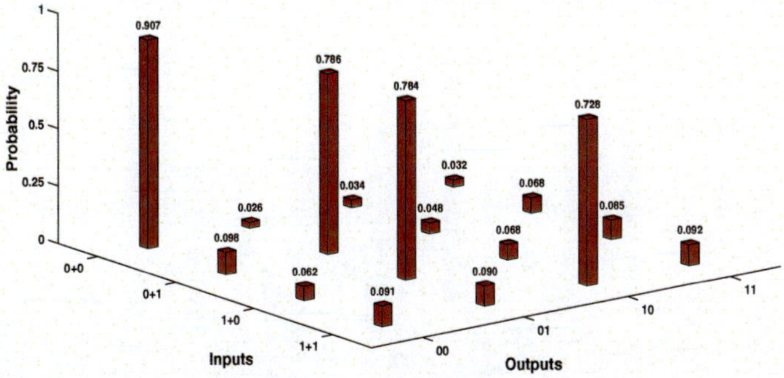

Fig. 6 Experimental results of one-qubit QFT adder on IBM quantum device: IBM Athens (5 qubits) [15]

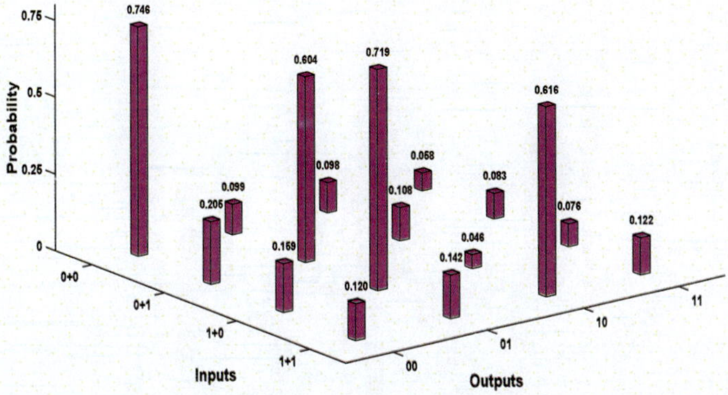

Fig. 7 Experimental results of one-qubit QFT adder on IBM quantum device: IBM Quito (5 qubits) [15]

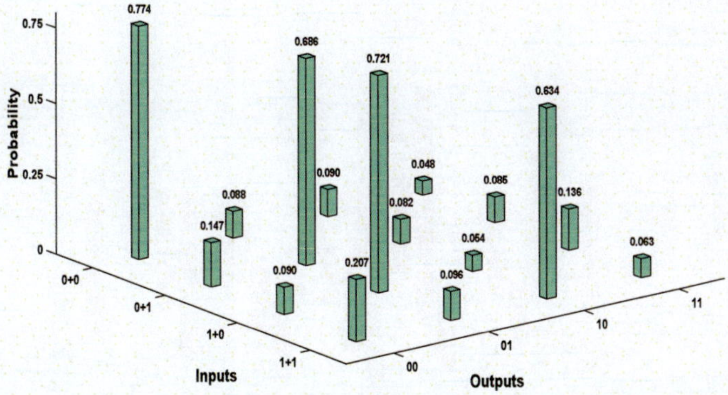

Fig. 8 Experimental results of one-qubit QFT adder on IBM quantum device: IBM Belem (5 qubits) [15]

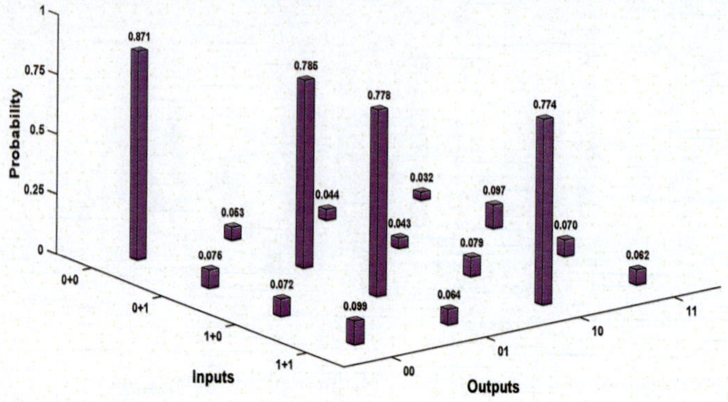

Fig. 9 Experimental results of one-qubit QFT adder on IBM quantum device: IBM Lima (5 qubits) [15]

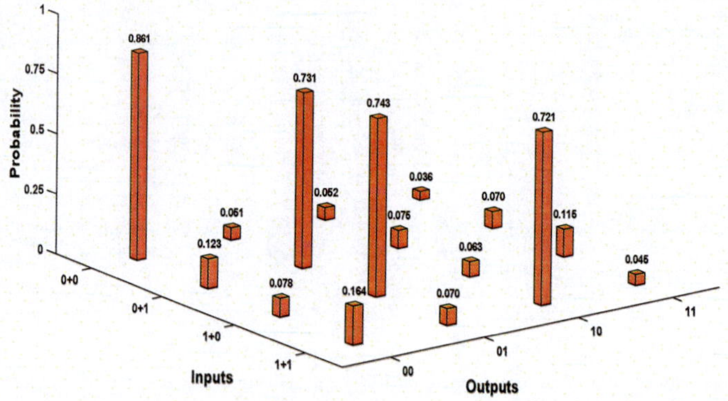

Fig. 10 Experimental results of one-qubit QFT adder on IBM quantum device: IBM Manila (5 qubits) [15]

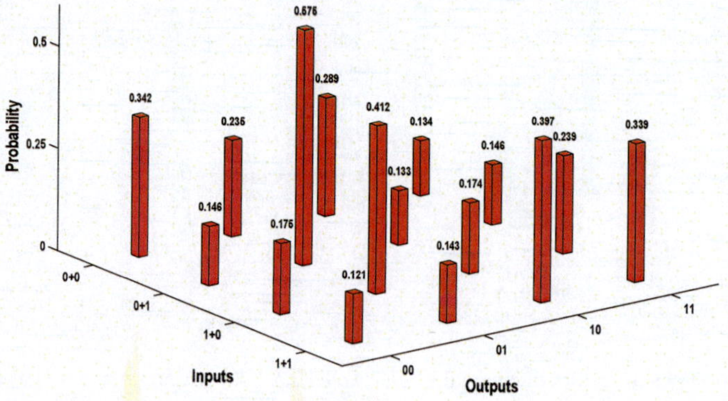

Fig. 11 Experimental results of one-qubit QFT adder on IBM quantum device: IBM Melbourne (15 qubits) [15]

Table 3 Accuracy rate of one-qubit QFT adder on real IBM quantum devices [15]

Sl. No.	IBM Quantum devices	Accuracy rate (in %)
1.	**IBM QASM Simulator (32 qubits)**	100
2.	**IBM YorkTown (5 qubits)**	58.725
3.	**IBM Santiago (5 qubits)**	63.025
4.	**IBM Athens (5 qubits)**	80.125
5.	**IBM Quito (5 qubits)**	67.125
6.	**IBM Belem (5 qubits)**	70.375
7.	**IBM Lima (5 qubits)**	80.2
8.	**IBM Manila (5 qubits)**	76.4
9.	**IBM Melbourne (15 qubits)**	43.15

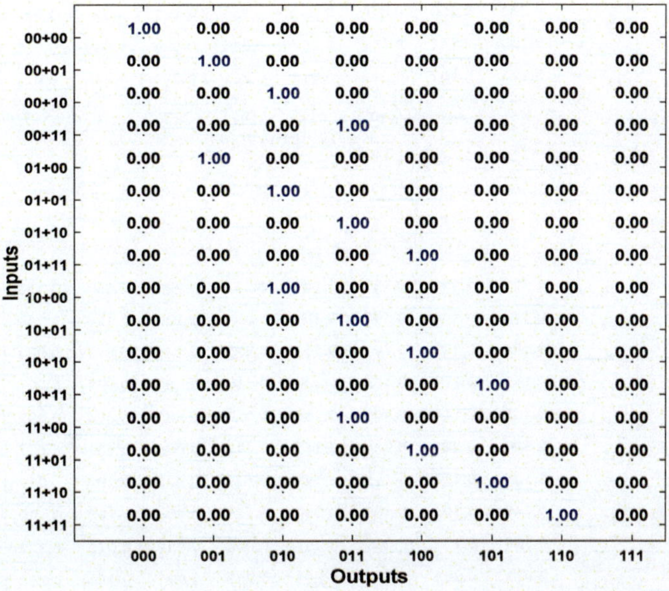

Fig. 12 Experimental results of two-qubit QFT adder on IBM quantum device: IBM QASM (32 qubits) [15]

An another differentiating factor of two-qubit QFT adder is that it requires only a minimum of five qubits and can run on any eight IBM systems readily available, whereas the conventional gate-based quantum adders requires a minimum of seven qubits and one ancilla qubit for computation possible for IBM Melbourne (e.g. ibmq_16_Melbourne) server only.

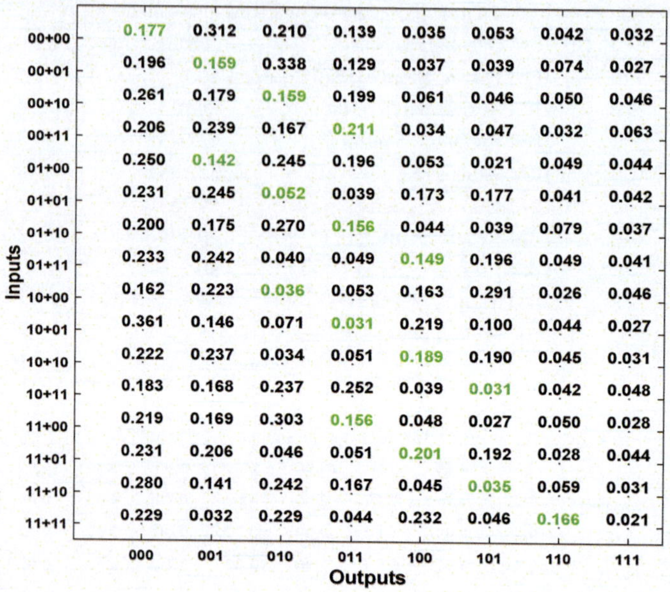

Fig. 13 Experimental results of two-qubit QFT adder on IBM quantum device: IBM YorkTown (5 qubits) [15]

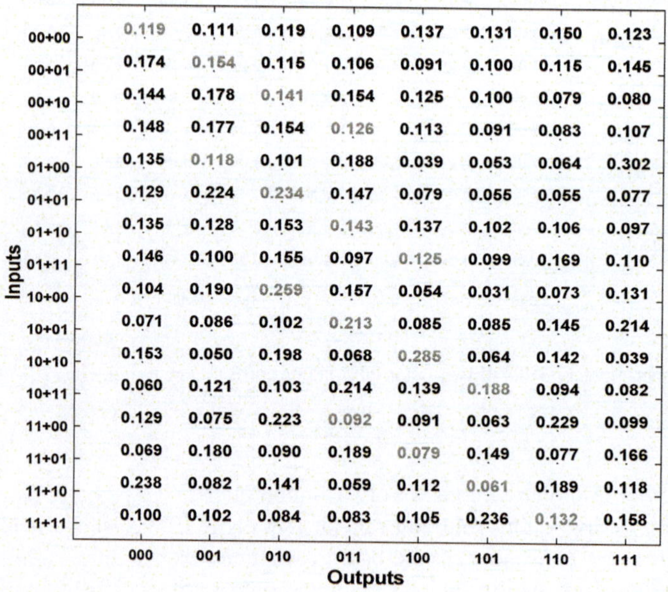

Fig. 14 Experimental results of two-qubit QFT adder on IBM quantum device: IBM Santiago (5 qubits) [15]

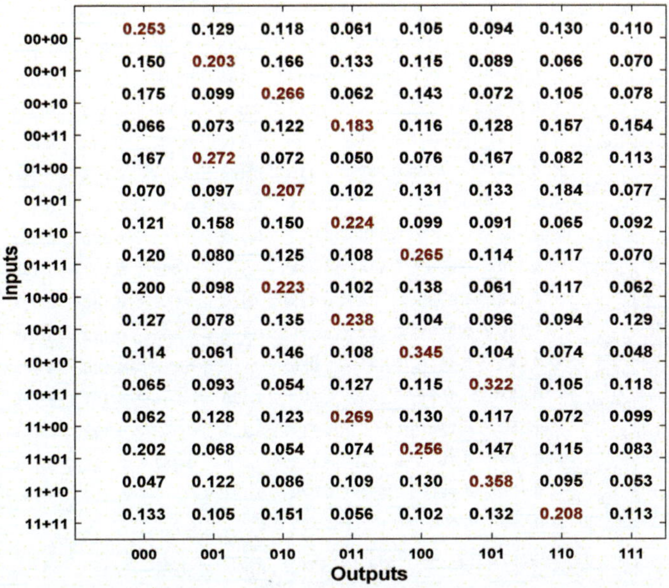

Fig. 15 Experimental results of two-qubit QFT adder on IBM quantum device: IBM Athens (5 qubits) [15]

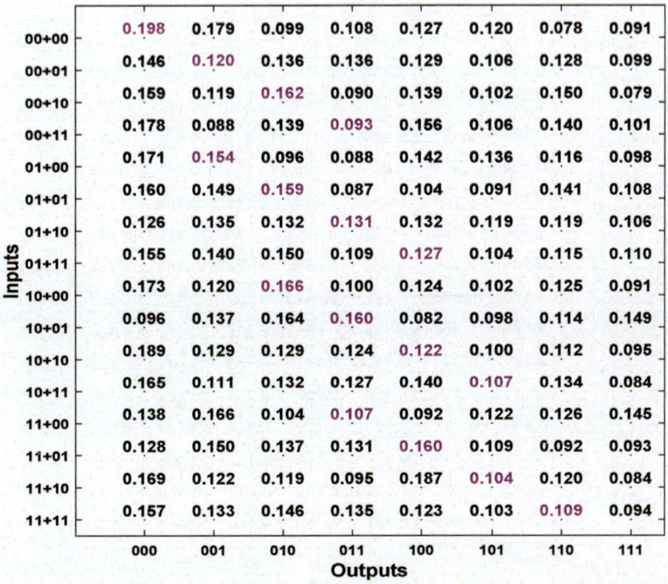

Fig. 16 Experimental results of two-qubit QFT adder on IBM quantum device: IBM Quito (5 qubits) [15]

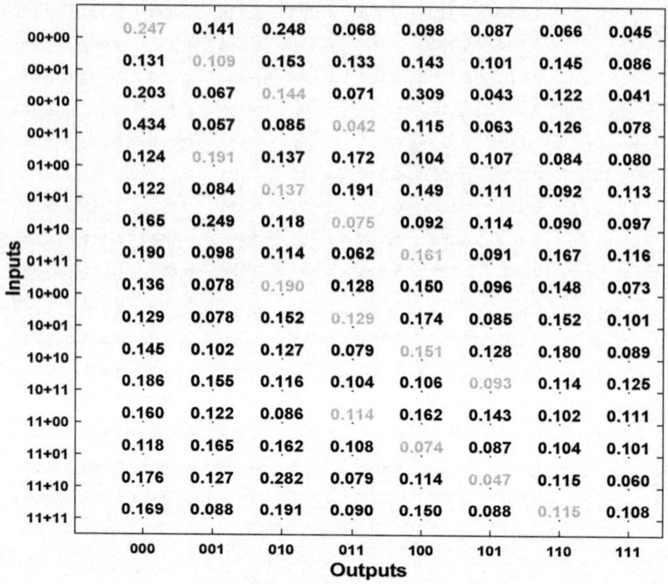

Fig. 17 Experimental results of two-qubit QFT adder on IBM quantum device: IBM Belem (5 qubits) [15]

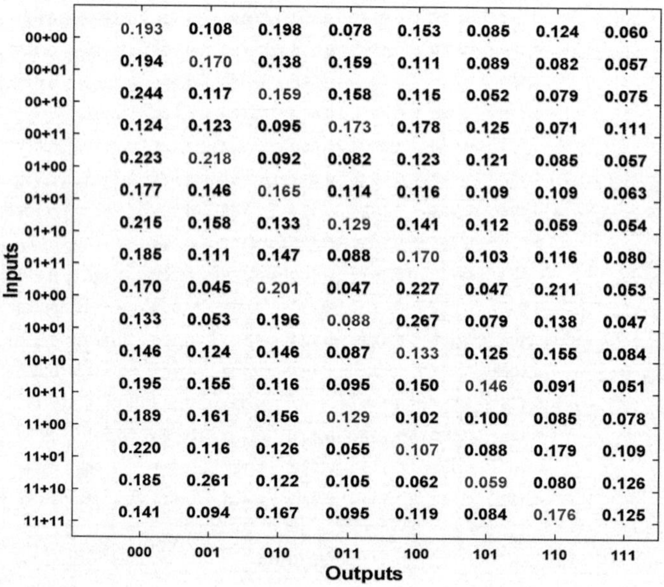

Fig. 18 Experimental results of two-qubit QFT adder on IBM quantum device: IBM Lima (5 qubits) [15]

Inputs \ Outputs	000	001	010	011	100	101	110	111
00+00	0.249	0.116	0.154	0.068	0.127	0.071	0.116	0.098
00+01	0.260	0.198	0.130	0.100	0.077	0.050	0.057	0.129
00+10	0.085	0.033	0.216	0.081	0.354	0.066	0.115	0.049
00+11	0.119	0.112	0.051	0.080	0.177	0.240	0.130	0.091
01+00	0.086	0.248	0.175	0.128	0.099	0.161	0.054	0.050
01+01	0.182	0.076	0.158	0.110	0.319	0.052	0.060	0.043
01+10	0.103	0.098	0.057	0.077	0.104	0.172	0.238	0.152
01+11	0.190	0.222	0.049	0.048	0.103	0.157	0.091	0.141
10+00	0.109	0.088	0.314	0.121	0.166	0.047	0.086	0.068
10+01	0.090	0.076	0.055	0.174	0.174	0.136	0.086	0.210
10+10	0.072	0.091	0.144	0.150	0.207	0.142	0.102	0.093
10+11	0.111	0.100	0.127	0.045	0.081	0.122	0.269	0.146
11+00	0.138	0.107	0.084	0.128	0.110	0.104	0.098	0.230
11+01	0.130	0.063	0.147	0.053	0.138	0.094	0.310	0.065
11+10	0.137	0.191	0.066	0.060	0.166	0.150	0.150	0.144
11+11	0.123	0.072	0.128	0.065	0.080	0.072	0.241	0.218

Fig. 19 Experimental results of two-qubit QFT adder on IBM quantum device: IBM Manila (5 qubits) [15]

Inputs \ Outputs	000	001	010	011	100	101	110	111
00+00	0.171	0.118	0.160	0.104	0.149	0.188	0.100	0.079
00+01	0.131	0.131	0.123	0.110	0.156	0.139	0.108	0.102
00+10	0.147	0.100	0.160	0.114	0.126	0.103	0.162	0.088
00+11	0.141	0.116	0.152	0.128	0.126	0.120	0.135	0.082
01+00	0.160	0.126	0.146	0.116	0.137	0.110	0.097	0.108
01+01	0.196	0.124	0.127	0.093	0.149	0.096	0.121	0.094
01+10	0.159	0.140	0.146	0.098	0.145	0.081	0.119	0.113
01+11	0.152	0.111	0.139	0.124	0.157	0.118	0.129	0.069
10+00	0.155	0.100	0.154	0.096	0.136	0.100	0.146	0.114
10+01	0.143	0.123	0.165	0.095	0.121	0.105	0.134	0.114
10+10	0.150	0.113	0.141	0.102	0.145	0.104	0.136	0.110
10+11	0.169	0.128	0.134	0.104	0.146	0.116	0.116	0.087
11+00	0.127	0.100	0.144	0.111	0.144	0.135	0.134	0.106
11+01	0.180	0.129	0.145	0.111	0.143	0.095	0.113	0.085
11+10	0.151	0.090	0.138	0.101	0.152	0.097	0.162	0.109
11+11	0.176	0.093	0.185	0.086	0.170	0.077	0.134	0.080

Fig. 20 Experimental results of two-qubit QFT adder on IBM quantum device: IBM Melbourne (15 qubits) [15]

Table 4 Accuracy rate of two-qubit QFT adder on real IBM quantum devices [15]

Sl. No.	IBM Quantum devices	Accuracy rate (in %)
1.	**IBM QASM Simulator (32 qubits)**	100
2.	**IBM YorkTown (5 qubits)**	12.8125
3.	**IBM Santiago (5 qubits)**	15.43125
4.	**IBM Athens (5 qubits)**	25.575
5.	**IBM Quito (5 qubits)**	13.62
6.	**IBM Belem (5 qubits)**	12.62
7.	**IBM Lima (5 qubits)**	15.1
8.	**IBM Manila (5 qubits)**	17.52
9.	**IBM Melbourne (15 qubits)**	13.08125

5 Conclusions

This paper discussed an experimental evaluation of the performance of the nine IBM QX hardware (YorkTown, Santiago, Athens, Quito, Belem, Lima, Manila and Melbourne) for quantum simulations of one-qubit and two-qubit QFT adder circuits. Quantum circuits were developed to represent quantum additions using Qiskit for simulating quantum computation and considered 1024 shots in each of the ten runs performed on IBM devices. The authors have considered accuracy rate as a metric to comparison the performance of various real quantum devices.

Future works include development of quantum circuits for higher-qubit QFT adders on real hardware. Moreover, error mitigation techniques to manage various types of hardware noise (readout error, gate error and state preparation error) and improve the simulation performance of quantum additions will also be investigated.

References

1. Nielsen MA, Chuang IL (2011) Quantum computation and quantum information, 10th edn. Cambridge University Press, USA
2. Shaik EH, Rangaswamy N (2020) Implementation of quantum gates based logic circuits using IBM Qiskit. In: 2020 5th International conference on computing, communication and security (ICCCS), pp 1–6
3. Sohel MA, Zia N, Ali MA, Zia N (2020) Quantum computing based implementation of full adder. In: 2020 IEEE International conference for innovation in technology (INOCON), pp 1–4
4. Beauregard S (2003) Circuit for Shor's algorithm using 2n+3 qubits. Quantum Inf Comput 3(2):175–185
5. Shor PW (1997) Polynomial-time algorithms for prime factorization and discrete logarithms on a quantum computer. SIAM J Comput 26(5):1484–1509
6. Draper TG (2000) Addition on a quantum computer. Available at https://arxiv.org/pdf/quant-ph/0008033.pdf

7. Asaka R, Sakai K, Yahagi R (2020) Quantum circuit for the Fast Fourier Transform. Quantum Inf Process 19(8)
8. Methachawalit W, Chongstitvatana P (2020) Adder circuit on IBM universal quantum computers. In: 2020 17th International conference on electrical engineering/electronics, computer, telecommunications and information technology (ECTI-CON), pp 92–95
9. Zhou SS, Loke T, Izaac JA, Wang JB (2017) Quantum Fourier Transform in computational basis. Quantum Inf Process 16(3)
10. Ruiz-Perez L, Garcia-Escartin JC (2017) Quantum arithmetic with the Quantum Fourier Transform. Quantum Inf Process 16(6)
11. IBM quantum experience. Available at https://quantum-computing.ibm.com/. Accessed on 5 June 2021
12. Sutor RS (2019) Dancing with qubits, 1st edn. Packt Publishing
13. Singh D, Jakhodia S, Jajodia B (2021) Experimental evaluation of adder circuits on IBM QX Hardware. In: International conference on inventive computation technologies (ICICIT). Springer, Berlin, pp 1–13
14. Cross AW, Bishop LS, Smolin JA, Gambetta JM (2017) Open quantum assembly language. Available at https://arxiv.org/pdf/1707.03429.pdf
15. IBM quantum backends. Available at https://quantum-computing.ibm.com/docs/manage/backends/. Accessed on 5 June 2021

Wireless Networks and Internet of Everything (IoE)

A Microcontroller Based Hardware Implementation to Detect and Protect a Transformer Against Various Faults

Shafqat Nabi Mughal and Mustansir Nizam

Abstract The objective of this paper is to provide a mechanism to detect and protect a transformer against various kinds of faults. The protection scheme for a transformer is highly desirable as the continuity of power supply is directly associated with it. A reliable transformer protection scheme is desirable for safe and economically viable operation in the power system. In order to protect a transformer against different kinds of faults many methods are in vogue, however, each of them is suitable for a particular kind of fault and is very complex in nature. This paper describes a design and execution of microcontroller-based system for protecting a transformer. In this research work, a microcontroller is used to detect electrical faults and send signal to a relay to disconnect the transformer from the system. At the time of fault occurrence, it generates a signal to the control room providing information pertaining to position of fault. The proposed methodology has been hardware implemented as well.

Keywords Transformer · Microcontroller · GSM · Relay · Temperature sensor

1 Introduction

An electric power system is a network containing different electrical instruments used in the generation, transmission, and distribution of electric power. In order to meet the specific needs of different types of consumers, special devices that change the voltage levels are required. Some of these devices comprise of transformers, capacitor banks, protection devices, etc. Protection of power system against various kinds of fault is necessity for better power quality. The occurrence of fault is dangerous for both electric power consumers and the electric network itself. The main function of a power system is consumer should receive a stable supply of electric power and also make sure that the electrical instrument does not get damaged. The transformer is a heart of electrical power system. It is the most important instrument to transfer

S. N. Mughal (✉) · M. Nizam
Department of Electrical Engineering, SoET, BGSB University, Rajouri, Jammu and Kashmir, India
e-mail: snmughal@bgsbu.ac.in

P. K. Singh et al. (eds.), *Emerging Technologies for Computing, Communication and Smart Cities*, Lecture Notes in Electrical Engineering 875,
https://doi.org/10.1007/978-981-19-0284-0_32

electrical energy from one circuit to another. A number of transformer protection schemes are available in the literature such as microprocessor-based relay systems, differential protection system, etc. However, in this research work, we are designing a mechanism using microcontroller and Global System Mobile (GSM) technology along with a step-down transformer. Here the system voltage is decreased from 220 to 12 V and an automotive localization system is developed using GSM-SMS services. This system communicates a message to predefined numbers about the status of different electrical parameters such as current, temperature, voltage, power etc.

Transformer being heart of the electric power system, hence its protection against various faults is very important. These are used in everyday life for different purposes [1, 2]. They are also used to isolate two circuits electrically [3, 4]. They are used to increase or decrease the alternating voltages in the various protective systems as well [5, 6]. They are used to decrease the voltage from high level to low level for safety [7]. They are used as rectifier also [8, 9]. They are used in power supplies, voltage regulators, voltage stabilizers, etc. [10].

With the increase in the demand of power, transformers become overloaded. The overloading of transformers arises owing to the increase in voltage/current levels than the specified ratings of a transformer. Due to overloading, a large amount of heat gets generated resulting in the breakdown of the insulation system leading to reduction in the life expectancy of a transformer. So a control mechanism is needed to overcome this overloading [11, 12]. Nowadays, microcontroller-based relay protection is used for overloading protection, however, due to harmonics present it becomes problematic to differentiate between different kinds of faults [13]. Earlier solid state relays were used and now a days there is a transition towards digital relays [14]. In the proposed method, we have provided microcontroller-based protection against overload and overheating. Moreover using GSM technology an SMS can also be generated to the control room indicating the nature and location of fault as well.

2 Problem Formulation

The safety of transformers is critical since they are critical components of the electrical power grid. Transformers can be damaged by the overcurrent phenomena. When a transformer is damaged, the electrical supply to customers is disrupted. Blackouts wreak havoc on the economy and throw customers' work schedules off. This illustrates why this system is required and how it may assist in mitigating the impact of a transformer malfunction. To function at maximum capacity, electrical transformers require the coordination of many systems as well as routine maintenance. A failure can be caused by a variety of sources, including mechanical or electrical failures, lightning, or even human mistake. Fortunately, a lengthy history of failure documentation and increased maintenance standards can assist in preventing or detecting failures before serious harm occurs. Insulation deterioration is a typical problem with older transformers. As a transformer warms up, the high temperatures

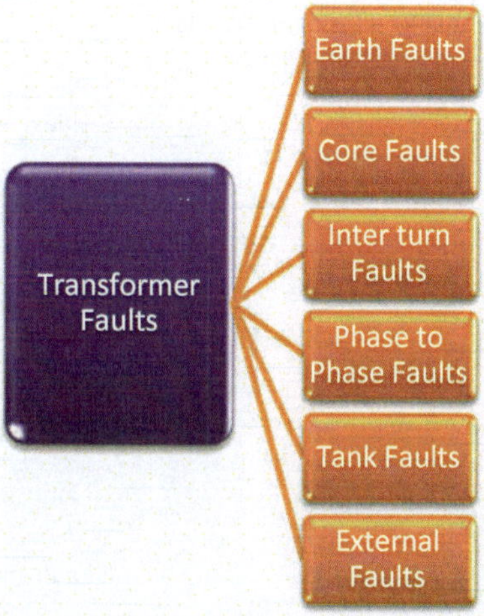

Fig. 1 Types of transformer faults

can cause the insulating system to degrade over time. If the transformer is repeatedly overloaded, the system will deteriorate faster. A compromised insulating system is the end outcome of minor, gradual improvements in loading capacity over time. Overloading leads to overheating and, eventually, thermal damage, which occurs through insulating fractures. Line surges, which are a major source of transformer failure, are a prevalent cause of electrical failure. Electrical failure can be caused by voltage spikes, switching surges, and line failures, to name a few. High-quality surge protection and frequent maintenance of the existing surge protection system are the key ways to avoid this Fig. 1 shows various types of transformer faults. These occur due to issues inside the transformer as well as due to external causes. Each type of fault causes interruption to power supply. Hence it becomes necessary to avoid them by providing suitable protections. Table 1 shows different types of faults and their protection mechanisms.

Modern power system demands highly accurate and efficient techniques/methods to locate and isolate the faulty section from the healthy section. This involves sensor-based daily monitoring operations for fast response. Since a typical microprocessor relay does not provide speedy response, so the research is shifted towards microcontroller-based relays. With the fast response of microcontroller-based systems, we have proposed a microcontroller-based setup to detect and isolate the fault in the transformer.

3 Proposed Model

The block diagram representation of our proposed setup is shown in Fig. 2. The main goal of this project is to create an automated overcurrent relay that employs a microcontroller to read transformer currents and isolates the transformer from the power

Table 1 Types of fault and its protection scheme

Type of fault	Type of protection
Phase-phase fault (primary side)	Differential relays and overcurrent protection
Phase-earth fault (primary side)	Differential relays and overcurrent protection
Phase-phase fault (secondary side)	Differential relay protection
Phase-earth fault (secondary side)	Differential relay and earth fault protection
Inter-turn faults	Differential relay, and Buchholz relay
Faults in the core	Differential relay and Buchholz relay
Faults in the oil tank	Differential relay, earth relay and Buchholz relay
Over flux	Over flux protection
Over heating	Thermal relays

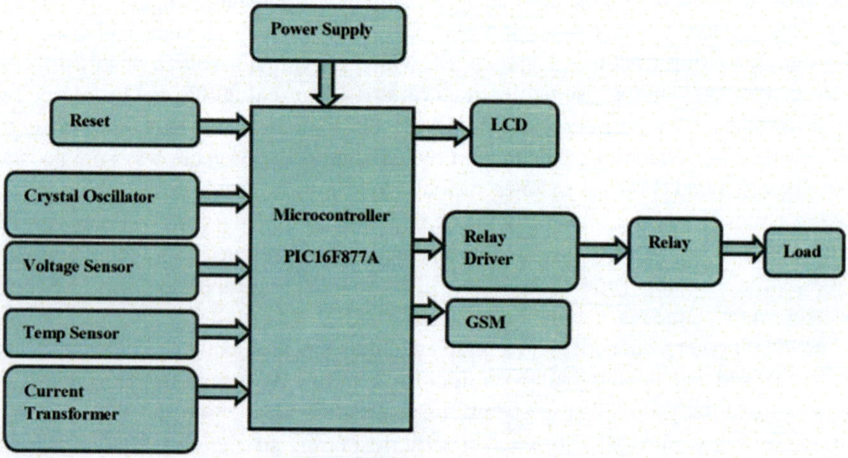

Fig. 2 Proposed setup for transformer protection

supply automatically in the event of a malfunction. As a result, the key control element in this system is the programmable Interrupt Controller (PIC microcontroller). This system's design is broken down into the following sections: Various components that are required in the design process of the hardware model are as follows.

- Transformer

  It is a device that is required to increase/decrease the voltage level of the system. In various stages of the power system network, a differently rated transformer is required. In our proposed setup we are using a stepdown transformer to change voltage level from 230 to 12 V.

- Electrical Load

  A load is something that consumes power. There are several types of loads like domestic, industrial, municipal, etc. Each of them are an amalgam of resistive, inductive, and capacitive types.
- Current transformer

  A current transformer (CT) is a special device used to measure current which is of higher magnitude. It is used for the protection of ammeter for measuring current in high power line. In our setup, the current rating of the device varies from 20A to 20 mA.
- Microcontroller

  A microcontroller PIC16f877a is used in the design setup to provide multiple options of use as it can accommodate many operations at one time owing to its variety of inputs and output pin configurations and flash memory technology. This controller being simple is easy to implement and provides a reliable setup as well.
- Relay Driver Circuit

  It is an integrated circuit comprising of NPN transistor in the design setup to control the operation of the relay. This is important to isolate the faulty section/transformer from the power supply.
- Relay

  A relay is a switch that causes the operation of the circuit breaker in the event of fault. It has a pickup current setting that initiates the operation of the circuit breaker if higher than predetermined value.
- Temperature Sensor

  In order to sense the change in the temperature, a temperature sensor is required which is basically a thermocouple. Here in our setup LM35 is chosen to sense the varying temperatures.
- LCD Display

  All the readings which are needed to be displayed for proper action are showcased using liquid crystal display panels. They provide information about various voltages, currents, etc.
- GSM

  The word GSM stands for global system for mobile communication. It has spread its use in all walks of life. Here in our system, we are using it to generate and transmit the signal in the form of SMS to the end user in the event of fault in the transformer.
- Power Supply

  A power supply is basically an energy source to make things work. In our setup 5 V supply was required which was generated using step-down transformer and rectifier/filter circuits, etc.

4 Hardware Implementation and Working

The hardware setup of the proposed design is shown in Fig. 3. It comprises various components as discussed in the Sect. 3.The design was fabricated in the laboratory.

The model was designed such that the system can be switched ON/OFF using a Personal Computer through GSM. By using all the described components above, a hardware setup was designed that disconnects the transformer from the power supply in the event of fault like overvoltage and overheating, etc. This setup will prevent the cascading effect of the damage of fault in the power system. The two cases are described as follows:

(a) Effect of temperature on transformer:

The heat produced during transformer operation raises the temperature of the transformer's interior structures. More efficient transformers have a smaller temperature increase, whereas less efficient transformers have a bigger temperature increase. A more efficient transformer produces less waste heat in the first place, but the temperature of the transformer rises as a function of both the amount of heat produced and the amount of heat removed. Make sure that a device with a minimal temperature rise isn't equally inefficient since it relies on fans to remove the surplus heat.

Dry-type transformers come in three temperature ranges: 80, 115, and 150 °C. Standard rises of 55 and 65 °C are available for liquid-filled transformers. These figures assume a maximum ambient temperature of 40 °C. For example, an 80C rise dry transformer operating at full-rated load in a 40C ambient environment will have

Fig. 3 Hardware implementation of the model

an average winding temperature of 120 °C. (Temperatures within the transformer may be greater than usual at so-called hot spots.). The 80C rise unit offers greater space for an occasional overload than the 150 °C rise unit, because most dry transformers employ the same insulation on their windings (usually rated at 220 °C), regardless of the design temperature rise.

Figure 4 shows the flowchart demonstrating the protection against rise in temperature. In the proposed setup the preset value of temperature is set to 40 °C. If the temperature becomes greater than this predefined value, relay operates isolating a transformer from the power supply thereby protecting it.

(b) Effect of over-voltage on transformer:

When electrical equipment is subjected to excessive voltage, the insulation becomes stressed, and the equipment is more likely to be damaged. Surges can induce sparking and flashing between phase and ground at the weakest point in the network, as well as the breakdown of gaseous, solid, and liquid insulation, and the failure of transformers and spinning machinery.

The iron losses (hysteresis and eddy current losses) in a transformer are affected by changes in voltage and/or frequency. Hysteresis loss (P_h) and eddy current loss (P_e) vary according to the following relationships as shown below as long as the flux

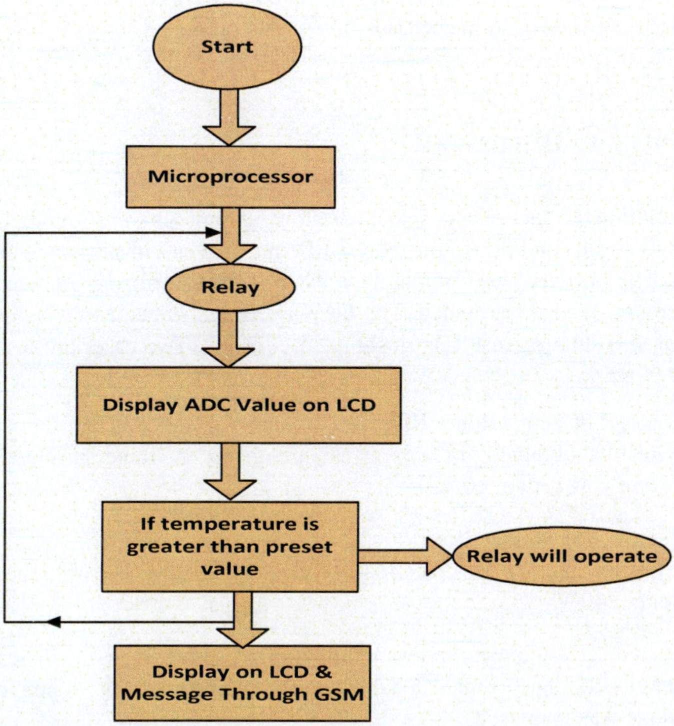

Fig. 4 Flowchart of circuit for temperature [15]

changes are sinusoidal with respect to the line.

$$P_h \propto f(\text{Ø}_{\text{maz}})^z \tag{1}$$

$$P_e \propto f^2(\text{Ø}_{\text{maz}})^2 \tag{2}$$

where x varies from 1.5 to 2.5 depending upon the iron grade used.

If the frequency and voltage of the transformer are changed in the same proportion, the flux density will stay unchanged and the no-load current will similarly appear to be unaffected. The transformer may be safely used at a frequency lower than the rated frequency, with a correspondingly lower voltage. Iron losses will be decreased in this circumstance. The core losses may become unacceptable if the transformer is operated with increased voltage and frequency in the same proportion.

When the frequency is increased while the supply voltage remains fixed, the hysteresis loss is reduced while the eddy current losses are unaltered. At higher frequencies, some voltage rise may be allowed, but how much depends on the relative degree of hysteresis and eddy current losses, as well as the quality of iron employed in the transformer core.

A working methodology during overvoltage scenario is depicted in the form of flow chart as shown in Fig. 5.If the voltage becomes greater than the preset value (300 V), relay operates and hence transformer gets protected.

5 Results and Discussion

After completing the hardware design, the following findings were obtained. Tables 2 and 3 illustrates the condition of the relay, LED, and buzzer with respect to the change in current. The primary side circulating current of the transformer increases owing to the decrease in load resistance. The decrease in resistance could be due to the breakdown of insulation resulting in shorts circuits etc. Two cases are discussed in this research work.

Case. I Effect of Temperature Rise

The results obtained for this effect are given in Table 2. The reference temperature was kept at 40 °C.

Case. II Effect of Voltage Rise.

The results obtained for this effect are given in Table 3. The reference voltage was kept at 240 V.

It has been discovered that the sensor output increases owing to the increase in the transformer circulating current. The table above demonstrates this. When the sensor output voltage was increased.

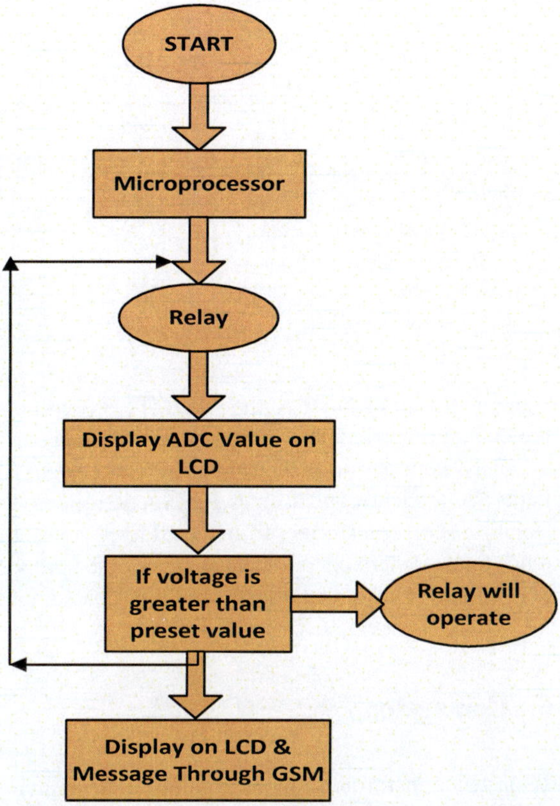

Fig. 5 Flowchart of circuit for overvoltage [15]

Table 2 Effect of temperature rise and state of system

Loading	Temperature (°C)	Sensor output (V)	Relay state	Buzzer state	LED	LCD
Low	25	0.05	OFF	OFF	Green	–
Normal	30	0.95	OFF	OFF	Green	–
Medium	35	1.89	OFF	OFF	Green	–
High	45	2.5	ON	ON	Red	Temperature fault
Very high	50	3	ON	ON	Red	Temperature fault

When the transformer current supersedes the rated value kept, the system, as illustrated in Tables 2 and 3 activates the LCD display. The LCD shows "OVERCURRENT FAULT" and the buzzer sounds to inform the system operator about a fault in a specific transformer. To disconnect the transformer from the electricity system, a

Table 3 Effect of voltage rise and state of system

Loading	Voltage (V)	Sensor output (V)	Relay state	Buzzer state	LED	LCD
Low	150	0.05	OFF	OFF	Green	–
Normal	200	0.95	OFF	OFF	Green	–
Medium	220	1.89	OFF	OFF	Green	–
High	300	2.5	ON	ON	Red	Overvoltage fault
Very high	320	3	ON	ON	Red	Over voltage fault

signal is also transmitted to the relay. The procedure of transformer separation from the system is carried out in this type using a relay and a contactor.

The system performs the role of an over-current relay, which detects overcurrent faults automatically and utilizes a microcontroller to keep the fault record, sometimes isolation of the transformer from the afflicted zone, and communicate the fault occurrence to the system operator. The benefit of the proposed methods over conventional over current relay is that after the fault is cleared, it automatically recloses.

6 Conclusion

To protect transformers against failures, many relaying principles have been devised and deployed. Over current, over flux, and overheating relays protect transformers against overloads and other externally imposed situations. Transformers are protected against internal faults by differential relays. By doing this research work, a novel method of fault detection and protection for important devices like Transformer has been established. This method will be able to protect costly equipment like transformer etc. This method will not only detect the fault but also initiate the required action to isolate transformer from the power supply. The system works by sending an SMS to the end user who can switch off the system remotely.

References

1. Bashi SM, Marium N, Rafa A (2007) Power transformer protection using microcontroller-based relay. J Appl Sci 7(12):1602–1607
2. Thiyagarajan V, Palanivel TG (2010) An efficient monitoring of substations using microcontroller based monitoring systems. Int J Res Rev Appl Sci 4(1):63–68
3. Benmouyal G, Guzman A, Zocholl S, Altuve HJ (2002) A current-based solution for transformer differential protection. II. Relay description and evaluation. IEEE Trans Power Delivery 17(4):886–893

4. Ahamed SS (2009) The role of zigbee technology in future data communication system. IEEE J Theor Appl Inf Technol 5(2):129–135
5. Thote PB, Daigavane MB, Bawane NG (2015) MATLAB simulink of based digital protection of transformer. Int J Res Eng Technol 3(2):484–487
6. Jardini JA, Brittes JL, Magrini LC, Bini MA, Yasuoka J (2005) Power transformer temperature evaluation for overloading conditions. IEEE Trans Power Delivery 20(1):179–184
7. Grainger JJ (2003) Power system analysis. Tata McGraw-Hill Publishing Co ltd.
8. Mao H (2010) Research of wireless monitoring system in power distribution transformer station based on GPRS. In: The 2nd international conference on computer and automation engineering (ICCAE), vol 5, pp 386–389
9. Nath S, Chakrabarti A, Mukhopadhyay AK (2006) A microcomputer-based unique digital fault diagnosis scheme of radial transformer feeders. IEEE Trans Power Delivery 21(4):1824–1829
10. Perera N, Rajapakse AD (2012) Design and hardware implementation of a modular transient directional protection scheme using current signals. IET Gen Trans Distrib 6(6):554–562
11. Mughal SN, Thaker R (2018) A novel approach for harmonic analysis using MATLAB. Int J Adv Eng Res Dev Emerg Trends Innov Electron Commun Eng 5(1)
12. Malik H, Mughal SN, Sood YR, Jarial RK (2011) Application and implementation of artificial intelligence in electrical system. In: International conference on advances in computing and communication (ICACC-2011). NIT Hamirpur (Himachal Pradesh)
13. Wadhwa CL (2005) Electrical power systems. New Age International (P) Ltd., New Delhi, pp 656–657
14. Malik H, Azeem A, Mughal SN, Singh S (2011) Artificial intelligence techniques for incipient fault diagnosis and condition assessment in transformers. In: International conference on emerging trends in engineering (ICETE-11). Karnataka, pp 5–8
15. Salunkhe K, Shaikh S, Kunure BS (2016) Transformer protection using microcontroller based relay & monitoring using GSM technology. Int Eng Res J (IERJ) 2(2):813–817

Fuzzy—KNN-Assisted Vehicular Localization for Bluetooth and Wi-Fi Scenario

Shubhangi Verma, O. P. Singh, Sachin Kumar, and Sumita Mishra

Abstract The exponential expansion of intelligent and smart communication, infrastructure, and other frontier technology has restructured many domains of day-to-day life and has given birth to new opportunities for managing cities more effectively and holistically. Technologies such as high-speed internet, 5G mobile networks, the internet of things, and big data play an increasingly important role in transition towards "smart cities". Technological innovations such as autonomous vehicles among them have taken the lead. Number of automobiles companies are offering different variants of autonomous vehicles with innovative solutions for issues related to safety, connectivity, transportation, and more. In terms of data transfer with optimal connection and speed, VANET technology is around here but with changing trends in technology limitations have surfaced. The work proposed performs vehicular localization employing fuzzy systems, for localization the proposed algorithm uses k-nearest algorithm with fuzzy decision system. System suggested employs two distance estimating techniques—Euclidian distance and physical distance between test nodes and reference nodes. RSSI values were evaluated for BLE and Wi-Fi scenarios for approximately 100 different vehicle positions in a 2 × 2 km simulated environment. BLE due range constraints would require a setup with V2V and V2I network. Simulated MATLAB environment was used for testing results. RSSI values were then fed to Neural Network to estimate MSE and R values for both scenarios. Experiment results demonstrate the effectiveness and merit of the two proposed scenarios, and it was found, localization through Bluetooth outperforms slightly as compared to Wi-Fi.

Keywords VANET · Fuzzy systems · RSSI · Localization · K-NN

S. Verma · O. P. Singh · S. Kumar (✉) · S. Mishra
Amity University, Lucknow Campus, Lucknow, India
e-mail: skumar3@lko.amity.edu

O. P. Singh
e-mail: opsingh@amity.edu

S. Mishra
e-mail: smishra3@lko.amity.edu

P. K. Singh et al. (eds.), *Emerging Technologies for Computing, Communication and Smart Cities*, Lecture Notes in Electrical Engineering 875,
https://doi.org/10.1007/978-981-19-0284-0_33

1 Introduction

Vehicular communication systems or VANET are networks in which roadside units and vehicles are the nodes that communicate with each other and share information like traffic status and safety information. Figure 1 depicts the vehicle to vehicle (V2V) and vehicle to infrastructure (V2I) communication in VANET architecture. Sharing of information in real-time communications between infrastructure and vehicles has become possible owing to innovations in wireless communications. It is important for one to find solutions based on issues with vehicular communication like accidents, visibility, and safe environment. Other important applications include monitoring systems, routing problems, traffic information for vehicular networks, and vehicle localization systems. With an increase in vehicle density, it has become extremely important to improve vehicle's navigation and localization systems. Due to the random and fast mobility of vehicles and weak signals, it is difficult to get accurate results of location. Hence various localization techniques have been developed and are under research.

The focus is on widely integrating with other hybrid technologies to improve the localization system on parameters of accuracy, connectivity, timing constraints, and other factors associated with VANETs. A recent survey was published on VANET localization techniques [1] where the authors discussed about the improvements in the localization and classified the VANET localization use case techniques on the basis of 1.Non-Cooperative Localization techniques 2.Cooperative Localization techniques as shown in Fig. 2. The cooperative technique is further classified as GPS and non-GPS based-techniques. The various techniques use filters based on the application and for non-cooperative techniques various communication based on cellular towers, zigbee, or other communication protocols is used.

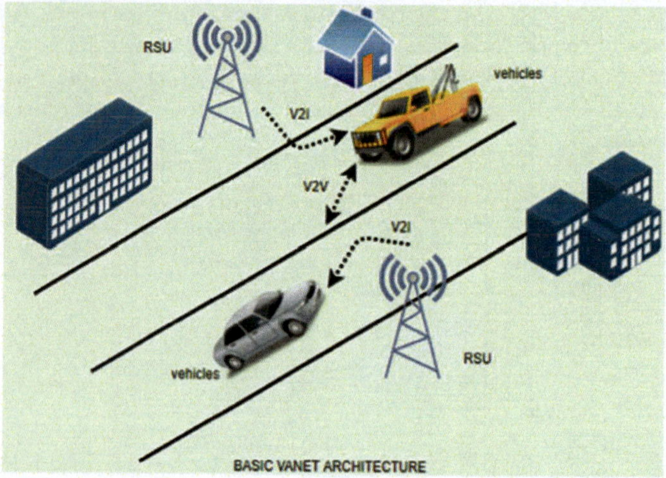

Fig. 1 VANET architecture

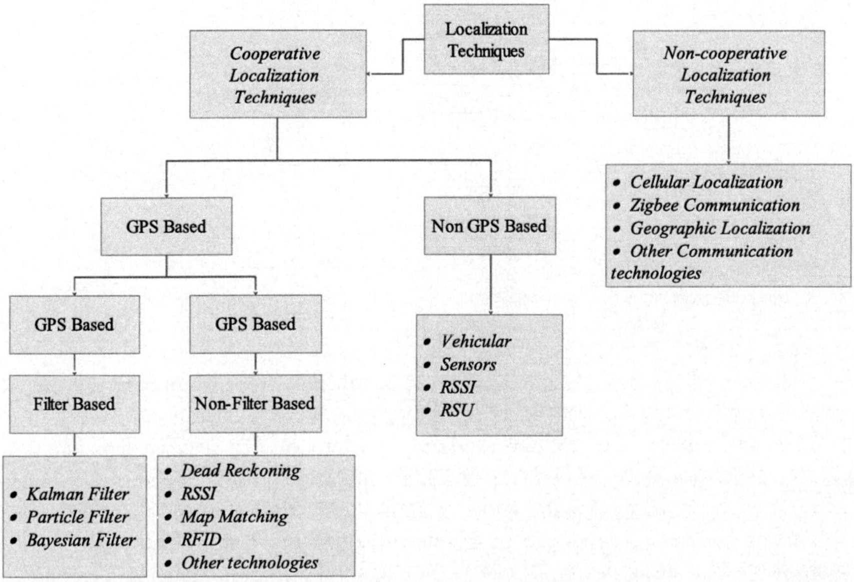

Fig. 2 VANET localization techniques

Table 1, presents a comparison of related technologies employed for entity-to-entity communications. Bluetooth Low Energy (BLE) offers several advantages as compared to Bluetooth classic (BC) for vehicle to vehicle and vehicle to infrastructure communications. BLE offers lower latency and with less bit rate a range beyond 100m is possible, further BLE for exchanging data does not require devices to be paired. Wi-Fi on the other hand offers significantly higher data rate not necessarily required for V2V applications.

Wi-Fi has extremely high energy requirements which is a major disadvantage. DSRC, has been specifically designed for Vehicle-to-vehicle communication and it provides a better performance compared to all existing ones employed. One of

Table 1 Related V2V technologies [3]

	BC	BLE	WiFi	DSRC
Band	2.4	2.4	2.4 / 5	5.9
Channel BW	1	2	20/40	10
Number of Ch.	79	40	11	7
Modulation	GFSK	GFSK	OFDM	OFDM
Max. bit rate (Mbps)	3	1	600	27
Max. power (mW)	100	10	100	2000
Approx. range (m)	10	100	100	1000
Approx. latency (ms)	100	6	50	1

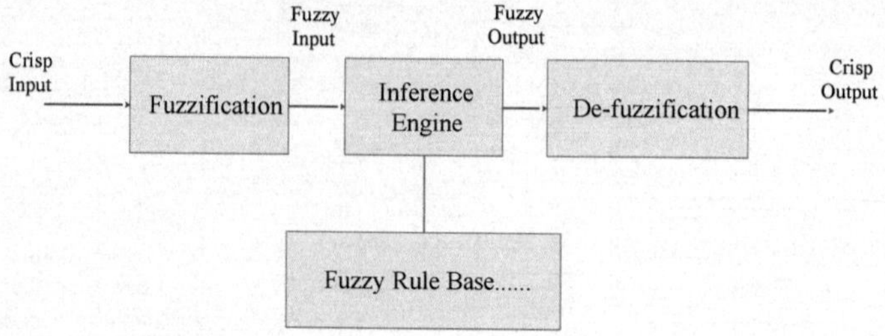

Fig. 3 Fuzzy logic system

the major disadvantages of DSRC is that it is still not widely deployed. The work presented compares BLE and Wi-Fi technology for V2V localization.

Work survey based on location-based routing protocols, [2] discussed about issues and their solutions in the VANET communications. Here the authors presented issues and challenges associated with factors including local optimum problems, accurate positioning, broadcast overhead problem, and optimum routing selection. Hence location-based routing protocols can be one of the best solutions as this technique does not rely on pre-destination routes.

Another interesting work on fuzzy logic in wireless localization [4] explains the approach as a feasible and attractive to overcome some uncertainties and challenges in wireless localization systems. The authors reviewed various fuzzy-related methodologies and techniques. They also evaluated the techniques based on four factors, i.e., environmental metrics, system metrics, fuzzy system, and positioning metrics. They classified fuzzy-based localization methods in two ways, i.e., fuzzy system will be incorporated as core estimation process in positioning system and another method is where fuzzy system will assist the overall positioning system. The fuzzy system can be further classified as fuzzy clustering techniques, fuzzy optimization method, fuzzy arithmetic and analysis, and hybrid-based approach.

The fuzzy inference system has three stages which includes Fuzzification—where the fuzzy sets using the linguistic variables are constructed, then Fuzzy rule base and Defuzzification—which produce a non-fuzzy output. The fuzzy rules work on the pre-defined rule regarding the problem statement as shown in Fig. 3. Paper is organized as follows section II addresses related work, section III presents proposed methodology in detail. Simulation and results are discussed in section IV and finally, section V concludes the paper with discussion about future direction.

2 Related Work

Jiyao An et al explained about the vehicle location problem in urban road conditions using the GPS, WiFi, and Cellular Network (CN) positioning systems. The fuzzy-based hybrid location algorithm is developed with some advantages like enhancing the positioning features by new fuzzy-weighting location mechanism and also can adjust by using fuzzy Kalman filtering method. The experiment results are evaluated to show effectiveness and merit of the proposed approach [5].

Lina Altoaimy et al proposed the localization method based on fuzzy logic and location information of neighbors. The objective of the proposed methodology is to estimate the vehicle's location by considering the neighbor's vehicle location information. In order to achieve accurate localization, the authors modeled vehicles' weights with fuzzy logic system, by utilizing the distance and heading information to obtain the weight values. They also considered the concept of centroid localization. The results obtained from the simulation are promising and shows the effectiveness of the proposed methodology in different traffic densities [6].

Sudhir Kumar et al discussed about the distributed and range-free algorithm to track vehicles. Connectivity with audible anchor nodes is used to track vehicles for range-free. The method is described as low cost and saves the cost of spending on GPS systems in the cars. The results of the proposed system in a simulation show the improved performance when compared with the other conventional methods [7].

Azat Rozyyev et al presented the results of the study where they combined KNN with fuzzy logic technique which improved the accuracy of KNN. In comparison with other localization techniques in terms of high accuracy, multilateration, and fuzzy logic indoor positioning system fuzzy KNN got better results in terms of accuracy and algorithm complexity [8].

Juan Cota-Ruiz et al proposed a hop-weighted scheme which can be useful for distributed multi-hop localization based on distance. This localization approach generates spatial locations and estimates position of unknown nodes and then computes local functions in order to minimize distance errors. The proposed methodology achieves root mean square error values which are less than the radio range of the unknown nodes with less number of iterations [9].

Kaur. A et al proposed another weighted centroid DV-Hop algorithm. This methodology uses weights that use different factors like number of anchors, radius, and to determine location of nearest anchor of unknown node. With the help of simulation and theoretical analysis the proposed methodology y outperforms the traditional algorithm of DV-hop in terms of power consumption and localization error [10].

Kang S et al presented the algorithm based on a target localization using an approximated error covariance matrix based weighted least squares solution, which integrates the received signal strength and angle of arrival data for wireless network communication. The proposed methodology does not need any prerequisite knowledge of the target true position and noise variance. Simulations show the superior performance of the method [11].

Shi X et al presented the HPWB algorithm in which unknown nodes are divided into three categories according to their number of neighboring anchors. For category one, weighted method based on RSSI was proposed. For category two and three the exclusion method was presented. The simulation results show that the proposed method HPWB got high precision, in comparison with the traditional bounding box method and weighted centroid algorithm [12].

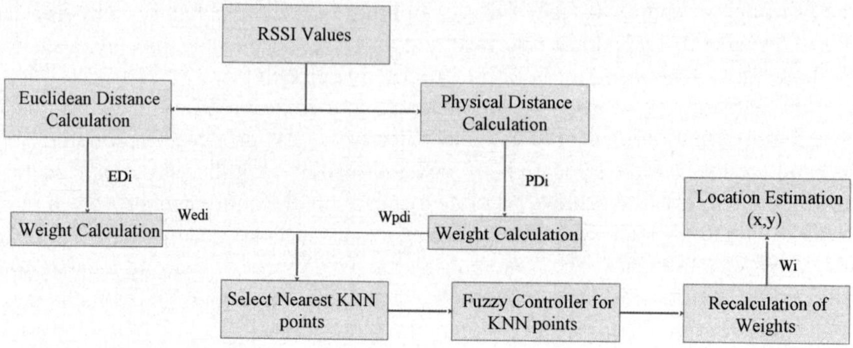

Fig. 4 Proposed architecture

3 Methodology

3.1 Architecture

Paper presents a methodology for estimating location of vehicles in a vehicular network, the technique parallelly evaluates Euclidean and Physical distance, performs weight calculation to minimize error then selects *k* nearest neighbors with Fuzzy logic recalculates the weight for accurate localization. Figure 4 depicts the architecture of the proposed methodology.

3.2 RSSI Values

There are many radio propagation models known for wireless communications that predict signal-strength loss with distance—path loss. These models are basically used in those simulations simulating WSN. When a practical approach is required there are two main variables measured when defining received-power and link-quality: RSSI stands for Received Signal Strength Indicator, It is the measured power of a received radio signal. It is implemented and widely used in 802.11 standards. Received power can be calculated from RSSI. The other one is LQI stands for Link Quality Indicator. LQI estimates how easily the received signal can be modulated when considering noise in the channel. Practical examples of RSSI and LQI values can be summarized as:

1. Weak signal and noise presence will give low RSSI and low LQI.
2. Weak signal with noise absence will give low RSSI and high LQI.
3. Strong signal and without noise gives high RSSI and high LQI.
4. Strong signal in a noisy environment will return high RSSI and low LQI.
5. And finally, strong noise my give high RSSI and low LQI.

RSSI is estimated calculated by the given Formula 1:

$$\text{RSSI} = \text{Pt} - \text{Pl(d)} \quad (1)$$

Where Pt is signal transmission power and Pl(d) is path loss when the distance is d. The unit for both power factors is dBm.

$$\text{Pl(d)} = \text{Pl(d0)} + 10n\log\left(\frac{\text{d}}{\text{d0}}\right) + X \quad (2)$$

Here 'X' is the random number of the Gaussian distribution and 'd' is distance between receiver and transmitter and 'n' is weakness factor of the environment.

Therefore,

$$\text{RSSI} = P_\gamma + G_\gamma - \text{Pl(d)} \quad (3)$$

Where P_γ is the wireless signal power and

G_γ is gain of transmitting antenna.

3.3 Distance Calculation

The two distances employed for estimation here is Euclidean distance and the physical distance based on the RSSI values.

Euclidean distance is calculated using Formula 4.

$$\text{EDi} = \sqrt{\sum_{j=1}^{n}\left(\text{RSSj} - \text{RSS}_i^j\right)^2}, \quad i = 1, 2, 3, \ldots N \quad (4)$$

The weights are calculated using Formula 5.

$$\text{WEDi} = \frac{\frac{1}{\text{EDi}}}{\sum_{i=1}^{k}\frac{1}{\text{EDi}}} \quad (5)$$

Physical distance is calculated using Formula 6.

$$\text{PDi} = \sum_{j=1}^{n}\text{PD}_i^j$$
$$\text{where} \quad \text{PD}_i^j = d^j - d_i^j, \quad i = 1, 2, 3, \ldots N \quad (6)$$

Where $\text{d}^j = \text{d}_0 10^{\frac{\text{RSS(d0)}-\text{RSSj}}{10n}}$ $\quad \text{d}_i^j = d_0 10^{\frac{\text{RSS(d0)}-\text{RSS}_j^i}{10n}}$

The weights are calculated using Formula 7.

$$\mathrm{WPDi} = \frac{\frac{1}{\mathrm{PDi}}}{\sum_{i=1}^{k} \frac{1}{\mathrm{PDi}}} \quad (7)$$

3.4 K-NN Points

K-nearest neighbors have been widely used to address localization issues, however, traditional KNN uses only Euclidean distance to measures the similarity between two vectors.

The K-NN algorithm working can be summarized as:

Step 1: Select the number K of the neighbors.
Step 2: Calculate the Euclidean distance of K number of neighbors.
Step 3: Take the K nearest neighbors as per the calculated Euclidean distance.
Step 4: Among these k neighbors, count the number of the data points in each category.
Step 5: Assign the new data points to that category for which the number of the neighbor is maximum.

3.5 Fuzzy Controller and Weights Calculation

Inserting the model in the fuzzy controller and get the common k-points and then based on the weighted formula we can get location points.

Recalculation of weights can be done using Formula 8.

$$\mathrm{wi} = \frac{\mathrm{Wpdi} + \mathrm{Wedi}}{\sum_{i=1}^{k}(\mathrm{Wpdi} + \mathrm{Wedi})} \quad \mathrm{wi} = \frac{\mathrm{Wpdi.Wedi}}{\sum_{i=1}^{k}(\mathrm{Wpdi.Wedi})} \quad (8)$$

Location can be estimated using Formula 9 (Figs 5, 6 and 7).

$$(\hat{x}, \hat{y}) = \sum_{i=1}^{k} \mathrm{wi}(\mathrm{xi}, \mathrm{yi}) \quad (9)$$

Root mean square value is estimated employing following relationship:

$$\delta = \sqrt{(\hat{x} - x)^2 + (\hat{y} - y)^2}$$

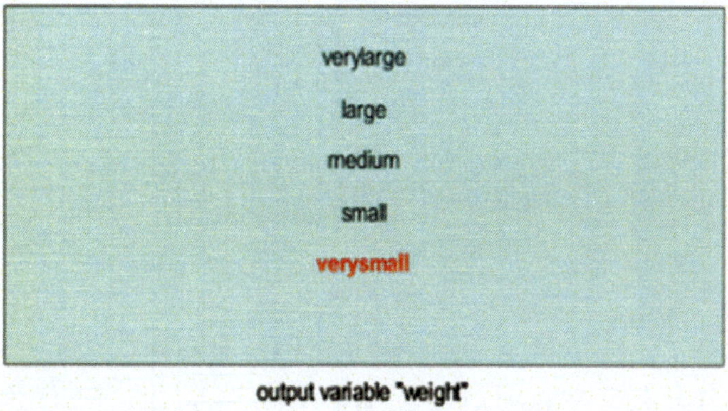

Fig. 5 Input variable as distance

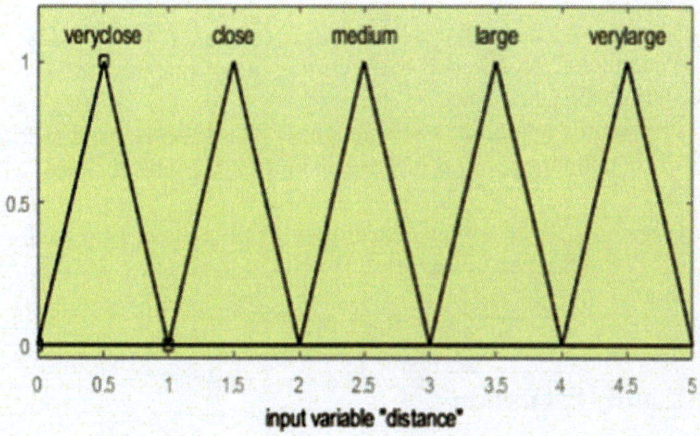

Fig. 6 Output variable as weights

Fig. 7 Rules evaluation

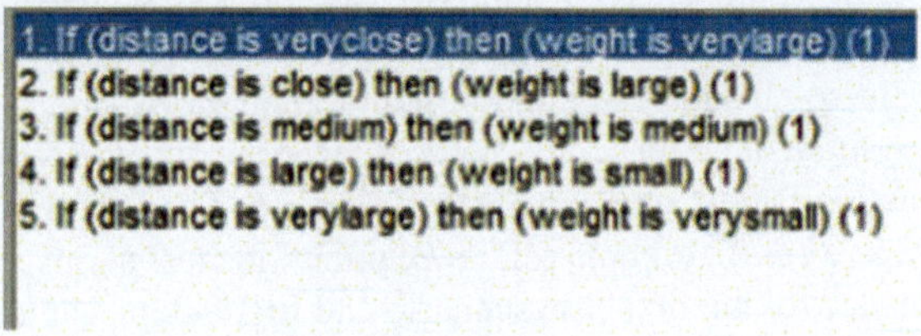

$$\text{ME} = \frac{1}{n}\sum_{i=1}^{n}\delta_i$$

$$\text{RMSE} = \sqrt{\frac{1}{n}\sum_{i=1}^{n}\delta_i^2}$$

Where δ represents mean error, ME represents cumulative mean error and RMSE represents root mean square error [13–18].

3.6 Algorithm

Algorithm estimates RSSI values based on reference nodes and test nodes, i.e., target node: (x, y), simulation of the mathematical equations will perform the following steps.

Step 1. Start.
Step 2. Measuring RSSI values from reference nodes to target nodes.
Step 3. Calculation of distance between target and reference nodes: Euclidean distance and Physical distance.
Step 4. Finding the common nearest neighbors based on KNN algorithm.
Step 5. The common points inserted into fuzzy controller and weights are calculated.
Step 6. According to the weights location of target node is estimated.
Step 7. MSE and R are estimated through NN tool.
Step 8. Stop.

4 Result and Discussion

To test the proposed algorithm, the experiment was carried out in simulation environment in MATLAB 2019. Simulation model was used to generate target and reference nodes for the area (2 × 2 km) for two different scenarios in an approximated simulated environment. Two different wireless techniques are used for the different scenarios Bluetooth and Wi-Fi and based on their RSSI values are calculated to estimate the position of target nodes. We compared the two wireless techniques based on the evaluation factors mentioned in section 3.6. The sampling rate of the RSSI acquisition was set as 2 s for both reference points and test points. Fig. 8 shows the distribution of nodes for the first case scenario and Fig. 9 depicts the second scenario respectively.

The weights calculated in FIS and the reference coordinates of k NN are used to calculate the coordinates of the target, RSSI was estimated employing reference and test nodes, approximately 100 vehicle locations were selected for both Bluetooth and Wi-Fi scenarios for RSSI estimation.

The accuracy of RSSI values of each model was evaluated using Neural Network model. Figs. 10 and 11 depicts the scatter plot of RSSI estimate for Bluetooth and Wi-Fi scenarios.

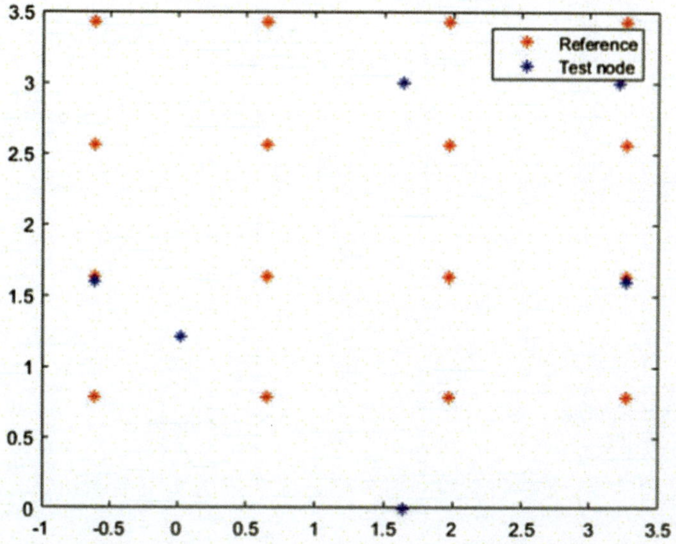

Fig. 8 Scenario 1—Bluetooth

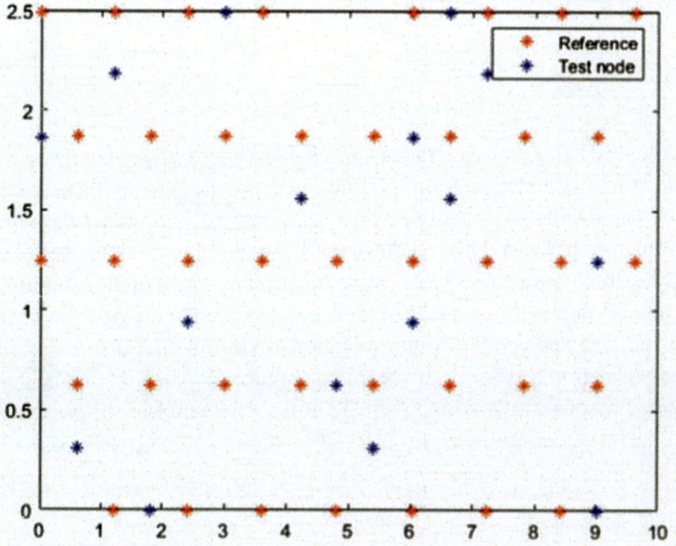

Fig. 9 Scenario 2—WiFi

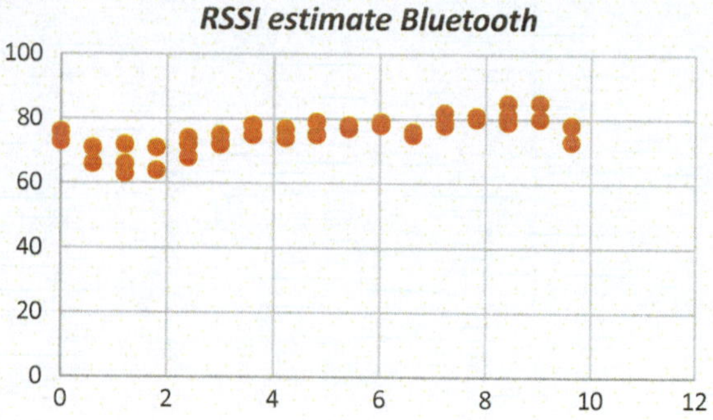

Fig. 10 RSSI estimates scenario1—Bluetooth

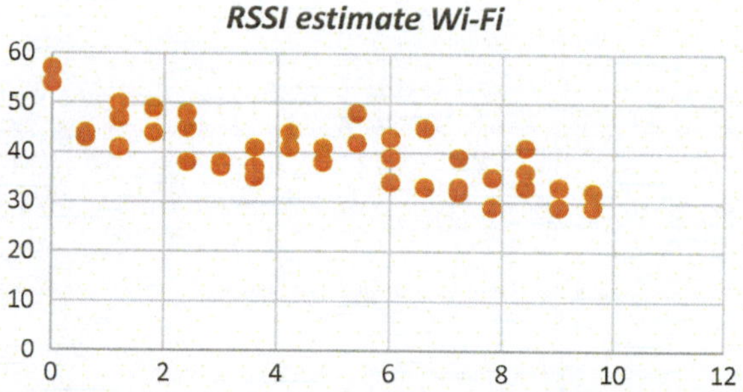

Fig. 11 RSSI estimates scenario1—Wi-Fi

Mean Squared Error is the average squared difference between outputs and targets. Lower values are better. Regression *R* Values measure the correlation between outputs and targets. An *R* value of 1 means a close relationship, 0 a random relationship. Zero means no error. The Mean Square Values (MSE) and *R* values estimated for Scenario 1 and 2 with *x* and *y* coordinates of vehicle location are presented in Table 2 [19–24].

MSE values obtained suggested increase in error in scenario 2 as compared to scenario 1 for the same vehicle communication environment. Further, Figs 12 and 13 depict best validation test for both scenarios. Figure 12 presents that best validation performance for scenario 1: which is 55.37 at Epoch 7 and Fig. 13 presents that best validation performance for scenario 2: Wi-Fi which is 59.58 at Epoch 3.

Table 2 MSE and *R* values for scenario 1: Bluetooth and scenario 2: Wi-Fi

MSE and *R* values		Scenario 1: Bluetooth		Scenario 2: Wi-Fi	
	Samples	MSE	*R*	MSE	*R*
Training	73	33.29196e-0	6.10186e-1	24.03158e-0	7.24842e-1
Validation	16	55.36769e-0	4.65164e-1	59.58184e-0	4.32347e-1
Testing	16	41.50336e-0	3.30810e-1	49.95423e-0	6.01166e-1

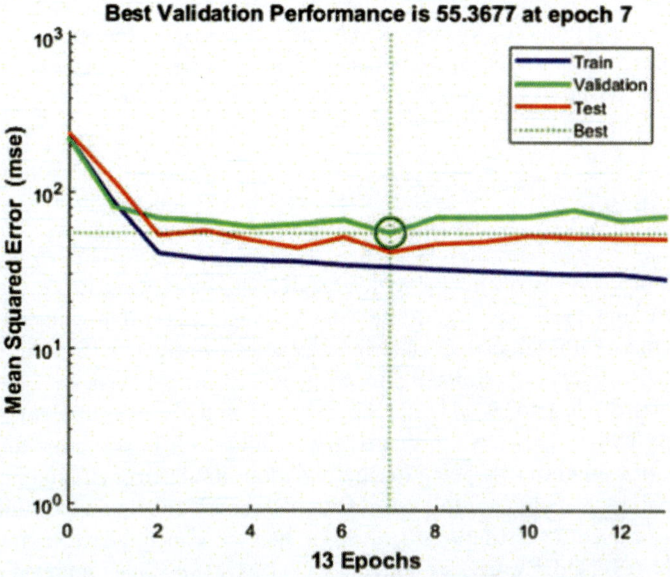

Fig. 12 Present that best validation performance for scenario 1: Bluetooth

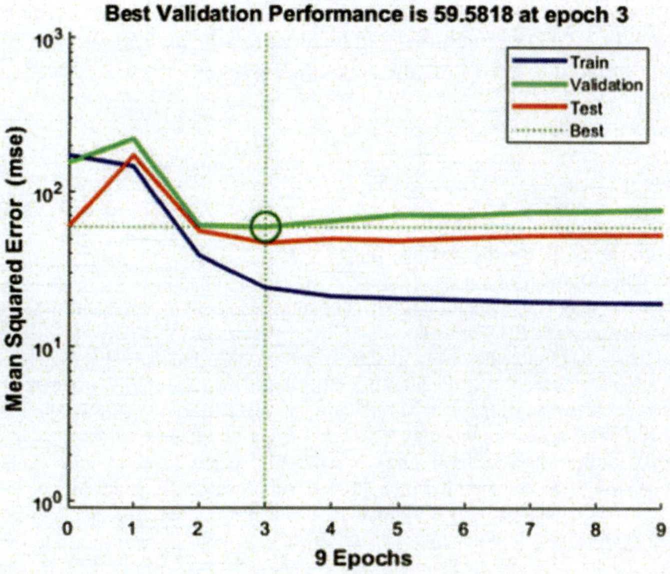

Fig. 13 Presents that best validation performance for scenario 2: Wi-Fi

Methodology proposed compares vehicular localization for approximately 100 different target points for a simulated environment, in which Bluetooth localization outperforms Wi-Fi.

5 Conclusion

The work proposed performs vehicular localization employing fuzzy systems, for localization the proposed algorithm uses k-nearest algorithm with fuzzy decision system to form the cluster of vehicles under investigation. System suggested employs two distances estimating techniques—Euclidian distance and physical distance between test nodes and reference nodes. Simulation was implemented using MATLAB. RSSI values were evaluated for BLE and Wi-Fi scenarios for approximately 100 different vehicle positions in a 2 $\times$ 2 km simulated environment. BLE due range constraints would require a setup with V2V and V2I network. Simulated environment was used for testing results. RSSI values were then fed to Neural Network to estimate MSE and R values for both the scenarios. MSE values obtained suggested increase in error in scenario 2 as compared to scenario 1 for same vehicle communication environment. Validation performance for scenario 1: Bluetooth comes as epoch 7 which is 55.37 and for scenario 2: Wi-Fi comes as epoch 3 which is 59.58. From the results, it can be concluded that localization through Bluetooth outperforms slightly as compared to Wi-Fi owing to positional losses.

References

1. Günay FB, Öztürk E, Çavdar T, Sinan Hanay Y, KhanAR (2020) Vehicular Ad Hoc Network (VANET) localization techniques: a survey. In: Computational methods in engineering
2. Srivastava A, Prakash A, Tripathi R (2020) Location based routing protocols in VANET: issues and existing solutions. Elsevier
3. Kelarestaghi KB, Foruhandeh M et al (2019) Survey on vehicular Ad Hoc networks and its access technologies security vulnerabilities and countermeasures. arXiv e-prints
4. Alakhras M, Oussalah M, Hussein M (2020) A survey of fuzzy logic in wireless localization. EURASIP J Wireless Commun Netw
5. An J, Yu Y, Tang J, Zhan J (2019) Fuzzy-based hybrid location algorithm for vehicle position in VANETs via fuzzy kalman filtering approach. Adv Fuzzy Syst, Hindawi
6. Altoaimy L, Mahgoub I (2014) Fuzzy logic based localization for vehicular Ad Hoc networks. In: IEEE symposium on computational intelligence in vehicles and transportation systems (CIVTS)
7. Kumar S, Kislay K, Singh MK, Hegde RM (2014) A range-free tracking algorithm in vehicular Ad-Hoc Networks. In: Twentieth national conference on communications (NCC)
8. Rozyyev A, Hasbullah H, Subhan F (2012) Combined K-nearest neighbors and fuzzy logic indoor localization technique for wireless sensor network. Res J Inf Technol
9. Cota-Ruiz J, Gonzalez-Landaeta R, Diaz-Roman JD, Mederos-Madrazo B, Sifuentes E (2019) A weighted and distributed algorithm for multi-hop localization. Int J Distrib Sens Netw
10. Kaur A, Kumar P, Gupta GP (2017) A weighted centroid localization algorithm for randomly deployed wireless sensor networks. J King Saud Univ Comput Inf Sci

11. Kang S, Kim T, Chung W (2020) Hybrid RSS/AOA localization using approximated weighted least square in wireless sensor networks. Sensors 20(4):1159
12. Shi X, Zhang L (2013) High-precision weighted Bounding Box localization algorithm for Wireless Sensor Network. In: IEEE third international conference on information science and technology (ICIST)
13. Altoaimy L, Mahgoub I (2016) Mobility data verification for vehicle localization in vehicular Ad Hoc Networks. IEEE Wireless Commun Netw Conf
14. Altoaimy L, Mahgoub I (2014) OWL: optimized weighted localization for vehicular ad hoc networks. In: International conference on connected vehicles and expo (ICCVE)
15. Altoaimy L, Mahgoub I (2016) Mobility data verification for vehicle localization in Vehicular Ad Hoc Networks. IEEE Wireless Commun Netw Conf
16. Sun L et al (2012) An RSU-assisted localization method in non-GPS highway traffic with dead reckoning and V2R communications. In: 2nd international conference on consumer electronics, communications and networks (CECNet), pp 149–152
17. Khattab A, et al (2015) High accuracy GPS-free vehicle localization framework via an INS-assisted single RSU. Int J Distrib Sens Netw. https://doi.org/10.1155/2015/795036
18. Qafzezi E, Bylykbashi K, Ikeda M, Matsuo K, Barolli L (2020) Coordination and management of cloud, fog and edge resources in SDN-VANETs using fuzzy logic: a comparison study for two fuzzy-based systems. Internet of Things 11:100169. https://doi.org/10.1016/j.iot.2020.100169
19. Balta M, Özçelik I (2020) A 3-stage fuzzy-decision tree model for traffic signal optimization in urban city via a SDN based VANET architecture. Future Gen Comput Syst 104:142–158. https://doi.org/10.1016/j.future.2019.10.020
20. Ghaffari A (2019) Hybrid opportunistic and position-based routing protocol in vehicular ad hoc networks. J Ambient Intell Human Comput. https://doi.org/10.1007/s12652-019-01316-z
21. Purkait R, Tripathi S (2019) Fuzzy logic based multi-criteria intelligent forward routing in VANET. Wireless Personal Communications
22. Hu X, Ma L, Ding Y, Xu J, Li Y, Ma S (2019) Fuzzy logic-based geographic routing protocol for dynamic wireless sensor networks. Sensors 19(1):196. https://doi.org/10.3390/s19010196
23. Peng X, Chen R, Yu K et al (2020) An improved weighted K-nearest neighbor algorithm for indoor localization. Electronics 9:2117. https://doi.org/10.3390/electronics9122117
24. Khanna A, Rodrigues JJPC, Gupta N, Swaroop A, Gupta D (2020) Local mutual exclusion algorithm using fuzzy logic for flying Ad hoc networks. Comput Commun 156:101–111

A Proposal of Iot Based Evapotranspiration Irrigation System for Water Conservation

Simrat Walia and Jyotsna Sengupta

Abstract Water conservation is a prime concern in arid and semi-arid regions. The farming sector consumes a lot of water in irrigation and other activities. IoT (Internet of Things) and WSN (Wireless Sensor Network) are extensively used in precision agriculture to control fundamental activities like Irrigation Scheduling, Data Acquisition, Data Storage and Analysis. These systems involve sensors inputs to be evaluated against some threshold values and decisions are taken. The proposed automated irrigation system involves monitoring environmental parameters and controlling the irrigation process remotely, involves short-range communication from sensors to the base station and long-range communication from the base station to the internet. The focus of the proposed system is to conserve water and other resources. The system incorporates use of sensor data, evapotranspiration of region and precipitation information from online weather forecast. The hardware components of system are microcontroller, 5v DC motor, L293 motor driver module, resistive soil moisture sensor and breadboard. The soil moisture sensor continuously records the moisture of the soil and transmits readings to the microcontroller. The DC submersible motor connected to the microcontroller through L293 motor driver module, supply water when soil moisture level is below the threshold. The system generates irrigation schedules are based on evapotranspiration of the region and sensor feedback.

Keywords IoT (Internet of Things) · Automated · Irrigation · Evapotranspiration · GPRS · ZigBee

1 Introduction

Water is a fundamental resource for life and having importance in many sectors such as agriculture, industry, recreation and residential. The freshwater resources are

S. Walia (✉)
Department of CSE, Baba Banda Singh Bahadur Engineering College, Fatehgarh Sahib, India

J. Sengupta
Department of Computer Science, Punjabi University, Patiala, India

P. K. Singh et al. (eds.), *Emerging Technologies for Computing, Communication and Smart Cities*, Lecture Notes in Electrical Engineering 875,
https://doi.org/10.1007/978-981-19-0284-0_34

groundwater and surface water. These resources are insufficient as demand for freshwater increase with population growth. In human civilization, agriculture consumes approximately seventy percent of freshwater [1]. The continuous population growth requires more crop production to meet increasing food requirements of the population. The effective utilization of water in agriculture is need of hour owing to limited resources, climate change and depletion of ground water resources. Water conservation is a major concern in developing countries where economies are primarily driven by agriculture. The Indian irrigation mainly dependent on groundwater, facing several challenges like over exploitation, poor water supply infrastructure and depletion of groundwater resources. The country has a huge area of 39 mha irrigated with groundwater, followed by China with 19 mha and USA with 17 mha. The country has a requirement to conserve current water resources to meet urban, industrial and agricultural needs of present and future [1].

Farmers are an integral part of society as 50% of the workforce in India depend on agriculture for their livelihood. There are several challenges from routine activities to harvest in this sector. The current era is of IT and varied technology provides solutions for these challenges encompassing sensors, autonomous vehicles, automatic devices, web services, mobile devices and IoT. In modern age, automation of the irrigation processes using sensors and machine learning techniques is an eminent idea for research.

1.1 Motivation

The IoT and WSN technologies link real-world objects to the internet through wireless and wired technologies. These objects are on the network can interact and exchange the data aiming at enhancement in service, cost cutting in procedures and saving manpower. Farmers have started to use computer and internet to organize data, monitor crop cultivation and to automate routine tasks. IoT is a dynamic field to due rapid advancements in technology of sensors, communication and micro-controllers. Therefore, this is an eminent area for research and gaining more attention from the researchers. There are many automated irrigation systems based on evapotranspiration and sensors readings based on crop and geography of location.

Punjab is an agricultural state where those crops are grown which consumes a lot of water. As a result, the water resources like surface water and groundwater are not properly utilized. The automation of irrigation system using IOT techniques is proposed in this paper to conserve groundwater according to the climate of state.

1.2 Literature Review

Irrigation is an accurate supply of water to plants for optimum plant growth in terms of quantity and frequency. The plant growth and yield are influenced by the amount of water supply as the excess water harms the crops by making roots inefficient in taking nutrients from the soil while lower supply causes slow seed germination [2]. The different ways to supply water are flood, drip and sprinkler irrigation. Many of these procedures are manual and involve human labor to ON/OFF supply based on estimation. The various pitfalls of these methods are: less supply, more supply and late supply. These issues can be addressed in automated irrigation system. The automation of irrigation systems is categorized as open loop and closed loop based on control. The various inputs like length of watering, volume of water, start time, end time and watering periods are provided by the operator of the system in open loop. In a closed loop, a computer program recommends time and length of irrigation [3].

Gutierrez et al. [4] proposed automated drip irrigation in Canada that saved 90% more water as compare to manual irrigation. The system had two components such as WSU (Wireless Sensor Units) to obtain sensor information and a WIU (Wireless Information Unit) to transmit that information for analysis. The WSU consist of an RF transceiver, sensors and a microcontroller. It was a wireless network of soil moisture and temperature sensors situated in the roots of plants. The WIU had a GPRS module that used public mobile network to transmit the sensor observations to a web server. Giusti and Libelli [2] proposed a fuzzy decision support system based on soil water predictions to improve a web based irrigation system. The system took the decisions based on climate data (rain, temperature and solar radiation) and agricultural data (soil composition, crop characteristics and site location).

Ghosh et al. [5] proposed an Arduino based automated irrigation system with remote control from an Android smartphone. The system had user interface which displays sensor data on smartphone and irrigation frequency and volume were recorded through the android app. The results showed that the automated fields consumed less water as compared to the manually irrigated fields. Hu and Shao [6] provided a remote irrigation system with a cloud platform. The various parameters such as soil temperature, humidity and CO_2 concentration considered to take the irrigation decisions. Bandara et al. [7] presented an artificial sprinkler irrigation mechanism for green roof vegetation in Sri Lanka. It had predicted evapotranspiration to determine crop water need.

Sivagami et al. [8] proposed an automatic irrigation system in a greenhouse for both drip and sprinkler, which calculated the irrigation requirements based on soil moisture and environmental conditions. Kwok and Sun [9] developed a machine learning based irrigation system that utilized information like crop water requirements and sensor inputs. The system has two main modules: plant recognizing mobile app to detect plant type by taking a photograph and an Arduino based irrigation controller. Aydin et al [10] presented a technique to receive sensor data withy WeMos D1 Mini microcontroller and sent it to MQTT server. Murthy et al. [11] proposed an Irrigation Control that predicted the Maximum Allowable Runtime (MAR) for each sprinkler zone with weather data acquired from imminent weather station and soil surface runoff. An et al. [12] presented a sensor based automated irrigation system by monitoring substrate temperature. The performance of system measured by considering the total number of leaves, the length and width of the leaves, chlorophyll content in leaves and roots length. Munir et al [13] presented an approach where decision dependent on two things KNN based ontology and sensor data.

1.3 Objective

The literature review presents many methods to automate irrigation systems like IoT based systems, fuzzy decision support system, evapotranspiration based irrigation scheduling and machine learning based irrigation system. Following a review of the pros and cons of previous studies, the objective of this paper is to propose an automated system using IoT that will use sensor inputs, evapotranspiration and precipitation information from internet to schedule irrigation and to take decisions about irrigation.

The paper consists of five main sections. The second section presents the background of the paper. Third section describes the proposed system. The fourth section consists of results, fifth section contains conclusion and sixth section provides future scope.

2 Background

This section presents main activities and a common framework of automated irrigation system. The main activities of system are sensor data collection, data transfer, data storage and analysis. The Fig. 1 describes the common framework of the system having sensor nodes, base station and server. The base station collects data from

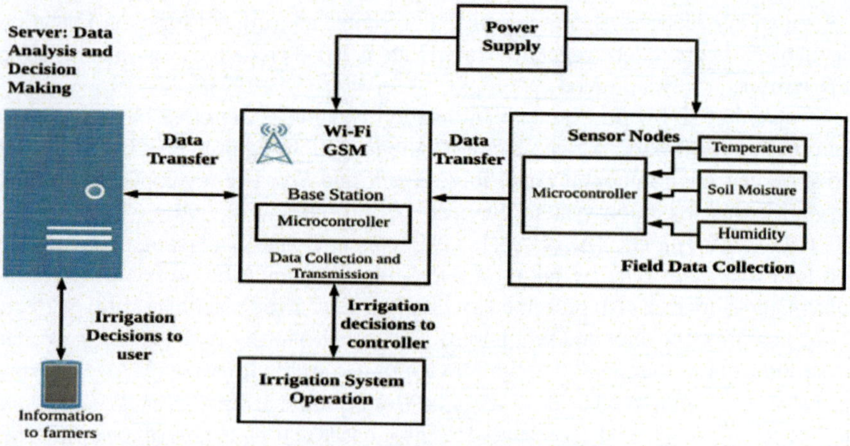

Fig. 1 Common framework of automated irrigation system

sensor nodes and then transmits the obtained information to the server. The server side has a remote user or web application to take the decisions based on analysis of collected sensor data.

The automated system corresponds to various factors such as sensors used, number and types of sensor nodes, communication, data storage and power sources. The main activities and hardware components are discussed as follows:

2.1 Data Acquisition

On a base station, the data is collected from in-field sensor nodes made up from various sensors such as soil moisture sensor, humidity and temperature sensor. The base station uses appropriate internet technologies to send it to a cloud network or web server. Gutierrez et al. [4] employed the use of web application for data collection and analysis. Ghosh et al. [5] proposed the use of a cloud computing platform to store and analyze data. Murthy et al. [11] used MQTT protocol to transfer zone wise sensor data on web server implemented on AWS platform.

2.2 Data Storage and Analysis

After data collection, the various values obtained from sensors are analyzed against some threshold values and then decisions are taken accordingly. Lenka and Mohapatra [14] developed a hybrid system for crop-specific irrigation based on a feed forward neural network and fuzzy logic-based decision support system (DSS). The fuzzy logic-based DSS predicts soil moisture, which is delivered to the farmer with SMS notification. The real-time agriculture data becomes the input and training data to the feedforward neural network, which trained to take the decisions in fully automated irrigation system.

2.3 Sensors Used

Sensors are used to collect readings of soil temperature, soil moisture, soil pH and humidity, which are important for irrigation scheduling. These parameters are determined by the type of crop and irrigation technique.

2.4 Microcontroller

A microcontroller is an essential component of an automated irrigation system that collects data from sensor nodes, converts that data into digital form and then transmits that data to cloud or web server. The various characteristics, such as low cost, versatility to connect with sensor nodes and low power consumption taken into account while selecting a microcontroller. Gutierrez et al. [4] used single chip (PIC24FJ64GB004) 16-bit microcontroller programmed in C compiler to transfer data.

2.5 Communication Protocols

The automated irrigation system needs two types of communication technologies: short-range communication from sensors to the base station and long-range communication from the base station to the internet. The Internet Protocol Version 6 (IPv6), ZigBee and Near Field Communication (NFC) are used for low range private area network communication while GPRS and Wi-Fi are technologies used for communicating over the internet.

2.6 Power Sources

The power requirements of sensor nodes and base station are met by rechargeable batteries, dry cell batteries and solar power. The low power consumptions are always preferred to design a modest system, which can be used in real life situations. Gutierrez et al. [4] used solar powered photovoltaic cell to supply power to a wireless sensor unit (WSU) and rechargeable battery to supply power to wireless information unit (WIU).

3 Proposed System

In this section, the system is proposed to automate irrigation processes using IoT infrastructure and sensor network by considering the influencing factors such as weather, surface runoffs, deep percolation and precipitation. The system takes advantages of feedforward and feedback control mechanisms. The feedforward control is focused on crop evapotranspiration to determine crop water requirements and feedback control takes sensor readings to inquire about soil moisture in the fields. The evapotranspiration, sensor feedback and precipitation information from online weather forecasts are used to produce irrigation schedules. Figure 2 illustrates

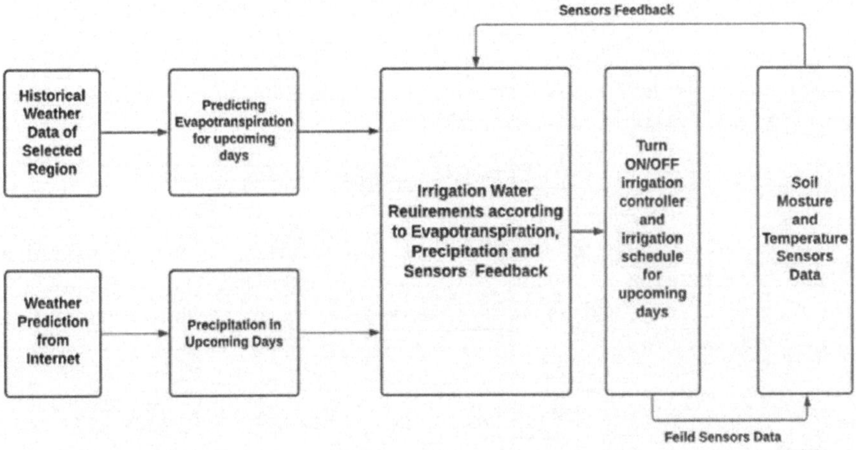

Fig. 2 Schematic diagram of proposed system

schematic diagram of proposed system where evapotranspiration, precipitation and sensor readings are used as input parameters, and the outputs is irrigation schedules.

The system involves use of historical weather data, online weather forecast and sensor feedback in various modules. The various modules of proposed system are discussed as follows:

3.1 Evapotranspiration Prediction

The crop water requirements are determined from crop evapotranspiration (ET_c) that in turn calculated with reference crop evapotranspiration (ET_o). The ET_o is estimating the evapotranspiration on the reference hypothetical crop of grass with height 0.12_{m}. It depends on many factors such as humidity, temperature, atmospheric pressure, solar radiation and wind speed. The present method is to predict ET_o of upcoming days from historical weather data of the region involves machine learning. The dataset is in JSON format having hourly values of previous five years. The training part consists of calculating reference evapotranspiration is calculated with the Hargreaves Samani empirical method that uses the air temperature as main parameter. The equation of method as follows:

$$ET_o = \alpha * (T_{\text{mean}} + 17.8)(T_{\max} - T_{\min})^{1/2} R_a \tag{1}$$

where ET_o is the reference evapotranspiration [mm day^{-1}] is a measure of maximum removal water from the surface with the processes of evaporation and transpiration by the atmosphere. T_{mean} is the mean daily temperature [°C] as $T_{\text{mean}} = (T_{\max} +$

T_{min})/ 2. α is empirical constant having value 0.0023 and R_a is extraterrestrial solar radiation [15].

The reference evapotranspiration is used to calculate the crop evapotranspiration. The crop evapotranspiration is calculated from Eq. 2.

$$ET_c = K_c * ET_o \quad (2)$$

ET_c is crop evapotranspiration. K_c is the crop coefficient of evapotranspiration depends upon the type of crop and geographical location [15]. The system predicts ET_o of upcoming days with an artificial neural network which trained and tested on past weather data [16].

3.2 Precipitation Information from Online Weather Forecast

It is important factor to decide timings, number of irrigation cycles, duration and interval between successive irrigation. Punjab has semi-arid climate and having deep water tables. It requires regular irrigation at small intervals to minimize the effect of percolation and evaporation. The system collects weather forecast information of next five days from internet.

3.3 Data Collection from Sensors

The hardware of proposed system consists of sensor nodes having soil moisture and soil temperature sensors connected to ESP32 microcontroller. The ESP 32 has 32-bit LX6 microprocessor that can operate at 160 or 240 MHz having Bluetooth and Wi-Fi modules for data transfer act as base station to send data on Thingspeak cloud platform. The sensor data is collected over cloud platform with Wi-Fi module for analysis and decision making.

3.4 Data Analysis

It involves designing a system capable of generating irrigation schedules for crops based on evapotranspiration, weather forecasts and sensor inputs. The irrigation time and duration will be decided by the system. The steps of data analysis as follows:

1. Collect historical weather data and precipitation information of selected region.
2. Maintain the information in a dataset.
3. Using the Hargreaves Samani empirical method, calculate the crop evapotranspiration in dataset [15].

4. Train and Test the model for prediction of evapotranspiration using Artificial Neural Network [17].
5. Make a prediction of potential evapotranspiration for upcoming days.
6. Find out the crop irrigation requirements from potential evapotranspiration.
7. Classify the day as rainy, cloudy, partly cloudy or sunshine based on the amount of precipitation.
8. The intervals between sensor readings and the duration of irrigation cycles are determined by the type of day.
9. Make decisions using a rule-driven structure based on sensor inputs and potential crop evapotranspiration.
10. Supply water according to need and generate irrigation schedule.

This system addresses irrigation water requirements in open fields and experimental test will be conducted using flood irrigation in a small garden in Fatehgarh Sahib, Indian Punjab.

4 Results

The results of various modules are presented this section. The hardware unit for sensor data capturing and watering system is proposed. The main components of system are Microcontroller, 5v DC Motor, L293 Motor Driver Module, Resistive Soil Moisture Sensor and Breadboard. The soil moisture sensor continuously monitors the moisture content of the soil and transmits readings to the microcontroller every minute. The device is attached to the DC submersible motor through the L293 motor driver module. When the moisture content is less than the threshold value motor pumps water.

Figure 3 shows moisture data collected over a serial port which are represented in graphical form in Figs. 4 and 5. The soil moisture sensor data retrieved in minute (Time interval of 3 min) and second (Time interval of 3 s) intervals are plotted as a graph in Figs. 4 and 5 respectively. The sensor reading was obtained with various

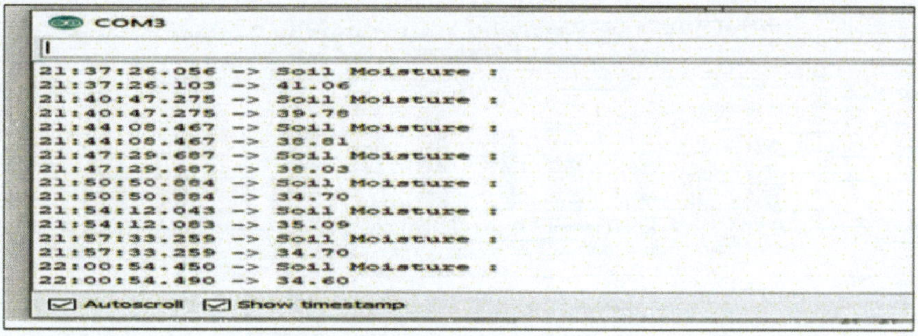

Fig. 3 Snapshot of soil moisture data from serial port

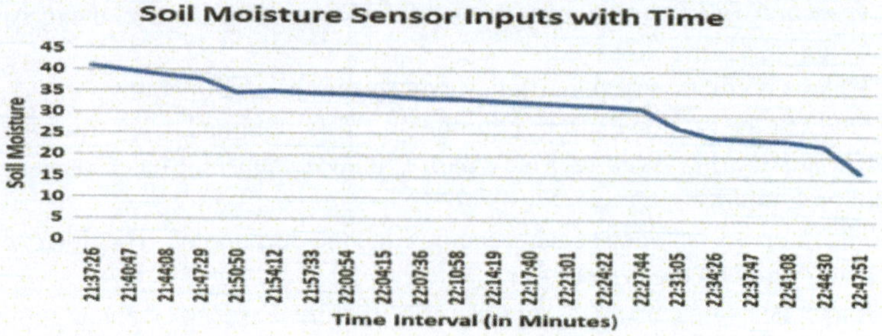

Fig. 4 Data retrieved from soil moisture sensor in time interval in minutes

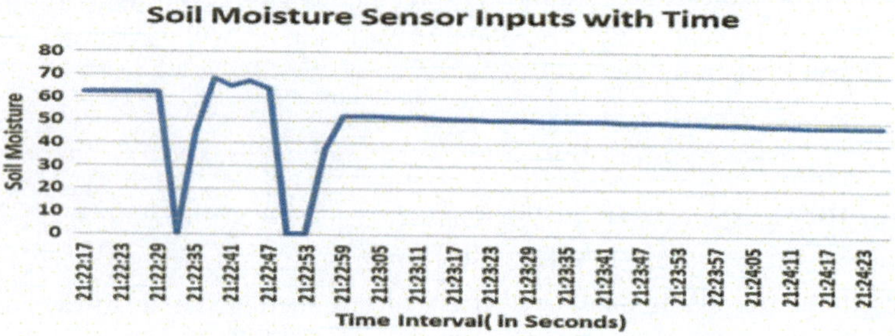

Fig. 5 Data retrieved from soil moisture sensor in time interval in seconds

soil moisture conditions such as highly, moderately and low. The maximum and minimum values of moisture content of soil in minute interval are 41.06 and 16.62% respectively. The maximum and minimum values of soil moisture in second interval are 68.9 and 0.29% respectively.

The weather forecast module retrieves weather forecast data from internet through API. The data has various parameters like city_name, latitude, longitude, temp_minimum, temp_maximum, wind speed, wind degree (direction), cloudiness and rainfall. Figure 6 displays maximum and minimum temperatures in Celsius

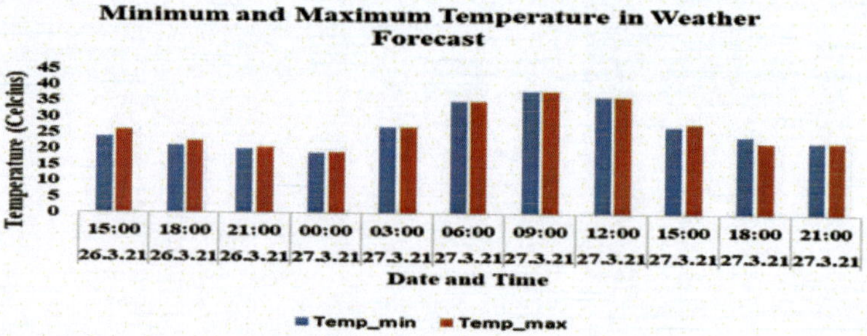

Fig. 6 Minimum and maximum temperature from online weather forecast

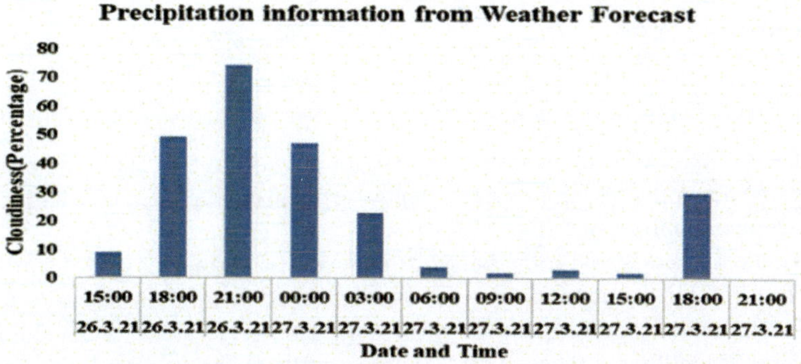

Fig. 7 Cloudiness percentage in online weather forecast

against the date and time received from online weather forecast. Figure 7 displays the cloudiness in percentage against the date and time received from online weather forecast.

Characteristics of Dataset to Predict ET0: The dataset mentioned in Sect. 3 for prediction of ET0 has the following characteristics:

1. The dataset contains Historical Weather data of Punjab.
2. It includes parameters: city_name, temp_minimum, temp_maximum, pressure, humidity, cloudiness, weather_description (Rainy, Clear, Foggy etc.), rain_fall.
3. The data set is hourly data the period of 2016 to sept 2020.
4. The value of ET_0 in data is manually calculated with Hargreaves Samani formula.
5. The dataset is divided into testing and training data.

5 Conclusion

Automated irrigation systems can save energy and human labor while also contributing significantly to water conservation. In this paper, various irrigation techniques available from literature have been explored in based techniques of hardware deployment, sensors usage and background of study. The fundamental activities and common infrastructure are presented in the consecutive section. By considering the benefits and limitations of various techniques, an automated irrigation system based on irrigation scheduling is proposed to obtain robust high performance system that utilizes precipitation forecast, evapotranspiration and sensor information. The main challenge for automated irrigation system is high deployment cost, power sources and effected by uncertainties and disturbances such as climate change, weather uncertainties, type of soil and salinization. In order to minimize these factors and improve the response of the system are areas for future work.

6 Future Scope

The few advancements for automated irrigation system are listed as follows:

1. Cost: The cost involves in wireless sensor network set up and maintenance of those networks. To minimize the deployment and maintenance cost will increase scope of applications.
2. Control approaches in the presence of model uncertainties and disturbances: These factors can interfere in system performance. To minimize these factors and improve the response characteristics of the system are areas for development.
3. User Friendliness: These users of these systems are farmers, therefore user friendly interface and closed loop operations thrust areas.

References

1. Water Ethics and the World Commission on the Ethics of Scientific Knowledge and Technology, & Appelgren, B. (2004). Water in agriculture (5)
2. Giusti E, Marsili-Libelli S (2014) A fuzzy decision support system for irrigation and water conservation in agriculture. Environ Model Softw 63:73–86
3. Obota ME, Inyama HC (2013) Soil moisture based irrigation control system for rice cropping using wireless sensor network. Int J Eng Sci (IJES) 2(3):37–43
4. Gutierrez J, Villa-Medina JF, Nieto-Garibay A, Porta-Gandara MA (2014) Automated irrigation system using a wireless sensor network and GPRS Module. IEEE Trans Instrum Meas 63:166–176
5. Ghosh S, Sayyed S, Wani K, Mhatre M, Hingoliwala HA (2017) Smart irrigation: a smart drip irrigation system using cloud, android and data mining. In: Proceedings of IEEE International conference on advances in electronics, communication and computer technology, ICAECCT, pp 236–239
6. Hu F, Shao L (2017) Design of remote irrigation system in farmland based on the cloud platform. In: Proc. 29th Chinese Control Decis. Conf. CCDC 2017, pp 1125–1129
7. Bandara AGN, Balasooriya BMAN, Bandara HGIW, Buddhasiri KS, Muthugala MAVJ, Jayasekara AGBP, Chandima DP (2017) Smart irrigation controlling system for green roofs based on predicted evapotranspiration. In: Proceedings of electrical engineering conference (EECon), pp 31–36
8. Sivagami A, Hareeshvare U, Maheshwar S, Venkatachalapathy VSK (2018) Automated irrigation system for greenhouse monitoring. J Instit Eng (India) 99(2):183–191
9. Kwok J, Sun Y (2018) A smart IoT-based irrigation system with automated plant recognition using deep learning. In: ACM international conference proceeding series, pp 87–91
10. Aydin O, Kandemir CA, Kirat U (2019) An artificial intelligence and Internet of things based automated irrigation. In: International conference on computer applications in food and agriculture, Konya
11. Murthy A, Green C, Stoleru R, Bhunia S, Swanson C, Chaspari T (2020) Machine learning-based irrigation control optimization. In: BuildSys 2019—proceedings ACM international conference syst. Energy-Efficient Buildings, Cities, Transport, pp 213–222
12. An SK, Lee HB, Kim J, Kim KS (2021) Efficient water management for cymbidium grown in coir dust using a soil moisture sensor-based automated irrigation system. Agronomy 11(41):2021
13. Munir MS, Bajwa IS, Ashraf A, Anwar W, Rashid R (2021) Intelligent and smart irrigation system using edge computing and IoT. Wiley Complex. https://doi.org/10.1155/2021/6691571

14. Lenka SK, Mohapatra AG (2016) Neuro-fuzzy-based smart DSS for crop specific irrigation control and SMS notification generation for precision agriculture. Int J Conver Computing 2(1):3–22
15. Cruz-Blanco M, Lorite IJ, Santos C (2014) An innovative remote sensing based reference evapotranspiration method to support irrigation water management under semi-arid conditions. Agric Water Manag 131:135–145
16. Adeyemi O, Grove I, Peets S, Domun Y, Norton T (2018) Dynamic neural network modelling of soil moisture content for predictive irrigation scheduling. Sensors (Switzerland) 18(10):3408. https://doi.org/10.3390/s18103408
17. Zanetti SS, Sousa EF, Oliveira VP, Almeida FT, Bernardo S (2007) Estimating evapotranspiration using artificial neural network and minimum climatological data. J Irrig Drainage Eng 133(2):83–89

A Comprehensive Study of "etcd"—An Open-Source Distributed Key-Value Store with Relevant Distributed Databases

Husen Saifibhai Nalawala, Jaymin Shah, Smita Agrawal, and Parita Oza

Abstract Distributed systems overcome various limitations of a centralized system and offer several advantages like high performance, increased availability, and extensibility at a low cost. Their rise is unprecedented, and it is only going to increase hereon. But for a distributed system to be effective, it needs a consistent data store, which would store all its necessary metadata and also serve as a single source of truth for the entire system. These data changes need to be stored and to be communicated quickly in a consistent manner across all the nodes in the cluster. It should have fault tolerant capability and should be able to handle failures without any manual intervention. One such open-source data store is etcd. In this paper, we presented a comprehensive analysis of etcd along with its internal working and characteristics. Comparative analysis of etcd with other relevant data stores is also presented in the paper. The paper also sheds light on the application domains and research areas related to etcd.

Keywords Distributed systems · Key-value store · etcd · Kubernetes · High availability

1 Introduction

A database is a collection of data in a well-defined structure and should serve organizational purposes. Databases can be classified based on different factors like type

H. S. Nalawala · J. Shah · S. Agrawal (✉) · P. Oza
Computer Science and Engineering Department, Nirma University, Gujarat, India
e-mail: smita.agrawal@nirmauni.ac.in

H. S. Nalawala
e-mail: 19mca005@nirmauni.ac.in

J. Shah
e-mail: 19mca035@nirmauni.ac.in

P. Oza
e-mail: parita.prajapati@nirmauni.ac.in

P. K. Singh et al. (eds.), *Emerging Technologies for Computing, Communication and Smart Cities*, Lecture Notes in Electrical Engineering 875,
https://doi.org/10.1007/978-981-19-0284-0_35

of content, application area, and underlying data structure. Some examples include operational databases, data warehouses, cloud database, distributed database, graph database, document-oriented database, spatial database, and temporal database. A key-value store is one such database which contains a list of records, and each record in turn contains different fields with data. Every record is identified by a unique key which is used for storing and retrieving the data. Distributed system is a system where the components are located on different computers attached through a network. Components interact with one another towards a common goal, for solving a large computational problem [1]. Kubernetes is a famous open-source container orchestration platform of distributed systems. It helps to manage containerized workloads as they scale, by taking care of configuration, deployment, load balancing, service discovery, and health monitoring across all the clusters. In order to achieve coordination amongst all its clusters and pods, Kubernetes needs a single source of information which stores the current state of the entire system. To fulfil this need, etcd comes into play. etcd serves as a single source of information and stores all the critical information that are needed to keep Kubernetes or other similar distributed systems running.

This paper provides a detailed review on etcd, which includes its characteristics, internal working, and security mechanisms. This paper also compares etcd with other open-source databases like ZooKeeper and Consul which serve the same purpose and are open-source in nature as well. We discuss the pros and cons of each database, which would ultimately help in deciding the appropriate tool as the situation demands. Next, we draw our attention back to etcd and provide application domains where etcd played a pivotal role.

The rest of the paper is structured as follows: Sect. 2 introduces etcd, its characteristics, and security mechanisms, which is followed by a comparative study between etcd and its contemporaries in Sect. 3. In Sect. 4, we present the research opportunities and application domains of etcd and finally conclude the paper in Sect. 5.

2 Overview of etcd

etcd is an open-source, key-value database which holds critical information required for proper functioning of distributed systems in a reliable manner. Fig. 1 represents an etcd interface and its key-value database. The name etcd is derived from the "/etc." directory in the UNIX system, where all the configuration files are stored. "d" here in etcd stands for "distributed" [2].

etcd can have a wide range of applications, but its most notable use case is as a backbone to Kubernetes cluster. To ensure data consistency across all the distributed nodes, Raft algorithm [4] is used. Managing the state of each node in a distributed cluster is the most challenging task too. During state management, problems like race conditions and network partitions may arise which need special care. In essence, a distributed system needs a reliable coordinator who tracks the changes in each node and communicates the same to the entire cluster in a timely and reliable manner [4]. It

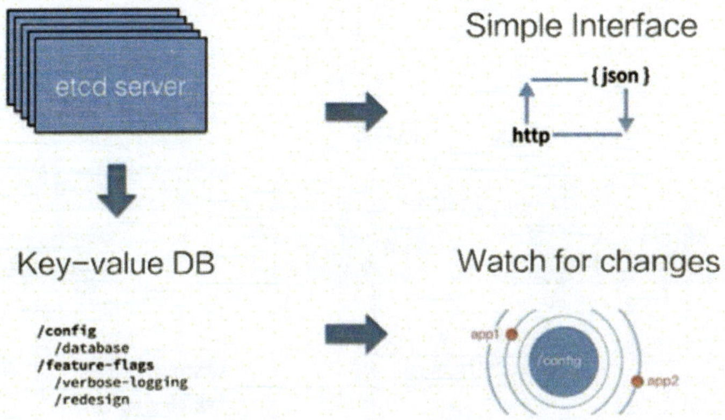

Fig. 1 etcd Key-value database [3]

stores information like data cluster configuration, state of the cluster, node existence in the cluster, service state (running or not), pods state (running or not), and node state (running or not) at any given point of time [5]. Other notable adopters of etcd technology include ROOK—the open-source distributed storage system manager, CoreDNS—an open-source DNS server, Openstack—which provides a set of software components for cloud, and M3—which serves as Uber's metric platform [2]. etcd is written in Go programming language, and its backbone is Cloud Native Computing Foundation. Go has been around for quite some time and thus has excellent community support behind it [2]. Some characteristics of etcd which make it desirable for managing distributed workload are [4]:

- **Highly Available—**It manages hardware failures and network partitions easily and is designed in such a way that there is no single point of failure.
- **Fully Replicated—**Each node in the etcd cluster can access the entire data store.
- **Fast—**Can perform up to 10,000 writes per second.
- **Consistent—**Returns the latest data value during each read operation.
- **Secure—**It supports automatic TSL and optional SSL for security. Additionally, it is recommended to implement role-based access and apply the principle of least privilege to limit the access to secure configuration data.
- **Simple—**Be it a simple Web application or a highly distributed system like Kubernetes, any application can read–write using HTTP/JSON tools.

2.1 Working of etcd

This section discusses working of etcd and its functionality. The working of etcd is presented using three aspects that are RAFT Consensus Algorithm, high availability systems, and etcd clusters which graphically represent in Fig. 2.

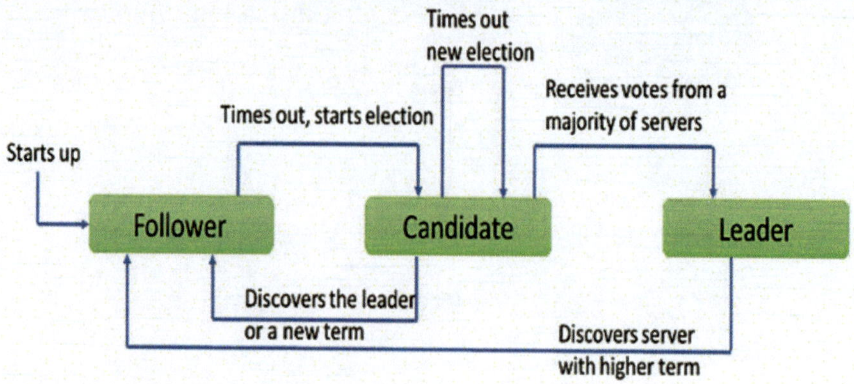

Fig. 2 etcd stages [1]

RAFT Consensus Algorithm

As discussed earlier, etcd is a highly available database system which uses RAFT algorithm to give uptime of 99.99% [6–8]. etcd uses RAFT Consensus Algorithm for communicating between etcd machines. There are three stages of a server in the RAFT cluster—leader, follower, and candidate. By this, all the machines are fully replicated, if the leader node goes down, then other machines or the follower nodes can become candidate, and then, one of them is elected as leader node. Any candidate receiving a majority vote becomes a leader. If the leader finds another leader or a server with a higher term id, it becomes a follower again. Leader election is one of the main aspects of RAFT. There are three scenarios which can occur during this process mentioned as follows.

a. Winning the Election:
   - Each node votes for only one candidate in a single term. The leader requires a vote from the majority of nodes to win.
b. Receiving an < Append Entries Message > :
   - If the term is > = to the current term of that node, recognize the other node as the leader node. Transferal to the follower state.
c. No one is elected split votes (TIE):
   - Each node times out and starts a new election. Timeout period is randomized. This reduces the chances of split votes happening again.

High Availability Systems

Computing environments configured to provide nearly full-time availability are known as high availability systems. This type of system typically has redundant hardware and software that makes the system available regardless of failures. Well-designed high availability systems circumvent having single points-of-failure. Any hardware or software component that can fail has a redundant component of the same type. When failure occurs, the failover process moves processing performed by the failed component to the backup component. This process remasters system wide resources, recovers partial or failed transactions, and restores the system to normal, preferably within a matter of microseconds. The more transparent that failover is to users, the higher the availability of the system.

etcd clusters

etcd is an open-source distributed key-value database used to hold and manage the critical information that distributed systems need to keep running. Most notably, it manages the configuration data, state data, and metadata for Kubernetes, the popular container orchestration platform. Starting an etcd cluster statically requires that each member knows another in the cluster as shown in Fig. 3. In a number of cases, the IPs of the cluster members may be unknown ahead of time. In this case, the etcd cluster can be bootstrapped with the help of a discovery service [9, 10]. In a number of cases, the IPs of the cluster peers may not be known before. This is common during the usage of cloud providers or when the network is running through DHCP. In this scenario, rather than stating a static configuration, it uses a present etcd cluster to bootstrap a new one, called "discovery." If no existing cluster is available, it uses the public discovery service hosted at discovery etcd.io.

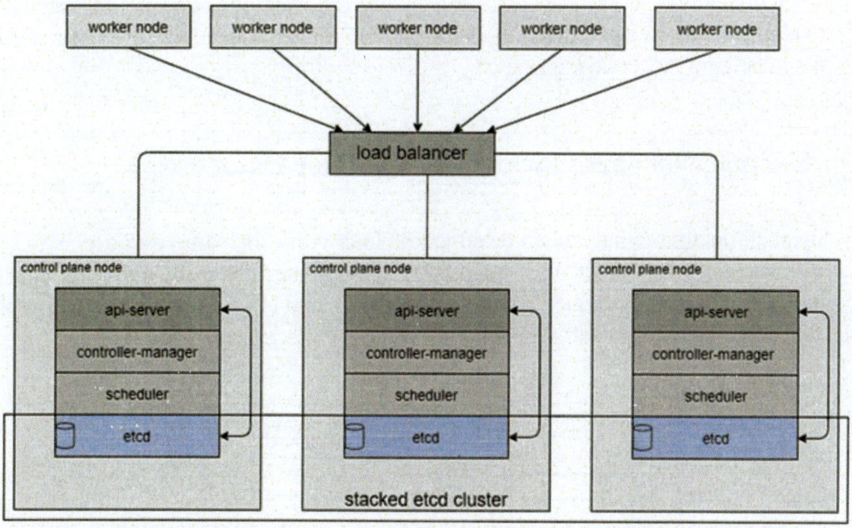

Fig. 3 etcd cluster service discovery

2.2 *Security Model of etchd*

This model comes into effect when we want to allow our etcd cluster to communicate directly to different clusters, servers or the clients. We will understand better with four different scenarios.

First we have to create a CA certificate and a signed key pair for each member in a cluster.

Scenario 1: Client-to-server transport security with HTTPS.

For this connection, we only need to have a CA certificate and a signed key pair.

Scenario 2: Client-to-server authentication with HTTPS client certificates.

For this scenario, we have allowed client interaction directly with HTTPS client certificates, clients will provide their certificates to the server, and then, the server will validate that the certificate is signed and supplied by the appropriate CA authority and then will decide whether to serve the request.

Scenario 3: Transport security & client certificates in a cluster.

etcd supports the same method as mentioned above for peer communication in the same cluster, i.e. that all the communication between the members of the cluster will be encrypted and authenticated using the client certificates.

Scenario 4: Automatic self-signed transport security.

etcd supports self-signed certificates where encryption is needed, but authentication is not needed. This simplifies the deployment because there is no need for managing keys outside of etcd.

3 Comparison of etcd with Other Similar Databases

Many such databases have been developed which serve the same purpose as etcd, i.e. manage information across clusters in a distributed environment. Two most commonly used databases are Apache's ZooKeeper and HashiCorp's Consul. These databases are also compared with etcd in this section (see Table 1).

Table 1 Comparative study of etcd with relevant distributed databases

Properties	etcd	Zookeeper	Consul
Latest version	3.4.0	3.6.2	1.9.1
Maximum database size	Several gigabytes	Few megabytes to few gigabytes	Hundreds of megabytes
HTTP/JSON API support	Yes	No	No
Multi-version concurrency control	Yes	No	No

3.1 ZooKeeper

ZooKeeper is the oldest technology compared to others, and it is created with the aim to enable coordination and control between Apache Hadoop Clusters. Hadoop is an open-source framework which allows the storage and processing of large volumes of data [11, 12]. It stores the data in a hierarchical key-value store, much like a file system. It is used by companies like Reddit, Yelp, Yahoo!, Facebook, and Twitter. Biggest advantages of ZooKeeper are that it is mature, robust, supports high availability, reliable, and it has a rich feature set. ZooKeeper performs much better in read-heavy environments, where there are more read operations than write. ZooKeeper is written in Java, which proves it is robust and reliable. ZooKeeper is quite complex compared to its contemporaries, and as a result, it requires a deeper knowledge than expected for the use case. The rich set of features it provides turns out to be more of a liability than an asset. Further, ZooKeeper supports limited language binding because it implements Jute RPC protocol, which is unique.

ZooKeeper is older than etcd, and it gave many lessons to the latter. Some of the improvements etcd made over ZooKeeper include performing stable read and write operation even under high load, a robust key monitoring mechanism which always gives a notification prior to dropping any event, dynamic reconfiguration of cluster membership, a multi-version data model for concurrency control, and a wider range of language and framework support. It can be concluded that ZooKeeper paved the path for services like etcd, which are considerably simpler and lightweight [13, 14].

3.2 Consul

Consul is a service discovery framework. Unlike etcd and ZooKeeper, consul implements service discovery framework within it (embedded), and so there is no need to use a third party tool. Users only need to register their services and discover them using HTTP or DNS. Similar to etcd, it implements a key-value database which is also based on RAFT algorithm and APIs which can be used with HTTP/JSON. Consul provides built-in facilities for failure detection, health check-up, and services related to DNS. Data is stored in key-value format and can be retrieved using RESTful APIs with HTTP. Some out-of-the-box features are service discovery, health checking, multi-data centre, and secure service communication. etcd and Consul have different aims and solve different problems. While etcd is a good choice for a distributed key-value store with reliability being a chief characteristic, Consul is more aimed towards end-to-end service discovery of a cluster. Also, etcd is preferred when the system needs to be fault tolerant, while Consul provides an overall good infrastructure system for service discovery. In case of etcd, the fault tolerance mechanism increases latency, while Consul is more prone to network failures [13, 14].

4 Application Domains and Research Opportunities

Earlier section talked about etcd as a single source of information for the entire cluster of nodes like Kubernetes. An analogy to understand this could be of a job scheduler, which is currently executing some work. After finishing the work, the scheduler would want to communicate the status of finished tasks with the entire cluster. Here, it is important that this communication happens as quickly as possible to keep the entire system in a stable state. etcd, being fault tolerant as well, presents itself as an ideal choice for backend Kubernetes. Another popular use case of etcd is serving as a backend of Chinese giant Alibaba Group. Since 2017, Alibaba Group has been using etcd to store its key metadata information [15]. Note that they have taken advantage of the open-source nature of etcd and optimized it to make it more efficient and serve its need. Some of the modifications include using a dedicated machine to run etcd, improving memory index, optimizing boltdb (It is used by etcd to store key-values; it can be thought of as a database for databases.), and also client side best practices like not creating key-values which change constantly, avoiding large values when performing calculations and limiting number of lease objects created at a single time [16]. They also created a maintenance and management platform called Alpha, which manages etcd clusters effectively, and is a stop solution for different tasks. It is based on an existing platform called etcd operator, but is designed to cater Alibaba's needs. The result of all these optimizations is that etcd could withstand even the most extremes of network traffic, like the Double 11 shopping event of Alibaba, which generated around 325,000 orders/second. We can see how the open-source nature of etcd can be used to tweak its performance and serve the business needs [17].

As we know that etcd was developed to provide organizations with superior control over their microservice architectures, but there have been some gaps lately as all the focus of the development is shifted towards optimizing the etcd for working with Kubernetes. Some aspects of developing the etcd for organizational work and personal uses are lagging. We would like to point out some fields where it can be improved, but before that we should know that etcd now is an open-source project available freely on GitHub, so one can change it according to their needs. etcd was initially created so that smaller companies could get data centre facilities like bigger companies, but looking at the current time it provides open-source service all thanks to Kubernetes.

- Optimization of RAFT for better latency.
- When there is a large amount of data pumped in directly to etcd node.

5 Conclusion

This study focused on the importance of etcd as a backbone of various distributed systems. We started our discussion with a brief introduction to distributed systems, and in particular, we discussed Kubernetes. In the following section, we introduced etcd, laid out its characteristics, and stated the reasons which make it suitable as a backbone of distributed systems. The internal working of etcd was also presented in this section. We also have compared etcd with other existing systems that serve the same purpose. We also pointed out the research opportunities and areas where etcd could improve itself. In a nutshell, the open-source nature of etcd makes it highly flexible, and with proper configurations, it can withstand even the heaviest of the loads. In future, we would like to undertake a detailed performance analysis of etcd and other similar databases which serve the same purpose as etcd.

References

1. Tanenbaum AS, Steen M (2002) Distributed systems: principles and paradigms. In: Upper Saddle River. Pearson Prentice Hall, NJ. ISBN 0-13-088893-1
2. etcd Authors (n.d.). etcd https://etcd.io/
3. AlibabaCloud. https://alibaba-cloud.medium.com/getting-started-with-kubernetes-etcd-a26cba0b4258
4. Mizerany B (2014) About Etcd, the Distributed key-value store used for kubernetes, Google's cluster container manager the newstack. https://thenewstack.io/about-etcd-the-distributed-key-value-store-used-for-kubernetes-googles-cluster-container-manager/
5. IBM Cloud Education (2019) etcd IBM. https://www.ibm.com/cloud/learn/etcd
6. Juggery L (2019) A Closer Look at Etcd: The Brain of a Kubernetes Cluster Medium. https://medium.com/better-programming/a-closer-look-at-etcd-the-brain-of-a-kubernetes-cluster-788c8ea759a5
7. Nangare S (2019) A Guide to Kubernetes Etcd: all you need to know to set up Etcd clusters super user. https://superuser.openstack.org/articles/a-guide-to-kubernetes-etcd-all-you-need-to-know-to-set-up-etcd-clusters
8. Kelverson E (2015) Getting started with etcd compose. https://www.compose.com/articles/getting-started-with-etcd/
9. Agrawal SS, Patel A (2019) CSG cluster: a collaborative similarity based graph clustering for community detection in complex networks. Int J Eng Adv Technol 8(5):1682–1687. https://www.ijeat.org/wp-content/uploads/papers/v8i5/E7475068519.pdf
10. Agrawal S, Patel A (2020) Clustering algorithm for community detection in complex network: a comprehensive review. Rec Patents Comput Sci 13(1)
11. Desai R, Gandhi A, Agrawal S, Kathiria P, Oza P (2020) IoT-based home automation with smart fan and AC using NodeMCU. In: Proceedings of ICRIC 2019. Springer, Cham, pp 197–207
12. Yadav S, Verma J, Agrawal S (2017) SUTRON: IoT-based industrial/home security and automation system to compete the smarter world. Int J Appl Res Inf Technol Comput 8(2):193–198
13. Farcic V (2015) Service discovery: zookeeper vs etcd vs Consul technology conversation. https://technologyconversations.com/2015/09/08/service-discovery-zookeeper-vs-etcd-vs-consul/
14. etcd Authors (2021) etcd versus other key-value stores etcd. https://etcd.io/docs/v3.3.12/learning/why/
15. AlibabaCloud. https://www.alibabacloud.com/blog/fast-stable-and-efficient-etcd-performance-after-2019-double-11_595736
16. Masani KI, Oza P, Agrawal S (2019) Predictive maintenance and monitoring of industrial machine using machine learning. Scalable Comput Pract Exp 20(4):663–668
17. Xingyu C (2020) Fast, stable, and efficient: etcd Performance After 2019 Double

AgriBot: Smart Autonomous Agriculture Robot for Multipurpose Farming Application Using IOT

Hari Mohan Rai, Manish Chauhan, Himanshu Sharma, Netik Bhardwaj, and Lokesh Kumar

Abstract Internet of Things (IoT) is used all around the globe for connecting things with each other. IoT is a term broadly used for devices that are connected to each other via embedded sensors or with the use of wireless networks may be cellular or Wi-Fi. The proposed system is basically an Agricultural Robot or "AgriBots" used for increasing the productivity and quality of the crop and also to reduce the time and labor cost. The system explains about the network of sensors and the applications of different sensors in the agricultural fields. There are number of Agricultural Robots that already exist at present but they are used at small scale only. In the existing system, the monitoring of the parameters such as the soil moisture and temperature are done by using the manual method. In the proposed system, the IoT is integrated with the Arduino UNO to improve the efficiency of the agricultural fields. The details collected by the Robot from the agricultural field will be stored on cloud and can be monitored without any human interaction.

Keywords Agriculture robot · Seeding · Ploughing · Autonomous · Farming · Sensors

1 Introduction

Encouraging the modernization of the Agricultural industry has become more important in the recent years. The agricultural industry has many challenges to overcome, such as the increasing demands of food due to the rapidly increasing population across the globe, to maintain and keep a check on the quantity and quality of the food items and also the lacking interest of the young people in the Agriculture sector. The development of robotics is spreading in every sector all around the world, as robotics technology provides better implementation and improved quality for several purposes [1]. In the upcoming years, there will be a boost in the number of "AgriBots" or Agricultural Robots. Although the driverless tractors have already developed by

H. M. Rai · M. Chauhan (✉) · H. Sharma · N. Bhardwaj · L. Kumar
Krishna Engineering College, Ghaziabad, Uttar Pradesh 201007, India
e-mail: manish.chauhan@krishnacollege.ac.in

P. K. Singh et al. (eds.), *Emerging Technologies for Computing, Communication and Smart Cities*, Lecture Notes in Electrical Engineering 875,
https://doi.org/10.1007/978-981-19-0284-0_36

the Engineers but this cannot replace the use of the human operated tractors fully as this process is something which requires human assistance [2].

There are several application of IoT and it is being used in almost every application such as, industrial internet of things (IIoT) for industries [3], internet of medical things (IoMT) for medical sector [4], internet of vehicles (IoV) for vehicles [5], and internet of everything (IoE) [6]. So, IoT application in agriculture sector also required for its advancement. As we need more grains on the same agriculture land because the population is increasing with very rapid rate, and it will be approximately 9.7 billion at the end of 2050 [7].

The objective is to design an automated IoT enabled robot which can perform all the farming operations without any manual intervention. Designing, implementing, and employing a Robot (Agricultural Robot) in farm are considered as building a whole Digital farming project [8]. Digitalization of farming refers to the automation of each and every aspect of Farming. With the help of Robot, one can monitor each and every activity going on in the field. There can be different Robots for a particular activity or a Robot who can do multi-tasking. The implementation of Robots in Agriculture is not at large scale in present. The robotics market is still at its early stages of Agricultural development [9]. While the development of these AgriBots (Agricultural Robots) the key points that are to be kept in mind are—the speed, energy efficiency, accuracy [10], light weight, and the autonomous behavior of the Robot.

In India, almost 60% of people are indulged in farming as their occupation [11]. The farmers tend to be the backbone of the Food production [12]. Till date, the most common method used by farmers for farming is the Conventional method of farming which is a time taking process and requires more manpower in the fields. The Robots enable the farmers to monitor their fields virtually with the help of IoT. The Robot can perform various processes like Seed sowing, spraying pesticides, ploughing, harvesting, irrigation, and many other activities. During harvesting, the Robot can pick up the right fruit at the right time based on the inputs provided to it and the data is continuously passed to the farmer to monitor it [13]. Like in the proposed system, the data is stored on the Cloud server and for its continuous monitoring. The deployment of Robots in the farm will provide great help and ease to the farmers and also provide right analysis in various aspects of farming like maintaining seed planting uniformly will help in increasing the efficiency of seed sowing [14].

Agriculture is emerging as the high-tech industry today and many big companies are involving into it [15]. The application of these Agricultural Robots is increasing day to day because this provide effective solution with return of investment and also this is very important in case of doing the activities that might result in risk of health of the farmers. This will provide safety to farmers. Apart from various activities such as Seed sowing, ploughing, irrigation, and others, the proposed system will check the soil moisture with the use of moisture sensor and will water the plants whenever required.

In this work, we have designed and implemented a robot which is capable of performing several farming operations such as Seed sowing, ploughing, irrigation, fertilizer spreading, and harvesting. The developed robot can be operated using Bluetooth, remote control, and also by feeding automatic data. The robot can be operated through Wi-fi or using GSM module remotely and the data can be stored, analyzed, and controlled through cloud.

2 Literature Survey

In recent past, several research has been presented by various researchers to improve the agriculture sector by introducing smart vehicles or IoTs. Auat Cheein and Carelli [8] presented an agricultural robot that provides unmanned service for a variety of agricultural operations. In this paper, automated applications such as terrain leveling, irrigation, weed detection robotics are discussed and analyzed. Ayaz et al. [16] presented the use of IoT with wireless sensors to make agriculture smart and the problem faced in integrating these devices with the classical farming process. In this work, the authors also discuss the use of types of sensors to suit a variety of agricultural applications such as seeding, harvesting, packing, transportation, irrigation, etc. LeVoir et al. [17] proposed chip and high accuracy adaptive location sensing technique for the self-governing farming robot in precision agriculture. Based on low cost camera and global positioning system, the real time location is traced of the rover. Bodkhe et al. [18] have presented a detail study of blockchain based irrigation system, its challenges, and opportunity. In this paper, authors have surveyed the advantages, disadvantages, application of blockchain in precise irrigation and also regarding security related to the payment transaction between stakeholders. They have also surveyed and proposed the integration of blockchain with precision irrigation which will secure the sensors of the connected IoT devices.

Aashish et al. [19] reviewed the overall planning, development, and the fabrication of the Agrobot or the Agriculture Robot that works with the help of battery or any other alternate energy supply. The Robot is able to dig the soil, level the soil, water spray the field and fertilizer separately. The autonomous vehicle has gained the rapid interest in the agricultural field. Almost 40% of the population chooses Agriculture as their fir occupation. The system consists of a Relay switch and Bluetooth. As this language is familiar to most people so it will be easy to interact with the Robot. The Robot uses solar energy to work. The advantage of the Robot is that it is automatic hence, hands-free, and fast input operations of the data. The Robot intends to execute all the essential functions required to be conducted within the farms. Gonzalez-De-Santos et al. [7] proposed a smart farming vehicle called Unmanned Ground Vehicle (UGV). The work discusses the design and development of automated convention vehicle robots and the design and development of mobile platforms. Two types of robots analyzed, wheeled and wheeled, and their adaptability to crops and terrain are discussed. Cihan Yurtseveri et al. [12] developed and presented a robot in farming with row-like seeding operation. It has four different types of sub modules for digging, irrigation, wheel movement, and seeding. All of the Robot's mechanical and electrical characteristics are determined by the architectural requirements, while the device control and interface are determined by stakeholder preferences. In their proposed model, they have assembled and integrated all four subsystem into the robot. They have verified the model using test result and practical performance based on all proposed four types of sub modules. Oktarina et al. [20] have proposed a harvesting robot in their pilot project for tomato plant. In their project, they used to select the tomatoes based on their color and sizes. They have used image processing technique for classing and proving command based on that to robot for selecting the tomatoes. In their agriculture robot, they have also classified the tomatoes based on their sizes. So in their pilot, project they have built and design a robot which can automatically detect the, picked and classify the types of tomatoes based on their color and sizes.

Poonguzhali and Gomathi [21] implemented and designed a IoT based agriculture robot for seeding and ploughing application. In this work, Wi-Fi technology has been used to control robotic vehicles and farming is also done without human intervention of manpower. Three main functions have been implemented, automatic sowing, tillage, and sprinkling of water based on time intervals. Bhupal Naik et al. [14] presented automated IoT based smart farming methods, such as highly rough yield control and data collection task. In this work, various types of sensors are used to extract information about the crop and farm and crop monitoring are also done using these sensors. The information extracted through the sensor is stored on the Internet and based on that, crop and field forecasts are made. Patel et al. [22] have presented a method using 5G network for improving the supply chain of agriculture food based on block chain technology. The proposed method is a credit based system which helps farmers and also guarantees transparency and trust in the application of food chain supply. The presented model provides better result as compared to traditional schemes.

Although a lot of research has been done on agricultural robots in recent years, there is a lot of room for improvement. Very little work has been done on automated agricultural land selection and updation using Land Rover. Most agricultural robots are not fully automated and at the same time they are not capable of doing all the agricultural tasks alone without human intervention.

In this paper, we have developed an agricultural robot (AgriBot) which is capable of performing all the agricultural task such as seed sowing, harvesting, irrigation, soil digging, fertilizer spraying, and many more. The proposed AgriBot is fully automated with multiple operation and can be handled without human intervention using Bluetooth, remote, and wi-fi module.

The paper is structured as: Sect. 1 presents the motivation, objective, and general discussion. The literature survey is conducted on the recently published work in Sect. 2. Section 3 deals with modeling and analysis where the hardware components used are described. The proposed methodology used for this work is presented in Sect. 4. Section 5 describes the results and discussion on the proposed work and conclusion and future work are summarized in Sect. 6.

3 Modeling and Analysis

3.1 Hardware Components

Arduino UNO, NodeMCU ESP8266, General Purpose PCB, DC Motor, Motor Pump, Moisture Sensor, Vibrator, Relay, Capacitor, Batteries, Connecting wires, Water tubes, Platform and wheels (for Structure).

Arduino UNO. Arduino UNO is the backbone of our prototyped model which has been implemented for multiple agricultural operations. The power supply, connection of different types of motors for various operation has been connected in Arduino Uno. It is used to control all the operation of proposed robotic vehicle with software embedding. It is an open-source and flexible electronic platform prototype which enables easy use of both hardware and software. The microcontroller Arduino UNO is based on ATmega328P. It consists of 14 digital Input/Output pins, 6 analog pins, a USB connection, and a reset button. Table 1 shows the technical specifications and components used for implementation of this work.

NodeMCU ESP8266. The Arduino UNO board has been connected on the NodeMCU ESP8266 is a development board which is used to combine control of Inputs/Outputs and the Wi-Fi capabilities. It is affordable to use and an open-source

Table 1 Technical specifications of Arduino Uno

Name of components	Description
Microcontroller	*ATmega328P*
Operating voltage	5 V
Input voltage (recommended)	7–12 V
Input voltage (limit)	6–20 V
Digital I10 Pins	14
PWM digital 1/0 Pins	6
Analog input pins	6
DC current per I/0 Pin	20 mA
DC current for 3.3 V Pin	50 mA
Flash memory	32 KB (ATmega328P)
SRAM	2 KB (ATmega328P)
EEPROM	1 KB (ATmega328P)
Clock speed	16 MHz
Length	68.6 mm
Width	53.4 mm
Weight	25 g

(a) (b)

Fig. 1 **a** Arduino Uno. **b** NodeMCU ESP8266

IoT platform. We have used Node MCU and Arduino Uno for connecting and implementing all other components. Figure 1 shows the pictorial view of Arduino Uno and Node MCU ESP8266 used in this work.

General Purpose PCB. Since our developed AgriBot is a prototype model, for its implementation instead of designed PCB we have used general purpose PCB. Printed Circuit Boards (PCBs) are used to support the electronic components mechanically and connect them electrically using the conductive pathways, tracks from copper sheets laminated onto the non-conducting substrate.

DC Motor. There are several DC motor that has been utilized for different agricultural operations such as for digging, for harvesting, for seed sowing, for ploughing, etc. DC Motors converts the electricity into motion by utilizing electromagnetic induction. By using a variable supply of voltage or by changing the strength of current in its field windings, it is possible to control the speed of the DC Motor.

Motor Pump. A Motor pump is used to transfer or move any type of liquid/gases from one place to another with the help of mechanical action. The pump converts the motor's energy into mechanical flow of liquid.

Moisture Sensor. The moisture sensor is used to measure the volumetric content of water present in the soil. The Moisture sensor works as the resistance is inversely proportional to the moisture present in the soil. If there is more water present in the soil better will be the conductivity, hence, results in low resistance and vice-versa.

Vibration Sensor. Vibration sensor is a coreless motor and is compact in size. It has magnetic properties and is small sized. Based on these features, its performance is highly consistent. Here, it is used for seed sowing, the vibrator vibrates and the seeds fall on the desired place.

Relay. A Relay is a switch that connects or disconnects two circuits and is electrically operated. The Relay or switch has a set of input terminals for controlling single or multiple signals and also a set of operating contact terminals. Small electrical stimuli are converted into larger currents with the help of Relay. A Relay may a different number of contacts in many different forms, such as make contacts, break contacts, or combinations thereof.

Batteries. Batteries are used to give power supply to the Robot. They convert chemical energy to electrical energy. In this project, we have used rechargeable lithium ion battery which is used to enable the Node MCU, Arduino Uno, and other necessary components. Connecting wires are used for the connections between the different components of the Robot. As the project is a prototype device hence we have used connection wires for connecting all components with each other. Also, there are the structural components used in designing the Robot.

3.2 Software Components

Software is one the important section of this project, some software used in this project are; Arduino IDE, Embedded C, OrCAD design, and Cloud server.

Arduino IDE. The Arduino Integrated Development Environment (IDE) is used to program Arduino in Java programming language. It is common to all boards and can run both offline and online. It contains a text editor where the code is written, a message area, console, common functions on toolbar with buttons and series of menus. The Arduino IDE connects the Arduino and the hardware to get the programs uploaded and communicate among them.

Embedded C. Embedded C is the name given to the programming language C associated with hardware architecture. Embedded C is the extended version of the C programming language with some additional header files.

OrCAD design. OrCAD software is used to create electronic semantics, to perform mixed signal simulation and electronic prints to manufacture printed circuit boards. It is used by electronic engineers or technicians for various projects.

Cloud Server. The data monitored is then sent to the cloud server with the help of the Wi-Fi module embedded on the Arduino UNO. There are many cloud platforms that are free source and are available on the internet to which the data is sent to the cloud using the Wi-Fi module and the users can monitor the output i.e., the volumetric water content present in the soil.

4 Proposed Methodology

The main problem in the modern era in the field of Agriculture is that it lacks the manpower i.e., the availability of farm labor, lack of knowledge about the field and soil, increase in the labor wages, and the wastage of resources and raw material required for farming. The innovation of the Agricultural robot is used to overcome

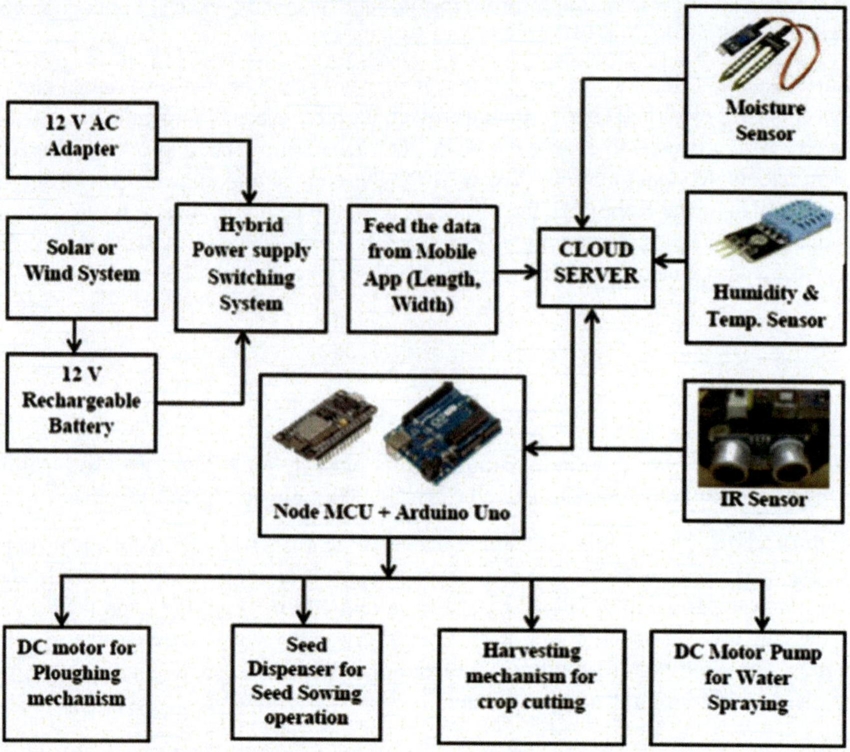

Fig. 2 Block diagram of the proposed system

these disadvantages. The use of Robotics technology is new in the field of Agriculture, the opportunities have enhanced the Robots. The Robots perform various operations such as Ploughing, Seed sowing, Irrigation, and Harvesting. Figure 2 shows the block diagram of the proposed system. This has enabled the farmers to reduce the environmental impacts, increase the accuracy and efficiency and to manage each individual plant in an efficient manner.

This autonomous Robot can go the Agricultural land through the crop lines and can easily perform the tasks that require more energy and are really tiring, also might be unsafe for the farmers. The structure of the Robot is designed in such a way that the Robot can perform different tasks such as seeding, plug system easily even on the uneven land. The wheels of the Agribot are designed so it can travel easily even when the soil is wet or uneven.

The measurements of field collected are provided to the robot through the farmer's mobile, data from mobile will be sent to the cloud server from which the data will be passed to the Arduino with the help of NodeMCU and based on these readings or measurements the Robot will follow the path in the field.

4.1 Ploughing Operation

Ploughing is done using cultivator, the farmer will provide field measurements as Input through his/her mobile phone and that data will go to the cloud server, then from the cloud server the data will come to the Robot via NodeMCU and cultivator is attached to the Robot. Robot will then follow the path according to the measurements given by farmer.

4.2 Seed Sowing Operation

Farmer will fill the seed dispenser with seeds (whichever he/she wants to grow in the field), again the Robot follows the path according to measurement provided by farmer, there is a narrow hole at the bottom of seed dispenser and then when the vibrator vibrates the seeds will fall on the desired location.

4.3 Irrigation Operation

Irrigation is done based on the Output given by moisture sensor, the data from moisture sensor goes to NodeMCU ESP8266. NodeMCU will then communicate to the cloud server and will send the data to the cloud, from cloud the data will come to the water pump controller which will then take a decision based on the previously feed values to decide whether to Switch ON the pump or not. The type of Irrigation method used in the proposed system is Drip irrigation system which will conserve water. The water flows through the water tubes in the whole field. The motor will Turn OFF after some fixed amount of time. The robot possesses a humidity sensor and temperature sensor which will record the temperature and humidity of the field and send the data using the NodeMCU to the cloud server and from the cloud that data will go to the mobile phone of farmer. Based on this data the farmer can decide which crop will be good for the field as different crops require different necessary conditions to grow.

4.4 Harvesting Operation

Harvesting is done using Straw walker for wheat like crops and Corn head for core like crops. A DC motor is used to rotate the Straw walker and Corn head work simply. In this mode also the robot will follow the path based on the measurements only.

5 Results and Discussions

The proposed system performs the following functions—Ploughing, seed sowing, harvesting, and irrigation. All these functions are performed with the help of different sensors. The temperature and moisture sensors are used to collect the data from the surroundings and send it to the cloud server which then goes to the Robot and then the further action takes place and the Robot sprinkles the water in the field. Then comes the Ploughing mechanism, especially before the seed sown are loaded into the module. Some factors are pre-defined according to the type of seeds that are to be sown into the field such as the depth of the ground to be dug and the distance to be maintained. Figure 3 shows the designed agricultural robot with ploughing mechanism.

The DC motor is used for the movement of the Robot which is controlled by the software modules. The Agricultural robots can easily and efficiently perform the tasks like Ploughing, seeding, and ground leveling. Figure 4 visualizes the seed dispenser used in AgriBot for performing the seeding operation.

The traditional method of seed sowing is much time taking process. So, agricultural robots are useful for sowing the seeds in the field over the traditional method. With the help of the AgriBot, seed can be sown throughout the field at the right place with much more accuracy. For this, geographical mapping also plays an important role when used for large scale purposes. Both the Robotics and geographical mapping work well together in developing an autonomous Agriculture Robot. AgriBot with harvesting mechanism for crop cutting operation can be seen in Fig. 5.

Harvesting is an important activity of Farming or Cultivation and can be automated with the help of Robotics technology and using a combine harvester. Although in some cases there may be a need of using the manual harvesting process but for collecting the apples, grapes, and some other fruits and crops, the autonomous Agriculture Robot is a great approach.

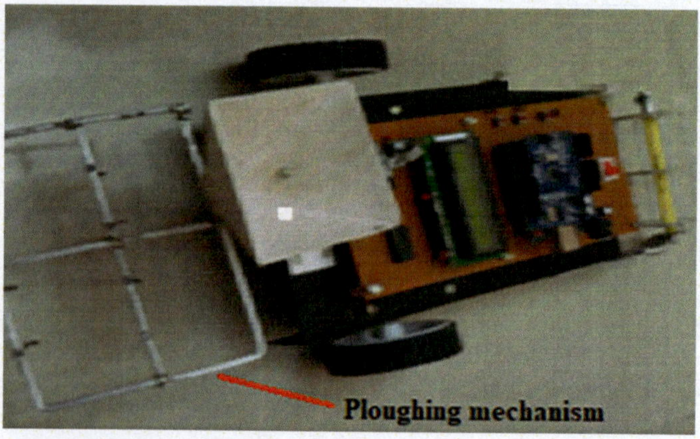

Fig. 3 Designed AgriBot with Ploughing mechanism

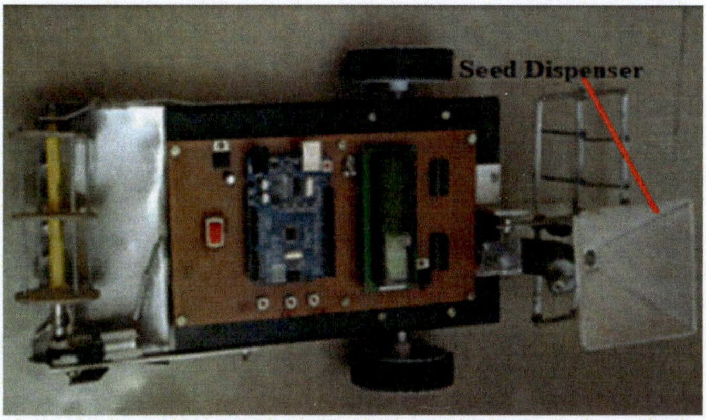

Fig. 4 (e) AgriBot with seed dispenser for seed sowing operation

Fig. 5 AgriBot with harvesting mechanism for crop cutting operation

In the Irrigation process, the moisture sensor is used to observe the environmental conditions. Based on the observations given by the moisture sensor, the Robot takes the further action that whether to spray water to the plants or remain in steady state. The irrigation process is somewhat same as that of spraying the fertilizers, but the only difference is that every crop has its own requirement of fertilizers, as some need in the initial phase when the seed is germinating and some when the seed starts developing in the field.

6 Conclusion and Future Scope

The proposed system provides an innovation in the field of Agriculture and Agribusiness. Also, in the money related angle it is very useful for the farmers. This also reduced certain kinds of monotonous work in the Agriculture sector and thus emphasize that many more young individuals to opt for farming as their occupation. When comparing the work done by the Robot and the manpower used in doing the same work, it can be concluded that the average cost for the Robot's maintenance is less than the total average cost in maintaining the human being including all the fringes. Also, the productivity and the quality of the food increased. The use of Robots has improved the management control as they carry pre-planned procedures with great accuracy. The Robots meet all the Occupational safety and the Health Administration standards. In the proposed system, we constructed a multipurpose Robot or Agriculture vehicle that can perform different mechanism like Ploughing, seed sowing, irrigation, harvesting, and many more other mechanisms can also be added it in. To increase the use of Agricultural Robots, the old barriers of limiting their use should break up now to move to a more advance level in the field of Agriculture.

In future scope, these Agriculture based Agribots can be enhanced in many other ways like Nursery bot, Herder bot, Wine bot, Bee bot, and Hamster bots. This would qualify the standards from the current precision to autonomous farming methodologies. Apart from the Agricultural sector, the Agricultural Robots are also used in the Horticulture and also for Managing and monitoring of the environment.

References

1. Sowjanya KD, Sindhu R, Parijatham M, Srikanth K, Bhargav P (2017) Multipurpose autonomous agricultural robot. In: 2017 International conference of electronics, communication and aerospace technology (ICECA). IEEE, pp 696–699 https://doi.org/10.1109/ICECA.2017.8212756
2. Krishnan A, Swarna SS, BH (2020) Robotics, IoT, and AI in the automation of agricultural industry: a review. In: 2020 IEEE Bangalore humanitarian technology conference (B-HTC). IEEE, pp 1–6. https://doi.org/10.1109/B-HTC50970.2020.9297856
3. Boyes H, Hallaq B, Cunningham J, Watson T (2018) The industrial internet of things (IIoT): an analysis framework. Comput Ind 101:1–12. https://doi.org/10.1016/j.compind.2018.04.015
4. Khan SR, Sikandar M, Almogren A, Ud Din I, Guerrieri A, Fortino G (2020) IoMT-based computational approach for detecting brain tumor. Futur Gener Comput Syst 109:360–367. https://doi.org/10.1016/j.future.2020.03.054
5. Priyan MK, Devi GU (2019) A survey on internet of vehicles: applications, technologies, challenges and opportunities. Int J Adv Intell Paradig 12:98. https://doi.org/10.1504/IJAIP.2019.096957
6. Majeed A (2017) Developing countries and Internet-of-Everything (IoE). In: 2017 IEEE 7th annual computing and communication workshop and conference (CCWC). IEEE, pp 1–4. https://doi.org/10.1109/CCWC.2017.7868459
7. Gonzalez-De-Santos P, Fernández R, Sepúlveda D, Navas E, Armada M (2020) Unmanned ground vehicles for smart farms. In: Agronomy—climate change and food security. IntechOpen. https://doi.org/10.5772/intechopen.90683

8. Auat Cheein FA, Carelli R (2013) Agricultural robotics: unmanned robotic service units in agricultural tasks. IEEE Ind Electron Mag 7:48–58. https://doi.org/10.1109/MIE.2013.2252957
9. Ponnusamy V, Natarajan S (2021) Precision agriculture using advanced technology of IoT, unmanned aerial vehicle, augmented reality, and machine learning. Presented at the (2021). https://doi.org/10.1007/978-3-030-52624-5_14
10. Barman A, Neogi B, Pal S (2020) Solar-powered automated IoT-based drip irrigation system. Presented at the (2020). https://doi.org/10.1007/978-981-13-9177-4_2
11. Arjun KM (2013) Indian agriculture- status, importance. Int J Agric Food Sci Technol 4:343–346
12. Yurtsever C, Ertas Y, Sustam O, Ulu C (2020) Development of an Agriculture Robot for Row-type Seed Sowing Applications. J Selcuk Spec ICAT' 20:228–242
13. Apalak MK, Karaboga D, Akay B (2014) The artificial bee colony algorithm in layer optimization for the maximum fundamental frequency of symmetrical laminated composite plates. Eng Optim 46:420–437
14. Bhupal Naik DS, Ramakrishna Sajja V, Jhansi Lakshmi P, Venkatesulu D (2021) Smart farming using IoT. Presented at the (2021). https://doi.org/10.1007/978-981-15-9516-5_34
15. Baylou P (1987) Agricultural robots. IFAC Proc 20:111–119. https://doi.org/10.1016/s1474-6670(17)55251-9
16. Ayaz M, Ammad-Uddin M, Sharif Z, Mansour A, Aggoune EHM (2019) Internet-of-Things (IoT)-based smart agriculture: toward making the fields talk. IEEE Access 7:129551–129583. https://doi.org/10.1109/ACCESS.2019.2932609
17. LeVoir SJ, Farley PA, Sun T, Xu C (2020) High-accuracy adaptive low-cost location sensing subsystems for autonomous rover in precision agriculture. IEEE Open J Ind Appl 1:74–94. https://doi.org/10.1109/ojia.2020.3015253
18. Bodkhe U, Tanwar S, Bhattacharya P, Kumar N (2020) Blockchain for precision irrigation: opportunities and challenges. Trans Emerg Telecommun Technol. https://doi.org/10.1002/ett.4059
19. Kumar AAA (2020) Review on multipurpose agriculture robot. Int J Res Appl Sci Eng Technol 8:1314–1318. https://doi.org/10.22214/ijraset.2020.5209
20. Oktarina Y, Dewi T, Risma P, Nawawi M (2020) Tomato harvesting arm robot manipulator; a pilot project. In: Journal of physics: conference series. Institute of Physics Publishing, p 12003. https://doi.org/10.1088/1742-6596/1500/1/012003
21. Poonguzhali S, Gomathi T (2021) Design and implementation of ploughing and seeding of agriculture robot using IOT. Presented at the (2021). https://doi.org/10.1007/978-981-15-7394-1_58
22. Patel N, Shukla A, Tanwar S, Singh D (2021) KRanTi: Blockchain-based farmer's credit scheme for agriculture-food supply chain. Trans Emerg Telecommun Technol. https://doi.org/10.1002/ett.4286

Mobile Application Voting System: A Means to Achieve a Seamless Election Process in Developing Countries

Abidemi Emmanuel Adeniyi, Roseline Oluwaseun Ogundokun, Sanjay Misra, Jonathan Oluranti, and Ravin Ahuja

Abstract Voting is a concept used to describe the part of the election process. It is a means by which the citizens choose who to lead them for a designated period. There is various type of manual and electronic voting processes currently in use. Manual voting processes have become a tool by which government bodies in Nigeria and other African countries at considerable take advantage of to push unworthy people into power. The Nigeria voting system is a typical example of this misfortune, where voters are subjected to long queues before they can perform their legal duty as a citizen. This existing system is faced with numerous challenges such as hooliganism where glorified thugs snatch ballot boxes and disrupt the peace and tranquillity of the voting process. Therefore, a loyal citizen who is bound to vote is unable to perform their legal duty, leading to the manipulation of results and other voting crises. This research proposed a mobile voting platform to deal with the challenges as mentioned earlier associated with a manual voting system that is ineffective and inconvenient for citizens. The proposed system will improve how the election is being conducted in Nigeria and other countries that are faced with similar challenges in the voting process. The scheme aims to allow eligible voters with registered voters card (PVC) in Nigeria and diaspora to cast their votes in their respective places of residence as long as the mobile application is accessible on their mobile devices which will be

A. E. Adeniyi · R. O. Ogundokun
Department of Computer Science, University of Ilorin, Ilorin, Nigeria
e-mail: adeniyi.emmanuel@lmu.edu.ng

R. O. Ogundokun
e-mail: ogundokun.roseline@lmu.edu.ng

S. Misra (✉)
Department of Computer Science and Communication, Ostfold University College, Halden, Norway
e-mail: sanjay.misra@hiof.no

J. Oluranti
Centre for ICT/ICE Research, Covenant University of Technology, Otta, Nigeria
e-mail: jonathan.oluranti@covenantuniversity.edu.ng

R. Ahuja
Shri Viswakarma Skill University, Gurgaon, Hariyana, India

P. K. Singh et al. (eds.), *Emerging Technologies for Computing, Communication and Smart Cities*, Lecture Notes in Electrical Engineering 875,
https://doi.org/10.1007/978-981-19-0284-0_37

available on various versions such as Android, iOS, Windows operating system. Each voter's details will be secure through the use of various cryptographic techniques and verified with the use of one-time password during the voting process. This process will make the election process flawless, efficient, convenient, secured and timely in the area of result compilation and final verdict. Also, the system will eliminate violence and result in manipulation.

Keywords Voting system · Election process · Mobile application · Mobile voting system · PVC

1 Introduction

In a democratic form of government, voting is the legal right of the citizen to choose who they want to lead then for a specific period. The event which takes place every four years known as the electoral process is a period preceded by active preparation by members of a registered political party under the rule of law of the government. In most developing countries, many people rally around prominent people who can afford a ticket to contest for office; these individuals gain the support of a political party which gives them the opportunity to be further showcased to the entire country. Unlike the military government where commanders are appointed based on ranks and professional achievement, people are ruled based on a set of principles that may be somewhat difficult or unbearable for members of society. This gave birth to democracy which gives the people the authority to choose their governing legislation, it is a pure representation of which people are and how power is distributed among them are the base to the democratic development and constitution.

In most cases, how people choose who governs them is based on some apparent factors which span through correctness, fairness, a sense of purpose and achievement, and more importantly, the future, and this is an essential requirement to integrity in the electoral process. However, many individuals are biased in the opinion, and this set of people mostly wants their candidate to emerge triumphant in an election process. Voting technology has essentially remained stagnant for long [1]. Nevertheless, certain aspects stayed the same even longer than anticipated, such as a public ballot where voters signed their signatures on a general vote under the name of one candidate or another. This approach had the benefit of being easy to count and difficult to falsify, but the procedure was best suited for smaller elections. Then came along the printed ballot as the government grew and parties in Nigeria became more organized, and printed ballots became more popular. Each group will distribute multi-page ballots with the names of the officials running for the different election offices. Voters would take the vote to be counted from the party they chose and drop it in the ballot box.

The cost of modern voting machines is a recurrent barrier for states trying to boost [2]. For instance, in Nigeria, the price for acquiring a card reader is higher, and most of these machine is not adequately designed for electoral purposes which makes it more challenging to get an accurate compilation of results at the end of an election. It is quite evident that this method has only made the voting process less secured, as most of these devices fail on the field when people are trying to cast their votes. The findings of the analysis carried out on the existing voting system in Nigeria using the Hypertext processor Web platform and Google Android mobile application in the native language of people show impressive outcomes. In future elections in Nigeria, it will ensure a high level of citizen participation and free, equal, open, convenient and confidential electoral processes [3]. The need for a multifactorial authentication solution, such as one-time passcode (OTP) for security authentication, plays an essential role in this project, and it is our main focus. Using mobile as voting mechanisms can resolve most of the above worries. If properly implemented, the remote votes on handheld phones could improve voter turnout, minimize election management costs and enable voters to engage with recognizable technologies.

Although voters may not be able to visit voting booths physically for several reasons, they may need to vote remotely, for example, from home or when travelling abroad. Hence, there is a great demand for simple, straightforward, and, most importantly, safe remote voting procedures. Today, the most popular way to vote remotely is by postal where electors cast their ballots by mail. This ignores adequate authentication, however, and requires a time-consuming process. Internet voting has been implemented to allow for more versatility [4]. The voting process should remain secrete when voting takes place, and no individual vote or authority should be connected to the voter who casts a vote. This is essential even if the voter is not aware of the security concerns, this may raise. In this work, we look to improve the way people vote seamlessly in Nigeria and hope this method can be adopted in various sectors of the world. The current system consists of techniques such as paper ballots, voting machine booster, punch card and voting machine visual. The biggest problem with the new system was a time-intensive one that used to require a lot of time to register. In the new scheme, the paper-based polling process was used, which also yielded the results of bogus polling [5].

The elector gets a blank ballot and uses a pen or marker to indicate that he wishes to vote for which candidate. Hand-counted polls are a time-consuming and labour-consuming process, but paper ballots are easy to make, and the votes can be retained for verification, this type is still the most common way of voting. Appropriate training will need to be provided to the polling duty supervise associates [5]. During polling day, the staff fellows concerned are requisite to be present half an hour before the polling booth/station opening to check that all arrangements are being made correctly. After voting, another group of officers will take care of counting the ballots [5]. The process can prove tedious, error-prone and costly with all these phases, groups and protocols involved. However, some technological implementation currently in the Jamaican environment makes the process semi-manual, but this is far from what a completely ICT-driven process might accomplish. The semi-manual method only requires the government to retain information about electors in a database that can be accessed on a computer at the voting day to enable quicker searches [5]. Electronic voting (EV) means using a machine instead of using a conventional vote at voting centres or via postal mail [6]. It includes the process of making a decision or voicing a view about a group such as an assembly of an electorate, usually after consultations, arguments, or election campaigns [1]. This combines several forms of polling, such as kiosks, the Internet, telephones, punch cards and mark-sense or optical ballots. India, as the largest democracy in the world with a 1.1 billion [7] population, developed electronic voting machines (EVMs). Embraced and sponsored by-election electors will address issues related to the conventional paper-based voting system. In 2003, Estonia's experience with the implementation of online voting in the USA and about 3.4% of electors were able to use mobile e-voting, and by 2007 mobile e-voting elections were proven healthy amid questions about hacker attacks, identity theft and abuse of voting counts [8].

This study therefore proposed a mobile voting platform to deal with the challenges as mentioned earlier associated with a manual voting system, that is, ineffective and inconvenient for citizens. The proposed system will improve how the election is being conducted in Nigeria and other countries that are faced with similar challenges in the voting process. The scheme aims to allow eligible voters with registered voters card (PVC) in Nigeria and diaspora to cast their votes in their respective places of residence as long as the mobile application is accessible on their mobile devices which will be available on various versions such as Android, iOS, Windows operating system. Each voter's details will be secure through the use of various cryptographic techniques and verified with the use of one-time password during the voting process.

The paper is structured as follows: Sect. 2 discusses the related research to the proposed study. Section 3 discusses the material and method used in this study while Sect. 4 discusses the experimentation results and discussion on the findings discovered from the study. Section 5 concludes the study, and future works were suggested in the section as well.

2 Literature Review

Sontakke et al. [9] proposed the development of an e-voting platform which is an online voting system that allows people to cast their ballots through their gadgets or via the use of an e-voting website. The study used the one-time-password (OTP) technique that is used more commonly on the web to tell the difference between a human using a web server and an automated bot. If the results of the matching algorithm match three points, it verifies the voter ID of this individual and then checked with Aadhaar ID if he is entitled to vote and then be given a voting form which will lead to the person to the third stage of the authentication using the one-time password (OTP) method. Kohno et al. [10] explained automated voting system security functionality, and e-voting is more comfortable than physical voting. The author also reveals that electors will cast total ballots without insider rights without any processes being found inside the voting terminal program. Ciprian Stănică-Ezeanu [11] analysed the system of e-voting by explaining its benefits and drawbacks. His research focused mainly on security mechanisms including firewalls or SSL communications, which are necessary but not sufficient to ensure the specific security of e-voting specifications. The author also explains the additional layer of quality improvement in online voting process using pseudorandom one-time password (OTP) and specializing information systems to tackle the unique threats posed by automated voting and maintaining essential security standards such as the safety of voters, election fairness and voter identification. Equally, the author proposed using biometrics and smartcard to authenticate people. One big concern discussed by the author is the disparity between biometric and "classical" verification, such as smart cards. The suggested e-voting program does not interfere in any manner with the real user's biometric features but also authenticates the user on the smart card with the aid of the user's certificate of authentication.

Rossler [12] proposed using remote Internet voting to improve the accessibility of voters, raise the confidence of voters and increase turnout. Authors proposed remote poll-site online voting as the next move forward in the survey because it offers greater voter comfort but does not risk protection at the same time. Avi [13] discussed the security precautions required for the interactive electronic voting method by reflecting on two situations where electors cast their votes over the web–Arizona Democratic Primary 2000 and Student Council Votes at the University of Virginia. The author states that there are four critical criteria for a secure voting system: security, affordability, secrecy and honesty. Shin-Yan et al. [14], this paper proposes a new, oblivious signature scheme with a proxy signature feature to fulfil security criteria, such as identity confidentiality, safety and message authentication, and to ensure that no one can deceive other users (including the server). They suggest an online voting system based on the existing oblivious and proxy signing scheme and incorporate this scheme in a mobile framework to enable users to vote safely and conveniently. Safety analyses and efficiency analyses are presented to show the functionality and efficacy of the method proposed. Ullah et al. [15] proposed a hybrid cryptosystem-mobile phone voting system (MPVS) with an electronic mobile voting process. In the event of casting votes first from the cell phone and then pooling booth, the device avoids double voting. The suggested method is more effective and accurate, given that data would be transmitted via secure SMS to the election commission server. Our machine required no Internet and no unique hardware computer that reduced the cost. The device only needed a cell telephone and SIM card.

Patil et al. [16] successfully introduced our smartphone voting program in this project and have used Firebase as the database. This program helps the user to build and vote for the candidates without being on the polling booth physically present. The application is encrypted and does not require the person to vote again or register with the same telephone number or the same Aadhaar card number also. As Microsoft Azure with cognitive software face API' is not a free source, the face authentication feature was not built into this program. But for this, a separate module has been developed. The various reviewed work proposed different techniques [17, 18] of the electronic voting process and government systems [19–21], but there are still some lapses in the process. The first reviewed work does not put security concern into consideration, Ullah et al. [15] lacked timely delivery of data to the service since it was not Internet-enabled, and this makes cause delay of the voting process or casting of votes by the voters. The last reviewed work uses a facial authentication feature; this feature can consume more space and memory usage of the user device. This proposed system is user-friendly and consumes less memory and space on the user device. It also deals with the security concern by providing a one-time password login authentication.

3 Material and Method

The research aims to develop an accessible and usable system allowing voters to cast a vote instantly without the time and place limit. Mobile development tools are used to design the mobile voting application. There are native mobile development tools, as well as cross-platform mobile development tools. The cross-platform tool is used in this research for the application to be able to run on Windows, Android and iOS platforms. The proposed application will be easily installed on consumer mobile phones, with minimal images or graphics for cheap mobile devices. Adobe XD was used to model the system while the ionic framework was used because it allows for quick prototyping and the pleasant environment when integrated with angular. Mobile phones are portable devices; so far, they can be moved easily from one place to another [22–27]. The mobile phone as proposed in this research will be the best tool to achieve a seamless voting process in this part of the world and as well eliminate any form of chaos during and aftermath of the election process in Nigeria. This will increase our integrity as a country in electing who to lead the country for a given period.

There are techniques available that facilitate the registration and voting process of this proposed system to ensure seamless voting process, to secure the votes and details of each voters from the third party and to tackle the issue of multiple voting by a voter. The details of each voters will be secured using various cryptographic techniques and save of the server. Digital information will be used to ensure the identity of voters and that each voter votes only once. Each voters will be sent an OTP code for the authentication of their digital information before the voting processing.

3.1 *Security Techniques Used*

The Rivest–Shamir–Adleman (RSA) cryptographic algorithm was used to secure the voters details because of it large factorization of large prime numbers. The RSA cryptographic is widely used for data and messages security in network and computer system which involves two keys, public and private keys. The public key is known to everyone and is used to encrypt data. The private is kept secret from knowing everyone except owner, and it is used to decrypt data. The RSA cryptographic algorithm goes thus:

`Select two large prime numbers` ***p and q***

Compute n such that $n = p * q$
Compute $phi(n)$ such that $phi(n) = (p - 1)(q - 1)$
Choose a random integer $e, e < phi(n)$ and $gcd(e, phi(n)) = 1$
that $e * d = 1 mod\, phi(n)$
Private key is (n, e)
Public key is (n, d)
Encryption is $C = M^e mod n$
Decryption is $M = C^d mod n$

OTP is often used as an extra element for multivariate verification and authentication applications. They are only relevant for a particular order for official approval or identity verification. To prevent password lists, it is easy to send an OTP to the voters via short message service. The voter's mobile number must be registered with a provider that offers SMS OTPs for verification or authorization. OTPs are very common as an added permission or authentication element for mobile and web-based services.

3.2 Hardware Setup for the Proposed System

The voters will have to secure the facility of multimedia phone with Android 4.0 Operating System compatible.

The user mobile SIM must be a registered SIM from the telecommunication company.

The process will require the user mobile phone to be connected to the Internet of at least the second generation of network.

The voter's phone must be able to receive message from the third party in order to receive the OTP authentication code for voting process.

Features of the proposed system include:

Availability: the proposed application is available for use as and when required.

Accuracy: the result computation is done automatically and accumulated as the vote is being cast appropriately. This reduces errors encountered during manual calculations.

Flexibility: the application can populate different political parties and candidates' identity to conduct the election efficiently.

Mobility: the application is convenient and possesses high usability for portable mobile devices.

Security: the proposed system provides anonymity of the votes being cast and as well ensures data security from the third party (Fig. 1).

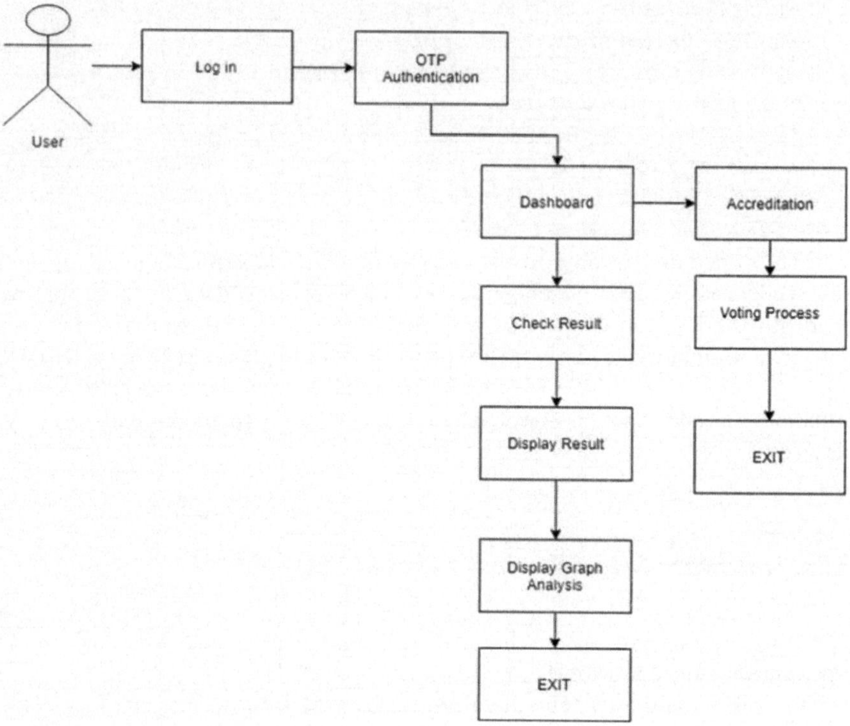

Fig. 1 Proposed system model

4 Experimentation Results and Discussion

The proposed system is a mobile-based voting system. This application operates on two modules; the first module is modelled on registration, accreditation of voters, and the voting process. In contrast, the second module is modelled on result collation, result checker, and graph analysis of the result. Each registered voter is to download the application on their mobile phones which come in a different versions (Android, Windows, iOS) and register their devices with the Permanent Voters Card (PVC) pin; after this has been done, they will be granted access to use the application to vote for their preferred candidate. The voting process begins by each user login into the application and online time password (OTP) will be sent to the registered voter's number to continue the process, after which it has been authenticated, the user can cast his/her votes and wait for the allotted time before he/she can go back to check the result and graph analysis of the result.

This section, therefore, describes the various performance of the proposed system. It explains how each component works together to achieve a seamless mobile voting process. The goal of the proposed approach is to simplify the voting process in Nigeria whereby automate the counting and result generation process of the casted vote, thereby eliminating any form of doubt which is the central issue in the existing system.

According to Fig. 2, if the user card number is valid, a feedback OTP message interface will be displayed for the user to proceed with the voting process.

Figure 3 displays the registration page for the first-time user of the application. After registration, the mobile voting application will be activated on the user's mobile phone.

When the user tries to log in, a token is activated, which sent a one-time password to the user to authenticate and continue the process. As shown in Fig. 4, if the user enters the correct OTP within the stipulated time, the application opens the users' details page where the necessary details of the user can be confirmed and then proceed to the voting page menu else the application will take the user back to the main login page if the OTP is not correct or it has expired.

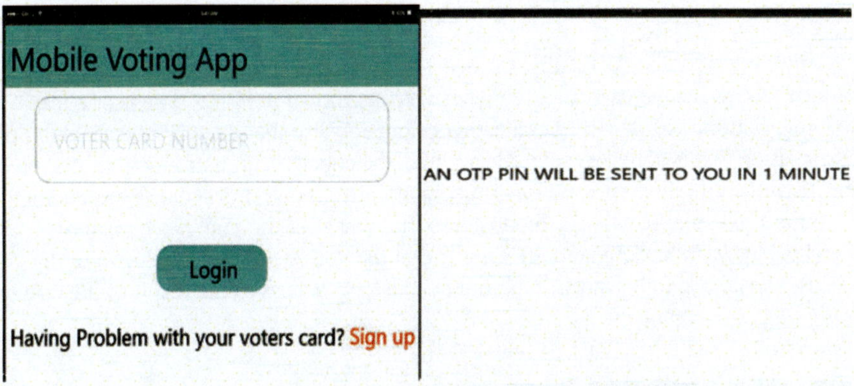

Fig. 2 Mobile voting application login page with feedback message

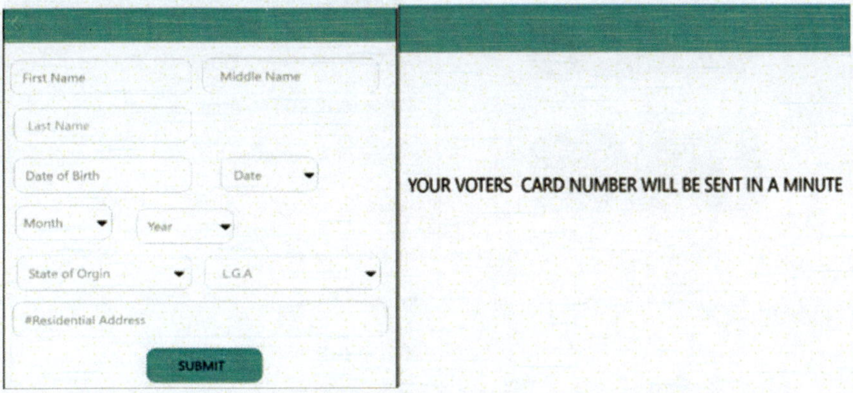

Fig. 3 Registration page with feedback message

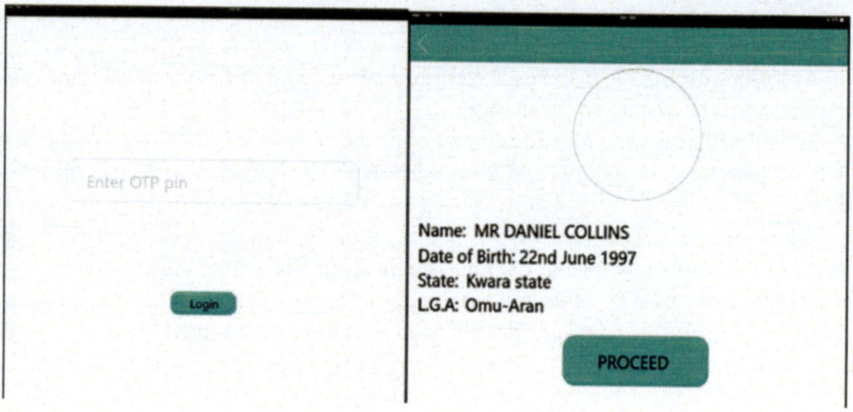

Fig. 4 OTP interface with voter's details page

Upon login and successfully verified the user's essential detail, the voting page will be loaded for the user to cast his/her vote, as shown in Fig. 5. The user clicks on the vote button, and it will load a category page where the user can select which of the position that is active to vote. Meanwhile, the check result button will be disabled during the voting and be enabled immediately after the voting process has ended.

Figure 6 displays message feedback to the user after the vote has been cast.

This page will be activated after the end of the voting process, as shown in Fig. 7; the user will have to login again before he can have access to the result of the casted vote.

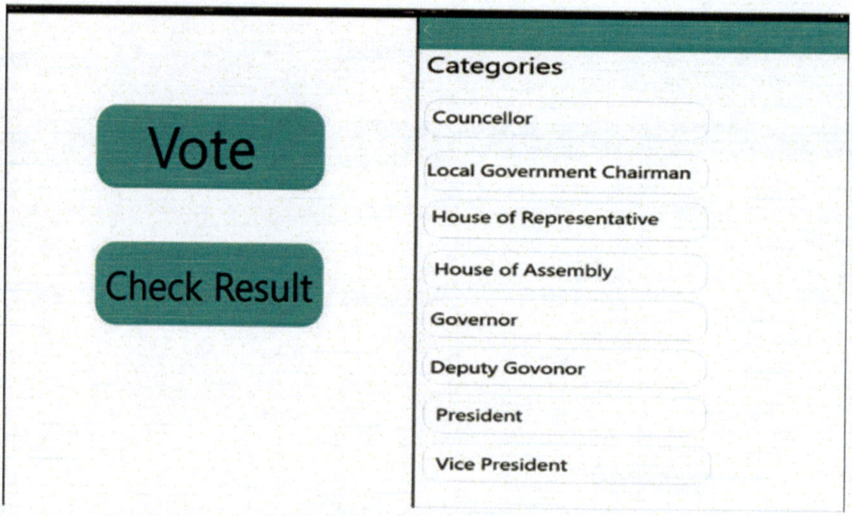

Fig. 5 Voting and category selection menu

Fig. 6 Message feedback

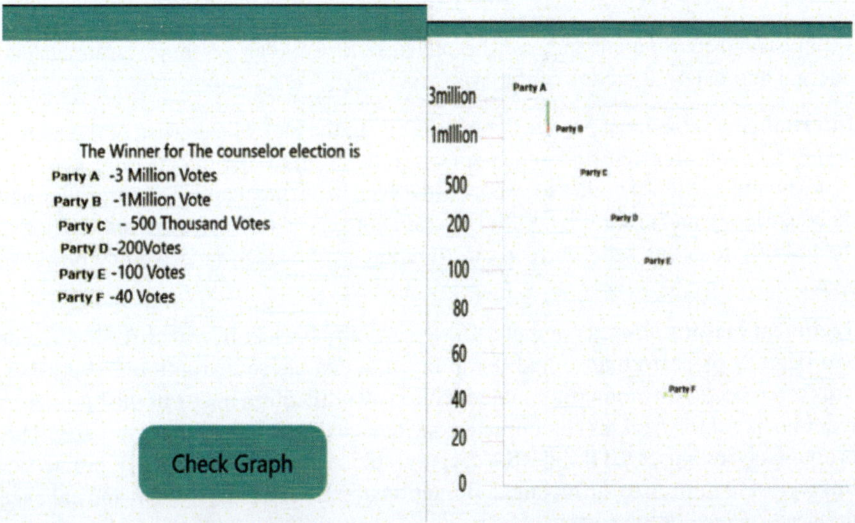

Fig. 7 Result page and graph analysis

This page will be displayed when the users try to vote when the voting process has been ended as shown in Fig. 8, while the second error message will be displayed when the users try to view the result when the voting process is ongoing.

5 Technical Limitation of Mobile Voting Applications

The following are the most common concerns of mobile voting application:

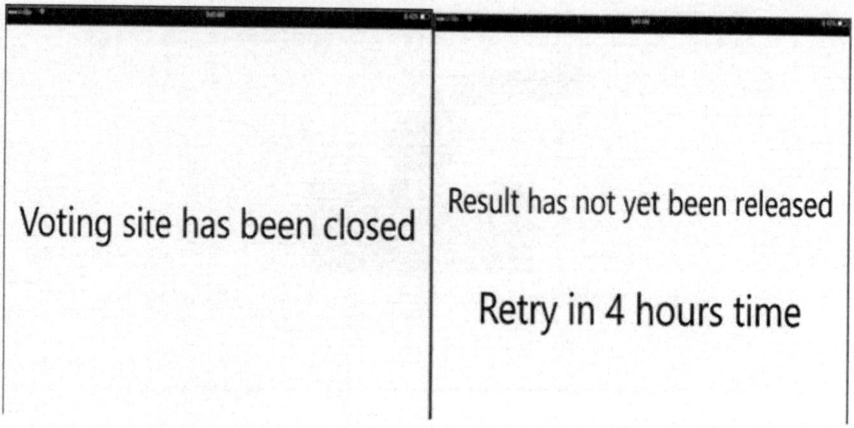

Fig. 8 Error feedback message

Physical Environment factors: the physical factors can be major grouped into internal and external factors of the firm.

Internal: this is the most pressing concerns due to the lack of expertise of the mobile voting application usage and management of the digital voter information.

External: electoral bodies may not be involved in the mechanism of enrolment and its consistency and credibility. Officials can gain access to the system and comprise their ability to count votes correctly or to influence the election, driven by power or wealth.

Technical factors: several computers have currently been built with differing speeds and degrees of performance and durability that can cause disruption and crashes. This often occurs in biometrics authentication. Modification in any of the biometrics features like skin wrinkles can influence the devices from functioning properly. This proposed system uses OTP and RSA cryptographic techniques in place of biometrics fingerprint to minimize the technical limitation of mobile voting system and enhance the performance of the system.

6 Conclusion

The main focus of this proposed system is to provide an easy means of conducting a free and fair election in the country (Nigeria). This proposed system will help the eligible citizen to cast their vote for the candidate of their choice at the convenient of their house, thereby eliminating all form of physical stress (queue and favouritism) that usually leads to commotion during the existing voting process. The system was built in a way that promotes one person one vote's law and also eliminates fraud voters in the sense that each user has a unique voter's number and instant one-time password to activate their mobile application during the voting process. Once a vote is being cast, it will be automatically added up to the database server for automatic summary and result generation. This feature makes the proposed system to be seamless, saves time and voters life during the voting process. The proposed approach is user-friendly, which makes it easy to use and efficient in its operation. The authors recommend that the mobile voting system should be used in various organizations, public and private sectors, even in the education system when trying to choose their leader.

References

1. Bellis M (2007) The history of voting machines. Retrieved 9th November 2016, from http://inventors.about.com/library/weekly/aa111300b.html Another reference
2. Okediran OO (2019) Mobile phones: a panacea for the implementation of E-voting in Nigeria. Asian J Res Comput Sci 1–15
3. Falade A, Adebiyi AA, Ayo CK, Adebiyi M, Okesola O (2019) E-voting system: the pathway to the free and fair election in Nigeria. Electron Government an Int J 15(4):439–452
4. Kulyk O, Volkamer M (2016) Efficiency comparison of various approaches in e-voting protocols. In: International conference on financial cryptography and data security. Springer, Berlin, Heidelberg, pp 209–223
5. Jumb V, Martin J, Figer P, Rebello A (2015) Mobile voting using finger print authentication. Int J Eng Adv Technol (IJEAT) 4(4):141
6. Anderson C (2006) A timeline of electronic voting in the United States. Retrieved 28th November 2016, from http://www.indypendent.org/?p=608
7. Das A (2015) Usability of the electronic voting system in India and innovatory approach. Int J Appl Sci Eng Res 4(5):633–642
8. Heiberg S, Laud P, Willemson J (2011) The application of voting for Estonian parliamentary elections of 2011. In: International conference on E-voting and identity, Springer, Berlin Heidelberg, pp 208–223
9. Sontakke C, Payghan S, Raut S, Deshmukh S, Chande M, Manowar DJ (2017) Online voting system via mobile. Int J Eng Sci Comput 7(5):12176–12178
10. Kohno T, Stubblefield A, Rubin A, Wallach DS (2004) Analysis of an electronic voting system. In: Proceedings of IEEE symposium on security and privacy 2004. pp 1–23
11. CiprianStănică-Ezeanu (2008) E-voting security. Buletinul Universităţii Petrol–Gaze din Ploieşti LX (2):93–97
12. Rossler TG (2011) E-voting: a survey and introduction. Available at http://wiki.agoraciudadana.org/ images/ 5/5 6/An%2 BIntroduction%2Bto%2BElectronic%2BVoting%2BSchemes.pdf Retrieved on 15th June 2012
13. Rubin A (2001) Security considerations for electronic remote voting over the internet. AT&T Labs–Research Florham Park, NJ. Available at http://avirubin.com/e-voting.security.html. Date Accessed 7th July 2012
14. Shin-Yan C, Tsung-Ju W, Jiun-Ming C (2017) Design and implementation of a mobile voting system using a novel oblivious and proxy signature. Secur Commun Netw. https://doi.org/10.1155/2017/3075210
15. Ullah M, Umar AI, Amin N, Nizamuddin (2016) An efficient and secure mobile phone voting system. IEEE, pp 332–336. 978-1-4799-0615-4/13/$31.00 ©2013
16. Patil H, Barot H, Gawhale K, Mhaisgawali A, Chaudhari S (2019) Mobile based voting application. Int J Res Appl Sci Eng Technol (IJRASET) 7(5):2181–2185

17. Abayomi-Zannu TP, Odun-Ayo I, Tatama BF, Misra S (2020) Implementing a mobile voting system utilizing blockchain technology and two-factor authentication in Nigeria. In: Proceedings of first international conference on computing, communications, and cyber-security (IC4S 2019). Springer, Singapore, pp 857–872
18. Anagha H, Chetana A, Jyothi B (2019). Mobile voting system. Int J Sci Eng Technol Res (IJSETR) 6(4). ISSN: 2278–7798
19. Jonathan O, Ayo CK, Misra S (2014) A comparative study of e-Government successful implementation between Nigeria and Republic of Korea. In: Asia-Pacific World congress on computer science and engineering, November. IEEE, pp 1–7
20. Edikan E, Misra S, Ahuja R, Sisa FP, Oluranti J (2019) Data acquisition for effective E-Governance: Nigeria, a case study. In: International conference on recent developments in science, engineering and technology, November. Springer, Singapore, pp 397–411
21. Okewu E, Misra S, Fernandez-Sanz L, Maskeliunas R, Damasevicius R (2018) An e-environment system for socio-economic sustainability and national security. Problemy Ekorozwoju/Problems of Sustain Developm 13(1):121–132
22. Jonathan O, Ogbunude C, Misra S, Damaševičius R, Maskeliunas R, Ahuja R (2018) Design and implementation of a mobile-based personal digital assistant (MPDA). In: International conference on computational intelligence, communications, and business analytics, July. Springer, Singapore, pp 15–28
23. Adeniyi EA, Awotunde JB, Ogundokun RO, Kolawole PO, Abiodun MK, Adeniyi AA (2020) Mobile health application and Covid-19: opportunities and challenges. J Critical Rev 7(15):3481–3488. https://doi.org/10.31838/Jcr.07.15.473
24. Sadiku PO, Ogundokun RO, Habib EAA, Akande A (2019) Design and implementation of an android based tourist guide. Int J Modern Hospital Tourism 1(1):1–33
25. Emmanuel AA, Adedoyin AE, Mukaila O, Roseline OO (2020) Application of smartphone qrcode scanner as a means of authenticating student identity card. Int J Eng Res Technol 13(1):48–53
26. Sowunmi OY, Misra S, Omoregbe N, Damasevicius R, Maskeliūnas R (2017) A semantic web-based framework for information retrieval in E-learning systems. In: International conference on recent developments in science, engineering and technology, October, Springer, Singapore, pp 96–106
27. Adewumi A, Omoregbe N, Misra S (2016) Usability evaluation of mobile access to institutional repository. Int J Pharmacy Technol 8(4):22892–22905

Adoption of Blockchain for Data Privacy in 6G-Envisioned Augmented Reality: Opportunities and Challenges

Umesh Bodkhe, Ashwin Verma, Deepti Saraswat, Pronaya Bhattacharya, and Sudeep Tanwar

Abstract The widespread adoption of real-interactivity, fuelled with developments in wireless communication infrastructures, such as sixth generation (6G), made augmented reality (AR) a possibility in a range of vertical applications. AR integrates the digital visuals into a virtual space that emulates the real-world feedback for the user. It supports a range of haptic and interactive communication and thus has attracted the interest of the research community to provide useful AR frameworks. However, as the user-private data is communicated through open channels, it raises privacy and security concerns. In parallel, decentralized computing has shifted resource provisioning to near device interactions that support the real-time connectivity of AR. To preserve privacy and induce trust among AR stakeholders, blockchain (BC)-based AR use cases are presented by researchers. However, the existing surveys have not focused on reference architectures and implementation challenges. Motivated by the research gap, the paper presents a systematic survey of the integration of BC to induce data privacy in AR ecosystems. A comparative analysis is presented, and the potential application of BC-based AR is discussed. A case study on *VoRtex* enterprise is presented, and a supporting architectural framework is discussed. The survey intends to present the readers deep insights about BC adoption in AR that would motivate industry practitioners, academicians and researchers to propose scalable BC-integrated solutions in AR/VR-based ecosystems.

U. Bodkhe (✉) · A. Verma · D. Saraswat · P. Bhattacharya · S. Tanwar
Department of Computer Science and Engineering, Institute of Technology, Nirma University, Gujarat, India
e-mail: umesh.bodkhe@nirmauni.ac.in

A. Verma
e-mail: ashwin.verma@nirmauni.ac.in

D. Saraswat
e-mail: deepti.saraswat@nirmauni.ac.in

P. Bhattacharya
e-mail: pronoya.bhattacharya@nirmauni.ac.in

S. Tanwar
e-mail: sudeep.tanwar@nirmauni.ac.in

P. K. Singh et al. (eds.), *Emerging Technologies for Computing, Communication and Smart Cities*, Lecture Notes in Electrical Engineering 875,
https://doi.org/10.1007/978-981-19-0284-0_38

Keywords Augmented reality · Blockchain · 6G · Data privacy · Security

1 Introduction

The advancements in holographic interactions in the beyond fifth-generation (5G) and emerging sixth-generation (6G) infrastructures have driven industry projects towards the design of scalable and interactive augmented reality (AR) designs. AR expands the physical world objects and interactive elements into a virtual space and can be processed through information layers, such as audio, visual graphics and cognitive sensing elements. Developing complete artificial world by replacing real things with virtual things is not possible in AR, while virtual reality (VR) has the capability to do the same. AR can be visualized by incorporating voices, pictures and graphics in the real world. By superimposing the computer-generated images, actual real-world environment can be viewed differently through AR by changing the reality [1]. In the same environment, AR allows users to experience the coexistence of real and virtual entities.

Several corporate companies amount to huge investments in the AR market. Development and launch of new hardware and software accessories, involvement in a variety of applications and increased funding in market sectors are fuelling AR growth. AR provides a variety of applications starting from entertainment, health care, advertising, real-estate, virtual experiences market, art, monetization of games, tourism, e-commerce to education and training, and many more. AR devices collect data from various sources but experience security and privacy issues due to the scale, scope and sensitivity of the latter. The information collected by AR devices can be classified as *Observed* (location data, motion tracking and user-provided information/generated data), *Observable* (like virtual person/object, avatar and real-time world/in-application), *Computed* (recommendation/advertising and biometric information) and *Associated* (login/contacts/payment information, virtual assets and IP address). These data in one or another form may be attacked by adversaries. There are various possible attacks on AR data that are private for a person or an organization. The breaches mainly include the following attack scenarios.

- A malicious attacker can utilize the identity of another person for whom the desired immersive experience is not created, enabling fraud and theft.
- Revealing biographical, health information, web-browsing, way of living, age/gender/race, etc., can cause significant harm if disclosed.
- Misuse and outflow of derived sensitive information may lead to potential damage of identity of the intended user.
- Linking of user data (IP address, username/password with other descriptive information results in access to private/public and social accounts. This may lead to economic and reputational harms.
- Data, location services, sound interaction, surveillance information collected from AR device, which includes user information can be used for various purposes, viz.

selling, marketing, overlaying, spying and third-party interaction which imposes privacy threats.

There are several risks associated with developing immersive technologies or content like consumer privacy, data security, infringement, copyright and malicious attack restoration. Recently, there are developments of distrust, scepticism among consumers with corporate firms due to misuse of data. A privacy agreement between users and the company is required to address the lack of transparency and accountability. Moreover, the data demands security to avoid fraudulence and end-to-end delivery, processing, exchange and delivery to the consumer-oriented application to ascertain particular behaviour. These challenges can be catered through BC which acts as a ledger and stores all the transactions in a decentralized way.

1.1 Benefits of Adopting BC in AR

Figure 1 depicts the benefits of BC adoption in various AR applications as follows.

1. Requirement of fast image/video display and processing enables AR assets to be stored locally. Moreover, AR devices exchange data to and from the local server because of memory and processing limitation.
2. AR-oriented applications are data and bandwidth-hungry. Overloading of the central server may occur in the case of multiple AR devices asynchronous communication. BC can provide data decentralization in such a situation.

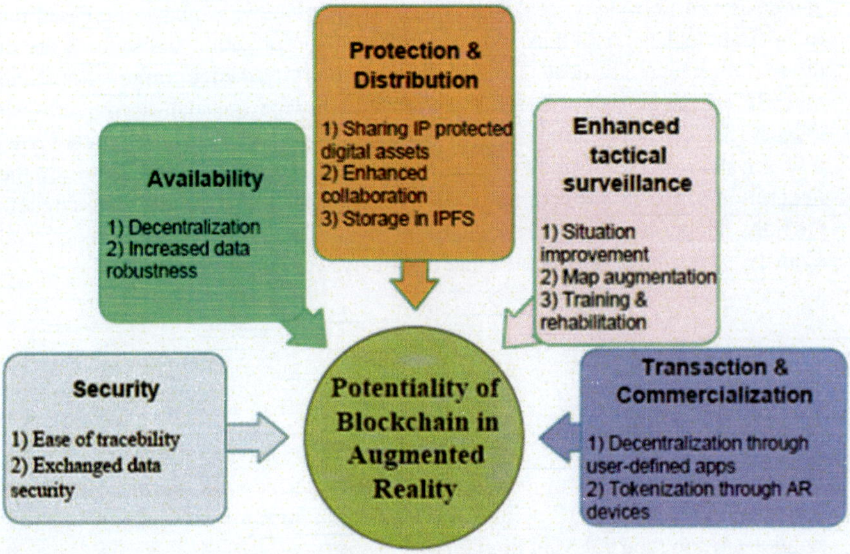

Fig.1 Potential applications of BC in AR

3. BC can help in the commercialization of AR devices through the creation of a user-defined marketplace for storing and uploading AR content in a decentralized fashion. Moreover, the data can be stored and peered using an interplanetary file system (IPFS) that allows authorized stakeholders to access records with IPFS key, and hashed IPFS references are stored as external references in the BC ledger. This improves the scalability of BC, as more transactions can be fitted in each block due to the small transaction size.
4. BC-based tokenization enhances financial transactions where AR devices act as hardware and perform peer-to-peer transactions on digital tokens.

1.2 Motivation

The motivations for the paper are as follows.

- The shift towards beyond 5G and emerging 6G communications has transformed networks to support terahertz (THz) bands, sub-millimetre wave communications and photonic communications, at extremely low latency. Moreover, 6G allows services to be provided as network-in-a-box (NIB) requirements that simplify network and management operations [2]. Due to this, a range of holographic technologies is designed that would break the distance barriers and support real-time interactions to support three-dimensional (3D) imagery, automated cars, AR, digital-twin technology, scalable industry 4.0, virtual reality (VR) and massive Internet of Things (IoT) communications. In the design of massive and scalable AR frameworks, however, the simplicity and flexibility of network configurations, and design of network virtual functions to drive 6G devices, have increased the risk of data leakage, and thus, user data privacy, authorization and security are of prime importance. The need is intensified in decentralized edge computations, and thus, BC leverages trust, chronology and auditability in stored data as transactional ledgers [3]. Recently, with rising privacy regulations and data sharing concerns, research studies have shifted towards security in AR applications [4]. However, the integration of BC in AR communications is not explored in detail by researchers globally. Thus, the survey presents a systematic discussion of data privacy and security issues in BC-envisioned and responsive AR applications.

1.3 Key Takeaways

The major key takeaways of the survey are as follows.

- The novelty of this survey is to emphasize on privacy and security concerns achieved by blending BC with AR which help researchers to include these technologies in their relevant domains.

- We present a comparative analysis of existing surveys and based on identified research gaps have proposed key discussions on emerging BC-based AR/VR projects, and the open issues and challenges of the integration.
- A real-world case study *VoRtex* presents that amalgamates BC in scalable AR/VR space to present an interactive education platform. The user-centric design and access restrictions are presented as part of the case study.

This paper is organized as follows. Section 2 presents the related work. Section 3 illustrates the key discussions on the data privacy issues in single and networked AR applications. Section 4 presents the integration of BC in AR verticals. Section 5 presents some real-world BC-assisted AR/VR deployments. Section 6 presents the open issues and challenges in deploying BC in AR applications. Section 7 presents the case study of *VoRtex* Enterprise Platform, a BC-based AR interactive platform. Finally, Sect. 8 presents the concluding remarks and future scope of the work.

2 Related Work

Till date, surveys have majorly focused on secured techniques and approaches to envision data privacy in AR applications. The authors have analysed some parameters such as privacy, integrity, security and privacy-preserving algorithms. Table 1 presents a comparative analysis of state-of-the-art approaches in AR/VR.

Nguyen et al. [5] discussed the importance of BC-based projects in AR domain through the use of digital ledger. Authors also highlighted the potential Table 1: Comparative analysis of existing state-of-the-art surveys in AR/VR with the proposed survey.

benefits of BC which can improve a security in AR. Carames et al. [8] presented a BC-based holistic approach in AR. Mewes et al. [13] presented projector-based AR visualization concepts, but not considered any of the security issues. Abbas et al. [9] performed in-depth survey on variety of BC applications including AR. They presented BC as the most efficient security scheme to resolve the privacy issues in intelligent mobile augmented reality (IMAR) project. Cannavo et al. [10] outlined a comprehensive survey on the integration of VR, AR and BC. They discussed different use cases for AR/VR with virtual experience marketplace and monetization of games. BC is the key solution for the betterment of user's interaction with various digital content by using gestures, gaze and various natural interfaces. Authors in [14] that, similar service is currently offered by the company named 'CEEK' [14], which offers its user to participate in live music events with AR/VR by enhancing the entertainment experience. According to [15], the opportunities, use cases and challenges of AR are frequently studied and illustrated based on the review of 5G developments the data movement and speed of the communication still exist challenges in the AR/VR in terms of security too. Hence, researchers integrated 6G with AR/VR for a better response as well an edge of artificial intelligence.

Table 1 Comparative analysis of state-of-the-art approaches in AR/VR

Authors	Year	Objective	Pros	Cons
Nguyen et al. [5]	2018	BC-based projects were identified and optimized the security of the project	Use of digital ledger	Experimental analysis was not found
French et al. [6]	2018	Integration of BC, 5G and AR	Discussed several models to increase BC-assisted manufacturing capabilities, with 5G network orchestration and management support	Attack scenarios about user authorization, and data confidentiality is not discussed
Werner et al. [7]	2018	Detailed study on AR and its implementation in analogue and digital maps and images	Location-based system with digital maps and images	Data tampering
Carames et al. [8]	2019	Focused on the use of BC in the cyber-resilient automotive industry	BC can benefit automotive industry and ensure data security, anonymity and accountability	Interoperability issues of BC adaption to different AR and IoT verticals are not discussed
Abbas et al. [9]	2019	An overview of BC to provide security to objects of intelligent mobile AR and BC	Efficient security scheme was proposed	Privacy of the mobile data in augmented reality was not discussed
Cannavo et al. 10]	2020	Author has drawn attention from the traditional use of BC to AR and VR	Integrate AR and BC within unified platforms	Privacy during data exchange was maintained between the users on different platform
Lim et al . [11]	2021	Identified BC as a key solution on the vulnerability of the VR/AR software platform	Proposed use of BC to secure the AR/VR system	Un-secure data exchange between the users on different platforms
Liao et al. [12]	2021	proposed novel consensus mechanism as Proof-of-Cache-Offloading (PoCO)	Quality of Service (QoS)	Deployment was possible in 5G and yet to finalize on 6G nodes

(continued)

Table 1 (continued)

Authors	Year	Objective	Pros	Cons
Proposed	2021	Integration of BC in AR/VR space, along with possible industry deployments, and potential challenges are discussed	Security and privacy issues of emerging AR applications, along with possible benefits of mitigation of attack vectors through BC are discussed. A suitable case study is discussed	-

3 Data Privacy Issues in AR

In today's new era, AR is becoming more and more popular. Though there are various security and privacy risks present in AR systems, for example, eye tracking systems that mainly employ biometric identification, AR can enhance the accuracy for authentication and improves the user experience, but the same would require intricate and explicit details of user identification to be stored in released tables, and that might hinder the privacy of the customer. An adversary can spoof the identity by using fake identities as well data privacy can be spoiled in AR systems.

3.1 Issues in Multiple Applications

Multiple applications in AR, this category is having more than one single application in a single AR system.

- *Data Access*: There is a lot of shared data available in multiple application environments. Access for this cross-application sharing data needs to be given properly else it can lead to the use of one application data accessed by some malicious application in the network [16].
- *Input*: Usually users interact with the AR system using voice or embedded gloves, etc. In multiple application scenarios when a user tries to interact with one application, might be brought to other applications' focus. And malicious applications might steal the input of another application and by registering that input, later be used to access that application.
- *Output*: In multiple applications of AR, applications have more than one output device. Conflict among them might be created when more than one application attempts to use that output device, it could lead to security concerns. For instance, a malicious application tries to access content presented by another application.

3.2 Issues in Single Application

- *Data Access*: AR applications are having various data such as sensor data, GPS data, video and audio feeds, temperature, and accelerometers reading. Whom to provide how much access to these data is difficult in AR single application, which might not be able to secure data privacy.
- *Input*: Input in single AR applications also faces similar input validations issues as a conventional application such as integrity is not maintained.
- *Output*: In a single application, a malicious application might use some techniques which can sensory overload the users. Output attacks in a single application are more serious in the AR domain, as it is difficult to distinguish feedback of users as virtual or real and due to that users might not be able to shut down the system which leads to loss of data in running mode. BC-based solution for data privacy in AR.

4 Integration of BC with AR

In this section, we presented the role of BC and its integration with AR to preserve the privacy of user data. BC technology guarantees security and authenticates transactions and records by cryptography [17–19]. Theoretically, to boost the efficiency of AR platforms, decentralized systems such as BC can provide instant and unchanging data transfer [20]. The network of BC is used to store data retrieved from client devices as well as data providers for protection of integrity at which hashed data is posted [21, 22]. Data accessibility from the provider, each request for data access should be handled using a decentralized permission management procedure to get approval from the data holder. A decentralized protocol for the control of permissions is a key part of a decentralized network [23]. BC technology is an emerging technology. Due to this, companies have started to integrate BC technology and AR [24, 25]. There is a centralized system to store or retrieve information from a huge volume of data. But BC technology is having distributed ledger technology which ensures data security at a higher level.

BC-based AR applications are usually having two chains—one is for basic data of the application and another is for data used by transactions. Some AR applications need to be more secure than other ones such as AR shopping applications. Application built on BC technology is providing high security as it is immutable and transparent. Therefore with the changing time, it needs to be adapt faster-growing technologies together to provide a way for securing important data over the Internet. BC is advantageous in the AR field in many ways. For instance, deciding whether a digital photo is an original photograph or a copy of a digital photograph is very difficult. But BC can handle this issue easily by allowing the developers to create unique digital assets that cannot be copied by other developers or users [26–28]. Thus, the integration of BC technology in AR, for sure gives trust to customers for relying on such real-time applications.

5 Real-Life Project Deployments of BC-Assisted AR/VR Space

This section describes BC-integrated AR/VR projects which are excellent instances of BC technologies being used beyond the financial sector. Table 2 is the feature table based on the software specifications and standard scenarios for the development of BC-virtual reality platform. Here label is used to represent essential feature.

- *CEEK*: CEEK virtual reality is a producer of immersive VR content which provides multiple VR channels with social and 3D experience. Virtual reality experiences involve 360 music videos, charities and sports events. The CEEK service platform is compatible with all devices including smartphones, desktops and smart TVs.
- *Decentraland*: Decentraland (MANA) describes itself as an Ethereum BC-virtual reality platform empowering users to develop, access, monetize content and applications. Users in this virtual world buy plots of land which they can explore, build on and monetize later [29].
- *Vibehub*: VibeHub is a ground-breaking platform specializing in new AR technologies are designed to revolutionize how people communicate and monetize content on the internet. Build a cutting-edge network that uses both AR and VR, allowing users to attend a variety of events being conducted such as live concerts, develop new skills, and in virtual reality connect with friends and family.
- *MARKSPACE*: MARK.SPACE is a platform that is open source for the development of 3D / AR / VR-compatible spaces and entities of any type, as well as for their rapid launch and incorporation into a specific environment. This platform is BC-based and offers a crypto-economy.

Table 2 Comparison of BC-virtual reality projects

Criteria/features	VoRtex	CEEK	Decentraland	Vibehub	Virtual universe	Mark space
3D computer graphics	•	•	•	•	•	•
360 degree videos	°	•	°	°	°	°
VW and source or SDK tools	•	°	•	•	•	•
Open source or SDK tools	•	•	•	°	°	°
AI and story-telling	•	°	°	•	°	°
Crypto-currency support	•	•	•	•	•	•

• Parameter is considered, °Parameter is not considered

6 Open Issues of Integration of 6G and BC in AR and Challenges During Deployments

This section gives glimpses for meaningful insights to the research community and readers about the potential of decentralized BC in the AR domain to overcome security and privacy-related issues. BC has also some limitations in terms of scalability, network measurements and privacy attack vulnerability in the smart contract. In this section, we presented open issues of BC in AR and challenges during the implementation phase in detail. Figure 2 presents the open issues and challenges of mainstream and widespread adoption of AR/VR in industry projects.

In the figure, we have presented the challenges and have proposed the possible key directions in blue-boxes. Inclusion of 5G and 6G services, like ultrareliable low-latency services (uRLLC), 6G-based enhanced reliable low-latency services, 5G-enhanced mobile broadband (eMBB) and 6G-further enhanced mobile broadband (FeMBB). 5G-supports a peak data rate of 20 Gbps, user-experienced data rate of 0.1 Gbps and connection density of 10^6 devices/per sq. km. The round-trip time latency of 5G tactile and haptic communications is < 1 ms. Thus, 5G can address some of the key challenges of AR deployments. 6G services, on the other hand, offers a user-experienced data rate of 1 Tbps, and a user-experienced data rate of 1 Gbps. The connection density is $10 \times$ times of 5G and is 10^7 devices/per sq. km. The round-trip latency over the air (OTA) interface in 6G varies from $10 - 100$ microseconds (μ s). Thus, 6G supports a range of AR verticals and addresses the quality-of-interaction (QoI), further than 5G. For security and privacy considerations, users can design databases to follow K-anonymity and differential, privacy-based models. Similarly, BC-based issues can be addressed through decentralized storage via interplanetary file systems (IPFS), permissioned chain structures that improve the mining latency and improve bandwidth. For smart contract verification, we can

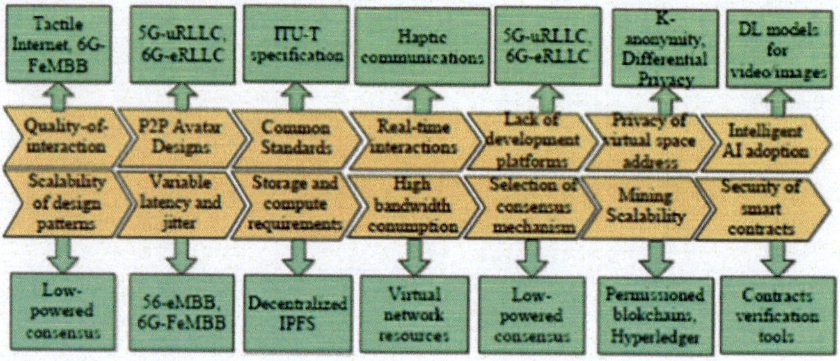

Fig. 2 Open issues and possible research directions of integration of 6G and BC in AR/VR deployments

predict the possible attack vectors to contracts like gas flaws, injection dependence, transaction re-entrancy and many more.

- *Peer-to-peer bandwidth*: The first is downloading speed since it has historically been too slow to get a file from distributed peer-to-peer (P2P) storage network.
- *Avatar Designs*: The other challenge for P2P designs, however, is to include updated information for each avatar (profile) on which other avatars (profiles) are its neighbours.
- *Lack of uniformity in standards*: Space with a central location and transparent boundaries where users can move and communicate with objects and control materials, with interactive features such as role play and participation in teamwork.
- *Variable application service provisioning*: Enables users to interact in a network or an environment using variable rates of audio depending on the average distance among users.
- *Scalability and high latency*: Sometimes the use of BC for this network will enhance the scalability because of its decentralized nature and it provides less throughput with high latency.
- *High requirement of Storage and Computation power*: BC-based consensus algorithms require the high computation power and high storage facility to process and store the data.
- *Vulnerability of smart contracts*: Smart contracts are susceptible to various attacks such as re-entrancy, timestamp dependence, short address and delegate call.

7 Case Study: *VoRtex Enterprise*: BC-Assisted Decentralized VR Platform

VoRtex is an educational platform that is developed using BC technology integrated with VR. This platform enterprise artificial industrial environment which is similar to reality to employ training from various fields like health care, automotive and space. To provide trust and security about the privacy of data in the virtual world (VW). Jovanovic *et al.* [20] proposed a solution platform with combination of BC technology. This platform is having a user-centric design and allows the processing of data using a BC distributed environment, by which data ownership and integrity functionality can be provided. The decentralized BC-based model works as a back-end, where data is represented in the form of blocks and is synchronized among connected users. Figure 3 represents the high-level software architecture of *VoRtex* platform. The proposed solution architecture is listed as follows.

8 Concluding Remarks and Future Scope

Technical advancements in wireless communication infrastructures and the emergence of connected things made the realization of real-time visual communications a reality. With the emergence of beyond 5G and 6G-driven applications, AR/VR

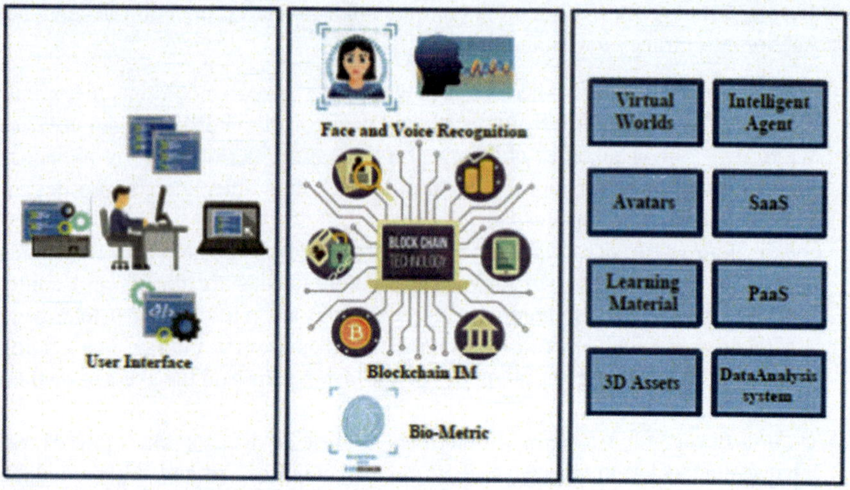

Fig. 3 Component-based software architecture of VoRtex

has transitioned data-driven applications with virtual and haptic connectivity, with massive device-connectivity. AR/VR space has become a complex ecosystem, with high ingested data, which has made security and privacy issues more paramount. Thus, the adoption of BC in AR/VR-based ecosystems handles the inherent issues of trust, security and privacy among AR stakeholders. The survey systematically highlights the potential BC-envisioned AR use cases and presents the key drivers and benefits. As part of the future scope, the authors intend to investigate the highlighted open issues, and possible frameworks that would help practitioners in the design of scalable decentralized AR solutions, with a view of addressing the trade-offs of required security and privacy counterparts.

References

1. Forescout (2020) What is augmented reality (AR) and how does it work. https://thinkmobiles.com/blog/what-is-augmented-reality/ (2021), [Online; Accessed 03 –December 2020]
2. Hsu CH, Manogaran G, Srivastava G, Chilamkurti N (2021) 6g-enabled network in box (nib) for industrial applications and services. IEEE Trans Indust Inform 1–1. https://doi.org/10.1109/TII.2021.306770
3. Bodkhe U, Tanwar S, Bhattacharya P, Verma A (2021) Blockchain adoption for trusted medical records in healthcare 4.0 applications: a survey. In: Singh PK, Wierzchoń ST, Tanwar S, Ganzha M, Rodrigues JJPC (eds) Proceedings of second international conference on computing, communications, and cybersecurity. Springer Singapore, Singapore, pp 759–774
4. Yamakami T (2020) A privacy threat model in xr applications. In: Barolli L, Okada Y, Amato F (eds) Advances in internet, data and web technologies. Springer International Publishing, Cham, pp 384–394

5. Nguyen QK, Dang QV (2018) Blockchain technology-opportunities for emerging economies. In: 2018 4th international conference on green technology and sustainable development (GTSD). IEEE, pp 478–482
6. French A, Risius M, Shim JP, Bordnick P, Parsons J, Holtman D (2018) Virtual reality and blockchain technology in the era of 5g: Disrupting business and society
7. Werner PA (2019) Review of implementation of augmented reality into the georeferenced analogue and digital maps and images. Information 10(1):12
8. Fraga-Lamas P, Fernández-Caramés TM (2019) A review on blockchain technologies for an advanced and cyber-resilient automotive industry. IEEE Access 7:17578–17598
9. Abbas QE, Sung-Bong J (2019) A survey of blockchain and its applications. In: 2019 International conference on artificial intelligence in information and communication (ICAIIC). IEEE, pp 001–003
10. Cannavo A, Lamberti F (2020) How blockchain, virtual reality and augmented reality are converging, and why. IEEE Consumer Electron Magazine
11. How can blockchain accelerate the virtual reality revolution? https://www.entrepreneur.com/article/365482, note = Accessed: 2021–02–03
12. Liao S, Wu J, Li J, Konstantin K (2020) Information-centric massive IOT based ubiquitous connected vr/ar in 6g: A proposed caching consensus approach. IEEE Internet of Things J
13. Mewes A, Heinrich F, Hensen B, Wacker F, Lawonn K, Hansen C (2018) Concepts for augmented reality visualisation to support needle guidance inside the mri. Healthcare Technology Letters 5(5):172–176. https://doi.org/10.1049/htl.2018.5076
14. Ceek AV (2020) Ceek:.http://www.ceek.com/ (May 2020)
15. Tang F, Kawamoto Y, Kato N, Liu J (2019) Future intelligent and secure vehicular network toward 6g: Machine-learning approaches. Proc IEEE 108(2):292–307
16. Ladha A, Bhattacharya P, Chaubey N, Bodkhe U (2020) Iigpts: Iot-based framework for intelligent green public transportation system. In: Singh PK, Pawłowski W, Kumar N, Tanwar S, Rodrigues JJPC, Obaidat MSO (eds) Proceedings of first international conference on computing, communications, and cyber-security (IC4S 2019), vol 121. Springer International Publishing, pp 183–195
17. Bodkhe U, Tanwar S (2020) A taxonomy of secure data dissemination techniques for IOT environment. IET Software 1–12. https://doi.org/10.1049/ietsen.2020.0006
18. Forescout (2021) "privacy and blockchain. https://en.wikipedia.org/wiki/ Privacyand-blockchain. [Online; Accessed 13 December 2020]
19. Tanwar S, Bhatia Q, Patel P, Kumari A, Singh PK, Hong WC (2020) Machine learning adoption in blockchain-based smart applications: the challenges, and a way forward. IEEE Access 8:474–488. https://doi.org/10.1109/ACCESS.2019.2961372
20. Jovanović A, Milosavljević A (2019) Decentralized virtual reality blockchain-based platform. Facta Universitatis Series: Autom Control Robot 18(1):057–077. https://doi.org/10.22190/FUACR1901057J
21. Bhattacharya P, Tanwar S, Bodke U, Tyagi S, Kumar N (2019) Bindaas: blockchain-based deep-learning as-a-service in healthcare 4.0 applications. IEEE Trans Netw Sci Eng 1–1. https://doi.org/10.1109/TNSE.2019.2961932
22. Bodkhe U, Bhattacharya P, Tanwar S, Tyagi S, Kumar N, Obaidat MS (2019) Blohost: blockchain enabled smart tourism and hospitality management. In: 2019 international conference on computer, information and telecommunication systems (CITS), Shanghai, China. IEEE, pp 1–5
23. Bodkhe U, Tanwar S (2021) Secure data dissemination techniques for IOT applications: Research challenges and opportunities. Software: Practice and Exper n/a(n/a) 1–23. https://doi.org/10.1002/spe.2811, https://onlinelibrary.wiley.com/doi/abs/https://doi.org/10.1002/spe.2811
24. Bodkhe U, Tanwar S, Bhattacharya P, Kumar N (2020) Blockchain for precision irrigation: opportunities and challenges. Trans Emerg Telecommun Technol e4059. https://doi.org/10.1002/ett.4059, https: //onlinelibrary.wiley.com/doi/abs/https://doi.org/10.1002/ett.4059

25. Dhalor M (2021) Blockchain in augmented reality. https://www.ommzi.com/blockchain-augmented-reality/ [Online; Accessed 13- December 2020
26. Bodkhe U, Tanwar S, Parekh K, Khanpara P, Tyagi S, Kumar N, Alazab M (2020) Blockchain for industry 4.0: a comprehensive review. IEEE Access 8:79764–79800
27. Mehta D, Tanwar S, Bodkhe U, Shukla A, Kumar N (2021) Blockchainbased royalty contract transactions scheme for industry 4.0 supply-chain management. Inform Process Managem 58:1–20. https://doi.org/10.1016/j.ipm.2021.102586
28. Price (2021) Next reality "the future of augmented reality and blockchain technology. https://medium.com/@NextRealityNews/the-future-of-augmented-reality-blockchaintechnology-9a24b8a0a1ec. Online; Accessed 13 Dec 2020
29. Coinmarketcap (2021) Decentraland (mana. https://coinmarketcap.com/currencies/decentraland/. [Online; Accessed 13 December 2020]

Integration Blockchain for Data Sharing and Collaboration in Mobile Healthcare Applications

Inaam Zia, Pawan Singh, Anil Kr. Tiwari, and Amit Pandey

Abstract Enabled by way of cellular and wearable era, private fitness record provides huge and growing importance of healthcare, which is beneficial for both care vendors and medical studies. The secure method and handy sharing of personal health statistics is vital to the improvement of the interplay and collaborating with the healthcare enterprise. Confronted with the capability privacy troubles and vulnerabilities present in current private health statistics garage and system of sharing, which also include the concept of self-sovereign facts ownership, we advocate an modern person-centric fitness data sharing solution via making use of a decentralized and permissioned blockchain to guard privateness the usage of channel formation scheme and beautify the identification management the use of the club service supported with the aid of the blockchain. A cellular utility is deployed to accumulate fitness data from non-public wearable devices, guide input, and scientific gadgets, and synchronize information to the cloud for records sharing with healthcare insurance agencies and healthcare companies. To maintain the integrity of fitness statistics, inside each document, an evidence of validation and integrity is permanently retrieved from database stored in cloud location and is very important to the blockchain network. Furthermore, for performance and scalable and considerations, we undertake a tree-based totally processing statically and batching method to handle big facts units of personal fitness statistics captured and uploaded by means of a mobile platform.

Keywords Health care · e-health · Privateness · Permissioned blockchain · Get right of entry to manage · Scalability · Integrity · Wearable devices · Cell platform

I. Zia (✉) · P. Singh · A. Kr. Tiwari
Amity University Uttar Pradesh, Lucknow Campus, India

A. Pandey
College of Informatics, Bule Hora University, Bule Hora, Ethiopia

P. K. Singh et al. (eds.), *Emerging Technologies for Computing, Communication and Smart Cities*, Lecture Notes in Electrical Engineering 875,
https://doi.org/10.1007/978-981-19-0284-0_39

1 Introduction

The healthcare system is one of the most demanding and evolving sectors in the world, any countries growth is directly proportional to the how good the health system is, and it constitutes to certain amount of GDP of a country [1]. The industry may include companies which mainly consist of generalization and commercialization of the products sometimes without taking into consideration of the factor of healing and improving of the sector. It is considered as one of the most complex system in the world which contain counterfeiting of drugs without being able to discriminate the bad ones from the good one, not being able to break the heavy barriers such as distributed patient record and try to fetch each and every record of the patient from different hospitals is a very difficult task, achieving a transparency during data sharing is a very difficult task too and if there is not proper sharing of data among different parties it may lead to disastrous impact on the health of a patient. There has been cases where it has been seen that the record has been sold for money by the middle man so we have to produce something which may lead to a system where there is no middle man and there is security and transparency of data which may lead to trustworthy system.

There are many research papers which has discussed about the use blockchain in health care [2–6].

2 Blockchain

Bitcoin is one of the famous cryptocurrencies, and the technology on which it is based on is known as blockchain [7] because of the release and use of bitcoin the blockchain became very famous, people started talking about it and taking interest in it. Bitcoin works same as other currency like Dollar and Pounds, but it uses different concept to maintain and handle these cryptocurrencies. Due to its volatile nature and anonymous founder of various cryptocurrencies, the government of various country is banning or regulating the laws to monitors these cryptocurrencies, one of the features of these cryptocurrencies is that it is very hard to track transactions of these cryptocurrencies, and there are hardly any experts present with the government of the countries to do so which makes it the most difficult for government to legalize it. But there are various positives sides of blockchain in which are interested which are decentralization of data which is maintaining a copy of record at every node making the data secure from losing it and there cannot be changes done in the previous node as there is concept of proof-of-work. So, it can be concluded that with the help of blockchain we can develop tools which can be helpful in various aspects of life like in the medical field or in the shipment industries. As we know any cryptocurrencies can be mined only it matches the hash function [8] of it.

2.1 Consensus

The consensus refers to a network of blockchain on which mechanism should be used to build a particular network, there are various factors which are taken into consideration before going for the model which determines which transaction should be accepted or rejected based on the model selected for the network. Merely a regulation of this protocol deals with the regulation of transaction whether it is valid or not, adding the newly created node in the network or rejection of node based on the mechanism. Basically, there are categorized into following:

1. In the case of global consensus model, there is a main block which is basically called as genesis block, and every node which is similar to the genesis block will be stored after validation. Most popular example is Bitcoin and Ethereum.
2. In the early of consensus, it only deals with the concerned nodes, and there is ignorance of other nodes. For example, blockchains are nano [9] and trust chain [10].

One of the most common and popular mechanisms is known as Nakamoto consensus. The Nakamoto consensus deals with the concepts of proof-of-work (POW) for accepting or rejection of new node in the block. If a miner is mining complete a task and find a hash of the new node it will be added to the network as it is validated by the miner this concept is known as proof-of-work. the stake [11].

2.2 Smart Contracts

A smart contract is a kind of contract which does not require any manual adjustment, everything is done automatically, it is verified automatically, it is executed automatically, and there cannot be any changes done in the code which is programmed for the logic which later on runs on the blockchain. A custom logic or rules or guidelines can be written on the blockchain; on the satisfaction of these rules, some actions are triggered Mostly all the cryptocurrency are adapting these smart contracts.

2.3 Blockchain Types

Blockchain can be classified into three categories which are as follows:

- Public blockchain as the name suggests it is open to all and anybody can take part into it. It is basically based on the trust of decentralization one of the most common example it.
- Private blockchain is personal to an organization or group; it is made for public use and only used by the persons in the organization, and a personal blockchain is a permissioned blockchain centralized to 1 governing organization.

2.4 Challenges and Future

In spite of having so much capacity, there are various obstacle in the blockchain. The concept of blockchain is that it maintains a copy of transaction at every node which makes the makes the transaction very sluggish; it is fine for very a small transaction but not in the case of very large transaction. Due to the previous point, it is hard to enhance the scalability of blockchain in the software. In case of bitcoin and Ethereum also, there it is very hard to enhance the scalability though it provides decentralization and consistency. In the future, we should be working on improving the feature of scalability. As the scalability of the software is increased, it opens a door for various other features which may increase the efficiency of the software.

3 Device Layout

3.1 System Evaluation

Non-public health information sharing. 6 entities which are included, particularly consumer, wearable gadgets, healthcare company, insurance employer, the blockchain network and cloud database (Fig. 1).

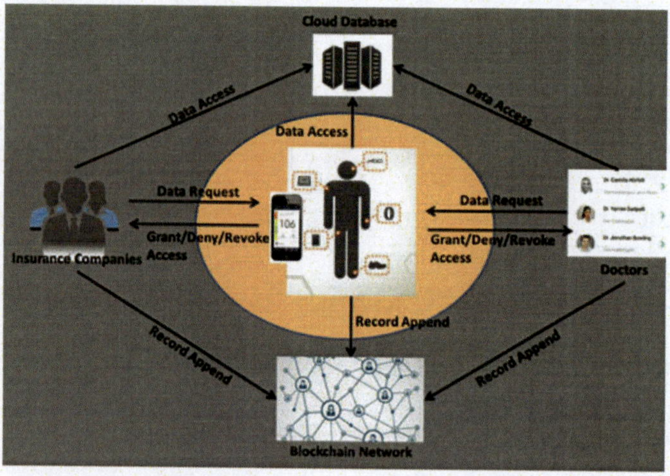

Fig. 1 Consumer-centric personal fitness facts sharing

3.2 Machine Entities

- **User.** System customers gather record from wearable gadgets which screen users' health facts which include strolling distance, napping situations, and heartbeat. The ones records are uploaded to the database on cloud hosted on trustworthy platform through a mobile application. Consumer is the sole owner of private fitness statistics and is liable for allowing, revoking, and denying facts get right of entry to from another parties, which are healthcare vendors and insurance corporations. If the person is in the search of scientific treatment, the user would percentage the health facts with the desired doctors. If the treatment is completed, the information gets right of entry to is revoked to disclaim similarly access from the doctors. Identical state of affairs applies to user-insurance organization relations. Except, user also can report normal activities in line with a specific scientific remedy along with medication utilization to proportion with the remedy issuer for adjustment and higher improvement.
- **Wearable devices.** Wearable devices serve to rework authentic health data into user-friendly layout, and then, the information is sync with the aid of the user to their active account. Each account is related to a fixed of wearable gadgets and viable clinical gadgets. While a chunk of fitness facts generated, it is going to be uploaded to the blockchain community for report preserving and integrity protection.
- **Healthcare company.** In healthcare companies, doctors are appointed by a positive person to perform medical check, provide a few recommendations, or provide clinical treatment. In the meantime, the medical remedy records may be uploaded to the blockchain network for sharing statistics with different healthcare carriers below the user's permission. And the contemporary healthcare provider may request access to previous fitness data and clinical remedy from the person. Every information request and the records corresponding get right of entry to are recorded at the blockchain.
- **Medical insurance enterprise.** User might also request a quotation of health insurance from medical insurance businesses or marketers to choose a right health coverage plan. To provide higher insurance regulations, insurance companies request facts get admission to from customers consisting of user health facts from wearable gadgets and scientific remedy history. Customers with preceding clinical remedy(s) may also want to pay a better rate, and the history cannot be denied by way of customers to prevent insurance fraud. Users can pick out not to percentage workout facts because of privacy problems, but commonly they would preference to percentage because everyday workout can bring down the insurance pay price. But customers cannot conceal or adjust medical treatment history facts considering the fact that the once data is permanently recorded on the blockchain community and the integrity and trustworthiness is ensured. Furthermore, the coverage claims also can be recorded at the blockchain.
- **Blockchain community.** The blockchain network is used for three functions. For fitness records accumulated from each wearable gadgets and healthcare carriers,

each of the hashed information entry is uploaded to the blockchain community for integrity safety. For private fitness information, get admission from healthcare company and medical insurance business enterprise, and every of the information that gets right of entry to request must be processed to get a permission from the owner with a decentralized permission control protocol. The received entry to manipulate policies must be saved in a dispensed manner at the blockchain which ensures balance. Besides, each of the get entry to request and access pastime ought to be recorded on the blockchain for in addition auditing or investigation.

- **Cloud database.** The cloud database stores consumer health associated statistics, records requests from the healthcare company and coverage businesses, records get entry to document and records access manipulate policy. Statistics get admission to is responsible and traceable. Once information leakage is detected, the malicious entity may be diagnosed.

3.3 Device Implementation

The device captures a personal health data using a gadget which is wearable which can be a smart watch or any tracker. It can be used in the tracking of medical utilization. The private information can also be stored in cloud where the user can buy a subscription of the cloud services with sufficient storage which can store abundant amount of data (Fig. 2).

Private health data requires integrity and validation of the data. All the public data which are not that private information is moved from the user computer or tools to the cloud server for the validation of the data. The wearable gadget which is in the form of smart watches or tracker is used to collect the daily motion data of the user which are maintained as a statistic. Merkle tree model is used for the scalability of the product. Merkle tree is a tree structure which is in a form of binary; it contained an information in the form hash. Based on the time it is generated, the information is ordered. In Merkle, a pair is maintained of every two nodes, and the time hash

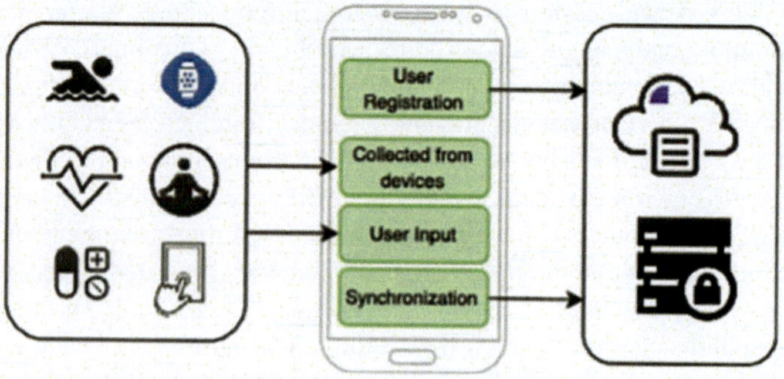

Fig. 2 Personal health data collection

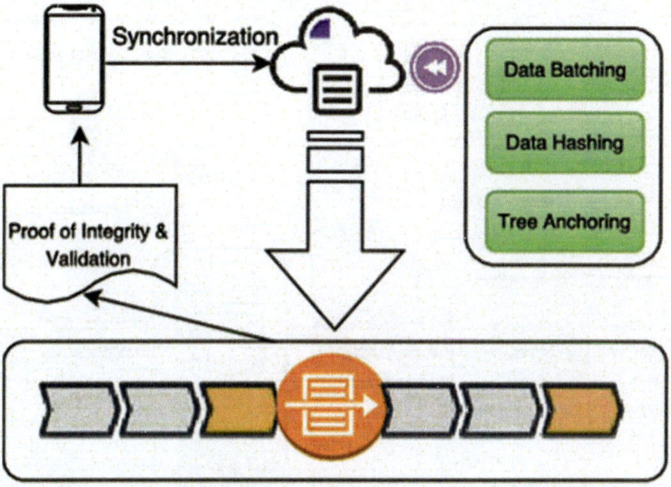

Fig. 3 Personal fitness information integrity safety

becomes the leaf of the node. The following step is repeated until an unused hash is found out which then becomes the tree of the node (Fig. 3).

It is a process to improve the record validation and authenticity of it which later on help in the scalability of the product. Merkle algorithm becomes the root of the transaction, and verification of the data is 1/3rd dependent on the other parties. Storing all the hash brings a lot of advantages. The hash is very effective to find any changes or modification it can be detected very easily (Fig. 4).

Information sharing is done using a Hyperledger. The data is stored in the cloud server, firstly the data is given by the user, and some information is also shared by the HealthCare provider, Insurance Company. If there is any fabrication by the client, it is checked by the peers, it is seen from the ordering services, and there are various channels maintained to carry out transaction. There is a Hyperledger fabric network maintained which contains a CA, peers, ordering service and channels.

3.4 System Assessment

Our system adopts a consumer-centric version for processing non-public health data the usage of blockchain network, ensuring the statistics possession of people, as well as records integrity. The operations at the records facts are fantastically inter-operable and like-minded with cutting-edge structures. Via implementing access manipulate policies, customers can cope with their non-public records without demanding about the privateness problems. With all of the security targets proposed in segment i accomplished, it is far crucial to assess the machine performance, concerning to the scalability and efficiency of the information integrity proof generation and facts

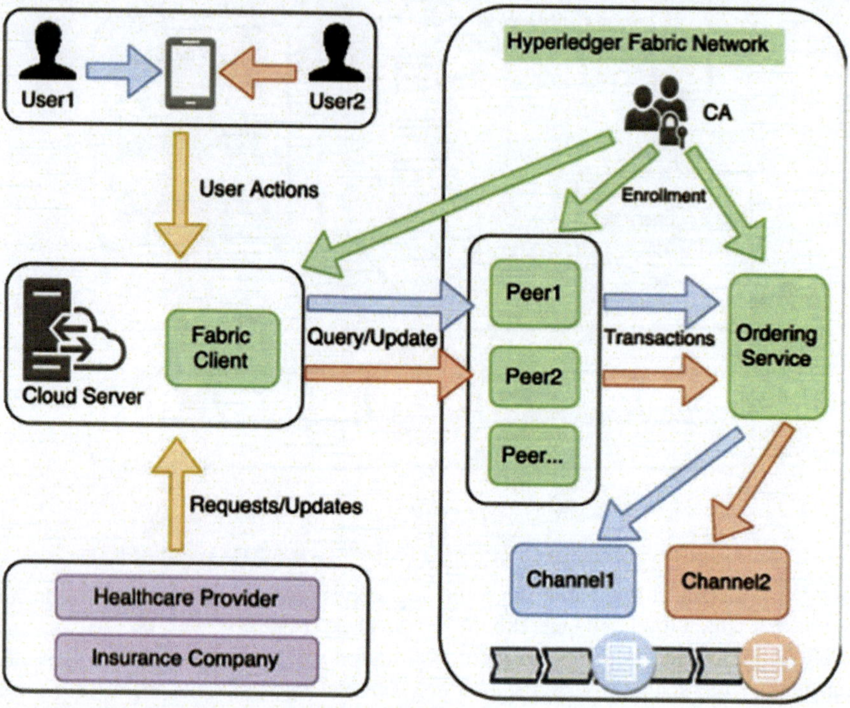

Fig. 4 Information sharing and collaboration the usage of Hyperledger fiber and channel for cellular customers

validation method. We take a look at distinctive numbers of concurrent statistics with a variety from 1 to 10,000. Parents 5 and 6 suggest the common time value, respectively (Figs. 5 and 6).

From these two figures, we are able to conclude that the device can deal with a huge dataset at low latency, which indicates the scalability and efficiency of the facts

Fig. 5 Average time for integrity evidence era

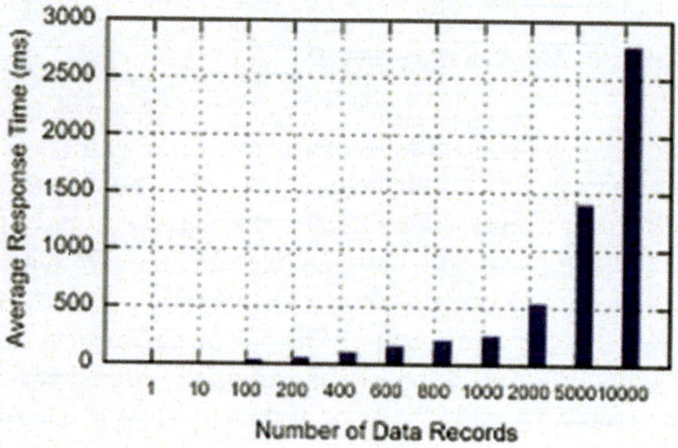

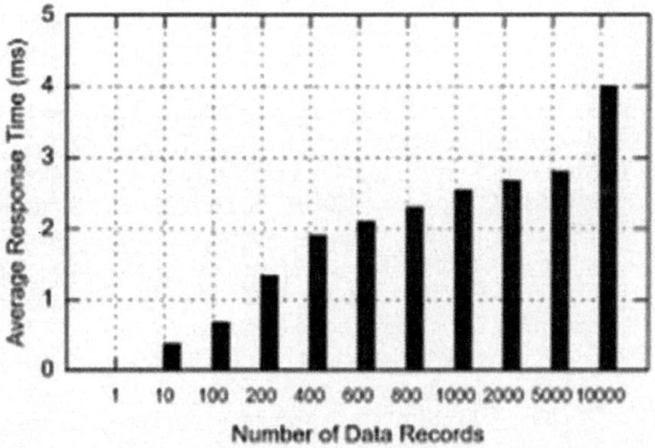

Fig. 6 Average time for integrity evidence validation

process. Through adopting Merkle tree method to batch statistics, we implement a set of rules with the computation complexity of o(log2n). That is a crucial gain when the information facts are accumulated at a high frequency.

4 Conclusions

This research paper deals with fact sharing of medical records of patient and record of the doctor. The paper deals with the removing the extra cost for the patient. It uses the blockchain to implement the concept. It does not deal with all the various other aspects of the blockchain. Medrec is a medical management system which deals with the smart contract which increases the privacy of the system. In this paper, we implement a mobile healthcare system for collection and monitoring of the system and providing an efficient result according to it. The system deals with the privacy, integrity, and validation of the data.

References

1. OECD.Health.spending (2018) https://doi.org/10.1787/8643de7e-enhttps://www.oecdilibrary.org/content/data/8643de7e-en
2. Kuo T-T, Kim H-E, Ohno-Machado L (2017) Blockchain distributed ledger technologies for biomedical and health care applications. J Amer Med Info Assoc 24(6):1211–1220. 1093/jamia/ocx068. URL https://doi.org/10.1093/jamia/ocx068
3. Dhillon V, Metcalf D, Hooper M (2017) Blockchain in health care. Apress, Berkeley, CA, pp 125–138. ISBN 978–1–4842–3081–7. https://doi.org/10.1007/978-1-4842-3081-7_9. URL https://doi.org/10.1007/978-1-4842-30817_9
4. Engelhardt MA, Espinosa D (2017) Hitching healthcare to the chain : an introduction to blockchain technology in the healthcare sector. An Introduct Blockchain Technol Healthcare Sect 7(10):22–35

5. Randall D, Goel P, Abujamra R (2017) Blockchain applications and use cases in health information technology. J Health and Med Info 8(3):8–11. https://doi.org/10.4172/2157-7420.1000276
6. Sharma R (2018) Blockchain in Healthcare. Accessed 18 Oct 2018. http://www.fccco.org/uploads/publications/Blockchaininhealthcare_FCCCO_RS.pdf
7. Nakamoto S (2008) Bitcoin: a peer-to-peer electronic cash system
8. Rogaway P, Shrimpton T (2004) Cryptographic hash-function basics: Definitions, implications, and separations for preimage resistance, second-preimage resistance, and collision resistance. In: International workshop on fast software encryption. Springer, pp 371–388
9. LeMahieu C (2018) Nano: a feeless distributed cryptocurrency network. 2018 Accessed 10 November 2018. https: //nano.org/en/whitepaper
10. Otte P, de Vos M, Pouwelse J (2017) Trustchain: a sybil-resistant scalable blockchain. Future Gener Comput Syst
11. Saleh F (2018) Blockchain without waste: Pproof-of-stake
12. Bell EA, Ohno-Machado L, Adela Grando M (2014) Sharing my health data: a survey of data sharing preferences of healthy individuals. In: AMIA annual symposium proceedings. American Medical Informatics Association, pp 1699
13. Behner FWP, Hecht M-L (2018) Fighting counterfeit pharmaceuticals. Accessed 15 Nov 2018. https://www.strategyand.pwc.com/reports/counterfeit-pharmaceuticals
14. Script—Bitcoin Wiki (2018) Accessed 15 Nov 2018. https://en.bitcoin.it/wiki/Script
15. Wood G (2014) Ethereum: a secure decentralised generalised transaction ledger. Ethereum project yellow paper 151:1–32
16. Peterson J, Krug J, Zoltu M, Williams A, Alexander S (2018) Augur: a decentralized oracle and prediction market platform. Accessed 18 October 2018. https://www.augur.net/whitepaper.pdf
17. 2018 (accessed October, 2018). URL https://0xproject.com/pdfs/0x_white_paper.pdf. 0x: An open protocol for decentralized exchange on the Ethereum blockchain
18. LeMahieu C (2018) NEM–distributed ledger technology. Accessed 10 Nov 2018. https://nem.io/
19. Androulaki E, Barger A, Bortnikov V, Cachin C, Christidis K, De Caro A, Enyeart D, Ferris C, Laventman G, Manevich Y et al (2018) Hyperledger fabric: a distributed operating system for permissioned blockchains. In: Proceedings of the thirteenth eurosys conference. ACM, pp 30
20. Iris (2018) Develop the world's new financial system. Accessed Nov15 2018. https://www.stellar.org/
21. Neo-project. *NEO*. https://neo.org/
22. Zheng Z, Xie S, Dai H, Chen X, Wang H (2017) An overview of blockchain technology: architecture, consensus, and future trends. In: Big data (BigData Congress), 2017 IEEE international congress on IEEE, pp 557–564
23. Kish LJ, Topol EJ (2015) Unpatients-why patients should own their medical data. Nat Biotechnol 33(9):921–924
24. Clippinger JH (2017) Why self-sovereignty matters. https: //idcubed.org/chapter-2-self-sovereignty-matters/ Online; Accessed 7 March 2017
25. Liang X, Shetty S, Tosh D, Kamhoua C, Kwiat K, Njilla L (2017) Provchain: a blockchain-based data provenance architecture in cloud environment with enhanced privacy and availability. In: International symposium on cluster, cloud and grid computing. IEEE/ACM
26. Tosh DK, Shetty S, Liang X, Kamhoua CA, Kwiat KA, Njilla L (2017) Security implications of blockchain cloud with analysis of block withholding attack. In: Proceedings of the 17th IEEE/ACM international symposium on cluster, cloud and grid computing, ser. CCGrid '17. Piscataway, NJ, USA: IEEE Press, pp 458–467
27. T. O. of the National Coordinator for Health IT (ONC) (2016) the National Institute for Standards, and T. (NIST). Use of blockchain in healthcare and research workshop
28. Cachin C (2016) Architecture of the hyperledger blockchain fabric. In: Workshop on distributed cryptocurrencies and consensus ledgers

29. Author F, Author S (2016) Title of a proceedings paper. In: Editor F, Editor S (eds) CONFERENCE 2016, LNCS, vol 9999. Springer, Heidelberg, pp 1–13
30. Author F, Author S, Author T (1999) Book title, 2nd edn. Publisher, Location
31. Author F (2010) Contribution title. In: 9th international proceedings on proceedings. Publisher, Location, pp 1–2
32. LNCS Homepage (2016). http://www.springer.com/lncs. Last Accessed 21 Nov 2016

Communication Technologies, Security and Privacy

Reusability Estimation of Object-Oriented (O–O) Systems Using SOM Technique

Bharti Bisht and Parul Gandhi

Abstract Due to recent changes and advancements in software sector, there was an alarming need to cut down time as well as efforts required during complete software development process. While constructing and designing product and services, assuring the quality of software product is very important aspect in order to boost market value of that product. To attain both quality as well as productivity objectives, there is an urge for software reusability. Reusability is an essential measure that can be used to improve overall software quality with lesser cost and efforts. Data Mining is a process to extract useful data from a larger set containing raw data. It also means analyzing patterns from large collection of data. In Object-Oriented (O–O) Software Systems, class reusability can be easily obtained by inspecting metrics values. Inspecting metric values would save time to develop software components from initial phase. In this paper various software metrics related to software reusability of Object-oriented (O–O) software systems has been outlined with the help of MATLAB environment. SOM Technique was applied to create clusters of metrics values that were derived from two Java based system that were a part of COMET dataset that was easily available online. The goal of this study was to identify relationship among metrics values and different level of reusability of particular class. This study shows how SOM technique can be applied on metrics with SOM grids of different sizes. The study depicts elimination of two of the software metrics i.e. NOC and DIT due to poor distribution depicted by both of them.

Keywords Object-oriented software system · Software metrics · Reusability · SOM

1 Introduction

Reuse based engineering [7] is an approach in which the software development process is organized in order to reuse existing software components. This approach

---

B. Bisht · P. Gandhi (✉)
FCA, Manav Rachna International Institute of Research and Studies Faridabad, Faridabad, India
e-mail: parul.fca@mriu.edu.in

P. K. Singh et al. (eds.), *Emerging Technologies for Computing, Communication and Smart Cities*, Lecture Notes in Electrical Engineering 875,
https://doi.org/10.1007/978-981-19-0284-0_40

comes with many advantages such as reduced cost as well as time and even contributes in improving software quality. Metrics represents standard of measure of particular property that a software system possess. These also act as a indicator of quality of software system that will help managers of particular project to control software development process and will also help software developers to analyze software quality [15]. Metrics plays an important role in measuring reusability of OO (Object-Oriented) systems. OO (Object-Oriented) approach is even contributing towards construction of software in component form [7] and also plays an important role in reuse process. In current scenario, there is much demand for OO (Object-Oriented) techniques that would directly contribute in improving reusability level of software [10]. Software Reusability is one of the remarkable approach [8] to reveal all those artifacts [10] from extant components for formulating new software systems.

According to [11], software reusability is distinctive approach for a software component that portrays software's proficiency of reuse [3]. Reusability also points out the reuse proficiency of particular software component [3]. It means that [5] if reusability of particular software component is low, then that component would be less reusable. From [12], reusability means the employment of hitherto software [20] written in the form of design together with its code. This mechanism has been already inspected generally during development evolution of different software projects. The most imperative advantage showcased by software reusability is that it contributes in reducing the total bugs present in software [10].

The main objective behind this study was as follows:

(1) This study attempts to use metrics as an indicator for measuring the capability of the reusable components.
(2) This study would help to predict those components that can be reused early during software development process.
(3) SOM technique has been used in this study would help to categorize reusable components using OO (Object-Oriented) metrics.

2 Reusability Concepts

Researcher through its study [11] was able to explore that a efficient as well as effective software is required which can be produced in short span of time. This study also helped to find the profitable contribution [11] of reusability in improving productivity as well as quality of software. Authors in [9] has mentioned about the main purpose of Component based approach that this creates only reusable modules not the complete system that can be easily reused. So through their study they highlighted about the richness of O–O (Object oriented).In past there have been lot of studies which have contributed towards introduction of various approaches [15] to find out various software characteristics using different metrics. Authors through their study [19] measured the reusability level of software using two metrics i.e. fan-in metric [19] and fan-out metric [19]. Kayarvizhy and Kanmani [21] developed their own metrics to compute software modules quality characteristics such as complexity as

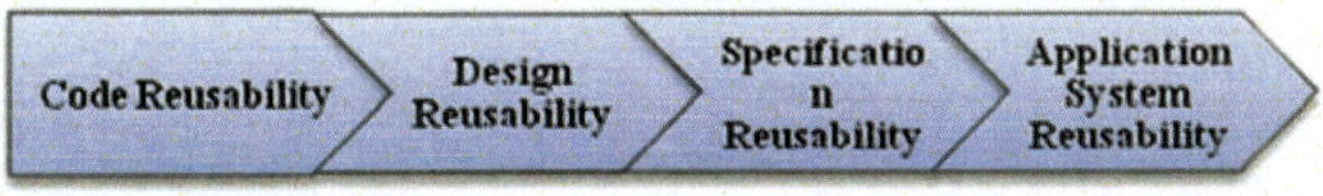

Fig. 1 Levels of reusability. (*Source* Self)

well as reusability. Further in this study they used two methods. In first method using methods present in a software component, CR (Component Reusability) [21] was calculated whereas in second method, level of reusability was computed using metric known as CRL (Component Reuse level) [21].

Reusability is a process of amending software [9] through extant reusable units or modules rather than formulating a very brand new software system. These reusable modules have been precisely tested and also being verified [22], which make them suitable enough for contributing in the improvement of software quality [22]. Reusability helps to increase productivity and improve quality of software. It is not only dependent on code but also envelopes software development cycle valuables like software components [8], test suites [8] documentations [8] and designs [8]. Reusability can also be attained by adopting software metrics approach. Metrics play an important role during software development process and as well as management of software. Organization uses different metrics to apply for their software development phase which includes various reuse metrics, Object-oriented metrics as well as quality metrics. These metrics also helps in shaping reuse practices applied during entire software development life cycle [11]. Software components which are developed within any organization are not only confined to be used in that particular environment but also can be dispersed as object code and can be easily reused by any other environments also.

Figure 1 represents different Reusability levels in Software Development lifecycle:

2.1 *Code Reusability [8]*

This is the most common level of software reuse approach. This is applied during implementation stage of software development life-cycle.

2.2 *Design Reusability [8]*

This one is considered as the highest reusability level. In this, software design miniature is reused.

2.3 Specification Reusability [8]

In this, issues at the specification phase which occurs due to inefficient reusable code [8] is removed.

2.4 Application System Reusability [8]

This is the special case of reusability as this helps to reuse complete software system. They are combined and even their accuracy is also assessed.

The different benefits provided by Software Reusability are [18]:

- Dependability is increased
- Improves Productivity
- Accelerated Development [8]
- Effectiveness Increased [8]
- Reduces Operational costs [15]

In this study SOM technique has been used through which it is very easy to categorize the components that are easily reusable.

3 SOM Technique

This is one of the familiar pattern recognition as well as classification algorithm. It is one of the Artificial Neural Network model that represents competitive learning [7] and is also unsupervised technique. SOM uses arranged collection of neurons in the form of two-dimensional grid. This reduction of high-dimensional grid into two-dimensional grid is very useful in data visualization technique which in turn helps to determine similar patterns [6]. It helps to transfer incoming input signals into 2-D map [8] and even explore underlying structure of input.

SOM contains 2 neuron layers i.e. input and output layers. Every input layer is connected to every output layer. The degree of minutiae's of findings helps to determine SOM grid size [9], more generalized findings requires less sized grid. Similar kinds of clusters are represented as neighbors in grid but different kind of clusters are at very greater distance to each other.

Rotaru and Dobre [29] used SOM technique to details of various components, for characterizing these software modules present in repositories that are extracted from manual of software and the results obtained during the study were propitious. In [31], authors proposed new technique i.e. SSOM technique in order to classify software components. In [25], authors have applied SOM technique in analyzing software components with the help of different metrics values for detecting quality dimensions. This approach helped designers to investigate and have better understanding

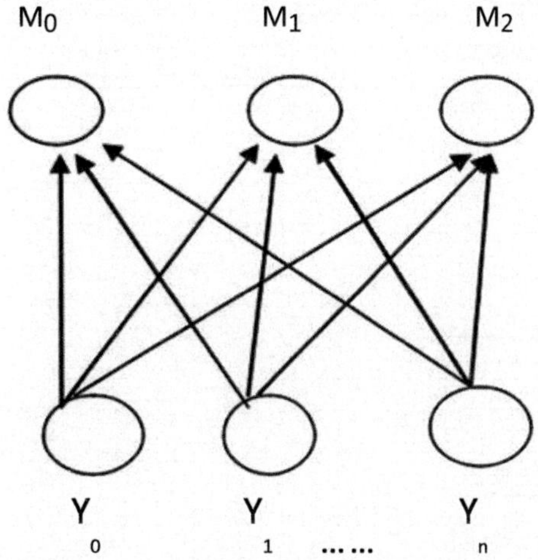

Fig. 2 SOM architecture [10]

of software clusters. To describe the self-organizing mechanism, the collection of input variables can be defined as real vector [7] $y = \{a_1, a_2 \ldots a_n\}^T \in R^n$. SOM array contains I neurons, each neuron is associated with real vector $n_i = \{b_{i1,}\ b_{i2\ldots}.b_{\mathrm{in}}\}^T \in R^n$. Each input variable has an image on SOM array, which can be described as an element of an array n_c that meets with y. This matching is calculated with the help of distance function i.e. d(y, n_i).Applying this function to all input variables is defined as a mapping of n-dimensional input space to 2-D SOM array [7]. During learning phase, when these real vectors are amended, the nodes that are near in the array to the present matching node will stimulate other to grasp values of attributes obtained from input vector y.

Figure 2 represents architecture of SOM containing 3 clusters and n input vectors: M_0.

Figure 3 represents different steps involved in SOM.

In this study SOM technique has been used as it is very easy to interpret as well as understood the data. It is also an easy method to find out similarities among various software modules.

4 Methodology Used

The main aim of conducting this study is to apprehend, analyze and project effects of values of software modules metric on level of reusability. In this study, SOM technique has been used to define various reusability levels. This study contains following stages shown below by Fig 4.

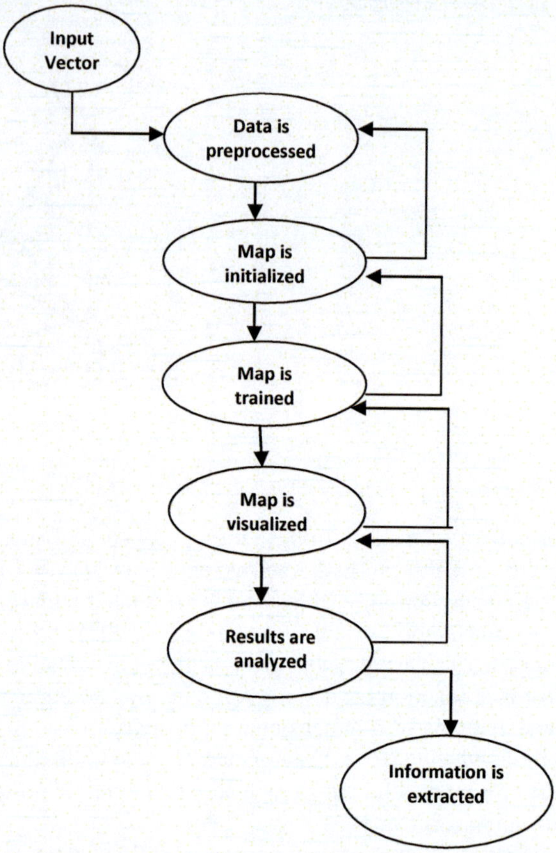

Fig. 3 Steps involved in SOM. (*Source* Self)

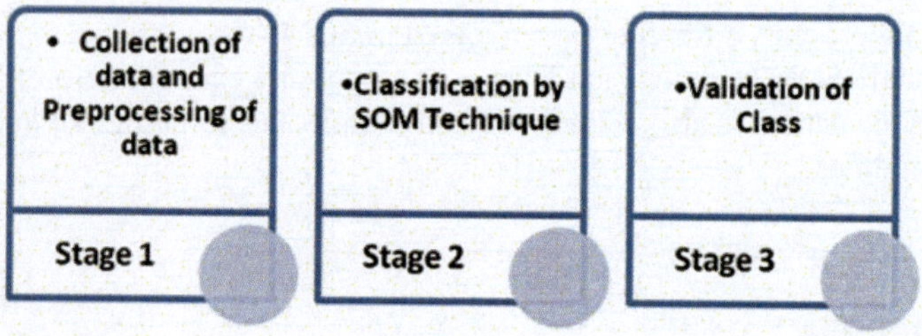

Fig. 4 Stages of study

4.1 Collection and Preprocessing of Data

For this study, online available COMET dataset [10] available online is taken as input dataset. Semi-automated approach [7] was applied on datasets in order to measure values of software metrics which are used in this study. Complete information of classes with their attributes and the methods used by the particular class were used as input vector.

Class level metrics are used for experimenting with SOM technique. The various metrics used for study are:

- Weighted Methods per Class (WMC)
- Response for a Class (RFC)
- Lack of Cohesion in Methods(LCOM) [4]
- Depth of Inheritance Tree of a Class (DIT) [13]
- Number of Children (NOC)
- Coupling between Objects(CBO) [15]

The values of above used metrics were collected [6] from 2 selected systems i.e. Eclipse JDT [5] and Hibernate [5].

Table 1 provides the minimum, maximum, standard deviation, mean, median values for different metrics used in study (NOC, DIT, LCOM, WMC, CBO and RFC [19] for Eclipse JDT system [5] which is used during study [10].

Table 2 provides the minimum, maximum, standard deviation, mean, median values for different metrics used in study (NOC, DIT, LCOM, WMC, CBO and RFC [19] for Hibernate system [5] which is used during study [10].

The above analytical descriptions of both Python programs revealed that NOC and DIT metric have low median value as well as mean value. This results shows that inheritance is not a part of these programs which were used as input dataset.

Table 3 represents the threshold value for class level metrics used in this study [19].

Table 1 Analytical description (Eclipse JDT system)

	WMC	NOC	DIT	RFC	CBO	LCOM
Minimum value	1	0	1	0	0	0
Maximum value	2076	24	9	664	227	80,987
Standard deviation value	158.26	1.96	1.77	65.87	21.38	2765.67
Mean value	64.61	0.52	2.68	42.55	17.88	234.37
Median value	24	0	2	25	7	6

Table 2 Analytical description (Hibernate system)

	WMC	NOC	DIT	RFC	CBO	LCOM
Minimum value	0	0	1	0	0	0
Maximum value	227	106	7	337	128	1374
Standard deviation value	26.72	3.85	1.51	35.63	16.34	82.64
Mean value	24.59	0.54	2.27	37.19	16.10	18.73
Median value	14	0	2	27	15	1

Table 3 Threshold value of class level metrics

	WMC	NOC	DIT	RFC	CBO	LCOM
Threshold value	21	3	3	45	14	21
Reference used	[18]	[18]	[21]	[21]	[18]	[16]

Threshold value for selected metrics used as quantitative approach to specify qualitative for software quality [20] and this in turn useful to find out classes with high risk [20].The threshold values helps developers in complete review process of classes of particular software. Exceeded threshold upper bound value gives a sign of problem and helps to find out error in the design of class of particular software [34].

During data preprocessing stage, there should be preparation of data that can easily available to be used by any of the clustering techniques [18].The dataset which has been used contains different Object-Oriented classes with metrics value -1 which shows that particular class don't exist in a particular time.

In this study, the used dataset is normalized by the Min-Max normalization technique [12]

$$v_i' = \frac{v_i' - \min}{\max_A - \min_A}(\text{new\_ } \max_A - \text{new\_} \min_A) + \text{new\_} \min_A \tag{1}$$

4.2 *Classification by SOM Technique*

This study uses Silhouette score technique to validate clustering process. This method can be used to analyze the distance separating different resulting clusters [11]. Silhouette graph shows how close different points in a cluster to the different points of neighboring cluster to previous one. If the coefficient value is very near to + 1 then this means that the selected cluster is very far from the neighboring one. Value of 0 tells that selected cluster is very near to or on boundary of neighboring one and negative one tells that the selected clusters are not correct one to be selected for clustering process.

The steps which are followed to find out Silhouette score of data set S containing m objects are as follows:

- Find the average distance between object $j \,\varepsilon\, S$ and other objects part of that cluster to which j belongs. Let this value be $p(j)$
- Now calculate minimum average distance [4] from j to all clusters of which j is not a part. Let this value be $q(j)$
- Silhouette score $s(j)$ is calculated using below Eq. 2 [20]:

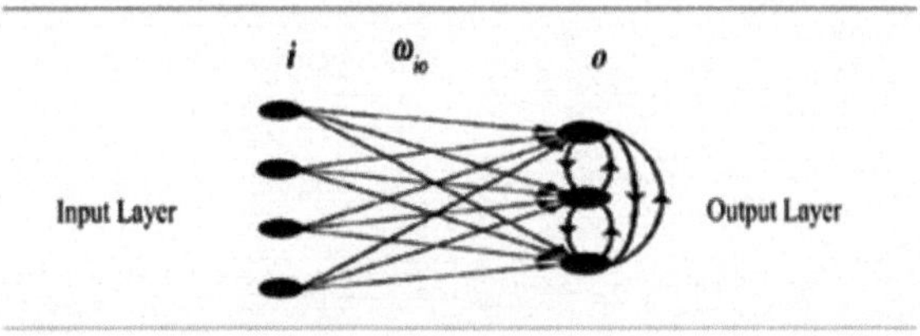

Fig. 5 Simple learning network [11]

$$s(j) = \frac{q(j) - p(j)}{max\{q(j), p(j)} \quad (2)$$

Figure 5 represents Simple learning network that has been taken into the consideration for this study:

5 Experimental Analysis

In this section experiments were analyzed separately for each Java based systems [6] which was used as the dataset for the study. Size of SOM Grid and the epoch's number was updated during experiments. Values of learning rate used during experiment are:

Learning rate value of Ordering stage = 0.7

Steps of Ordering stage = 1005

Learning rate value of Tuning Stage = 0.05 Neighborhood distance of Tuning Stage = 2.

It was observed that as epoch's number was increasing during experiment, the average values of Silhouette coefficient [24] was also increasing. The epoch's number [5] was then modified to 1000 because Silhouette coefficient became constant [8] for all the input dataset. Highest value of Silhouette coefficient was selected and then the value of metrics used in study was analyzed for every cluster in order to find the connection with different reusability level of classes.

After evaluating every cluster during experiment it was observed that the process of clustering was dominated by NOC metric as well as DIT metric. The values of different metrics were also examined in every cluster and were not able to find common connection between classes that are a part of same cluster [5]. The results found during experiment showed that if NOC metric as well as DIT metric are eliminated, then the results can be enhanced. After eliminating both metrics i.e. NOC and DIT metric [5] again the experiments were applied.

All the experiments done on Eclipse JDT [5] are shown below by Table 4.

It can be found from above table that Silhouette Coefficient Average is changed whenever there is a change in Cluster Size.

Figure 6 represents Silhouette Coefficient plot of Experiment 7 of Eclipse JDT which was performed during this study:

Table 4 Experiments done on Eclipse JDT system [5]

Number of experiment	Cluster size	Silhouette coefficient average
1	[9]	0.4938
2	[8, 9]	0.4601
3	[7]	0.4588
4	[6, 7]	0.5412
5	[6]	0.5456
6	[4, 5]	0.6345
7	[2]	0.7080
8	[2, 3]	0.6181

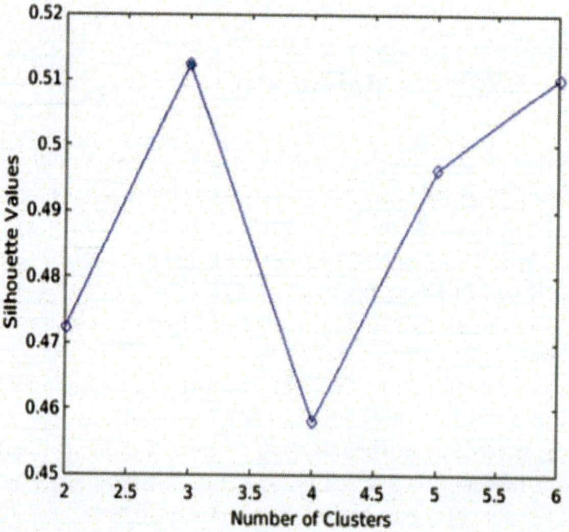

Fig. 6 Silhouette coefficient plot of eclipse JDT [12]

Negative value of Silhouette Coefficient is not preferred because it indicates that the object is much related to other cluster more than the cluster to which it belongs. The desired values are the values which approach to 1.

The classes which are a part of different clusters can be categorized into 3 divisions. These divisions summarize the relationship between class level metrics and reusability factor [8]. The divisions are as follows:

- Division 1: Highly reusable cluster [9] consists those classes which are not exceeding threshold values and have low values of LCOM, WMC, CBO, and RFC metrics [11]
- Division 2: Medium reusable cluster [9] consists of those classes that are near threshold values and even values of LCOM, WMC, CBO, and RFC metrics is medium

- Division 3: Low Reusable cluster [9] consists those classes that are exceeding threshold values and values of LCOM, WMC, CBO, and RFC metrics is highest

Based on above divisions, the complete analysis of each and every cluster vectors [28] used in experiment 7 of Eclipse JDT is represented in Table 5.

In the above table, if the Avg. of any one of the metric part of particular cluster is high then it can be concluded that the classes that are included [3] in this cluster have metrics with higher values, so they have less reusable ability. When there is higher percentage value of Ex. %, it can be concluded that there are lot of classes in this particular cluster that are outrunning the threshold value, so this also becomes less reusable [11].

From above table, it can be concluded that experiment 3 and 4 [11] show low reusability level because almost every classes in both experiments exceeds threshold values [6] and the average of Silhouette coefficient is very high. Experiment 2 [11] consists of every classes that are reusable and metrics have minimum value, so this one is Highly Reusable. Experiment 1 [11] has medium reusability level because almost every classes in this experiment exceeds threshold value [6] and average of Silhouette coefficient is lesser than Experiment 3 and 4 but more than experiment 2 [11].

Hibernate as well as Eclipse JDT systems were analyzed many times [3] to find out the highest average value of Silhouette coefficient. In Eclipse JDT, value of NOC (Number of Children) metric is 0 in 81.2% of the classes and 80.5% classes have values 4, 5 and 6 for DIT (Depth of Inheritance Tree of a Class) metric [4]. In Hibernate, value of NOC (Number of Children) metric is 0 in 82.3% of the classes [6] and 76.5% classes have values of 4 and 5 for DIT (Depth of Inheritance Tree of a Class) metric [5]. Therefore it can be concluded that DIT as well as NOC metric had poor distribution of classes, so these 2 metrics were removed from input vector used in experiments done in Eclipse JDT [7].

All the experiments done on Hibernate [5] are shown below by Table 6.

Figure 7 represents Silhouette Coefficient plot of Experiment 7 of Hibernate which was performed during this study:

The complete analysis of each and every cluster vectors used in experiment 7 of Hibernate is represented in Table 7.

Table 5 Analysis of experiment 7 done on eclipse JDT system [7]

Cluster number	WMC		RFC		CBO		LCOM		No of classes
	Avg.	Ex. %	Avg.	Ex. %	Avg.	Ex. %	Avg.	Ex. %	
1	77.5	95.2	63.5	82.4	23.6	95.2	102.1	45	223
2	14.6	27.5	14.4	2.4	5.8	7.8	14.4	27.8	703
3	1090.5	100	5	100	135.2	100	9054	100	15
4	254.3	100	152. 3	100	52.5	97.3	1211.2	76.7	74
Threshold	22		46		15		22		

Table 6 .

Number of experiment	Cluster size	Silhouette coefficient average
1	[10]	0.4555
2	[9, 10]	0.4458
3	[8]	0.4257
4	[7, 8]	0.5299
5	[6]	0.5301
6	[5, 6]	0.6137
7	[2]	0.8011
8	[2, 3]	0.7097

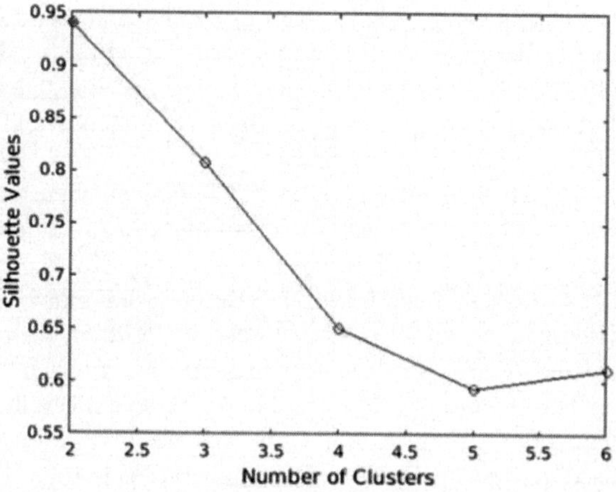

Fig. 7 Silhouette coefficient plot of hibernate [12]

Table 7 Analysis of experiment 7 done on hibernate [10]

Cluster number	WMC		RFC		CBO		LCOM		No of classes
	Avg.	Ex. %	Avg.	Ex. %	Avg.	Ex. %	Avg.	Ex. %	
1	43.7	82.5	72.6	79.4	28.4	99.5	253.8	57	185
2	7.8	5.6	12.7	0.9	7.7	12.7	10.4	11	969
3	463	100	588	100	128.9	100	12958	100	2
4	221.7	100	275.6	100	85.7	100	2825.9	89.6	19
Threshold	22		46		15		22		

From the above table it can be concluded that Cluster number 3 has high values with two vectors for all class level metrics used in this study [11]. Cluster Number 3 as well as Cluster Number 4 have low reusability value. Cluster Number 2 has high reusability value because all metrics have minimum values [10]. Cluster number 1 has medium reusability level as the classes in this cluster exceeds the threshold value and average of silhouette coefficient is lesser than cluster number 3 and cluster number 4 but greater than cluster number 2 [13].

6 Conclusion and Future Work

In this paper, SOM Technique is used to cluster class level metrics. This clustering was build on threshold values of class level metrics. Through this paper we have tried to show that for clustering metrics, SOM technique is applied and even this technique helps to visualize relationship between metrics and reusability level of software.

SOM technique is used due to its capability of clustering data and even its special property of spatial autocorrelation [14]. This technique also helped in finding out patterns in class level metrics and the relation with different reusability level of software. Validity of clustering was done on basis of highest average value of silhouette coefficient after applying different sizes of grid and different epoch's number. At first stage, we applied SOM technique on all class level metrics. We found that DIT and NOC metric [27] influenced results of clustering process due to their poor distribution [11]. The results of experiments revealed that clustering process becomes more meaningful [13].

The future study would investigate SOM technique on more metrics to get much better results than the current results also. We would use same dataset of both systems with elimination of three metrics i.e. NOC,DIT and LCOM due to poor distribution [12].

Acknowledgements I would like to acknowledge Dr Parul Gandhi, Professor and Ph.D. Coordinator in Faculty of Computer Applications, MRIIRS, Faridabad for her continuous support and helpful guidance during the preparation time of this article.

References

1. Abreu B, Carapuca R (1994) Candidate metrics for object-oriented software within a taxonomy framework. J Syst Softw 26:87–96
2. Aggarwal K, Singh Y, Kaur A, Malhotra R (2006) Empirical study of object-oriented metrics. J Object Technol 5(8):110–115
3. Arifa B, Mohamed N, Archana K (2013) Study of software reusability in software components. IJET 5:2455–2460
4. Amin F, Mahmood A, Oxley A (2011) Reusability assessment of open source components for software product lines. Int J New Comput Archit Appl (IJNCAA) 1(3):519–533
5. Alexandre A, Almeida E, Meira S (2006) A software component quality model: a preliminary evaluation. IEEE 4:1–8
6. Barnard J (1998) A new reusability metric for object-oriented software. Software Qual J 7(1):20–25
7. Basili J, B. V.R, Melo W (1996) How reuse influences productivity in object-oriented systems. Commun ACM 30(10):104–114

8. Bhatia P, Mann R (2009) An approach to measure software reusability of OO design. COIT-2009, pp 26–30
9. Budhija N, Singh B, Ahuja S (2013) Detection of reusable components in object oriented programming using quality metrics. Int J Adv Res Comput Sci Softw Eng 3(1)
10. Caldiera G, Basili V (1991) Identtifying and quantiffying reusable software components. IEEE pp 61–70
11. Chahal K, Singh H (2008) A metrics based approach to evaluate design of software components. IEEE 269–272
12. Chidamber R, Kemerer F (1994) A metrics suite for object-oriented design. IEEE Trans Softw Eng 20(6):476–492
13. Deeba K, Amutha B (2016) Classification algorithms of data mining. Indian J Sci Technol 9:2–5
14. Gandhi P, Bhatia K (2010) Reusability metrics for object-oriented system: an alternative approach. IJSE 1(4):63–72
15. Gandhi P, Bhatia PK (2011) Estimation of generic reusability for object-oriented software: an empirical approach. ACM 36(3):1–4
16. Gill SN (2006) Importance of software component characterization for better software reusability. In ACM SIGSOFT Softw Eng Notes 31:1–3
17. Goel B, Bhatia P (2013) Analysis of reusability of object-oriented systems using object-oriented metrics. ACM 38:1–5
18. Hudaib A, Huneiti A (2016) Software reusability classification and prediction using self-organizing map (SOM). Commun Netw 8:179–192
19. Bhagwan J, Oberoi A (2011) Software modules clustering: an effective approach for reusability. J Inform Eng Appl 1(4)
20. Kakkar P, Sharma M, Sandu P (2012) Modeling of reusability of procedure based software components using Naïve Bayes classifier approach. Int J Comput Appl 55:12–17
21. Kayarvizhy N, Kanmani S (2011) Analysis of quality of object-oriented systems using object-oriented metrics. ICECT 2:25–32
22. Maheswari G, Chitra K (2019) Enhancing reusability and measuring performance merits of software component. Int J Innov Technol Explor Eng 8:1577–1583
23. Manhas S, Vashisht R, Sandhu P, Neeru N (2010) Reusability evaluation model for procedure based software systems. Int J Comput Electri Eng 2:1107–1110
24. Mateen A, Kausar S, Sattar A (2017) A software reuse approach and its effect on software quality, an empirical study for the software industry. Int J Managem IT Eng 7(2):266–279
25. Morisio EM, Tully C (2002) Success and failure factors in software reuse. IEEE Trans Softw Eng 28:340–357
26. Nyasente S, Mwangi P, Kimani S (2014) A metrics-based framework for measuring the reusability of object-oriented software components. J Inform Eng Appl 4:71–84
27. Olague H, Etzkorn HL, Gholston S, Quattlebaum S (2007) Empirical validation of three software metrics suites to predict fault-proneness of object-oriented classes developed using highly iterative or agile software development processes. IEEE Trans Software Eng 33:402–419
28. Padhy N, Singh R, Satapathy (2018) Software reusability metrics estimation: algorithms, models and optimization techniques. Elsevier 69:653–66
29. Rotaru P, Dobre M (2014) Reusability metrics for software components. IEEE 21:24–29
30. Sandhu P, Singh H (2006) A reusability evaluation model for OO-based software components. Int J Comput Sci 1(4):259–264
31. Sharma A, Grover PS, Kumar R (2009) Reusability assessment for software components. ACM SIGSOFT Softw Eng Notes 34:1–6
32. Shatnawi A, Abdelhak DS (2013) Mining reusable software components from object-oriented source code of a set of similar software, information reuse and integration. In: 2013 IEEE 14th international conference on IEEE, pp 2–18
33. Singh G (2013) Metrics for measuring the quality of object-oriented software. ACMSIGSOFT 38:1–5

34. Subramanyam R, Krishnan S (2003) Empirical analysis of CK metrics for object-oriented design complexity: implications for software defects software engineering. IEEE Trans 29:297–310
35. Tully C (2018) Success and failure factors in software reuse. IEEE Trans Software Eng 28:340–357

The First Corpus for Detecting Fake News in Hausa Language

Sukairaj Hafiz Imam, Abubakar Ahmad Musa, and Ankur Choudhary

Abstract The rapid spread of deceptive news especially in Africa has become a global issue in last decade. This triggers the attention of the research community to develop efficient and reliable classification approaches for fake news detection so as to prevent its spread in the community. It has been explored that fake news in regional languages spread with a faster pace as compare to English language in local regions. Hausa is a very common language in Nigeria and some West African countries. So, it opens the challenge to detect the fake news in Hausa language. This paper presents the first corpus for the detection of fake news in Hausa. A dataset has been formed by collecting the labeled real and fake news consists of 2600 articles. In order to classify the fake news in Hausa, six different classifiers have been utilized. The performance of these approaches is then evaluated over different metrics and compared to determine the best model on the Hausa language dataset. The experimental results indicate that support vector machine (SVM) outperformed the other classifiers used by achieving 85% accuracy while AdaBoost happens to emerge as the fair model with 70% accuracy.

Keywords Hausa · Corpus · Machine learning (ML) · Natural language processing (NLP) · LSTM · Bi-LSTM · CNN · RNN

1 Introduction

It is not wrong to say that deception is the act of creating and spreading information across the globe with the sole aim of misleading people or achieving a negative goal [1–3]. Nowadays, the emergence of internet and online social media platforms has speed up the pace of news spreading, which has its own pros and cons [4]. The positive side of this spreading is the important news reached to the population very easily but the negative side is fake news. Fake news is a good example of this deception which has been popular these days, as several fake news contents have been initiated and

S. H. Imam (✉) · A. A. Musa · A. Choudhary
Department of Computer Science and Engineering, Sharda University, Greater Noida, India

P. K. Singh et al. (eds.), *Emerging Technologies for Computing, Communication and Smart Cities*, Lecture Notes in Electrical Engineering 875,
https://doi.org/10.1007/978-981-19-0284-0_41

propagated, thereby creating a lot chaos and abnormalities within the environment [5, 6]. This is a global problem that requires an immediate attention.

Hausa is adherent of the Afro-asiatic and is the most broadly spoken language within the Chadic branch of that family. Ethnologies estimated that it was spoken as a first language by some 47 million people and as a second language by another 25 million, bringing the total number of Hausa speakers to an estimated 72 million. According to more recent estimations, Hausa would be spoken by 100–150 million people.[1]

In line of this huge population, the spread of unauthentic news is a great threat to the people speaking Hausa language because it is ease to dismay the society.

Taking Nigeria as a case study, where most of the northern parts of its populace are Hausa speakers, fake news contents have been dispersed through social media platforms like Facebook, Twitter. This leads to the occurrence of violence which destroys the lives and properties of thousands of people. For example: A fake news broadcasted by Nigerian Television Authority (NTA) in November 1989, related to the death of the first Governor General and President of Nigeria, Dr. Nnamdi Azikwe.

After listening this news, many newspapers had published it on their front page. It takes much effort and time before Azikwe proved his health status and convinced the world that president was still alive and the wrong news was managed. In 2019 (Thirty years later), rumors spread that President Buhari had died during one of his prolonged nonappearances from Nigeria in 2017 due to medical issues. The imaginary death of President Buhari spread like wildfire on top social media portals like Twitter, Facebook and WhatsApp. Most recently was the fake news on vital issues like: religious matters and ethics, security issues, coronavirus, damaging the reputation of respected personnel and government officials, as such, the negative impact, danger and harmful challenges of fake news written in Hausa language cannot be overemphasized.

Literature revealed that lots of techniques have been employed to identify fake news so as to limit its danger. This ranges from traditional based approaches where individuals try to distinguish between real and fake news, considering some basic features of the news content, and modern artificial intelligence-based approaches. It also comprises of natural language progressing techniques, machine learning-based classifiers, deep learning-based models, recurrent learning-based approaches, and data mining-based techniques [1, 3, 4, 7–20].

These technology-driven solutions proposed in the literature have proven their performance mainly on English language-based content [10–13, 15–21] but the approaches utilized for other language news content are relatively very a few as compare to English language [1, 4, 7–9]. However, currently no effort has been made public on the automatic detection of fake news written in Hausa language despite the large size of its speakers and the danger of misleading its huge populace.

This paper aims to fill this gap by presenting a corpus for detecting fake news written in Hausa language. The contribution proposed by this paper can be summarized as follows:

---

[1] https://en.wikipedia.org/wiki/Hausa_language.

- The paper is solely on Hausa Language.
- The first labeled dataset has been created for detection of fake news in Hausa language. The dataset consists of 2600 news articles comprising of real and fake news selected from key topics like: Business, health, entertainment, sports, politics as well as religion.
- Furthermore, we studied and employed some of the most used machine learning classifiers to examine and compare their performance on this dataset. Hoping that this work will serve as a baseline approach for Hausa-based fake news detection systems.
- The different parts of this paper comprise of Sect. 2—related literature review, Sect. 3—the Hausa corpus and the methodology used, Sect. 4—presentation and discussion of results, and Sect. 5 which concludes the whole process.

2 Related Work

In the past years, many researches have been done for detecting fake news. Supervised and unsupervised machine learning is the most popular algorithms used by the researchers. Furthermore, few researchers used deep learning and semi-supervised algorithms.

Most of the work done for detecting fake news were English-based [1, 10–13, 15–25]. Few researches were carried out using non-English medium. Al-yahya et al. [26] used NN and transformer-based language models for Arabic fake news detection and compared their performance to each other where transformer-based outperform NN. Pratiwi et al. [27] built an automatic method for detecting fake news in Indonesian language where PHP-ML library was utilized to classify between real and fake news content. Mansur and Onan et al. [28] presented a dataset for satirical text classification in Turkish language in which nine different surveys carried out on non-identical text classification. Naïve Bayes, LR, C4.5, and SVM were employed and SVM got the best accuracy of 89.70%.Vogel and Jiang et al. [22] presented the first dataset in German language for fake news detection. SVM and CNN resulted an outstanding performance of 72 and 89%, respectively. Alves et al. [29] carried out a survey on the 2018 Brazilian election record shared over the social media platform. LSTM (Naïve), bidirectional LSTM (Bi-LSTM), and deep Bi-LSTM achieved the highest accuracy. Duran et al. [30] trained different algorithms on lexical features such as bag of words (BOW), part of speech (POS tags), n-grams, and n-grams combination for confirming the authenticity of news content in Spanish language. The outcome revealed that SVM outperformed the rest of the algorithms used. Alzanin and Azmi [31] utilized supervised Gaussian Naïve Bayes (GNB) and semi-supervised expectation–maximization (E–M) to detect fake news in Arabic tweets. The supervised approach excelled the semi-supervised resulting up to 78.6% accuracy. Vicario et al. [32] proposed a scheme that identifies a polarizing news content on social media to forecast the authenticity of news in Italian Facebook dataset with the aid of different classifiers like LR, SVM, KNN, NN. Kim et al. [33] used unified key sentence information in Korean article dataset for distinguishing the genuineness of the news content by using bilateral multi-perspective matching model using five layers. Liu et al. [34] presented a French dataset for detecting sarcasm news using machine learning algorithms in which logistic regression outpaces the rest with 92.17%. Silva et al. [1] presented an extensive survey on how to automatically classify different news contents as fake or real in Portuguese. SVM and logistic regression performed better in the different experiments conducted. Hossain et al. [4] created the first dataset in Bangla and compare between human and machine performance in classifying fake and real news content. Different models like SVM, LR, RF, LSTM, CNN, and Bidirectional Encoder Representations from Transformers (BERT) were employed. Amjad [7] provided the first dataset in Urdu language for confirming news authenticity. SVM, LR, RF, multinomial Naïve Bayes, Bernoulli Naïve Bayes, DT, and AdaBoost were used. However, AdaBoost was resulted to be the most accurate compared to others.

3 The First Corpus for Detecting Fake News in Hausa Language

This section provides an overview of the data collection process as well as the compilation statistics. We collected thousands of real news articles from different reliable sources, a special thanks to Abubakar Rimi Television, Kano, for providing us the news scripts contains more than one thousand authentic news. This dataset compiled news content from six different domain: (i) Business domain, (ii) health domain, (iii) entertainment domain, (iv) sports domain, (v) politics as well as (vi) religious. The next subsections discuss dataset details such as: news sources, news domains, etc.

3.1 *Authentic and Fake News Collection*

Different authentic and reputed news channels, radio stations and newspaper sites have been chosen for collecting the data and creating the news dataset. The major points keep in to consideration in handling and collecting news data is as follows:

- The dataset was prepared by collection and annotation of data manually.

The news was considered as authentic if any of the condition is satisfied:

- If the news was published in trustworthy newspaper and news agency.
- If the same news was published by different newspapers with evidence (Fig. 1).

Table 1 shows the different chosen sources for authentic news.
Table 2 described the categories of news.

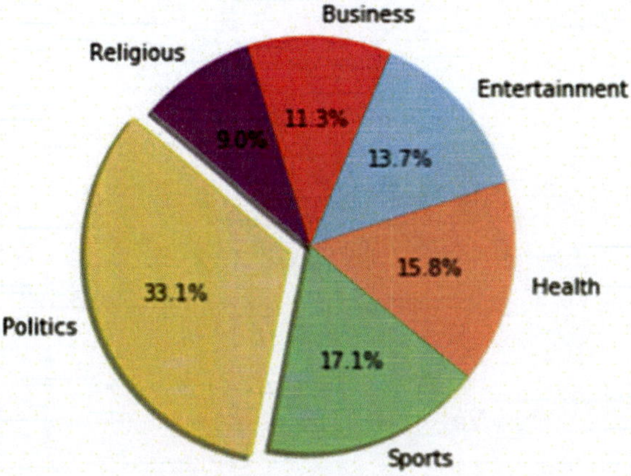

Fig. 1 Frequency of news by category in Hausa corpus

Table 1 Sources of authentic news

S. no	Source	Origin	Url
1	Abubakar Rimi television, Kano	Nigeria	www.artvkano.org
2	Freedom Radio, Kano	Nigeria	freedomradionig.com
3	Aminiya	Nigeria	aminiya.dailytrust.ng.com
4	Guarantee Radio	Nigeria	guaranteeradio.com
5	Daily Nigerian	Nigeria	dailynigerian.com
6	Legit Hausa	Nigeria	hausa.legit.ng
8	Sahara Hausa	Nigeria	sahara.hausang.com
9	Premium Times Hausa	Nigeria	hausa.premiumtimesng.com
10	Rariya	Nigeria	rariyajarida.com
11	Leadership Hausa	Nigeria	hausa.leadership.ng
12	Radio Kano	Nigeria	radiokano.com
13	Rahama radio	Nigeria	rahamradio.com
14	DW Hausa	Germany	www.dwhausa.com
15	RFI Hausa	France	www.rfi.fr
16	CRI Hausa	China	hausa.cri.cn
17	BBC Hausa	England	www.bbchausa.com
18	VOA Hausa	America	www.voahausa.com
19	Mikiya	Nigeria	mikiyahausa.com

Table 2 News categories

S. no	Categories	Real news count	Fake news count
1	Business	125	100
2	Entertainment	151	200
3	Health	175	260
4	Politics	366	500
5	Religious	100	94
6	Sport	189	340
7	Totals	1106	1494

3.2 Pipeline: Text Data

An overview of the pipeline composed of five phases: News collection, data preprocessing, feature extraction, models selected and verification of effectiveness of algorithms Selected. This process described how news is collected until the detection of final result (Fig. 2).

3.2.1 Data Preprocessing

In preprocessing, different approaches were used to remove undesired characteristics from the collected dataset.

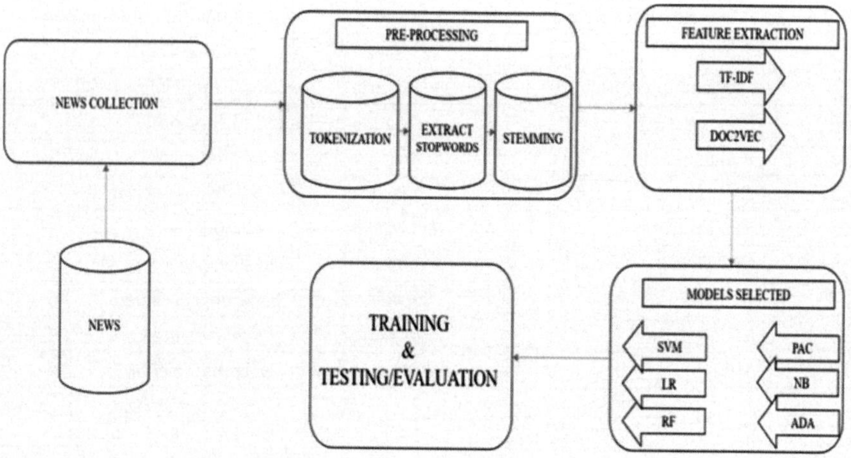

Fig. 2 Text pipeline

3.2.2 Tokenization

It is the initial stage of NLP. The main advantage of this phase is to translate a word into phrases, symbols, or other meaningful meaning known as tokens [35].

3.2.3 Extraction of Stop Words

They are the English words which are inessential and ignoring it has no any impact to a sentence meaning? NOTE: only 22 languages have stop words in Python[2] and Hausa is not among them. Stop words were used only because few words in Hausa resembles that of English language.

3.2.4 Stemming

It is the method of eliminating suffixes to get to the origin form of the word. It refers to heuristics that split suffixes [36].

3.2.5 Tf-Idf

It is an advancement from IDF which was proposed by Spärck Jones [37]. The term frequency (TF) was used in calculating the number of times a word appears in the dataset divided by the total number of words in the dataset. The demonstration of the equation represented by

$$tfi, j = \frac{ni, j}{\sum kni, j} \tag{1}$$

TF-IDF weight of term i in document j in a corpus of N documents is calculated as:

$$\text{Weight}\, i,\ j = tfi,\ j^{*}\log(\frac{N}{dfi}) \tag{2}$$

(where (*tfij*) is a number of times term *i* appear in document *j* and (*dfi*) is a number of documents containing term *i*). The main idea behind the TF-IDF is from the theory of language modeling where the terms in given documents are categorized into words with relevancy and those without relevancy for a given document. Furthermore, TF-IDF can be used to measure the effectiveness of a term in in the document collection [38].

[2] https://pypi.org/project/stop-words/#available-languages.

3.2.6 Doc2vec

It is an extended version of word2vec introduced by Le and Mikolov. It is applied to a document as whole instead of individual words. Doc2Vec works on the logic that the denotation of a word is solely depends on the document that it is present in. The vectors computed by Doc2Vec can be utilized for finding resemblances between documents [39–41]. In some sources, Doc2Vec referred to paragraph2vec and its modified version of word2vec algorithm [42].

3.2.7 Performance Evaluation Metrics

Evaluation metrics are constantly used in selecting relevant algorithms, and it gives us chance to test the effectiveness of the algorithms. It helps in comparing/evaluating the performances of different types of algorithms for detecting fake news. The most common metrics are true positive (TP), false positive (FP), true negative (TN), false negative (FN), precision (Pr), accuracy (Acc) and recall (Re). The performance was evaluated using the following criteria:

$$\text{Acc} = \frac{TP + TN}{TP + TN + FP + FN} \tag{3}$$

$$\text{Pr} = \frac{TP}{TP + FP} \tag{4}$$

$$\text{Re} = \frac{TP}{TP + FN} \tag{5}$$

$$\text{F\_score} = \frac{2^{*} Re^{*} Pr}{Re + Pr} \tag{6}$$

The selection of this metrics is in line with the work in [3, 15].

3.3 Algorithms

This contains the detail explanation of the proposed algorithm used for detecting fake news in Hausa language. Initially, we started by preprocessing our dataset by removing redundancy and characters such as numbers. The application of feature extraction helps in reducing the dimension of feature state. The last but not the least was choosing the models that fits our dataset and provide good performance or high accuracy. Six different algorithms were selected due to their high predictive performance and broad use in related research papers. These include: NB [7, 9, 11, 12, 20], SVM [1, 4, 7, 9, 12, 20], AdaBoost [7, 9, 12], RF [1, 4, 7, 9, 20] and LR [7, 9, 12], PAC [43].

3.3.1 Logistic Regression (LR)

It is used in categorizing text base on an extensive feature set, in which the paired output is (fake (0) /real (1)). It is an algorithm which is used in classifying data base on binary or numerous sections. Hyperparameters tuning is applied for generating an excellent and an accurate outcome on the datasets. Mathematically, the LR can be defined as follow:

$$h_{\theta}(\mathrm{X}) = \frac{1}{1 + e^{-(\beta_o + \beta_1 X)}} \tag{7}$$

3.3.2 Support Vector Machine (SVM)

This also used for paired categorization task and is accessible via several kernels tasks. The objective causes the use of SVM algorithm which is to evaluate a marginal boarder on the origin of attribute set to classified data points. The length of the decision boundary differs depending on the quantity of attributes. Different chances of the decision boundary occurring in N-dimensional planetary. The charge is detecting that the planetary separating the two types of data points has a strong margin. A mathematical illustration of the SVM can be defined as:

$$\mathrm{J}(\theta) = \frac{1}{2}\sum_{\mathrm{j}=1}^{\mathrm{n}} \theta_{\mathrm{j}}^{2} \tag{8}$$

Such that

$$\theta^{t} x^{(i)} \geq 1, \quad y^{(i)} = 1 \tag{9}$$

$$\theta^{t} x^{(i)} \leq -1, \quad y^{(i)} = 0 \tag{10}$$

3.3.3 Random Forest (RF)

The model was introduced as improvement technique of decision trees. It comprises huge amount of decision trees, dedicated for predicting the outcome of the category, and concluding estimate is grounded on the category with the most votes. The rate of mistakes/errors in RF is quiet less due to little correlation among trees. There are different algorithms that can determine the division in the decision tree based on regression or classification problems. For categorization problems, this paper invites the Gini index as a cost function to guess the division of the dataset. The mathematical equation is:

$$G_{ind} = 1 - \sum_{i+1}^{c} (p_i)^2 \tag{11}$$

3.3.4 K-Nearest Neighbors (KNN)

It is a learning mode in which an experimental variable does not need to guess the result of the precise data. The KNN algorithm evaluates the distance between a new data point and its closest neighbor; the advantage of K is to calculate the preponderance of votes from its neighbors, if the value of K is 1, and amount of data point is assigned to the closest class. The mathematical equation to evaluate the interval between 2 dots can be considered as:

$$\text{Euclidean distance} = \sqrt{\sum_{i=1}^{k} (x_i - y_i)^2} \tag{12}$$

3.3.5 Passive-Aggressive Classifier (PAC)

This classifier is used for extensive learning. This classifier became aggressive any erroneousness or unsuitable outcome but remain reflexive if exact result is achieved while categorizing.

The general steps for passive-aggressive classifier are as follows:

Step 1: Initialization of weights $wt = [0, 0, 0, \ldots]$.

Step 2: Select the document for classification $\mathrm{dc} = (\mathrm{dc1}, \mathrm{dc2} \ldots, \mathrm{dcn})$.

Step 3: Perform TF-IDF on document and then normalize dc.

Step 4: Apply conditions if $\mathrm{dc}^{\mathrm{T}}\mathrm{wt} > = 0, +1$ and -1. Based on that fix the output as $+1$ or -1.

Step 5: Compute Loss.

4 Comparison and Discussion of Result

4.1 Machine Learning Approach

The dataset was splitted into 80% for training set and 20% for testing on the selected models using scikit-learn [44]. The implementation was carried out using DOC2VEC and TF-IDF in which two different word embedding techniques utilized to map the news content and compare the performance of the selected models on each technique. The concluded result is summarize below (Figs. 3 and 4):

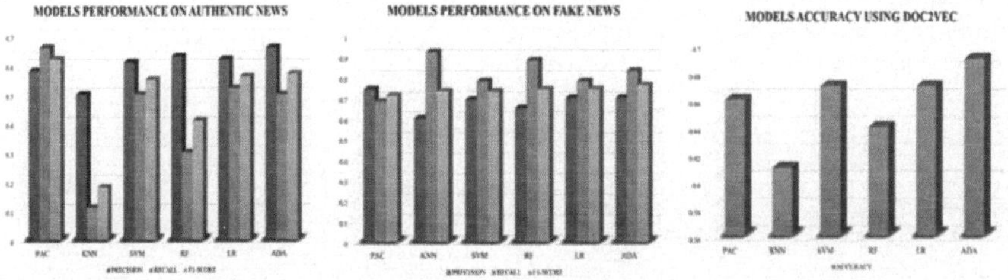

Fig. 3 Model performance using DOC2VEC

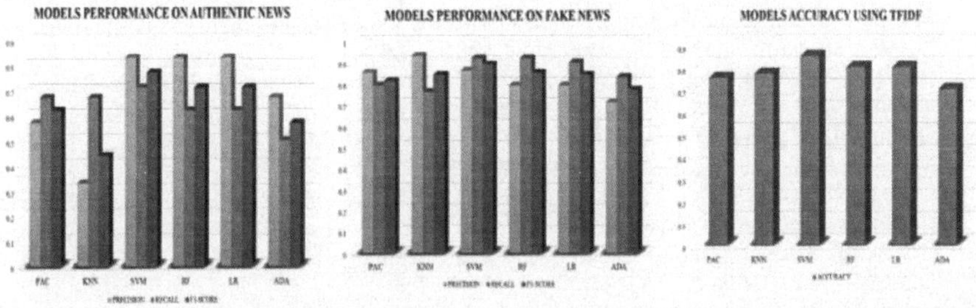

Fig. 4 Model performance using TF-IDF

It is clearly shown from the above classification of results that models trained and tested with TF-IDF outperform the models trained and tested using Doc2Vec. The results unveiled that some classifiers did a magnificent work in differentiating authentic news content. However, some classifiers showed fair performance. It was observed that support vector machine (SVM) indicated an impressive performance in detecting fake news in Hausa language with 85% using TF-IDF and AdaBoost 69% using Doc2Vec. Classifiers such as AdaBoost to be the fair classifier using TF-IDF and multinomial Naïve Bayes using Doc2Vec.

5 Conclusion

The first labeled dataset is presented in Hausa language; the real news contents were generated from the internet and other legitimate sources, while the fake news contents were drafted by the expert journalist of ARTV Kano. The dataset was annotated manually cleaned and processed and used for this work. Six different classifiers were employed for this experiment; however, it was observed that SVM outperformed all the classifiers used to achieve 85% accuracy. Due to several factors, we could only annotate 2600 news contents; we look forward to the improvement of this dataset in the future, also the viability of incorporating other languages like the main three Nigerian languages (Hausa, Yoruba & Igbo) and other African languages like Swahili.

References

1. Silva RM, Santos RLS, Almeida TA, Pardo TAS (2020) Towards automatically filtering fake news in Portuguese. Expert Syst Appl 146:113199. https://doi.org/10.1016/j.eswa.2020.113199
2. De keersmaecker J, Roets A (2017) 'Fake news': Incorrect, but hard to correct. The role of cognitive ability on the impact of false information on social impressions. Intelligence. 65:107–110. https://doi.org/10.1016/j.intell.2017.10.005.
3. Zhang X, Ghorbani AA (2020) An overview of online fake news: characterization, detection, and discussion. Inf Process Manag 57:102025. https://doi.org/10.1016/j.ipm.2019.03.004
4. Hossain MZ, Rahman MA, Islam MS, Kar S (2020) BanFakeNews: a dataset for detecting fake news in Bangla
5. Kim J, Tabibian B, Oh A, Schölkopf B, Gomez-Rodriguez M (2018) Leveraging the crowd to detect and reduce the spread of fake news and misinformation. WSDM 2018—proceedings 11th ACM international confernce web search data Mining 2018 Feburary, pp 324–332. https://doi.org/10.1145/3159652.3159734.
6. Sadiku M, Eze T, Musa S (2018) Fake news and misinformation. Int J Adv Sci Res Eng 4:187–190. https://doi.org/10.31695/ijasre.2018.32728.
7. Amjad M, Sidorov G, Zhila A, Gómez-Adorno H, Voronkov I, Gelbukh A (2020) Bend the truth: benchmark dataset for fake news detection in urdu language and its evaluation. J Intell Fuzzy Syst 1–13. https://doi.org/10.3233/jifs-179905
8. Abonizio HQ, de Morais JI, Tavares GM, Junior SB (2020) Language-independent fake news detection: english, Portuguese, and Spanish mutual features. Futur Internet 12:1–18. https://doi.org/10.3390/FI12050087
9. Faustini PHA, Covões TF (2020) Fake news detection in multiple platforms and languages. Expert Syst Appl 158:113503.https://doi.org/10.1016/j.eswa.2020.113503
10. Ozbay FA, Alatas B (2020) Fake news detection within online social media using supervised artificial intelligence algorithms. Phys A Stat Mech its Appl 540:123174.https://doi.org/10.1016/j.physa.2019.123174
11. Mahabub A (2020) A robust technique of fake news detection using ensemble voting classifier and comparison with other classifiers. SN Appl Sci 2:1–9. https://doi.org/10.1007/s42452-020-2326-y
12. Singh VK, Ghosh I, Sonagara D (2020) Detecting fake news stories via multimodal analysis. J Assoc Inf Sci Technol 1–15. https://doi.org/10.1002/asi.24359
13. Kumar S, Asthana R, Upadhyay S, Upreti N, Akbar M (2020) Fake news detection using deep learning models: a novel approach. Trans Emerg Telecommun Technol 31:1–23. https://doi.org/10.1002/ett.3767
14. Kong SH, Tan LM, Gan KH, Samsudin NH (2020) Fake news detection using deep learning. In: ISCAIE 2020—IEEE 10th symposium computing application industry electronics, pp 102–107. https://doi.org/10.1109/ISCAIE47305.2020.9108841
15. Kaliyar RK, Goswami A, Narang P, Sinha S (2020) FNDNet—a deep convolutional neural network for fake news detection. Cogn Syst Res 61:32–44. https://doi.org/10.1016/j.cogsys.2019.12.005
16. Kaliyar RK, Goswami A, Narang P (2020) DeepFakE: improving fake news detection using tensor decomposition-based deep neural network. J Supercomput.https://doi.org/10.1007/s11227-020-03294-y
17. Huang YF, Chen PH (2020) Fake news detection using an ensemble learning model based on self-adaptive harmony search algorithms. Expert Syst Appl 159:113584.https://doi.org/10.1016/j.eswa.2020.113584

18. Antoun W, Baly F, Achour R, Hussein A, Hajj H (2020) State of the art models for fake news detection tasks. In: 2020 IEEE international conference informatics, IoT, enabling technology ICIoT 2020. pp 519–524. https://doi.org/10.1109/ICIoT48696.2020.9089487
19. Wang Y, Yang W, Ma F, Xu J, Zhong B, Deng Q, Gao J (2020) Weak supervision for fake news detection via reinforcement learning. Proc AAAI Conf Artif Intell 34:516–523. https://doi.org/10.1609/aaai.v34i01.5389
20. Meneses Silva CV, Silva Fontes R, Colaço Júnior M (2020) Intelligent fake news detection: a systematic mapping. J Appl Secur Res 0:1–22.https://doi.org/10.1080/19361610.2020.1761224
21. Zhang J, Dong B, Yu PS (2020) FakeDetector: effective fake news detection with deep diffusive neural network. In: Proceedings–international conference data engineering 2020-April, pp 1826–1829. https://doi.org/10.1109/ICDE48307.2020.00180
22. Vogel I, Jiang P (2019) Fake news detection with the new German dataset "GermanFakeNC". https://doi.org/10.1007/978-3-030-30760-8
23. Song C, Ning N, Zhang Y, Wu B (2021) A multimodal fake news detection model based on crossmodal attention residual and multichannel convolutional neural networks. Inf Process Manag 58:102437. https://doi.org/10.1016/j.ipm.2020.102437
24. Scott W, Bansal R, Kaushik A, Sengupta S (2021) Cross-SEAN : a cross-stitch semi-supervised neural attention model for COVID-19 fake news detection. Appl Soft Comput J 107:107393. https://doi.org/10.1016/j.asoc.2021.107393
25. Jiang TAO, Li JP (2021) A novel stacking approach for accurate detection of fake news. 9. https://doi.org/10.1109/ACCESS.2021.3056079
26. Al-yahya M, Al-khalifa H, Al-baity H, Alsaeed D, Essam A (2021) Arabic fake news detection : comparative study of neural networks and transformer-based approaches
27. Pratiwi IYR, Asmara RA, Rahutomo F (2018) Study of hoax news detection using naïve bayes classifier in Indonesian language. In: Proceedings 11th international conference information communications technology systems ICTS 2017, 2018-January, pp 73–78. https://doi.org/10.1109/ICTS.2017.8265649
28. Mansur TA, Onan A (2019) Satire detection in turkish news articles: a machine learning approach. Mansur Alp Toçoğlu1 and Aytuğ Onan2(&) 1 Faculty of Technology, Department of Software Engineering, Manisa Celal Bayar University, 45400 Manisa, Turkey mansur.tocoglu@cbu.edu.tr 2 Faculty of Engineering and Architecture, Department of Computer Engineering
29. Jairo Alves L, Leila W, Paulo Q, Carlos Cardoso E, Luan C (2019) Brazilian presidential elections in the era of misinformation: a machine learning approach to analyse fake news. Fluminense Federal University, Rio de Janeiro, Brazil jairo.luciano@gmail.com, {leila_weitzel,carloseac, luanpereiracunha}@id.uff.br 2 Universidade de Évora, Évora 17, Portugal pq@uevora.pt. https://doi.org/10.1007/978-3-030-33904-3
30. Posadas-Durán JP, Gomez-Adorno H, Sidorov G, Escobar JJM (2019) Detection of fake news in a new corpus for the Spanish language. J Intell Fuzzy Syst 36:4868–4876. https://doi.org/10.3233/JIFS-179034
31. Alzanin SM, Azmi AM (2019) Rumor detection in Arabic tweets using semi-supervised and unsupervised expectation–maximization. Knowledge-Based Syst 185:104945. https://doi.org/10.1016/j.knosys.2019.104945
32. Vicario MD, Quattrociocchi W, Scala A, Zollo F (2019) Polarization and fake news. ACM Trans Web 13:1–22. https://doi.org/10.1145/3316809
33. Kim N, Seo D, Jeong CS (2019) FAMOUS: fake news detection model based on unified key sentence information. In: Proceedings IEEE International conference software engineering services sciences ICSESS. 2018 Nov, pp 617–620. https://doi.org/10.1109/ICSESS.2018.8663864
34. Liu Z, Shabani S, Balet NG, Sokhn M (2019) Detection of satiric news on social media: analysis of the phenomenon with a French dataset. In: Proceedings—international conference computing communications networks, ICCCN. 2019-July, 1–6. https://doi.org/10.1109/ICCCN.2019.8847041

35. Verma T, Renu R, Gaur D (2014) Tokenization and filtering process in RapidMiner. Int J Appl Inf Syst 7:16–18. https://doi.org/10.5120/ijais14-451139
36. Balakrishnan V, Ethel L-Y (2014) Stemming and lemmatization: a comparison of retrieval performances. Lect Notes Softw Eng 2:262–267. https://doi.org/10.7763/lnse.2014.v2.134
37. Spärck Jones K (2004) A statistical interpretation of term specificity and its application in retrieval. J Doc 60:493–502. https://doi.org/10.1108/00220410410560573
38. Zhang W, Yoshida T, Tang X (2011) A comparative study of TF*IDF, LSI and multi-words for text classification. Expert Syst Appl 38:2758–2765. https://doi.org/10.1016/j.eswa.2010.08.066
39. Bilgin M, Şentürk İF (2017) Sentiment analysis on Twitter data with semi-supervised Doc2Vec. In: 2nd international conference computing science engineering UBMK 2017. pp 661–666. https://doi.org/10.1109/UBMK.2017.8093492
40. Shuai Q, Huang Y, Jin L, Pang L (2018) Sentiment analysis on chinese hotel reviews with Doc2Vec and classifiers. In: Proceedings 2018 IEEE 3rd advanced information technology electronics automation control conference IAEAC 2018. pp 1171–1174. https://doi.org/10.1109/IAEAC.2018.8577581
41. Chen Q, Sokolova M (2018) Word2Vec and Doc2Vec in unsupervised sentiment analysis of clinical discharge summaries
42. Huang G, Bhayani (2009) Twitter sentiment analysis (final project results). J Inform
43. Saloni G, Priyanka M (2019) Fake news detection using passive-aggressive classifier
44. Buitinck L, Louppe G, Blondel M, Pedregosa F, Mueller A, Grisel O, Niculae V, Prettenhofer P, Gramfort A, Grobler J, Layton R, Vanderplas J, Joly A, Holt B, Varoquaux G (2013) API design for machine learning software: experiences from the scikit-learn project. pp 1–15

A Tour Towards the Security Issues of Mobile Cloud Computing: A Survey

Vaibhav Kalra, Shubhankar Rahi, Neerav pawar, Poonam Tanwar, and Ms Shweta Sharma

Abstract The rapid growth of mobile devices has seriously challenged mobile cloud computing. It has become one of the biggest issues in the IT world as growth of these smart phone devices and tablets has increased web-based malicious activities around the world. Such as there is a lot of data and security issues like risk of data theft and violation of privacy rights. Securing data is more hard and critical in mobile cloud computing. The purpose of this study is to present major security and privacy challenges in this field which has been of great interest to academics and the research community.

Keywords Mobile cloud computing · Mobile computing · Cloud computing · Research directions

1 Introduction

Lastly, organized processing and wanted applications have encouraged unstable growth of use system, such as distributed computing, programming as an assistance, interpersonal organizations, online stores, and so on. As an important application model in the Internet age, cloud computing has been an important research topic for scientific and industrial communities since 2007. Generally, cloud computing is defined as a wide range of malware provided by an online system program. These bundled programs may have a set of economical servers or PCs, which organize various computing resources according to a specific management plan and provide customers with secure, consistent, quick, transparent and simple services such as data warehouse, computer access and use. According to the topmost technologies provided by Gartner [1–3], ubiquity and mobility are two important characteristics of the next-generation network that offers various personalized network services through multiple methods of communication and network access. The primary technologies of cloud computing include computing, resources and specific applications

V. Kalra · S. Rahi · N. pawar · P. Tanwar (✉) · M. S. Sharma
Manav Rachna International Institute of Research and Studies Faridabad, Faridabad, Haryana, India

P. K. Singh et al. (eds.), *Emerging Technologies for Computing, Communication and Smart Cities*, Lecture Notes in Electrical Engineering 875,
https://doi.org/10.1007/978-981-19-0284-0_42

such as things that can be sold, such as water, gas or electricity to users. Therefore, the combination of the ubiquitous network and cloud computing creates a new computing model, namely mobile cloud computing. Although mobile cloud computing contributes a lot to our daily life, it will also bring many problems. In short, at the heart of these challenges and difficulties is the question of how to combine the two technologies to make them smaller: on the one hand, to ensure that mobile devices take full advantage of the benefits of cloud computing to improve and expand their operations [4].

2 Literature View

In this paper, we have given extensive data in regards to the cloud security issues. The cloud board secuity issues and cloud access technique security issues are likewise featured. We additionally proposed the portable distributed computing application and the difficulties which have existing arrangements and ways to deal with the beat these difficulties. This paper gives us the key open exploration issues related with the portable utilization of the distributed computing which assists us with for the most part perusing and have an outline of the MCC, their issues, existing arrangements, and the distinctive approaches. The creator inform the point that branch of software engineering proposes and carried out that at whatever point another client verification component of versatile distributed computing utilizing the unique finger impression acknowledgment framework in which a global diary of computer science and designing innovation present us the idea of MCC and told the new wording Mobile Cloud and other different MCC applications, significant concern and security worries with a portion of the anticipations. We discussed the engineering of the MCC with these various administrations required by their customer and the worker in MCC. We additionally discussed the major distinctive security issues which emerges about how much safe the portable distributed computing climate is left [3, 5–21].

3 Security Issue

Security and protection are significant issues in versatile distributed computing applications and still face some tremendous difficulties. Client protection and the honesty of information or applications are one of the fundamental issues in accomplishing versatile distributed computing. It is a mix of distributed computing and portable organizations. For this, security issues are isolated in to two classes: cloud security and portable organization client security. A mobile network user's security and protection are significant issues in portable distributed computing applications and still face some colossal difficulties. Client protection and the honesty of information or applications are one of the fundamental issues in accomplishing versatile distributed computing. It is a mix of distributed computing and portable organizations. For this, security issues are isolated into two classes: cloud security and portable organization clients security [1–25].

3.1 Security for Mobile Applications

The requesting way to deal with identifying security issues is low to present and executing security programming and antivirus programs on portable devices. All versatile contraptions are fixed with upkeep and force limits. These dangers can be additionally undermined with PCs that are viable with ensuring the contraption. A few techniques are intended to look for security apparatuses in the cloud. Prior to utilizing a particular portable application on versatile, it might encounter some degree of hazard evaluation. To start with, on the off chance that it is checked and not noxious, the record will be shipped off the customer's portable contraptions. Versatile contraptions are a gentle exercise, for instance, running on a cloud security worker rather than plague programming or hazard ID [3, 5–11].

4 Security

It creates situations for protection issues by determining the current area and giving private data, including significant data from the client the utilization of area-based administrations (LBS) given by global positioning system (GPS) gadgets. Different dangers to revealing individual information can be limited by choosing and examining the requirements of the endeavor, and just the predetermined administrations ought to be acquired and carried out locally through enemy of infection programming or danger location programs. Programs should move to the cloud networks.

4.1 Getting Information on the Cloud

Individuals and associations can impact the cloud to store a lot of information or applications. In any case, care should be taken to guarantee the respectability, confirmation, and computerized privileges of the information or application during handling.

5 Integrity

The whole portable cloud client should guarantee the respectability of their data put away on the cloud organization. All entrance should be validated and confirmed.

It is proposed to give various ways to deal with keep up the respectability of the information put away in the cloud. For instance, all the data put away by every individual or association in the cloud network is labeled or empowered with them, in which they just move, refresh, or erase information.

6 Authentication

Unlike normal electronic applications, distributed computing applications offer numerous extraordinary types of assistance to clients. Subsequently, the validation cycle in distributed computing is totally different contrasted with an ordinary web application. Getting to various cloud-based applications utilizing a username and secret word is a solitary symbol. For instance, all the data put away by every individual or association in the cloud network is labeled or empowered with them, in which they just move, refresh, or erase information.

7 Digital Rights Management

Digital rights the board (DRM) is a deliberate way to deal with copyright insurance for advanced media. DRM is typically carried out by inserting code that forestalls duplicating, indicating the time span for which substance can be gotten to, or restricting the quantity of gadgets that can be introduced on the media.

Mobile Cloud Computer: -

Mobile computing has become a popular word and is increasingly popular in today's computer field. Advances in mobile computers, network connectivity, and security technologies such as mobile phones, smart phones, PDAs, GPS navigation, and laptops have shown impressive growth. Alternatively, with wireless technologies such as WiMAX, connectivity networks, and WiFi, users can display the Internet successfully but are not limited to wires as before. Therefore, those mobile devices are accepted as their first choice for work and have fun in their daily lives.

8 Characteristic

Characteristics are as follows: -

8.1 *Travel*

Mobile nodes on mobile computer networks that communicate with others, including random nodes in the cable network via mobile help (MS) channels while in operation.

8.2 *Different Types of Network Conditions*

The network functions commonly used for mobile nodes are not specified; such networks may be high-bandwidth wired network networks or low-bandwidth broadband (WWAN) networks, or they may be phased out.

8.3 *Frequent Termination*

Mobile nodes do not always maintain connections, such as wireless communication, charging network status, but wireless network remains idle or disconnected due to low battery power.

8.4 *Disk-Symmetric Network Connection*

Servers and access points and other MSS provide powerful access, but such capabilities weaken call centers. Therefore, there are connection bandwidths with high variability between downlink and uplink.

8.5 *Low Dependability*

Network system terminals that signify mobile phones, networks, database platforms, and security application development improvements should be considered as signs of disruption and recurrence may occur.

9 Characteristics

Cloud square measurement options are as follows:

9.1 Performance

'Cloud' is often thought of as a resource pool wherever all underlying platforms are used. Complete users access the desired resources through the browser and get information from cloud computing providers while not maintaining their information centers. In addition, some virtual machines (VMs) square measure are usually installed on the server too much to increase the utility capacity; and VMs support upload migration in the event of server overload.

9.2 Reliability and Extension

Cloud computing provides a secure mode for storing user information and users are not concerned about issues such as software system changes, leak repairs, virus attacks, and loss of information. If a failure occurs on a server or VM, cloud computing systems transfer and store that information to different machines and then remove those nodes that fail on the system systems to create the whole system operating normally. Meanwhile, the cloud is usually expanded from horizontal and overhead to a very large network, making a variety of applications from thousands of nodes and strangers.

9.3 Independence

The associate in nursing cloud system is self-contained, repairs itself and assigns hardware, software and storage resources to the most demanding consumers, so administrators are clear to eliminate users.

(3) Provocation:

First, cloud computing requires the associate in nursing mechanism to be developed to produce a secure and high-powered service because of the variety of complaints used and the infrastructure used in computer use. In addition, due to the resource-intensive resource information centers, a resource management plan and measurement strategies are required to avoid wasting energy. Lastly, a small, straightforward, and easy-to-use square integration is very important for service providers in the cloud computing, so the exact same standard as normal | equal level} required dry.

10 Architecture of Mobile Cloud Computing

Versatile distributed computing, in its easiest way, alludes to the foundation that happens outside the cell phone, both information stockpiling and information preparing. Portable cloud applications remove processing force and information stockpiling from cell phones and the cloud, bringing applications and MCs not exclusively to cell phone clients yet to a more extensive scope of versatile supporters. Cell phones assume a significant part in our day-by-day lives as it has gotten very accommodating in a compelling specialized instrument paying little heed to time and place. The size of cell phones is restricted to a base size; there are limits on handling abilities, battery life, information stockpiling, and show limit. Asset concentrated applications are huge information that require maximum usage of central processing unit (CPU), random access memory (RAM) to run these applications. These require a more drawn out battery life, which is absent in current cell phones, and incredible applications with high information use, for example, games and discourse acknowledgment are genuine instances of this. To beat these impediments, portable distributed computing has been presented with versatile administrations, with the advancement of (mobile cloud computing (MCC) and the expanding number of cell phone applications in cloud moving. This lessens the specialized equipment and processing force of cell phones. A very much organized construction proposes an absence of coordination among IT and business objectives. Many MCC structures have been created from various stages, and at present, four kinds of designs are being utilized for the current MCC. This is the overall construction of the MCC [25] (Fig. 1).

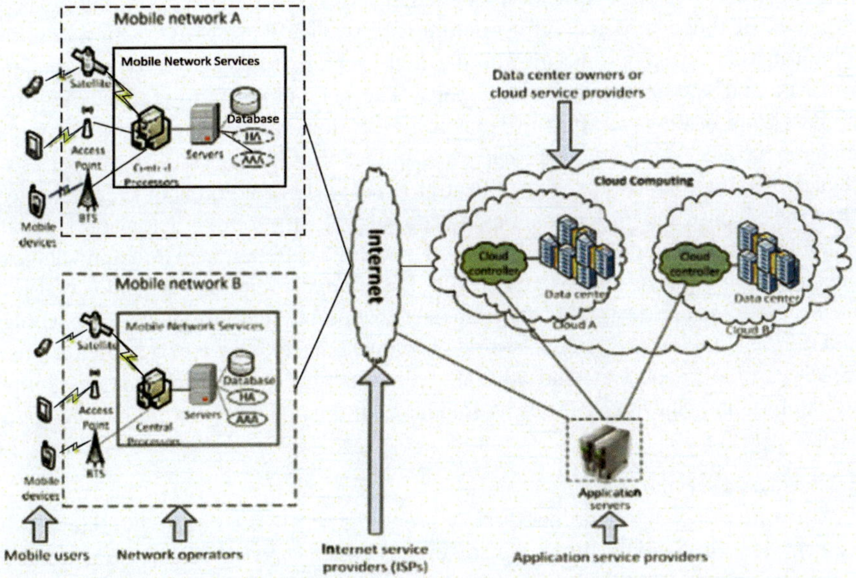

Fig. 1 Mobile cloud computing (MCC) architecture of [25]

This structure consists of mobile devices connected to the mobile network via basic channels (base transceiver channels, access points, or satellites) that connect and control the active interaction between the network and the mobile device. Medium user data requests and information are transmitted via a central processor from a server connected to mobile network services. Mobile users receive authentication, authentication, and accounting (AAA) services from mobile network providers based on user information on the home agent and data server. User requests are sent to the cloud via the Internet. These cloud applications are managed by cloud administrators and provide services at their request. These services are available through utility computing, virtualization and service-oriented architecture. The current structures below apply to all cloud-based development models.

Cloud services are usually categorized based on layer configuration. Within the higher levels of this paradigm, infrastructure as a service (IaaS), platform as a service (PaaS), and software as a service (SaaS) are set [1–3, 5, 12–17].

Data center layer

This layer provides hardware comfort and cloud infrastructure. In the data center layer, many servers connected to networks are very fast to provide services to customers. Typically, data centers with high power supply and low risk of disaster are built in densely populated areas. Data center layer (DCL) plays an important role in the data center because it connects all data center resources.

Infrastructure as a service (IaaS).

IaaS as a service is a distributed computing administration where organizations lease or rent workers for registering and capacity in the cloud. Clients can run any working framework or application on a leased worker with no support and upkeep expenses of those workers. Infrastructure as a service is a provisioning model in which an association rethinks gear used to help tasks, including capacity, equipment, workers, and systems administration parts. The IaaS server farm is situated at the highest point of the layer. It gives stockpiling, workers, organizing parts and equipment on a 'installment premise' that you use. IaaS rapidly increases and down with request, paying you just for your utilization. This will assist you with keeping away from the expense and intricacy of buying and keeping up your own actual worker and other datacenter foundation. Every asset is given as a different help unit and you just need to employ one as long as you need it. Distributed computing specialist co-ops, for example, Azure deals with the framework when buying, introducing, designing and keeping up your own product working framework, middleware and applications. Instances of Amazon EC2 and S3 IaaS. Infrastructure as a service gives essential stockpiling and figuring abilities as standard administrations over an organization [4, 18–25] (Fig. 2).

Software as a service (SaaS).

Software as a service (or SaaS): How programs are distributed on the Internet—as a service. SaaS applications are sometimes called web-based software, wanted software, or hosted software. No matter the name, SAS applications run on a SAS provider server.

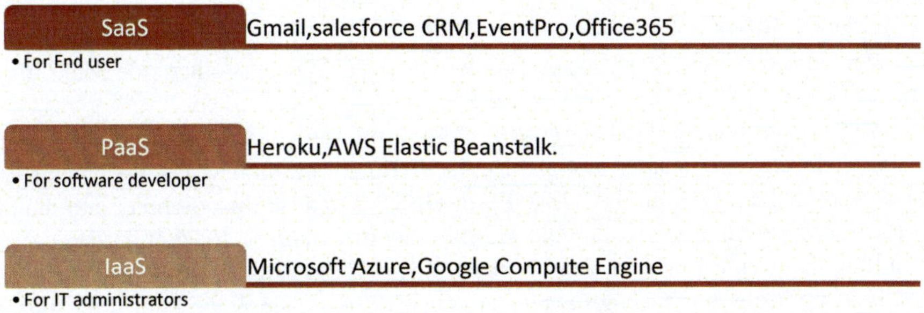

Fig. 2 Cloud service model [25]

The most well-known example of Saleforce is Salesforce.com, but many other examples have hit the market, including providing Google apps for basic business services, including email and word processing (Google Microsystems, 2009). Saleforce is a leader in promoting this service model. Microsoft Mesh Live lets you share files and folders on multiple devices at the same time.

Platform as a Service (PaaS)

A platform as a service (PaaS) cloud is full of development and distribution environments, with resources to help you deliver everything from simple cloud-based applications to advanced, cloud-enabled business applications. Like IaaS, PaaS includes infrastructure—servers, storage and communications—but at the same time middleware, development tools, enterprise intelligence services (BI), data management systems and more. PaaS is designed to support the entire web application life process: build, test, deployment, storage and development. PaaS allows you to avoid the costs and difficulties of purchasing and maintaining software licenses, built-in app infrastructure and middleware etc.

The platform integrates software (PaaS) layer and provides it as a service that can be used to build high-quality services. The client interacts with the platform through the API and does what it takes to maintain and measure the platform to provide a certain level of service (Sun Microsystems, 2009). Examples of PA are Google App Engine, Microsoft Azure and Amazon Map Reduction / Simple Service.

In this way, on a portable cloud computing, data storage and integration is embedded in the cloud, and the user gets seamless, desired service without worrying about battery life or the ability to process mobile devices.

11 Idea and System

Similar to cloud computing, there are local unit loads but no consistent descriptions of what mobile cloud computing is. With this paper, we tend to think of a completely different computer mode consisting of MC and CC that provides cloud-based client services using web and mobile devices. Whereas other side MCC may be the advancement of a portable computer.

MCC is simply divided into cloud computing and mobile computing. Those mobile devices are usually laptops, PDAs, smart phones and so on. It connects to a hotspot or base station via 3G, WiFi, or GPRS. Because the computer and major processors are migrated to the 'cloud,' the potential demand for mobile devices is limited, and some low-cost mobile phones or perhaps non-smartphones can even deliver mobile goods through a central platform. Al- even if the consumer uses a cloud that is converted from PCs or installed devices to mobile devices, the main idea remains the computer of the cloud. Mobile users send cloud service requests via an Internet browser or desktop application, then cloud management assigns resources to the application to ensure integration, and mobile computing and visualization activities can be forced to ensure QoS integration is complete [1–24].

12 Challenges with Solutions

The main purpose of using cloud computing is to create a simple and fast way for users to access and retrieve information from the cloud, which is a simpler and faster way to access cloud computing services effectively through the exploitation of contextual devices. The biggest challenge of mobile cloud computing comes from the features of mobile devices and wireless networks, in addition to its limitations and limitations, and such challenges make application, editing and installing on mobile devices and additional distribution more difficult than cloud-mounted devices. In the case of mobile cloud computing, mobile device issues, wireless communication quality, application forms, and support from a computer cloud to mobile square measure all important factors that contribute to testing from a computer. The second table provides a summary of the planned challenges and a few solutions related to computer cloud computing [1–24].

12.1 Ratings for Mobile Devices

Talking about cloud-based devices is a key factor in resource pressure. Although smart phones are clearly developed in various fields such as hardware and memory, storage, screen size, wireless communication, audio technology, and applications, there are still major limitations such as restrictions on computer capabilities and power resources, using complex systems. In this regard, PCs and laptops are in a state of overcrowding, and these smartphones such as the iPhone 4S, humanoid serials, Windows Mobile serials are three times less processed, eight times memory, five to ten times in the final capacity and ten times the network data rate. Normally, a smartphone should be charged daily such as making calls, triggering messages, aquatic web, public access, and various network applications. In line with previous development trends, computer duplication capabilities and rapid development of screen technology can create more and more applications installed on smartphones. If battery technology cannot be improved in a short period of time, then how to save battery power on a smartphone can be a major problem we often encounter these days.

The process efficiency, storage, battery time, and communication of these smartphones are systematically improved in the event of a portable computer. However, such large differences could continue at a concert of major challenges in mobile cloud computing.

12.2 Communication Quality

Separated from the cable operator using physical integration to ensure data integrity, the data transmission in mobile cloud sites remains dynamic and therefore the organization continues due to the existing network overlay permissions. In addition, the information center for a large business and web service provider service is often a way to complete users, especially for mobile device users. In a wireless network, the network delay is two hundred ms in the 'last mile' but only five ms in the old wireless network.

13 Open Analysis Issues

Although some come with a cloud-based computer already shipped around the world, there is still a long way to go for business use, and a few aspects of analysis should be considered in any project.

13.1 Delivery of Information

Due to the nature of the resources, mobile devices have potential challenges in cloud access, static access, data transfer, etc. These type of issues may be resolved by the application (service) and a central location that can provide a policy for all cloud-based computer programs [25].

13.2 *Classification of Tasks*

Investigators differentiate tasks/applications from mobile devices to multiple subtasks and bring their number to work in the cloud, which can be a reasonable decision for mobile devices that have been denied resources. Though the accurate strategy or rules on how to differentiate these functions are not defined and to be processed by the cloud and those by device.

13.3 *Higher Service*

The actual purpose of MC is to offer PC-friendly platform for mobile terminals. On the other hand, because of completely different options available between mobile devices and PCs and user are often unable to directly integrate services from PC platforms to mobile devices. Therefore, any analysis should attempt to develop a strategy in such a way as to provide appropriate and friendly communication services for mobile devices.

References

1. Hewitt C (2008) Orgs for scalable, robust, privacy-friendly client cloud computing. IEEE Internet Comput 12(5):96–99
2. Buyya R, Yeo C, Venugopal S (2008) Market-oriented cloud computing: vision , hype, and reality for delivering it services as computing utilities. In: High performance computing and communications 2008, HPCC'08. 10th IEEE international conference on IEEE, pp 5–13
3. Youseff L, Butrico M, Da Silva D (2008) Toward a unified ontology of cloud computing. In: Grid computing environments workshop, GCE'08. IEEE, pp 1–10
4. Mobile cloud computing subscribers to total nearly one billion by 2014, 2009, Sept). [Online]. Available http://www.abiresearch.com/press/1484
5. Shankar S (2009) Amazon elastic compute cloud
6. Zahariev A (2009) Google app engine. Helsinki University of Technology
7. Microsoft azure homepage (2011) [Online] Available http://www.windowsazure.com/en-us/
8. McCarthy J (1961) Speech givento celebrate mits centennial [Online] Available http://en.wikipedia.org/wiki/JohnMcCarthy (computerscientist)
9. The customer relationship management (crm) (2009) [Online]. Avail- able: http://en.wikipedia.org/wiki/Customerrelationshipmanagement
10. Rochwerger B et al (2009) There servoir model and architecture for open federated cloud computing. IBM J Res Develop 53(4):1–11
11. Boss G et al (2007) Cloud computing. IBM white paper, Version vol 1.
12. Mei L, Chan W, Tse T (2008) A tale of clouds: paradigm comparisons and some thoughts on research issues. In: Asia-pacific services com- puting conference, 2008. APSCC'08, IEEE, pp 464–469
13. Cohen R (2010) The cloud computing opportunity by the numbers. 2010 [Online] Available http://www.elasticvapor.com/2010/05/cloud-computing-opportunity-by-numbers.html
14. Marrapese B (2010) Google ceo: a few years later, the mobile phone becomes a super computer Dec [Online] Available http://www.itnews-blog.com/it/21320.html

15. Chetanet S et al (2010) Cloud computing for mobile world. Available at chetan. ueuo.com
16. Chunet et al (2011) Elastic execution between mobile device and cloud. Published in Proceedings of the sixth conference on computer systems. ACM, pp 301–314
17. Zhang X et al (2011) Towards an elastic application model for augmenting the computing capabilities of mobile devices with cloud computing. Mobile Netw Appl 16(3):270–284
18. Li L, Shen (2011) Virtualized screen: a third element for cloud-mobile convergence. Multimedia IEEE 18(2):4–11
19. Giurgiu et al (2009) Calling the cloud: enabling mobile phones as interfaces to cloud applications. In: Proceedings of the ACM/IFIP/USENIX 10th international conference on middleware. Springer, pp 83–102
20. Rellermeyer A, Roscoe (2011) R-osgi: distributed applications through software modularization. In: IFIP lecture notes in computer science(LNCS) vol 4834(4834). pp 1–20
21. Jeong S et al (2010) Towards an elastic application model for augmenting computing capabilities of mobile platforms. In: Mobile wireless middleware, operating systems and applications, pp 161–174
22. Satyanarayanan M, Bahl P, Caceres R, Davies N (2009) The case for vm-based cloudlets in mobile computing. Pervasive Comput IEEE 8(4):14–23
23. Marinelli E (2009) Hyrax: cloud computing on mobile devices using mapre-duce. DTIC Document, Tech. Rep
24. Cooney M (2011) Gartner: The top 10 strategic technology trends for 2012. [Online]. Available http://www.networkworld.com/news/2011/101811-gartner-technology-trends-252100.html
25. Hoang et al (2011) A survey of mobile cloud computing: architecture, applications, and approaches. Wireless Mobile Commun. https://doi.org/10.1002/wcm.1203

Subpixel Image Registration Using FFT-Based Technique

Swati Jindal, Bhaskar Dubey, and Jitali Patel

Abstract This paper presents an efficient image matching technique with translation, subpixel translation. Fast Fourier transform technique is the most powerful area-based technique that involves translation, rotation and other operation in frequency domain. FFT and its inverse are used to convert a two-dimensional image from the spatial domain to the frequency domain. The benefit of expressing an image in frequency domain is that executing certain operations on the frequencies is far more efficient than doing so in spatial domain. In this paper, we discussed different methods of subpixel estimation. Also, we validate the performance of our method under several different noise conditions. Our main goal to develop efficient technique to estimate translation and subpixel both by using FFT techniques. Our implementation is based on Python programming.

Keywords Image registration · Phase correlation · FFT

1 Introduction

Image registration is a crucial step in most of the image processing task such as remote sensing, weather forecasting and satellite image. Image registration is used to overlay two or more images with the same scene taken from different sensors having different orientation, the same sensor at different time and from different locations. The main goal of image registration is to find coordinate relationship between images.

S. Jindal (✉) · J. Patel
Institute of Technology, Nirma University, Ahmedabad, India
e-mail: 19mcei09@nirmauni.ac.in

J. Patel
e-mail: jitali.patel@nirmauni.ac.in

B. Dubey
Image Analysis and Quality Evaluation Division, Signal and Image Processing Group, Space Applications Centre, Ahmedabad, India
e-mail: bhaskard@sac.isro.gov.in

P. K. Singh et al. (eds.), *Emerging Technologies for Computing, Communication and Smart Cities*, Lecture Notes in Electrical Engineering 875,
https://doi.org/10.1007/978-981-19-0284-0_43

Image registration is divided into two categories: feature-based techniques and area-based techniques.

1. Feature-based techniques: This technique works on the object or feature point of an image. It can be further divided into algorithm such as an algorithm that uses high-level and low-level features, e.g. graph-theoretic method and feature-based method, respectively [1–3]. The low-level feature means edges and corners, while high-level feature means detect object or relation between features.

2. Area-based techniques: This technique works on intensity of images. It can be further divided into algorithm which uses frequency domain, e.g. FFT-based & algorithm which uses direct pixel value, e.g. cross-correlation [1–3].

Mathematical model: Let f and g be two images such that f is translated, rotated and scaled replica of image 'g'. That is,

$$f(x, y) = g(\frac{1}{|a|}(x \cos\theta + y \sin\theta - x_0, -x \cos\theta + y \cos\theta - y_0)) \tag{1}$$

where x_0 and y_0 are the translation in the x and y directions, respectively. $\theta$and a are the rotation angle and scale factor, respectively. Thus, the aim of any automatic image registration model is to estimate the parameters (x_0 , y_0 ,θ, a).

The fast Fourier transform(FFT) is the most powerful method of area-based technique. The registration methods use the Fourier domain method to match the images.

Geometric defects such as translation, rotation and scaling between images are corrected using FFT-based registration algorithms. This is accomplished by applying techniques from theory of Fourier transform to image data sets.

The FFT ratio is calculated using phase correlation, which is based on the translation property of the Fourier transform. The strongest correlation is represented by the sharp peak of the inversion of this ratio. Before phase correlation, the reference and sensed images are converted to log-polar images from Cartesian coordinates, i.e. the rotation in Cartesian coordinate is equivalent to translation in polar coordinates. After that, phase correlation computes similar ratio, scaling & rotation error which are represented as shift.

2 Literature Review

The image registration sector is highly capital-intensive sector and has attracted attention of researchers. A researcher from the world used different techniques of image registration to match images. Study examining efficiency of different algorithms includes work of Reddy et al.(1996), Samritjiarapon et al.(2008), Xie et al.(2003), Tong et al.(2019), Foroosh et al.(2002) and Ezzeldeen et al.(2010).

Ezzeldeen et al. [4] analysed different techniques, i.e. FFT-based techniques, contour-based, wavelet-based, Harris-PCNN and Harris moment-based technique. These techniques have been evaluated based on root mean square error(RMSE) and running time. Landsat Thematic Mapper and SPOT remote sensed image are used.

The sensed images are rotated with 1–20 angle in counterclockwise direction and, furthermore, compared referenced image and rotated images. In the end, it is concluded that most stable and best process is FFT, but FFT has higher RMSE value, and the RMSE value of wavelet technique is least, and smaller running time technique is contour based and FFT technique.

Foroosh et al. [5] discuss the subpixel image registration method with interpolation such as phase correlation, intensity interpolation and correlation interpolation. The accuracy of methods depends on the quality of the interpolation algorithm, also discusses subpixel image registration without interpolation such as differential properties of an image sequence and defines subpixel registration as an optimization problem. So the main motive is to find subpixel image registration using phase correlation. To estimate non-integer translation between images, author describes various methods such as 2D Dirichlet kernel function. The algorithm is tested on different set of images.

Samritjiarapon et al. [1] explore the Fourier phase correlation which is used to measure the translation, and best first search is used to analyse the final translation of angle or scaling parameters. Satellite image is used to implement data. The best technique is FFT-based technique with phase correlation and best first search as its error is less as compared to other techniques. Moreover, the computation time of FFT-based technique with the phase correlation technique is less. So FFT-based technique with phase correlation and the best first search techniques are observed as more stable.

Reddy et al. [6] evaluated the efficiency of FFT-based techniques. Phase correlation is used to get translation movement. To find scaling and rotation movement, use Fourier scaling and rotation properties. The proposed algorithm is tested on set of different images and different amount of noise added to the image. The advantage of the algorithm is that it computes result in fixed amount of time and low computational cost.

Xie et al. [7] conducted a study on FFT-based technique and implemented using Interactive Data Language (IDL) and added ENvironment for Visualizing Images (ENVI). As the image size increases, overall accuracy increases, and the accuracy of the algorithm is good compared to manual method. The ENVI user functions are very useful because function is inbuilt and it also reduces time.

Patel et al. [8] explore the challenge of estimating the rotation, and scaling is reduced to one estimating a 2D translation by resampling the Fourier magnitudes on the log-polar grid. Rather than standard normal correlation, the authors use phase correlation to convert from Cartesian to log-polar using bilinear interpolation algorithms. As a result, when it comes to registering aerial photographs, this method is extremely effective.

According to Juan et al. [9], SIFT is the slowest and least effective at lighting changes. SURF is quick, but it is not resistant to rotations or variations in lighting. PCA- SIFT's blur quality should increase. The approaches are chosen in accordance with the application.

From depth of published paper, we learned several methods for image registration such as contour based, wavelet based, FFT and many more, but the FFT-based technique is very efficient and accurate compared to other methods.

3 Methodology

3.1 *FFT-Based Image Registration Approach*

FFT-based image registration problems use the properties of Fourier transforms of the two images (often called master and slave, or original and replica) to estimate the translation in X and Y directions. The technique is highly efficient and robust and gives accurate results under the presence of high noise as well. The Fourier transform method basically is used to convert the image from spatial domain into frequency domain, it decomposes result into sine and cosine waves [10], and each point in the Fourier domain image is a particular frequency. Phase correlation method can be explained as follows: let f_1 be sensed image and f_2 be translated replica of f_1, so they differ by (x_0, y_0) displacement [6]:

$$f_2(x, y) = f_1(x - x_0, y - y_0) \tag{2}$$

their corresponding Fourier transform is [6]:

$$F_2(x, y) = \exp^{-j2\pi(ux_0+vy0)} * F_1(u, v) \tag{3}$$

where F_1 & F_2 are Fourier transform of f_1 and f_2 . Now phase correlation of two image f_1 and f_2 with their Fourier transform F_1 and F_2 is [6, 11]:

$$\frac{F_1(u, v)F_2^*(u, v)}{|F_1(u, v)F_2^*(u, v)|} = \exp^{j2\pi(ux_0+vy0)} \tag{4}$$

As shown in Eq. 4, F_2^* is complex conjugate of F_2. Now inverse Fourier transform of Eq. 4 results in an impulse at the location (x_0, y_0). Thus by locating the position of impulse, we get the required translation in image pairs.

3.2 *Subpixel Image Registration*

Our main aim is to find a method for subpixel translation with the benefits of the phase correlation method. In subpixel image, registration images are upsampled by scaling factor using bicubic interpolation. There are various interpolation techniques such as nearest neighbour, bilinear and bicubic interpolation. Bicubic interpolation is the best technique as compared to nearest neighbour and bilinear interpolation. Bicubic interpolation technique is generally used in professional image editing software like adobe photoshop and corel photoshop [2]. In this, we consider closest 4×4 neighbour of points for a total of 16 points or pixels. These 16 points are determined using the below equation [2] :

$$p(x, y) = \sum_{i=0}^{3} \sum_{j=0}^{3} a_{ij} x^i y^j \tag{5}$$

Hence, images are upsampled by bicubic interpolation. The scaling factor calculates the computation time of the interpolation technique.

Now upsampled images are shifted by an integer value, and upsampled shifted images are downsampled as the same scaling factor. After that, the result of phase correlation is observed as single peak value, while in the case of subpixel image shift, there is coherent peak in which one is largely adjacent to each other. So, further computing is required for subpixel shift.

Generally, subpixel image registration based on Fourier transform is divided into two categories:

(i) Spatial domain
(ii) Frequency domain

Spatial domain means by using correlation peak, while frequency domain means by using phase difference. To estimate subpixel shift, three techniques are used

(i) Centroid shift estimation
(ii) Parabolic fit
(iii) Dirichlet function

Centroid Shift Estimation In this, we simply calculate or estimate weighted average of local neighbourhood around phase correlation value (i.e. peak value)[12]. Weighted average is calculated using below equation:

$$x = \frac{x_0 pc[0] + x_1 pc[1] + \cdots + x_n pc[n]}{\sum_{i=0}^{n} pc[i]} \tag{6}$$

$$y = \frac{y_0 pc[0] + y_1 pc[1] + \cdots + y_n pc[n]}{\sum_{i=0}^{n} pc[i]} \tag{7}$$

where pc[0] pc[1] and ..., pc[n] are phase correlation value at point and $x_0, x_1, \ldots, x_n, y_0, y_1, \ldots and y_n$ are nearest neighbour of peak point. Here x and y are estimated values of subpixel displacement image registration.

Parabolic Fit In this, we used parabolic function [3, 11, 13], i.e.

$$PC(x, y) = a_0 x^2 + a_1 y^2 + a_2 xy + a_3 x + a_4 y + a_5 \tag{8}$$

First estimate peak value and then estimate subpixel value using six coefficients $(a_0, a_1, a_2, a_3, a_4, a_5)$ and nine neighbourhood values.

$$AX = B \tag{9}$$

where A is a rectangular matrix of size 9×6, X is coefficient matrix which contains unknown values and B is phase correlation matrix which contains phase correlation value as shown in below equation [11, 13]:.

$$A = \begin{bmatrix} x_0^2 & y_0^2 & x_0y_0 & x_0 & y_0 & 1 \\ x_1^2 & y_1^2 & x_1y_1 & x_1 & y_1 & 1 \\ . & . & . & . & . & . \\ . & . & . & . & . & . \\ . & . & . & . & . & . \\ . & . & . & . & . & . \\ . & . & . & . & . & . \\ . & . & . & . & . & . \\ x_8^2 & y_8^2 & x_8y_8 & x_8 & y_8 & 1 \end{bmatrix}, X = \begin{bmatrix} a_0 \\ a_1 \\ a_2 \\ a_3 \\ a_4 \\ a_5 \\ a_6 \\ a_7 \\ a_8 \end{bmatrix} and B = \begin{bmatrix} PC_0 \\ PC_1 \\ PC_2 \\ PC_3 \\ PC_4 \\ PC_5 \\ PC_6 \\ PC_7 \\ PC_8 \end{bmatrix} \tag{10}$$

Here $PC_0, PC_1, \ldots$and PC_8 are phase correlation values, and we have to determine coefficient matrix, i.e. X.

$$X = A^{-1}B \tag{11}$$

Now, peak value at x and y can be determined using below equation [3, 11–13]:

$$\Delta x = \frac{a_2a_4 - 2a_1a_3}{4a_0a_1 - a_2^2}, \Delta y = \frac{a_2a_3 - 2a_0a_4}{4a_0a_1 - a_2^2} \tag{12}$$

Hence, estimated values of x and y are determined as given in Table 4.

3.3 Dirichlet Function

Once we get peak value at point (x_0, y_0) using phase correlation. In subpixel image, displacement coherent peak occured which is one adjacent to each other, i.e. (x_h, y_0) & (x_0, y_h), where $x_h = x_0 \pm 1$, $y_h = y_0 \pm 1$.

Explain Dirichlet function by taking example: Consider peak which occurs at location (x_0, y_0) and two side peaks at $(x_0{+}1, y_0)$ and $(x_0, y_0{+}1)$. By applying Eq. (4), we will find phase correlation pc(0,0), pc(1,0) and pc(0,1). After applying below equation [5], we obtained value of $(\Delta x, \mathbf{\Delta y})$.

$$\Delta x = \frac{pc(1,0)}{pc(1,0) \pm pc(0,0)} \tag{13}$$

$$\Delta y = \frac{pc(0,1)}{pc(0,1) \pm pc(0,0)} \tag{14}$$

Here, we will obtain two solutions. So correct solution is to identify that the result is in an interval of (0,1) and the same sign as x_h - x_0 and the same for y-axis.

4 Implementation and Results

4.1 Algorithm

The main goal is to develop an efficient technique using Python that can estimate translation and rotation both. We first do our simulation experiment on Lena image; subsequently, the technique can be applied to any image in general. The implementation steps are shown in Fig. 1.

Repeat the steps under different noise conditions and carry out a sensitivity analysis with respect to noise.

Experiments on Bidirectional Translated Lena Image The matching of proposed algorithm is tested on a different set of images. Here Lena image is taken as test case. All the images are 8-bit greyscale image of size 512 × 512. In this, we give integer

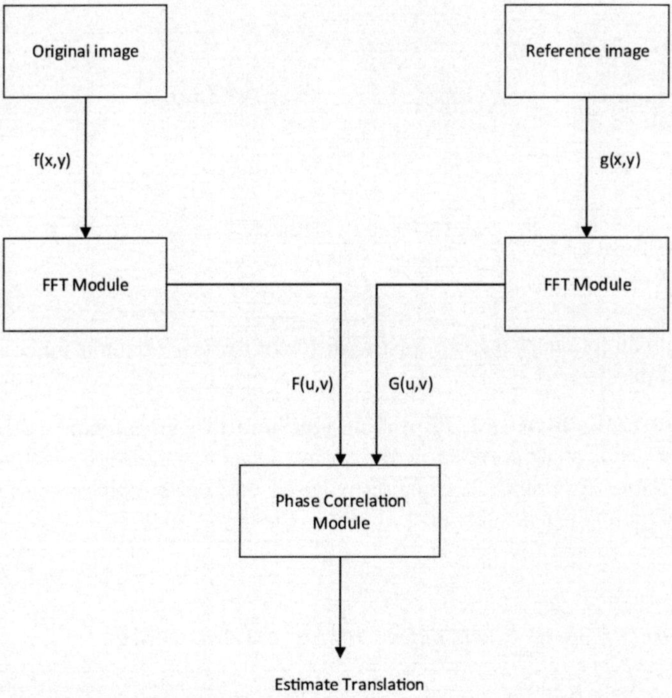

Fig. 1 Block diagram of algorithm

Fig. 2 **a**Original Lena image, **b** shifted Lena image by (−20, −20), **c** shifted Lena image by (−20, 10), **d** shifted Lena image by (−10, −10), **e** shifted Lena image by (20, 20)

Table 1 Simulated and translated Lena images

Fig	X-Shift	Y-Shift
a	0	0
b	−20	−20
c	−20	10
d	−10	20
e	20	20

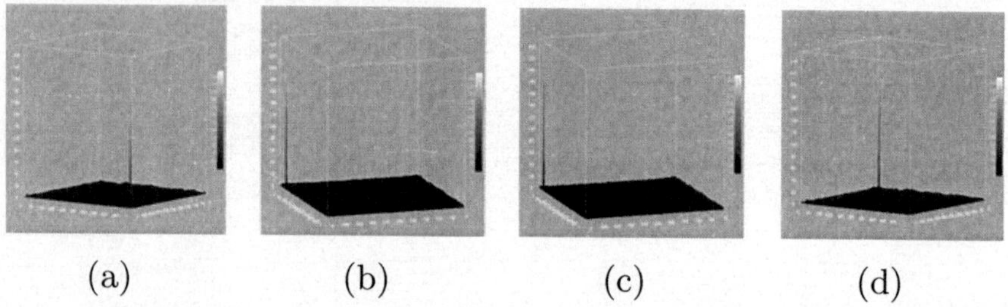

Fig. 3 **a** Surface plot of phase correlation, **b** surface plot of phase correlation, **c** surface plot of phase correlation, **d** surface plot of phase correlation

shift in both directions using Eq. 2. Now, different shifts are demonstrated as shown in Fig. 2 (Table 1).

Translation Estimation using Phase Correlation In this, we estimate the translation of the image using phase correlation. Here phase correlation means basically compute their FFT ratio. And phase correlation is based on Fourier shift property. Due to circular wrapping, $-x$ is equivalent to $512 - x$ (Fig. 3; Table 2).

4.2 Generation of Subpixel Translated Lena Image

In this section, we are explaining the process of subpixel translation. Lena image was used to demonstrate the result. For example, Lena image is amplified by scaling factor

Table 2 Estimated translation using phase correlation

Fig	Estimated X-Shift	Estimated Y-Shift
(a)	492(−20)	492(−20)
(b)	492(−20)	10
(c)	502(−10)	20
(d)	20	20

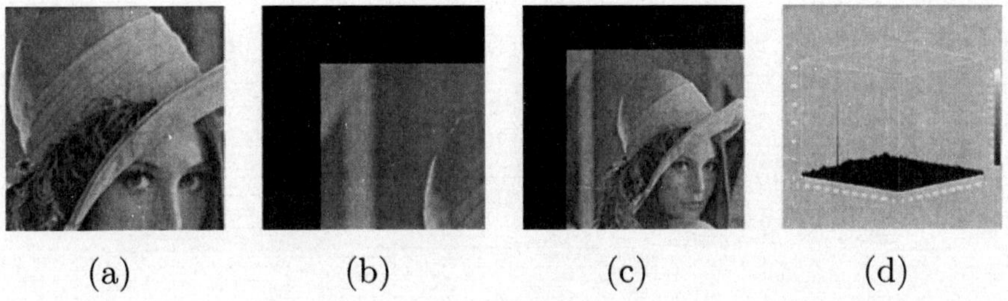

Fig. 4 **a** Zoomed image by factor 2, **b** integer shifted Lena image by (201, 201), **c** downsampled shifted Lena image by (100.5, 100.5), **d** surface plot of phase correlation of Lena image & (**c**)

2 using bicubic interpolation, and then image is translated by integer value(201,201). Some portion of amplified Lena image and the translated image are shown in Fig. 4a, b, respectively. Then translated zoomed Lena image is subsequently downsampled by the same factor, as shown in Fig. 4c.

Subpixel Image Registration Results The algorithm has been tested on 48 test cases of Lena image. The algorithm was applied to a different shift, which has been shown in Fig. 5. The surface plot of phase correlation of Lena image, which is translated in both axes, is shown in Fig. 5b, e, h, k. In subpixel shift, peak is not located at single point but located at various points which are adjacent to each other. So, further processing of subpixel image is required.

The simulation results of different algorithm are shown in Tables 3, 4 and 5 by examining error between the estimate and original shift. By comparing the method, it is found that the Dirichlet function is more efficient because the error is less as compared to other methods, and error is in an interval of (±0.1, ±0.25).

Subpixel Translation Under Noise Condition We tested our algorithm under different noise conditions. To obtain noisy images, Gaussian noise has been added to the normal image. Here 20db noise is added to the images. Due to the subpixel nature, there is the peak which is the one adjacent to each other. To estimate subpixel shift, Dirichlet function is used, and result shows the exact shift is recovered even under noise conditions as shown in Fig. 6.

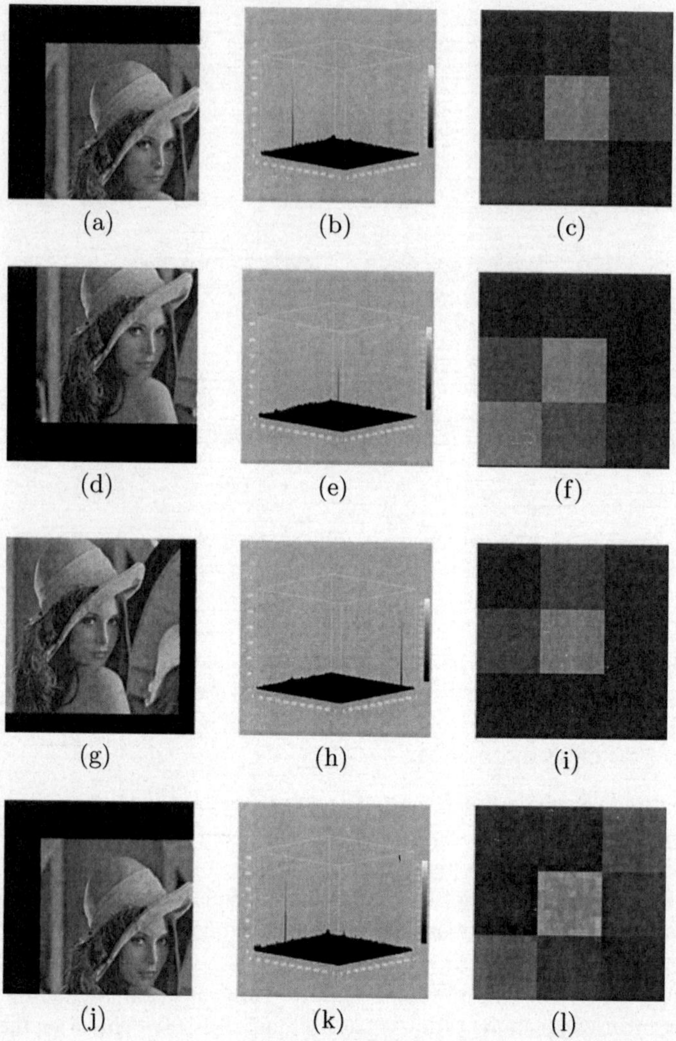

Fig. 5 **a** Shifted Lena image by (100.3, 100.3), **b** surface plot of phase correlation, **c** zoomed image of phase correlation, **d** shifted Lena image by (−100.5, 80.5), **e** surface plot of phase correlation, **f** zoomed image of phase correlation, **g** shifted Lena image by (−50.3, −40.3), (h) surface plot of phase correlation, **i** zoomed image of phase correlation, **j** shifted Lena image by (100.25, 100.25), **k** surface plot of phase correlation, **l** zoomed image of phase correlation

Table 3 Estimated subpixel displacement value using centroid function

S. No.	Original (x, y) shift	Estimate (x, y) shift	Error
1	(0,0)	(0,00	(0,0)
2	(100.5,100.5)	(100.23, 100.51)	(0.27, −0.01)
3	(100.33,100.33)	(100.18, 100.06)	(0.15,0.27)
4	(411.5,80.5)	(411.48, 80.59)	(0.02, −0.09)
5	(461.67,471.67)	(461.81,471.84)	(−0.14, −0.17)
6	(100.25,100.25)	(100.10,100.02)	(0.15,0.23)

Table 4 Estimated subpixel displacement value using parabolic fit

S.No.	Original (x, y) shift	Estimate (x, y) shift	Error
1	(0,0)	(0,00	(0,0)
2	(100.5,100.5)	(100.39, 99.96)	(0.11,0.54)
3	(100.33,100.33)	(100.13,100.01)	(0.2,0.23)
4	(411.5,80.5)	(411.98, 80.32)	(−0.48,0.18)
5	(461.67,471.67)	(461.91,471.93)	(−0.24, −0.26)
6	(100.25,100.25)	(100.81,99.36)	(−0.56,0.89)

Table 5 Estimated subpixel displacement value using Dirichlet function

S. No.	Original (x, y) shift	Estimate (x, y) shift	Error
1	(0,0)	(0,0)	(0,0)
2	(100.5,100.5)	(100.25,100.56)	(.25,−0.06)
3	(100.33,100.33)	(100.31,100.29)	(−0.01,0.01)
4	(411.5,80.5)	(411.40,80.68)	(0.1, −0.18)
5	(461.67,471.67)	(461.69,471.69)	(−0.02, −0.02)
6	(100.25,100.25)	(100.19,100.18)	(0.06, 0.07)

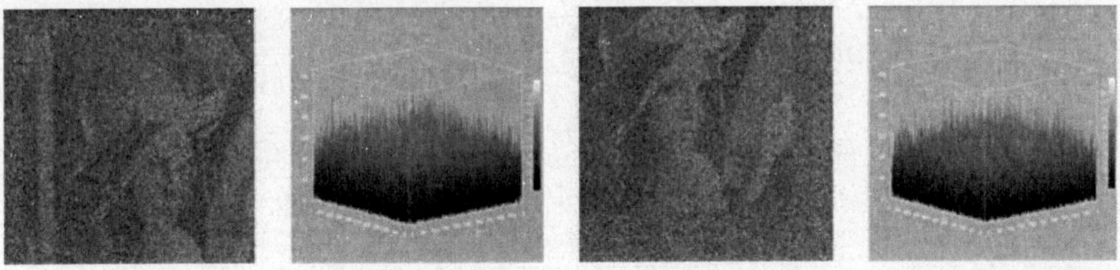

(a) (b) (c) (d)

Fig. 6 **a** Noisy image shifted by (50.5, 40.5), **b** surface plot of phase correlation of shifted noisy and original Lena image, **c** noisy image shifted by (−100.33, −80.33), (d)surface plot of phase correlation of shifted noisy and original Lena image

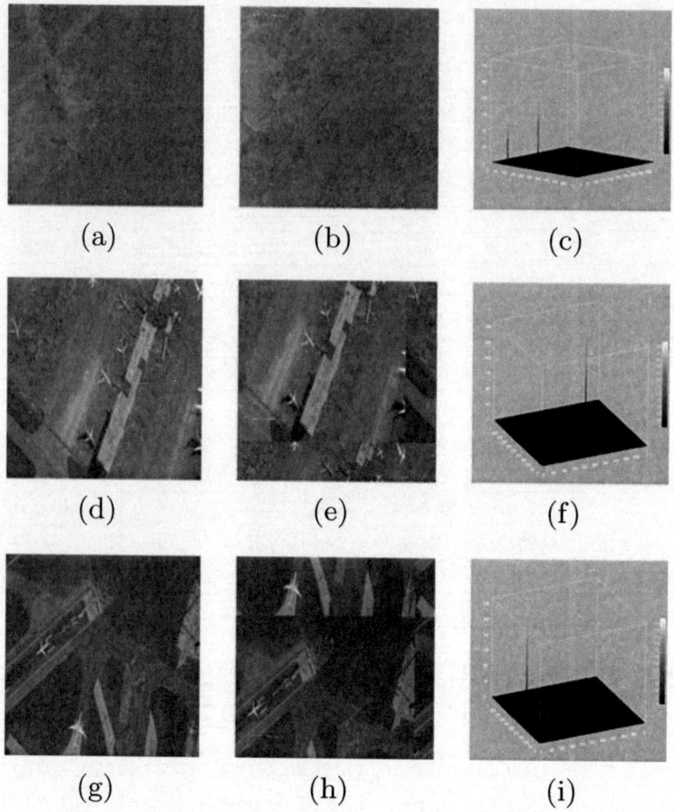

Fig. 7 **a** Landsat satellite image, **b** shifted image by (151,151), **c** surface plot of phase correlation of (**a**) & (**b**), **d** Atlanta Rareplane image, **e** shifted image by (−100, −80), **f** surface plot of phase correlation of (**d**) & (**e**), **g** London Rareplane image, **h** shifted image by (150, −50), **i** surface plot of phase correlation of (**g**) & (**h**)

4.3 Result of Satellite Image

The technique was also used to perform a separate set of satellite images from different bands of Landsat pictures. We have taken a satellite image from Earth Explorer, the latitude is 23.02, and the longitude is 72.57 of Ahmedabad space. The Landsat 8 data have 11 spectral bands, and image is of size 7731 × 7881. In this experiment, band 3 Landsat 8 image was taken as original image, and shift image by (151,151) in both axes was taken as sensed image. The FFT technique works only on image of size 2^n. The original landsat satellite image is of size 512 × 512 after that shift image by 151, 151 in both direction. Even under case of satellite image, algorithm provides accurate result. The result of phase correlation is impulse function, i.e. everywhere zero except at displacement point. It was discovered that comparing two bands of the same satellite image yielded precise results.

5 Future Work

Following is the work that needs to be done in future:

- The technique can be implemented on rotated and scaled images as well by transforming to FFT images in the polar coordinates.
- We plan to test the performance of rotation estimation by giving different degree of angle and under similar noise condition.
- Subsequently, the technique can be tested for any combination of translation, rotation and both.
- We plan to simulate studies with highly noised image pairs which are rotated, translated and scaled replica of each other.

6 Conclusion

An efficient FFT-based technique is implemented for automatic image registration. The accuracy of the FFT-based image registration method is very high, even under noise condition. We carefully designed simulation experiment that shifted the Lena image from -2 to 3 in both directions to generate 25 test cases. Moreover, add 10db PSNR noise to all the images and perform phase correlation. It is shown that phase correlation estimates the translation in images very accurately in more than 80 per cent case. So, this method is capable of estimating translation under noise condition. The advantage is that even if image size increases, its computation time is less and gives accurate results. The bicubic interpolation technique produced smoother result as compared to nearest neighbour and bilinear interpolation. In comparison with the centroid and parabolic functions, the Dirichlet function gives more reliable results when estimating subpixel translation. The Dirichlet function produces reliable results with an error range of (± 0.1,± 0.25). In the case of noise, the method produces an accurate result, and in simulation experiments, the error is less than one pixel. To determine the translation between satellite images taken over a three-month period provided a reliable result.

References

1. Samritjiarapon O, Chitsobhuk O (2008) An fft-based technique and best-first search for image registration. In: International symposium on communications and information technologies. IEEE, pp 364–367
2. Gonzalez RC, Woods RE (2008) Digital image processing. Prentice Hall
3. Jurgen L (2019) Sub-pixel image translation estimation on a nanosatellite platform, Ph.D. thesis. Stellenbosch University, Stellenbosch
4. Ezzeldeen R, Ramadan H, Nazmy T, Yehia MA, Abdel-Wahab M (2010) Comparative study for image registration techniques of remote sensing images. Egyptian J Remote Sensing Space Sci 13(1):31–36

5. Foroosh H, Zerubia JB, Berthod M (2002) Extension of phase correlation to subpixel registration. IEEE Trans Image Process 11(3):188–200
6. Reddy BS, Chatterji BN (1996) An fft-based technique for translation, rotation, and scale-invariant image registration. IEEE Trans Image Process 5(8):1266–1271
7. Xie H, Hicks N, Keller GR, Huang H, Kreinovich V (2003) An idl/envi implementation of the fft-based algorithm for automatic image registration. Comput Geosci 29(8):1045–1055
8. Patel D, Gandhi V, Patel V (2014) Image registration using log polar transform and fft based scale invariant. Int J Eng Res Appl 4(7):79–84
9. Juan L, Gwun O (2009) A comparison of sift, pca-sift and surf. Int J Image Process (IJIP) 3(4):143–152
10. Sarvaiya JN, Patnaik S, Kothari K (2012) Image registration using log polar transform and phase correlation to recover higher scale. J Pattern Recogn Res 7(1):90–105
11. Jingying J, Xiaodong H, Kexin X, Qilian Y (2002) Phase correlation-based matching method with sub-pixel accuracy for translated and rotated images. In: 6th international conference on signal processing, vol 1. IEEE, pp 752–755
12. Tong X, Ye Z, Xu Y, Gao S, Xie H, Du Q, Liu S, Xu X, Liu S, Luan K et al (2019) Image registration with fourier-based image correlation: a comprehensive review of developments and applications. IEEE J Selected Topics Appl Earth Observations Remote Sensing 12(10):4062–4081
13. Gleason SS, Hunt MA, Jatko WB (1991) Subpixel measurement of image features based on paraboloid surface fit. In: Machine vision systems integration in industry, vol 1386. International Society for Optics and Photonics, pp 135–144

Impact of Green Communication and Technology System

Sumit Dhariwal and Avinash Raipuria

Abstract Green communication implies inspecting out for the examination of viability as for biological conditions, imperativeness capability, and the purposes behind correspondence on mobile phones generally. The commitment to a green exchange to enhance business environmental responsibility and encourage an ecological period of framework types of materials and buildings. This paper attempts to introduce the most recent examination on green connections and frameworks organization for state of the art wired and prompt and underhanded consequences for nature. Ongoing exploration on convenient advancement incorporates the improvement of the quantity of pinion wheels manhandled yearly that has set off the requirement for upgrade in the field of essentialness successful correspondence. The paper is a composing study concerning the shows to improve essentialness adequacy in green correspondence frameworks. It clarifies the various pieces of examination, plan, scattering, and advancement of shows, just as constructions of green communications and frameworks organization.

Keywords Green communication · ICT · D2D · MIMO · GIOT and HTr network

1 Introduction

I have contributed tremendously to the communication of time, and considering that, I have given attention to the environment by using the energy from the work of energy through this paper. The use of the communication equipment increases the likelihood

---

The original version of this chapter was revised: A affiliation to an earlier paper was omitted. The correction to this chapter was available at
https://doi.org/10.1007/978-981-19-0284-0_58

S. Dhariwal (✉)
Manipal University Jaipur, Rajasthan, India
e-mail: sumit.dhariwal@jaipur.manipal.edu

A. Raipuria
Malaysia University of Science and Technology, Kota Damansara, Selangor, Malaysia
e-mail: avinash.raipuria@phd.must.edu.my

P. K. Singh et al. (eds.), *Emerging Technologies for Computing, Communication and Smart Cities*, Lecture Notes in Electrical Engineering 875,
https://doi.org/10.1007/978-981-19-0284-0_44

of the spread of radiation, which is fatal for the environment. The importance of green communication emphasized by this paper offers a refined solution to all the problems the researcher has attempted to solve.

It has been noticed that presently, the total 4% energy is spent by the ICT framework that reasons across three percentage (3%) of the commonly CO_2 expenditures, once the contrasted with the generally CO_2 releases through planes is a total CO_2 quarter of the productions by different automobiles. Pursuant to ongoing study, versatile endorsers comprise the majority of the global populace. The supposition made by the media transmission market is that to reach that cutting edge, an expansion in endorsers will be needed, as well as per supporter's information rate, and the rollout of extra base stations if there is an occurrence of portable systems. The move of versatile systems is primarily centered on diminishing the vitality utilization of terminals, whose battery power forces necessities in a similar respect.

As a reality, late thoughts of versatile innovation incorporate the development various hardware abused each year that has introduced the importance of progressing in the field of green communications. Get necessities the number of subscribers and base stations have both been increased, which have inspired an expansion in communication movement.

It talks about the problem of energy proficiency and utilization in interchanges systems. Ericsson's ongoing test, report says half of versatile administrator effective costs is equal to the expenses of the energy. In this manner, media communications applications can have an immediate, economical effect on bringing down ozone depleting substance discharges, control utilization, and energy proficient remote transmission strategies (Fig. 1).

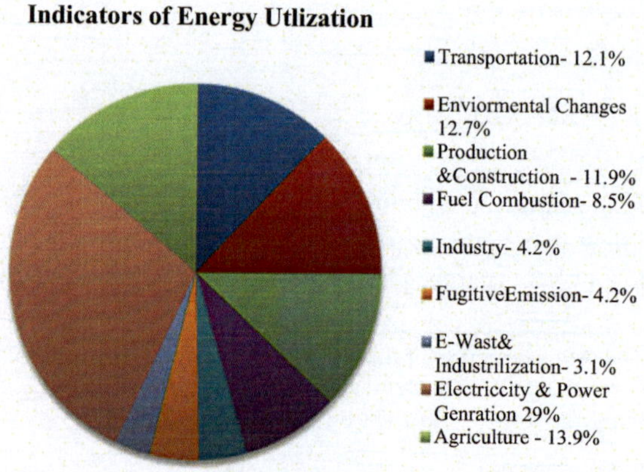

Fig. 1 Energy utilization indicators

2 What is Green Communication?

Green Communication is a communication system in which we communicate using work using radiation, thereby eliminating the risk of diseases mainly including nervous systems as well as the prevention of nature.

Green correspondence is the act of choosing energy proficient interchanges and systems administration advancements and items, limiting asset utilized at whatever point conceivable in all parts of correspondence. The number of portable membership in the data and correspondence innovation the number of subscribers and the number of base stations has been expanded. According to current analyses, worldwide portable memberships have increased dramatically from 500_million in_2000 to 5_billion in the year of 2012. Also, tend to achieve worldwide infiltration of hundred percentage (100%) after the year of 2020.

3 Techniques

There's a compelling impulse for green communication in the improvement of 5G. Part of the progress is less scrutinized because of the green communication used in 5G. It is stated below.

3.1 [D2D] Device-To-Device Communication

D2D indicates to communication between devices, which might be phones or vehicles. This innovation opens up new gadget driven interchanges that regularly require no immediate correspondence with network framework, consequently to address part of the organization limit issue as 5G permits more gadgets to be associated in quicker, more solid organizations [11]. Guarantees are normal shown in Fig. 2.

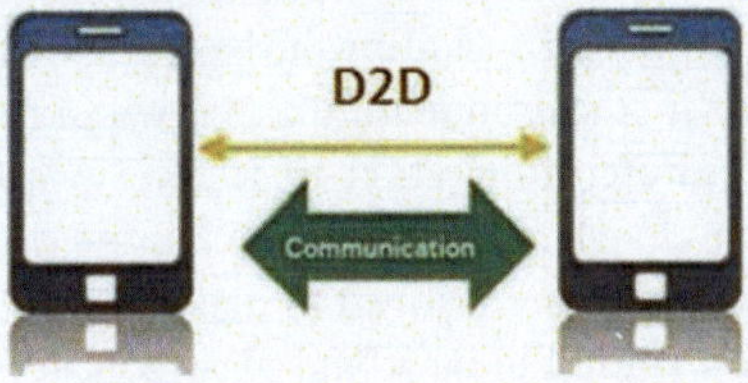

Fig. 2 Device-to-Device Communication

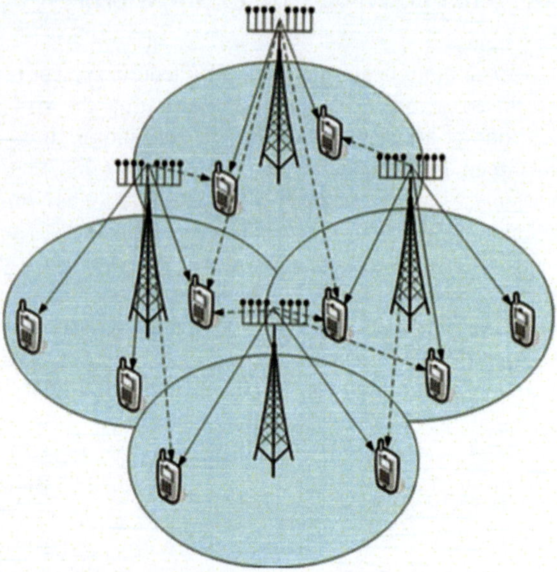

Fig. 3 Multiple input multiple outputs Malik et al. (2017)

3.2 *Multiuser Multiple Input Multiple Output*

Multiuser Multiple Input Multiple Output (MIMO) frameworks are on the ascent. In such frameworks, a base station having different receiving wires all the while serves various clients having single radio wire gear [12]. All clients in such situations can share multiplexing advantages, as delineated in Fig. 3.

3.3 *Hiterogeneous Networks*

One includes the utilization of a heterogeneous organization of macro cells and some more modest cells (e.g., microcells, pico cells, and femtocells). A backhaul network is finished by adding the base station to the core network through wired [13], wireless, or mixed architecture as shown in Fig. 4.

Force utilization in HetNet can be constrained by placing little cells into rest mode when there is a low burden or no heap [12]. In [13], advancement and logical methods are utilized to make the general framework proficient and to control power utilization in backhaul networks.

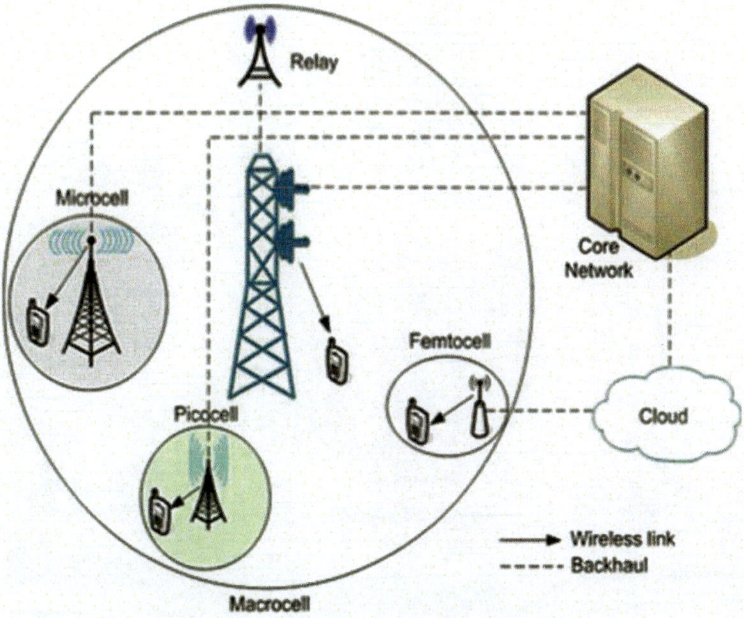

Fig. 4 A 5G heterogeneous network system design Malik et al. (2017)

3.4 *Green Internet of Things*

Green Internet of Things 5G is another possible element of green correspondence as shown in Fig. 6. Energy Efficiency in IoT has been pointed toward giving joining of various fields, that is concentrated in [14] (Fig. 5).

4 Types of Environment Impacts

The environmental impact of the terms is used and regularly used when examining green progress in communication. The carbon discharges as of now get the most consideration because of the issue of a worldwide temperature alteration and the adjustment in the atmosphere. When thinking about the ecologically friendly arrangement, air contamination, water contamination, and earth characteristic, security of the ozone layer, common property use, squander decrease, and so on must be considered as the principle sway.

Media communications hardware typically contains a lot of scarce resources and heavy metals, and the biggest test is the primary test on earth to isolate these materials through mining and waste treatment. With this technique, the amount of waste created with and without the material can be determined. We see that the reuse is 93 kg,

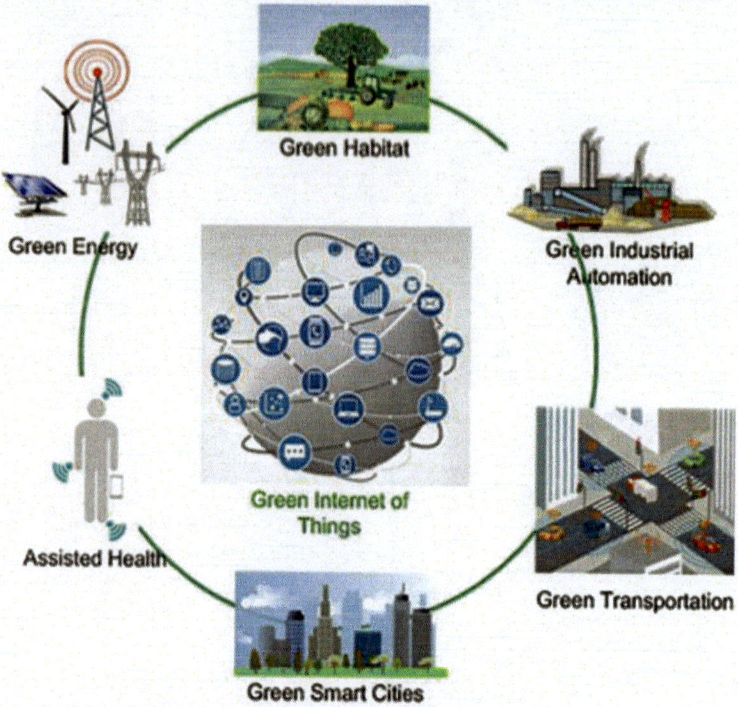

Fig. 5 The Internet of Things is going green Malik et al. (2017)

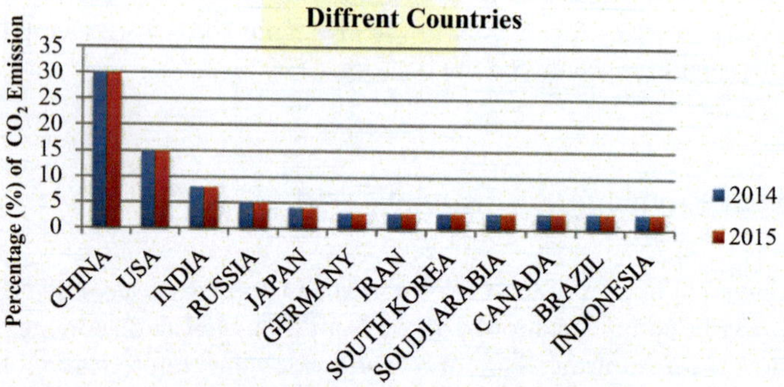

Fig. 6 CO_2 production by the most important providers Sejja et al. (2017)

whereas, without the reuse of 12 kg of auxiliary material for PC, there would be about 500 kg of mining waste. So, to have a total perspective on the natural impact of an object, each of the five phases must be considered, for the most part, material extraction, manufacture, use, transportation, and end of life.

Energy use and materials used in data communication innovation, which is identified as a carbon discharge, must also be considered. When examining such an equivalent, we need to consider the discharge of a substance that damages ozone, which incorporates the Kyoto convention, for example, CH_4 (methane), N_2O (nitrous oxide), CO_2 (carbon dioxide), PFCs, HFCs, and SF_6 (sulfur hexafluoride). Greenhouse gases have an earth-wide temperature raise potential that is assumed for a hundred-year horizon. These temperatures are most important for the EWT of carbon dioxide. CO_2 emissions are mainly associated with off-lattice locations that include remote areas. Diesel-control generators fuel most such destinations. The ozone harming substance outflows for atomic power are extremely low and atomic power impacts affect nature, for example, the treatment of atomic waste.

5 Express and Secondary Influences

When breaking down the immediate and backhanded effects ecological inviting arrangements are to be considered. For instance, actualizing an answer, which diminishes the energy utilization of an administration, brings about an immediate effect. Aberrant effects of arrangements are identified with the more extensive ideas and selection of the arrangement. Composition of letter supplying attendance, issue of vehicle, use of paper, and so forth. Backhanded effect decrease ordinarily constrains the ecological issues. Since the varieties in the political, money related, educational elements are subject to these, those decreases are normally harder to anticipate. Expanding proficiency is a fundamental methodology in restricting ecological effects. On the off chance that we think about the email address to which we are by replacing each letter sent by email to a great extent lessening the effect of those letters. This case exhibits that instating the circuitous natural effects of arrangements is a troublesome errand, which must be done with extraordinary consideration.

5.1 Mobile Devices

Different portable specialized gadgets, for example, cell phones, smart watches, personal wearable specialty gadgets, human service gadgets, and AI glasses have led us to a sensible society. With the rapid progress of remote correspondence, the absence of co-activity between portable hubs affects the nature, as well as results in an imbalance in the use of the environment, creating unnecessary energy use of cell phones.

The classes coming into the system can be broadly divided into three types, for the most part, standard cell phones, PCs, and cell phones. Ongoing testing shows performance is estimated at 1.5 kwh per year, depending on each 50th hour of charge. That equates to a backup state of 30% of the battery's limit each day and 40% of the remaining time, as well as versatile and green communication between systems for example, data sharing, the range of energy mindfulness, directing adjustment, and information reserving empowers giving potential advantages to enhancing and adjusting the asset utilization and recoveries the energy of altogether portable and remote systems, consequently green.

5.2 *Carbon Footmark Models and Conventions*

Communication turns into the most extreme significant and promising exploration point for future portable systems. The point of the exceptional issue is to spur inquires about to distribute their most recent research cutting-edge problems and difficulties in the green communication field for mobiles. Assessing versatile systems, which incorporate sorts of hardware of the client, utilizing systems, projects of industry of endorsers working. Use of appropriate systems as a result of the information traffic performed by portable supporters [1, 1].

Carbon impressions used for portable interchange models can be divided into five types:

(1) Mobiles production which is comparable to the creation including minimal effort mobiles, Smartphone's, and frameworks, on brief deals in the year and covering all administrators recently year.
(2) Mobile gadget's activities such as producing and then charging its battery and hook. For this situation framework task for frameworks, additional screens and different administrators are incorporated.
(3) RAN destinations task, which incorporates the productivity if there should be an occurrence of the power to base station locations, control sites, and center destinations.
(4) The operator exercises incorporate workplace tasks, supplies exercises, tasks in the vehicle armada, and business venture out activity identified with all clients' business exercises.
(5) Mobile system clients create server farms and transport. That is a versatile system client creates the allotment of different systems [2] (Fig. 7).

5.3 *Radio Approach Systems*

Development for the radio access organize relies upon system type hardware overall growth test. The measure of RAN power utilization in 2007 was about 20 kwh. The ongoing examination gives us the thought regarding the development of new locales and the expulsion of old site hardware. From the study, it is noticed that on a normal there is a 10% diminishing of new base station when contrasted with the earlier year. The base station model can be seen as a mix of the items offered.

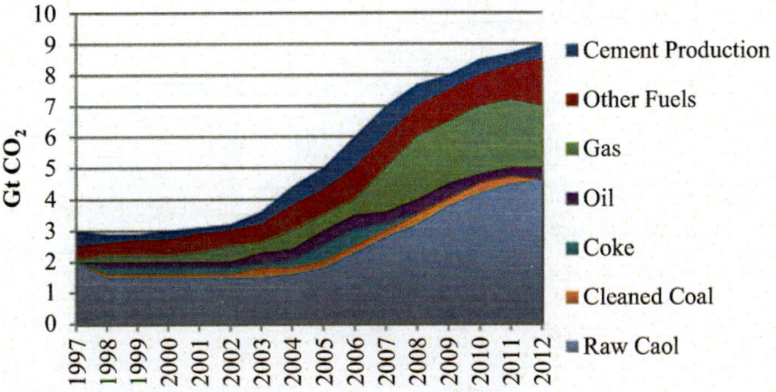

Fig. 7 China's main foundations Sejja et al. (2017)

We estimate that by 2022, the pattern progresses by as much as eight percent each year over the study time and is considered to be in steady progress. The worldwide normal of base station site power is anticipated to increase to approximately 1.8 kW in 2007 and decrease to approximately 1.2 kw in 2020, underestimates for the installation of new and removal of old gear each year. Investigation accepts a takeoff model expecting somewhere in the range of 600,000 and 675,000 destinations recently conveyed and up to 300,000 locales removed from administration every year [1].

6 Energy Competence Matrices

Energy proficiency measurements give data that can be utilized to survey and look at the energy utilization of different segments of a cell organize and of the system all in all. These measurements additionally help us to set a long haul to investigate objectives for lessening energy utilization. With the expansion in research exercises relating to green correspondence and because of the inherent contrasts and significance of different correspondence frameworks and execution measures, it is hard for one single measurement to get the job done.

While the meanings of vitality proficiency measurements at the part and gear levels are genuinely direct, it is all the more testing to characterize energy effectiveness measurements at the framework or system level. Measurements from the base of the system evaluate energy effectiveness at the system level, thinking about the system's limitations and properties.

Objectives related to the green cell system:

- Enhancement of energy proficiency.
- Improving knowledge of the system through interchanges between energy utilization and outer environments, that is a circulation capacity.

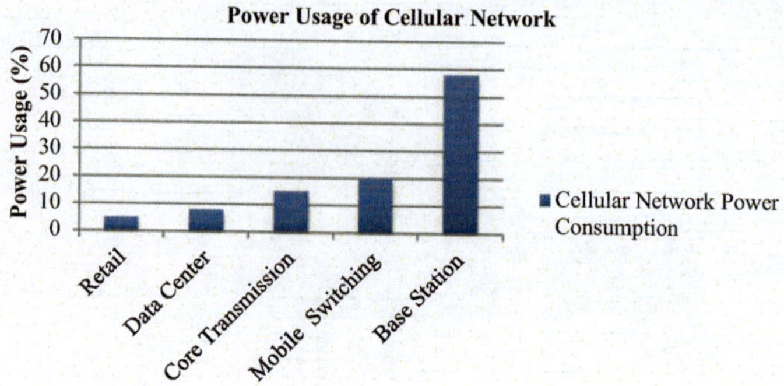

Fig. 8 Effect of cellular networks power consumption

- Incorporation of system foundation and system administrations to empower the system to remain gradually reactive what's more, to require less capacity to work.
- Low carbon emissions [2, 3].
- Energy savings in base stations [9, 10].
- Improvements in PA: Linear PAs → 90% wastage.

  DPD, Doherty, GaN based PA.

- Power saving mode: Sleep mode, discontinuous Tx/Rx.
- Optimization: BS placement, cell size (Fig. 8).

  Renewable energy:

- Sustainable bio
- Fuel
- Solar energy
- Wind energy

7 Conclusion

This paper shows a diagram of energy utilization issues in green correspondence systems and depicts organize energy sparing strategies. It is recognized that the regular energy utilization issue green correspondence arranges and portrays the systems to improve the energy efficiency of the system is used. This issue can be arranged illuminated by time and repetition spaces energy sparing issue crosses the various frameworks or systems is less comprehended. More endeavors are required from the displaying to specific arrangements. Radio access systems are additionally incorporated into this paper. Other significant issues incorporate the variety of portable administrator inclusion and the kinds of administrations starting with one territory then onto the next. As correspondence systems experience exponential development around the world, it is important to design as well as improve the remote access system.

References

1. Thirumaran JJ, Dhinakaran S (2018) A review on green communications. Int J Eng Res Technol (IJERT) 6(13):1–3
2. Alsharif MH, Nordin R, Ismail M (2013) Survey of green radio communications networks: techniques and recent advances. J Comput Netw Commun, Article ID 453893, 13
3. Strinati EC, Domenico AD, Herault L (2011) Green communications: an emerging challenge for mobile broadband communication networks. J Green Eng, CEA, LETI, MINATEC, 38000 Grenoble, France 267–301
4. Samdanis K, Rost P, Maeder A, Meo M, Verikoukis C (2015) Green communications: principles, concepts and practice. 1st edn. Wiley, Ltd.
5. Zhang H, Gladisch A, Pickavet M, Tao Z, Mohr W (2010) Energy efficiency in communications. IEEE Commun Mag, IEEE Xplore. 48(11):48–49
6. Mauri L, Rodrigues J (2012) Green communication and networking. In: Joel (eds) Second international conference, GreeNets 2012, Gaudia, Spain, Oct 25–26
7. Chen T, Harjula I, Kiviranta M (2011) Energy saving techniques in layered structure of LTE systems. In: Mobile VCE Wksp. Green Radio: In: conjunction with SDR'11-WInnComm-Europe, Brussels, Belgium, June 2011
8. Chen T, Yang Y, Zhang H, Kim H, Horneman K (2011) Network energy saving technologies for green wireless access networks. IEEE Wireless Commun 18(5):30–38
9. Hasan Z (2011) Green cellular networks: a survey, some research issues and challenges. IEEE Commun Surv Tuts 13(4):524–40
10. Mancuso V (2011) Reducing costs and pollution in cellular networks. IEEE Commun Magz 49(8):55–62
11. Gandotra P, Jha R, Jain S (2017) Green communication in next generation cellular networks: a survey. IEEE Access (99):1–1
12. Zhou X, Bai B, Chen W (2015) Antenna selection in energy efficient MIMO systems: a survey. China Commun 12(9):162–173
13. Mowla MM, Ahmad A, Habibi D, Phung QV (2017) A green communication model for 5G systems. IEEE Trans Green Commun Netw (99):1–1
14. Shaikh FK, Zeadally S, Exposito E (2017) Enabling technologies for green internet of things. IEEE Syst J 11(2):983–994
15. Zhang D, Zhou Z, Mumtaz S, Rodriguez J, Sato T (2016) One integrated energy efficiency proposal for 5G IoT communications. IEEE Internet of Things J 3(6):1346–1354

Recent Advancement in Fine-Grained Access Control and Secure Data Sharing Scheme for Distributed Environment

Vidhi Pandya, Atufaali Saiyed, and Khushi Patel

Abstract The demand for Big Data also increases with the increasing rate of usage of Internet. The storage, usage, and management of Big Data are the biggest challenge in today's world. Cloud computing and fog computing are new eras of IT industries which provide various computing services, distributed storage and resources over the internet which is available any-time-anywhere in the world using Internet but still, the security and privacy of these data is the major concern for business enterprises. The data shared among different users over distributed storage must be configured with fine-grained access control mechanisms. The distributed platform can use various encryption standards like DES, AES, Blowfish, Elliptic curve, RSA, etc. but they only offer confidentiality and integrity but for the fine-grained access control policies the mechanisms like identity-based encryption and attribute-based encryption algorithm are the best possible solution for authentication and confidentiality. Another option is homomorphic encryption which allows the processing of encrypted data. In this paper, the identity-based encryption, attribute-based encryption, and homomorphic encryption techniques are analyzed with their use cases and limitations which can be useful in the field of Industrial IoT, Industrial cloud, Big Data generated manufacturing industries, etc.

V. Pandya
Department of Computer Science & Engineering, Devang Patel Institute of Advance Technology and Research (DEPSTAR), Faculty of Technology and Engineering (FTE), Charotar University of Science and Technology (CHARUSAT), CHARUSAT Campus, Changa 388421, India

A. Saiyed
Department of Information Technology, Devang Patel Institute of Advance Technology and Research (DEPSTAR), Faculty of Technology and Engineering (FTE), Charotar University of Science and Technology (CHARUSAT), CHARUSAT Campus, Changa 388421, India

K. Patel (✉)
Department of Computer Engineering, Devang Patel Institute of Advance Technology and Research (DEPSTAR), Faculty of Technology and Engineering (FTE), Charotar University of Science and Technology (CHARUSAT), CHARUSAT Campus, Changa 388421, India
e-mail: khushipatel.ce@charusat.ac.in

P. K. Singh et al. (eds.), *Emerging Technologies for Computing, Communication and Smart Cities*, Lecture Notes in Electrical Engineering 875,
https://doi.org/10.1007/978-981-19-0284-0_45

Keywords Identity-Based Encryption · Attribute-based encryption · Homomorphic encryption · Cloud computing · Distributed computing

1 Introduction

With the advancement of Internet Technology, millions of computing devices are operated and connected through internet which generates 2.5 quintillion bytes of data each day [1]. According to the statistics of Forbes [2], "the amount of the newly created data in 2020 was predicted to grow 44X to reach 35 zettabytes (35 trillion gigabytes) and it is predicted that 175 zettabytes of new data will be created around the world in 2025". Today every industry like finance, health care, retail, government, etc. generates a large amount of data which is Big Data. Big data improves efficiency of a business model. It is a core of Industry 4.0. Still, the storage, usage, and management of Big Data are the biggest challenge in today's world [3]. As many IoT devices like sensors, actuators, RFID cards, Mobile devices are not able to store the data on the local storage; it increases the demand of third-party storage for sharing and processing the data. Nowadays many large companies like Amazon, IBM, Microsoft provide various services through cloud platforms [3]. Currently, cloud computing is the best option for the storage, management, and processing of Big data. However, the cloud computing offers many benefits [4] like outsourcing of storage, computation to servers using Internet using several fees, flexibility, scalability, less maintenance, access anywhere–anytime, etc. major issues [5] associated with cloud environment are (i) control and transparency rights of the data will be in hand of service provider and (ii) Trust and dependency. Other security concerns with respect to cloud computing are related to security end user machines and data, security of data during communication, and file system security.

Secure Interface and storage, secure data transfer, separation of data, and user access control are the major requirements for solving the issues related to trust and transparency in cloud environments. The use of encryption and digital signature at data producer side are the best ways to provide data confidentiality. Many strong and efficient schemes like DES, Triple DES, AES, Blowfish, IDEA, RSA, Elgamal, ECC, SHA-512, MD5, etc. are available for data encryption, digital signature, and hashing which can be used to provide confidentiality and integrity but it can't handle the fine-grained access control policies.

In order to provide legal access to authorized users, various access control mechanisms can be used but the main problem occurs when a server got compromised or a secret key got revealed by any of the users. Moreover, in order to accomplish the needs of resources and computing power with respect to the current scenario, most of the applications need to be migrated to the distributed and dynamic environment in which during the lifetime new users need to be added, some access control rights need to be changed and/or removing the current users. The cloud service providers need to adopt efficient access control and encryption schemes which solve both the issues related to authentication and confidentiality.

In order to solve the issues associated with conventional Public Key Infrastructure (PKI) and access control schemes, Identity Based Encryption was presented by Shamir in 1984. Attribute-Based Encryption (ABE) is another public key encryption technique which can be used to provide accurate access control to data stored on a third-party server by restricting privacy leakage of data producers and users. Attribute-based encryption works on the attribute the person has to manage access control policies. Another scheme which can be used to share the data securely on a third-party remote server is homomorphic encryption which leads to significant growth in the security domain and specifically in cloud computing. Homomorphic encryption provides the transmission securely.

In this paper, Sect. 2 covers Encipherment and Digital Signature schemes and Sect. 3 covers fine-grained access control mechanisms suitable for cloud, fog, and IoT-based applications which requires fine-grained access control, confidentiality, and integrity.

2 Encipherment and Digital Signature Schemes

The encryption techniques are used to protect the private data from unauthorized users. The user can simply encrypt the data before storing it in a cloud platform and at the time of downloading, the data can be decrypted before usage but it is not suitable at the time of data sharing among multiple users. In case of symmetric encryption, the owner needs to share the secret key if data needs to be shared [6]. In case of asymmetric encryption, data duplication will be the major issue as data needs to be converted into ciphertext using user's public key and shared through cloud and if data is shared among multiple users multiple copies will be generated to share the same data because data must be encrypted using receivers' public key and then need to be shared on cloud storage [6]. In this section, various techniques are covered for offering confidentiality and integrity.

2.1 Data Encryption Standard (DES)

DES falls under the category of symmetric key algorithm which is used to encrypt or decrypt the message/information, developed by IBM in 1977. DES algorithm is developed based on Horst Fiestel [7]. It is made up of fiestel network which divides the input into two parts—left and right part of 32-bit [8]. Besides 16 fiestel rounds, it includes two permutations, i.e., initial permutation which is held at the beginning, and the final permutation which takes place at the end. In the DES algorithm, round key generator is used to generate a key for each round. To create round keys, it takes input of 64 bits which is known as cipher key.

In the early days, DES was a comprehensively used algorithm in various applications like military, commercial, and communication systems [7]. According to the avalanche principle, strength of an algorithm is directly dependent on key size. As DES algorithm uses 56 bits key, the key length has become very small with respect to the super computers or workstations [9] In addition to this, DES is vulnerable to brute force attack, differential cryptanalysis attack [8] and man in the middle attack.

To overcome problems associated with DES, NIST tried to develop a new and more secure DES algorithm [9] which is later on known as Multiple—DES [8]. Triple DES, the most commonly used multiple-DES algorithm is proposed by IBM to overcome the disadvantages of DES algorithm [8, 9]. It is the same as DES but the only change is that it is used thrice on the same data [7] and used 168-bit key. To provide more confusion or to improve the security, later various modes of 3DES is introduced, i.e., DES-EDE2, DES-EDE3, DES-EEE3, DES-EDE2, and DES-EEE2 [9]. However, the brute force attack can be applied on 3DES to threaten security goals provided by 3DES [8].

2.2 Advanced Encryption Standard (AES)

As a trade-off of DES and Triple DES algorithms, National Institute of Standard and Technology (NIST) built up an algorithm called Advanced Encryption Standard [8]. AES was evolved in 2001 by Vincent Rijmen and Joan Daeman [10] based on Rijndael cipher [7]. The primary issue in the DES algorithm is the small key size. AES defeats this issue by giving three distinct key-size dependent on the quantity of rounds utilized in the encryption procedure. In the view of the key size, AES is isolated into three modes that are known as AES-128, AES-192, and AES-256 which uses 10, 12, and 14 rounds, respectively. A single round is comprised of four transformations, i.e., SubBytes, ShiftRows, Mixing, and Add Round Key.

Instead of Feistel network, AES is a substitution—permutation network [10]. Unlike DES, AES is not vulnerable to brute force attack and has a strong avalanche effect [8] so that it is more secure. There are many applications where one can use AES such as monetary transaction, security applications, and in a small device to send encrypted message or data over a network [7].

2.3 Blowfish

Blowfish is the symmetric algorithm that was developed in 1993 by Bruce Schneier. Blowfish is used as a stand in for DES or IDEA. The key size of blowfish is 32–448 bits. The data is encrypted at a rate of eighteen clock cycles per byte. Blowfish algorithm is divided into 2 phases. The key is break is divided into subkeys in the expansion process. A key size is not more than 448 bits [11]. There are 32 S boxes and P arrays. In the blowfish, 16 round Feistal and large keys are used. CAST-128 is similar to blowfish where fixed S boxes are used. The input is of 8 bit and output is of 32 bits. The input of the p array is used by the rounds, then after the last round half of the block is XOR [9]. The function divides the input of 32 bit into four eight-bit quarters and then this is used as input in the S boxes. After XOR final output is generated [11]. Considering it is a Feistal network, it can be reversed simply by XORing the *p* array into a Ciphertext block, then the p entries are used in reverse order [11]. In many applications blowfish is used because it consists of strong encryption.

2.4 RSA (Rivest–Shamir–Adleman)

The RSA [7, 12] algorithm is one of the first widely adopted public key encryption and digital signature algorithms which uses positive integers theory. This algorithm can be used for encrypting the data without exchanging the secret key and also used for digital signature. The modular exponential theorem is used for converting plaintext into ciphertext and vice versa in RSA. To achieve confidentiality, public key can be utilized for converting plaintext to ciphertext and a secret key which is known to user only, can be used to decrypt the data while in order to provide authentication, keys will be used in reverse order where the document/message can be encrypted by using the secret key which is only known to user that proves the identity of any user [12, 13]. Key generation process and encryption are slower. It is only suitable for encryption of small data units, not scalable, memory usage is high, vulnerable to factorization attack, timing attack and chosen plaintext attack are the limitations of RSA.

2.5 Elliptic Curve Cryptography (ECC)

Elliptic Curve Cryptography was developed in 1985 by Neal Koblitz and Victor S. Miller [14] which is based on Elliptic curve theory. It is like RSA but it is faster and provides many ways to the researcher of the algorithms. ECC is applied in wireless communication like RFID, mobile phones, etc. Currently, it is accepted commercially and many professional bodies have also adopted the algorithm. It provides security with a smaller key size than other cryptography algorithms. It covers algorithms like key exchange and digital signature. The building blocks of ECC is point multiplication [15]. In ECC the smaller parameters in the algorithm are used in comparison with RSA. The advantages of storage space, electrical power, and bandwidth are important in smart devices. The efficiency of ECC depends on the computations and fast algorithm for the multiplication.

3 Fine-Grained Access Control Mechanisms for Distributed Environment

Security is the major concern because of digitalization over the world. The data is store nowadays in the cloud while leads to security issues. The data which is stored is encrypted in any storage medium but if the algorithm or the operation is to be performed then the decryption is required. Moreover, if encryption is performed with symmetric or asymmetric encryption techniques, keys must be shared with cloud service providers which lead to privacy issues. Here the user must be authenticated before accessing data on the cloud as well as data must be secured from cloud service providers to solve the issues related to secure data sharing among multiple users. To overcome these challenges some techniques have been introduced which are covered in detail here.

3.1 Homomorphic Encryption

The homomorphic encryption was introduced to perform directly operation on the data which is encrypted. The homomorphic encryption concept can be explained by the example of the diamond shop. The shop owner wanted her workers in the shop to access and fixed things such gold, diamond, ring, etc. However, the owner did not trust the workers so she doesn't want the worker to directly access any of the things and so the owner decided to make the transparent impenetrable glovebox [16]. The box was transparent so things inside it are clearly visible but the door is lock with key. And only key access was with owner. Worker can only access the things with gloves. Thus, the jewelry shop analogy represents the function of homomorphic.

First time in 1978 Ronald Rivest, Leonard Adleman, and Michael Dertouzos [17] proposed the Homomorphic Encryption. In Homomorphic encryption specific type of computation is performed on the ciphertext which gives encrypted result and the result is also in the cipher text. If the enduser stores the private data on the unknown party server and wanted that no one other user can access the data. But in the traditional encryption, the data is protected when it is transmitting but data is not secured at the time of computation is being performed. In comparison, homomorphic encryption provides the data security from starting to end even while in the computations. Network intrusion, browser attack, and DDOS attacks become risks to cloud computing. Some of the issues in cloud computing are availability, data security, third-party control, privacy, and legal issues. To solve these security issues in the cloud computing, the solution is homomorphic encryption algorithm [18]. The solution is fit for the data in cloud computing and secure transmission in the cloud.

Example: In Fig. 1, the set of the integers are used. The first operation is performed is addition and two further the halving and doubling of number is performed. In the addition number 3 and 5 are taken as plaintext. The inputs are encrypted by doubling

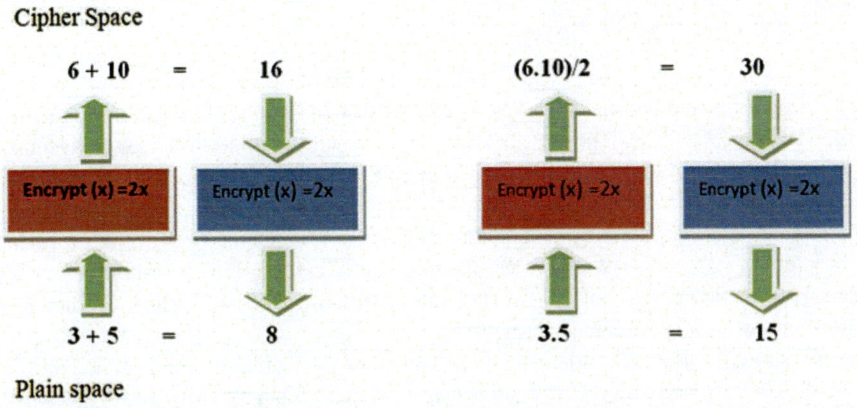

Fig. 1 Numeric value example using homomorphic encryption

the values 6 and 10, respectively. Then in the cipher text addition is performed between 6 and 10, which resulting in 16. This result is decrypted by dividing 2 and the value obtained is the actual value of plaintext. The Table 1 brief out the recent research remarks on homomorphic Encryption.

Table 1 Recent research directions in homomorphic encryption

Source	Year	Technique	Remark
[19]	2016	Homomorphic encryption	To reduce the size of cipher text for efficient data processing
[20]	2019	Fully homomorphic encryption	To enhance the Capability for the data by proposing the secure distributed model
[21]	2019	Fully homomorphic encryption	Main Focus on novelty of privacy preservation during the service selection process in an untrusted cloud service
[22]	2019	Homomorphic encryption and fully homomorphic encryption	Comparison of performance and efficiency of simple Homomorphic Encryption and Fully Homomorphic Encryption
[23]	2019	Fully homomorphic encryption	With the different key sizes, the experimental analysis is done and it gives the execution in millisecond for security level

3.2 *Identity-Based Encryption (IBE)*

IBE was proposed by Shamir in 1984 [24] but Bohen and Franklin has implemented the IBE in 2001 [25]. The IBE, public-key encryption suitable for large systems, solves the issues related to key management and certificate authority of conventional cryptosystem. The idea behind IBE was to protect important data from unauthorized access. In IBE, Private Key Generator (PKG) provides secret keys to the users and the identity of users like email, IP addresses, etc. can be used as a public key for encryption. Basically, IDE is divided into four sections, i.e., Setup, Private Key Extraction, Encryption, and Decryption.

As shown in Fig. 2, sender Alice uses receiver Bob's identity and pK_{PKG} to hide the message. After encrypting the message, Alice sends a message to receiver Bob via insecure channel. The receiver Bob can decrypt the message using sK_{IDBob}. Thus, IBE eliminates the need for PKI which was the heart of conventional approach [26] and makes certificate free system [27]. For the small framework, this methodology is exceptionally productive yet it isn't valuable for a large system to give the private key to all the users by utilizing secure channel. In this manner, to evacuate the utilization of secure channel, Tseng and Tsai changed the algorithm and included the idea of time update key with identity key (private key) [26]. PKG produces time update keys which are mapped with the client's private key occasionally and sends it to clients by means of public channel [26]. Other scenarios where IBE can be applied is on the cloud to provide confidentiality, fine-grained access control, backward secrecy, and forward secrecy [28]. To provide forward and backward secrecy, PKG must be online constantly and a secure channel must be set up among PKG and user while giving the private key [25]. One downside is that for an extremely enormous number of clients, PKG can be a bottleneck [25, 27] (Table 2).

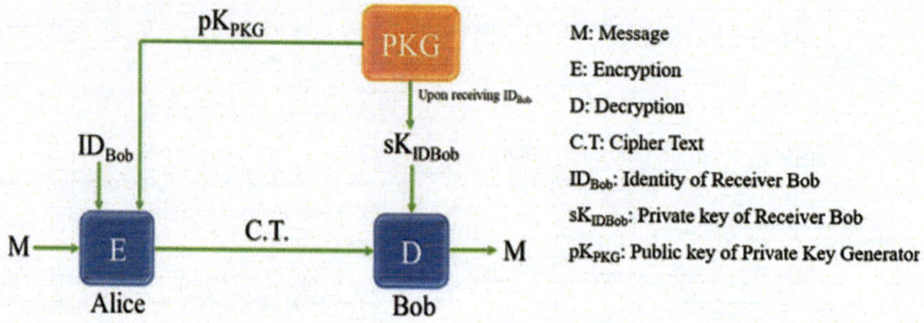

Fig. 2 Identity-Based Encryption [29]

Table 2 Recent research directions in area of Identity-Based Encryption

Ref	Year	Technique	Remark
[30]	2020	Self-Updated Encryption (SUE)	RS-IBE Scheme is used to encrypt the text but the problem occurs with this technique when a user wants to decrypt the ciphertext to get the original message. The proposed scheme of this paper uses RUE technique with RS-IBE to overcome the problem
[31]	2020	RIBE with SD method	In this paper, the proposed approach is made up of IBE Scheme, IBR Scheme, and two-level HIBE Scheme. As user's credentials changes dynamically in IBE, there is the requirement to provide a technique or method to handle key revocation. Sub-Set Difference Method is used by authors to solve the problem. It provides the security against chosen-ciphertext attack
[32]	2020	IBE under RSA (IBEET-RSA)	Security is the major concern in any of the fields. In this paper, security in Healthcare is discussed. To provide the basics goals of security, IBE under RSA mechanism is used in proposed system of paper for WBAN (Wireless Body Area Network). The proposed scheme is efficient and secure against chosen identity and chosen ciphertext attacks
[33]	2020	HIBAF	In this paper, Hierarchical Identity based encryption mechanism is used to provide security in fog computing. The burden of PKG is also reduced in the proposed approach but does not remove it completely
[34]	2019	IBE with FBAC (Function-Based Access Control)	This paper mainly focuses on the authentication issue in IoT. Devices must be authenticated to prevent over privilege access. To overcome this kind of problem, IBE with FBAC mechanism is proposed in this paper. This paper also provides the research direction for dynamic access control schemes and to verify the validity of IoT devices

3.3 Attribute-Based Encryption

Attribute-Based Encryption (ABE) [4, 35] is the public key encipher which was first invented by Sahai and Waters in 2006 that provides both confidentiality and fine-grained access control. In ABE, keys and ciphertext are based on the identity of the consumer or set of attributes inherited with the consumer. The ciphertext is generated in such a way that the decryption of the ciphertext is only possible when the attributes associated with key and the attributes associated with ciphertext matches. In ABE, attributes will be used at the time of encryption and secret keys corresponding to access policy will be used at the time of decryption to achieve revocation, collusion resistance, and scalability. The ABE also reduces the problem related to data duplication on cloud storage and owners can simply modify the access control policies to change the data sharing rights without using extra resources. The actors [4] involved in ABE are Data Owner (DO), Cloud Storage Server (CSS), Data User (DU)/Data Consumers (DC), and Trusted Authority (TA) (Key Generation Center). The authority uses attributes to generate the user's secret key along with the master key. The authority distributes public keys along with attributes to the Data Owner and private keys to Data users. For successful data decryption, at least some m components of attributes in encrypted text should match with attributes in secret key. In order to add a new system or change in an existing system, it is required to redefine the keys. The basic architecture of the ABE scheme is as shown in Fig. 3.

There are different four algorithms [35] involved in ABE are Initial Arrangement, Key generation process, Encryption, and Decryption. Initially, by using two random numbers p and q from the Zp* domain, the third-party trusted authority generates Public Parameters (PP) and master secret key (MSK). During the key generation, the trusted authority selects a distinct random number r from Zp* for each consumer and a distinct random number ri from Zp*for each attribute *X* from a set of attributes (AS) to generate each consumer's private key Kpt. The trusted authority generates the attribute keys for each consumer by utilizing the Master Key (MSK), Consumer's indices, and set of attributes (AS). The TA communicates the attribute group AGi for every attribute *X* from AS to CSS. The cloud storage generated encryption keys for each Consumers and Access structure tree T. The data owner can use the encryption

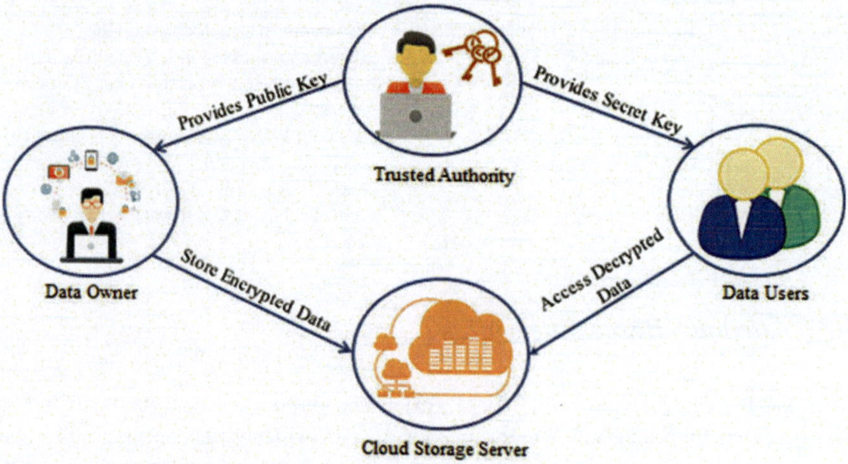

Fig. 3 Architecture of Attribute-based Encryption [4]

algorithm with public parameters (PP), set of attributes, and an access structure tree. The plaintext can be retrieved by using secret key and access structure parameters only if the set of attributes associated with ciphertext matches with the set of attributes associated with the secret key.

The data privacy, accurate access control rights, prevention of combination of different user's attributes, prevention of accessibility after certain amount of period elapsed, prevention of accessing the resources once user left the system are the functional parameters to evaluate any of the ABE Scheme and the total amount required for computing resources and storage are performance parameters [6, 36]. One limitation of ABE is the data owner needs to encrypt the data using each user's public key.

Generally, schemes can be divided into different classes: one is Key-Policy Attribute-based Encryption (KP-ABE) and another is Ciphertext- Policy-based attribute-based encryption (CP-ABE).

In the KP-ABE scheme, the public key received from the trusted authority and different attributes are utilized to convert the plaintext to ciphertext. The encrypted data can be decrypted only if the attribute associated with the user's private access policy matches with the attributes associated with ciphertext [37]. The KP-ABE scheme is less expressive as compared to others. Moreover, it is not suitable for synergic systems because access rights are incorporated with private keys instead of ciphertext. In the CP-ABE, the encrypted data needs to be attached with access policy and the user's set of attributes can be used to describe a private key. It is like a role-based access control scheme. The public key received from the key generation authority is used by the data owner to specify the access plan which can be used to encrypt the data. The data can be decrypted by the consumer using a private key received from the trusted authority if the attributes associated with the private key match with the associated attribute of ciphertext [37].

Many extended versions of ABE are Multi-authority, proxy re-encryption and accountability, revocation, etc. Key-Policy ABE can be further categorized into two different categories [4] like (i) Multi-Authority (ii) Revocable. Ciphertext-Policy ABE (CP-ABE) can be categorized into categories like (i) Hidden- Policy (ii) Hierarchical (iii) Revocable (iv) Multi- Authority (Table 3).

4 Use Cases: Health Care

The smart healthcare is the most promising area which uses cloud infrastructure to share the data among stakeholders like hospitals, doctors, patients, etc. The privacy of shared smart healthcare data is the most crucial requirement in current scenario [44]. As shown in Fig. 4, the patient data must be shared among multiple stakeholders through the cloud environment to offer the services like remote tele surgeries and diagnosis, remote consultancy, intrahospital monitoring. The hospital authorities store the patient data on the cloud infrastructure which can be accessed by stakeholders anywhere–anytime but it attracts different security threats [45]. In order

Table 3 Recent research directions in the area of Attribute-based Encryption

Ref	Year	Type of Paper/Type of Proposed Scheme	Remark
[3]	2019	Key Generation	Secure ABE key generation scheme for Industrial cloud is proposed to solve the issues related to privacy of users in terms of attributed associated with some products by separating attribute auditing and key generation functions
[38]	2018	Authorization Framework for Electronic Health Records	Cloud-based Electronic Heath record scheme which uses the concepts of semantic web like Web Ontology Language (OWL) and Semantic Web Rule Language (SWRL) is proposed. *Limitations*: The proposed scheme may open the door for another category of risk which is not considered
[39]	2020	Survey on KP-ABE, CP-ABE, anti-quantum ABE, generic ABE constructions, and their variations	In this Paper, ABE schemes are analyzed on the basis of security and performance. *Future directions*: (i) Existing Fully secure ABE schemes are not up to the mark in terms of efficiency, access policy a usage of random oracle model. (ii) Complexity of Bilinear pairing is higher. (iii) anti-quantum ABE is the new direction
[40]	2019	Linear Secret sharing based policy update and file update CP-ABE Scheme	The Proposed scheme uses proxy cloud service provider for policy updates and secret value used for policy update and file update is invariable for security enhancement. *Future Direction*: Time Complexity can be reduced for file update. Blockchain technology can be integrated for transparent policy update
[41]	2018	ciphertext-policy attribute-based encryption with equality test	The authors proposed the equality tests scheme which is useful for delegating a cloud server for equality test without having access to plain text. *Future Direction*: Space Complexity can be focused. Moreover, the model is tested for indistinguishability- chosen plaintext attack but still more security level can be achieved
[42]	2019	Survey: Revocation in CP-ABE	*Future Directions*: Revocation problem needs to be solved. Fault tolerance, single point of failure problem and user accountability can be resolved by adopting Decentralized Blockchain based environment. Reducing the computation overhead is another area which attracts researchers

(continued)

Table 3 (continued)

Ref	Year	Type of Paper/Type of Proposed Scheme	Remark
[43]	2020	Survey on CP-ABE for access control, authentication, Signcryption and authorization	*Future Direction*: Revocation problem and adoption of CP-ABE in the area of fog computing and Edge computing is in demand

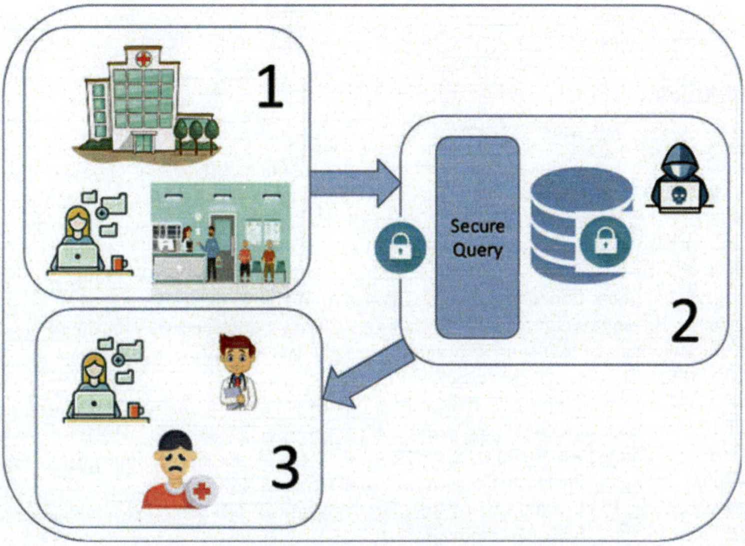

Fig. 4 Requirement of fine-grained access control in healthcare [46]

to offer the fine-grained access control, the IBE and ABE are the best available options while to process the encrypted data without decrypting the data homomorphic encryption is the best available option.

5 Conclusion

This paper mainly focuses on confidentiality of data and the access control when the data is stored at a third-party server. The various encryption technologies which fall under broadly in two categories, i.e., traditional and modern are discussed in this paper. Basically, the traditional encryption technologies like DES, AES, RSA, ECC, etc. work with confidentiality and integrity but according to the current scenario efficient access control is required in case of sharing data among multiple users in a distributed environment. The IBE and ABE consider the same and instead of hiding the data to provide confidentiality, it avoids letting out the stored information to unauthorized parties by controlling the access to the data. Many researchers have already proposed different versions of IBE and ABE but still limitations are there which opens doors for attackers. Currently, IBE and ABE are the emerging area for researchers for privacy preservation in cloud computing. Homomorphic is another technique which is used to provide end-to-end communication security in distributed environments by performing specific type of computation. Extended versions of Homomorphic encryption schemes are also available for secret sharing but still doors are open for researchers.

References

1. Schultz J (2019) How much data is created on the internet each day? 08 June 2019
2. Press G (2020) 6 predictions about data in 2020 and the coming decade, 6 Jan 2020
3. Song Y, Wang H, Wei X, Wu LH (2019) Efficient attribute-based encryption with privacy-preserving key generation and its application in industrial cloud. Secur Commun Netw
4. Kumar P, Alphonse P (2018) Attribute based encryption in cloud computing: a survey, gap analysis, and future directions. J Netw Comput Appl 108:37–52
5. Charanya R, Aramudhan M (2016) Survey on access control issues in cloud computing. In: International conference on emerging trends in engineering, technology and science (ICETETS)
6. Liu C-W, Hsien W-F, Yang CC, Hwang M-S (2016) A survey of attribute-based access control with user revocation in cloud data storage. Int J Netw Secur 18(5):900–916
7. Maqsood F, Ahmed M, Ali MM, Shah MA (2017) Cryptography: a comparative analysis for modern techniques. Int J Adv Comput Sci Appl 8(6):442–448
8. Verma A, Guha P, Mishra S (2016) Comparative study of different cryptographic algorithms. Int J Emerg Trends Technol Comput Sci (IJETTCS) 5(2):58–63
9. Bhanot R, Hans R (2015) A review and comparative analysis of various encryption algorithms. Int J Secur Appl 9(4):289–306
10. Harini M, Pushpa Gowri K, Pavithra C, Selvarani MP (2017) Comparative study and analysis of various Cryptographic Algorithms. Int J Sci Eng Res 8(5):2229–5518
11. Forouzan BA, Mukhopadhyay D (2015) Cryptography and network security. McGraw Hill Education (India) Private Limited
12. Arora R, Parashar A, Transforming CC (2013) Secure user data in cloud computing using encryption algorithms. Int J Eng Res Appl 3(4):1922–1926
13. Bhardwaj A, Subrahmanyam GV, Avasthi V, Sastry H (2016) Security algorithms for cloud computing. In: International conference on computational modeling and security (CMS 2016)
14. Li X, Shen X, Chen H (2011) Elgamal digital signature algorithm of adding a random number. J Netw 6(5):774
15. Amara M, Siad A (2011) Elliptic curve cryptography and its applications. In: International workshop on systems, signal processing and their applications
16. Martins P, Sousa L, Mariano A (2017) A survey on fully homomorphic encryption: an engineering perspective. ACM Comput Surv (CSUR) 50(6):1–33
17. Parmar PV, Padhar SB, Patel SN, Bhatt NI, Jhaveri RH (2014) Survey of various homomorphic encryption algorithms and schemes. Int J Comput Appl 91(8)
18. Zhao F, Li C, Liu CF (2014) A cloud computing security solution based on fully homomorphic encryption. In: 16th International conference on advanced communication technology
19. Ojha S, Rajput V (2017) AES and MD5 based secure authentication in cloud computing. In: 2017 International conference on I-SMAC (IoT in social, mobile, analytics and cloud) (I-SMAC)

20. Patel J (2017) Secure hashing algorithm and advance encryption algorithm in cloud computing. Int J Comput Inform Eng 11(6):753–757
21. Kaur N, Singh H (2012) Efficient and secure data storage in cloud computing through blowfish, RSA and hash function. Int J Sci Res (IJSR) 4(5)
22. Geng Y (2019) Homomorphic encryption technology for cloud computing. Procedia Comput Sci 154:73–83
23. El-Yahyaoui A, ECH-CHERIF EL KETTANI MD (2019) A verifiable fully homomorphic encryption scheme for cloud computing security. Technologies 7(1):21
24. Li F, Khan MK (2011) A survey of identity-based signcryption. IETE Tech Rev 28(3):265–272
25. Boldyreva A, Goyal V, Kumar V (2008) Identity-based encryption with efficient revocation. In: Proceedings of the 15th ACM conference on computer and communications security
26. Tseng Y-M, Tsai T-T, Huang S-S, Huang C-P (2016) Identity-based encryption with cloud revocation authority and its applications. IEEE Trans Cloud Comput 6(4):1041–1053
27. Li H, Dai Y, Tian L, Yang H (2009) Identity-based authentication for cloud computing. In: IEEE international conference on cloud computing
28. Wei J, Liu W, Hu X (2016) Secure data sharing in cloud computing using revocable-storage identity-based encryption. IEEE Trans Cloud Comput 6(4):1136–1148
29. Baek J, Newmarch J, Safavi-Naini R, Susilo W (2004) A survey of identity-based cryptography. In: Proceeding of Australian Unix users group annual conference
30. Lee K (2020) Comments on "secure data sharing in cloud computing using revocable-storage identity-based encryption". IEEE Trans Cloud Comput
31. Lee K (2020) A generic construction for revocable identity-based encryption with subset difference methods. PLoS ONE 15(9):0239053
32. Ramadan M, Liao Y, Li F, Zhou S, Abdalla H (2020) IBEET-RSA: Identity-based encryption with equality test over RSA for wireless body area networks. Mobile Netw Appl 25(1):223–233
33. Farjana N, Roy S, Mahi M, Nayeen J, Whaiduzzaman M (2020) An identity-based encryption scheme for data security in fog computing. In: Proceedings of international joint conference on computational intelligence
34. Yan H, Wang Y, Jia C, Li J, Xiang Y, Pedrycz W (2019) IoT-FBAC: Function-based access control scheme using identity-based encryption in IoT. Futur Gener Comput Syst 95:344–353
35. Rajeev A, Kenu TS, Arun R, Babu SS (2016) A survey on attribute based encryption schemes for data sharing. J Comput Eng 19–23
36. Garg H, Dave M (2019) Securing user access at IoT middleware using attribute based access control. In: 2019 10th International conference on computing, communication and networking technologies (ICCCNT)
37. Edemacu K, Park HK, Jang B, Kim JW (2019) Privacy provision in collaborative ehealth with attribute-based encryption: survey, challenges and future directions. IEEE Access 7:89614–89636
38. Joshi M, Joshi KP, Finin T (2019) Delegated authorization framework for EHR services using attribute based encryption. IEEE Trans Serv Comput
39. Zhang Y, Deng RH, Xu S, Sun J, Li Q, Zheng D (2020) Attribute-based encryption for cloud computing access control: a survey. ACM Comput Surv (CSUR) 53(4):1–41
40. Li J, Wang S, Li Y, Wang H, Wang H, Wang H, Chen J, You Z (2019) An efficient attribute-based encryption scheme with policy update and file update in cloud computing. IEEE Trans Ind Inform 15(12):6500–6509
41. Wang Q, Peng L, Xiong H, Sun J, Qin Z (2017) Ciphertext-policy attribute-based encryption with delegated equality test in cloud computing. IEEE Access 6:760–771
42. Al-Dahhan RR, Shi Q, Lee GM, Kifayat K (2019) Survey on revocation in ciphertext-policy attribute-based encryption. Sensors 19(7):1695
43. Belguith S, Kaaniche N, Hammoudeh M (2019) Analysis of attribute-based cryptographic techniques and their application to protect cloud services. Trans Emerg Telecommun Technol 3667
44. Patel K, Vyas S, Pandya V, saiyed A (2019) IoT: leading challenges, issues and explication using latest technologies. In: 2019 3rd International conference on electronics, communication and aerospace technology (ICECA), pp 757–762.https://doi.org/10.1109/ICECA.2019.8821970

45. Patel A, Shah N, Ramoliya D, Nayak A (2020) A detailed review of cloud security: issues, threats & attacks. In: 2020 4th International conference on electronics, communication and aerospace technology (ICECA), pp 758–764
46. Archer D, Chen L, Cheon JH, Gilad-Bachrach R, Hallman RA, Huang Z, Jiang X, Kumaresan R, Malin BA, Sofia H (2017) Applications of homomorphic encryption. HomomorphicEncryption.org, Redmond WA, Tech. Rep

Context-Based Multi-System Integration for Authorization Using Dynamic Privilege Escalation

Nidhi Sinha, Meenatchi Sundram, and Abhijit Sinha

Abstract Since the inception of the Internet of Things (IoT), the question of Security had been pertaining all along. The initial research was focused on encryption. Then, focus on speed and something that takes less space over the network and storage; multiple study findings also tried to address application-level Security like Authentication. Despite advancements in Authentication, Authorization, and application of context-aware Security, much of the research focus is still needed to address concerns democratic Security, applicability, generic integration, and interoperability of heterogeneous systems. This paper addresses integrating with heterogeneous systems and how the solution can be enhanced to implement more comprehensive interoperable. The paper finds a trusted way to authenticate the device and ascertain its identity, how to trust it, and how much to charge—Authorize all for the entity outside the ecosystem keeping context as a crucial decision point.

Keywords Internet of Things · Authentication · Nonce · Privilege Escalation · Grant · Permission · Dynamic authorization · Context · Context-aware security

1 Introduction

Whenever transmission occurs, the central concern to prioritize is Security and privacy [1]. When the talk is about the devices that interact, it does not know what kind of security threat may occur before, during, and post connection. Humans have intuitions about something wrong that can happen whenever they enter an unknown and unsafe zone; however, as devices are non-living, they never fear getting hacked or destroyed. They either keep working as programmed or stop, and in many scenarios, they may slow down. Here, the concept is instrumented to make the device understand or have knowledge about the threats that may cause physical damage to the machines/devices and affect the transmission and compromise the network. Security in IoT is one of the most important aspects to be handled with care, and all

N. Sinha (✉) · M. Sundram · A. Sinha
Garden City University, Bengaluru, India

P. K. Singh et al. (eds.), *Emerging Technologies for Computing, Communication and Smart Cities*, Lecture Notes in Electrical Engineering 875,
https://doi.org/10.1007/978-981-19-0284-0_46

the precautions should be taken care of when there is talk about overall technology. Security can be a cause of concern and treated with appropriate Authentication & Authorization [2, 3]. It can be more fine-tuned when granting permission will be set even when devices are packed together. The priority [4] setting of devices can be changed, and the Authorization is also be shifted to more prior or less prior conditions. From the context perspective, the level of approval can differ depending on its situation (context). The device or back-end involved in decision-making behaves smarter and is aware of the overall context. The main objective of all these activities is to make wise (smart) and intelligent devices like humans. The application of the proposed solution will ensure smooth communication where the escalation of privilege is required (a) emergency time calling an ambulance. (b) alerts by govt agencies about some calamity like a tsunami (c) traffic clearance for VVIP movement, (d) broadcasting health alerts to nearby people on their IoT enabled devices, and similarly many more use-cases.

First comes Authentication, which indicates whether the operator/requesting entity (a person or another device) is the one it claims to be or not. The credentials provided by the organization play a vital role when the Authentication of a person or device. Everything has its unique identity and critical password in every case of accessing things or data, or network. Regarding Authorization, the entity has permission to access a particular space. Therefore, it can be taken care of as authorization states (access levels) where the entity can perform specific operations. Based on context, the absolute priority of the device gets change to bring out the appropriate action by the interconnected devices.

2 Review of Literature

2.1 *Review of Literature Done Based on Authentication survey Summary*

Context-Aware Scalable Authentication (CASA) [5] makes Authentication easier or harder based on some user parameters (e.g., a user's location or time since the last log in) comparatively than making it equally rigid for all cases. However, the core idea of this paper is based on the current user location, and typing pin makes Authentication much complex for the unauthorized person on run time, which brings more Security at run time. The paper does not discuss Authorization and setting permission dynamically.

The author discussed the context-awareness [6] techniques for authentication and access control mechanisms. Context-awareness and context-security concepts highlight contextual attributes used to support and enhance IoT's authentication and access control mechanism. An authentication framework was described by Goel et al. [7]. To create a context-aware environment and encourage role-based decision-making, the authors used a combination of destination access control and a user's background, authentication policies, and other factors tagging with a lightweight a provision for extending is included in the framework. Malek et al. [8] proposed a context-aware paradigm. For Context-aware environment and computing authentication services, the proposed design can allow users to take action in context-aware computing. Domains establish trust and share secrets between parties based on their ideal level of assurance. The context-aware authentication service uses Context-data.

2.2 Authorization Survey Summary

Hu and Weaver [9] proposed a flexible and dynamic context-aware system. For healthcare applications, there is a security infrastructure. The control model extends the accessibility the role-based access control mechanism by tying access privileges to context-sensitive data restrictions. They listed their model's capabilities as follows, demonstrating how authorization decision-making is based on context Concerning tasks details. Wullems et al. [10] applied context-aware computing to the environment to improve. The architecture of Authorization the architecture suggested facilitates the role-based access control process. Access control policy that is mindful of the context. They explained the use of dynamic architecture to incorporate the architecture. However, the study not addressed run-time Authorization and priority setting.

2.3 Context-Based Security Survey Summary

Matos et al. [11] proposed a system that stated context information about any system, applications, and devices' (information) environment and events related to integrating the different vertical domains. The proposed architecture in that paper supports Edge to fog-centric context sharing, which reduces unwanted data exchange. The paper deals with two challenges; the first is the heterogeneity, and the second is scalability and real-time sharing. However, the paper stated that dynamic behavior is missing in real-time sharing. Therefore, Context-aware Security is introduced but not used. In this survey, Perera et al. [12] observations are based on context-aware on the perspective Internet of Things. The paper analyzed the in-depth context lifecycle over the IoT effect of context. The whole paper states that the challenges in context-aware are handled in desktop, web mobile, sensor network, and pervasive computing. Authors Filho et al. [13] stated that security breaches in a private network by IoT edge devices are an extensive concern to IoT's comprehensive adoption.

The author discussed four different technologies named Trust execution environment (TEE), Security by separation, Blockchain for trusted communication, Context-aware Security to cover various vulnerabilities and threats in the same machine. The author stated the research limitation is overhead imposed by the cryptographic operation required by CoT and Blockchain; to overcome this, run allotment and discard run play a crucial role in overcoming the overhead problem. Author Ghali et al. [14] proposed a new access control scheme, CATRAC: Context-Aware Trust- and Role-Based Access Control for Composite Web Services' higher-level control and Security. In this semantic key and public key, cryptography is used for Authentication. CATRAC addresses the threats used for both single or composite web service scenarios. Both role-based and trust-based access controls combine to provide the enhanced scheme to provide a better framework. The role-based access control method restricts the IoT environment limitation is going to be overcome in our proposed design of run-time security consideration.

Author Perera et al. [15] proposed the Context-Awareness for Internet of Things (CA4IOT) architecture to help users by automating selecting the sensors according to the problems/tasks at hand. The main focus is automated filtering, fusion, and reasoning applied to the collective data stream using sensors. The CA4IOT architecture consists of four layers: Data, Semantics, and Context Dissemination Layer (DSCDL), Processing and Reasoning Layer (CPRL), Context and Semantic Discovery Layer (CSDL), and Sensor Data Acquisition Layer (SDAL) to provide context-aware Security however paper lacking of context-aware Security. Author Jagadamaba et al. [16] focuses on adaptive security schemes which use the security parameters like context and trust. Many other security parameters are applied to the different applications and evaluated by security credentials like control, privacy, and context awareness. The findings are that context acquisitions make the system adaptive to new changes context and trust are essential to developing an adaptive and accurate security framework. Therefore, security schemes must be attended to carefully by incorporating trust and context-aware middleware. The paper focuses on Authentication, Authorization, and Access management; however, dynamic Security was not addressed.

Jason et al. [17] identified three benefits for building context-aware applications. First, the infrastructure can be neutral of hardware platform, operating system, and programming language, so many different devices and applications can access the infrastructure. Second, middleware decouples the individual pieces of infrastructure toes upgraded independently and dynamically while the system is still running. Third, devices use sensors, processing power services, and data in the infrastructure. Said et al. [18] proposed a context-aware security controller. Using a context-aware security controller to reduce security risks stimulates protection frameworks in response to the current situation details such as the type of application and the device's capabilities. Context-aware access control was proposed by Zhang et al. [19]. A dynamic role model for ubiquitous applications is a scheme for access control dependent on the user's location and access. However, it is restricted. Authentication processes that are feasible to protect applications that are widely used. Thiago et al. [3] proposed an extension of UPnP Universal plug-and-Play, which aims to connect consumer electronics and smart devices from different makes and models (Hetrogenity and Interoperability). UpnP-UP follows the Authentication and Authorization mechanism for UpnP standards, maintaining backward compatibility with earlier versions of UpnP. However, no talk about context-aware security dynamically.

2.4 Problem Introduction

As we see in the scenario depicted in Fig. 1 the ambulance needs to connect with multiple devices to request for green signal/corridor from traffic signaling systems. Requesting pass or warning vehicles those moving ahead to provide access and warning to the hospital to ensure the emergency ward is ready, despite the underlying fact that the ambulance in question is possibly not associated with any systems. Other systems in the above description need to ensure that the request urged by ambulance can be authenticated as someone it can trust and authorized to allow operation. It cannot be done with typical certificate/signature validation alone. It can establish identity (authenticate), but what if the system is hacked or the intended activity is not approved, which must be validated authorization check.

This paper attempts to solve an IoT device trying to communicate with different devices with their control (IAM) system—overall, in a heterogeneous environment. Allowing, at the same time, successfully authenticate and authorize to allow execution. As this paper purely focuses on security aspects, few assumptions are made: The IoT devices can connect using a shared medium like Bluetooth/GATT (Bluetooth Low Energy Generic Attributes)/ZigBee. The protocol to communicate is also familiar, similar to CoAP. All devices implement and provide a standard interface for communication using CoAP like authenticating. To do an operation like asking to open the gate, send an alarm, or in the case of a vehicle, it will flash a message to the driver, and a hospital will ring warning for medical staff to be ready. The system will provide a green signal blocking other traffic to ensure a fast pass across the crossing for the ambulance.

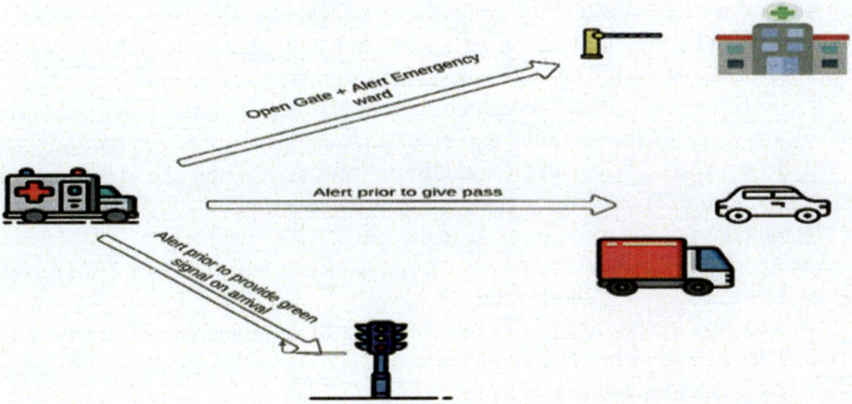

Fig. 1 The ambulance needs to connect to multiple heterogeneous devices/systems to get authorized and communicate [18, 20]

3 Solution

Summarizing the problem statement—this paper intends to solve Authentication and authorize the action to be performed by requesting an IoT device. It needs to be achieved under-listed constraints such that the devices can be heterogeneous, for example: of different makes and models. Moreover, the ecosystem they are controlled in is heterogeneous and need not know each other or be connected. Example: traffic system, hospital.

3.1 Overview

Figure 2 below depicts a high-level sequence of action. In the below scenario, the ambulance hereafter will also be called Requestor. It needs privileged access to send messages or requests, attention to various Requestees like vehicles moving ahead to get past, or requestee like traffic signal to get green signal blocking others for speedier transport or requestee like a hospital for alerting staff be ready and open the gates. The challenge for the Requestee devices is two-fold, viz:

- Find a trusted way to authenticate the device and ascertain its identity.
- How to trust and how much to charge—Authorize all for the entity outside the ecosystem.

The high-level flow is explained in steps at this moment.

Requestor like an ambulance connects with its own IAM server (belonging to the same ecosystem to which the ambulance belongs, like a hospital or ambulance control system like 108 in India) to get a unique token. It needs to be passed to other devices to establish identities and request authorization.

The IAM system where the ambulance belongs registers the token and sets parameters (detailed further in the paper) to help establish identity and purpose.

At the same time, the device (Requestor) on the move (or not) can try to connect with other devices in the vicinity (radius of reach like in case of Bluetooth 10 m, or in case of Zigbee up to 100 m) and share the token.

The Requestee (device receiving the request) connects with their own IAM server for Authentication and Authorization.

The IAM server, upon receiving a request, the IAM server identifies that the token does not belong to their system, instead of from Central Identity and Access Registry hence requests it to establish identity.

Once identity is established, and purpose is resolved, the Requestee IAM server evaluates the context and may authorize the action. Once approved, the Requestee device allows acting as a message to the driver or raise an alert in the hospital emergency ward to be better prepared.

The solution assumes the ambulance indiscriminately keeps broadcasting about the emergency. The context is that the ambulance ecosystem evaluates the need to raise the mode to the emergency level. At the same time, it tries to connect to relay its need of emergency, the other systems (within their ecosystem) try to evaluate the condition with fitment in their context. For example, the moving vehicles on receiving distress signal post successful Authentication and Authorization will assess' the context. They may deem fit to show the message to the driver (which can be intrusive at other times). On a comparable basis, a security camera-based IoT device in a nearby cafe getting distress signal from ambulance even post successful Authorization of device, when evaluating the context of the fitment of the message, may deem unfit to process and act any.

3.2 Details

The detailed steps involved are broken into two significant steps. However, before the actual Step kicks in. First, the role of Central Identity and Access Registry and Requestor IAM needs to be clarified. The Central Identity & Access Registry acts as a central governing and authorizing identity system, managed by a government body or consortium of industry bodies as a self-regulated and trusted party. Any organization or industry that needs to have its devices with escalated privilege must be registered with the central identity system. The registration and trust granted need to be mutual and require a manual approval process.

Step 1. In Step 1 above, the device needs complete access. In this case, the "Requestor device" comets with its own IAM server. The IAM server validates the context (evaluate need), generates a Nauthorization pass, and registers the same with the Central Identity server. Then, the Central Identity server initiates a mock test to validate the Nauthorization pass with the recorded IAM service before approving for requesting from other systems. Once registered successfully with the Central Identity server, the generated pass is returned to a device for further usage (Figs. 3 and 4).

Step 2. The steps involved are detailed below, where the requesting device (Requestor) needs access while broadcasting or trying to connect to multiple devices en-route. Requestor devices keep broadcasting or connecting to multiple devices in route, indicating its intention and passing its identity embedded in Nauthorization Pass to all devices trying to connect to and get validated.

It is more of async communication, where the Requestor generally does not wait for success or response (however, it can if needs to be). Instead, the Requestee devices can react once it finds action authorized in ways it can, like show distress message on driver's head-mounted display screen to give pass or traffic signal can block other ways except for the requestor device's path is arriving.

The Requestee device, upon receiving the request, the Requestee device would validate the token for two purposes, identify the device, and get it authorized of action that the device needs to perform. In this paper, we intend to have a genuine intention of action that can be taken wisely by the Requestee device itself (as in case of grabbing attention during an emergency). However, specific activities can also be chipped in.

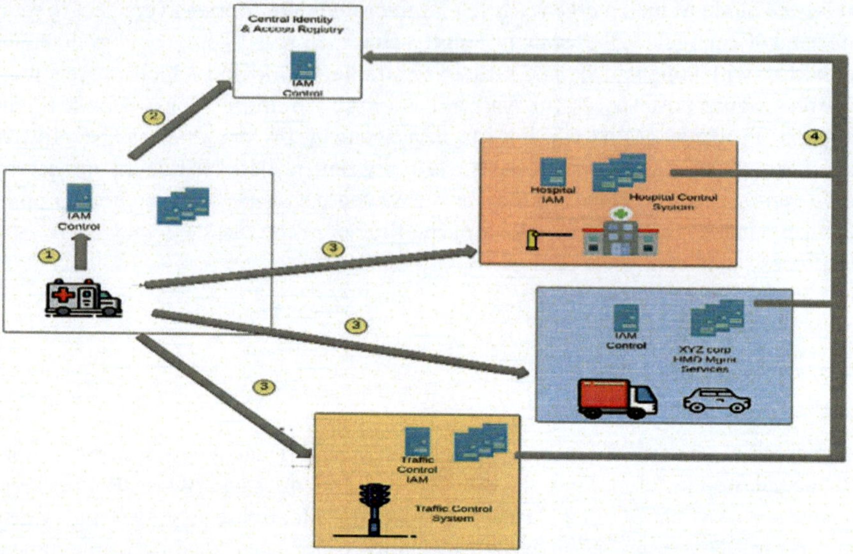

Fig. 2 Solution overview [20] depicts the high-level flow of devices connecting authenticate and authorize

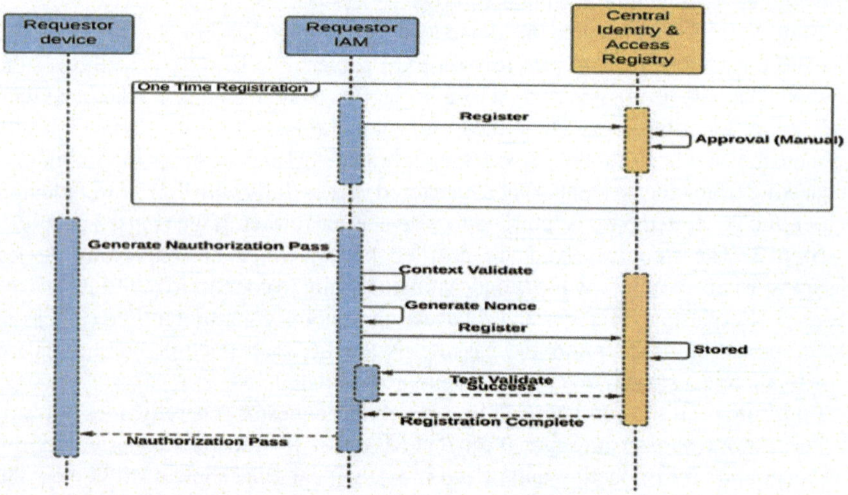

Fig. 3 A sequence of steps involved in generating Nauthorization pass

Requestee IAM, in turn, would try to validate the token, which is not registered locally and hence would fail. It would, in turn, get it authenticated from Central Identity and access registry. The Central Identity and Access Registry system is authenticated locally. However, it double checks with issuing organizations regarding

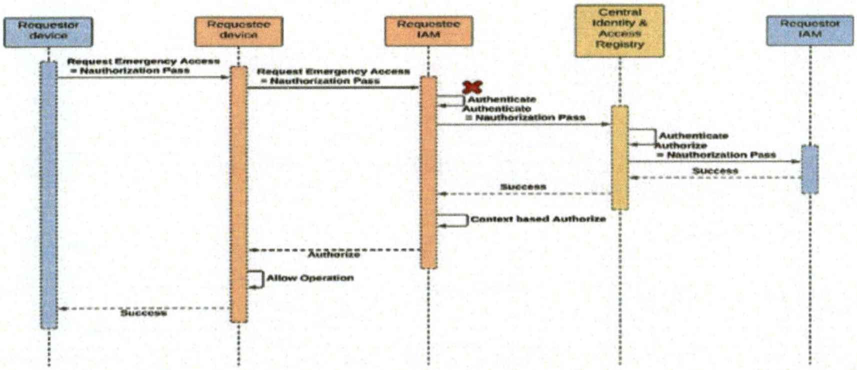

Fig. 4 The sequence depicting validation of requesting device and getting authenticated through a chain of trust mechanism

validity. The issuing organization can indicate revocation of Nauthorization pass as context (need) is no longer valid or purpose solved.

Once approved, the requestee IAM can re-evaluate the context of what it can do with the emergency request pending, like display message or give signal or ring alarm.

The optional last success message from the Requestee device to Requestor can be async and need not necessarily indicate the success or failure of the request. However, it can be devised to return a message for better communication.

4 Implementation and Evaluation

This section will detail the implementation (setup) and evaluation against various parameters.

4.1 Implementation

Two server setups were programmed: (i) the Central Identity and Access Registry and (ii) the Requestee/Requestor IAM. The code for Requestee and Requestor IAM are clubbed as one web service since reason in the real-world that the same organization can act like Requestee or to evaluate Requestor Identity server and are essentially bare IAM server which integrates with Central Identity and Access Registry to either register it is a device or being called upon for evaluation.

The web services are written in Java 8, run within a docker container supporting 1 CPU and 1 GB of RAM. As in many IAM server cases, the database used is relational SQL DB, which we used PostgreSQL. The container runs on a desktop with a processor Intel i5-1020U with four cores with a base processing speed of 1.6 GHz. The desktop has an overall RAM of 16 GB DDR4, which clocks at a speed of 2666 MHz. The host Operating system used is desktop Ubuntu 20.04, while the docker container uses CentOS using Sun JAVA. The Requestor and requestee device are simulated using Postman as their primary job is to hit the API endpoint exposed by either Requestor/Requestee IAM.

4.2 Evaluation

Below we look at various parameters and evaluate how the solution is against them. Finally, we evaluate against selected OWASP TOP 10 security threats applicable to evaluate Security.

Broken Authentication. The Nauthorization pass generated is stored in a central server. Any device passing the Nauthorization Pass to the Requestee device (and hence Requestee IAM), the pass is re-evaluated with Requestor for validity apart from local verification in Central Identity and Access Registry. It makes the solution grim for being hacked of manipulation or extended use.

Sensitive data exposure. The communication assumed is encrypted as in SSL over COAP or ZigBee, making it indifferent to raw text reading or sniffing. Also, the Nauthorization Pass critical systems; this just a token that does not reveal anything. However, it can also be implemented based on JWT with signed content.

Broken Access Control. The access control (Authorization) cannot be used in a prolonged fashion. Whenever a Requestee gets a request for operation, it connects to Central Identity and Access Registry, which connects with the originator to re-evaluate the applicability. Thus, it Makes it secure against prolonged unwarranted use and keeps the usage under check.

Action Replay. Action Replay is possible by hackers gaining the entire packet stream in transit and replaying the request payload. The net result would be overwhelming for requestee device sets (devices & IAM) and may eventually lead to system failure or serve other genuine requests. However, the duration of usage is limited and hence the impact. It can also be mitigated by intelligent monitoring systems employed by the current IAM system to distinguish between genuine and hacked repeat offenses.

Storage Overhead. The device needs to store the Nauthorization Pass and discover the device while passing its pass alone. The Nauthorization Pass, like other tokens, is lightweight and contains UUID in a gene. However, it can be extended to have JWT like structure for extended use. The UUID recommended solution does not need more than few additional bytes and is hence very lightweight to store and manage.

Communication Overhead. The suggested solution is like token passing over HTTP(S) or COAP and no different. The back-end integration does the heavy lifting or authenticating and Authorization. In this case, the frontend is devices with absolutely no additional protocol or bytes to transfer other than authorization access

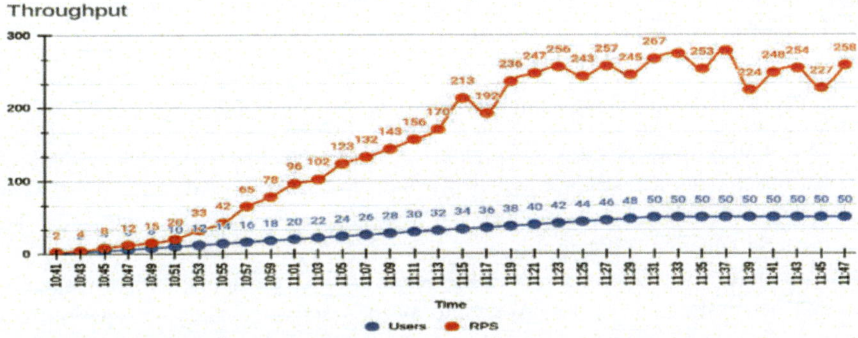

Fig. 5 Graph showing throughput with 50 concurrent users

token, which is Nauthorization Pass, making this solution compatible with existing IoT device limitations (Fig. 5).

Performance. The observed performance is noted below when run using JMeter calling endpoint to grant access to Requestee IAM (assuming being called from Requestee device). As we see from Fig. 3 and Fig. 4, the throughput hovers around 250 requests/sec on average. Furthermore, the latency stays pretty swift to respond within ~ 60 ms, keeping it highly responsive and scalable at par with any IAM solution (Fig. 6).

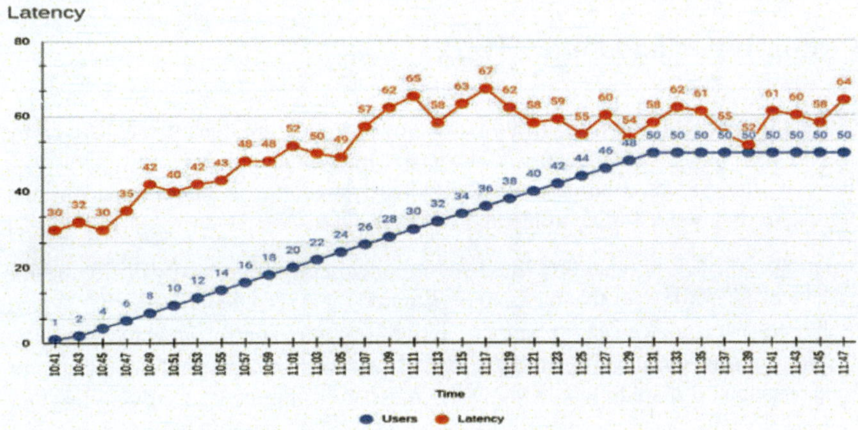

Fig. 6 Graph depicting latency observed with increased concurrent users over time

4.3 *Benefits of This Proposal*

Benefits of this proposal are summarized below.

- The solution provides a way to integrate the various heterogeneous systems to trust each other and still not with leaky doors to ensure those devices of different origin and make from the altogether different ecosystem (IAM control) can be trusted to operate and gain access.
- The context is evaluated numerous times to validate the applicability repeatedly and, hence, the chance to retain or withdraw the emergency status of the pass issued to the Requestor device. The context is also evaluated by giving the organization (origin) and the requestee system suitable applicability and rejecting the request if deemed unfit.
- The fundamentals of token for authenticating and Authorization remain intact and used in the same philosophy. The back-end does all the integration and problem solving with the trust-based mechanism.
- The Nauthorization pass can be revoked easily. The requestee system is under no obligation to fulfill the request despite authenticated and authorized giving autonomy of operation at its discretion.
- As we see in Latency and throughput comparison charts, the result data sets worked well compared to any standard IAM solutions in the market like Key cloak.

5 Conclusion and Future Work

The solution serves to Authentication and Authorization are dynamically driven by context and chance for heterogeneous devices to interconnect and execute operations. However, the current solution does not address few concerns that are evident and summarize below for further study and solutions:

- The specific action is not presented in this paper being requested by the Requestor device. Therefore, for this paper, it is assumed that it is a generic emergency and the requestee device can take care of action it needs to perform, which would be primarily monotonous like traffic signal giving way for a particular direction, or car dashboard display would show the message to the user to provide brighter side.
- The trust between Central Identity and Access Registry and Requestor organization is high and implicit. The current solution does not consider if the Requestor organization is compromised or hacked. There can be separate monitoring systems built into Central Identity and Access Registry to block or get the organization re-trusted temporarily.

We conclude our current work with acknowledgment to extend this solution further for better integrity and fine-grained control. The current solution addresses connectivity limited to general emergency and action mostly leftover to Requestee device rather than Requestor identifying and needing specific action to be performed.

References

1. Zhang G, Parashar M (2004) Context-aware dynamic access control for pervasive applications. In: Proceedings of the communication networks and distributed systems modelling and simulation conference, pp 21–30
2. Hong JI, Landay JA (2001) An infrastructure approach to context-aware computing. Human-Comput Interact 16(2–4):287–303
3. Icons used in Fig. 1 and Fig. 2 viz: truck, CT-scan, gate, hospital building and server made by https://www.flaticon.com/authors/smashicons. Last Accessed 19 May 2021
4. Malek B, Miri A, Karmouch A (2008) A framework for context-aware authentication. In: 4th International conference on intelligent environments, pp 1–8
5. Thaker R, Yusera Farooq Khan YF, Mughal S (2018) Fog approach in the internet of things: a review. IJSRCSEIT, 4(1b)
6. Sinha N, Sundaram M, Sinha A (2020) Authorization secure dynamic privileged escalation. RTEICT, Bangalore India, pp 110–117
7. Sales T, Sales L, Almeida H, Perkusich A (2021) Towards UPnP-UP: developing context-aware UPnP applications (position paper)
8. Sinha N, Sundaram M, Sinha A (2021) Instinctive and effective authorization for internet of things. In: Proceedings of second (IC4S). Springer, Singapore, pp 567–579
9. Hayashi E, Das S, Amini S, Hong J, Oakley I (2013) Casa: context-aware scalable authentication. In: Proceedings of the 9th SOUPS '13. ACM, vol 3. NY, USA, pp 1–3:10
10. Habib K, Leister W (2015) Context-aware authentication for the internet of things
11. Goel D, Kher E, Joag S, Mujumdar V, Griss M, Dey A (2010) Context-aware authentication framework. In: Mobile computing, applications, and services, vol 35. Springer, Berlin, Heidelberg, pp 26–41
12. Hu J, Weaver AC (2004) A dynamic, context-aware security infrastructure for distributed healthcare applications. In: Proceedings of the (PSPT), pp 1–8
13. Wullems C, Looi M, Clark A (2004) Towards context-aware security: an authorization architecture for intranet environments. In: Proceedings of the (PERCOMW'04), pp 132–137
14. De Matos E, Tiburski R, Amaral L, Hessel F (2018) Context interoperability for IoT through an edge-centric context sharing architecture. In: 2018 IEEE symposium on computers and communications (ISCC). Natal, Brazil, pp 439–440
15. Perera C, Zaslavsky A, Christen P, Georgakopoulos D (2014) Context-aware computing for the IoTs: a survey. IEEE Comm Surv Tutorials 16(1):414–454
16. Filho S, Tiburski R, Moratelli C, de Matos E, Hessel F (2019) Privacy and security of internet of things devices
17. Ghali C, Chehab A, Kayssi A (2010) CATRAC: context-aware trust- and role-based access control for composite web services. In: 2010 10th IEEE ICCIT, pp 1085–1089
18. Perera C, Zaslavsky A, Christen P, Georgakopoulos D (2020) CA4IOT: context-awareness for internet of things. In: Presented at: IEEE international conference on internet of things (iThings 2012). Besancon, France, pp 20–23 Nov 2012. Green Computing and Communications (GreenCom), IEEE/ACM Int'l Conference on & Int'l Conference on Cyber, Physical and Social Computing (CPSCom). IEEE, 775–782, (2020)
19. Jagadamaba G, Babu BS (2017) Adaptive security schemes based on context and trust for ubiquitous computing environment: a comprehensive survey. Indian J Sci Technol

20. Said SBH, Guillouard K, Bonnin J-M (2013) On the benefit of context-awareness for security mechanisms in LTE/EPS networks. In: Proceedings of the (PIMRC), pp 2414–2428

Next Generation Computing Technologies

Machine Learning-Based Security Solutions for Healthcare: An Overview

Pallavi Arora, Baljeet Kaur, and Marcio Andrey Teixeira

Abstract The eruption of chronic diseases such as COVID-19 has re-emphasized the need for people all over the world to have access to urgent healthcare. The latest pandemic has shown the flaws in the conventional healthcare system, namely that hospitals and clinics alone are incapable of dealing with such a crisis. Smart and integrated wearables are one of the main technologies that favor modern healthcare solutions (Godi, B., Viswanadham, S., Muttipati, A.S., Prakash Samantray, O., Gadiraju student, S.R.: E-Healthcare Monitoring System using IoT with Machine Learning Approaches. In: 2020 International Conference on Computer Science, Engineering and Applications (ICCSEA). pp. 1–5. IEEE, Gunupur, India (2020). https://doi.org/10.1109/ICCSEA49143.2020.9132937). These wearables can now gather data on an unparalleled scale thanks to advancements in the Internet of Things (IoT). Healthcare is among many fields that have been transformed by IoT, with the Internet of Medical Things (IoMT) systems being introduced as an IoT branch. Patients with chronic diseases can be monitored remotely using IoMT systems. As a result, it can offer prompt diagnostics to patients, potentially saving their lives in the event of an emergency. However, protection in these vital systems is a major issue that has arisen as a result of their widespread use. This paper presents an overview of the technologies that are being used on IoMT as well as some security problems found in the literature. This survey will give an insight to the readers regarding the importance of security in healthcare and the different machine learning methods used to address that issue.

Keywords Machine learning · Healthcare · Security · Internet of Things

P. Arora (✉)
IK Gujral Punjab Technical University, Kapurthala, Punjab, India

B. Kaur
Guru Nanak Dev Engineering College, Ludhiana, Punjab, India
e-mail: baljeetkaur@gndec.ac.in

M. A. Teixeira
Department of Informatics, Federal Institute of Education, Science, and Technology of São Paulo, Catanduva, SP 15808-305, Brazil
e-mail: marcio.andrey@ifsp.edu.br

P. K. Singh et al. (eds.), *Emerging Technologies for Computing, Communication and Smart Cities*, Lecture Notes in Electrical Engineering 875,
https://doi.org/10.1007/978-981-19-0284-0_47

1 Introduction

IoT has become an essential part of our daily lives in the twenty-first century. IoT has many applications in various fields ranging from healthcare, Smart Homes, Smart Grids, and Connected Cars to Industrial Internet. Healthcare is one of the most basic human needs, but with a growing population and an increasing number of diseases, the cost of healthcare facilities is also skyrocketing. In such cases, the advancement of IoT has paved a new path in the healthcare industry that can aid in the discovery of new treatment options [1]. IoT has enabled people to provide and receive healthcare services through a network of interconnected networks and IoT devices. This paper provides a thorough review of the literature on the various types of models proposed by researchers over the years involving machine learning and deep learning [2, 3] in IoT healthcare.

1.1 Contribution

The following are the primary contributions of this paper:

- We gave an insight into IoMT system architecture and different types of IoMT devices.
- We emphasized the traditional healthcare system's security and privacy issues.
- We discussed the usage of machine learning algorithms in healthcare security.

1.2 Organization

The remainder of the paper is organized as follows: Sect. 2 acquaints with the principles of IoT in healthcare. Section 3 provides a comprehensive description of IoMT-related assaults at different stages. Section 4 provides a review of the literature concerning the use of machine learning techniques for IoMT device security. Finally, Sect. 5 provides the paper's concluding thoughts.

2 IoT in Healthcare

2.1 IoMT Types

IoMT systems focus on providing essential or enhanced assistance for a wide range of health conditions [4]. Ingrafted devices, such as pacemakers treating cardiovascular disease, are required for particular healthcare disorders. In contrast, helping gadgets are mostly wearables for better healthcare experiences, such as smartwatches. Because of these distinctions, IoMT systems are divided into two categories: Implantable medical devices (IMDs) and Internet of wearable devices (IoWDs) [5].

- **Implantable Medical Devices (IMDs)**: An IMD itself is embedded equipment that is used to substitute, augment, or reinforce a tissue function. A pacemaker, for instance, seems to be an IMD that regulates irregular heartbeats by assisting the heart to pump at a standard rhythm whether pumping very quickly or slowly. IMDs are tiny as well as own a decent energy duration. As a result, enabling such gadgets to remain within the living organism over a longer period of time, minimal energy usage little memory capability with compact cells which endure a prolonged time was indeed required. Instances of embedded gadgets include cardiac pacemakers, implantable cardiac defibrillators (ICDs), coronary stents, hip implants, interocular lenses, and implantable insulin pumps (Fig. 1).
- **Internet of Wearable Devices (IoWDs)**: Such are wearable gadgets that monitor biomarkers such as pulse rate and might even positively affect people's general wellbeing. Some examples include wristbands, telemonitoring bands, electrocardiogram (ECG) screens, and bloodstream pressure screens. Wearables however are presently the most recognized IoWDs for monitoring biomarkers such as pulse rate and mobility. So, when the individual is not active, the monitoring can identify slow and fast heartbeats. They are now commonly used for non-critical patient monitoring. However, due to detector precision and energy power are both limited, these devices are unlikely to replace IMDs in critical situations.

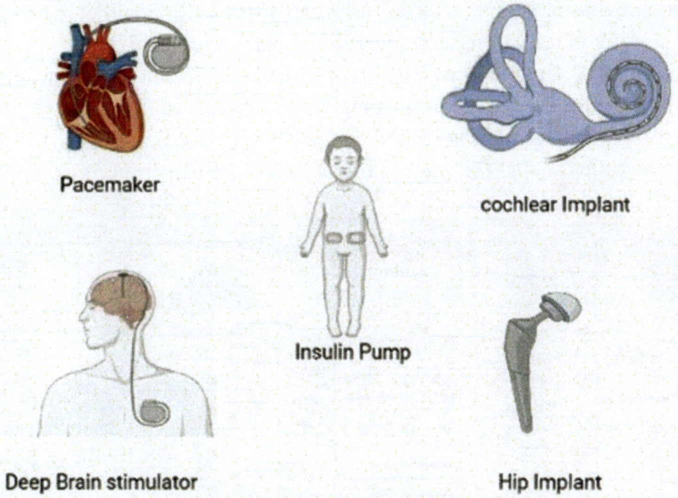

Fig. 1 Examples of IMDs and their associated body parts

2.2 IoMT System Architecture

The vast majority of existing IoMT setups are organized into four tiers [6, 7], as illustrated in Fig. 2. Such levels encompass various information phases, beginning well with individual's biomarkers collection, and continuing with information keeping as well as interpretation through the doctor enabling examination. Furthermore, any client may access individual medical status via the website. IoMT architecture is organized in various layers as follows:

- **Sensor layer**: Such a stratum has a collection of tiny sensors that are embedded or donned by the patient that gather biometric data. Communication protocols such as Wireless networking, Bluetooth, or the MedRadio available bandwidth, which is reserved for IMDs, are being utilized towards transmitting data to the next level [8].
- **Gateway layer**: IoMT sensors have storage and processing constraints and hence transfer the data forward to the next layer, i.e., the gateway layer, without being processed. The devices in this layer might have been the client's cellphone or indeed a dedicated access point (AP), which seem to be usually greater powerful than perhaps detectors. It can carry out some data preparation tasks including verification, relatively brief storing input, and basic artificial intelligence-based interpretation. Devices also use the Internet to communicate biosensor information to the server.
- **Cloud layer**: The cloud layer is in charge of receiving data from the gateway and storing, analyzing, and securely accessing it. Information analysis may detect variations arising in a person's condition and display these for healthcare professionals as well as clients for further action could be part of the analysis.
- **Visualization/Action Layer**: This layer provides information to physicians and clients in order for them to take care of their wellness. Additionally, it encompasses the doctor's recommendations based on the patient's medical issues. Prescriptions and dosage adjustments for various medications are illustrations of actions.

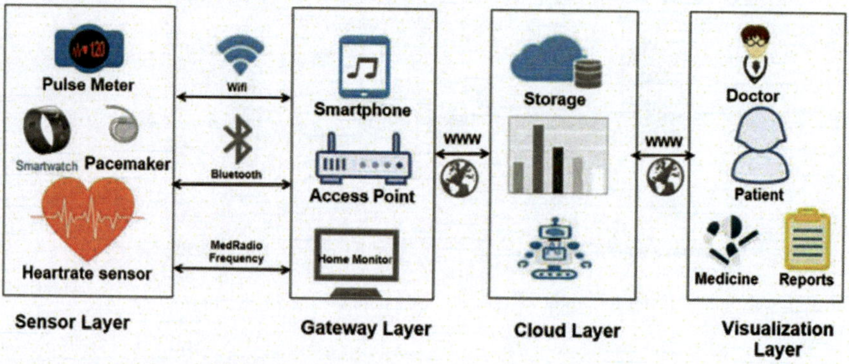

Fig. 2 IoMT system architecture

3 IoMT Security Model

3.1 Different Stages of IoMT Threats

IoMT platforms shall safeguard medical information throughout various phases of gathering, transfer, as well as storing [5].

(1) **Attack during data collection**: The initial phase of an IoMT protocol is the collection of patient data in the sensor layer. At such a point, assaults could either be software-based (information manipulation) or hardware-based (i.e., detector instrument alteration). Unless the detector equipment, as well as programming, is compromised, these attacks can endanger patients' lives. As a result, for maintaining the mechanism operating, it is vital to secure information from such threats.

(2) **Attack during data transit**: It involves equipment communication across all four levels, such as IoMT devices with in device tier and APs in the gateway levels. At this step, threats can change or prevent sensing information from being communicated. While a result, safeguarding from data in transit threats could prohibit information being getting distorted as it was transmitted between the four levels.

(3) **Attack during data storage**: The client's information is stored in the cloud layer after it has been gathered and disseminated between both sensor and gateway layers. Attacks in this layer range from account theft to DoS (denial of service) attacks. It is very important to protect data at this stage since most of the time data is resting thereby imposing a greater risk (Fig. 3).

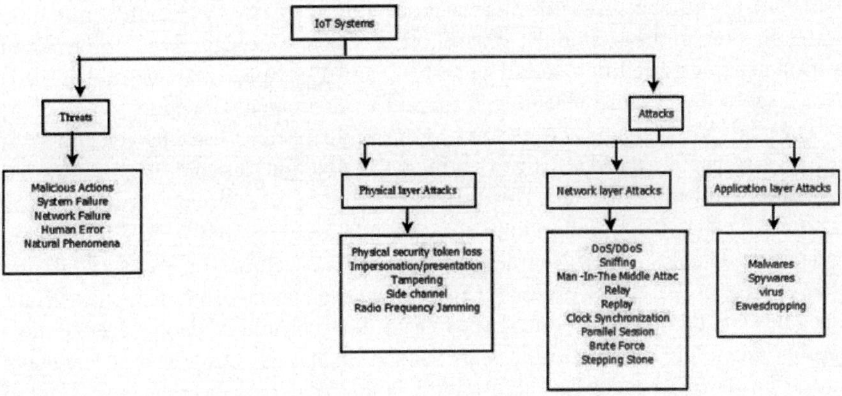

Fig. 3 IoMT security threats and attacks at various layers

4 Literature Review

4.1 Machine Learning in Healthcare

Machine learning is used to find trends in medical data and has excellent disease prediction capabilities. Authors in [9] evaluate a variety of machine learning algorithms for developing effective decision support for healthcare systems. This paper discusses various machine learning methods for predicting diseases such as heart disease, breast cancer, diabetes, and thyroid disease. It contributes to addressing the research gaps in the development of effective decision support systems for medical applications.

The research in [10] seeks to develop a model capable of properly predicting the probability for hyperglycemia in people. This project incorporates three machine learning classification methods to diagnose hyperglycemia during a preliminary phase: Decision Tree, SVM, and Naive Bayes. The studies make use of the Pima Indians Diabetes Database (PIDD), which is provided through the UCI machine learning repository. Precision, Accuracy, F-Measure, and Recall are all used to test the performance of the three algorithms. The classification system's precision is measured by the multitude of incidents that have been actually observed vs those that are mistakenly categorized. According to the data, Naive Bayes outperforms with a 76% precision.

In [11] the authors proposed wearable IoT for health tracking. In order to enable real-time health tracking, it uses the BASN (body area sensor network) platform. Among the worn sensors that have been incorporated are pulse, body heat, and arterial pressure detectors. Secondly, most of the worn fitness tracking communication is based on a smartphone as an information processing, display, as well as dissemination portal, that has an impact on the smartphone's typical everyday usage. Information is instantly transferred into web data servers, and a compact wrist-worn display is being placed like a substitute strategy for real-world content viewing.

The study showed in [12] does not stop at simple machine learning algorithms; it also digs deeper into the idea of deep learning. Deep Neural Networks were contrasted to Gradient Boosting Decision Tree, logistic regression, and support vector machine methods for diagnosing infarction based on a dataset acquired from a large-scale population database. Deep Neural Networks had an accuracy of 87.3%, while the others performed similarly, except for SVMs. Out of all the algorithms, support vector machines had the lowest accuracy. As a result, Deep Neural Networks are the most appropriate, while SVMs are the least suitable. Although Deep Neural Networks have been shown to be the best algorithm, it is still more complex than the others, so one must consider the algorithm's complexity before using it in real life.

Authors [13] emphasized the conventional healthcare system's security and privacy issues and also spoke about how blockchain can be used and integrated into the healthcare system. They proposed a healthcare architecture for remote patient monitoring and also stressed the usage of ML in blockchain to improve security.

Table 1 A brief comparison of the works described in Subsection 4.1

Author	Objective	ML technique used	Dataset	Disease
[9]	Developing effective decision support for healthcare systems	J48, SVM, Naïve Bayes, Decision Tree	UCI Diabetic Research Institute in Chennai Swami Vivekananda Diagnostic Centre Hospital	Heart disease, breast cancer, diabetes, and thyroid disease
[10]	To create a model that can accurately predict the risk of diabetes in patients	Decision Tree, SVM, and Naive Bayes	The Pima Indians Diabetes Database (PIDD)	Diabetes
[11]	To develop wearable IoT for real-time health monitoring	SVM and neural network	Data from wearable sensor to cloud Server directly	Predicts disease that will emerge in near future
[12]	To investigate the viability and efficacy of Deep Neural Networks (DNN) and other ML approaches for predicting stroke	Deep Learning and Gradient Boosting Decision Tree, logistic regression, and support vector machine algorithms	Customized	Stroke
[13]	To boost safety and confidentiality of patient's data as well as usage of ml with blockchain technology for healthcare	Blockchain along with Naive Bayesian, k-Nearest Neighbor, and Decision Tree	Real-time data from cloud server	Earlier prediction of disease

In the end, they discussed various research issues regarding blockchain and also suggested solutions for those challenges (Table 1).

4.2 Security in Healthcare

Despite the various advantages of using mobile devices, there are significant privacy and security issues with mobile health [14]. The use of successful encryption techniques to prevent data leakage is crucial for the growth of mHealth, which is fundamentally based on trust. Researchers have looked at the privacy and security concerns surrounding mHealth from a variety of angles. Because of its current inadequate security support, health care data security has a loophole in that it is seldom supported by direct implementation of Database Management System (DBMS) security features.

Healthcare technology has the potential to protect, prolong and upgrade someone's life [15]. Techniques range beyond storing electronic health records (EHRs) to tracking wellbeing and delivering medicines, as well as telemedicine technology that allows treatment to be delivered remotely—even across countries. Because of interconnected, easily accessible access points, obsolete systems, and a lack of focus on cybersecurity, healthcare technologies are vulnerable to security threats. While the focus has appeared to be on patient care, healthcare innovations contain large quantities of useful and sensitive information. Because healthcare recognition has been of greater significance unlike alternative forms of identification, personal profit is often the motivation behind assaults. Certain attacks, such as information warfare, could well be driven by sociopolitical considerations. People's lives are jeopardized if important health systems are targeted.

The advancement of information technology has sparked a transformative phase in all sectors in terms of integrating these technologies in order to take advantage of new functionality and capabilities [16]. Healthcare is a target because of the lack of security mechanisms, as well as the importance of the data collected or stored. We believe it is necessary to start on the procedural front preferably only at the country's scale having the goal of guaranteeing that the distributors of medical systems satisfy specific consented security standards. The security standards for computers, or similar devices used in the industry, maybe implemented and governed by national regulations.

The aim of the paper proposed in [17] is to examine different forms of security threats, as well as their depth and effect on organizations. Various IoT attacks are discussed in this article, along with current solutions. Furthermore, these IoT attacks are characterized based on the attack's vulnerability to compromise the network. Since IoT employs a network architecture that is similar to traditional network architecture for communication among various devices, it inherits the flaws of conventional network architecture. To combat these attacks, researchers have suggested a variety of solutions. Implementing all of these protection measures and strategies at the same time consumes computer computation and battery power, which is incompatible with IoT technology and its devices.

The Internet connects us to the actual world through individual medical sensors, vicinity systems, connected residences, smart automobiles, as well as automation networks [18]. All of this cutting-edge IoT creation creates new security problems and research gaps that must be tackled. Despite the fact that TCP/IP (Transport control protocol/Internet protocol) is the most common Internet communication protocol, to connect to a centralized module through which information is transmitted to the database, IoT devices may consider a brief network communication. There are research opportunities for developing, implementing, and testing new lightweight algorithms to secure data in IoT networks cluster information is encrypted in the server, on the internet, and perhaps even on smartphones. Even secured connections will also not secure customer information if an attacker acquires accessibility of the database through the use of the internet, storage, or portable devices. The safety of IoT secret information, on the internet, and even on smartphones is a difficult problem to solve. The use of artificial intelligence and machine learning to safeguard IoT devices is a relatively latest field of research. Another growing subject is network automation and intent-based networking, in which AI-based and ML-based technology handle network access and components. To protect your network, you can use network function virtualization (NFV) and software-defined networking (SDN). Usage of intend-based networking in conjunction with SDN for IoT devices as well as security is still a work in progress.

Authors in [19] presented a cellphone established medical procedure for the Healthcare 4.0 period in this study. Patients may self-identify and use personal smartphones and smartwatches, generating a communication channel that can be used across owned devices. Following mutual authentication, the web gateway checks individual clients. The AVISPA tool is used to describe the approach's security. They investigated the inherent security and performance qualities that improve health care and calculated the associated connection as well as computing expenses. The findings indicate the proposed methodology delivers higher protection than competing cutting-edge alternatives.

In [20], the authors designed a Blockchain-based architecture for the effective storage and management of electronic health records (EHRs). This also allows patients, providers, and third parties to have safe and efficient access to medical data while protecting the patient's private information. The objectives of this study are to examine how their suggested framework serves the demands of patients, providers, and third parties, as well as to comprehend how the framework addresses security and safety problems in healthcare 4.0 (Table 2).

Table 2 A brief comparison of the works described in Subsection 4.2

Author	Objective	Merits	Demerits
[14]	To conduct a comprehensive exploration of the security and safety elements of mHealth gadgets	Theoretically discussed importance of safety of mHealth gadgets	No practical examples of types of attacks on these gadgets
[15]	Assuring the safety of healthcare services and patient data	A systematic analysis of patterns, dangers, and solutions for electronic healthcare	Because of the nature of this research, only English papers were included for assessment
[16]	To investigate cyberspace security concerns in healthcare industry	Insisted on setting bare minimum security standards for healthcare gadgets	No discourse on healthcare security solution
[17]	Examines numerous IoT attacks, comparison and their defenses	In-depth survey of attacks on IoT and comparing them on basis of their effectiveness and severity of damage	Lacks discussion on usage of techniques such as ML, blockchain, etc.,

(continued)

Table 2 (continued)

Author	Objective	Merits	Demerits
[18]	To highlight the problems and open concerns around IoT security	Security solutions using emerging technologies are discussed	Didn't mention about implementation of these techniques in IoT
[19]	Offer a mobile-based solution for protecting electronic medical records	Suggested methodology ensures a high level of security while incurring lower computing and transmission costs	The topic of privacy is not discussed in depth
[20]	To present a Blockchain-based solution for effective EHR storage and management	Discussed framework and it's implementation for preserving security and privacy of all related to health care	–

5 Conclusion and Future Scope

We provided an overview of IoT technological advances in healthcare in this paper. A number of research challenges have been identified, with the expectation that they will become major research trends in the coming years. We would like to highlight the importance of protecting IoMT gadgets and medical technologies as a whole (hereafter referred to as IoMT systems) is, however, a significant challenge. IoMT systems are distinct from other systems as they have the ability to impact patients' lives and raise privacy issues if their identities are disclosed. We hope that this study will be beneficial for researchers and professionals in the field, assisting them in grasping the tremendous potential of IoT in the medical domain and recognizing major IOMT challenges. Based on the IoMT system architecture shown in Fig. 2, as future work, we are developing a security framework to strengthen the security of the IoMT system between the Gateway Layer and the Cloud Layer. Our framework will be composed of an intrusion detection system (IDS) based on machine learning algorithms.

References

1. Godi B, Viswanadham S, Muttipati AS, Prakash Samantray O, Gadiraju student SR (2020) E-healthcare monitoring system using IoT with machine learning approaches. In: 2020 International conference on computer science, engineering and applications (ICCSEA). IEEE, Gunupur, India, pp 1–5. https://doi.org/10.1109/ICCSEA49143.2020.9132937
2. Durga S, Nag R, Daniel E (2019) Survey on machine learning and deep learning algorithms used in internet of things (IoT) healthcare. In: 2019 3rd International conference on computing methodologies and communication (ICCMC). IEEE, Erode, India, pp 1018–1022. https://doi.org/10.1109/ICCMC.2019.8819806

3. Qayyum A, Qadir J, Bilal M, Al-Fuqaha A (2021) Secure and robust machine learning for healthcare: a survey. IEEE Rev Biomed Eng 14:156–180. https://doi.org/10.1109/RBME.2020.3013489
4. Joyia GJ, Liaqat RM, Farooq A, Rehman S (2017) Internet of medical things (IOMT): applications, benefits and future challenges in healthcare domain. JCM. National University of Sciences and Technology, Islamabad, Pakistan. https://doi.org/10.12720/jcm.12.4.240-247
5. Ghubaish A, Salman T, Zolanvari M, Unal D, Al-Ali AK, Jain R (2020) Recent advances in the internet of medical things (IoMT) systems security. IEEE Internet Things J 1–1. https://doi.org/10.1109/JIOT.2020.3045653
6. Mawgoud AA, Karadawy AIE, Tawfik BS (2020) A secure authentication technique in internet of medical things through machine learning. 1538814 Bytes. https://doi.org/10.6084/M9.FIGSHARE.13311479.V2
7. Hathaliya JJ, Tanwar S (2020) An exhaustive survey on security and privacy issues in Healthcare 4.0. Comput Commun 153:311–335. https://doi.org/10.1016/j.comcom.2020.02.018
8. Medical Device Radiocommunications Service (MedRadio), https://www.fcc.gov/medical-device-radiocommunications-service-medradio. Last Accessed 13 Apr 2021
9. Shailaja K, Seetharamulu B, Jabbar MA (2018) Machine learning in healthcare: a review. In: 2018 Second international conference on electronics, communication and aerospace technology (ICECA). IEEE, Coimbatore, pp 910–914. https://doi.org/10.1109/ICECA.2018.8474918
10. Sisodia D, Sisodia DS (2018) Prediction of diabetes using classification algorithms. Procedia Comput Sci 132:1578–1585. https://doi.org/10.1016/j.procs.2018.05.122
11. Wan J, AAH Al-awlaqi M, Li M, O'Grady M, Gu X, Wang J, Cao N (2018) Wearable IoT enabled real-time health monitoring system. EURASIP J Wireless Commun Netw 298. https://doi.org/10.1186/s13638-018-1308-x
12. Hung C-Y, Chen W-C, Lai P-T, Lin C-H, Lee C-C (2017) Comparing deep neural network and other machine learning algorithms for stroke prediction in a large-scale population-based electronic medical claims database. In: 2017 39th Annual international conference of the IEEE engineering in medicine and biology society (EMBC). IEEE, Seogwipo, pp 3110–3113. https://doi.org/10.1109/EMBC.2017.8037515
13. Hathaliya J, Sharma P, Tanwar S, Gupta R (2019) Blockchain-based remote patient monitoring in healthcare 4.0. In: 2019 IEEE 9th International conference on advanced computing (IACC). IEEE, Tiruchirappalli, India, pp 87–91. https://doi.org/10.1109/IACC48062.2019.8971593
14. Bhuyan SS, Kim H, Isehunwa OO, Kumar N, Bhatt J, Wyant DK, Kedia S, Chang CF, Dasgupta D (2017) Privacy and security issues in mobile health: current research and future directions. Health Policy Technol 6:188–191. https://doi.org/10.1016/j.hlpt.2017.01.004
15. Coventry L, Branley D (2018) Cybersecurity in healthcare: a narrative review of trends, threats and ways forward. Maturitas 113:48–52. https://doi.org/10.1016/j.maturitas.2018.04.008
16. Buzdugan A (2020) Integration of cyber security in healthcare equipment. In: Tiginyanu I, Sontea V, Railean S (eds) 4th International conference on nanotechnologies and biomedical engineering: proceedings of ICNBME-2019, Sept 18–21, 2019, Chisinau, Moldova. Springer International Publishing, Cham, pp 681–684. https://doi.org/10.1007/978-3-030-31866-6
17. Deogirikar J, Vidhate A (2017) Security attacks in IoT: a survey, vol 6
18. Sadique KM, Rahmani R, Johannesson P (2018) Towards security on internet of things: applications and challenges in technology. Procedia Comput Sci 141:199–206. https://doi.org/10.1016/j.procs.2018.10.168
19. Hathaliya JJ, Tanwar S, Evans R (2020) Securing electronic healthcare records: a mobile-based biometric authentication approach. J Inform Secur Appl 53:102528.https://doi.org/10.1016/j.jisa.2020.102528
20. Vora J, Nayyar A, Tanwar S, Tyagi S, Kumar N, Obaidat MS, Rodrigues JJPC (2018) BHEEM: a blockchain-based framework for securing electronic health records. In: 2018 IEEE Globecom Workshops (GC Wkshps), pp 1–6. https://doi.org/10.1109/GLOCOMW.2018.8644088

Fake News Detection Techniques: A Survey

Sahitya Diwakar, Divyansh Bansal, Raghav Jagya, Poonam Tanwar, and Shweta Sharma

Abstract Lately, on account of the thriving development of online relational associations, fake news for various businesses, what is more, political purposes have been appearing in tremendous numbers what is more, extensive in the online world. The development of beguiling information in standard access news sources, for instance, electronic media channels, news web diaries, and online papers have made it attempting to perceive reliable news sources, therefore growing the prerequisite for computational contraptions prepared to give pieces of information into the faithful nature of online substance (Poovaraghan et al. (2019) Fake news accuracy using naive bayes classifier. Int J Recent Technol Eng). This paper considers past and current systems for fake news distinguishing proof in text-based associations while determining how and why fake news exists regardless.

Keywords Fake news · News media · Social community

1 Introduction

As most of the time in our lives is spent interacting online through social media platforms like Instagram, Facebook, more and more people around the world prefer to see news via social media rather than traditional news platforms. The reason for this change in platforms is easy to understand as it is faster and cheaper to spread the news through social media rather than news platforms like newspapers and television. Moreover, it is easy to debate some news on social media among other readers and discuss it. In the US 62% of readers get news from social media in 2016, while in 2012 only 49% of readers got news from social media [1]. It is also said that social media now outperforms television as the major news source. However, despite the advantages of social media, the quality of news on social media is lower than traditional news organizations because it is cheap to provide news online and

S. Diwakar · D. Bansal · R. Jagya · P. Tanwar (✉) · S. Sharma
Computer Science and Engineering, FET, Manav Rachna International Institute of Research and Studies, Haryana, Faridabad, India

P. K. Singh et al. (eds.), *Emerging Technologies for Computing, Communication and Smart Cities*, Lecture Notes in Electrical Engineering 875,
https://doi.org/10.1007/978-981-19-0284-0_48

much faster and easier, large volumes of fake news are produced online for a variety of purposes.

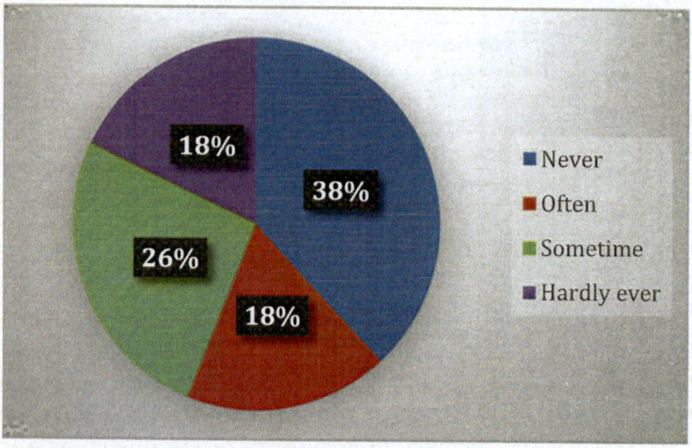

News Use Across Social Media Platforms 2016. *Source* Survey conducted Jan 12 Feb, 2016

The extensive spread of fake news can have a serious negative impact on individuals and society. It has a major impact on locals who have no idea how to distinguish between fake or real news. For example, the most popular real news spread through television did not spread as widely as the most popular Fake news was spread on Facebook. Fake news is mostly manipulated by protagonists with a hidden agenda of conveying some political message. Fake news also changes how people see real news. To help reduce the negative impact caused by fake news people must know how to differentiate between fake or real news and we must develop algorithms that automatically detect fake news on social media. Detecting fake news on social media opens a variety of challenges and research problems which make it very challenging for automated detection. Fake news is written with an intention of confusing readers and misleading them, which makes it not so simple to detect fake news based on its content.

The key motivation of this survey is that Fake news detection on social media is still at the early age of development, and there are still many challenging issues that need further attention. It is necessary to discuss potential research directions that can improve fake news detection and its capabilities. To guide the future of fake news detection research, appropriate clarifications are necessary. We give an overview of three existing fake news detection methods Style-based, Content-based, Naïve Bayes Classifier in detail from various research till now.

2 Literature Survey

This research seeks to bring together a series of similar works in the field of social media identification of fake news. As a result, I conducted a survey using different platforms like Facebook, Twitter, and others. To determine the potential of worldly experiences to be relied upon. Style-based methods focus on capturing the writing style of the news content and try to classify articles based on it [2]. Most of the works were based on the fact that there were several similarities between fake news and spam emails like both having a huge number of grammatical mistakes, trying to manipulate the reader's opinions, and using a somewhat similar word set. Although these approaches can be very effective at times, they fail to explain what exactly is fake in the news content. While on the other hand, content-based methods for detecting fake news, also known as fact-checking [3], take in use of the content of fake news rather than the style. It is the most straightforward way and aims at checking the truth value of the claims in the news content. The contemporary approaches majorly focus on finding a path in an existing knowledge graph for a certain triple. We will focus more on content-based approaches in this paper since it looks more promising in the current status quo. Because there has been a significant growth in the distribution of fake news via social media, there have been changes in the percentage of efficiency of algorithms in detecting fake news. Changes are good for some approaches and negative for others. For example, the Naive Bayes method has an efficiency of 0.92, which is 80% efficient in the market, while others have an efficiency of 0.73–0.82, which is 60% effective [4].

2.1 Domain

We learn from the papers that there are several social media domains, such as Facebook (Tucker J and Savyan), Twitter, and others. People will spread false news using this method. Automatic real-time monitoring of harmful posts on Facebook and using opinion mining to consider user activity on Facebook.

2.2 Dataset

The datasets (Zhou and Monti) that were used to propose new models that aid in the analysis of false news included Snopes, PolitiFact, BuzzFeed, and many others. For the identification of false news in their respective works, datasets from models such as the PHEME dataset [5], BuzzFeed [2], Epinions, and others were gathered.

2.3 Issues

Owing to shortcomings in the databases and the length of the news story, the key problems illustrated include difficulty in classification accuracy with inaccurate content on social media. The article would demonstrate how the model failed to work well in a particular data case, but it will not say that their dataset is representative of the Facebook population (p. Diwan).

3 Previous Method

Different types of methods to detect fake news:

- Fake News Detection from Style Based
- Fake News Detection from Content-Based
- Naïve Bayes Classifier

3.1 Fake News Detection from Style Based

The style-based fake news detection method analyses the news report based on the writing style of people who spread fake news. In this method, it is easily detectable that is this news is fake or not because in spam emails and fake news there are so many grammatical errors that are used to manipulate the ideology and thinking of people. Fake news authors used to use self-oriented pronouns rather than oriented pronounce more sensory-based words [6]. It generally uses deep syntax, data representation, sentiment analysis, and semantic analysis to understand fake news. The main assumption is that the fake news author has no previous experience regarding an event that is why they end up contradicting themselves or may leave important facts which are very important for the investigation [6]. But the biggest problem with this method is it does not tell us which part of the news is fake.

Fake news teller is generally using unethical plans to spread useless baseless fake news by misdirecting the data and impact enormous networks of buyers, requiring specific composing styles important to interest and convince a wide extent of shoppers that is not found in evident news stories. Style-based methods can easily identify fake news by identifying the composition style of news. Style-based method is further divided into two main categories: Deception-arranged and Objectivity situated.

Stylometric strategies of fraud type capture articulation or beguiling cases from news content. Inspiration of the pre-signing location begins from legal brain research (i.e., the undeutsch hypothesis) [7], and various criminological instruments including content-based content investigations [8] and scientific-based content analysis (Tony Lesce, 1990) has been made. Even more than the end-model of late normal language handling, developing applied at the duplicate spot stage from the perspective accompanying: deep grammar and rhetorical construction. The profound grammar model has been done utilizing the probabilistic settings for free grammar (PCFG), with which sentences can be converted into decisions that describe the grammar structure. Considering PCFG, various guidelines can be produced for scaly locations, for example, the rules of creation that are not lexicalized and grandparents' rules [9]. The expository design hypothesis can be used to capture the contrast between misleading and honest sentences [10]. In-depth organizational models, for example, convolutional nerve organizations (CNN), have also been applied to the truth of fake newsgroups [11].

• Methodologies arranged to objectivity the signs of capture style that can demonstrate the decrease in the objectivity of the content of the information and, therefore, the possibility of deceiving buyers, such as Hyper Partisan styles and sensationalist reports. Hyper partisan styles address an outrageous behavior for a specific ideological group, which corresponds regularly with solid inspiration to make counterfeit news. Featured based on linguistics can be applied to distinguish hyper partisans [12]. Sensationalist reports address those items that do not contain well-informed news, but depending on the characteristics of obtaining eyes (ie, deceptive content) with an inclination for beautification, sensationalization, alarm achievement, etc. Frequently, the news titles will summarize the significant prospects of the article that the writer needs to transmit, and consequently deceptive and cheated the deceptive content titles can complete as a decent marker to recognize counterfeit items [13, 14].

3.2 Fake News Detection from Content-Based

The content-based method of finding fake news is not like the style-based method. As in the style-based method, it does not check the styling, but it checks the content of the news report. It checks the content of the news report with previous research, report, content and checks the knowledge graph, and then tells if the news is fake or not. But here is also a problem that if there is no knowledge graph available on the internet or there are no previous reports on the internet.

There are some key features that tell us about the basic language and sentence structure of news. The semantic features of the content are acceptable pointers and help to discover suspicious news, which is focused to change the mindset of the reader. The writing style feature tells us different writing methods of the author of Counterfeit news to change the mindset of the reader [15].

Counterfeit news makers expect to emulate a similar method of composing of genuine news writers to delude the pursuers and urge them to accept their cases yet, there are a few loopholes that are still left out. These loopholes help to help to differentiate between the genuine news writer and fake news writers such as the use of several syntactical structures and keystroke features that gives clarity on the sentence structure of the news, we cannot predict the semantics of the sentences from the writing style which plays an important role in recognizing nature of the news (fake or real) [16].

Semantic features are vital highlights which are used to uncover fake news. There are few studies which had used this semantic feature to detect counterfeit news [17]. Results of studies have shown that after using this semantic feature we can easily recognize counterfeit news, but it needs more hands-on and examination [18].

It is a necessity to use semantic features in counterfeit news detection, just a couple of research have used these highlights in the research [17]. The aftereffects of the research have shown that utilizing semantic features to recognize counterfeit news is vital and successful and wants more examination [18]. To uncover the fake client accounts and suspicious clients, sentimental features are a way to deal [19]. Authors who write about these kinds of news generally overdraw the facts and mislead the reader. With the help of sentiments, it is easy to detect false and the psychological keywords can uncover the feelings, judgements [20, 21]. For the utilization of the sentiment of news so many methods have been proposed some of them are arousal valence, happiness detection and analysis, and many more [15, 22].

It has been demonstrated as an adequate technique to identify counterfeit news by joining a few scores of sentiments and can recognize and can differentiate fake bot records and genuine human records [19]. For the analysis of visual content such as images, videos, popular images, multi-image frequency, and so on, the visual-based feature analysis is proposed. These are the cues for detecting fake information [3, 16, 14, 23, 24, 25].

3.3 Naïve Bayes Classifier

This method is proposed from the mathematics theorem that is known as the Bayes theorem. This theorem is used to calculate the "probability that something will happen, given that something else has already occurred" [2]. It is a machine learning class, so it is predicted by "membership probabilities" and for each record, or data that is belong to that one class. And the class with maximum chances is considered as "most likely class" [23]. In many cases, this method was proven wrong because sometimes when two things are dependent on each other it takes them to be independent of each other. Let us say there is news related to Donald Trump but if in the scanning of Naïve Bayes classifier, it reads Donald in many cases and the same article there could also have Trump, which Naïve Bayes classifier treat as two different people which change the result of news flakiness. So, the biggest flaw of method is that it considers every one of the features to be discrete, which may not generally be the situation [24].

Naive Bayes classifiers are part of straightforward machine learning. For checking whether any news is fake utilizing pipelining ideas, or it is genuine Naïve Bayes is a well-known method. There are many calculations that zero in on regular standard, so that is not the solitary calculation for preparing such classifiers. For checking whether the news is fake or genuine Naive Bayes can be utilized. Naive Bayes classifiers are part of straightforward machine learning. Naive Bayes is well-known calculation which is utilized to discover the exactness of the news regardless of whether its genuine or phony utilizing multinomial NB and pipelining ideas. There are many calculations that zero in on basic standard, so Naïve Bayes is not the only algorithm for preparing such classifiers. This method is derived from the mathematics theorem which is known as the Bayes theorem. This theorem is used to calculate the "probability that something will happen, given that something else has already occurred" [6].

Naïve Bayes is a probabilistic classifier based on independent classification of features.

$$P(A|B) = \frac{P(B|A).P(A)}{P(B)} \tag{1}$$

These Naïve Bayes classifiers are the pairs of clear probabilistic classifiers that depend on applying Bayes theory with strong independence doubts between the features. For the development of models that are used to appoint the class marks to issue occasions, in which the name of class is taken by some limited sets, this method is used [2]. It is a singular computation for the preparation of these kinds of classifiers, however, the pairs of calculations are dependent on a regular rule: all Naive Bayes classifiers are expected to be that the approximation of the specified component is undertaken of another component where the class variable is given.

Naive Bayes classifiers are a mainstream measurable procedure of email filtering. They arose in the 90 s and were one of the principal endeavors to handle spam separating issues. Naive Bayes normally use pack of words highlights to distinguish spam email; a methodology regularly utilized in text order. Innocent Bayes Classifiers works by connecting the syntactic or not syntactic development with the spams and without spam messages and after applying this method it figures out whether the email is a spam message or not.

As it is a part of machine learning class, it can be performed by predicting "membership probabilities" for every class, data, or any of the record that belongs to that class. The class which has the maximum chances shall be determined as the "most likely class" [6]. In many cases, this method was proven wrong because sometimes when two things are dependent on each other it takes them to be independent of each other. Let us say there is news related to Donald Trump but if in the scanning of Naïve Bayes classifier, it reads Donald in many cases and the same article there could also have trump, which Naïve Bayes classifier treat as two different people which change the result of news flakiness [24]. This method does not always classify the data, so there is a higher risk of matching with unrelated data, which ultimately affects the reliability of the news [26]. The advantages and disadvantages of all are shown in Table 1.

Table 1 Comparative analysis of Fake news detection methods

S.no	Methods	Advantages	Disadvantages
1	Fake News Detection from Style Based	This process studies the writing style of the author and then gives results It generally uses deep syntax, data representation, sentiment analysis, and semantic analysis to understand fake news	This process doesn't study the content of the author and verify from online sources and then gives results. This process crashes when there is no previous papers or data on the internet regarding a particular topic
2	Fake News Detection from Content-Based	This process studies the content of the author and verifies from online sources and then gives results	This process crashes when there is no previous papers or data on the internet regarding a particular topic
3	Naïve Bayes Classifier	This process involves counting the number of times a word appears in the headline or stanza, converting that to a probability, and then calculating the probability that it is a fake or not	Sometimes in this domain people use hybrid cloud algorithm to detect fake news, which is less accurate and required large amount of space. Sometimes, it takes human input

4 Challenges

News is always provided by some journalist or blog writer via websites or articles. The problem arises in 3 different ways:

- Misinformation—This is the interface between worlds and journalists. Disinformation- This is the interface between journalist and their content.
- Misleading—This is the interface between content provided by journalists and readers or viewers.

Misinformation arises when a journalist does not have a clear understanding of the target system that he or she is discussing. Misinformation is characterized by a lack of veracity, that is, Information about the target system is not accurately conveyed in the journalist's message. Fake news writer uses minor mistakes done by honest journalists to create a big issue.

Disinformation is a bit different from misinformation, disinformation is characterized by the journalists' wrong intentions and arises when a journalist wants to change the mindset or mislead the audience via their message Wardle [27]. A journalist like this promotes false information, perhaps even true information that implies something false. Disinformation consists of deception of both the content and authorship. They can, for example, present content as though it were the product of genuine journalistic practices when, in fact, it was not [28, 29]. Misleading content is characterized by its ability to mislead the audience. News can an audience for several reasons. A vague or omitted article may mislead the audience. Sarcastically writing, for example, misleads readers who do not realize the author was trying to entertain rather than informing them about something [30].

5 Conclusion

In this paper, we have examined the fake news story, the maker also, subject identification issue. Because of the news increased heterogeneous informal organization, a bunch of expressed and dormant highlights can be removed from the text-based data of information articles, makers, and subjects individually. Fake news interferes with the capacity of a client to observe valuable data from the internet benefits particularly when news gets basic for dynamic. Thinking about the changing scene of the advanced business world, the issue of fake news has gotten something other than a showcasing issue as it warrants genuine endeavors from security scientists. Fake news identification has many open issues that require the consideration of scientists. For example, to lessen the spread of fake news, distinguishing key components associated with the spread of information is a significant advance.

References

1. Gottfried J, Shearer E (2016) News use across social media platforms
2. Granik M, Mesyura V (2017) fake news detection using naive bayes classifier. UKRCON IEEE
3. Shu K, Sliva A, Wang S, Tang J, Liu H (2017) Fake news detection on social media: a data mining perspective. Computer Science & Engineering, Arizona State University, Tempe, AZ, USA
4. Sharma U, Saran S, Patil SM (2021) Fake news detection using machine learning algorithms. Department of Information Technology Bharati Vidyapeeth College of Engineering Navi Mumbai, India, IJERT
5. Zhang X, Ghorbani AA (2020) An overview of online fake news: characterization, detection, and discussion. Inform Process Manage
6. Stahl K (2018) Fake news detection in social media. In: Candidate BS (ed) Department of Mathematics and Department of Computer Sciences, California State University Stanislaus, 1 University Circle, Turlock, CA 95382
7. Undeutsch U (1967) Beurteilung der glaubhaftigkeit von aussagen. Hand- buch der psychologie
8. Vrij A (2005) Criteria-based content analysis: a qualitative review of the first 37 studies. Psychol Public Policy Law
9. Wang WY (2017) Liar, liar pants on fire: a new benchmark dataset for fake news detection. arXiv preprint
10. Potthast M, Kiesel J, Reinartz K, Bevendorff J, Stein B (2017) A stylometric inquiry into hyper partisan and fake news. arXiv preprint
11. Rubin VL, Lukoianova T (2015) Truth and deception at the rhetorical structure level. J Assoc Inform Sci Technol 66(5):905–917
12. Chen Y, Conroy NJ, Rubin VL (2015) Misleading online content: recognizing clickbait as false news. In: Proceedings of the ACM on workshop on multimodal deception detection, ACM
13. Saxena R (2017) How the naive bayes classifier works in machine learning. Retrieved 2017
14. Kaustav et al (2020) A facial expression recognition system to predict emotions. In: International conference on intelligent engineering and management (ICIEM)

15. Zhang J, Dong B, Philip SY (2019) FAKEDETECTOR: effective fake news detection with deep diffusive neural network. IFM Lab, Department of Computer Science, Florida State University, FL, USA BDSC Lab, Department of Computer Science, University of Illinois at Chicago, IL, USA
16. Hussein A, Ahmad F, Kamaruddin S (2019) Content-social based features for fake news detection model from Twitter. Int J Adv Trends Comput Sci Eng
17. Jia S et al (2018) Fake reviews detection based on LDA. In: 4th International conference on information management (ICIM). IEEE
18. Xu K et al (2019) Detecting fake news over online social media via domain reputations and content understanding. Tsinghua Sci Technol
19. Dickerson JP, Kagan V, Subrahmanian V (2014) Using sentiment to detect bots on twitter: Are humans more opinionated than bots? In: IEEE/ACM international conference on advances in social networks analysis and mining
20. Thakur P, Shrivastava DR (2018) A review on text based emotion recognition system. Int J Adv Trends Comput Sci Eng
21. Zhao J et al (2014) FluxFlow: visual analysis of anomalous information spreading on social media. IEEE Trans Vis Comput Graphics
22. Oueslati O, Khalil AIS, Ounelli H (2018) Sentiment analysis for helpful reviews prediction. Int J
23. Tanwar P, Rai P (2020) A proposed system for opinion mining using machine learning, NLP and classifiers. IAES Int J Artif Intell (IJ-AI) 9(4):726–733, ISSN: 2252-8938. https://doi.org/10.11591/ijai.v9.i4.pp726-733
24. Petkar PB, Sonawane SS (2020) Fake news detection: a survey of techniques. Int J Innov Technol Exploring Eng (IJITEE)
25. Konagala V, Bano S (2020) fake news detection using deep learning. IGI Global
26. Poovaraghan RJ, Priya MK, Vamsi PS, Mewara M, Loganathan S (2019) Fake news accuracy using naive bayes classifier. Int J Recent Technol Eng
27. Wardle C, Derakhshan H (2018) Thinking about 'information disorder': formats of misinformation, disinformation and mal-information. In: Ireton C, Posetti J (eds) Journalism, 'fake news', & disinformation: handbook for journalism education and training
28. Pepp J, Michaelson E, Sterkin RK (2019) What's new about fake news? J Ethics Social Philos
29. Fallis D, Mathiesen K (2019) Fake news is counterfeit news. Routledge Taylor and Francis group
30. Garrett RK, Bond R, Poulsen S (2019) Too many people think satirical news is real. The Conversation Academic rigour, journalistic flair

Impact of Image Classification in Dermatology: Automated Diagnosis

Anukriti Singh, Devam Kumar Sinha, and Namrata Dhanda

Abstract The implementation of AI and ML in health care sector is majorly in cardiology, radiology, and ophthalmology but its employment scope in dermatology is minimal. Early diagnosis of skin diseases may prevent the disease to grow worse in condition. Machine-Learning has the caliber to improvise the dermatologist's practices by making the diagnosis process less tiresome and enable people to get a more personalized treatment through the use of mobile applications etc. Numerous technologies based on machine learning algorithms have emerged in different areas of medicine. Dermatological diseases are usually easily visible and diagnosis are carried on by visual screening of the abrasion, recognizing, and comparing with the patterns of known skin diseases. Three commonly occurring skin diseases namely Acne Vulgaris, Lichen Planus, and SJS (Stevens-Johnson syndrome) & TEN (toxic epidermal necrolysis) have been studied and worked upon in this paper. The purpose is to classify correctly these three diseases by implementation of machine-learning algorithms such as SVM, CNN, logistic-regression, Random-Forest, and naïve-Bayes. After analysis, it was found that CNN gives the highest accuracy among the discussed models in the prediction of the disease.

Keywords Dermatologist · Lesions · Visual Screening · SVM · CNN · Logistic regression · Random forest · Image classification

1 Introduction

Being the largest organ of the body with an approximate area of 20 square feet, skin is vital to produce the sense of touch, heat, cold, regulates the temperature of the human body, and safeguards us from microbes present in the environment. Skin is susceptible to both external and internal conditions.

As it protects the vital organs of the body from harmful pathogens, its proper care is extremely crucial. The skin is exposed to an environment full of pathogens is prone

A. Singh (✉) · D. K. Sinha · N. Dhanda
Amity University Uttar Pradesh Lucknow, Lucknow, India

P. K. Singh et al. (eds.), *Emerging Technologies for Computing, Communication and Smart Cities*, Lecture Notes in Electrical Engineering 875,
https://doi.org/10.1007/978-981-19-0284-0_49

to various kinds of diseases which may range from something as harmless as acne to something as dreadful as skin cancer.

The dermatologist is required to track the symptoms accurately and create a proper diagnosis as any discrepancy in judgment regarding critical skin diseases might prove deadly for the patient. Thus, an efficient approach must be established for the clear diagnosis of the symptoms of skin diseases in the early stage. Since 1955, AI has been emerging at a fast pace, revolutionizing every major field. Artificial intelligence and machine learning cater to the need for automation, deeper insights, predictive analysis, and increased accuracy.

With this fast-paced advancement in AI tech, skin diagnosis procedures can be technically designed for the initial diagnosis of skin infections. A variety of innovative ideas can be accessed for imagery and pattern-based detection for various skin diseases [1]. Since the employment of machine learning in the field of dermatology is still relatively new, this research aims at introducing the readers to the prospective technological methods driven by machine learning to obtain a more accurate and easily accessible diagnosis.

Machine learning can be a benefactor in the early identification of distinguished classes of skin infections. Skin disease symptoms are easily visible and often exhibit patterns. Through image classification, these diseases might be categorized and diagnosis can be fast-tracked and be made more personalized. Image-classification is a supervised-learning model in which a bunch of objective classes is defined and models are trained for the detection of correct class [2]. For the emerging technologies to be successful in the desired motive, it is required that medical professionals have a basic understanding of machine learning. The research provides a comparative study of different algorithmic models to find the most accurate image classification model.

2 Literature Review

The inter-merge and relation development between the technology and health-care services have led to frequent and rapid development in the field of image processing methods for the benefits and progress in the field of medicines. For an accurate diagnosis, many applications of digital image-based mechanisms like Computed Tomography (CT), Magnetic-Resonance-Imaging (MRI) and Digital-Subtraction-Angiography (DSA) are in use and have gained a lot of popularity.

Many organizations are working to build more efficient mechanisms for benefiting the healthcare field with modern technical aspects. Quite a few researchers have done work to detect skin diseases through different algorithms. A vivid literature survey is mentioned as follows. Ercal et al. [3] adapted efficient color metrics using the RGB planes. That serves in differentiating the tumor from its background. Image-segmentations are carried out using the appropriate coordinate transformations. Boundary shadows are sketched out by taking out the tumor portion of the classified image. That's a very efficient method to detect tumors while diagnosis.

Sigurdsson et al. [4] arrayed skin laceration using the in vitro Raman spectroscopy. This technique made use of a nonlinear-neural-network classifier for the work.

Distinctive brands of diversity express unambiguous lipids & proteins that give the facts and figures and relevant facts for the detection of skin lesions. Aberg [5] applies electrical bio-impedance for the assessment of skin cancers and lacerations. Multi-frequency impedance spectrum is made use to differentiate the skin-cancer and benign-nevi. Wong et al. [6] put forward a novel looping stochastic-region merging technique for segmentation of skin lesion infected area out of the macroscopic images. Initially, in the approach mentioned, stochastic-region merging is applied on the pixel level.

After this pixelated level, it upgrades to the region level till the convergence takes place. Demyanov et al. [7] made use of deep-convolutional-neural-networks, for the image-classification algo combined with data-augmentation for the successful investigation to autodetection of the dermoscopic-patterns and skin abrasion analysis.

Grana [8] gave a proficient mathematical perspective for the lesion border assessment. This method takes illumination values alongside a normal way to any contour at any instance point in consideration. Sigurdsson and Wighton [9] published automated skin lesion diagnosis. It's a model-based on the usage of supervised learning and makes use of MAP-estimation for portraying the results for identification. Emre Celebi et al. [10] make use of ensembles from the thresholding techniques for the detection of lesion boundaries in dermoscopy pictures. Ganster et al. [11] devised a computer-based system for analyzing images that are acquired via ELM.

General segmentation algorithms using fusion approach are applied to obtain the binary mask of skin lesion. The danger of injury is determined dependent on the shape and radiometric highlights. The neighborhood and worldwide boundaries are additionally taken under consideration for better outcomes. The framework serves to improve the early recognition of harmful melanoma. The procedure determined psoriasis vulgaris to have high affectability and particularity. The strategy determined psoriasis vulgaris to have high affectability and particularity. An inventive approach for auto-segmentation and classification of skin lesions was published by Sumithra et al. [12].

The discussed approach makes use of SVM and k-N neighbor algorithm for the detection of lesion. Lu et al. [13] utilize 2D computerized picture division and readjusting the size to arrange smooth-pixels consolidating the discussed procedures with Markov arbitrary field_(MRF). A solid division procedure is set up. Salimi et al. [14] differentiated various skin diseases with the use of a pattern recognition technique. Kolkur et al. [15] presented a novel skin identification calculation that improves the discovery of skin pixels that includes RGB, HSV, and YCbCr shading models. Kotian and Deepa [16] read auto conclusion framework for skin illness. Strategies, for example, picture line ID and highlight information mining are carried out utilizing Matlab programming. Oyola and Arroyo [17] gathered and characterize a picture of varicella through Hough change and applied the shading change, adjustment, and edge discovery procedures of picture handling. It helps in a better conclusion of varicella identification. Hung and Sapiro [18] recommended a technique for skin injury recognition based on a halfway differential condition. In view of the morphological sifting through PDE, a form model of sores was figured out, that helps in precisely recognizing the illness. 3-dimensional computed tomography (CT) imago logical technique was applied by Zhong et al. [19]. Kumar and Singh [20] related skin-cancer imagery across various types of neural networks. A bunch of skin cancer images was trained then tested using MATLAB. This helps in the characterization of skin-malignant growth.

According to research done in Medline, Embase (in December 2019), and WoS DB of articles retaining AI & ML in the field of dermatology. Articles from 2000 to 2020 are analyzed and included to concentrate on newer and better methods. A great deal of examination and applications are as of now trending in the field of clinical imaging and determination. Still, we need to focus on giving skin checking logically definite systems, minimal effort, and dependability.

3 Overview of Skin Diseases

The skin safeguards us from dangerous bacteria and other pathogens present in the environment. It's quite common for it to get infected and develop symptoms. The symptoms may be harmless but they can be as life-threatening as skin cancer. The reasons can be external environment-based or even internal because of the gene abnormality or any hormonal misbalance.

Major studies making use of ML in dermatology create have a clear focus on classification of skin lesions for a different disease. These studies generally use CNN for image recognition and classification. First, an already trained CNN (i.e., AlexNet) was employed to take out the workings, and these characteristics were then categorized by a simpler Machine-Learning algorithm like k-nearest neighbors or SVMs. In the current time, most of the CNNs take out characteristics and classify images by end-to-end learning [2]. This paper classifies aims at categorizing three kinds of skin diseases namely lichen planus, acne and Stevens-Johnson syndrome (SJS), and toxic epidermal necrolysis (TEN).

3.1 Acne

Acne is usually called acne-vulgaris. It is an immedicable inflaming infection. It happens over the body parts like neck, face, upper chest and back, and so forth. As per examines, acne is one of the most common and predominant sicknesses on the planet. It influences the patient both mentally just as psychosocially. The irritation of the skin happens because of the oil emitted from the sebaceous organ or that is usually known as oil-organs of the skin [21]. Skin inflammation is rarely a perilous situation, yet it influences the confidence of an individual. Individuals with the age

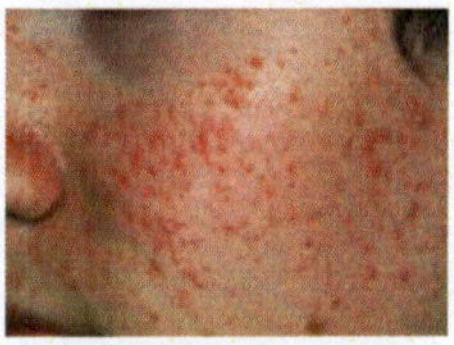

Fig. 1 Acne vulgaris

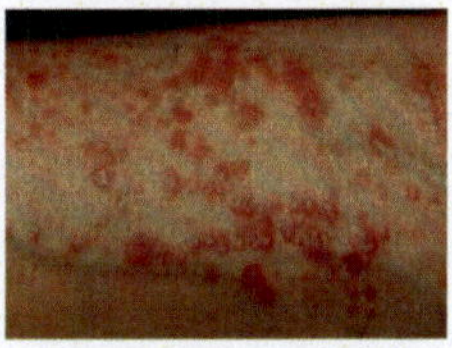

Fig. 2 Lichen planus

bunch going from 12–24 are more inclined to skin inflammation and around 85% of the populace is being influenced [21] (Fig. 1).

3.2 Lichen Planus

Lichen planus is a genuinely unmistakable, constant, immunologically interceded mucocutaneous sickness, of dubious etiology that was first depicted as an infection of the skin that can likewise influence mucosal surfaces, including the oral cavity lining. The commonness of LP in everyone is in the range of 0.9–1.2%. It is seen that roughly 40% of lesions happen on both oral and cutaneous surfaces, 35% on cutaneous, and 25% on mucosal surfaces alone. [22] Oral lichen planus is known to influence all genders between the ages of 30 and 70 years [22] (Fig. 2).

3.3 Stevens-Johnson Syndrome and Toxic Epidermal Necrolysis (SJS and TEN)

Stevens-Johnson syndrome and toxic epidermal necrolysis (SJS-TEN) are uncommon, possibly dangerous, extreme mucocutaneous antagonistic responses described by broad epidermal separation, the disintegration of mucosae, and serious indications. The medical highlights of SJS-TEN are trademark and the analysis is clinical. The first indications usually involve fever, conjunctivitis mimicking febrile-illness of infective origin and upper-respiratory-tract symptoms. This is accompanied by the disengagement of mucous-membranes (conjunctival, anogenital, oropharyngeal and nasal) [23]. Generally, more than two mucous-membranes are associated. Cutaneous-lesions, in the appearance of dusky erythematous-purpura/flat

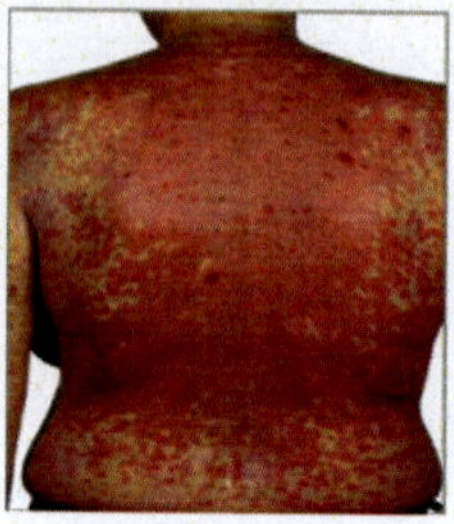

Fig. 3 SJS and TEN

typical-atypical target abrasions, break-out with ache and burning commotion (Fig. 3).

4 Proposed Work and Algorithms Used

In this comprehensive paper, the implementation of 5 machine-learning models that could detect 3-kinds of skin diseases namely lichen planus, acne and Stevens-Johnson syndrome and toxic epidermal Necrolysis (SJS-TEN) is discussed. The dataset for this research has been extracted through Bing Downloader. To further carry out the image classification process, the skin dataset is first preprocessed. While preprocessing, we first read the images using 'imread' function and loaded the folders containing the images into an array. Next, we resized them to the dimension (150, 150, 3) using 'resize' function. The images were segregated into three classes namely 0, 1, and 2. The dataset has been divided into the training–testing data: 80% is used for training purpose and rest 20% is utilized for testing. Thereafter, the models are trained using five variant classification algorithms namely naive Bayes classifier, logistic-regression, SVM, random-forest, and CNN for the classification of three mentioned skin diseases.

4.1 Logistic-Regression

Logistic-regression is statistical-model which utilizes employs logistic-functions to model up a binary-dependent variable, though plenty more complicated extensions exist. A logistic-model contains a dependent-variable which has 3-values which are possible that are indicated via an indicator-variable, where the 3-values are labeled as 0, 1 and 2. [24] In the logistic-model, the log-odds for the value tagged as "1" is a linear concoction of 1 or more than 1 predictor; the independent-variables can individually be a binary variable (2 classes, coded through the indicator variable) or a continuous-variable (any real value). The corresponding probability for the value tagged "1" can be valued somewhere from 0 to 1. The dependent variable in a binary-logistic-regression model has 2-levels (categorical) [24]. Outputs that have > 2 values are included in a model by multinomial logistic regression. The logistic-regression model plainly models the probable outcome of output in respective of input and doesn't undergo statistical-classification as this isn't a classifier, but it still can be employed well to use as a classifier, for example by selecting a cutoff-value and categorizing inputs which have the probability-index > the cutoff as 1 class, lesser the cutoff as the other, that's a different method to employ a binary-classifier.

4.2 Kernel SVM

The SVM is a supervised-learning algorithm that is usually implemented for classification although it can also be made use of for regression. The primary idea focuses on the labeled data (training data) by which the algorithm attempts to find the optimal hyperplane that can be employed to classify new data points. In two dimensions the hyperplane is considered a simple line.

Generally, a learning algorithm attempts at learning the most common characteristics, i.e., what distinguishes one class from another class, and the classification is based on those representative characteristics learned, so classification is based on differences between various classes. The SVM has its work in the other way around. It detects the most similar examples between classes. Those will be the support vectors.

4.3 Naive Bayes

Naive Bayes is an extremely easy approach used for the construction of classifiers/models that provide class labels to problems that are presented as vectors of feature values, in which case the class labels are extracted from the finite sets. There isn't any other algorithm that trains these types of classifiers, but a family-of-algorithms follows on a common principle that each and every naive Bayes classifier follows the assumption that the value of a specific feature isn't dependent over the value of any other feature.

For few kinds of probability models, naïve-Bayes classifiers can be effectively trained under a supervised learning atmosphere. In real-time uses, parameter estimation for naïve-Bayes models makes use of the way of max likelihood.

4.4 Random Forest

Random-forest, as the name infers, comprises an enormous number of individual choice trees that operate as a group. Each tree in the random forest provides a class prediction and the class having the largest number of votes takes the form of our model's outcome. In the language of data-science, the primary reason for the random forest model to work this well is that:

> "A large number of relatively uncorrelated models (trees) operating as a committee will outperform any of the individual constituent models [24]."

The low interrelation among models is the reason. Uncorrelated models are capable of producing ensemble outcome that is more correct than any other separate prediction outcome. The reason behind this is that the trees guard each other from their respective errors just as long as all of them don't persistently make errors in the same perception.

4.5 Convolutional Neural Networks

A CNN is a Deep-Learning algorithm that accepts an image as input, allocate importance, i.e., learnable weightage and biases to different prospect and objects of the image, and having ability to distinguish one from other. The pre-processing needed in CNN is much less in comparison to other classification algorithms. While in primeval methods, filters were hand-engineered, by proper training, Convolutional Networks possess the capability to accumulate these filters/characteristics. [12] The architecture of a CNN is homologous to the way of the association pattern of neurons and was greatly motivated by the establishment of the Visual Cortex. Each neuron replies to stimuli only in a constrained region of the ocular field known as the Receptive-Field. A cumulation of likewise fields overlaps to enfold the entire visual area.

5 Result Analysis

We had three different classes of skin diseases namely lichen planus, acne, and SJS and TEN in the suggested skin disease classification scenario. The data was then segregated into these three classes. We segregated the dataset between the training and testing dataset into the ratio of 80:20. We then employed the five models using Sklearn and Tensorflow and obtained the below-mentioned accuracies using r2_score metric. Among all these Naïve Bayes algorithms provides the least accuracy percentage, counting 47.4 and 49 as in testing and training accuracy respectively. The best mechanism that can be adopted for skin lesion detection is the convolutional neural network, with the most efficient working percentage. Its training accuracy is 86.77 and the training accuracy shows to be 89.05. So, the order of accuracy is as follows: Naïve Bayes < Kernel SVM < Random Forest < Logistic Regression < convolutional neural network (Fig. 4).

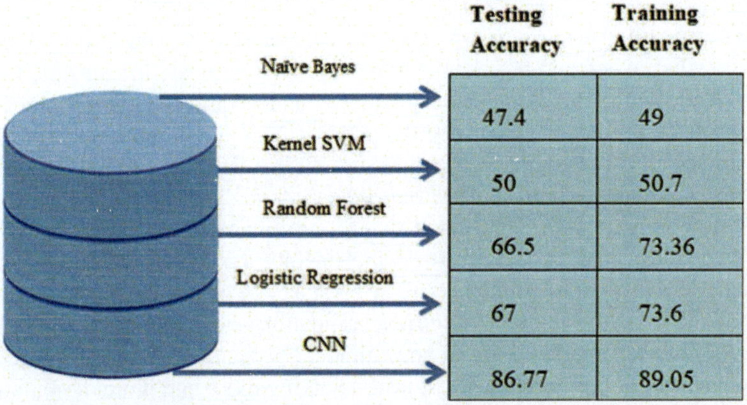

Fig. 4 Comparison of various classification models for skin disease diagnosis

6 Conclusion

Our early discussion started with brief factual information on skin diseases and their current treatments. In the field of skin disease treatment, the most crucial and major issue that the medical industry faces, is the detection of the lesion. The skin laceration can be diagnosed, retrieved, and cured efficiently, but only if analyzed beforehand at a primal stage. Literature survey clarifies that various skin disease inspection methods are being implemented. However, there is still a dire requirement to differentiate skin diseases at a real primary stage. Machine-learning algorithms hold the capacity to offer quite a few impactful methods for initial detection for skin laceration. It can also help people to take real-time measures for their skincare. If invested upon well for futuristic approaches, these techniques can provide pertinent assistance for certain and can also come up with many unified approaches for skin lesion issues prevention. The development in these mechanisms will also lead to assistance to patients and dermatologists to cure the skin lesion speedily and effectively. Many Machine Learning can be effectively used and applied to detect the skin lesion boundary very effectively. Naïve Bayes, Kernel SVM, Random Forest, Logistic Regression, and convolutional neural network are some of the used algorithms mentioned above in the discussion and have been given a mention in the literature survey. The CNN Model though had the most accurate among all the models.

Researches and implementations of medical information and figures are accessible in a very limited manner. We need more real-time info's to be made available in the coming future so that the detection of skin lacerations can be better tested and investigated with the latest advances in Artificial Intelligence and the pros of diagnosis assistance led with Artificial Intelligence. Machine learning offers an outstanding potential in the field of dermatology, ranging from efficient diagnosis to putting forward much better, effective, and safer treatments. As the Machine Learning and Artificial intelligence mechanisms advance, dermatologists will need to obtain at least a basic understanding of how these mechanisms work, and also to figure out that with what, when and how it should be applied effectively in real-time medical set-up. While ML methods are powerful, they still have quite a few similarities to the past clinical equipment, in which doctors' interaction is very important for the proper application of the tools in a real-world set-up. We are also required to be acquainted with the techniques of how budding biases can interact and make changes with the black-box feature of all the algorithms. One more crucial and important development needed is to make all these mechanisms which include the skin tones and color. Future researches in Machine Learning need to be crystal clear by providing algorithms and datasets to the public for more futuristic validation and implementations. Before launching into the market, careful and diligent peer-reviewed anticipated clinical proceedings should be conducted. At last, involving more and more dermatologists in the process development and trials of Machine Learning is of vital importance for producing efficient and clinically effective technologies.

References

1. Juyal P, Sharma S (2020) Machine learning algorithms based skin disease detection
2. Chan S, Reddy V, Myers B et al (2020) Machine learning in dermatology: current applications, opportunities, and limitations. Dermatol Ther (Heidelb) 10:365–386
3. Erçal F, Moganti M, Stoecker WV, Moss RH (1993) Detection of skin tumor boundaries in color images. IEEE Trans Med Imaging 12(3):624–626
4. Sigurdsson S, Philipsen PA, Hansen LK, Larsen J, Gniadecka M, Wulf HC (2004) Detection of skin cancer by classification of Raman spectra. IEEE Trans Biomed Eng 51(10):1784–1793
5. Aberg P, Nicander I, Hansson J, Geladi P, Holmgren U, Ollmar S (2004) Skin cancer identification using multifrequency electrical impedance-a potential screening tool. IEEE Trans Biomed Eng 51(12):2097–2102
6. Wong A, Scharcanski J, Fieguth P (2011) Automatic skin lesion segmentation via iterative stochastic region merging. IEEE Trans Inf Technol Biomed 15(6):929–936
7. Demyanov S, Chakravorty R, Abedini M, Halpern A, Garnavi R (2016) Classification of dermoscopy patterns using deep convolutional neural networks. In 2016 IEEE 13th International symposium on biomedical imaging (ISBI). IEEE, pp 364–368
8. Grana C, Pellacani G, Cucchiara R, Seidenari S (2003) A new algorithm for border description of polarized light surface microscopic images of pigmented skin lesions. IEEE Trans Med Imaging 22(8):959–964
9. Wighton P, Lee TK, Lui H, McLean DI, Atkins (2011) Generalizing common tasks in automated skin lesion diagnosis. IEEE Trans Inform Technol Biomed 15(4):622–629
10. Emre Celebi M, Wen Q, Hwang S, Iyatomi H, Schaefer G (2013) Lesion border detection in dermoscopy images using ensembles of thresholding methods. Skin Res Technol 19(1):e252–e258
11. Ganster H, Pinz P, Rohrer R, Wildling E, Binder M, Kittler H (2001) Automated melanoma recognition. IEEE Trans Med Imaging 20(3):233–239
12. R Sumithra R, Suhil M, Guru DS (2015) Segmentation and classification of skin lesions for disease diagnosis. Procedia Comput Sci 45:76–85
13. Juan Lu J, Kazmierczak E, Manton JH, Sinclair R (2012) Automatic segmentation of scaling in 2-d psoriasis skin images. IEEE Trans Med Imaging 32(4):719–730
14. Salimi S, Sabbagh Nobarian M, Rajebi S (2015) Skin disease images recognition based on classification methods. Int J Tech Phys Problems Eng 22(7):78–85

15. Kolkur S, Kalbande D, Shimpi P, Bapat C, Jatakia J (2017) Human skin detection using RGB, HSV and YCBCR color models. arXiv preprint arXiv:1708.02694
16. Kotian AL, Deepa K (2017) Detection and classification of skin diseases by image analysis using MATLAB. Int J Emerg Res Manage Technol 6(5):779–784
17. Oyola J, Arroyo V, Ruedin A, Acevedo D (2012) Detection of chickenpox vesicles in digital images of skin lesions. In: Iberoamerican Congress on pattern recognition. Springer, pp 583–590
18. Chung DH, Sapiro G (2000) Segmenting skin lesions with partial-differential-equations-based image processing algorithms. IEEE Trans Med Imaging 19(7):763–767
19. Zhong LS, Jin X, Quan C, Zhi-ping WEI, Yan-qun LIU (2011) Diagnostic applicability of confocal laser scanning microscopy in psoriasis vulgaris. Chin J Dermatovenereology 25(8):607–608
20. Kuila S, Dhanda N, Joardar S, Neogy S, Kuila J (2019) A generic survey on medical big data analysis using internet of things. In: First international conference on artificial intelligence and cognitive computing. Springer, Singapore, pp 265–276
21. Janani SK, Raman S (2019) A comprehensive review on acne, its pathogenesis, treatment, in-vitro and in-vivo models for induction and evaluation methods
22. Madalli V (2013) Lichen planus-a review. IOSR J Dent Med Sci 12:61–69. https://doi.org/10.9790/0853-1216169
23. Gupta LK, Martin AM, Agarwal N, DSouza P, Das S, Kumar R, Pande S, Das NK, Kumaresan M, Kumar P, Garg A, Singh S (2016) Guidelines for the management of Stevens–Johnson syndrome/toxic epidermal necrolysis: an Indian perspective. Indian J Dermatol Venereol Leprol 82:603–625
24. Patnaik SK, Sidhu MS, Gehlot Y, Sharma B, Muthu P (2018) Automated skin disease identification using deep learning algorithm. Biomed Pharmacol J 11(3)

IPFS-Based Image Sharing Using Perceptual Hash Code with Digital Right Management

D. Geethanjali, R. Priya, and R. Bhavani

Abstract In the current scenario, a large amount of information is exchanged over the Internet. An unauthorized user access may cause the information to be modified, disseminated, misused, detected, and used. Digital rights management (DRM) is a method of preserving copyrights in digital media. It often requires more elements than simply restricting or prohibiting copies of the content. This strategy includes the deployment of technologies that limit the copying and use of copyrighted works as well as proprietary software. The proposed approach has two modules. They are hash code generation of an image and Store and retrieve the data from IPFS. First, select the watermarked image. The different types of perceptual hashes are average hash, perceptual hash, difference hash, and wavelet hashing. The perceptual hash codes are generated for the selected watermarked image. Then, the watermarked image and created copyright text files are shared through the IPFS system. The goal of this proposed system is to compare the perceptual hashes by using the hamming distance, size of the image, and calculation time. Based on the evaluation, difference hash algorithm provides a better result. So, it is used to check whether the send and retrieve files of IPFS are similar or not. In the IPFS file system, hash codes are used to find the location of a file effectively and securely. This paper proposes, how IPFS and perceptual hashes are preventing unauthorized user access to the system while sharing the images over the Internet. Genuine image authors' copyrights are protected by using perceptually identical images on a decentralized blockchain IPFS store.

Keywords Perceptual hashes · Blockchain · IPFS · Digital right management · Watermark

D. Geethanjali (✉) · R. Priya · R. Bhavani
Department of Computer Science and Engineering, Annamalai University, Chidambaram, Tamilnadu, India

P. K. Singh et al. (eds.), *Emerging Technologies for Computing, Communication and Smart Cities*, Lecture Notes in Electrical Engineering 875,
https://doi.org/10.1007/978-981-19-0284-0_50

1 Introduction

The rapid advancement of technology has resulted in a significant increase in multimedia content, particularly in recent years. The total amount of digital content has risen at an exponential rate and continues to do so. We live in a world where almost everything can be digitalized and sent from one end of the globe to the other in seconds. On the Internet, modern technology allows for fast data transfer, data content alteration, and the creation and retransmission of unauthorized copies. However, in the network media, security, robustness, and content protections remain a difficult task. To solve those problems, the proposed system provides copyright verification with some concealed data.

The contribution of the work is the secure sharing of data with the help of hash algorithms. This may help for various applications like digital garment design works, research record-keeping systems, medical applications, online education, and so on. The work aims to discuss the applications of perceptual image hashing which is better than cryptographic hash functions. This proposed work is divided into two-step processes. First, generate and compare the perceptual hash code for the watermarked image. The authorized users' blockchain list was created. Next, the watermarked image and copyright text file are shared through IPFS. The following is a breakdown of the paper's structure. The related works of the existing systems are listed in Sect. 2. The proposed system is described in Sect. 3. The implementation results and design are presented in Sect. 4. The conclusion and future efforts are described in Sect. 5.

1.1 Hashing

The essential qualities of the perceptual hash algorithm are differentiability, collision tolerance, summarization, stability, and uni-direction. Hashing is a file-based algorithm for calculating a fixed-size bit string value. A file is essentially a set of data blocks. Hashing reduces the length of the data to a fixed-length value or key that represents the original string. When compared to the cryptographic hash, perceptual hashes are not affected by the avalanche effect and are less susceptible to perceptual changes in images such as rotation, cropping, re-sizing, grayscale conversion, salt and pepper noise, exposure or saturation changes, and so on. The hash function returned bit value. They are called hash values, hash codes, digests, or simply hashes. The perceptual hashes are used in various applications like digital garment design works, research record-keeping systems with digital right management (DRM), similarity, and dissimilarity of an image.

1.2 IPFS (InterPlanetary File System)

The images and hashes are stored using the IPFS architecture. It is a peer-to-peer hypermedia distribution protocol that is decentralized. By ensuring that files with identical content are only kept once, the IPFS system eliminates content duplication. This creates the Internet world as a faster, secure, and more open environment. Data warehoused on IPFS can be scrambled with any usual encryption techniques used to retrieve the stored information. The registered authenticated users can only access the system with the proper encryption keys. The diagrammatic representation of the IPFS system based on blockchain is shown in Fig. 1.

The IPFS system works based on the following three fundamental principles. They are (1) Unique identification through content addressing. (2) Content linking by using directed acyclic graphs (DAGs). (3) Content discovery through distributed hash tables (DHTs).

IPFS operates by tying all networked devices to the same file structure. To ensure the immutability, IPFS file structure combined with the Merkle trees. A hash tree, also called the Merkle tree, is used to efficiently and securely encrypt blockchain data. It enables the peer-to-peer blockchain network to quickly verify blockchain data as well as transfer vast volumes of data from one computer node to the next. The directed acyclic graphs are permitting the user to know the versions of content on IPFS. InterPlanetary version control is a distributed version control framework based on IPFS, similar to git (InterPlanetary File System). It can handle any type of data, not just human-readable text. It is also mainly suitable for versioning large files.

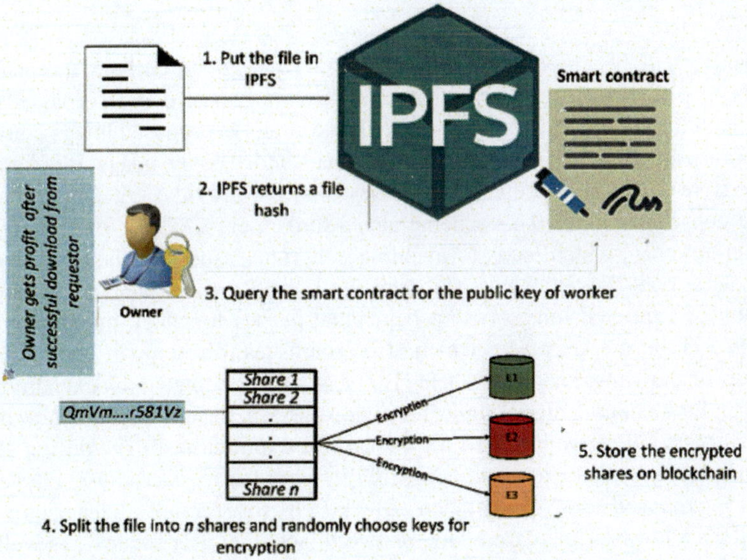

Fig. 1 Diagrammatic representation of IPFS system based on blockchain

Distributed hash table (DHT) serves as a catalog and navigation system, assisting the network in keeping track of and finding data. The IPFS system performs the file searching operation in two steps. First, IPFS identifies the file with content addressing through the hash value. Next, the network connection establishment of the corresponding node is identified; then, the file is retrieved.

The most significant benefit of employing IPFS in this manner is that nodes become individual servers capable of serving content to others. IPFS is widely utilized as a data-sharing storage platform. It offers great availability and performance, but it lacks the capacity to trace access and authentication, making it difficult to investigate illicit access and authorship.

1.3 Blockchain

A blockchain is a decentralized software network that serves as a digital record and allows for the secure transfer of property without the use of a third party. Blockchain is a technology that facilitates the digital exchange of units of value. It is the current technology that enables the digital flow of information. On a blockchain network, anything from money transfers to land titles to votes can be tokenized, stored, and traded.

2 Related Work

Geethanjali et al. [1] proposed a secure transaction through blockchain. It is achieved with the help of watermarked image and generated image hash code. It is passed to the blockchain for the secure sharing of data. Dedge et al. [2] proposed image copyright protection using blockchain by a digital watermark added on the image. Blockchain is used to store the data in a protected manner, which also provides proof-of-ownership for the copyright holder. This scheme also affords a cross-check functionality for commercial users which leads to avoiding copyright issues by checking whether the image is copyrighted or not. Mehta et al. [3] proposed automatic detection and rejection of tampered images using perceptual hashes for stock photo Web sites displayed in the market places which also establishes robust smart contracts with the help of the Ethereum blockchain. Lu et al. [4] presented a blockchain-based approach for managing digital rights for digital garment design works. A new proof-of-delivery technique is proposed for the sake of trade fairness. The design effects were displayed with privacy and compatibility with the registered users. Nyaletey et al. [5] proposed a novel approach (BlockIPFS) to generate a clear audit trail. BlockIPFS allows for increased data trustworthiness and user security, as well as a great solution to track all activity linked to a specific file utilizing blockchain as a service. Zhu et al. [6] proposed a combination of the perceptual hash algorithm and SIFT features, which increase the exactness of the image retrieval. It also eliminates the perceptual hash algorithm's influence on geometric attacks, imagescale changes, and other abnormal conditions. Rajalakshmi et al. [7] proposed a mixing of IPFS and blockchain technologies and the traditional encryption methods login id, biometric authentication, and secrete question are asked to check the authenticity. They used to maintain the academic research records in a secure manner with tamper-proof. This system improves the privacy, authenticity, and access control of the registered users. Steichen et al. [8] suggested an InterPlanetary File System (IPFS). It may be used to efficiently store and share enormous volumes of data. This work described a reformed version of the InterPlanetary File System (IPFS) that uses Ethereum smart contracts to enable access-controlled file sharing. The smart contract is used to keep track of the access control list. When a file is uploaded, downloaded, or moved, it also allows interaction with the smart contract. Tang et al. [9] proposed medical image sharing in a secure manner built through blockchain smart contracts and credit scores. This provides the solution for the following problems, for cross-domain distribution and patient privacy protection problem while sharing the medical image information. Meng et al. [10] offered a design scheme based on digital watermarking and blockchain to increase the security and transparency of information in the copyright management system. Zhang et al. [11] proposed a DRM mechanism based on blockchain technology. This blockchain technology is used to keep track of copyright transactions and authorizations in a transparent and secure manner. The smart contract establishes the copyright transaction's trustworthy automated licenses. Ma et al. [12] proposed a technique, based on watermark and blockchain, a comprehensive and high-level security artwork picture digital right management strategy for online abusing detection was presented. Zhaofeng Ma, et al. [13], proposed, a trustworthy, protected, effective and tamper resistance digital content services using DRM chain scheme. Li Weng, et al. [14], Image content authentication is achieved through wavelet watermarking with perceptual hashing.

3 Proposed System

The available technologies are becoming more accessible and open source in the current scenario. This has posed significant challenges in recent years in terms of data management, authentication, and security of sharing data. This proposed work was implemented in two steps.

They are (1) perceptual hash code generation. (2) datastore and retrieval from IPFS. Figure 2 illustrates the block diagram of the proposed system.

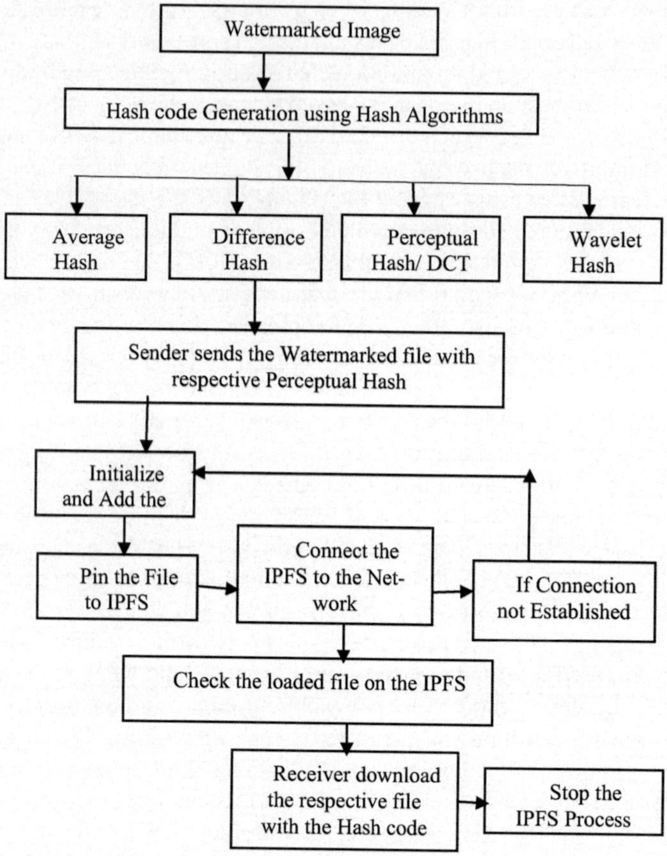

Fig. 2 Block diagram of the proposed system

3.1 Watermark Process

Get the cover image. Get the watermark image. The discrete wavelet transform technology was used to incorporate the watermark image on the cover image. Create a copyright text file for the watermarked image.

3.2 Perceptual Hash Code Generation

Perceptual hashing functions are not the same as cryptographic hashing functions. In contrast to cryptography hashes, especially, perceptual hashes are similar if their inputs have similar characteristics. The distinction (Hamming distance) metric of

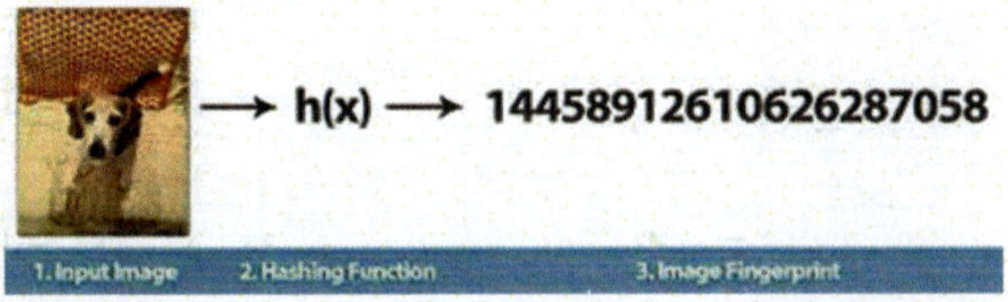

Fig. 3 Average hash code generation process

dissimilarity is used between perceptual hashes. The perceptual hashes give effective encryption and are more accurate and more compact.

To calculate the hash value, perceptual hashes usually follow the methods below. The input image was given to the hash function. The image hashing function calculates the image hash value and displays the hash code for the respective image. This is performed based on the visual content of an image. The image hashing is used to perform some of the following applications like content-based image retrieval, near-duplicate detection, and reverse image search engines. The four most popular perceptual hash algorithms are average hash (AHA), difference hash (DHA), perceptual/DCT hash (PHA), and wavelet hash (WHA). Each perceptual hash algorithm works based on the given below steps.

3.2.1 Algorithm Steps for Average Hash

(i) Resize the input image size into 8×8.
(ii) Convert the resized image into a grayscale image.
(iii) Calculate gray average value, and it is denoted as M.
(iv) According to the comparison with the gray average to get the binary map.

$$h(i) = \begin{cases} 0 \text{ if } A(i)pM \\ 1 \text{ otherwise} \end{cases} \quad (1 \le i \le 64) \tag{1}$$

where $h(i)$ is denoted as average hash, $A(i)$ represents the normalization of each pixel value based on the comparison of average value M.
(v) Combine binary maps into 64-bit hash values (fingerprints) in order.

Figure 3 shows the general hash code generation process for the average hash.

3.2.2 Algorithm Steps for Difference Hash

(i) Resize the image size into 9×8.
(ii) Convert the resized image into a grayscale image.
(iii) Calculate the difference value (compare the size of two adjacent values in each row) to get a binary map

$$d\text{px}_{r,c} = \text{px}_{r,c+1} - \text{px}_{r,c} \tag{2}$$

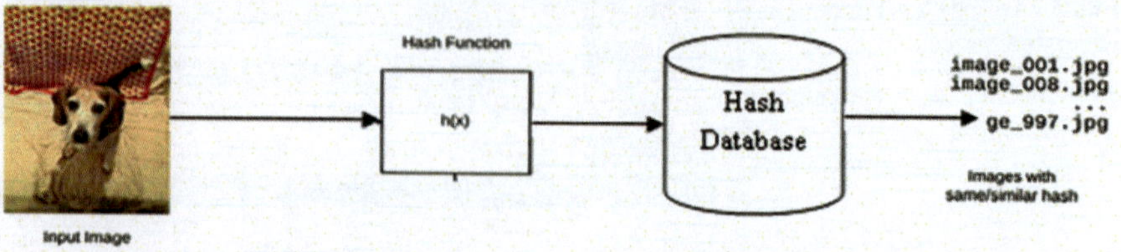

Fig. 4 Difference hash code generation process

where $px_{r,c}$ is denoted as the gray value of the resized image at r-row and c-column, $dpx_{r,c}$ specifies the difference value. Like this, the pixels per row yield 8 differences between adjacent pixels, and 8 differences of 8 rows become an 8×8 image.

(iv) Combine binary maps into 64-bit hash values (fingerprints) in order.

Figure 4 shows the general hash code generation process for difference hash.

3.2.3 Algorithm Steps for Perceptual Hash

(i) Resize the image size into 8×8.
(ii) Convert the resized image into a grayscale image.
(iii) Calculate the DCT transform of the image.
(iv) Reduce DCT.
(v) Calculate the mean (m) of the DCT coefficients.

$$m = \frac{1}{64}\sum_{r=1}^{8}\sum_{c=1}^{8} d_{r,c} \quad (3)$$

where r, c are rows and columns of the DCT coefficients. d refers to the mean difference between the DCT coefficients.

(vi) Compare and judge to get a binary map.

$$h_i = \begin{cases} 0 \; C_i pm \\ 1 \; C_i \geq m \end{cases}, \quad i = 0, \quad k \;\; 63. \quad (4)$$

where h_i denotes bit of the perceptual hash value at the i position; C_i is the array of DCT coefficients; m specifies the mean value of the DCT coefficients.

(vii) Binary graphs are combined into a 64-bit hash value (fingerprint) in sequence.

Figure 5 shows the general hash code generation process for perceptual/DCT hash.

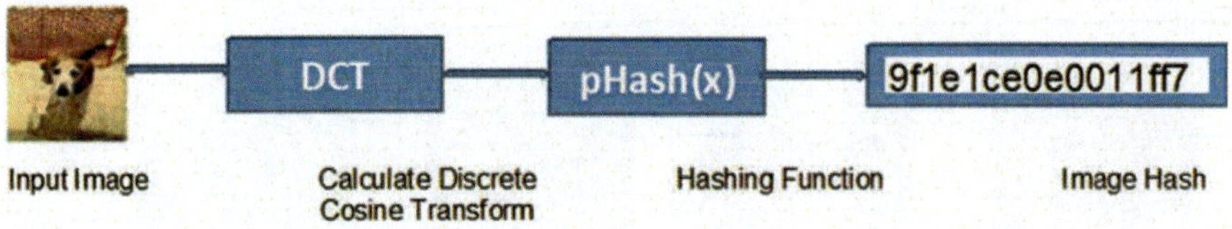

Fig. 5 Perceptual hash code generation process

3.2.4 Algorithm Steps for Wavelet Hash

(i) It generates a gray value image sized 8 × 8.
(ii) The image is then transformed into a two-dimensional wavelet transform.
(iii) Removes the lowest Haar frequency LL (max). The lowest frequency has only one data point/pixel, which shows the contrast image and is not very useful for hashing.
(iv) Binary graphs are combined into a 64-bit hash value (fingerprint) in sequence.

Figure 6 shows the general hash code generation process for wavelet hash.

3.3 Data Store and Retrieval from Interplanetary File System (IPFS)

The centralized servers used to store and retrieve the users data. It also resolves any issues that may arise in centralized server. The watermarked image was stored and distributed using IPFS. The following steps are performed to add and retrieve the data through IPFS.

(i) Download and install IPFS for your system.
(ii) Check whether the path of the file is stored in a proper location.
(iii) Initialize and add the file to IPFS. Pin the file to IPFS.
(iv) Connect the IPFS system to the network.
(v) Check the loaded file on the IPFS by using the https://ipfs.io/ipfs/Hashcode of a file/file directory

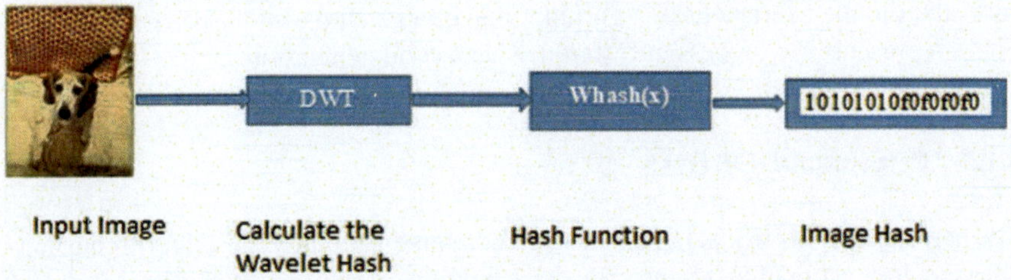

Fig. 6 Wavelet hash code generation process

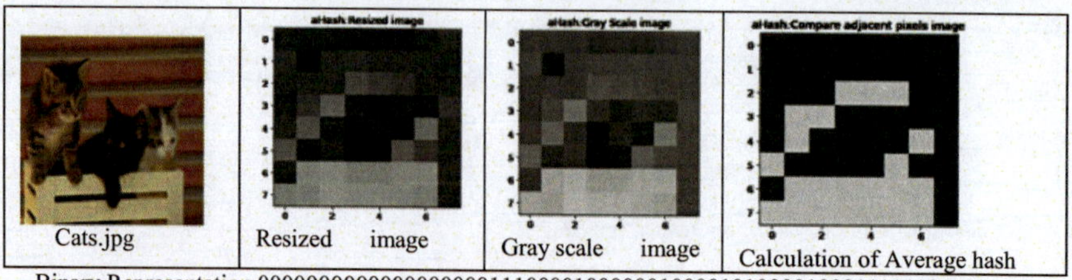

Binary Representation 0000000000000000000111000010000001000010100001000111111011111110
Hexadecimal Representation: 00001c6042847efe

Fig. 7 Representation of the generated average hash code for Cat.jpg image

(vi) Finally, the daemon is stopped by using the Ctrl + C keys.

4 Implementation Results and Discussion

4.1 Perceptual Hash Code Generation

4.1.1 Average Hash

To obtain an average hash, execute the following steps: First, the size of an image was reduced (for example, to an 8 × 8 image, to further shorten the number of computations); then, the image was converted to gray scale, 3rd average the resulting colors (for an 8 × 8 image, we average 64 colors), 4th calculate the bits by comparing if each color value is greater than or less than the mean, 5th build the hash. The following Fig. 7 represents the process of average hash code generation.

4.1.2 Differential Hash

The differential hash works in the following way: The difference hash algorithm primarily the image was scaled down to 9 × 8 pixels, then creates a grayscale image from the input image. Next, from each row, the first 8 pixels are reviewed serially from the left side to the right side and compared to their right hand neighbor, which, equivalent to the average hash algorithm, outcomes in a 64-bit hash. The following Fig. 8 represents the process of difference hash code generation.

4.1.3 Perceptual/DCT Hash

The perceptual hash works in the subsequent manner: Initially, this algorithm calculates the gray value of an image and scales it down. In this case, to desire a factor of 4, which is why we scaled down to 8 * 4 × 8 * 4, that is, a 32 × 32 image. Apply

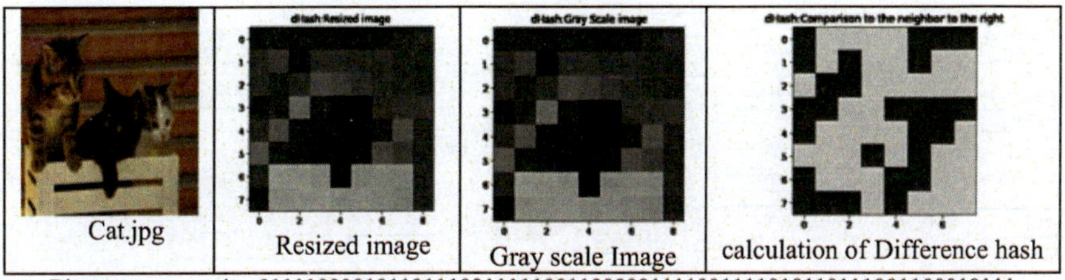

Binary representation 0111100001011011100111110011000001111001111010110111001100010111
Hexadecimal Representation: 785b9f3079eb7317

Fig. 8 Representation of the generated difference hash code for Cat.jpg image

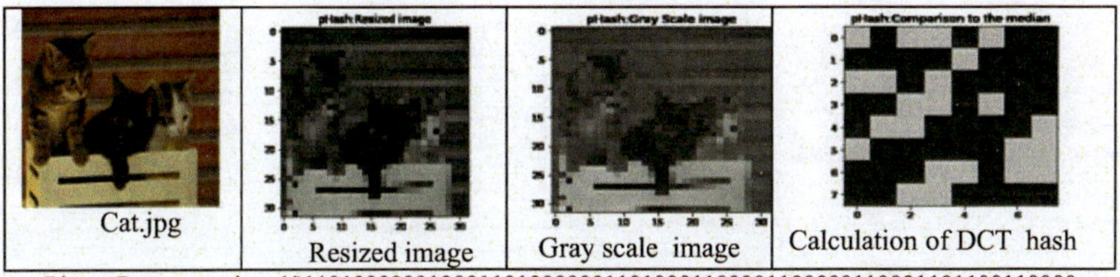

Binary Representation: 1011010000001000110100000011010001100001100000110001101100110000
Hexadecimal Representation: b408d03461831b30

Fig. 9 Representation of the generated perceptual hash code for Cat.jpg

a discrete cosine transform on this image, first per row and afterward per column. The pixels with high frequencies are now located in the upper-left corner. Because of this reason, the image cropped to the upper left 8 × 8 pixels. The following Fig. 9 represents the process of perceptual hash code generation.

4.1.4 Wavelet Hash

The wavelet hash function as follows: This wavelet hash algorithm also generates a gray value image sized 8 × 8. A two-dimensional wavelet transform is applied to the image. To improve the test results to set the top row to 0, that is, to black and re-apply the wavelet transform three times. It is implemented by using the image hash library. Next, similar to the perceptual hash algorithm, each pixel is compared to the median, and the hash is generated. The following Fig. 10 represents the process of wavelet hash code generation.

Perceptual hash codes are implemented based on their algorithms. The binary form hash code is generated and is converted to a hexadecimal representation. Usually, this hexadecimal format is used for further operations for easy understanding of humans. The following Fig. 11 shows the sample images are taken from the open database for testing the proposed system.

The following Table 1 shows the generated perceptual hash codes for sample images.

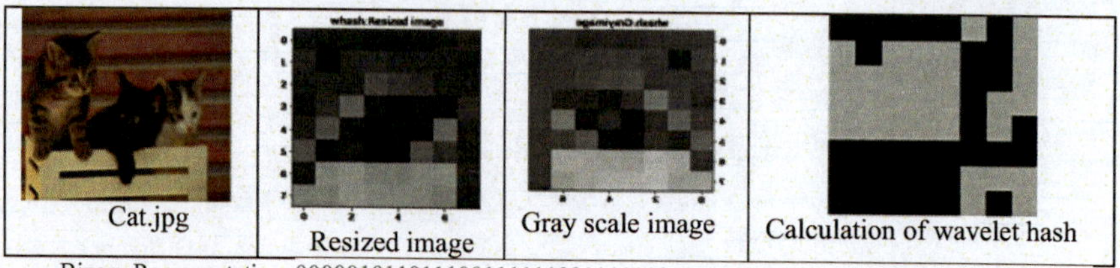

Binary Representation: 0000010110111001111110011111101111111010000000000000011100000101
Hexadecimal Representation: 05b9 f9fb fa00 0705

Fig. 10 Representation of the generated wavelet hash code for Cat.jpg image

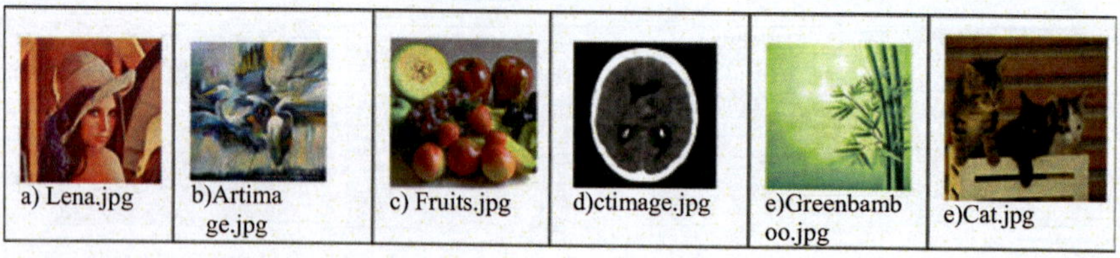

Fig. 11 Sample test images for the proposed system

Among the four perceptual hashes based on the hamming distance, calculation time, the difference hash (DHA) provides a minimal error and better result. So, DHA is used for further simulation process of IPFS.

4.2 Storing and Retrieval of an Image from IPFS

The go-ipfs is an open-source software. It is provided by the IPFS Web site. Figure 12 shows the downloaded IPFS installed into the system. And also, it was checked whether the path of the file is stored in a proper location.

The given Fig. 13 shows the initialization of the IPFS system.

The given Fig. 14 represents the creation of a new folder and adding the watermarked image file and copyright text file into that folder using IPFS.

Figure 15 represents that the file folder is pin to the IPFS, using the hash code

Figure 16 shows the IPFS system connection that is established to the network

File/File directory is searched by using the hash code of a respective folder. The file/file folder is retrieved from IPFS by using the following link. https://ipfs.io/ipfs/QmVPzqYHb16nc4iRJB62gv5WxYG5sMZK5YzP99R1ascKzE. This shows the folder file names with their and size of each file.

Figure 17 shows the loaded file on the IPFS by using the file/file directory.

Figures 18 and 19 shows the downloaded folder content of an image and copyright text file from an IPFS.

The following Fig. 20 shows the downloaded folder of a Lena image which contains the two files, watermarked-lena.jpg and lena.txt.

Table 1 Generated perceptual hash codes for sample images

Sample images	Average hash	Difference hash	DCT hash	Wavelet hash
Lena.jpg	ac6969eb61e0a080	58dbdbdbdb414832	fb819e56b0d55612	ae69e9ebe1e0a888
Artimage.jpg	eeed6721208bf300	c8c9cdc5c91a86c2	a392f5f946a439c2	eeef67352003fb00
Fruits.jpg	fff5e1c004032323	8d4989096df64e47	e3f2a51d1e318966	fff5e1c0040b23a7
Ctimage.jpg	c7c667e7e723c18	71cc8e96b696d471	c4d30b4c2dcb6c3e	387c667e56723c18
Greenbaboo.jpg	fcfcfcf4f8743400	e1f1c1edc1e5e4fc	fcfcfcf4f8743400	fcfcf8f478743000
Cat.jpg	00001c6042847efe	b408d03461831b30	b408d03461831b30	05b9 f9fb fa00 0705

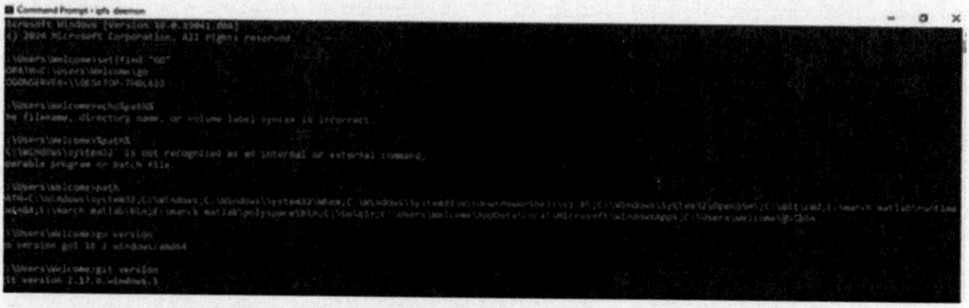

Fig. 12 Installation and file storing location of IPFS

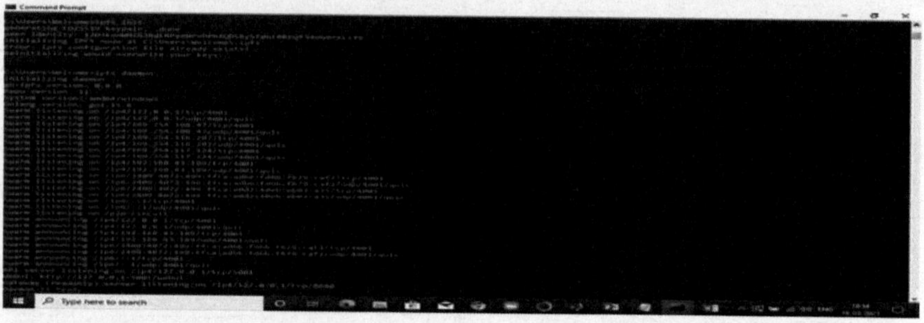

Fig. 13 Initialization of the IPFS

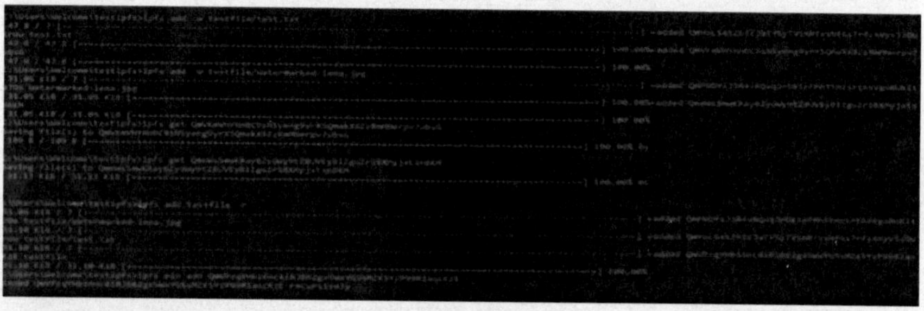

Fig. 14 Adding the image file and copyright file to IPFS

Fig. 15 Pin the file folder to IPFS

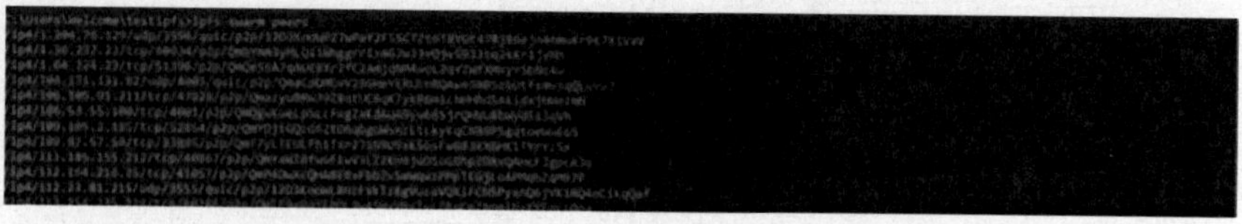

Fig. 16 Connect the IPFS system to the network

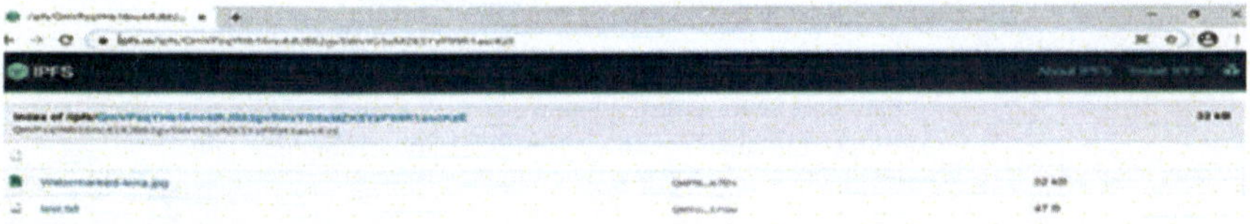

Fig. 17 Loaded folder of Lena image on the IPFS

Fig. 18 Downloaded watermarked-lena.jpg image

Fig. 19 Downloaded copyright text file of a watermarked image Lena from the IPFS

Fig. 20 Downloaded folder of a Lena image

The following Table 2 shows the similarity matches between the sent watermarked image and retrieved image from the IPFS.

Hamming distance is used to compare the two images. It refers to the number of dissimilar bits between two hashes. The hamming distance (HD) value is 0; then, the source and retrieved images are similar. If the HD value is between 1 and 10, then the two images are hypothetically variance, and the HD value is greater than 10; then,

Table 2 Similarity match between the sent and retrieval image of IPFS

Image name	Hash value of sent watermarked image	Hash value of retrieved image from IPFS	Hamming distance	Size of the image (in kilo bytes)	Result
Lena.jpg	898d82a4466ca583	898d82a4466ca583	0	30	Same
Artimage.jpg	372752a9364429f6	372752a9364429f6	0	42	Same
Fruits.jpg	709c34b6920528b0	709c34b6920528b0	0	40	Same
Ctimage.jpg	452b514949492b05	452b514949492b05	0	28	Same
Greenbamboo.jpg	0d061a13360a1b03	0d061a13360a1b03	0	25	Same
Cat.jpg	70539f3371eb7357	70539f3371eb7357	0	26	Same

the images are different. As per the result, the hamming distance value and size of the images are similar while sent and retrieved data. So, the hash code performs the better retrieval of data from IPFS with good protection.

4.3 *Performance Evaluation*

4.3.1 Peak Signal-To-Noise Ratio (PSNR)

It is a measure of the peak error between the two images. The ratio between the maximum possible power of a signal and the power of corrupting noise affects the fidelity. The formula for the PSNR value of an image is given in Eq. (5).

$$\text{PSNR} = 10\log_{10}\frac{255^2}{\text{MSE}} \tag{5}$$

4.3.2 Normalized Correlation (NC)

It evaluates the performance of the watermarking scheme. To estimate and check the quality of the extracted watermark image and the original watermark image. It is given in Eq. (6)

$$\text{NC} = \frac{\sum_{i=0}^{m}\sum_{j=0}^{n} W(i,j)W'(i,j)}{\sqrt{\sum_{i=0}^{m}\sum_{j=0}^{n} W(i,j)^2}\sqrt{\sum_{i=0}^{m}\sum_{j=0}^{n} W'(i,j)^2}} \tag{6}$$

Table 3 Attacked images

Salt and pepper noise (0.015)	Crop	Rotation (5°)	Rotation (10°)

4.3.3 Bit Error Rate (BER)

It is defined as percentage of bits that have errors relative to the total number of bits received in a transmission. It has been calculated by using Eq. (7)

$$\text{BER} = \frac{\text{No. of Error Bits}}{\text{Total no. of bits sent}} \tag{7}$$

4.4 Watermark Attacks

To check the robustness of the proposed system, is tested by applying the following attacks like salt and pepper, crop, and rotation attacks on the sent image. Table 3 shows the attacked image.

The PSNR of the embedding value of the Lena image is 36 dB. The retrieved image NCC values of all the test images are 1 for both before and after attacks. At the same time after the attack, the bit error value varies from 0.8816 to 0.9952. The retrieved images are recognizable and have reasonable correlation coefficients.

5 Conclusion and Future Enhancement

The proposed system can find the similarity between sent and retrieval images using the perceptual hash. As per the results, the identical data or image may be identified, with a different watermark images. The perceptual hashes have the following better properties. It is a useful technique for rapidly accessing data based on a key value. It also provides image protection. IPFS provides a better result of storing and retrieval of information from IPFS. In future, it can be used to evaluate the performance measure of the perceptual hashing and apply the transformation attacks on the image. Perceptual hash functions are commonly used in detecting cases of online copyright infringement as well as in digital forensics. The effectiveness of perceptual hashing and IPFS file storage will be applied in real-time application digital garment design work. The smart contract will be established through blockchain with the authenticated user.

References

1. Geethanjali D, Priya R, Bhavani R (2020) Blockchain-based protected digital copyright management with digital watermarking. In: Hemanth DJ et al (eds) Intelligent systems and computer technology. IOS Press. https://doi.org/10.3233/APC200113
2. Dedge O, Shingade R, Jagtap A, Yadav A, Kamble A (2020) Image copyright protection system using blockchain. Int J Future Gener Commun Networking 13(3s):37–43
3. Mehta R, Kapoor N, Sourav S, Shorey R (2019) Decentralised image sharing and copyright protection using blockchain and perceptual hashes. In: 11th international conference on communication systems and networks (COMSNETS). IEEE. 978-1-5386-7902-9/19$31.00 ©2019
4. Lu Z, Shi Y, Tao R, Zhang Z (2019) Blockchain for digital rights management of design works. IEEE. 978-1-7281-0945-9/19/$31.00©2019
5. Nyaletey E, Parizi RM, Zhang Q, Raymond Choo K-K (2019) BlockIPFS—blockchain-enabled interplanetary file system for forensic and trusted data traceability. In: IEEE international conference on blockchain (blockchain). IEEE. 978-1-7281-4693-5/19/$31.00 ©2019. https://doi.org/10.1109/Blockchain.2019.00012
6. Zhu W, Chang J (2019) Research and implementation of perceptual hash algorithm for image retrieval. AIP Conf Proc 2073(020083):2019. https://doi.org/10.1063/1.5090737PublishedOnline:13February
7. Rajalakshmi A, Lakshmy KV, Sindhu M, Amritha P (2018) A blockchain and IPFS based framework for secure research record keeping. Int J Pure Appl Math 119(15):1437–1442. ISSN: 1314-3395 (on-line version)
8. Steichen M, Fiz B, Norvill R, Shbair W, State R (2018) Blockchain-based, decentralized access control for IPFS. In: IEEE conferences on internet of things, green computing and communications, cyber, physical and social computing, smart data, blockchain, computer and information technology, congress on cybermatics. IEEE. 978-1-5386-7975-3/18/$31.00 ©2018. https://doi.org/10.1109/Cybermatics_2018.2018.00253
9. Tang H, Tong N, Ouyang J (2018) Medical images sharing system based on blockchain and smart contract of credit scores. In: Proceedings of 2018 1st IEEE international conference on hot information-centric networking (HotICN 2018). IEEE. 978-1-5386-4870-18/$31.00©2018
10. Meng Z, Morizumi T, Miyata S, Kinoshita H (2018)Design scheme of copyright management system based on digital watermarking and blockchain. In: IEEE 42nd annual computer software and applications conference (COMPSAC), Tokyo, pp 359–364
11. Zhang Z, Zhao L (2018) A design of digital rights management mechanism based on blockchain technology. In: Chen et al (eds) ICBC 2018. Springer International Publishing, pp 32–46. https://doi.org/10.1007/978-3-319-94478-4_3
12. Ma Z, Huang W, Gao H (2018) A new blockchain-based trusted DRM scheme for built-in content protection. EURASIP J Image Video Proc
13. Ma Z, Jiang M, Gao H, Wang Z (2018) Blockchain for digital rights management. Futur Gener Comput Syst. https://doi.org/10.1016/j.future.2018.07.029
14. Weng L, Darazi R, Preneel B, Macq B, Dooms A (2012) Robust image content authentication using perceptual hashing and watermarking. In: Lin W et al (eds) PCM 2012. pp 315–326

Mid-Price Prediction Using Online Kernel Adaptive Filtering

Shambhavi Mishra, Tanveer Ahmed, and Vipul Mishra

Abstract Stock price prediction is an important and popular area of time-series prediction. Although numerous methods have been proposed for stock price prediction in literature, a good model with excellent performance is still a long way to go. In addition, earlier prediction methods were based on offline classification and regression, thereby requiring independent and large batch-oriented samples. We would emphasize on one point: a financial time-series data is non-stationary in nature. Hence, traditional regression and classification-based techniques are not ideal for predicting stock prices. In this paper, we propose an "***online***" kernel adaptive filtering (KAF)-based approach for stock price prediction. We experiment with ten KAF methods and analyze their predictive capability on nine different time windows such as: one day, sixty minutes, thirty minutes, twenty-five minutes, twenty minutes, fifteen minutes, ten minutes, five minutes, and one minute. The process is repeated for all 50 stocks of the National Stock Exchange's main Index (NIFTY-50). The empirical results from the experiment show that KAF has several advantages over the current state-of-the-art. For instance, the probability that we can predict the next up or a down movement of a stock is 66%. Further, their low latency makes them a preferred choice for high-frequency trading.

Keywords Stock price prediction · Financial time-series · Kernel adaptive filtering · Online learning

S. Mishra (✉) · T. Ahmed · V. Mishra
Bennett University, Greater Noida, India
e-mail: sm7835@bennett.edu.in

T. Ahmed
e-mail: tanveer.ahmed@bennett.edu.in

V. Mishra
e-mail: vipul.mishra@bennett.edu.in

P. K. Singh et al. (eds.), *Emerging Technologies for Computing, Communication and Smart Cities*, Lecture Notes in Electrical Engineering 875,
https://doi.org/10.1007/978-981-19-0284-0_51

1 Introduction

In economics and investments research, time-series has always been a prominent area of investigation. Time-series has a multitude of applications, with stock price prediction being one of the popular domains. A good stock price prediction model can minimize the risk in trading and can optimize asset management. It can reduce overhead costs and mitigate market risks [1]. Although popular, stock price prediction at any instance of time has remained one of the non-trivial problems, earlier research has demonstrated that the stock price prediction is challenging [2]. Hence, significant efforts have been devoted in addressing the issue; however, the results in past research showed a little improvement in achieving the goal. A number of studies were conducted where various methods have been developed for the improvement in the stock price prediction. These methods vary from support vector machines (SVM) [3], fuzzy logic [4], particle swarm optimization [5], neural networks [6], and genetic algorithms [7] to name a few. Despite the vast body of work, we found that the authors have often overlooked the potential of kernel adaptive filtering (KAF) for stock price prediction. Though there are few studies [8, 9] that utilized KAF methods for stock price prediction, a large-scale evaluation of KAF methods for stock price prediction was limited from the previous research. It is also important to mention here that KAF methods can play a vital role in stock price prediction because they require fewer iterations to converge. Along with that, KAF methods have shown outstanding performance on non-stationary time-series prediction [10, 11]. Moreover, in time-series prediction, KAF methods have been applied extensively [10, 12, 13]. Hence, it is logical to apply the paradigm to predicting stock prices.

To address the challenges and as a potential solution, we utilized the paradigm of online-KAF for stock's mid-price prediction. Since, the issue of financial series forecasting is non-trivial, we work with 50 stocks of the main Index at the National Stock Exchange (NSE): Nifty-50. In contrast to the existing methods, the data was collected at multiple time windows such as one minute, five minutes, ten minutes, fifteen minutes, twenty minutes, twenty-five minutes, thirty minutes, sixty minutes, and one day. We then apply the prediction algorithms to each of these windows and try to show the predictive capabilities of KAF algorithms. In addition to this, we experiment with ten different algorithms and try to find the best algorithm that would produce good results.

The main contributions of the paper are:

- A novel KAF-based method for stock's mid price prediction. The mid price is calculated using two aspects; (high + low)/2 and (open + close)/2.
- Experimentation with ten KAF algorithms on nine different time windows and fifty stocks.

To the best of our knowledge, this article is the first wherein we apply and test the performance of ten different algorithms of the KAF class on multiple time windows.

The paper has been structured as follows. Section 2 summarized the details of the existing work. The proposed method for stock's mid price prediction using KAF-based methods is discussed in Sect. 3. The experiments performed and their results are included in Sect. 4. Finally, the conclusions and future directions are given in Sect. 5.

2 Related Work

This section elucidates the detail of the existing work in the domain of stock price prediction. Stock prediction is a tedious task, and the early methods were focused on past prices using indicators [14, 15]. It has also been shown that the stock prediction is noisy, chaotic, and follows nonlinearity [16–18]. The use of linear methods like correlations, autoregressive integrated moving average (ARIMA) models, and moving averages were also explored in the past [19, 20]. Due to the chaotic behavior of the stocks, in these methods, even short-time prediction needs heavy computation [17]. Much of the literature also focused on machine learning (ML) techniques due to their ability to handle complex patterns [21, 22].

Moreover, there is a great scope for sequential learning-based online methods as the time-series data follows nonlinearity. Sequential learning techniques have the ability of generating faster and acute results [23]. Neural network and support vector regression were also proposed in sequential learning techniques [11, 24]. However, support vector regression and neural network-based techniques have several disadvantages that make them less suitable for stock prediction task. In spite of the powerful generalization ability, support vector regression is unable to perform well on the large datasets. Neural network methods also showed slow convergence issue and needs high computational power during training and inference [25]. Another popular class of online learning-based method is kernel adaptive filtering (KAF) [10]. For sequential learning technique, kernel-based online learning is efficient and reduces the computational burden [12, 26]. In contrast to the other existing methods of the literature, KAF techniques have gain more popularity as these methods have several advantages like low computational complexity, non-parametric, and faster converges [10].

In one of the researches [26], the authors compared multi-filter neural network (MFNN) with random forest, SVR, and other NN-based approach for stock prediction. The results demonstrated that MFNN outperformed other methods. In [27, 28], long short-term memory (LSTM) was also used to predict stock prices. [29] combined SVR and kernel principal component analysis (KPCA) for improving short-term prediction performance. In addition to this, [30] proposed the idea of multi-step prediction using meta-cognitive recurrent kernel online learning. In one of the researches [31], the authors used a simple LSTM network with single layer and 200 nodes. In addition, the network used by [32] also uses a single layer LSTM with 140 nodes. In contract to [31–33] used a deep architecture with four LSTM layers and 96 nodes in the hidden layers. Each LSTM layer was further followed by a dropout layer.

Table 1 Time window of 60-min (Stock-Adaniports)

Day	High price	Low price	(High price + low price)/2	% Change in mid-price
1 d	387.65	384.05	385.85	−0.1555
2 d	386	384.5	385.25	0.0324
3 d	386	384.75	385.375	−0.0194
4 d	386.9	384.7	385.8	0.4671
5 d	388.9	385.3	387.1	0.5683
6 d	390.75	387.85	389.3	−0.4880
7 d	388.65	386.15	387.4	0.1484
8 d	388.95	387	387.975	NA

If we choose $M = 4$, then **Input** = {−0.1555, 0.0324, −0.0194, 0.4671}.**Output** = {0.5683}

The authors of another research [29] found that in high dimensional space, training ML-based methods for stock prediction requires longer processing time and high computing resources. To alleviate this, [34] proposed convolutional neural network (CNN) for automatic feature selection and market prediction. Recently, the use of deep reinforcement learning has also been explored for stock prediction [35] for adaptive stock trading strategy. The authors proposed gated deterministic policy gradient trading strategy (GDPG) and gated deep Q-learning trading strategy (GDQN) for extracting financial features.

Furthermore, there are some studies where the authors made use of KAF for stock prediction, i.e., [36]. In another method [37], the authors proposed two important aspects of kernel filtering; step-size and kernel-size. Whereas, [36] introduces the concept of the local model. Although the wok by Sergio et al. [36, 37] showed some preliminary work on KAF for stock prediction, the effectiveness of the on the large-scale dataset eludes literature. We also believe that the potential of the KAF methods has not been utilized and analyzed thoroughly for stock prediction. Moreover, in literature, no one thoroughly investigated stock price prediction with multiple time windows. To the best of our knowledge, we are the first to test multiple KAF algorithms on a large-scale dataset with various time windows.

3 Methodology

In this section, we have discussed the online KAF-based learning. KAF works with self-tuning, and input–output mapping is formulated according to an optimization criterion determined by an error signal. The main objective is to learn the input–output mapping $f : V \rightarrow R$, based on a well-known sequence of data $((v_1, d_1), (v_2, d_2), \ldots, (v_i, d_i))$, where $V \subseteq R^L$ is the input space, v_i, $i = 1, \ldots, n$ is the system input at sample time, and d_i is the desired response. The filter adaptively adjust weights, $\omega(i-1)$, where i denotes the discrete time interval. KAF has the following advan-

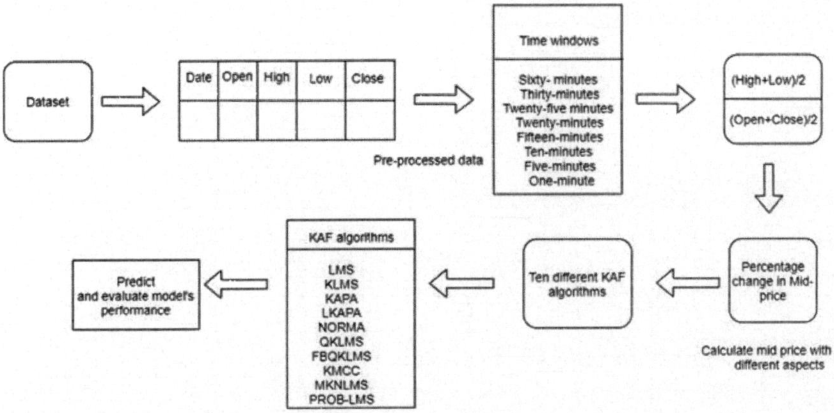

Fig. 1 Proposed stock price prediction framework

tages compared to the other existing techniques of regression and classification: KAF follows the idea of universal approximators. It does not suffer from local minima because it uses gradient decent and follow the concept of online learning and also fits on non-stationary conditions well.

$$\kappa < v, v' >= \exp\frac{(||v - v'||^2)}{\sigma^2}. \tag{1}$$

In stock price prediction, we predict stock's future values depending upon the historical values. In the proposed work, the objective is to predict stock's mid-price using online KAF. Here, we enforced the concept of auto-regression of order n to predict future change in the stock price. Table 1 shows the sample regression formulation and considering the 60-minute time window for one stock (Adaniports). We predict and analyze the stock's mid-price using two different scenarios (high + low)/2 and (open + close)/2 for all nine time window. The proposed stock price prediction framework is shown in Fig. 1. There are various methods following the real-time online prediction concepts [38]. Therefore, in the proposed work, we have worked with the following algorithms:

1. Least mean square (LMS)
2. Kernel least mean square (KLMS)
3. Kernel affine projection algorithm (KAPA)
4. Leaky kernel affine projection algorithm (LKAPA)
5. Normalized online regularized risk minimization algorithm (NORMA) [12]
6. Quantized kernel least mean square algorithm (QKLMS) [13]
7. Fixed budget quantized kernel least mean square algorithm (FBQKLMS) [39]
8. Kernel adaptive filtering with maximum crossentropy criterion (KMCC) [40]
9. Multi-kernel normalized least mean square (MKNLMS) [41]
10. Probabilistic least mean square filter (PROB-LMS) [42].

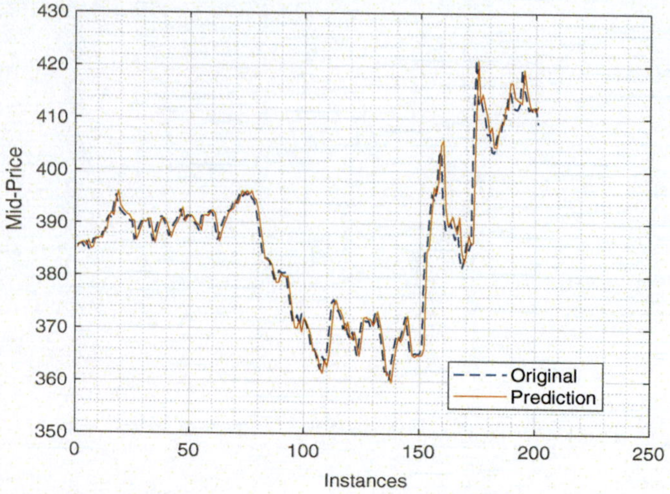

Fig. 2 Prediction for one stock (Adaniports) using KAPA (high + low)/2

4 Result

4.1 Dataset Description

To perform the experiments, the stock data from the National Stock Exchange of India was used. We archived data for one year between January 01, 2019, and December 31, 2019, from 9:15 a.m. to 3:30 p.m. The dataset used in the experiment is available at the link.[1] The original dataset contains only one-minute time window data and includes four different fields open, high, low, and close (OHLC) prices. In our research, data for different time windows was prepared by merging OHLC quotes from the one-minute data. More specifically, the dataset was created and preprocessed as per the nine prediction windows (one minute, five minutes, ten minutes, fifteen minutes, twenty minutes, twenty-five minutes, thirty minutes, sixty minutes, and one day). In addition to this, while creating the dataset, some practical assumptions were made: the trading hours were between 9:15 a.m. and 3:30 p.m., weekends data was not considered, and the values were normalized between zero and one. As discussed in Sect. 1, our aim was to predict the stock's mid-price by taking two different scenarios into consideration; (high + low)/2 and (open + close)/2. For this, we calculated the percentage change in mid-price. Then, we applied ten different KAF algorithms on the processed data. The empirical results obtained by applying KAF algorithms on the NIFTY-50 dataset shows the superiority of the proposed work and sets a new benchmark for future research.

[1] https://drive.google.com/drive/u/0/folders/0B8e3dtbFwQWUZ1I5dklCMmE5M2M.

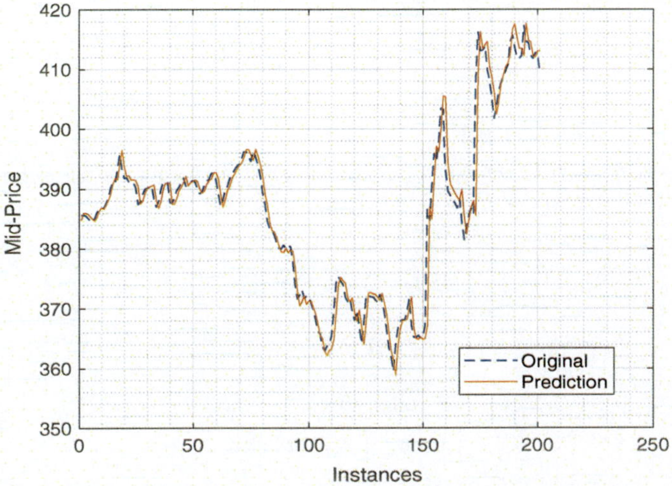

Fig. 3 Prediction for one stock (Adaniports) using KAPA (open + close)/2

Table 2 Parameter description for close-price using ten different KAF algorithms

Algorithms	σ	$\sigma_2 n$	$\sigma_2 d$	η	ϵ	Λ	$tcoff$	τ	$mu0$	P	nu	β
LMS	–	–	–	–	–	–	–	–	0.2	–	–	–
FBQKLMS	5	–	–	0.3	0.4	–	–	–	–	–	–	0.85
KLMS	7.0	–	–	1.7	–	–	–		–	–	–	–
KAPA	3.0	–	–	1.4	1E−2	–	–	–	0.2	20	–	–
LKAPA	5	–	–	0.03	–	1E−2	–	–	–	20	–	–
NORMA	3.0	–	–	–	1.5	1E−2	0	5000	–	–	–	–
QKLMS	4	–	–	0.2	0.5	–	–	–	–	–	–	–
KMCC	6.0	–	–	1.7	–	–	–	–	–	–	–	–
KNLMS	3.0	–	–	1.5	1E−4	–	–	–	2	–	–	–
PROB-LMS	–	2	3	–	–	0.9	–	–	–	–	–	–

σ= Kernel width , $\sigma_2 n$= Variance of observation noise , $\sigma_2 d$ = Variance of filter weight diffusion, η = Step-size, ϵ = Regularization parameter, Λ = Tikhonov regularization , tcoff = Learning rate coefficient, τ = memory size (terms retained in truncation), $mu0$ = Coherence criterion threshold, P = Memory length, nu= Approximate linear dependency (ALD) threshold

4.2 Evaluation Criterion

We used standard evaluation metrics and analyzed the performance of different KAF algorithms. These evaluation metrics use are mean squared error (MSE), and directional symmetry (DS). The different metrics are defined as:

Minimum Square Error (MSE)

It is also called the mean squared deviation (MSD), and it measures the average squared difference between the actual and predicted observation, i.e., the average prediction error, and it is given by:

$$\mathbf{MSE} = \sum_{i=1}^{n} \left(y_i - d_i\right)^2 \tag{2}$$

Directional Symmetry (DS)
In terms of time-series analysis, DS predicts the performance using positive and negative trends, from one time to the next.

$$\mathbf{DS} = \frac{100}{n} \sum_{i=1}^{n} D_i \tag{3}$$

where

$$D_i = \begin{cases} 0 & \text{otherwise} \\ 1 & (y_i - y_{i-1})(d_i - d_{i-1}) \geq 0 \end{cases}$$

In above equation, d_i represents the predicted output, y_i represents the actual values, and n is the time-step.

Measuring the evaluation metrics with Nifty-50

1. Table 2 shows the details of the different hyperparameters and their values for each of the algorithms.
2. With the help of standard evaluation metrics, we calculated error values and tried to quantify the best predictive performance. Overall, we get 50×2 (one for each stock) error values.
3. Finally, Tables 5 and 6 shows the predictive potential of the algorithms on all 50 stocks. We considered the average of all 50 error metrics to come up with the final number for a single time window and a single stock.

4.3 Prediction, Convergence Analysis

In this section, we discuss the predictive performance of the KAF class of algorithms using the convergence and prediction graphs. For this, we have shown the prediction graphs for one stock (Adaniports) with the KAPA algorithm. Figure 2 shows the prediction result for (high + low)/2 and that for (open + close)/2 is shown in Fig. 3. It should be noted here that the graphs show the result for an individual stock (Adaniports) and one prediction window (sixty minutes). It can be seen from the graphs that the predicted curve fits strongly against the original curve. Similar results were obtained for other stocks. We can clearly see that the predictions are not accurate; nevertheless, they are close. The prediction result that we achieved, therefore, shows the excellent capability of the proposed work.

Figures 4 and 5 show the convergence results for mid-price using (high + low)/2 and (open + close)/2, respectively. In this case, similarly to prediction graphs, we have taken individual stock (Adaniports) and one prediction window (sixty minutes)

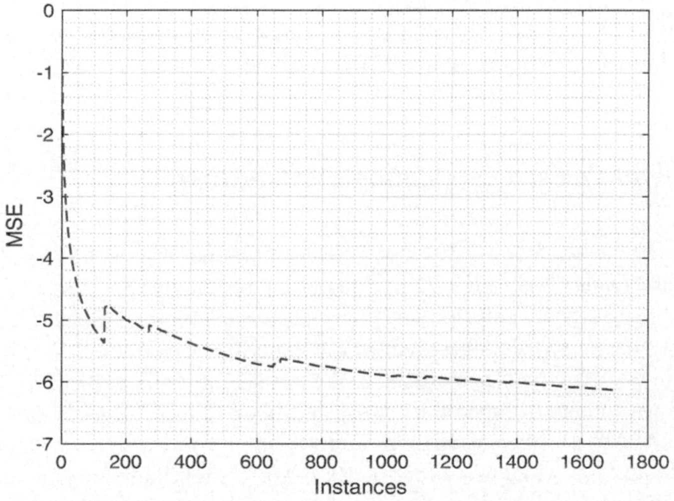

Fig. 4 Error convergence for one stock (Adaniports) using KAPA (high + low)/2

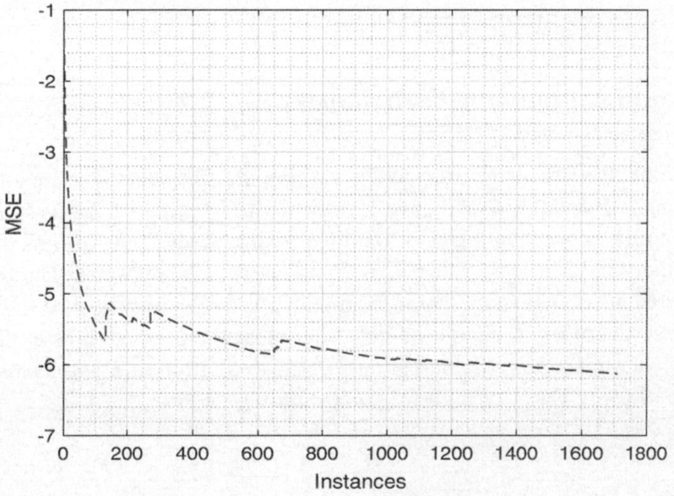

Fig. 5 Error convergence for one stock (Adaniports) using KAPA (open + close)/2

using the KAPA algorithm. Convergence graphs show that the algorithm converges quickly, i.e., at the 1000th data point. This is because algorithms adaptively adjusts its weight and learn quickly from the data. Sometimes fluctuations were present in the graphs, it is nevertheless acceptable because there will be some noise in the new data and minor changes are inevitable. We also analyzed the performance in terms of movement of a stock using directional symmetry. To do this, we used ten different algorithms listed in Sect. 3. Table 6 shows the probability that we can predict the next up or a down movement is 66%; using KNLMS, we obtained the best result at the twenty-minute time window in terms of directional symmetry.

4.4 Comprehensive Evaluation of Algorithms

We experiment with two different aspects using ten different aspects. First, the results are analyzed using (high + low)/2 to rule out the best algorithms. Second, we tried mid-price using (open + close)/2. The final analysis results are present in Tables 5 and 6. Since we have used ten algorithms in our experimentation; therefore, it becomes essential to compare their performance. We can see in terms of MSE, KAPA gives the best performance compared to other algorithms. From Tables 5 and 6, we can see that KAPA performed well in minimizing the error, the lowest error obtained at one-minute time window is order of 10^{-4}. We got a minimum error value at the one-minute time window. The DS results shows a contradiction. For DS, KNLMS performed superior on all time windows. Overall, the results shows the predictive superiority of KNLMS and KAPA algorithms.

4.5 Experimentation with Dictionary Size

The experiments were also performed by varying the dictionary size, and the results are summarized in Table 4. We have chosen KMCC because it is insusceptible to outliers. While experimenting with dictionary size, KMCC algorithms and sixty-minute time windows were selected with one stock (Adaniports). The dictionary size experiments demonstrate that if the dictionary size increases, then the system's performance degrades. It should be noted here that the prediction of an individual stock with dictionary size 1000 takes only 0.41 s execution time, and it shows the of advantage of using KAF for high-frequency trading.

4.6 Comparison with Other Methods

We compare our result with [31–33]. These are some of the popular deep learning-based stock prediction methods. All the methods were implemented based on the architecture details and hyperparameters setting found in the articles. Further, all authors of the articles used LSTM-based deep learning models to predict the stock prices. We trained all the methods on the Nifty-50 dataset. For experimentation, we selected fifty stocks for the sixty-minute time windows to have consistency across various methods. The result of all the methods were then compared with the proposed KAF method. Table 3 shows the comparison of the proposed work with other existing stock prediction methods. The 80:20 splits was used for training and testing the

Table 3 Comparison of the proposed work with other state-of-the-art stock prediction methods

Method	MSE	RMSE	Execution time (s)
[31]	0.4001	0.5626	351.31
[32]	0.0148	0.0985	259.72
[33]	0.0325	0.1253	556.92
[43]	0.0163	0.1102	6392.85
Proposed method	0.0043	0.0633	4.27

Table 4 Effect of dictionary size with execution time, algorithm chosen KMCC (60-min)

Dictionary size	MSE	DS	Execution time (s)
1000	0.00467	0.56	0.41
5000	0.00477	0.63	0.45

Table 5 Result in terms of MSE and DS for mid price (high + low)/2

Time window	MSE	Best algorithms out of ten discussed (According to MSE)	DS	Best algorithms out of ten discussed (According to DS)
1-Day	0.0156	KAPA	0.5638	NORMA
60 min	0.0043	KAPA	0.5667	KNLMS
30 min	0.0025	KAPA	0.5582	KNLMS
25 min	0.0022	KAPA	0.5537	KNLMS
20 min	0.0018	KAPA	0.5497	KNLMS
15 min	0.0014	KAPA	0.5446	KNLMS
10 min	0.0010	KAPA	0.54447	KNLMS
5 min	0.0005	KAPA	0.5299	KNLMS
1 min	0.00012	KAPA	0.5516	KNLMS

deep learning methods. Time taken to train & test was recorded, and the results are summarized in the last column of Table 3. The results shown in Table 3 show the superiority of the proposed work compared to the other stock prediction methods in literature.

5 Conclusion

In this paper, we proposed the idea of online learning-driven KAF algorithms to predict a stock's mid-price. The performance was analyzed on Indian National Stock Exchange (NIFTY-50) using ten different algorithms. The experiment was performed on nine time windows. Different time windows were considered to show the predictive capability and the superiority of the work for both interday and intraday trading.

Table 6 Result in terms of MSE and DS for mid price (open + close)/2

Time window	MSE	Best algorithms out of ten discussed (According to MSE)	DS	Best algorithms out of ten discussed (According to DS)
1-Day	0.0149	KAPA	0.6247	KMCC
60 min	0.00256	QKLMS	0.6546	KNLMS
30 min	0.00224	QKLMS	0.6636	KNLMS
25 min	0.00184	QKLMS	0.6629	KNLMS
20 min	0.00143	QKLMS	0.6670	KNLMS
15 min	0.0010	QKLMS	0.6666	KNLMS
10 min	0.0005	QKLMS	0.6638	KNLMS
5 min	0.00012	QKLMS	0.6533	KNLMS
1 min	0.00011	KAPA	0.5965	QKLMS

In the literature, the intraday time windows are generally overlooked. We therefore tried to fill the gap by proposing the online KAF method. Experimental results also show that the KAF algorithms outperformed similar methods in terms of execution time and error minimization.

References

1. Abu-Mostafa YS, Atiya AF (1996) Introduction to financial forecasting. Appl intell 6(3):205–213
2. Clements MP, Franses PH, Swanson NR (2004) Forecasting economic and financial time-series with non-linear models. Int J Forecast 20(2):169–183
3. Sagala TW, Saputri MS, Mahendra R, Budi I (2020) Stock price movement prediction using technical analysis and sentiment analysis. In: Proceedings of the 2020 2nd Asia pacific information technology conference, pp 123–127
4. Rafiul Hassan Md (2009) A combination of hidden Markov model and fuzzy model for stock market forecasting. Neurocomputing 72(16–18):3439–3446
5. Ecer F, Ardabili S, Band SS, Mosavi A (2020) Training multilayer perceptron with genetic algorithms and particle swarm optimization for modeling stock price index prediction. Entropy 22(11):1239
6. Moghaddam AH, Moghaddam MH, Esfandyari M (2016) Stock market index prediction using artificial neural network. J Econom Fin Adm Sci 21(41):89–93
7. Pimenta A, Nametala CAL, Guimarães FG, Carran EG (2018) An automated investing method for stock market based on multiobjective genetic programming. Comput Econ 52(1):125–144
8. Huang T-C, Zaeem RN, Barber KS (2019) It is an equal failing to trust everybody and to trust nobody: stock price prediction using trust filters and enhanced user sentiment on twitter. ACM Trans Int Technol (TOIT) 19(4):1–20
9. Shian-Chang H, Chei-Chang C, Jui-Te C, Wu C-F (2020) A novel intelligent option price forecasting and trading system by multiple kernel adaptive filters. J Comput Appl Math 369:112560
10. Liu W, Park I, Principe JC (2009) An information theoretic approach of designing sparse kernel adaptive filters. IEEE Trans Neural Netw 20(12):1950–1961

11. Min H, Shuhui Z, Xu M, Tie Q, Ning W (2018) Multivariate chaotic time series online prediction based on improved kernel recursive least squares algorithm. IEEE Trans Cybern 49(4):1160–1172
12. Liu W, Pokharel PP, Principe JC (2008) The kernel least-mean-square algorithm. IEEE Trans Sig Proc 56(2):543–554
13. Chen B, Zhao S, Zhu P, Príncipe CJ (2011) Quantized kernel least mean square algorithm. IEEE Trans Neu Netw Learn Syst 23(1):22–32
14. Ningning Z, Aijing L, Pengjian S (2017) Multidimensional k-nearest neighbor model based on EEMD for financial time series forecasting. Phys A: Stat Mech Appl 477:161–173
15. You-Shyang C, Ching-Hsue C, Wei-Lun T (2014) Modeling fitting-function-based fuzzy time series patterns for evolving stock index forecasting. Appl Intell 41(2):327–347
16. Abraham A, Philip NS, Saratchandran P (2004) Modeling chaotic behavior of stock indices using intelligent paradigms. arXiv preprint: cs/0405018
17. Ritika S, Shashi S (2017) Stock prediction using deep learning. Multimedia Tools Appl 76(18):18569–18584
18. Bezerra PCS, Albuquerque PHM (2017) Volatility forecasting via SVR-GARCH with mixture of gaussian kernels. Comput Manage Sci 14(2):179–196
19. Kumar M, Thenmozhi M (2014) Forecasting stock index returns using ARIA-SVM, ARIMA-ANN, and ARIMA-random forest hybrid models. Int J Banking Acc Finance 5(3):284–308
20. Ju-Jie W, Jian-Zhou W, Zhe-George Z, Shu-Po G (2012) Stock index forecasting based on a hybrid model. Omega 40(6):758–766
21. Gurav U, Sidnal N (2018) Predict stock market behavior: role of machine learning algorithms. In: Intelligent computing and information and communication. Springer, pp 383–394
22. Qiu Y, Yang H-Y, Lu S, Chen W (2020) A novel hybrid model based on recurrent neural networks for stock market timing. Soft Comput 1–18
23. Juan P, Francisco Z-M, Botella-Rocamora P (2015) Online learning algorithm for time series forecasting suitable for low cost wireless sensor networks nodes. Sensors 15(4):9277–9304
24. Vrablecová P, Ezzeddine AB, Rozinajová V, Šárik S, Sangaiah AK (2018) Smart grid load forecasting using online support vector regression. Comput Electr Eng 65:102–117
25. Marco L, Matteo B, Paolo F (2013) Short-term traffic flow forecasting: an experimental comparison of time-series analysis and supervised learning. IEEE Trans Intell Transp Syst 14(2):871–882
26. Wen L, Lu Z, Lingxiao C (2019) Deep learning-based feature engineering for stock price movement prediction. Knowl-Based Syst 164:163–173
27. Song Y, Lee JW, Lee J (2019) A study on novel filtering and relationship between input-features and target-vectors in a deep learning model for stock price prediction. Appl Intell 49(3):897–911
28. Chenjie S, Massimo DP (2019) Improving trading technical analysis with TensorFlow long short-term memory (LSTM) neural network. J Fin Data Sci 5(1):1–11
29. Anass N, Abdelouahid L (2018) Short-term stock price forecasting using kernel principal component analysis and support vector machines: the case of casablanca stock exchange. Proced Comput Sci 127:161–169
30. Liu Z, Loo CK, Pasupa K, Seera M (2020) Meta-cognitive recurrent kernel online sequential extreme learning machine with kernel adaptive filter for concept drift handling. Eng Appl Artif Intell 88:103327
31. Mojtaba N, Pooyan N, Hamed J, Amir M, Ely S et al (2020) Deep learning for stock market prediction. Entropy 22(8):840
32. Penglei G, Rui Z, Xi Y (2020) The application of stock index price prediction with neural network. Math Comput Appl 25(3):53
33. Adil M, Mhamed H (2020) Stock market prediction using LSTM recurrent neural network. Proced Comput Sci 170:1168–1173
34. Ehsan H, Saman H (2019) CNNpred: CNN-based stock market prediction using a diverse set of variables. Expert Syst Appl 129:273–285

35. Wu X, Haolei C, Jianjia W, Luigi T, Vincenzo L, Hamido F (2020) Adaptive stock trading strategies with deep reinforcement learning methods. Inf Sci 538:142–158
36. Sergio G-V, Xiao-Jun Z, John K (2020) Stock returns prediction using kernel adaptive filtering within a stock market interdependence approach. Expert Syst Appl 160:113668
37. Sergio G-V, Xiao-Jun Z, John K (2019) Learning from data streams using kernel least-mean-square with multiple kernel-sizes and adaptive step-size. Neurocomputing 339:105–115
38. Van Vaerenbergh S, Santamaría I (2013) A comparative study of kernel adaptive filtering algorithms. In: 2013 IEEE digital signal processing and signal processing education meeting (DSP/SPE). IEEE, pp 181–186
39. Zhao S, Chen B, Zhu P, Príncipe JC (2013) Fixed budget quantized kernel least-mean-square algorithm. Sig Proc 93(9):2759–2770
40. Zhao S, Chen B, Principe JC (2011) Kernel adaptive filtering with maximum correntropy criterion. In: The 2011 international joint conference on neural networks. IEEE, 2011, pp 2012–2017
41. Masahiro Y (2012) Multikernel adaptive filtering. IEEE Trans Sig Proc 60(9):4672–4682
42. Fernandez-Bes J, Elvira V, Van Vaerenbergh S (2015) A probabilistic least-mean-squares filter. In: 2015 IEEE international conference on acoustics, speech and signal processing (ICASSP). IEEE, 2015, pp 2199–2203
43. Mahla N, Gholamreza M, Jamshid B (2019) Stock price prediction using deep learning algorithm and its comparison with machine learning algorithms. Intell Syst Account Fin Manage 26(4):164–174

Blockchain-Based Model for Secure IoT Communication in Smart Healthcare

Gawhar Hameed, Yashwant Singh, Shahidul Haq, and Bharti Rana

Abstract Over the last decade, a dramatic growth has been witnessed in healthcare monitoring systems and wireless healthcare monitoring systems have been developed in many countries of the world. Smart healthcare is one of the key areas where IoT infrastructures and solutions are widely used to facilitate the best possible patient surveillance, accurate diagnosis, and timely operation of patients with existing diseases. Smart healthcare systems on the other hand, face various challenges including data transparency, traceability, immutability, and security. Also, most of the existing healthcare systems are often controlled by a centralized monitoring unit that poses potential risks of single-point failures during natural calamities. Being motivated by the aforementioned issues, this paper aims at leveraging blockchain technology for smart healthcare to overcome the issues and challenges that smart healthcare faces as a result of traditional security policies. We propose a theoretical Autonomous Healthcare model, which divides the whole healthcare scenario into hospital-patient collaborations. The proposed model uses multiple blockchains, a global blockchain (GBC), which is used for collaboration among hospitals. Thus, creating a global healthcare system in which a patient is not limited to a certain set of region-based hospitals and doctors. Other blockchains (RBC) are specific to the regions of the patients for problems that are minor and require less interaction and consultation of doctors. Global blockchain (GBC) is a permission-less blockchain while region blockchains (RBC) are permissioned ones. The proposed model promises a decentralized framework that facilitates sharing and exchanging, and integration of information across all the users and third parties.

Keywords Internet of things (IoT) · Smart healthcare · Security attacks · Blockchain

---

G. Hameed · Y. Singh · S. Haq · B. Rana (✉)
Department of Computer Science and Information Technology, Central University of Jammu, Jammu and Kashmir, Jammu 181143, India

Y. Singh
e-mail: yashwant.csit@cujammu.ac.in

P. K. Singh et al. (eds.), *Emerging Technologies for Computing, Communication and Smart Cities*, Lecture Notes in Electrical Engineering 875,
https://doi.org/10.1007/978-981-19-0284-0_52

1 Introduction

The growth of the "Internet of things (IoT)" and related technologies has resulted in the development of smart systems such as: smart cities, smart transport systems, smart energy, and so on, where all objects (living and non-living) are interconnected to form a physical network in which all processes such as sensing, processing, and communication are automatically controlled and managed without human intervention [1]. With a plethora of space for improvement, the healthcare industry has grown tremendously by gaining some level of automation. The current healthcare industry has advanced beyond hospitals by bringing together medical devices, healthcare providers, and patients [2].

The Internet of things has brought about a significant transformation in the field of healthcare by changing the conventional healthcare systems to the modern smart healthcare system (SHS) [3]. SHS are developed to deal directly with the patient health-related data [4]. SHS provides in-depth insights about disease symptoms and reduces the need for routine health check-ups, which can benefit elderly people, diabetes patients, and others with secure management.

Although SHS offers various benefits, there are also some issues that need to be addressed. Security is the major concern in smart healthcare monitoring systems because of the massive amount of sensitive patient data being transferred across the Internet. The data is vulnerable to several security attacks [5]. Secondly, smart healthcare systems are often controlled by a centralized monitoring unit. In case of system failures, the stored data is inaccessible and there might be a chance of data theft or corruption of data. The existing techniques are not sufficient in addressing these security issues [6]. The purpose of the proposed model is to create a secure blockchain-based global healthcare system in which a patient is not limited to a certain set of region-based hospitals and doctors. Thus, provides global access to patient information from any region at any time.

The acronyms used in the article are described in Table 1. The organization of rest of the paper is as follows Sect. 2 discusses the literature work. Section 3 describes the security requirements and challenges in smart healthcare. Section 4 describes blockchain technology, its working, and various benefits in smart healthcare. Section 5 describes the proposed Autonomous Smart Healthcare model. Section 6 presents the conclusion and future work.

The research contributions of the study are as follows:

- Security requirements and challenges in smart healthcare have been thoroughly discussed.
- Blockchain utilization with various features in smart healthcare systems has been explained.
- A theoretical Autonomous Smart Healthcare model using blockchain has been proposed.

Table 1 Acronyms and their meaning

Acronym	Meaning
SHS	Smart healthcare system
IoT	Internet of things
GBC	Global blockchain
RBC	Regional blockchain
CH	Cluster head
TX	Transaction
D	Doctor
P	Patient
PUK	Public key
K	Shared key
$H_{1,1}$	Hospital 1 of cluster 1
PK	Private key
EP	Encrypted policy
$P_{1,1,1}D_{1,1,1}$	Access details of patient 1 of hospital 1 of cluster 1 to doctor 1 of hospital 1 of cluster 1
TXID	Transaction identity
PHash	Previous hash

2 Literature Survey

In El Zouka and Hosni [7], a secure authentication scheme for remote health monitoring of patients is presented. The presented system enables doctors to monitor the vital signals of the patient. The authentication mechanism is categorized into three steps namely, registration, login, and authentication. The fuzzy logic inference method is used for the processing of sensor signals. The author proves the reliability and effectiveness of the system.

In Hossain and Muhammad [8], the authors devise a remote health monitoring system. Vital signals are collected by the smart objects from the body of a patient and transmitted to a module used for analyzing the signals and taking appropriate decisions. Several smart devices are utilized to collect the voice and other signals of a patient. The captured signals are processed individually and serve as an input for the emotion detection system. These signals are then combined to provide a final value for making decisions regarding emotion.

In Chen et al. [9], a remote health monitoring system based on any cast routing is presented. The system chooses the nearest receiver automatically as its destination from a group of any cast to minimize the transmission delay and the control overhead. The delay of the path recovery is also minimized by the new protocol using route recovery from the intermediary routers belonging to the initial path. According to the experimental results obtained, the authors believe that the scheme is fast and reliable.

In Shakeel et al. [10], the authors provide a secure framework for ensuring the security and privacy of the transmitted health data over the Internet. A learning-based system is developed for minimizing malware attacks on health data. Medical data is examined across different layers based on the Q-learning approach to reduce malware attacks. The performance of the suggested scheme is estimated based on experimental outcomes.

In Elhoseny et al. [11], a security framework has been presented that provides a way to securely transmit health data in healthcare systems. The working of this model is based on four steps: first, the health data is encrypted by making use of a hybrid encryption technique. This encrypted data is hidden behind a cover image to generate a stego image. To recover the original data, embedded information is then extracted and decrypted. This proposed system was implemented in the MATLAB setup using simulation and the results were satisfactory.

3 Smart Healthcare: Security Requirements and Challenges

The wireless sensors and wearables collect vital information about the patient for predictive analysis. An unauthorized access to patient's data can lead to a catastrophe, not only compromising the privacy of patients but also puts their lives at risk. Therefore the security challenges encountered in smart healthcare are discussed subsequently. Due to the emerging security vulnerabilities during data transmission, smart healthcare necessitates to satisfy the various security requirements [12] as shown in Table 2.

- **Centralized storage**

  Healthcare systems are often controlled by a centralized monitoring unit. In case of system failures, the stored data is inaccessible and there might be a chance of data theft or corruption of data [14].
- **Lack of synchronization**

  Maintaining synchronization among various entities of the Smart Healthcare systems is a difficult task [10]. Since all the modules work independently from

Table 2 Security requirements for smart healthcare monitoring systems [13]

Requirement	Description
Confidentiality	Guarantees that patient health data is protected from unauthorized access by third parties
Integrity	Ensures that the patient data that has been collected is not modified
Availability	Ensures healthcare data is available on-demand without delay
Anonymity	Ensures the confidentiality of patient identity
Secure data transit	Ensures the data is not modified while it is in transit

each other, and have their own standards. Therefore, combining the data is a challenging task [15].

- **Scalability**

  Smart healthcare systems expand or contracts depending upon the need of patients. The existing system must not be compromised when the system is modified [16].

- **Availability**

  Historical records and current signals of the patient must be available to the examiner at any time and location. Based on the current technology these operations are likely unattainable.

4 Blockchain

A blockchain is a synchronized, shared ledger that includes linked blocks of transactions. BC is a distributed and shared ledger that records all transactions without involving a third party [17]. The first block in blockchain is called the genesis block. Except the genesis block, every other block contains the previous block's hash. Each participating node in a blockchain network contains a copy of blockchain and whenever a new block is attached to the blockchain, changes are reflected in all other blocks. The copies of blockchain get updated, preventing an intruder from modifying the block [18]. Each node in the network possesses two keys: a public key used to encrypt the messages sent to a node and a private key to decrypt the messages and allow a node to read them. Therefore, blockchain uses public key encryption to ensure non-repudiability and consistency, and irreversibility. The encrypted messages can only be decrypted with an appropriate private key [19]. The whole structure of blockchain is shown in Fig. 1.

4.1 Working of Blockchain

Blockchain is based on four key components including [20]: (a) a *peer-to-peer network*: in which network nodes communicate with each other with the help of a pair of keys (private/public). Transactions are signed using a private key, and the public key is used as an address to be reachable on the network. (b) *Distributed ledger*:

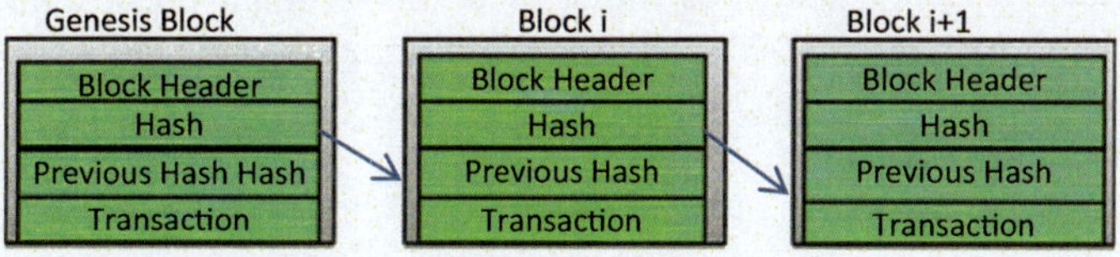

Fig. 1 Structure of blocks in a blockchain [19]

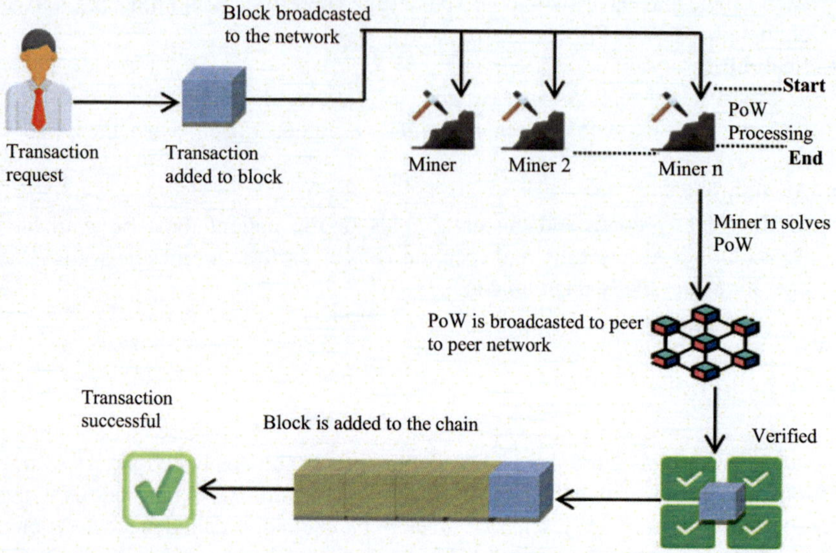

Fig. 2 Working of blockchain [23]

The ledger is open and public to everyone. (c) *Consensus mechanism*: Consensus is a method to ensure that the network nodes verify the transactions and agree with their order and existence on the ledger [21]. Proof of work (PoW) and proof of stake (PoS) are the most common ones. (d) *Public key cryptography*: To chain blocks together, the data in a block is subjected to a particular function called a 'cryptographic hash.' which creates a unique output for a specific input [22]. The various steps involved in the blockchain is demonstrated in Fig. 2.

1. A user requests a transaction.
2. The transaction is added to the block and broadcasted to the network.
3. Miner nodes perform mining to add the transaction into the network. The miner, who first solves the computation, broadcasts the newly created block into the network.
4. The block is distributed to peer-to-peer network consisting of computers called nodes.
5. The transaction is validated by network nodes with the support of algorithms. After validation, the transaction is verified with the help of cryptocurrency.
6. Now the transaction is merged with another transaction to create a new block.
7. This new block is attached to the already existing blockchain without being modified.
8. Transaction is successful.

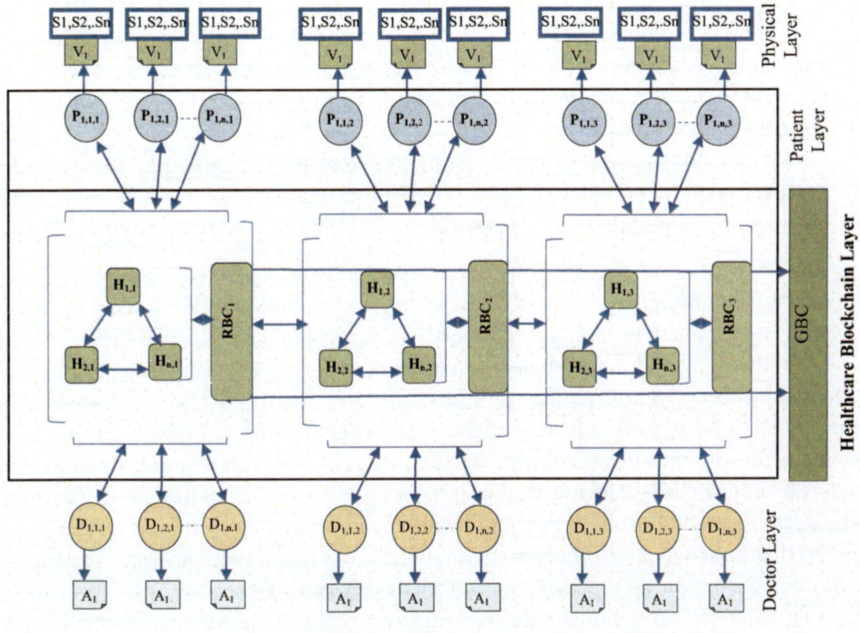

Fig. 3 Proposed autonomous model for smart healthcare using blockchain

4.2 *Smart Healthcare: Blockchain Benefits*

The smart healthcare sector is a perfect fit for blockchain technology because it has the ability to solve the security challenges that smart healthcare systems face [24]. Blockchain technology has numerous benefits that could be used in smart healthcare for secure communication as shown in Fig. 3.

- **Decentralized Storage**
  Decentralized storage is the key advantage of blockchain technology. It is the fundamental basis for improved security and authentication of data stored inside the network. The content that is added to the blockchain is not controlled by any central authority. Data storage is split between different servers, which allows quicker access to medical data, better data quality, and device interoperability [25].
- **Security**
  From various perspectives, blockchain is more trustworthy and secure than existing record-keeping systems. When a transaction is approved, it is encrypted and attached to the preceding transaction. Furthermore, rather than storing information on a single server, information is distributed throughout a network of computers, preventing hackers from accessing transaction data [26].
- **Persistency**

Persistency is the key feature of blockchain. Because of the distributed ledger, which is stored across multiple nodes, deletion of records after they have been added onto the blockchain is almost impossible [22].

- **Greater transparency**

  Blockchain transaction records are more transparent because they are available to all users of the network. As a result, all blockchain users have equal access to the network, allowing them to link, verify, and trace transaction activities [27].

Table 3 provides a comparative analysis of existing frameworks based on blockchain technology for providing security solutions in Smart Healthcare systems considering parameters such as the objectives, framework, merits, and limitations.

From the above table, we analyze that the existing smart healthcare monitoring systems lack in terms of patient mobility [30] and privacy of the patient's health data [31–34]. Also most of the existing systems relay on the centralized architecture which adds problems in case of natural calamities. Keeping these points in mind, we must develop smart healthcare system in which the patient is not limited to a certain set of region-based hospital but it should be a kind of global healthcare system [35–39] so that patients can connect from anywhere and based on their need, treatment is provided. In case they need to connect to a different doctor, the system must be capable of providing them the desired service without compromising their health data.

5 Proposed Autonomous Smart Healthcare Model Using Blockchain

Taking into consideration the need of an autonomous healthcare model, our proposed model divides the whole Healthcare scenario into hospital-patient collaborations as shown in Fig. 3. We have used multiple blockchains in our architecture, one is a global blockchain (GBC), which is used for collaboration of all the hospitals, thus creating a global healthcare system in which a patient is not limited to a certain set of region-based hospitals and doctors, other blockchains (RBC) are specific to the regions of the patients for problems that are minor and require less interaction and consultation of doctors. Global blockchain (GBC) is a permission-less blockchain while as region blockchains (RBC) are permissioned ones.

The proposed model consists of four layers physical layer, patient layer, healthcare blockchain layer, and doctor's layer. Physical layer consists of the sensors used to monitor the health of patients. Patient layer consists of all the patients of all the participating hospitals. Healthcare blockchain layer consists of all the participating hospitals in a blockchain and doctors layer consists of all the doctors of the participating hospitals. Every participating hospital in a region consists of Regional Blockchain (RBC). All the region-specific health clusters collaborate in a way to form a global healthcare model with the creation of a global blockchain (GBC).

Table 3 Analysis of existing blockchain-based frameworks for smart healthcare [27]

Author	Year	Objective	Framework/Model	Merits	Limitations
Mcghin et al. [25]	2020	To design and analyze a smart health monitoring system	BHEEM	The functioning of several subsystems of healthcare is defined and simulated with well-defined algorithms	Implementation of smart contracts are not defined formally
Bigini et al. [21]	2020	To propose various workflows for efficient data handling in the healthcare environment using blockchain technology	Smart contract	Improve the gathering, use, and sharing of patient data and data sub-processors	The cost linked with utilizing smart contracts for smart healthcare should be calculated
Hathaliya et al. [23]	2020	To design a blockchain-based decentralized eHealth architecture	eHealth architecture	The system is fault-tolerant and reliable	Lacks dynamic storage selection algorithm and patients privacy
Kumar et al. [1]	2020	To design, simulate and analyze the performance of blockchain-based smart health care system	IHIoT model	Various factors taken into consideration including design, security, and framework	Lack of implementation over different blockchain networks having different tools and techniques
Nasiri [28]	2019	Proposed a framework that integrates AI to make healthcare system smart and secure	Key agreement protocol	Access time and key generation time were reduced	Patient mobility has not been considered
Hathaliya and Tanwar [29]	2019	A patient monitoring system based on a wireless sensor network	Machine learning	Acceptable localization accuracy with low processing time	High information retrieval delay and redundant data

(continued)

Table 3 (continued)

Author	Year	Objective	Framework/Model	Merits	Limitations
Ray et al. [26]	2019	A detailed framework for a smart healthcare system is developed	S2SH framework	The application of blockchain to the Indian healthcare system is examined in-depth	The integration of various subsystems must be validated before approval
Al Karaki et al. [6]	2019	To develop a framework based on blockchain (DASS-CARE)	DASS-CARE	Provides real-time access and preserving integrity, confidentiality, and security of patient information	Validated is missing
Velliangiri and Karthikeyan Karunya [22]	2018	To develop an architecture that facilitates secure smart sensor management	Ethereum protocol	Overcomes a number of security vulnerabilities in smart healthcare monitoring	Patient privacy needs to be improved
Proposed model	2021	To develop a smart global healthcare monitoring system	Smart healthcare model using blockchain	Decentralized model with better security, and transparency of patient data	–

Let the collaborating hospitals in a region be identified as $H = \{H_{1,1}, H_{2,1}, \ldots .H_{n,1}\}$, where $H_{1,1}$ is identified as hospital 1 of cluster 1, $H_{2,1}$ is identified as hospital 2 of cluster 1, and so on. Let the patients of a specific cluster be represented by the set $P = \{P_{1,1,1}, P_{1,2,1}, \ldots .P_{1,n,1}\}$, where $P_{1,2,1}$ is identified as patient 1 of hospital 2 of cluster 1. The set of doctors in a specific cluster be represented by set $D = \{D_{1,1,1}, D_{1,2,1}, \ldots D_{1,n,1}\}$ whereas $D_{1,2,1}$ represents the doctor 1 of hospital 2 of cluster 1. The set V_1 of patient 1 represents the variables monitored by various wearable sensors such as ECG sensor, Oximeter sensor, Glucometer sensor, etc., and set A_1 represents the actions prescribed by the doctor after analyzing the V_1 set. These two sets can be described as $V_{1=}\{\mathrm{sd}_1, \mathrm{sd}_2, \ldots \mathrm{sd}_n\}$ and $A_1 = \{a_1, a_2, \ldots a_n\}$. We have described the three workflows of the model which include register patient, check patient, and delegate patient. These workflows are discussed below as:

1. **Register Patient**

A patient registers with a particular hospital through hospital portal or manually by a hospital receptionist by providing details which typically include patient's name, address, disease or problem and various other identifiers such as age, height, weight, etc. The transaction being encrypted by the hospital public key is represented as $\text{Encrypt}_{\text{HijPub}}(P_{\text{name}}, P_{\text{address}}, P_{\text{disease}}, P_{\text{gender}}, P_{\text{age}}, P_{\text{height}}, P_{\text{weight}})$. The particular hospital creates this transaction of the patient details and broadcasts to the regional hospital cluster. The cluster node adds the transaction to the blockchain through some consensus mechanism such as 'proof of work' or 'proof of stake'. After the transaction is added to the regional blockchain (RBC), RBC sends a notification to the particular doctor by a smart contract policy. In the end the doctor sends a confirmation message to the patient. All the subsequent transactions between the doctor and the patient are encrypted using the shared secret key provided by the concerned hospital. The encrypted transaction can be represented as EncryptSK($P_{i,j,k}$ $\text{D}_{i,j,k} = (P_{\text{name}}, P_{\text{id}}, P_{\text{puk}}, D_{\text{name}}, D_{\text{id}}, D_{\text{puk}})$). The proposed patient registration is illustrated using a workflow as shown in Fig. 4.

2. **Check Patient**

After registration the patient send a hello message to the doctor by providing a registration transaction Id. The doctor then issue a monitor vitals action on the patient. The vitals are monitored using various wearable sensors that are used by the patient. After a span of time, the patient's wearable sensors sends the vitals data to the concerned hospital. The participating clusters add the data to the blockchain in the form of a transaction. After the data is added to the blockchain, doctor is notified and gets an access to the patient's vital data. The doctor suggests further actions on the patient by issuing a transaction to the blockchain. The wearable sensors are represented as $V_i = \{s_1, s_2, s_3 \ldots\}$. The transactions are represented as $\text{Encrpyt}_{\text{sk}}(P_{i,j,k} D_{i,j,k}\ V_1 = \{\text{sd}_1, \text{sd}_2, \ldots \text{sd}_n\}, A_1 = \{a_1, a_2, \ldots a_n\})$, where sd_1 represents as sensor data one, $P_{i,j,k}$ patient

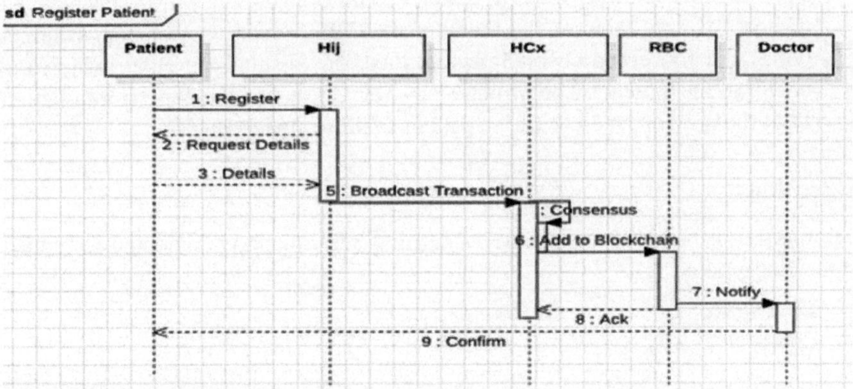

Fig. 4 Workflow of patient registration

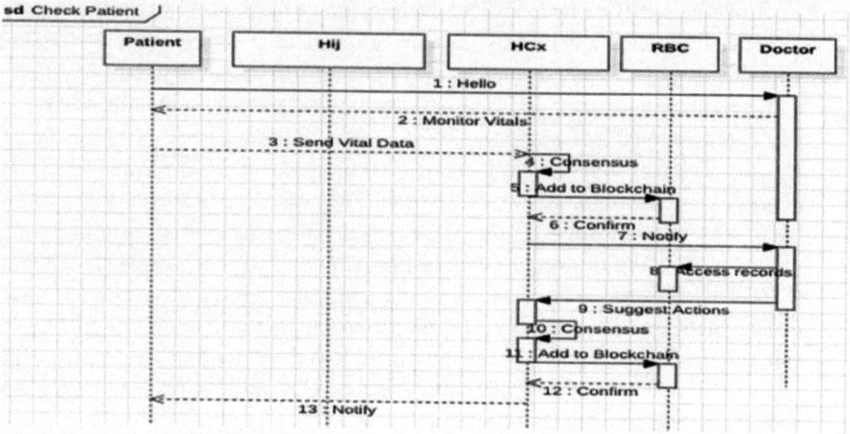

Fig. 5 Workflow of patient check

I of hospital *j* of cluster *k*. V_1 and A_1 represents first iteration of vital values and actions. These two sets get evolved continuously until a stable set of V_1 is obtained and the doctor is sure that patient does not need any more consultation. The whole scenario is illustrated using a workflow as shown in Fig. 5.

3. **Delegate Patient**

   If a doctor at some point feels to refer the patient to another expert doctor in the field. The doctor can create a referral transaction to RBC server as $\text{EncrpytH}_{\text{xPuk}}(P_{\text{gid}}, D_{\text{id}}, R, \text{Sk})$, where H_{xpuk} refers to the public key of current hospital, P_{gid} is the transaction id of the genesis block of patient. D_{id} is the doctors id, *R* is the reason for the referral, and Sk is the secret key of doctor and patient. The concerned hospital is notified about the referral transaction and delegate patient transaction is created in which the patient's records are added in a global blockchain (GBC). The concerned hospital cluster informs the referred doctor. After addition of the block to the GBC, based on the smart contract policy the concerned hospital gets notified, the hospital therefore assigns the doctor to the patient by providing a new secret key and the genesis transaction Id of the previous records of the patient as illustrated in Fig. 6.

Algorithms

Below discussed algorithms provide an idea how a transaction is validated and added to the blockchain.

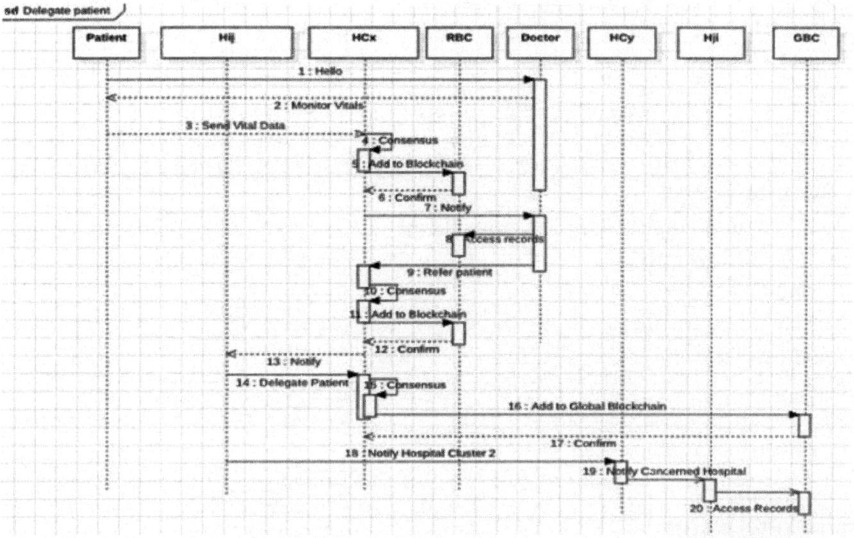

Fig. 6 Workflow of delegate patient

Algorithm 1: Validating a Transaction

ValTx
Begin
1. Patient P1 sends a request to H_{ij}
2: H_{ij} extracts Patients details and creates a transaction Tx as
3: Tx=(Pname, Ptype, PService, Ppuk, PHash),
4: Broadcast Tx to other Hospital Clusters H_{cx}
5: Set hash Difficulty=5
6: Tx.hash = Tx.compute_hash()
7: while not Tx.hash.Difficulty=5
8: Tx.hash = Tx.compute_hash()
9: if(addRBC(Tx)==1)
9: return True
End

Algorithm 2: Adding Transaction to Blockchain
addRBC(Tx)
Begin
1: Generate a block 'Blk' by adding Tx
2: PreviousHash = LastBlockHash // previous hash assigned to last block hash
3: if PreviousHash!= Blk.PreviousHash
4: return False
5: else
6: RBC.append(blk)
7: return True
End

6 Conclusion

Internet of things (IoT) has emerged as a key technology for the development of SHS over the past few years. SHS have been developed to provide real-time monitoring of patients. However, SHS are exposed to several security risks including information theft, security attacks and privacy of patient's data. In this study, we discussed various security issues that SHS faces as a result of traditional security policies. We also discussed the features of blockchain that can be utilized in healthcare systems for managing health data in a transparent, decentralized, and traceable manner. Finally, we proposed a theoretical blockchain-based Autonomous smart healthcare model which divides the whole Healthcare scenario into hospital-patient collaborations. The proposed model uses a permission-less global blockchain (GBC) for collaboration of all the hospitals, thus creating a global healthcare system in which a patient is not limited to a certain set of region-based hospitals and doctors. Permissioned regional blockchains (RBC) are used for specific regions of the patients. All the region-specific health clusters collaborate in a way to form a global healthcare model with the creation of a global blockchain (GBC).

In future, we will implement the proposed model using solidity framework. Also, we have described the three workflows of the model which include register patient, check patient, and delegate patient. In future, more workflows and functionalities will be added.

References

1. Kumar A, Krishnamurthi R, Nayyar A, Sharma K, Grover V, Hossain E (2020) A novel smart healthcare design, simulation, and implementation using healthcare 4.0 processes. IEEE Access 8:118433–118471. https://doi.org/10.1109/ACCESS.2020.3004790

2. Akkaş MA, Sokullu R, Ertürk Çetin H (2020) Healthcare and patient monitoring using IoT. Internet of Things 11:100173. https://doi.org/10.1016/j.iot.2020.100173
3. Hathaliya JJ, Tanwar S, Tyagi S, Kumar N (2019) Securing electronics healthcare records in healthcare 4.0: a biometric-based approach R. Comput Electr Eng 76:398–410. https://doi.org/10.1016/j.compeleceng.2019.04.017
4. El Zouka HA, Hosni MM (2019) Secure IoT communications for smart healthcare monitoring system. Internet of Things 100036. https://doi.org/10.1016/j.iot.2019.01.003
5. Aggarwal S, Chaudhary R, Singh G, Kumar N, Choo KR, Zomaya AY (2019) Blockchain for smart communities: applications, challenges and opportunities. J Netw Comput Appl 144(July):13–48. https://doi.org/10.1016/j.jnca.2019.06.018
6. Al-Karaki N, Gawanmeh A, Ayache M, Mashaleh A (2019) DASS-CARE: a decentralized, accessible, scalable, and secure healthcare framework using blockchain. In: 15th international wireless communications and mobile computing conference IWCMC 2019, pp 330–335. https://doi.org/10.1109/IWCMC.2019.8766714
7. El Zouka HA, Hosni MM (2019) Secure IoT communications for smart healthcare monitoring system. Internet of Things xxxx:100036. https://doi.org/10.1016/j.iot.2019.01.003
8. Hossain MS, Muhammad G (2018) Emotion-aware connected healthcare big data towards 5G. IEEE Internet Things J 5(4):2399–2406. https://doi.org/10.1109/JIOT.2017.2772959
9. Chen SK et al (2012) A reliable transmission protocol for zigbee-based wireless patient monitoring. IEEE Trans Inf Technol Biomed 16(1):6–16. https://doi.org/10.1109/TITB.2011.2171704
10. Shakeel PM, Baskar S, Dhulipala VRS, Mishra S, Jaber MM (2018) Maintaining security and privacy in health care system using learning based deep-Q-networks
11. Elhoseny M, Ramírez-González G, Abu-Elnasr OM, Shawkat SA, Arunkumar N, Farouk A (2018) Secure medical data transmission model for IoT-based healthcare systems. IEEE Access 6(c):20596–20608. https://doi.org/10.1109/ACCESS.2018.2817615
12. Farahani B, Firouzi F, Chang V, Badaroglu M, Constant N, Mankodiya K (2018) Towards fog-driven IoT eHealth: promises and challenges of IoT in medicine and healthcare. Futur Gener Comput Syst 78:659–676. https://doi.org/10.1016/j.future.2017.04.036
13. Moosavi SR et al (2015) SEA: a secure and efficient authentication and authorization architecture for IoT-based healthcare using smart gateways. Procedia Comput Sci 52(1):452–459. https://doi.org/10.1016/j.procs.2015.05.013
14. Tariq N, Qamar A, Asim M, Khan FA (2020) Blockchain and smart healthcare security: a survey. Procedia Comput Sci 175(2019):615–620. https://doi.org/10.1016/j.procs.2020.07.089
15. Islam SMR, Kwak D, Kabir MH, Hossain M, Kwak KS (2015) The internet of things for health care: a comprehensive survey. IEEE Access 3:678–708. https://doi.org/10.1109/ACCESS.2015.2437951
16. Lu Y, Sinnott RO (2019) Security and privacy solutions for smart healthcare systems. Elsevier Inc.
17. Eklund JM (2019) Blockchain technology in healthcare: a systematic review. https://doi.org/10.3390/healthcare7020056
18. Hussien HM, Yasin SM, Udzir SNI, Zaidan AA, Zaidan BB (2019) A systematic review for enabling of develop a blockchain technology in healthcare application: taxonomy, substantially analysis, motivations, challenges, recommendations and future direction
19. Griggs KN, Ossipova O, Kohlios CP, Baccarini AN, Howson EA, Hayajneh T (2021) Healthcare blockchain system using smart contracts for secure automated remote patient monitoring. 2018:1–7
20. Id AP, Tapas N (2018) Blockchain and IoT integration: a systematic survey
21. Bigini G, Freschi V (2020) A review on blockchain for the internet of medical things: definitions, challenges, applications, and vision. https://doi.org/10.3390/fi12120208
22. Velliangiri S, Karthikeyan Karunya P (2020) Blockchain technology: challenges and security issues in consensus algorithm. In: International conference on computer communication and informatics, ICCCI. https://doi.org/10.1109/ICCCI48352.2020.9104132

23. Hathaliya J, Sharma P, Tanwar S, Gupta R (2019) Blockchain-based remote patient monitoring in healthcare 4.0. pp 87–91
24. Tripathi G, Abdul M, Paiva S (2019) Healthcare S2HS—a blockchain based approach for smart healthcare system. Healthcare 100391. https://doi.org/10.1016/j.hjdsi.2019.100391
25. Mcghin T, Choo KR, Zhechao C, He D (2019) Blockchain in healthcare applications: research challenges and opportunities. J Netw Comput Appl 135:62–75. https://doi.org/10.1016/j.jnca.2019.02.027
26. Ray PP, Member S, Dash D, Salah K, Member S (2020) Blockchain for IoT-based healthcare: background, consensus, platforms, and use cases. 1–10
27. Yaqoob I (2021) Blockchain for healthcare data management: opportunities, challenges, and future recommendations. Neural Comput Appl 0123456789. https://doi.org/10.1007/s00521-020-05519-w
28. Nasiri S (2019) Security requirements of Internet of Things-based healthcare system: a survey study. 27(4):253–258. https://doi.org/10.5455/aim.2019.27.253-258
29. Hathaliya JJ, Tanwar S (2020) An exhaustive survey on security and privacy issues in healthcare 4.0. Comput Commun 153(February):311–335. https://doi.org/10.1016/j.comcom.2020.02.018
30. Ambarkar SS, Shekokar N (2020) Toward smart and secure IoT based healthcare system, vol 266. Springer International Publishing
31. Yeh KH (2016) A secure IoT-based healthcare system with body sensor networks. IEEE Access 4:10288–10299. https://doi.org/10.1109/ACCESS.2016.2638038
32. Newaz AKMI, Sikder AK, Rahman MA (2020) A survey on security and privacy issues in modern healthcare systems: attacks and defences. 1(1):1–40
33. Mawgoud AA, Karadawy AI, Tawfik BS A secure authentication technique in Internet of medical things through machine learning
34. Reena JK, Parameswari R (2019) A smart health care monitor system in IoT based human activities of daily living: a review. In: International conference on machine learning, big data, cloud parallel computing, pp 446–448
35. Koutras D, Stergiopoulos G, Dasaklis T, Kotzanikolaou P, Glynos D, Douligeris C (2020) Security in IoMT communications: a survey. pp 1–49. https://doi.org/10.3390/s20174828.
36. Redondi A, Chirico M, Borsani L, Cesana M, Tagliasacchi M (2013) An integrated system based on wireless sensor networks for patient monitoring, localization and tracking. Ad Hoc Netw 11(1):39–53. https://doi.org/10.1016/j.adhoc.2012.04.006
37. Luo E, Bhuiyan MZA, Wang G, Rahman MA, Wu J, Atiquzzaman M (2018) Privacyprotector: privacy-protected patient data collection in IoT-based healthcare systems. IEEE Commun Mag 56(2):163–168. https://doi.org/10.1109/MCOM.2018.1700364
38. Kaur PD, Chana I (2014) Cloud based intelligent system for delivering health care as a service. Comput Methods Programs Biomed 113(1):346–359. https://doi.org/10.1016/j.cmpb.2013.09.013
39. Yang G et al (2018) IoT-based remote pain monitoring system: from device to cloud platform. IEEE J Biomed Heal Informatics 22(6):1711–1719. https://doi.org/10.1109/JBHI.2017.2776351

Internet eXchange Points: A Systematic Case Study in Indian Context

Deepti Saraswat, Pronaya Bhattacharya, Samprat Bhavsar, Hardik Dave, Karan Patel, Shivam Punjani, Ashwin Verma, and Sudeep Tanwar

Abstract At present, Internet eXchange points (IXPs) drive the communication backbone of modern Internet, that allow the Internet to handle the high volume of data traffic. IXPs support critical infrastructures like clouds, data centers, and heterogeneous networks, and thus, it is predicted that the IXPs crossing traffic volume would reach 5016 exabytes by 2030. To cater to this high demand, researchers globally have addressed novel solutions of IXPs peering management, and service orchestration in clouds, data centers, and computational servers. IXP simplifies peering connections of different autonomous system (AS) by reducing the portion of Internet service providers (ISPs) traffic via upstream transit provider of latter, thereby minimizing data rate consumption cost. Although IXPs have been the resilient backbone of the modern Internet, research on the study of IXP is still in its early stages. In IXPs, the data packets follow an optimum path from source to reach to the destination through

D. Saraswat · P. Bhattacharya (✉) · S. Bhavsar · H. Dave · K. Patel · S. Punjani · A. Verma · S. Tanwar
Department of Computer Science and Engineering, Institute of Technology, Nirma University, Gujarat, India
e-mail: pronoya.bhattacharya@nirmauni.ac.in

D. Saraswat
e-mail: deepti.saraswat@nirmauni.ac.in

S. Bhavsar
e-mail: 19mcec02@nirmauni.ac.in

H. Dave
e-mail: 19mcec05@nirmauni.ac.in

K. Patel
e-mail: 19mcec10@nirmauni.ac.in

S. Punjani
e-mail: 19mcec19@nirmauni.ac.in

A. Verma
e-mail: ashwin.verma@nirmauni.ac.in

S. Tanwar
e-mail: sudeep.tanwar@nirmauni.ac.in

P. K. Singh et al. (eds.), *Emerging Technologies for Computing, Communication and Smart Cities*, Lecture Notes in Electrical Engineering 875,
https://doi.org/10.1007/978-981-19-0284-0_53

multiple hops via different AS networks. Thus, motivated by the aforementioned discussions, the article presents a systematic study of IXPs in the Indian context, and have analyzed the IXPs performance on publicly available data sets. We have presented comparative analysis in terms of connected AS average data rate, and user-level context bifurcations. Thus, the analysis intends to be a starting point for building smart solutions for the creation of robust IXPs that improves network services at a global scale.

Keywords Internet · Autonomous systems · Exchange points · Peering links · Communication services

1 Introduction

The term 'Internet' is defined as a group of interconnecting networks that is autonomous and provides a variety of resources to the end user. In this large pool of available networks, the efficient interconnection and architecture among them is a growing topic among researchers. Modern Internet has been spread across the globe in three tiers acting as pillars of Internet traffic transmission. Tier 1 network utilizes Internet protocol (IP) and provides capabilities to connect to any interconnect network (i.e., Internet) via a settlement-free interconnection also known as peering [1]. This means that two Tier 1 networks can exchange data or traffic without any fees. In the hierarchical structure of the Internet, tier 2 and tier 3 networks are responsible to pay the IP transit fees for propagating their traffic to the rest of the Internet. The backbone of internet constitutes a three-tier network architecture. Figure 1 shows the architecture of IXPs.

- At Tier 1, Internet backbone transmits the data traffic through the core routers. Tier 1 networks maintains the principal data routes between large computer networks [1]. These routers are maintained mainly by the government bodies, commercial and IXPs.
- Tier 2 ISP provides an interface between Tier 1 and Tier 3 networks. Tier 2 ISPs have regional or country reach.
- Tier 3 ISPs are closest to the end users and provides Internet connection by charging nominal amount. Based on data traffic distribution and generation, Tier 3 ISPs pays to Tier 2 ISPs.

IXPs are the fundamental part of the critical infrastructure and handles the peering links between different networks. It provides the platform for interconnection to ensure low-latency data transit via Internet backbone infrastructure. Nowadays, most of the business and consumer Internet data traffic is handled by ISPs. These ISPs form tier 2 or tier 3 level of Internet layer and have to pass through a large setup of physical layer connecting the continents.

These networks face long round-trip time (RTT) values (higher latency) and result in decreased performance of Internet connection. The main aspect here is the routing

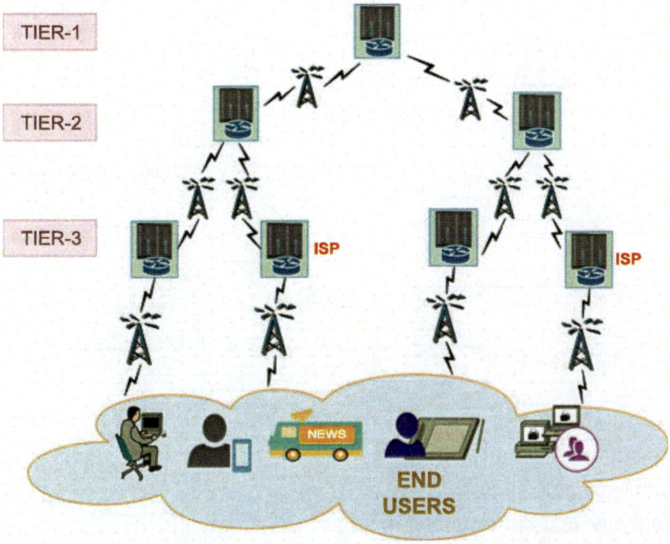

Fig. 1 Architecture of internet eXchange Points

capabilities of the network as well as the distance and most importantly, the number of hops it has to pass. Using IXPs, the latency can be reduced by bringing participating ISPs to the same platform. The routing performance can be further enhanced using the concept of peering links, which can be implemented using IXPs. Border gateway protocol (BGP) is an exterior routing protocol mainly used at the edge routers of the AS. BGP is used for exchanging the routing information between the routers of different AS. There are two variants of BGP protocol, viz., interior BGP (iBGP) and exterior BGP (eBGP). iBGP is used to exchange routing information within the AS, whereas eBGP or normal BGP is used for communication with external routers. The BGP routing configuration mainly depends on the routing and peering policies decided by the organization. The BGP is a path-vector routing protocol that utilizes path information and network policies to make routing decisions. BGP significantly maintains a graph that is revised based on the routing information transferred between different internal or external BGP routers. Table 1 presents the list of abbreviations used in the paper.

1.1 Motivation

The study of Internet backbone infrastructure and IXPs has been a topic of interest for researchers globally. In the current scenario, a large amount of data traffic is forwarded between different AS, and thus the overall analysis of IXP and its associated peering links is of paramount importance. To date, researchers globally have

Table 1 Abbreviations and their meanings

Abbreviation	Meaning
AS	Autonomous systems
BGP	Border gateway protocol
CAIDA	Center for applied internet data analysis
eBGP	External BGP
iBGP	Internal BGP
IXP	Internet eXchange points
ISP	Internet service providers
NIXI	National internet exchange of India

proposed solutions that optimize the bandwidth and link load management, routing control, and traffic distributions. We have focused on orchestration of different infrastructure, analysis, and their management as a coherent unit. To address the research, the paper presents a systematic analysis of IXPs and analyzes the relation of IXP and AS with iBGP and eBGP routing protocols. For the analysis part, we have considered the Indian IXP regions and have presented the analysis. The study is directed toward effective IXP management (load balancing of IXPs), and AS distribution with the consideration of traffic points.

1.2 Research Contributions

The research contributions of the article are as follows.

1. We provide the comparative analysis for different parameters of Internet backbone infrastructure, IXPs, and AS. The analysis helps to provide the current scenario of IXP ecosystems and the complexity of the Internet structure.
2. We analyze the relationship between the IXPs and AS and integrate the routing protocols such as iBGP and eBGP to study data flow through the Internet networks, utilizing Tier 1 networks and IXP infrastructure. We study the various properties of IXPs, AS, and routing in the Internet using BGP and provided a comparative analysis for the above parametric components on a global scale.
3. Considering Internet networking and current growth of IXP and its efficacy, we present a comparative analysis of the IXP data as well as IXP to AS relationship in the context of Indian regions for the past few years. The current IXP establishments at various cities in India based on IXP data and infrastructure based on database-driven resource data are also presented.

1.3 Article Structure

The article structure is as follows. Section 2 presents the studies carried out worldwide on IXP and AS in Internet networking architecture. Section 3 presents an overview of IXP and BGP systems. Section 4 presents the performance analysis of IXP and AS in the context of global as well as Indian regions. Section 5 presents a case study of analysis of IXP points in India located at heterogeneous locations, and finally, Sect. 6 concludes the article.

2 Related Work

A variety of studies have been carried out on IXPs and autonomous systems. The majority of them focus on the analysis of the current situation of IXPs and how the latter can enhance the current Internet networking more efficiently. In [1], the overall study of the topology of the Internet is carried out along with the economical analysis of field and routing overlays. The topology evolution study was carried out by constructing AS-level maps of the Internet topology from the datasets, viz., CAIDA and *RouteViews*. The routing performance was considered as the measurement parameter for the study. Lv et al. [2] studied the geographical distributions of the IXPs along with the creation time of the IXPs. The authors have analyzed the network traffic type from the dataset *PeeringDB* which is a large database of information about the IXPs and AS along with peer-to-peer network relationships formed by the IXPs. The authors in [3] study the performance parameters for IXP, the path latency, and propose a framework that compares the paths through IXPs with alternate paths available. The study concludes that the alternate paths have superior performance compared to another method, thus pointing to the need for improvement in the routing policies. Gavazza et al. [4] presented the reference architecture of the modern Internet, and presented software-defined networking (SDN) management planes to contribute to the load balancing of IXPs.

Briain et al. [5] proposed a case study on the eastern African coast and present a proof-of-concept system design for SDN-based IXP design that presents virtualized container designs to automate the IXP functions. Authors in [6] proposed solutions for IXP management through the identification of elephant flows. The elephant flows are traditionally large traffic flows with persistent high data ingestion. The paper presents a mechanism for the prediction of elephant flows through temporal patterns. For the same, the authors have applied a local weighted regression model that predicts the current flow behavior from the previous flows. The authors predict the volume and duration on 32 samples and accurately predict 80% of successful elephant flow prediction, with only 5% of false positives. Fanou et al. [7] have implemented a system for profiling IXPs in the region of interest and monitoring their growth. ARDA, a route-collector data analyzer tool provides insights into the status of the IXP interconnection at local, national, and regional levels. The same was tested in the

AfriNIC region. Subramani et al. [8] have developed *IXmon* an open-source system for detection of adversaries such as distributed reflective denial of service (DRDoS) for high-data IXPs and peering hubs.

In [9], the analysis of the IXP ecosystem in Brazil was carried out using BGP datasets obtained from looking glass (GS) server. The insights of the peering relationships between the autonomous systems per IXP were provided by inspecting the IPv4 and IPv6 prefix distribution. In [10], the authors study the inter-domain Internet topology and provide a country-specific analysis. The network science parameters were measured in a country-specific manner using routing information base (RIB) dataset of RouteViews project and routing information service (RIS).

The paper brings out analysis of IXPs conducted in terms of AS distribution based on network traffic (at global level and at Indian context), daily average data traffic at AS (at global level and at Indian context), utility of AS, and number of established peering links. The analysis also depicts the data rate distribution in terms of traffic type existing at IXP level that brings out advantages of future enhancement of IXPs in various regions. IXP boasts benefits to customers such as reduced cost, latency, and efficient local Internet connectivity. However, IXP still faces challenges such as lack of expertise and the setup costs, unavailability of reliable environment for local data exchange, back-haul connectivity, and high time-to-revenue. However, the challenges can be addressed though usage of efficient network platforms, increased automation, and scalable infrastructure.

3 Key Terminologies in Internet Infrastructure

The design of IXP involves an ISP router and connects to an Ethernet network provided at the IXP. The ISP peers with other participants at the IXP using BGP. Each participant in BGP communicates with all peers in a complete fashion, or either a subset of the peers via an AD provided in the IXP core. BGP makes traffic exchange routing decisions regarding network rules, hop counts, and other parameters controlled by network administrators.

3.1 Internet eXchange Points

The traditional layers of the Internet hierarchy consists of different transit providers or upstream providers that charges transit fees to propagate the data traffic to the rest of the Internet, either from the ISPs, or different AS. Because of this reason, ISPs/AS were brought to the same platform to improve the latency, RTT, routing performance, and save cost. IXP enhances the efficiency of transmission of data flowing through the Internet. The objective of establishing an IXP is to provide a platform for participating networks to be able to connect with external networks through a common switching mechanism rather than third-party networks for a transit connection. For mutual benefit, the transmission of data traffic is not billed by the participating AS, ISPs, or content delivery networks (CDNs).

IXPs consists of a collection of numerous network switches for different participating ISPs or AS. The former is normally established in the same city as the participating AS to reduce the latency. IXPs utilizes fiber-optic inter-repeater links (FOIRL), and fiber-distributed data interface (FDDI) rings when switches were not employed. The routing functionality of the data traffic between two participating ASs is provided using the BGP, an exterior routing protocol. Since the participating AS has established a peering relationship, the BGP routers announce the routes via the already established peering relationship. The other participant chooses whether to accept the routes or not. The routers at the edge of AS are called edge routers or eBGP peers and employ BGP to maintain information and facilitate information exchange with other participating systems' border or edge routers.

3.2 Border Gateway Protocol

Border gateway protocol (BGP) is an exterior routing protocol that facilitates the exchange of routing information between the edge routers of different ASs. It works internally as iBGP within AS and eBGP with external systems. Each participating network or AS is called a BGP speaker or peer; exchanges the routing information with the neighboring peers or systems, via network prefix announcements. Each peer maintains a table that contains information about all the routes to systems and exchanges this information to the neighboring peers. In this way, the BGP ensures that the participating AS collects all the routing information coming from the neighboring peers and propagate the information further. For reaching a particular destination, BGP can take n different available paths. The most suitable path is based on the organization's policies that depend on factors like reduced cost, increase reliability, and speed. BGP is vulnerable to security attacks; however, it can be secured using a combination of internet key exchange (IKE) and Internet protocol security (IPSec) [11].

4 Comparative Analysis of IXPs Performance

In this section, we present the comparative analysis for IXPs, ISPs as well as other existing AS. The analysis is carried out at two levels, viz., at the global level and Indian IXP ecosystem context. We also present the improvement measures for specific properties of IXP and AS in subsequent sections.

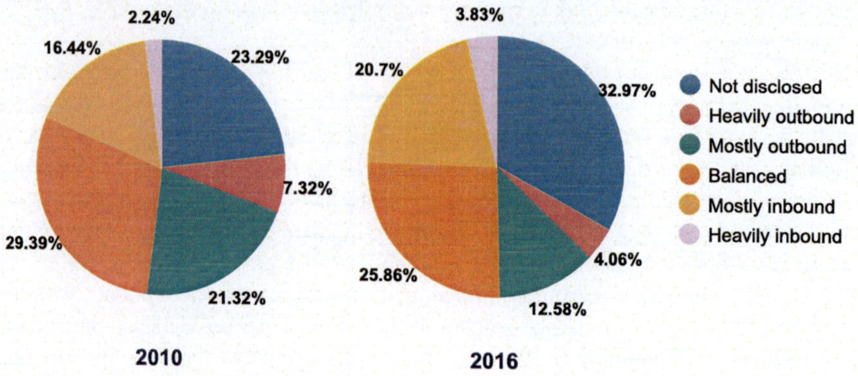

Fig. 2 Distribution of AS-based on network traffic type [12]

4.1 *Analysis of IXPs and as at Global-Level*

A variety of network traffic types exist based on the direction of traffic that moves between different AS, viz., mostly inbound, heavy inbound, balanced, mostly outbound, and heavy outbound. Here, inbound traffic indicates the data that is coming into the system or network, whereas outbound refers to the data traffic which is going outside from the system or network.

Fig. 2 shows the ratio comparison of AS with different network traffic types. The data is collected for autonomous systems for the year 2010 and 2016 from *PeeringDB* [12]. It can be inferred from the figure that majority of the traffic is either balanced or mostly inbound.

Figure 3 shows the comparison of data traffic rate in the AS from the data of the year 2010 to 2016. It can be seen that the majority of the AS face traffic in two data rates, viz., 100–1000 Mbps and 1–5 Gbps. From the above data, it is possible to derive that the capacity of autonomous systems in larger data traffic ranges has been increased. The impact of increased AS connection to IXPs enables more peering and connectivity and customer AS (network) directly exchanges traffic with one or several other networks.

Tables 2 and 3 showcases the distribution of data rate traffic information at AS connected to various IXPs in the year 2010 and 2016 according to their varying speed.

Table 4 shows the differentiation of AS in three different categories based on collected data. Three categories are defined, viz., transit/access AS, enterprise AS, and content AS. Transit/access AS are the ones that provide the functionality to route the data traffic from one AS to another. Based on the classification, it is clear that the majority of AS in the global scenario falls under transit/access or enterprise AS. The categorization of AS based on data collected from CAIDA [13]. It also depicts the comparison of the total number of AS having different types of links that provide access to the other network.

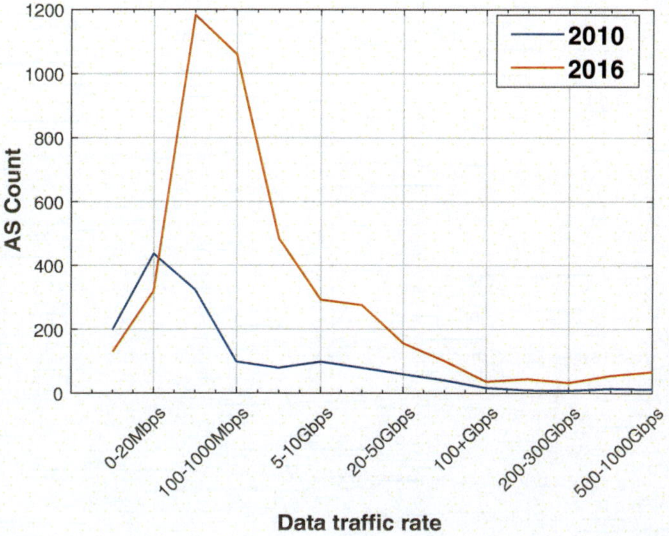

Fig. 3 Comparison of number of AS connected to IXPs and daily average data traffic rate [12]

Table 2 Total number of AS according to data traffic rate in the year 2010

Traffic at AS - 2010	
Traffic information	Total AS
Not disclosed	1057
0–20 Mbps	199
20–100 Mbps	436
100–1000 Mbps	324
1–5 Gbps	99
5–10 Gbps	80
10–20 Gbps	99
20–50 Gbps	80
50–100 Gbps	60
100+ Gbps	41
100–200 Gbps	17
200–300 Gbps	9
300–500 Gbps	8
500–1000 Gbps	14
1 Tbps+	12

Table 3 Total number of AS according to data traffic rate in the year 2016

Traffic at AS - 2016	
Traffic information	Total AS
Not disclosed	4160
0–20 Mbps	129
20–100 Mbps	319
100–1000 Mbps	1182
1–5 Gbps	1059
5–10 Gbps	484
10–20 Gbps	292
20–50 Gbps	275
50–100 Gbps	156
100+ Gbps	101
100–200 Gbps	36
200–300 Gbps	44
300–500 Gbps	32
500–1000 Gbps	54
1 Tbps+	66

Table 4 Number of AS of different types

Type of AS	Number of AS
Transmit/access	21,721
Enterprise	27,456
Content	2341

Table 5 Comparison of total number of P2P and P2C links

Year	P2P Link	P2C Link
1998	852	4921
2020	227,470	132,565

There are two types of links, viz., transit links and peering links. The transit links are also called provider-to-customer (P2C) where it charges fee from the individual ISP and AS for providing transit to rest of the Internet. The peering links are also called peer-to-peer (P2P) links where the participating peers (AS) share mutual traffic without any cost. This is also called settlement-free peering [2]. Based on the comparison of data and present insight, there is a massive exponential growth of P2P links which shows that there is a huge demand for settlement-free peering to save cost, reduce latency, and increase bandwidth.

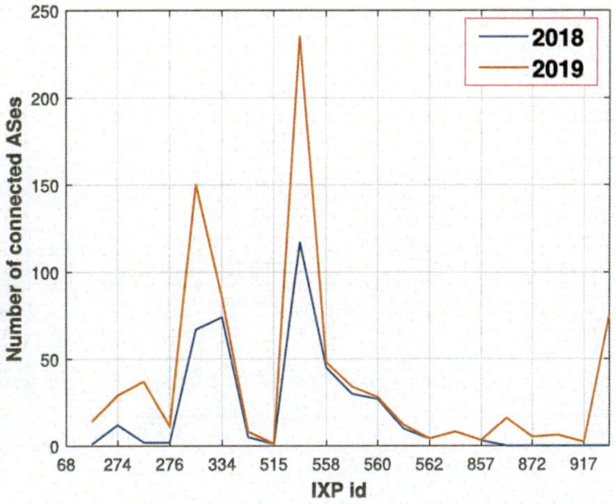

Fig. 4 Comparison of total AS connected to IXP [13]

Table 5 shows that there is an increase in peering links that exploit the benefits of peering. For the classification of links, 4395 links were classified using [12], 46,329 links were classified by CAIDA, and 783 were classified manually. The numbers of P2P links and P2C links from the year 1998 to 2020 have increased nearly 267 times and nearly 27 times, respectively.

5 Case Study: Analysis of IXP Eco-System in India

In our research work, there is a crucial emphasis on the analysis of IXPs in India, particularly due to the lack of any significant work that analyzes the potential and capacity of currently developing IXP infrastructure in India. As per statistics provided by VNI, IDC, and Ovum, the demand for Internet video in India has projected a growth of 640.1% during the year 2010–2015 which is second largest compared to any country in that duration. Australia tops the list with a growth of 1508.8%.

The analysis projects an increase in the majority of IXPs in India to achieve the benefit of peering. This is advantageous in providing larger bandwidths and increased efficiency to Internet users in remote areas. It was found that there are 22 active and working IXPs in India established in different cities and the majority of these IXPs have IPv6 deployment. Figure 4 shows the comparison of AS mapping to Indian IXPs IDs for two consecutive years (2018 & 2019) from data collected using [13] as mentioned in previous section. IXP IDs 515, 276, and 917 show a drastic increase in the number of AS. The trend shows a significant increase in the autonomous systems connected to particular IXP (with IXP IDs 515, 276 located at Mumbai and IXP ID 917 located at Chennai), leading to increased demand for peering for efficient Internet infrastructure establishment in Indian cities.

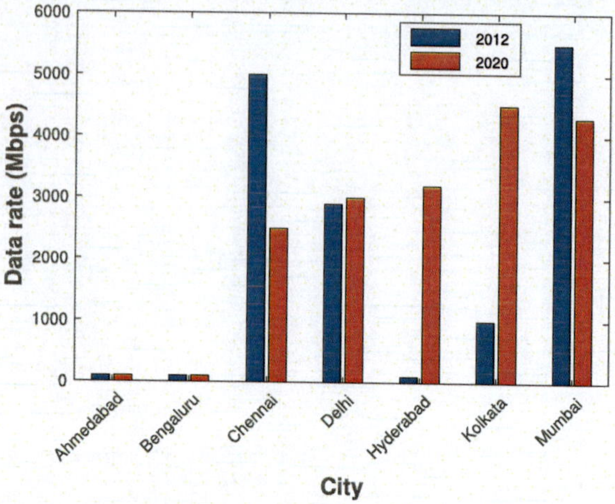

Fig. 5 Daily average data rate of IXPs in different cities of India [14]

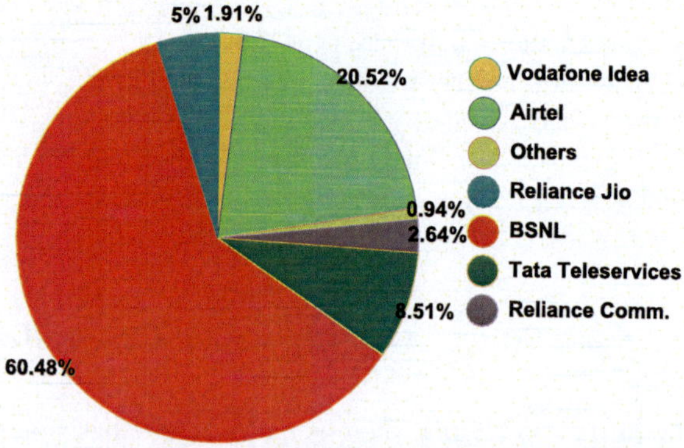

Fig. 6 Users ratio of Indian fixed-line operators as on 2019 [15]

Figure 5 shows the data rate and daily traffic exchange values averaged over multiple ISPs deployed at particular cities. The daily traffic exchange values are also averaged depending on load balance conditions of IXP; the exchange value varies from around 50,000 INR to 6,00,000 INR when port capacity is varied from 100 Mbps to 10 Gbps. This data is obtained from various ISPs registered with National Internet Exchange of India (NIXI) [14]. It is worth mentioning that the data represented is not static and varies as per every connection deal as the latter may vary on the basis of profit or loss incurred.

In India, the leading ISPs provide narrowband and broadband services based on user-oriented requirements like power consumption, transmitter power range, and data rate. The leading fixed-line operators provide services to users based on varied factors like reliability, cost, scalability, QoS, and communication throughput. Figure 6 lists the leading Indian fixed-line operators for the year 2019 that provides various services like Internet, landline connectivity, cellular, etc. Among these fixed-line operators, BSNL tops the list with 60.48%. According to Telecommunication Regu-

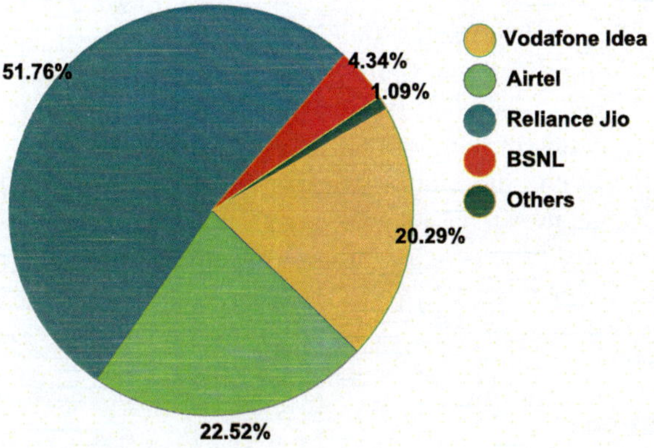

Fig. 7 Subscribers ratio of major ISPs in India as on 2019 [15]

latory Authority of India (TRAI), there were nearly 2.1 crores of wireline subscribers and 341 ISPs in India as of 2019 [15]. Figure 7 indicates ratio-wise subscribers of leading ISPs in India. These ISP provides both narrowband and broadband services as mentioned above.

6 Conclusion and Future Scope

The paper presents a comparative analysis of IXP, AS, and ISP that allows the exploration of IXPs for overall global and Indian scenarios. The ASs were bifurcated in buckets to observe and compare P2P and P2C network traffic links and average daily Internet data traffic. A case study is presented to analyze the IXP ecosystem concerning the Indian context. The IXPs from different cities were considered and compared the AS and average data rates connected to each IXP. The performance indicates that IXP can give much better results in terms of latency and bandwidth of individual AS when established at the correct location. It also encourages peering links which is beneficial mutually and provides better efficiency and performance for overall Internet infrastructure when implemented on large scale. Establishment of IXP reduces transit costs as the exchange of the data traffic for the participant is free. Due to efficient transmission of data and direct interconnection, bandwidth can be improved significantly between adjacent systems. As mentioned above, it reduces latency significantly as the participating systems connect directly without the intervention of third-party networks.

In the future, the authors would investigate the impact of network-coded CDN on IXPs that provide users with specified content type. For the same, SDN-enabled switch configurations would be studied that selectively bifurcates the flows based on content resource links.

References

1. Ahmad MZ, Guha R (2011) Internet exchange points and internet routing. In: 2011 19th IEEE international conference on network protocols, pp 292–294. https://doi.org/10.1109/ICNP.2011.6089065
2. Lv T, Qin D, Ge L (2018) Research and analysis of statistical characteristics of internet exchange points. In: 2018 sixth international symposium on computing and networking workshops (CANDARW), pp 558–560. https://doi.org/10.1109/CANDARW.2018.00110
3. Ahmad MZ, Guha R (2012) A tale of nine internet exchange points: studying path latencies through major regional ixps. In: 37th annual IEEE conference on local computer networks, pp 618–625. https://doi.org/10.1109/LCN.2012.6423683
4. Gavazza JAT, Melo JC, da Silva TB, Alberti AM, Rosa PF, de Oliveira Silva F, Verdi FL, Suruagy JA (2020) Future internet exchange point (fixp): enabling future internet architectures interconnection. In: Barolli L, Amato F, Moscato F, Enokido T, Takizawa M (eds) Advanced information networking and applications. Springer International Publishing, Cham, pp 703–714
5. Briain DÓ, Denieffe D, Okello D, Kavanagh Y (2020) Enabling models of internet exchange points for developing contexts. Dev Eng 5:100057. https://doi.org/10.1016/j.deveng.2020.100057
6. da Silva MVB, Jacobs AS, Pfitscher RJ, Granville LZ (2020) Predicting elephant flows in internet exchange point programmable networks. In: Barolli L, Takizawa M, Xhafa F, Enokido T (eds) Advanced information networking and applications. Springer International Publishing, Cham, pp 485–497
7. Fanou R, Sanchez-Agüero V, Valera F, Mwangi M, Coffin J (2019) A system for profiling the ixps in a region and monitoring their growth: spotlight at the internet frontier. Int J Netw Manage 29(2):e2056
8. Subramani K, Perdisci R, Konte M (2020) IXmon: detecting and analyzing DRDoS attacks at internet exchange points
9. Bucke Brito SH, Silva Santos MA, dos Reis Fontes R, Lachos Perez DA, Lourenço da Silva H, Esteve Rothenberg CR (2016) An analysis of the largest national ecosystem of public internet exchange points: the case of brazil. J Commun Inf Syst 31(1) (2016). https://doi.org/10.14209/jcis.2016.23. https://jcis.sbrt.org.br/jcis/article/view/371
10. Çakmak G, Aydin MN (2017) A country-specific analysis on internet interconnection ecosystems. In: 2017 9th international Congress on ultra modern telecommunications and control systems and workshops (ICUMT), pp 232–237. https://doi.org/10.1109/ICUMT.2017.8255151
11. Bakkali S, Benaboud H, Ben Mamoun M (2013) Security problems in BGP: an overview. In: 2013 national security days (JNS3), pp 1–5. https://doi.org/10.1109/JNS3.2013.6595458
12. Peeringdb, "the interconnection database" (2020). https://www.peeringdb.com/. Accessed on 15 Apr 2021
13. Caida, "caida data server" (2020). http://data.caida.org/datasets. Accessed on 15 Apr 2021

14. National internet exchange of India (nixi), "routing and tariff policy" (2019). https://nixi.in/en/routing-and-tarrif-policy. Accessed on 15 Apr 2021
15. Telecom subscription data (2019). https://www.trai.gov.in/release-publication/reports/telecom-subscriptions-reports. Accessed on 15 Apr 2021

Review on Security of Internet of Things: Security Requirements, Threats, and Proposed Solutions

Rayeesa Muzafar, Yashwant Singh, Pooja Anand, and Zakir Ahmad Sheikh

Abstract The Internet of Things (IoT) is a new technology that is quickly gaining popularity as a major research subject. The Internet of Things has begun to transform and restyle our lifestyle due to its rapid growth. As many things are linked to the Internet and these things correspond openly with one another with no human intercession, so the risk of cyberattacks is very high. Billions of connected things communicate with each other and can exchange sensitive information that may be leaked. Hence, strengthening IoT's security and preserving users' privacy are a major challenge. This paper seeks to provide a thorough examination of IoT security. IoT architecture with the taxonomy of security requirements depending on the attacks' aims is proposed after analyzing several IoT security threats. Furthermore, recent security solutions are explained and classified according to the application domains in which they are used.

Keywords IoT · Security · Privacy · IoT threats · Security requirements · IoT vulnerabilities · IoT attacks

1 Introduction

The Internet of Things (IoT) is an arrangement of gadgets that can communicate with each other and detect interactions to improve our lifestyle and focus on three key functions data transmission, data reception, and data processing [1]. The term IoT was given by Kevin Asthon in 1999 [2]. There are currently nine billion things linked to the Internet. The Internet of Things (IoT) envisions connecting objects all around the world. It is estimated that 75% of connected devices is exposed to cyber-assaults [3]. According to estimates, 25% of attacks on the industrial sector will be

R. Muzafar (✉) · Y. Singh · P. Anand · Z. A. Sheikh
Department of Computer Science and Information Technology, Central University of Jammu, Jammu, Jammu and Kashmir 181143, India

Y. Singh
e-mail: yashwant.csit@cujammu.ac.in

P. K. Singh et al. (eds.), *Emerging Technologies for Computing, Communication and Smart Cities*, Lecture Notes in Electrical Engineering 875,
https://doi.org/10.1007/978-981-19-0284-0_54

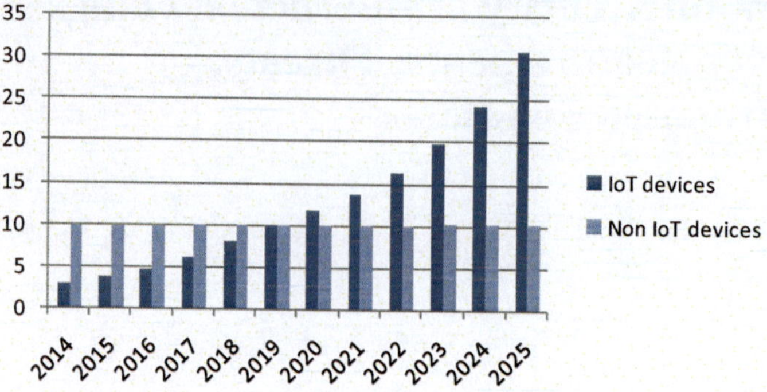

Fig. 1 Estimated devices for IoT and non-IoT [5]

ascribed to corrupt IoT systems by the end of 2020 [4]. Figure 1 shows an estimated graph of the predicted spike of IoT devices soon, based on a Statista report [5]. The complexity of controlling IoT is growing at an exponential rate. Advancement in a smart environment, smart homes, smart traffic, smart industries, smart farming, smart grids, etc., had made life very easier are some examples where IoT is strongly recommended.

As a transformational technique for delivering a plethora of services, the Internet of Things introduces a slew of security problems. The inbuilt vulnerabilities in IoT devices are the source of these issues. The market is filled with these insecure smart gadgets, which are simple to manipulate to gain access to an IoT system remotely. The exponential rise of IoT and its integration with other technologies has created a larger attack surface [6]. End-users and developers are unaware of the security threats that come with today's sophisticated smart applications. Cyberattacks like Mirai take use of these flaws in IoT devices [7].

Following are the contributions of the paper. To begin, it is necessary to have prior knowledge of the infrastructure with which we are dealing to comprehend IoT security; hence, we have covered IoT architecture; we have examined various threats and security requirements in IoT; we have presented various IoT security solutions.

The rest of the paper is organized as follows. Section 2 provides a brief literature review of recent developments in IoT, and the IoT architecture is discussed in Sect. 3. Section 4 presents a brief about IoT threats and requirements for security. Proposed security solutions are presented in Sect. 5. Finally, Sect. 6 concludes the paper.

2 Related Work

A study of protocols, access control models, and frameworks in IoT was undertaken by Ouaddah et al. [8]. The survey explored the objectives of scalability, flexibility, context awareness, interoperability, and granularity for preserving security and privacy against various existing methods of access control.

The authors described the perception, network, and application levels as well as some IoT security attacks. Then, they addressed security vulnerabilities and attacks at various tiers, as well as potential solutions. The basic shortcoming of this study is that various crucial security issues including access control, integrity, and data privacy are not included [9].

The security challenges in IoT were defined by Oracevic et al. Confidentiality, integrity, and authentication were the three key security needs they evaluated. After presenting some current IoT security solutions, they explored prospective routes for eventual study in the realm of IoT security. The survey's fundamental flaw is that the authors mainly focused on fundamental security needs, overlooking other critical security issues including access control, privacy, availability, and trust [10].

The authors addressed IoT security threats and vulnerabilities, as well as known ways to address these issues. The authors presented an analysis of the Internet of Things (IoT) by providing the distinction between the security challenges of IoT and that of a typical security network. The study discussed various layers of IoT (perception, network, and application), and various security threats related to each layer have been highlighted. The author then discussed various remedies for IoT threats and vulnerabilities. This study examines the security vulnerabilities posed by IoT devices in-depth. The IoT security categorization is still unclear, which is one of the work's limitations [11].

In Abdur et al. [12], IoT applications such as smart home, healthcare, and industry have been discussed. The fundamental security requirements in IoT including access control, confidentiality, authentication, and privacy have also been explored. The authors have oversimplified the security needs of IoT applications without considering any existing solutions which is a security flaw of this study.

In Mendez Mena et al. [13], the authors discussed current standards and security goals for IoT devices. The study specified various security criteria for IoT gadgets and data protection. The vulnerabilities of several technologies, such as WSN and RFID flaws, were determined, and several remedies were offered. They also discussed security issues and potential solutions. They did not, however, go into great detail about the weaknesses in the enabling technology.

The survey was conducted by Neshenko et al. that exposed a unique classification of IoT vulnerabilities, including layer-by-layer vulnerabilities, their remediation strategies, impact, and capabilities for situational awareness. The authors also offered a data-driven technique for evaluating IoT maliciousness empirically [14].

We have critically examined certain papers related to security solutions to IoT, dealing with issues at different layers of the protocol stack. None of the existing solutions are perfectly matching all the requirements, hence lagging with one or the other necessity. Some IoT security solutions and their challenges are mentioned in Table 1.

From the above table, we analyzed that IoT security needs to be taken into consideration. As IoT include more devices day by day, cyber-threat and cyberattack are a question. The privacy and security of the IoT system should be a major priority, and there should be no exposure to cyberattacks.

Table 1 Comparison of the proposed survey

Author	Year	Description	Challenges
Ouaddah et al. [8]	2017	The survey explored the objectives of scalability, flexibility, context awareness, interoperability, and granularity for preserving security and privacy against various existing methods of access control	Internet protocols were affected To create an access control model framework based on blockchain
Chahid et al. [9]	2017	A survey was done on IoT security for five layers	Confidentiality, authentication
Oracevic et al. [10]	2017	A survey for IoT security for physical and virtual things was presented	Authentication, integrity, confidentiality
Alaba et al. [11]	2017	A survey for IoT security for three layers was presented	Trust, privacy, authorization, authentication
Abdur et al. [12]	2017	A study for the security of IoT was presented	Confidentiality, privacy, access control, authentication
Mendez Mena et al. [13]	2018	A survey was done on IoT security and privacy for 3 layers	Privacy, availability, integrity, and confidentiality
Neshenko et al. [14]	2019	Classification of IoT vulnerabilities was presented, and also, a data-driven method for providing IoT-specific malicious signatures was presented	To create ways for detecting compromised and susceptible IoT devices, vulnerabilities in IoT software are automatically updated

3 IoT Architecture

The authors briefly describe the ITU-Telecommunication Standardization Sector Y. 2002s generic IoT design [15]. In this architecture, there are three layers, i.e., perception layer, network layer, and application layer.

Perception Layer: It is also called as device layer of IoT and gives a physical meaning to each thing. Data sensors are incorporated in many forms such as RFID tags, infrared sensors, or any other sensor devices that can sense the motion, temperature, humidity, and speed among other things [16]. This layer collects important information about things from the sensors which are connected to them and translates it into digital signals that are then transmitted for further action to the network layer. Threats common to this layer are relay attack, node capture, jamming attack, eavesdropping.

Network Layer: The function of this layer is to accept relevant information from the device layer in the form of digital signals and supports various communication networks such as Wi-Fi, GPRS network, Bluetooth, Zigbee, and GSM [17]. To limit the risks of threats and attacks, proper security tools such as any form of encryption or NIDS can be used. Threats common to this layer are routing attack, Sybil attack, the man-in-the-middle attack, and DDoS attack.

Application Layer: Depend on the analyzed data, this layer develops IoT applications for various types of industries. The application layer is particularly useful in the broadscale development of IoT systems because applications stimulate the development of IoT [18]. Smart homes, smart health care, smart transportation, smart grids, are some related applications for IoT. phishing attacks, DoS, session hijacking, malicious code injection are threats to this layer.

4 IoT Security Threats

The potentials of the cyberattack are a question, as more devices are included in the IoT every day [19]. Weak passwords, insecure firmware updates, improper mechanisms for authentication, ports for open debugging, and unsafe interfaces make the IoT system prone to cyberattacks [20]. This section analyzes different IoT security threats and their purposes. In Table 2, the analysis is summarized.

Table 2 IoT security attacks

Attack	Method	Purposes
Node tampering	Substitute sensor node physically	Service availability is affected, and intruders get access to sensitive information
Replay attack	Information is retransmitted by the malicious node	Trust and confidence of network are obtained
Wormhole attack	Intruder receives packets in the network at one point, tunnel them to a different part of the network, and then replay them into the network from there	Additional attacks are launched; data confidentiality is breached
Man-in-the-middle attack	Information between two nodes is secretly intercepted	The system is taken down, alter, update, eavesdrop the communication
Denial of service	Send data packets to IoT network	Data acquisition is compromised; the network is disabled; service provider resources are exhausted
Sinkhole attack	Traffic is diverted such that the base station does not receive complete data from nodes	Additional attacks are launched; data confidentiality is breached
Spoofing	The attacker pretends to be a legitimate user	Mac address is modified; intruder pretends to be someone else they are not

(a) *Spoofing*: Spoofing is a user's capacity to repudiate knowledge and seems to be something they are not.
(b) *Wormhole*: During a wormhole attack, the intruder receives packets in the network at one point, tunnel them to a different part of the network, and then replay them into the network from there. It is also called a tunneling attack.
(c) *Denial of services (DoS)*: It occurs when an attacker overloads a server with different types of packets, causing it to crash or begin rejecting legitimate users from the service. When a botnet strikes, it is known as a distributed denial-of-service attack (DDoS).
(d) *Node Capturing*: Attackers may attempt to capture or replace the IoT system's node with a malicious node. The attacker controls the new node, which seems to be part of the system.
(e) *Sinkhole attack*: This attack diverts traffic so that the base station does not receive complete data from nodes. As a result, the data provided by devices are jeopardized.
(f) *Replay attack*: In this type of attack, valid data transmission is repeated or delayed intentionally or fraudulently.
(g) *Man-in-the-middle attack*: Intruder secretly interrupts data transmission between two or more devices and can interrupt, edit, update, or replace the data without the knowledge of the victim.

4.1 Security Requirements

Security is a great test for the livelihood of IoT. For IoT, security requirements are both difficult and compulsory as noted by Alqassem [21]. Security requirements were described by three properties as coded by CIA Triad, i.e., confidentiality, integrity, and availability but was insufficient for the context of security. A list of requirements for security is given in Table 3.

Table 3 Requirements for security

Security requirement	Definition
Integrity [22]	To guarantee precision, wholeness, as well as the absence of unauthorized manipulation of data
Confidentiality [22]	To guarantee that data will be confidential by granting access to an authorized client
Privacy [23]	To guarantee that only the user has charged to disclose his personal information
Availability [24]	To ensure reliable access to all services in the system
Trustworthiness [25]	The ability to verify the identity and build trust.

5 IoT Security Solutions

This section describes recent solutions proposed for the security of IoT. The authors presented an intelligent security framework for IoT devices which assures data confidentiality and authentication [26]. It uses symmetric and asymmetric key encryption, with the key pair created using Regev's learning with errors (LWEs) mechanism [27]. In Song et al. [28], the authors introduce chaos-based privacy preservation (CPP), a lightweight communication protocol for safeguarding smart home systems. Agents and a server are part of the planned smart home concept. The agents communicate data to the central controller regularly, and the central controller responds with orders and responses. An asymmetric key, created via a chaos-based technique, is used to encrypt the data. For each data transmission, the symmetric key is updated. A secure smart shopping system was given by authors using ultra-high frequency RFID [29]. All objects in this system are RFID tagged, and RFID readers are installed in smart shelves and smart carts. The authors used ECC-based symmetric and asymmetric encryption/decryption to secure connection with the server. For healthcare system, author focused on distributed access control for the security of patients information [30]. A lightweight authentication scheme has been presented. Authors in [31] emphasized mutual authentication in Smart City applications. The paper describes a four-phase system setup, key generation, encryption, and decryption methodology for lightweight public-key cryptography. Sigfox provides security support with a variety of solutions for security, including public-key infrastructure, a robust firewall, hardware security module, and on-the-go security dispatching security solution; all of which are advantageous in the dynamic IoT environment setup [32]. We can employ honeypots to figure out what techniques and attack vectors attackers use to carry out harmful operations [33]. According to a review of per-day traffic, the IoT honeypot HIoTPOT discovers that the majority of attackers are interested in locating susceptible devices [34].

6 Conclusion

The Internet of Things (IoT) is a rapidly growing technology field that offers end customers a variety of cost-effective, efficient, and simple-to-use applications and services. However, security is one of the most important concerns as IoT systems become more widely used. We identified and studied the current state of security requirements in the IoT in this article. Common types of IoT attacks occurring in general and attacks occurring at different layers of IoT architecture in particular have been discussed in great detail. Furthermore, we have identified the necessity of intelligent object design for greater autonomy in terms of recognizing and recovering from threats.

Table 4 Security solutions

System	Technologies	Objectives	Domain
Intelligent security framework	Symmetric and asymmetric encryption using lattice-based cryptography	To address privacy and confidentiality in the IoT	Smart environment
CPP	Symmetric encryption using chaos-based cryptography, message authentication code	To achieve privacy-preserving in communication protocol	Smart home
Secure smart shopping system	Symmetric and asymmetric encryption based on ECC, message authentication code	To propose a secure smart shopping system	Smart environment
LDAC-KS	Lightweight encryption/decryption using pairing-based cryptography	To preserve medical data privacy and facilitate secured data retrieval	Health care
The lightweight mutual authentic cation protocol	Lightweight public-key encryption	To design a secure mutual authentication protocol for resource-constrained devices	Smart city
Sigfox	Public-key infrastructure, a robust firewall, hardware security module, and on-the-go security dispatching security solution	To provide security support with a variety of solutions for security	IoT environment
Honeypots	Hardware/software security module	It traps the adversaries by imitating real IoT assets but having no value for them, by calculatedly creating security vulnerabilities	IoT environment

References

1. Ud Din I et al (2019) The internet of things: a review of enabled technologies and future challenges. IEEE Access 7:7606–7640. https://doi.org/10.1109/ACCESS.2018.2886601
2. Dudhe PV, Kadam NV, Hushangabade RM, Deshmukh MS (2017) Applications. In: International conference on energy, communication data analytics and soft computing, pp 2650–2653
3. Naik S, Maral V (2017) Cyber security—IoT. In: RTEICT 2017—2nd IEEE international conference on recent trends electronics, information and communication technology proceedings, vol 2018, pp 764–767. https://doi.org/10.1109/RTEICT.2017.8256700

4. att-cybersecurity-insights-vol-5-the-ceos-guide-to-data-security @ www.business.att.com. [Online]. Available: https://www.business.att.com/learn/research-reports/att-cybersecurity-insights-vol-5-the-ceos-guide-to-data-security.html
5. Bhatt RS, Ragiri PR (2021) Security trends in Internet of Things: a survey. SN Appl Sci 3(1):1–14. https://doi.org/10.1007/s42452-021-04156-9
6. Anand P, Singh Y, Selwal A, Alazab M, Tanwar S, Kumar N (2020) IoT vulnerability assessment for sustainable computing: threats, current solutions, and open challenges. IEEE Access 8:168825–168853. https://doi.org/10.1109/ACCESS.2020.3022842
7. Kolias C, Kambourakis G, Stavrou A, Voas J (2017) DDoS in the IoT: Mirai and other botnets. Computer (Long Beach Calif) 50(7):80–84. https://doi.org/10.1109/MC.2017.201
8. Ouaddah A, Mousannif H, Abou Elkalam A, Ait Ouahman A (2017) Access control in IoT: survey and state of the art. In: International conference on multimedia computing and systems—proceedings, vol 0, pp 272–277. https://doi.org/10.1109/ICMCS.2016.7905662
9. Chahid Y, Benabdellah M, Azizi A (2017) 07934655.Pdf
10. Oracevic A, Dilek S, Ozdemir S (2017) Security in internet of things: a survey. In: International symposium on networks, computers and communications ISNCC 2017. https://doi.org/10.1109/ISNCC.2017.8072001
11. Alaba FA, Othman M, Hashem IAT, Alotaibi F (2017) Internet of things security: a survey. J Netw Comput Appl 88:10–28. https://doi.org/10.1016/j.jnca.2017.04.002
12. Abdur M, Habib S, Ali M, Ullah S (2017) Security issues in the internet of things (IoT): a comprehensive study. Int J Adv Comput Sci Appl 8(6). https://doi.org/10.14569/ijacsa.2017.080650
13. Mendez Mena D, Papapanagiotou I, Yang B (2018) Internet of things: survey on security. Inf Secur J 27(3):162–182. https://doi.org/10.1080/19393555.2018.1458258
14. Neshenko N, Bou-Harb E, Crichigno J, Kaddoum G, Ghani N (2019) Demystifying IoT security: an exhaustive survey on IoT vulnerabilities and a first empirical look on internet-scale IoT exploitations. IEEE Commun Surv Tutorials 21(3):2702–2733. https://doi.org/10.1109/COMST.2019.2910750
15. Zhang N et al (2017) Understanding IoT security through the data crystal ball: where we are now and where we are going to be. [Online]. Available: http://arxiv.org/abs/1703.09809
16. Bandyopadhyay D, Sen J (2011) Internet of things: applications and challenges in technology and standardization. Wirel Pers Commun 58(1):49–69. https://doi.org/10.1007/s11277-011-0288-5
17. dac @ onlinelibrary.wiley.com. [Online]. Available: https://onlinelibrary.wiley.com/doi/epdf/https://doi.org/10.1002/dac.3981
18. Fang WD, He W, Chen W, Shan LH, Ma FY (2016) Research on the application-driven architecture in internet of things. Front Artif Intell Appl 293:458–465. https://doi.org/10.3233/978-1-61499-722-1-458
19. Thakur K, Qiu M, Gai K, Ali ML (2016) An investigation on cyber security threats and security models. In: Proceedings—2nd IEEE international conference on cyber security cloud computing CSCloud 2015—IEEE international symposium smart cloud, IEEE SSC 2015, pp 307–311. https://doi.org/10.1109/CSCloud.2015.71
20. Internet_Of_Things_Iot_Security_Best_Practices_IEEE_Community_led_White_Paper @ www.academia.edu. [Online]. Available: https://www.academia.edu/32053241/Internet_Of_Things_Iot_Security_Best_Practices_IEEE_Community_led_White_Paper
21. 2591062 @ dl.acm.org. [Online]. Available: https://dl.acm.org/doi/abs/https://doi.org/10.1145/2591062.2591201
22. Patel VH, Patel S (2021) A review on IoT security: challenges and solution using lightweight cryptography and security service mechanisms offloading at fog. SSRN Electron J 640–648. https://doi.org/10.2139/ssrn.3769832
23. Misra S, Maheswaran M, Hashmi S (2017) Security challenges and approaches in internet of things. Springer, Cham
24. Roman R, Zhou J, Lopez J (2013) On the features and challenges of security and privacy in distributed internet of things. Comput Networks 57(10):2266–2279. https://doi.org/10.1016/j.comnet.2012.12.018

25. Abomhara M, Køien GM (2015) Cyber security and the internet of things: vulnerabilities, threats, intruders and attacks. J Cyber Secur Mobil 4(1):65–88. https://doi.org/10.13052/jcsm2245-1439.414
26. Sridhar S, Smys S (2017) Intelligent security framework for IoT devices. In: International conference on inventive systems and control intelligence, pp 1–5
27. Regev O (2009) On lattices, learning with errors, random linear codes, and cryptography. J ACM 56(6):1–37. https://doi.org/10.1145/1568318.1568324
28. Song T, Li R, Mei B, Yu J, Xing X, Cheng X (2017) A privacy preserving communication protocol for IoT applications in smart homes. IEEE Internet Things J 4(6):1844–1852. https://doi.org/10.1109/JIOT.2017.2707489
29. Li R, Song T, Capurso N, Yu J, Couture J, Cheng X (2017) IoT applications on secure smart shopping system. IEEE Internet Things J 4(6):1945–1954. https://doi.org/10.1109/JIOT.2017.2706698
30. Yang Y, Zheng X, Tang C (2017) Lightweight distributed secure data management system for health internet of things. J Netw Comput Appl 89:26–37. https://doi.org/10.1016/j.jnca.2016.11.017
31. Li N, Liu D, Nepal S (2017) Lightweight mutual authentication for IoT and its applications. IEEE Trans Sustain Comput 2(4):359–370. https://doi.org/10.1109/TSUSC.2017.2716953
32. Coman FL, Malarski KM, Petersen MN, Ruepp S (2019) Security issues in internet of things: vulnerability analysis of LoRaWAN, sigfox and NB-IoT. In: Global IoT summit, GIoTS 2019—proceedings. https://doi.org/10.1109/GIOTS.2019.8766430
33. Farivar F, Haghighi MS, Jolfaei A, Alazab M (2020) Artificial intelligence for detection, estimation, and compensation of malicious attacks in nonlinear cyber-physical systems and industrial IoT. IEEE Trans Ind Informatics 16(4):2716–2725. https://doi.org/10.1109/TII.2019.2956474
34. Gandhi UD, Kumar PM, Varatharajan R, Manogaran G, Sundarasekar R, Kadu S (2018) HIoTPOT: surveillance on IoT devices against recent threats. Wirel Pers Commun 103(2):1179–1194. https://doi.org/10.1007/s11277-018-5307-3

Emerging Security Issues in IoT

Umaira Ahad, Yashwant Singh, and Pooja Anand

Abstract The Internet of Things (IoT) is an innovative model that combines the physical world with the Internet and turns out to be one of the major hi-tech breakthroughs of computer technology. The low openness and absence of interoperability of a significant number of these gadgets in an enormous region of magnitude will make it extremely hard to plan explicit wellbeing techniques and actualize a particular security strategy for sustainable IoT. Moreover, IoT networks presented are helpless against assaults planned to disturb the organization. Thus, special measures are required to combat the upward security apprehensions in IoT. In this article, these security concerns with respect to different layers of basic architecture of IoT are studied in detail; with other emerging challenges, IoT is facing today.

Keywords Internet of Things · Intrusion detection system · Security in IoT

1 Introduction

The Internet of Things (IoT) is a rapidly emerging networking model that aims to link a wide range of devices to the Internet, collecting sensor-generated data, controlling appliances and devices remotely, monitoring environments, vehicles, and buildings, etc. [1]. IoT's vision is to use smart technology to link and control things at anytime, anywhere. IoT was first introduced in 1998, and Kevin Ashton invented the word in 1999 [2]. Moreover, the rise in the number of Internet-linked artifacts has prepared the IoT a progressively rising subject in recent years and is estimated to increase exponentially in the near-term years. According to some estimations, the numbers of associated gadgets before the end of 2025 is anticipated to hit 75 billion. The growth in IoT devices by 2040 is presented in Fig. 1. Additionally, in the coming years,

U. Ahad (✉) · Y. Singh · P. Anand
Department of Computer Science and Information Technology, Central University of Jammu, Jammu, Jammu and Kashmir 181143, India

Y. Singh
e-mail: yashwant.csit@cujammu.ac.in

P. K. Singh et al. (eds.), *Emerging Technologies for Computing, Communication and Smart Cities*, Lecture Notes in Electrical Engineering 875,
https://doi.org/10.1007/978-981-19-0284-0_55

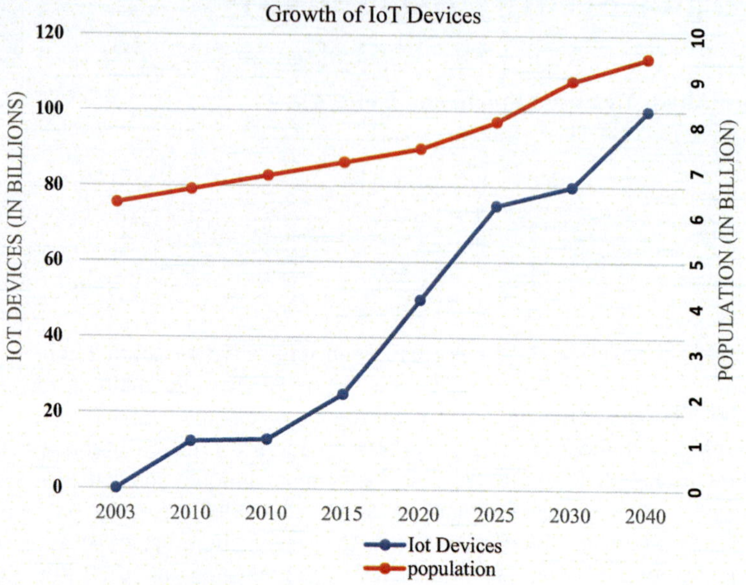

Fig. 1 Growth of IoT devices with population [6]

assaults that threaten resource-constrained IoT devices have multiplied manifold. Along with this, the inherent vulnerabilities have become the part and parcel of the IoT-based systems used in both industrial sectors and home environments [3].

Data security as per the trinity, i.e., data confidentiality, accessibility, and integrity are even a prey. To guarantee data protection, organizations must classify data according to its vulnerability and then allow access to people based on their authorization to access each class of data. In addition, data protection must not be disregarded in the current IoT period in such a way that personal details can be revealed publicly or to an individual not allowed to access it [4]. Based on the study of risk categories and IoT protection mechanisms, detection systems should be further configured. The intrusion detection technique evaluates assaults using data collected from various computer network collecting points. Integration of an active security approach capable of intercepting and responding to intrusions before they reach the network is required [5].

As the Internet is the heart and core of IoT, practically all security concerns that exist on the internet also affect IoT. In comparison to other traditional networks, the IoT's vital nodes are allocated to locations without administrative monitoring, with constrained capacity and resources, leaving the IoT's security concerns particularly challenging. However, the apparent benefits offered by the IoT paradigm are accompanied by major loopholes. The majority of data acquired by IoT devices is confidential and requires protection. These sensitive IoT data might be an open invitation for hackers to seize them and use them in a variety of ways [7]. These sensitive security issues are being stressed upon in this article. The paper is structured in the following sections: Sect. 2 presents the background of IoT in terms of the basic architecture of IoT followed by the applications of IoT; Sect. 3 delves deeper into security issues in IoT. Section 4 presents various challenges in IoT, and finally, we conclude in Sect. 5.

2 Background and Preliminaries

2.1 Architecture of IoT

As there is no unique architecture for the IoT, various architectural styles are suggested by different researchers. Researchers typically offer an IoT-layered architecture [8]. Three-layer architecture: This is the basic architecture of IoT, consisting of three layers: perception layer, network layer, and application layer as shown in Fig. 2.

(a) *Perception Layer*: Sensors are used in the perception layer to feel and collect knowledge about the world. The protocols for communication between devices in an IoT network are included in this layer. The number of homogeneous and heterogeneous devices linked to the Internet via communication protocols is increasingly growing. In terms of battery capacity, computing capabilities, and storage capital, IoT devices are limited. Wi-Fi, NFC, Zigbee, Sigfox, and long-range wide area networks (LoRaWAN) are IoT awareness layer technologies that link the physical world to the network [9].

(b) *Network Layer*: The network layer's most powerful role is to link the device and sensing layers. Fixed and mobile telephone networks, cable networks, the Internet, electrical communication networks, and private networks are examples of transmission networks. Fiber networking, cellular Internet, Ethernet access, and satellite connectivity are examples of network layer communication approaches [11].

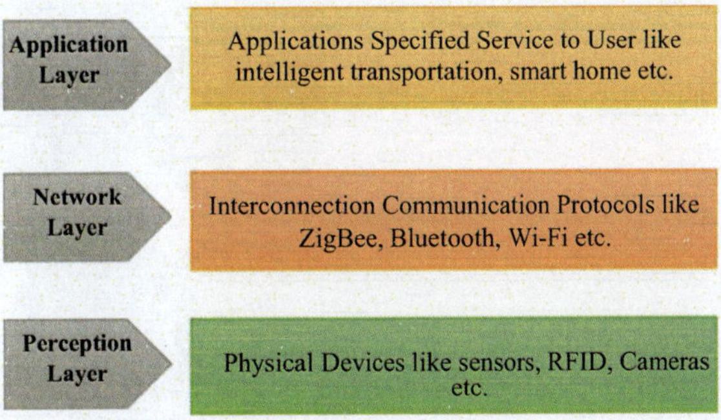

Fig. 2 Architecture of IoT [10]

(c) *Application layer*: Data formatting and presentation are handled by the application layer. On the Internet, HTTP is widely used as the application layer. Due to its verbose nature, which results in a large decoding overhead, HTTP is unsuitable for resource-constrained environments. Several alternative protocols, such as message queue telemetry transport (MQTT), and constrained application protocol (CoAP), have been proposed for IoT settings [12].

2.2 Applications of IoT

The IoT applications are widespread and expanded in all aspects of people's everyday lives that generally cover culture, sectors, and the environment. Some IoT applications are shown in Fig. 3. In the broad sphere of science and technology, IoT finds its application. It is used in computing, big data, smart city apps, smarter energy management systems, mobile charging apps, smart retail, automated transport, environmental monitoring, delivery of water, urban protection, smart surveillance, smart supply chain, etc. IoT devices allow hospitals to monitor their patients' health at home, reducing hospital stays while also providing real-time data that might save lives [13]. A smart home system has the potential to make our lives much easier. Starting from energy conservation, which includes the power controls feature of AC appliances and the thermostat, much of which is controlled to reduce the amount of energy used. IoT technologies have the potential to turn reactive care processes into constructive wellbeing systems. Current medical science relies on services that are deficient in vital real-world data. For medical review, it often uses leftover records, managed environments, and volunteers. Via research, real-time field data, and monitoring, the IoT opens doors to a sea of useful data [14].

Fig. 3 Applications of IoT

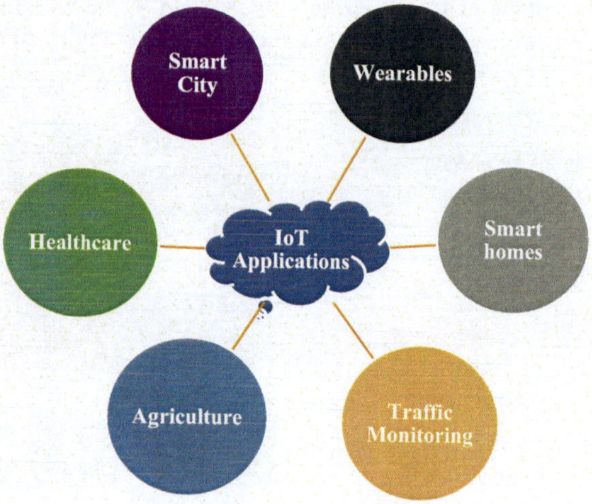

3 Security Issues in IoT

Different security risks in IoT implementations for the four layers are discussed in this section. The attacks that could be rendered on these four layers are depicted in Fig. 4.

(a) *Security issues at sensing layer*

- Node Capturing: A few low-power nodes, such as sensors and actuators, are used in IoT applications. These nodes are insecure against a variety of attackers. In the IoT framework, the assailants may effort to replicate or replace the hub with a malignant hub. The new hub may give off an impression of being essential for the framework is however constrained by the assailant. This may force the IoT system's cumulative protection to be negotiated [15].
- Malicious Code Injection Attack: The assault includes the aggressor infusing around noxious code into the memory of the hub. By and large, the firmware or programming of IoT hubs are overhauled broadcasting live, and this stretches a passage to the aggressors to infuse noxious code. Utilizing such malignant code, the assailants may compel the hubs to play out some unintentional capacities or may even attempt to get to the total IoT framework [16].
- False Data Injection Attack: When the aggressor seizes the hub, he or she can use it and inject false data into the system of IoT. This might cause

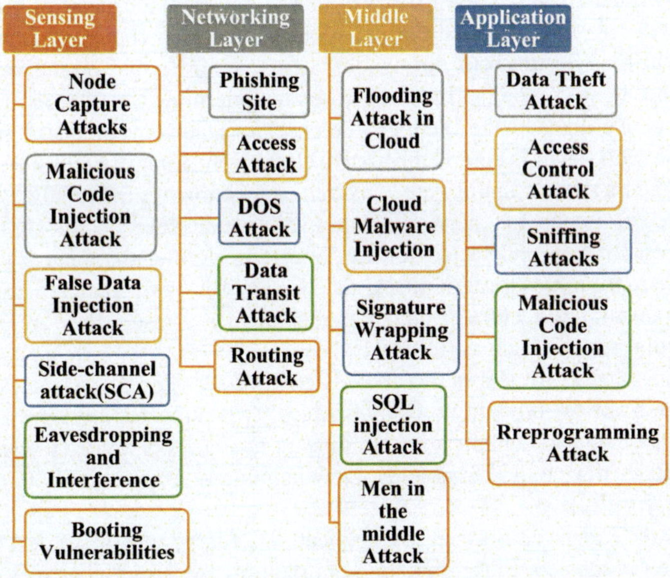

Fig. 4 Attacks in IoT [15]

erroneous consequences and the IoT application breaking down. A DDoS attack might also be launched using this method [17].

- Side-Channel Attacks (SCA): A variety of side-channel attacks can result in confidential data being leaked. Processor microarchitectures, electromagnetic emanation, and resource usage both expose confidential data to adversaries. Power use, laser-based attacks, pacing attacks, and electromagnetic attacks are all examples of side-channel attacks [18]. When integrating cryptographic modules, modern chips take different defense systems to avoid side-channel assaults.
- Eavesdropping and Interference: IoT systems are frequently made up of a variety of nodes that are distributed in open spaces. As a consequence, eavesdroppers can control those IoT applications. All through various stages along with transmitting data or authentication, attackers may listen in and collect information.
- Booting Attacks: During the boot operation, edge devices are susceptible to a variety of attacks. This is because the built-in security processes aren't allowed at that time. As a result of this weakness, when node devices are rebooted, attackers might try to target devices. Since gadgets drive through sleep–wake cycles and are usually low-powered, securing the boot phase in these devices is critical [19].

(b) *Security issues at network layer*

- Phishing Site Attack: These assaults regularly allude to attacks where a few IoT gadgets can be focused on by an insignificant exertion put by the assailant. The aggressors expect that at any rate not many of the gadgets will turn into a casualty of the assault. There is a chance of experiencing phishing destinations throughout clients visiting pages on the Web. When the client's record and secret phrase are undermined, the entire IoT climate being utilized by the client gets defenseless against cyberattacks. In IoT, the network layer is completely defenseless to phishing site attacks [20].
- Access Attack: In this type of assault, an unknown entity or adversary gains access to the IoT network. The assailant can keep on remaining in the organization undetected for the long term. Rather than causing network disruption, this form of attack aims to steal sensitive data. IoT frameworks transmit and receive critical data regularly; as a result, they are extremely vulnerable to such assaults [21].
- DDoS/DoS Attack: In distributed denial-of-service (DDoS) attack, when a Web site is "brought down by hackers," it is almost always the result of a DDoS attack. In plain words, this means that hackers wanted to make a Web site or device inaccessible by overwhelming or crippling it with too much traffic [22].
- Data Transit Attacks: In IoT applications, a large volume of data is shared and processed. Since data are important, hackers and other foes continue to pursue it. Cyberattacks can compromise data held in the cloud or on local computers, but data that are traveling from one place to another are

even more so. In IoT applications, there is a lot of data transfer between sensors, actuators, the cloud, and other devices. As a consequence of the various connection technologies being used in transforming, IoT apps are vulnerable to data attacks [23].

- Routing Attacks: During data transfer, malicious nodes in IoT applications might redirect routing routes. The sinkhole is another form of routing attack in which a fictitious routing path and recruits nodes to use it to route traffic by an adversary. A wormhole attack is a type of attack that, when combined with other types of attacks including sinkhole attacks, can pose a serious security risk. Another attack is a wormhole attack when this attack combines with others such as sinkhole attacks leads to a serious security issue [24].

(c) *Security issues at middleware layer*

- Man-in-the-Middle Attack: The MQTT protocol uses a publish–subscribe model for communication between clients and subscribers, with the MQTT broker serving as a middleman. By separating the subscribing and publishing clients from one another, messages can be transmitted without the recipient's awareness. If an attacker acquires control of the broker and act as a man-in-the-middle, the attacker will have complete control over all communication, even if the clients do not approve [25].
- SQL Injection Attack: A malicious SQL statement can be inserted into a program by an attacker. The attackers can then access the personal information of the user and even change database records. SQL injection attacks are caused by illegal access to private data such as credit card information, passwords, or personal user credentials [26].
- Signature Wrapping Attack: In the middleware of Web services, XML signatures are used. Through leveraging simple object access protocol (SOAP) vulnerabilities, the attacker may crack the algorithm's signature and accomplish operations or modify the message in a signature wrapping assault [27].
- Cloud Malware Injection: The adversary could gain access, insert a virtual machine, or malware using cloud malicious injection into the cloud. The attacker impersonates a genuine service by attempting to construct a malicious service module. Throughout this manner, the assailant can also get access to the victim's service requests, collect private records, and alter the data as needed.
- Flooding Attack in Cloud: Flooding attack is alike to a DoS attack in the cloud, and it affects service quality. The aggressors constantly send requests to service to deplete cloud resources. These assaults can hugely affect cloud frameworks by expanding the load on the cloud servers [28].

(d) *Security issues at the application layer*

- Data Thefts: IoT systems work with a lot of sensitive and personal information. Information on the move is much more vulnerable to attacks than information that is still, and there is a lot of information creation in IoT applications. When IoT apps are susceptible to data intrusion attacks, customers may be unwilling to register their sensitive information. Any of the methods and protocols being used to safeguard IoT applications against cyberattacks include encryption technology, device isolation, client and network authentication, privacy security, and so on [29].

- Access Control Attacks: A mechanism that restricts access to information or accounts to only approved users or processes is access control. In IoT applications, an access control attack is critical since granting access exposes the entire IoT architecture to attackers.
- Malicious Code Injection Attacks: Intruders frequently take the quickest or fastest route to gain access to a device or network. If the device is vulnerable to malicious scripts and misdirection as a consequence of insufficient code testing, it will be an attacker's first point of entry. Cross-site scripting (XSS) is a method used by attackers to inject malicious code into a trusted Web site. If an XSS attack is successful, an IoT account can be hijacked, and the IoT device can be rendered useless [30].
- Sniffing Attacks: Sniffer software can be used by foes to track IoT applications for network traffic. If there are not adequate defense mechanisms ready to prevent this attack, the foe will be capable to access sensitive data of the user [31].
- Reprogram Attacks on the off chance that the programming mechanism isn't ensured; the assaulters can modify IoT protests distantly. This may prompt the hacking of IoT organizations.

4 Emerging Challenges in IoT

Developing an IoT network alone would be a challenge, and this is a challenge. Developing IoT security algorithms allows us to gain a better knowledge of the entire challenges. IoT challenges can be summed up as and are depicted in Figure 5 but are not bound to these only [32].

- *Data mining challenges*: Data mining applications are becoming more relevant as more data becomes usable for processing and interpretation. Data mining software may be used to trigger disciplinary procedures to address urgent operating

Fig. 5 Challenges in IoT

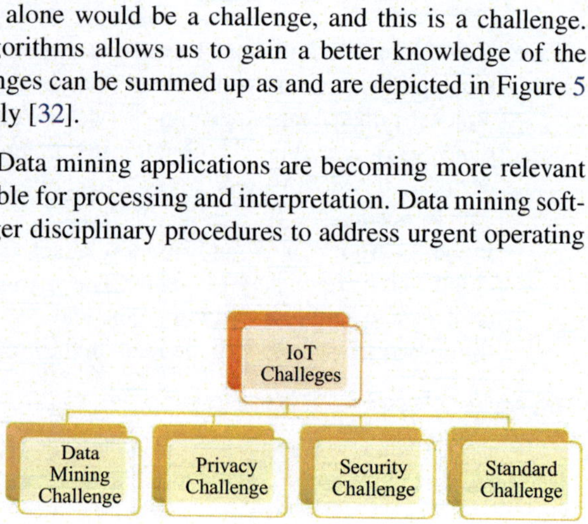

challenges or to alert managers about revelations about competitors' competitive moves and customer demand shifts that could affect their short- and long-term market operations. Using an algorithm and mathematical models, data must be tamed and interpreted. Unstructured photographs and video files do not lend themselves to traditional data processing techniques. There is a lack of qualified data scientists, which is compounded by the need for specialized data processing techniques to pit data streams from sensor networks and multimedia content data [33].

- *Privacy challenge*: In comparison to traditional networks, IoT privacy concerns are becoming increasingly prevalent. Because data involve confidential info, protection of privacy is a major security concern in the IoT. Security of IoT must shield wider management items and stages over traditional data security in line with the integration of things, networks, and services. The prevailing security style is based on human interactions and possibly will not be suited for usage in the IoT [34].
- *Security challenge*: The possible security vulnerability increases as the sheer variety of embedded devices incorporated into IoT systems grows. While the IoT increases business efficiency and improves human livelihoods, it also expands the attack surface for hackers and other data breaches. Each system had an average of 25 holes or risks of breaching the home network. Data encryption methods are seldom used by IoT devices. This tendency arises as a result of the difficulty posed by the widespread use of identical IoT devices. Apart from the capacity of certain gadgets to mechanically connect to other devices, this means that IoT users and developers alike must verify that they are not harming other users or the Internet itself [35].
- *Standard Challenge*: In the growth of IoT, standards are quite significant. A standard is required to ensure that all entities may use and access information in the same way. Professional design of IoT applications, infrastructures, devices, and services will be supported by standardization and cooperation. Overall, collaborating parties' standards, as well as the information models and protocols included in the standards, must be open. All members must be allowed to participate in the standard-setting process, and the resulting standards must be freely and publicly available [36].

There are numerous security concerns in the IoT, including confidentiality, integrity, availability, and authentication. We examined different methods that might be implemented to the IoT environment to address these concerns. The proposed framework is presented in Fig. 6 to mitigate these issues. Lightweight-based cryptographic solutions can overcome these issues; some learning-based solutions (ML, DL, transfer learning, federate learning) and blockchain-based solutions can also be implemented to maintain data privacy and integrity.

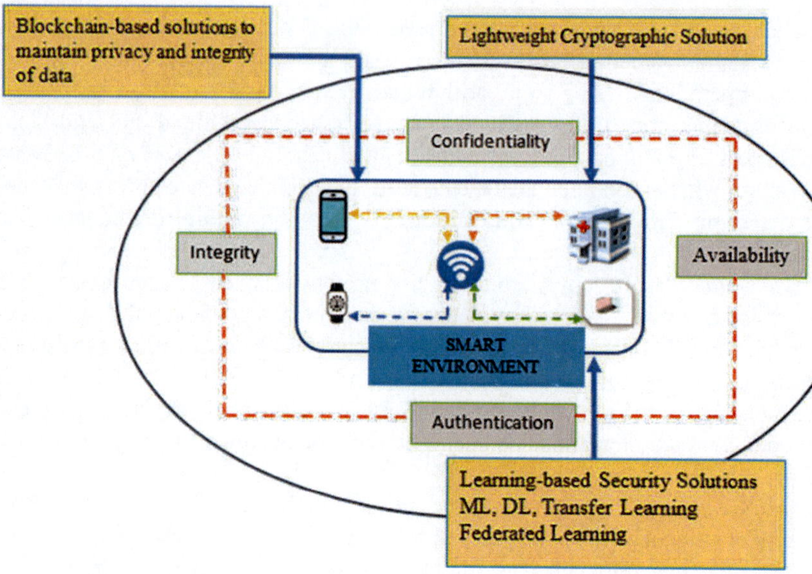

Fig. 6 Proposed solution framework for IoT

5 Conclusion

The IoT has a vital role in the advancement of high tech progress in the direction of automation and application design using machine learning and artificial intelligence. In both of these applications, data play a critical role. Data from devices or environments that are connected through networking interfaces must be transferred in a secure manner. The presence of the IoT worldview over the most recent couple of years has been releasing countless menaces and possible assaults against the security and protection of IoT articles and people. These dangers lead to hampering the acknowledgment of this worldview on the off chance that they have been left without appropriate countermeasures threatening its long-term viability. This article covered the background and preliminaries to IoT. Then, in next section, the security issues at different layers of IoT were critically analyzed and detailed. Additionally, the emerging challenges like privacy, security, data mining, and business challenges were highlighted. In the future, we aspire to build a secure blockchain-based IoT protocol that will safeguard IoT systems from the majority of integrity assaults.

References

1. Meneghello F, Calore M, Zucchetto D, Polese M, Zanella A (2019) IoT : internet of threats ? A survey of practical security vulnerabilities in real IoT devices. IEEE Internet Things J 1. https://doi.org/10.1109/JIOT.2019.2935189

2. Bhuvaneswari V, Porkodi R (2014) The internet of things (IOT) applications and communication enabling technology standards: an overview. In: Proceedings—2014 International conference on intelligent computing applications ICICA, pp 324–329. https://doi.org/10.1109/ICICA.2014.73
3. Stellios I, Kotzanikolaou P, Psarakis M, Alcaraz C, Lopez J (2018) A survey of IoT-enabled cyberattacks : assessing attack paths to critical infrastructures and services. IEEE Commun Surv Tutorials 1. https://doi.org/10.1109/COMST.2018.2855563
4. Makhdoom I, Abolhasan M, Lipman J, Liu RP, Ni W (2018) Anatomy of threats to the internet of things. IEEE Commun Surv Tutorials 1. https://doi.org/10.1109/COMST.2018.2874978
5. Liu L, Xu B, Zhang X, Wu X (2018) An intrusion detection method for internet of things based on suppressed fuzzy clustering. Eurasip J Wirel Commun Netw 1:2018. https://doi.org/10.1186/s13638-018-1128-z
6. Anand P, Singh Y, Selwal A, Singh PK, Felseghi RA, Raboaca MS (2020) IoVT: internet of vulnerable things? threat architecture, attack surfaces, and vulnerabilities in internet of things and its applications towards smart grids. Energies 13(18):1–23. https://doi.org/10.3390/en13184813
7. Hameed A (2019) Security issues in IoT : a survey. In: International conference on innovation and intelligence informatics, computing, and technologies, pp 1–5
8. Mrabet H, Belguith S, Alhomoud A, Jemai A (2020) A survey of IoT security based on a layered architecture of sensing and data analysis. Sensors (Switzerland) 20(13):1–20. https://doi.org/10.3390/s20133625
9. Sarkar C et al (2014) DIAT : a scalable distributed architecture for IoT. X(X):1–10. https://doi.org/10.1109/JIOT.2014.2387155
10. Li D, Deng L, Lee M, Wang H (2019) IoT data feature extraction and intrusion detection system for smart cities based on deep migration learning. Int J Inf Manage 49(April):533–545. https://doi.org/10.1016/j.ijinfomgt.2019.04.006
11. Dawoud A, Shahristani S, Raun C (2018) Deep learning and software-defined networks: towards secure IoT architecture. Internet of Things 3–4:82–89. https://doi.org/10.1016/j.iot.2018.09.003
12. Lv W, Meng F, Zhang C, Lv Y, Cao N, Jiang J (2017) A general architecture of IoT system. 659–664. https://doi.org/10.1109/CSE-EUC.2017.124
13. Mathew PS, Pillai AS, Palade V (2018) Applications of IoT in healthcare
14. Chaudhary S (2019) CRAIoT : concept, review and application (s) of IoT. In: 2019 4th International conference on internet things smart innovation and usages, pp 1–4
15. Deogirikar J, Vidhate A (2017) Security attacks in IoT: a survey. In: Proceedings of the international conference on IoT social mobile, analytics and cloud, I-SMAC, pp 32–37. https://doi.org/10.1109/I-SMAC.2017.8058363
16. Anand P, Singh Y, Selwal A, Alazab M, Member S, Tanwar S (2020) IoT vulnerability assessment for sustainable computing : threats, current solutions, and open challenges. 168825–168853. https://doi.org/10.1109/ACCESS.2020.3022842
17. Hassija V, Chamola V, Saxena V, Jain D, Goyal P, Sikdar B (2019) A survey on IoT security: application areas, security threats, and solution architectures. IEEE Access 7:82721–82743. https://doi.org/10.1109/ACCESS.2019.2924045
18. Tawalbeh L, Muheidat F, Tawalbeh M, Quwaider M (2020) IoT privacy and security : challenges and solutions. Appl Sci 1–17
19. Andrea I, Chrysostomou C, Hadjichristofi G (2015) Internet of things : security vulnerabilities and challenges. pp 180–187
20. Frustaci M, Pace P, Aloi G, Fortino G (2017) Evaluating critical security issues of the IoT world : present and future challenges. 4662. https://doi.org/10.1109/JIOT.2017.2767291
21. Mukherjee BA (2015) Physical-layer security in the internet of things : sensing and communication confidentiality under resource constraints. 103(10)
22. Galeano-Brajones J, Carmona-Murillo J, Valenzuela-Valdés JF, Luna-Valero F (2020) Detection and mitigation of DoS and DDoS attacks in iot-based stateful SDN: an experimental approach. Sensors (Switzerland) 20(3). https://doi.org/10.3390/s20030816

23. Sharma V, You I, Andersson K, Palmieri F, Rehmani MH, Lim J (2020) Security, privacy and trust for smart mobile-internet of things (M-IoT): a survey. https://doi.org/10.1109/ACCESS.2020.3022661
24. Yugha R, Chithra S (2020) A survey on technologies and security protocols: reference for future generation IoT. J Netw Comput Appl 169:102763. https://doi.org/10.1016/j.jnca.2020.102763
25. Tournier J, Lesueur F, Le Mouël F, Guyon L, Ben-Hassine H (2020) A survey of IoT protocols and their security issues through the lens of a generic IoT stack. Internet of Things 100264. https://doi.org/10.1016/j.iot.2020.100264
26. Ngu AH, Gutierrez M, Metsis V, Nepal S, Sheng QZ (2017) IoT middleware: a survey on issues and enabling technologies. IEEE Internet Things J 4(1):1–20. https://doi.org/10.1109/JIOT.2016.2615180
27. Ibitoye O, Shafiq O, Matrawy A (2019) Analyzing adversarial attacks against deep learning for intrusion detection in IoT networks. arXiv
28. Pal S, Hitchens M, Rabehaja T, Mukhopadhyay S (2020) Security requirements for the internet of things: a systematic approach. Sensors (Switzerland) 20(20):1–34. https://doi.org/10.3390/s20205897
29. Nebbione G, Calzarossa MC (2020) Security of IoT application layer protocols : challenges and findings. 1–20. https://doi.org/10.3390/fi12030055
30. Obaidat MA, Obeidat S, Holst J, Al Hayajneh A, Brown J (2020) A comprehensive and systematic survey on the internet of things: security and privacy challenges, security frameworks, enabling technologies, threats, vulnerabilities and countermeasures. Computers 9(2). https://doi.org/10.3390/computers9020044
31. Benkhelifa E, Welsh T, Hamouda W (2018) A critical review of practices and challenges in intrusion detection systems for IoT: toward universal and resilient systems. IEEE Commun Surv Tutorials 20(4):3496–3509. https://doi.org/10.1109/COMST.2018.2844742
32. Chen S et al (2020) A vision of IoT : applications, challenges, and opportunities with China perspective. 1(4):349–359
33. Challenges S (2021) Internet of things: evolution, concerns and security challenges
34. Farhan L, Alissa AE (2017) A survey on the challenges and opportunities of the internet of things (IoT)
35. Sengupta J, Ruj S, Das Bit S (2019) A comprehensive survey on attacks, security issues and blockchain solutions for IoT and IIoT. J Netw Comput Appl 102481. https://doi.org/10.1016/j.jnca.2019.102481
36. Panchiwala S, Shah M (2020) A comprehensive study on critical security issues and challenges of the IoT world

Robust and Imperceptible Multiple Watermarking Using Transform Domain Algorithm for Digital Media

Namita Agarwal, Amit Kumar, and Pradeep Kumar Singh

Abstract This paper discussed a novel robust multiple watermarking method with three transform domain techniques, discrete wavelet transforms (DWT), discrete cosine transforms (DCT) and singular value decomposition (SVD) for color images. For security reasons, secret media is inserted into same multimedia items to offers an additional level of security and to achieve two important performance metrices of watermarking. Firstly, the original media (image) is disintegrated into first level DWT and generate DWT coefficients then select the lower frequency sub-band for applying DCT then SVD. The secret image is also altered by DCT decomposition then SVD transformation is applied on DCT coefficients in embedding process. Then after, a watermarked image is produced by applying converse of all transform domain SVD then DCT and DWT. The secret media can be extracted with recovery algorithm. This procedure has been generously tried and evaluated against various attacks of watermarking and it is found that it achieves better robustness and imperceptibility.

Keywords Digital image watermarking · Discrete wavelet transforms (DWTs) · Discrete cosine transforms (DCT) · Singular value decomposition (SVD) · Robustness

1 Introduction

The Internet becomes an attractive fortune of information to many billion consumers in the era of twenty-first century. The instant growth of media information technologies has shined the way for transmission and data circulation worldwide. This digital disruption has nurtured the escalation of prohibited actions like copyright protection, illegal distribution, unauthorized copying and rewriting of multimedia information. In such a scenario security of senders, receivers, and content of electronic

N. Agarwal (✉) · A. Kumar
Jaypee University of Information Technology, Solan, Himachal Pradesh, India

P. K. Singh
KIET Group of Institutions, Ghaziabad, Uttar Pradesh, India

P. K. Singh et al. (eds.), *Emerging Technologies for Computing, Communication and Smart Cities*, Lecture Notes in Electrical Engineering 875,
https://doi.org/10.1007/978-981-19-0284-0_56

media is of serious significance [1]. Cryptography, steganography and watermarking several data secure techniques have been designed to protect multimedia data. Digital watermarking guarantees to shield the digital media from malicious attacks, piracy, interfering and distribution of information. Digital watermarking is the process of embedding the sensitive data into the accepting image without humiliating image visual quality. After recovery process, this sensitive information or the watermark is useful for content identification and copyright protection. Some believable applications of watermarking are electronic voting system, chip and hardware protection, smart healthcare and numerous additional are shown in Fig. 1 [2–9].

Robustness, imperceptibility, embedding capacity and security are various performance parameters for any watermarking algorithm [10]. Robustness can be described as ability of an algorithm to recovered and authenticate a watermark after being exposed to different attacks. Imperceptibility is defined as keeping the visual quality of original image after it is exposed to many distortions in the presence of watermarking algorithm. The amount of secret data that is feasible to embed into an original image is called embedding capacity. These watermarking performance parameters are independent to one another. Security is described by perseverance of watermarking algorithm against intended or unintended attacks to recover or alter secret information.

Based on the working are of watermarking, it is classified into two parts spatial and transform domain [11]. Spatial domain methods found easy to embed, information is directly embedded in pixel strengths of the original image. However, this domain offers low embedding capacity and powerless to watermarking attacks. Spatial domain techniques are least significant bit, patch work and local binary pattern

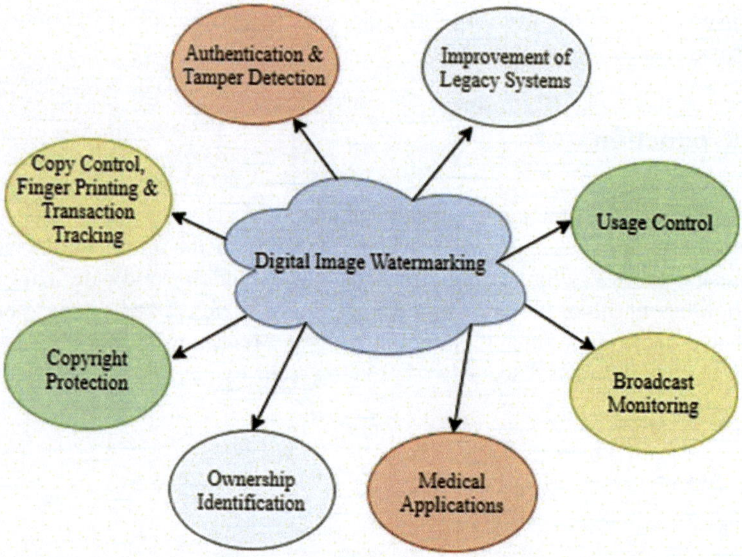

Fig. 1 Applications of watermarking [7]

and many more. In contrast to spatial domain, transform domain technique offers better robustness, imperceptibility, higher embedding capacity. Transform domain algorithms are DCT, DWT, SVD, redundant DWT (RDWT), lifting wavelet transform (LWT) and divisive normalization transform (DNT) and many more.

2 Related Work

Some watermarking methods have been discussed for security of data in past decades. Some motivating watermarking procedures for color images are discussed.

The author of [12] designed a robust and imperceptible watermarking based on DWT and SVD. Two distinct watermarks were inserted in color image matrix applying continuous, segmented and combined watermarking techniques. The author of [13] proposed a Quaternion discrete cosine transform (QDCT), SVD with computer-generated hologram-based watermarking method for color images. The discussed technique founds secure and robust against many attacks. According to Mahajan and Patil [14] they discussed an efficient hybrid watermarking technique with SVD and DWT. Firstly, the first secret information is fixed by second secret mark then joint watermark is inserted into the host media. In Kumar et al. [15] author proposed a watermarking algorithm based on RDWT, non-subsampled contourlet transform (NSCT), SVD and set partitioning in hierarchical trees (SPIHT) is described to achieve a robustness and imperceptibility. Arnold transform is also discussed for more security. Designed algorithm gained robustness, imperceptibility as well as high capacity. Wavelet-based watermarking is very much responsible for highest robustness and are presented in [16]. The maximum functioning of wavelet constructed watermarking scheme significantly lies on embedding and recovery method. Some hybrid watermarking algorithm that uplifts the performance metrices are presented by researchers [17–19].

A robust hybrid multiple watermarking schemes based on DWT, DCT and SVD are discussed in this paper. This algorithm offers the satisfactory performance measures such as PSNR and NC.

3 Theoretical Background

In this paper, a multiple image watermarking technique is proposed, grounded on DWT, DCT and SVD. Firstly, an original image is decomposed by DWT and generate a DWT coefficient then approximation band is selected for DCT then SVD. The secret media (image) will be transformed by DCT. The reason behind DCT decomposition of image watermark as it contains low frequency information and as long as no loss of data while extracting the secret image from watermarked image. A concise narration of these techniques is discussed in given below sections.

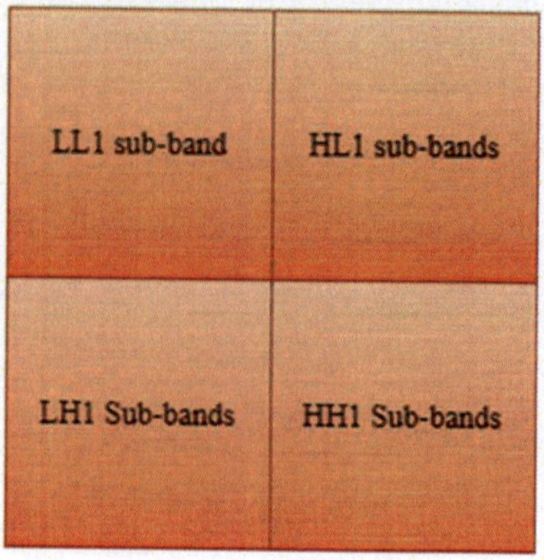

Fig. 2 DWT illustration [22]

4 Discrete Wavelet Transform (DWT)

The main concepts behind DWT it includes decaying of a media (image) in frequency conduits of continual bandwidth on logarithmic gage. The 2-D signal like image is decaying by DWT into a fixed number of four sub-bands (LL, LH, HL, HH) that are non-overlapping multi resolution, approximation band (LL), horizontal sub-band (LH), vertical sub-bands (HL) and diagonal sub-bands (HH) [20]. This procedure can be repetitive to obtain a level of disintegration as shown in Fig. 2. Approximation/lower frequency sub-bands are more delicate to human eyes [21].

5 Discrete Cosine Transform (DCT)

This transform divides an image matrix into three different parts of dissimilar regularities low, high and middle frequency constants. In middle frequency, it is easy to enclose the data that offers an extra challenge to the lossy compression techniques. Due to energy compaction property, it makes DCT very familiar to use in image watermarking [23]. This transform is used in both spatial and transform domain watermarking system. Hence, 2-D transform can be depicted by following Eq. 1.

$$F(x, y) = c(x)c(y)\frac{2}{N}\sum_{p=0}^{N-1}\sum_{q=0}^{N-1} f(p, q)\cos\frac{(2p+1)x\pi}{2N}\cos\frac{(2q+1)y\pi}{2N} \tag{1}$$

where $x, y, p, q = 0, 1, 2\ldots, N-1$, and

$$c(x) = \begin{cases} \frac{1}{\sqrt{2}} \text{ for } \in = 0 \\ 1 \quad \text{Otherwise} \end{cases} \tag{2}$$

The evaluation formula for inverse DCT (IDCT) is given by following Eq. 3:

$$F(p,q) = \frac{2}{N}\sum_{x=0}^{N-1}\sum_{y=0}^{N-1} c(x)c(y)F(x,y)\cos\frac{(2p+1)x\pi}{2N}\cos\frac{(2q+1)y\pi}{2N} \tag{3}$$

6 Singular Value Decomposition (SVD)

The Eq. 4 shows the SVD transformation of an image in rectangular matrix, denoted by P is as follows:

$$\mathrm{SVD}(P) = USV^{\mathrm{T}} \tag{4}$$

where U and V are orthogonal matrices and S represents the diagonal matrix of same size. SVD is usually working on the concept of finding singular value decomposition of an image and then update the singular values to embed the watermark [8]. Nowadays, SVD can be merged with other techniques like DCT and DWT and getting very trendy in watermarking applications.

7 Performance Procedures

Robustness and imperceptibility are two performance parameters for watermarking algorithm [24]. A higher PSNR value denotes that watermarked media is more analogous to original media. It also means that watermark is more imperceptible.

$$\mathrm{PSNR} = 10\log_{10}\frac{(\mathrm{I_{max}})^2}{\mathrm{MSE}} \tag{5}$$

wherever $I_{\max}$ is maximum pixel rate of an image and mean square error (MSE) is defined in Eq. 6

$$\mathrm{MSE} = \frac{1}{M\times N}\sum_{m=0}^{M-1}\sum_{n=0}^{N-1}\left(P(m,n) - \hat{P}(m,n)\right)^2 \tag{6}$$

where P is cover image matrix with size $M \times N$ and $\hat{P}$ is watermarked media (image) matrix with size of $M \times N$. Good imperceptibility is received if cover and watermarked image looks similar.

Normalized correlation is another performance parameter is used to calculate the robustness, by estimating the resemblances between cover and secret media [8].

$$\text{NCC} = \frac{\sum_i \sum_j W\ \text{original}(i, j) W\ \text{recovered}(i, j)}{\sum_i \sum_j W\ \text{original}^2(i, j)}$$

Here, W(original) and W(recovered) depicts the cover and extracted watermarks.

8 Proposed Algorithm

The recommended procedure is the grouping of DWT, DCT and SVD, which increases the robustness without substantial deprivation of the multimedia object (image) quality in contradiction of the signal processing attacks. The discussed method is the combination of two different process one is embedding and second is extraction. Figures 3 and 4 shows the block diagram for embedding and extraction procedures.

9 Experimental Results and Analysis

The size of cover and the secret image is 512×512 for testing the proposed method. In Barbara image the embedding algorithm is grounded on DWT, DCT and SVD. In this experiment PSNR and NC values are calculated at different gain factors (Fig. 5).

In Table 1 algorithm tested the PSNR and NC values at varying scale factor for Barbara image. The proposed algorithm attains maximum PSNR is 37.69 dB at gain factor 0.01 and maximum NC is 0.9998 at gain factor 0.1 without any attack. This table shows that increase in gain value, imperceptibility of images is degraded while improvement in robustness.

The Table 2 shows the outcome of proposed method for four different cover images with different secret image at same gain value 0.05. The maximum and minimum PSNR value achieved from image Lena is 37.65 and 35.98 from image Mandrill. The maximum NC value attained is 0.9948 from image Lena and lowest NC is 0.9889 from Mandrill image.

The Table 3 shows the robustness achieved by proposed algorithm for different attacks at different noise densities. The uppermost NC value 0.9995 is achieved at JPEG-compression (QF 90). The lowest NC is 0.6998 obtained by Salt and Pepper attack at noise density 0.05.

10 Conclusion

In this paper a novel robust multiple watermarking using transform domain techniques such as DWT, DCT and SVD instead of utilizing DWT, DCT and SVD

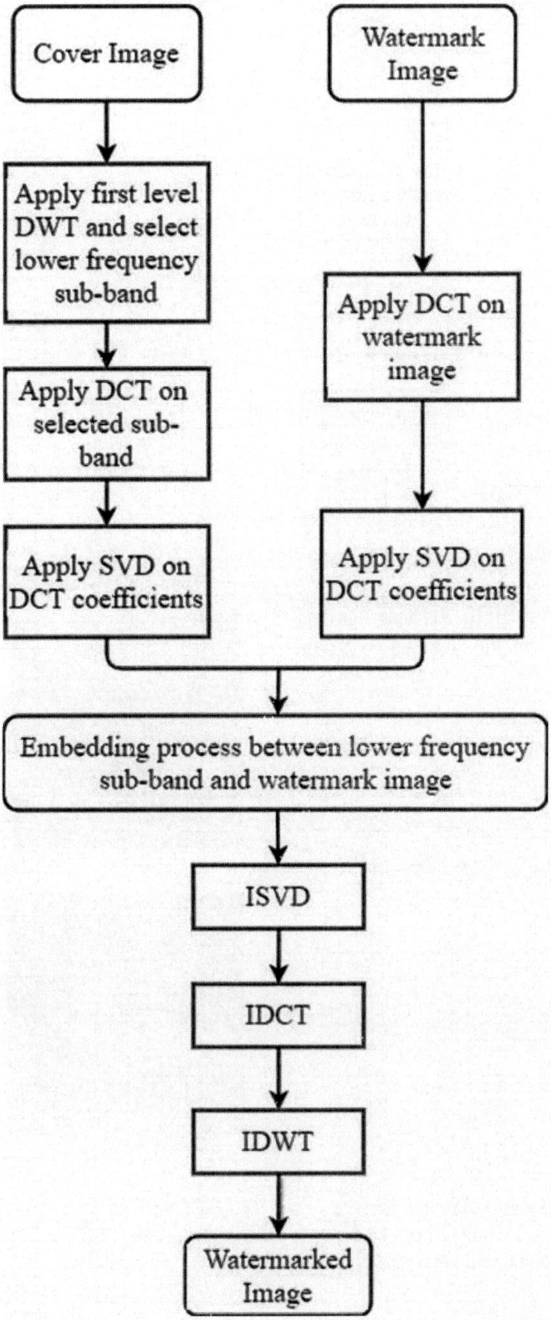

Fig. 3 Embedding algorithm

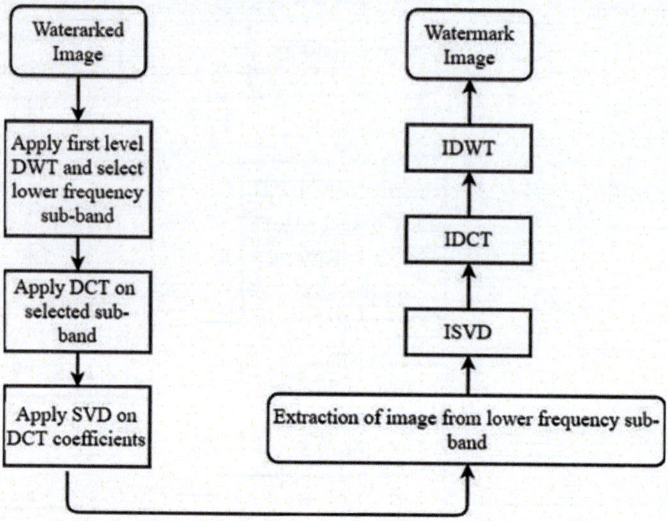

Fig. 4 Extraction algorithm

Fig. 5 (**a–d**) Lena, Barbara, Mandrill and Peppers images used as cover and secret media

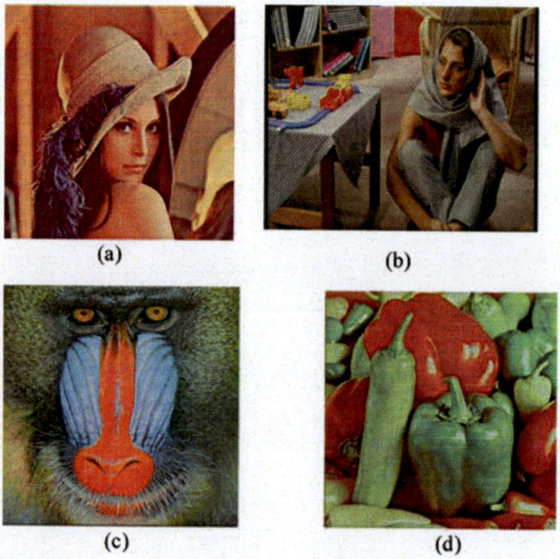

Table 1 Performance evaluation of proposed method at varying gain value

Image	Gain factor	PSNR	NC
Barbara	0.01	37.69	0.9936
Barbara	0.05	37.58	0.9958
Barbara	0.1	36.12	0.9998

Table 2 Performance evaluation for different images at same gain value

Cover image	Watermark image	Gain factor	PSNR	NC
Lena	Lena	0.05	37.65	0.9948
Peppers	Peppers	0.05	36.27	0.9938
Mandrill	Peppers	0.05	35.98	0.9889
Barbara	Lena	0.05	37.55	0.9942

Table 3 Performance evaluation of watermarking algorithm against attacks from image Barbara

S. No.	Attacks	Noise density	NC value
1	JPEG-compression	QF = 10	0.9613
2	JPEG-compression	QF = 50	0.9728
3	JPEG-compression	QF = 90	0.9995
4	Salt and pepper	0.01	0.7989
5	Salt and pepper	0.05	0.6998
6	Salt and pepper	0.001	0.9978
7	Gaussian noise	0.01	0.7883
8	Filtering	1×1	0.9992
9	Scaling	1.1	0.7651

in combination is presented for color images. For security reason a secret image is embedded into the cover media to produce a watermarked image, it achieves an appropriate performance measure in terms of robustness and imperceptibility. Designed algorithm may discover possible application in healthcare systems, electronic voting systems, video transmission and many more.

References

1. Ray A, Roy S (2020) Recent trends in image watermarking techniques for copyright protection: a survey. Int J Multimedia Inf Retrieval 1–22
2. Sharma A, Singh AK, Kumar P (2018) Combining Haar wavelet and Karhunen Loeve transforms for robust and imperceptible data hiding using digital images. J Intell Syst 27(1):91–103
3. Mousavi SM, Naghsh A, Abu-Bakr SAR (2014) Watermarking techniques used in medical images: a survey. J Digit Imaging 27(6):714–729
4. Horng SJ, Rosiyadi D, Fan P, Wang X, Khan MK (2013) An adaptive watermarking scheme for e-government document images. Multimedia Tools Appl 72(3):3085–3103
5. Mohanty SP, Sengupta A, Guturu P, Kougianos E (2017) Everything you want to know about watermarking: from papers marks to hardware protection. IEEE Consum Electron Mag 6(3):83–91
6. Singh AK, Dave M, Mohan A (2014) Wavelet based image watermarking: futuristic concepts in information security. Proc Natl Acad Sci, India, Sect A 84(3):345–359

7. Agbaje MO, Awodele O, Ogbonna AC (2015) Applications of digital watermarking to cyber security (cyber watermarking). In: Proceedings of informing science and IT education conference (InSITE), pp 1–11
8. Singh D, Singh SK (2017) DWT-SVD and DCT based robust and blind watermarking scheme for copyright protection. Multimedia Tools Appl 76(11):13001–13024
9. Singh AK, Kumar B, Dave M, Mohan A (2015) Robust and imperceptible dual watermarking for telemedicine applications. Wireless Pers Commun 80(4):1415–1433
10. Bhatnagar G (2012) A new facet in robust digital watermarking framework. AEU—Int J Electron Commun 66(4):275–285
11. Begum M, Uddin MS (2020) Digital image watermarking techniques: a review. Information 11(2):110
12. Mohananthini N, Yamuna G (2016) Comparison of multiple watermarking techniques using genetic algorithms. J Electr Syst Inf Technol 3(1):68–80
13. Li J, Lin Q, Yu C, Ren X, Li P (2018) AQDCT and SVD based color image watermarking scheme using an optimized encrypted binary computer-generated hologram. Soft Comput 22(1):47–65
14. Mahajan LH, Patil SA (2013) Image watermarking scheme using SVD. Int J Adv Res Sci Eng 2(6):69–77
15. Kumar C, Singh AK, Kumar P, Singh R, Singh S (2020) SPIHT—based multiple image watermarking in NSCT domain. Concurrency Comput Pract Experience 32(1)
16. Lai CC, Tsai CC (2010) Digital image watermarking using discrete wavelet transform and singular value decomposition. IEEE Trans Instrum Meas 59(11):3060–3063
17. Lin WH, Horng SJ, Kao TW, Fan P, Lee LC, Pan Y (2008) An efficient watermarking method based on significant difference of wavelet coefficients quantization. IEEE Trans Multimedia 10(5):746–757
18. Awasthi M, Lodhi H (2013) Robust image watermarking based on discrete wavelet transform, discrete cosine transforms and singular value decomposition. Adv Electr Electron Eng 3(8):971–976
19. Golshan F, Mohammadi K (2013) A hybrid intelligent SVD-based perceptual shaping of digital image watermark in DCT and DWT domain. Imaging Sci J 61(1):35–46
20. Langekaar GC, Setyawan I, Lagendijk RL (2000) Watermarking digital image and video data. A state-of-the-art overview. IEEE Signal Process Mag 17(5):20–46
21. Ali M, Ahn CW (2018) An optimal image watermarking approach through cuckoo search algorithm in wavelet domain. Int J Syst Assur Eng Manage 9(3):602–611
22. Daphal SD, Aher PK, Galande SB, Jagtap SK (2018) Comparison of key transform domain watermarking methods based on performance measures. In: International conference on communication, information and computing technology (ICCICT). IEEE, pp 1–6
23. Abdulrahman AK, Ozturk S (2019) A novel hybrid DCT and DWT based robust watermarking algorithm for color images. Multimedia Tools Appl 78(12):17027–17049
24. Khan M, Masood F (2019) A novel chaotic image encryption technique based on multiple discrete dynamical maps. Multimed Tools Appl 78(18):26203–26222

Blockchain-Driven and IoT-Assisted Chemical Supply-Chain Management

Pronaya Bhattacharya, Ashwin Verma, and Gulshan Sharma

Abstract Blockchain (BC) technologies have gained prominence in the scientific and industrial sectors, as they have the potential to benefit a wide range of sectors. The shift is attributed owing to the key benefits of immutability, chronology, and ownership at every point in the supply chain. Thus, through BC, and smart contracts (SCs), every point in the supply chain can capture and exchange transactional data, and thus, it enables efficient and automated chain operations. Thus, BC adoption introduces traceability and transparency and lowers costs. Owing to the potential benefits of BC, promising solutions can be adopted in the chemical industry sector, where BC-based solutions can be potentially used to manage inventory management, to logistics, and ensure that fake chemicals are not sold by malicious entities, without the requirement for third-party validation or authorization. Moreover, chemical products can be assigned low-powered Internet-of-Things (IoT) digital tags that simplify the chemical inventory operations. Thus, in this paper, we propose a BC-based IoT-assisted reference architecture for chemical industries and present the entire supply-chain operation. To support the claim, a practical case study, *ChemBlocks* is presented, that presents the potential benefits of integration of IoT tagging on chemical products, with ledger state information managed via BC. At every supply-chain point, SCs are executed that automated the transfer of funds, and asset transfer is achieved. The proposed case study eliminates cost and effort duplication and increases the return of investment (ROI) of chemical industries.

Keywords Blockchain · Chemical industries · Internet-of-Things · Manufacturing supply chain · Smart contract · Smart tagged objects

---

P. Bhattacharya (✉) · A. Verma
Institute of Technology, Nirma University, Ahmedabad, Gujarat, India
e-mail: pronoya.bhattacharya@nirmauni.ac.in

A. Verma
e-mail: ashwin.verma@nirmauni.ac.in

G. Sharma
Durban University of Technology, Durban, South Africa

P. K. Singh et al. (eds.), *Emerging Technologies for Computing, Communication and Smart Cities*, Lecture Notes in Electrical Engineering 875,
https://doi.org/10.1007/978-981-19-0284-0_57

1 Introduction

In smart manufacturing ecosystems, industries have shifted toward smart and automated supply-chain builds, where the industrial processes are automated through cyber-physical processes (CPS) that connect industrial stakeholders over heterogeneous networked applications. In such supply-chain builds, trusted supply relationships have gained prominence owing to the increased challenges of large distances, product genuineness, and quality of the product. In the chemical industries, chemicals are produced in mass quantities, where the raw chemicals are provided by basic chemical suppliers and are then processed and produced in chemical plants. These raw chemicals can be forged, owing to the different quantities of processed chemicals, and sold to manufacturers at a higher cost. Thus, fake chemical generation and detection is a timely process, and this affects the entire supply-cycle [7].

Once the chemicals are packed, they are assigned low-powered Internet-of-Things (IoT)-based tokens, that are barcodes, quick-response (QR), or near-field communication (NFC) codes assigned on packaged chemicals, and sent to manufacturer plant [5]. Figure 1 represents the generic chemical supply-chain cycle and highlights the main challenges in the ecosystem. From there, the final deliverable chemicals are produced, and products are packed, assigned IoT digital tokens, and sent to the warehouse. Once the chemicals reach the warehouse, they are sold to small-scale

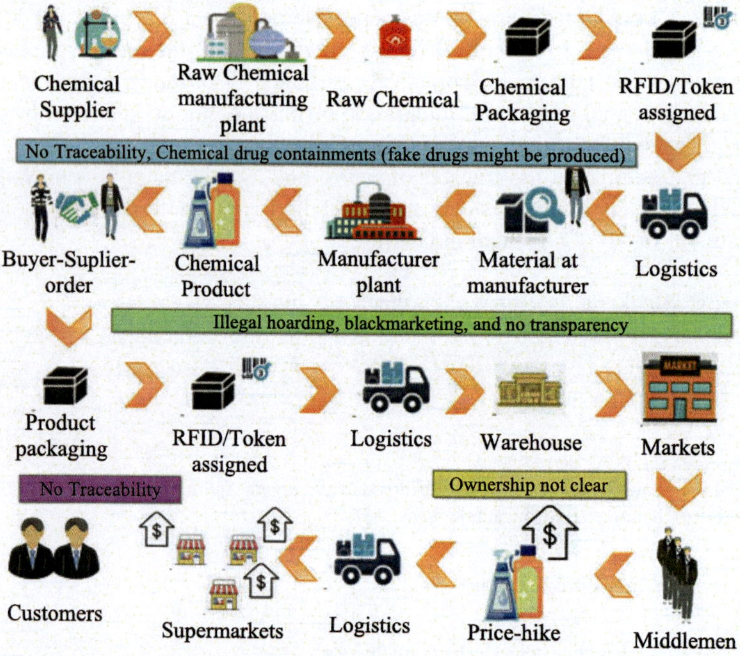

Fig. 1 Challenges in distribution in the generic supply chains in chemical industries

industries from open markets. Here, illegal hoarding and black-marketing of chemicals are possible, by reducing the supply of chemicals to open markets, the prices are hiked by third-party vendors. Once the hiked chemicals are bought by small-sized industries by such third-part vendors, they are not assured of the manufacturer, and ownership is not clear, in case of inferior quality chemicals. The prices are further hiked by small-sized vendors, to cover their logistic cost, and are sold at very high prices at supermarkets to general customers [8]. Thus, the global supply-chain process has to assure trust, transparency, and ownership through the entire chain, so that the entire process is visible to end customers.

Blockchain (BC) technology provides a means of building trust between suppliers and customers along the chemicals supply chain. BC allows verified provenance of the source of raw materials, tracking and tracing of processed chemicals from suppliers, through assures the quality control and inventory details to be added in the immutable transaction ledgers [10]. Thus, it allows that the desired product quality is maintained over the entire supply chain points, and counterfeit chemicals are detected, and the tracing and ownership to the malicious stakeholders are also identified [1]. In the chemical supply-chain ecosystems, we can document the entire process, and transactional payments between any two supply points can be done via smart contracts (SCs). SCs include the supplier, manufacturer, and packaging origin and create a single truth point across the entire ledger [2].

At each point of the supply chain, we consider that on every chemical product, IoT tags are attached. IoT tagging allows contactless reading and writing of information. Data that is written onto an IoT tag can be encrypted and published in BC, and stakeholders can determine the proper access control on the data sharing, despite the information being stored at distributed and heterogeneous networked locations [14].

1.1 Motivation

In chemical industries, drug composition to prepare the chemical in the manufacturing plant is a critical process. Currently, chemical industries are suffering from the issue of the genuineness of raw materials, high costs due to illegal hoarding and distribution. Owing to these limitations, BC is a preferred choice, and the integration of BC in chemical industries is still in its infancy. Thus, the paper discusses a reference architecture that integrates BC at different supply points in the distribution ecosystem and assures trust and transparency in the entire supply chain. At any supply-chain point, SCs assure that the required operating conditions of chemical storage, distribution, and collection are preserved. The paper also proposes a real-time case study that presents an industry use-case deployment of BC in the chemical industry. The potential benefits and issues and challenges are discussed for practical realization of BC-driven chemical industry supply-chain mechanisms.

1.2 Research Contributions

The following are the research contributions of the paper.

- A BC-based and IoT tag-driven reference architecture is proposed. Through IoT tag objects, we assure a low-cost solution of digitizing the chemical packaging records, and information of the source and destination endpoints. Once the chemical records (CR) are fetched, the meta-information is stored as a transactional ledger in BC.
- A case study of BC-driven chemical industry, named as *ChemBlocks* is discussed that presents the practical deployment conditions and the SCs that are proposed at different points of the supply chain.

1.3 Article Structure

The chapter is divided into five sections. Section 2 presents the proposed reference architecture that integrates BC and IoT to assist low-powered traceable supply-chain operations. Section 3 presents the open issues and potential challenges of the deployment. Section 4 presents the discussion of a proposed case study, *ChemBlocks* for chemical supply-chain ecosystems. Section 5 concludes the chapter.

2 BC-Driven and IoT-Assisted Chemical Supply-Chain Management: The Reference Architecture

In this section, we present the reference architecture of BC-driven and IoT-assisted chemical distribution and supply-chain management. The details are presented as follows.

2.1 The Entity Model

In the propose ecosystem, we consider the following entities.

- *Chemical supplier*: The chemical supplier, denoted by E_{CS}, distributes the raw chemicals that are mixed in appropriate proportions in the chemical manufacturing plant.
- *Chemical manufacturer*: The chemical manufacturer, denoted as E_{CM}, collects the raw chemicals from E_{CS}, processes them in desired proportion, and prepares the final raw chemical, denoted by F_{C}. F_{C} is then packaged into chemical containers, and a RFID token $T_{F_{\text{C}}}$ is attached. $T_{F_{\text{C}}}$ contains the attributes like $\{F_{\text{id}}, F_{\text{com}}, F_{\text{DOM}}, F_{\text{DOE}}, F_{\text{cost}}\}$, where F_{id} denotes the product ID, F_{com} denotes

the chemical composition information, F_{DOM} denotes the date of packaging, F_{DOE} denotes the expiry date, and F_{cost} denotes the chemical cost.

- *Logistics head*: Once E_{CM} prepares $T_{F_{\text{C}}}$, the packages are collected and transferred for shipments through the land, water, and air. The entire shipment process is handled by the logistics head, denoted by E_{LH}. E_{LH} is also responsible to handle cross-border financial settlements, duty and toll-taxes, and payment settlement to intermediaries who assist in the shipment process. E_{LH} delivers F_{C} to the manufacturer, who is responsible for preparing the chemical product.
- *Suppliers and buyers*: Once the chemical product is ready from F_{C}, denoted as CP, SCs are initiated depending on the product orders. Seller denotes E_{CM}, from which buyers (market buyers), purchase the product. The buyers, denoted as E_{B}, purchases the final product that includes the total cost C as $T(E_{\text{CS}} + C(E_{\text{LH}} + C(M_{\text{CP}}) + T$, where $C(E_{\text{CS}})$ denotes the raw chemical packaging cost, $C(C_{\text{LH}})$ denotes the logistics cost, and $C(M_{\text{CP}})$ denotes the manufacturing cost of CP, and T denotes the incurred taxes and other operational costs. In BC, this cost C is transparent between E_{CS}, E_{LH}, E_{CM}, E_{B}.
- *End-customers*: End-customers, denoted by E_{CU}, purchases CP from open markets, or supermarkets. The final cost, FC can be computed as $\text{FC} = C + E_{\text{LH}} + C(S) + T$, where $C(S)$ denotes the cost of product maintenance at supermarkets. As C is transparent to E_{CU} via BC, and the cost of E_{LH}, T, and $C(S)$ is entered by respective stakeholders in BC, the FC is transparent.

2.2 The Layered Reference Architecture

In this subsection, we discuss the reference architecture as a layered model between the entities. In the reference architecture, we consider three layers, namely the IoT layer, the supply-chain layer, and the contract layer. Figure 2 presents the architecture details.

IoT Layer—At this layer, we consider two supply-chain points. One point is between E_{CS} and E_{CM}, and the second one is between E_{CM} and E_{B}. At both supply points, we consider the raw chemical is packaged and processed into containers and are assigned a digital token identity, denoted as $T_{F_{\text{C}}}$. The digital identity can be an RFID barcode, QR, or NFC code, that can be processed through a low-powered reader device R_{d}. The containers are shipped through E_{LH} to destination points. To assist low-powered networking, we consider an IoT protocol stack S that consists of a communication module (ZigBee, WiFi, or 6LowPAN-assisted addresses). We consider IPv6 addressing module that communicates through low-powered personal area networks like ZigBee, WiFi, and 6LowPAN addresses, with information of source IPv6 and destination IPv6, for packet routing and delivery. At the transport layer, we consider real-time barcode bytes streaming assisted through either TCP or UDP, depending on the connection approach. The packet payload information is encrypted through lightweight symmetric ciphers, like elliptic curve cryptography (ECC), or advanced encryption standard (AES), or implementors might prefer trusted execu-

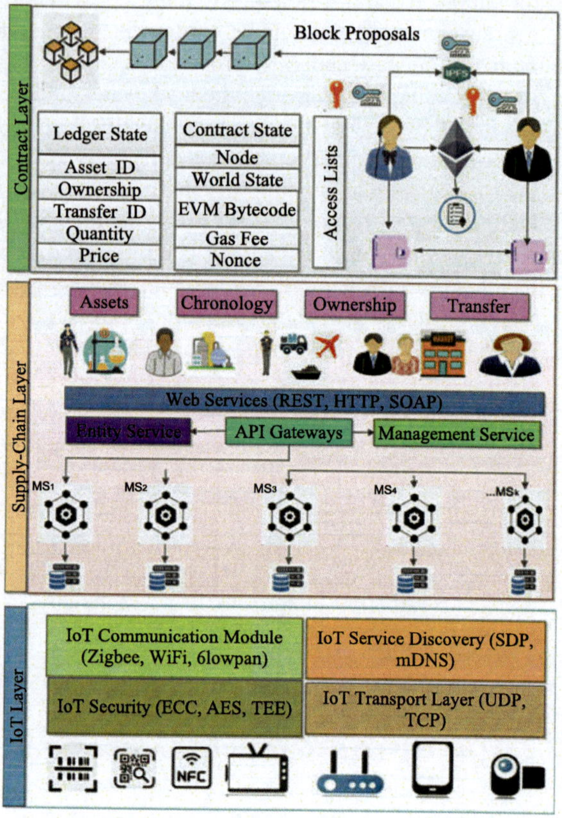

Fig. 2 Reference architecture of IoT-assisted BC-driven chemical supply-chain ecosystem

tion environments (TEE). For service discovery, we consider the service discovery protocol (SDP), or multicast DNS (mDNS) approach [6].

Supply-Chain Layer—At this layer, we consider a microservice functional environment, for specific functionalities like asset ownership functionality, chemical meta-information $\{F_{\text{id}}, F_{\text{com}}, F_{\text{DOM}}, F_{\text{DOE}}, F_{\text{cost}}\}$, logistics cost functionality that includes shipments via land, water, and air cargo facilities, over international land, air, and water boundaries. Here, the microservice handles the internal payment information through digital cryptocurrencies. This allows the payments to be traceable, and in the case of multi-currency scenarios, an exchange server ES is formed that converts one form of cryptocurrency to other, based on the current conversion rates. Another microservice involves the chemical stock inventory service. It includes information $\{O_S, T_S, F_S\}$, where O_S denotes the ordered quantity of stock, T_S denotes the delivered quantity, and F_S denotes the bad-stock quantity. Overall, we consider that k micro-services are operational, denoted as $\text{MS} = \{\text{MS}_1, \text{MS}_2, \ldots, \text{MS}_k\}$.

Each microservice has an associated local database D_{MS_k} that maintains the state and ledger information [4]. MS communicates with each other through event-driven application programming interface (API) gateways, which involves two event han-

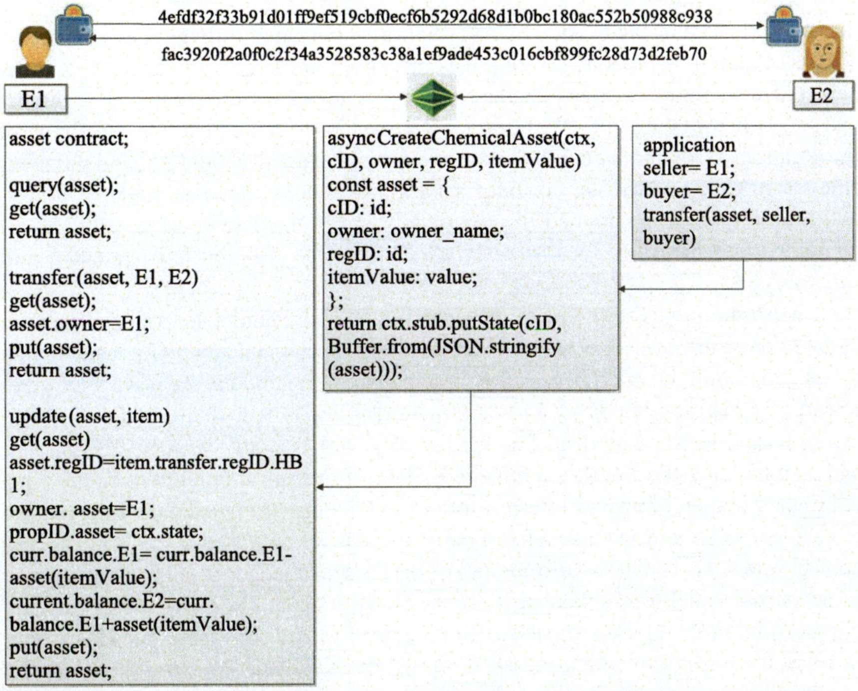

Fig. 3 SCs interface among the communicating entities

dlers, one for entity service handler that allows chemical stakeholders to loosely interact with MS via the web gateways. The other event handler is the management service handler, which allows event segregation, and web-event data handling, formatted exchange, and information caches. The implementations might adopt the representational state transfer (REST) web-interface communication points that allow stateless client–server HTTP request–reply paradigm. The session state information is cached at the client, which allows faster query response for frequently selected objects [12]. REST allows the functionality of transferable executable codes to the server end. The data exchange is normally preferred in JavaScript object notation (JSON) files. In simple object access protocol (SOAP), the web-communication interface follows the transactional states (ACID properties) and presents security interface endpoints through web-service security (WS-security), and web services description language (WDSL) [3].

The Smart Contract Layer—At this layer, we consider the transactional ledgers T_l are maintained between the chemical stakeholders. We maintain the ledger state information $I_l = \{A_{id}, O_d, T_{id}, Q, P\}$, where I denotes the ledger information, A_{id} denotes the asset-identifier, O_d denotes the ownership records, T_{id} denotes the asset transfer information that contains the information of final ownership, Q denotes the final quantity of F_C, or CP transferred between different supply-chain points, and P

denotes the price of the asset. The information is managed through a permissioned contract-chain information between two entities E_1 and E_2, as depicted in Fig. 3. As depicted in the figure, we consider the contract between E_1 and E_2, depicted as buyers and sellers in the contract C. We consider the transfer of chemical product CP as asset information. The contract conditions are finalized between E_1 and E_2, and payment channel mode is setup. We consider a permissioned SC transfer mode of assets, and information is maintained through ledger states L_s. The contract deployment contains three functionalities, *assetContract* interface, *CreateChemicalAsset*, and *transferAsset*.

The information is stored in world-state as a response, and read–write buffers are updated once the contact conditions are fulfilled. A channel setup C is established for secure communication between E_1 and E_2, and it maintains the privacy of asset transfer. The appropriate gas fee for contract execution is deducted, and the chemical asset ownership is transferred. The wallets W_{E_1} and W_{E_2} are, respectively, debited and credited once the contract is finalized. The contract information is published in a distributed ledger, known as interplanetary file systems (IPFS), and the information is visible to all authorized stakeholders, whose entries are considered in the access control lists [11]. The meta-information of the recorded ledger in IPFS is then stored as an unconfirmed transaction in Mempool [13]. Miners E_m proposes the mining phase, to initiate a block proposal B to the added to BC. To assess a record stored in IPFS, the hash of the meta-information is referenced from BC, and the IPFS key is fetched. Any authorized stakeholder then assesses the IPFS through two sets of keys, one is the IPFS key and the other is the users private key.

3 Open Issues and Challenges

In this section, we discuss the potential issues of BC adoption in chemical supply-chain ecosystems.

1. *Node scalability*—In chemical supply-chain ecosystems, owing to huge amount of data transfer among different stakeholders, the BC-governance model has to assure lightweight resource requirements for node validations. Recently, lightweight consensus approaches like IOTA, Merkle-DAG, and others are proposed, but there is no uniformity of the consensus protocols with the data governance model of heterogeneous applications.
2. *Privacy requirements*—In BC, the data is transparent and auditable. Thus, authorized stakeholders, based on access-list permissions, can view the ledger data. However, the access control mechanisms in BC are not universal, and most SC implementations do not have access control privilege. Thus, the privacy of data is not controllable in most implementations. Moreover, user anonymity is not guaranteed, as in public chains, the transaction ledger is visible through a public wallet address. Recently, industry practitioners have shifted toward controlled lock mechanisms in BC that proposes lock mechanisms on data transfer, access,

and ownership. However, the open-source implementation is still in its infancy, and most solutions are proprietary.

3. *Operational cost*—The next issue is the overall operational cost (OPEX) of the BC-based supply-chain ecosystems. In the chemical industries, the supply chain might consist of a lot of intermediate supply-chain points, and many stakeholders are involved in the entire process. The distribution of chemicals transparently, with assured chronology among all supply points, would involve high transactional costs, due to block mining and validations. Moreover, the block mining time is proportional to the mining difficulty, or the miners with a high amount of stakes. Miners with high computational resources might collude with other stakeholders in the chain, to monopolize the mining process, thereby increasing the price of the chemicals.
4. *Security attacks*—BC is considered secure, but it is vulnerable to security attacks like mining attacks, power attacks (51% attack), validator-based attacks, and collusion attacks. Moreover, SCs are programmable codes that are vulnerable to flaws in code logic that might induce vulnerabilities like gas-attacks, contract flow-attack, injection-attack, dependency-injection, name-space pollution attacks, re-entracy, and many more. Thus, formal verification of SCs against the possible attacks should be considered before the final deployments. However, the identification of attack vectors is complex in BC ecosystems.
5. *Data management*—In the chemical supply chain, a chemical supplier (sender) may transact in Bitcoin wallets, and the receiver wallet might be accepting Ethereum. Thus, the data management among the transactional ledgers would require effective exchange server integration in the payment system to perform the interconversion in real-time. At present, the conversion exchanges are limited and are not supported in all supply chain ecosystems.
6. *Consensus approach*—With the emergence of assisted technologies like IoT, artificial intelligence, and big data ecosystems, the design of generic consensus approaches that operate with the collected data from these platforms is in the early stages. In such ecosystems, BC leverages low transactional throughput and mostly is not suitable for public-chain systems. Alternative designs involve the design of permissioned consensus, but it reduces the BC-openness and transparency flow among all supply-chain stakeholders. Thus, the design of optimized public consensus schemes, that manages the node throughput is an open issue.
7. *Cross-boundary legislation*—Due to international regulations and government legislation, cryptocurrency is not considered as a monetary medium of transfers in many countries. To fully realize the potential of global supply-chain operations, logistics would require cross-boundary uniformity in legislative rules that involves payments via cryptocurrency tokens.

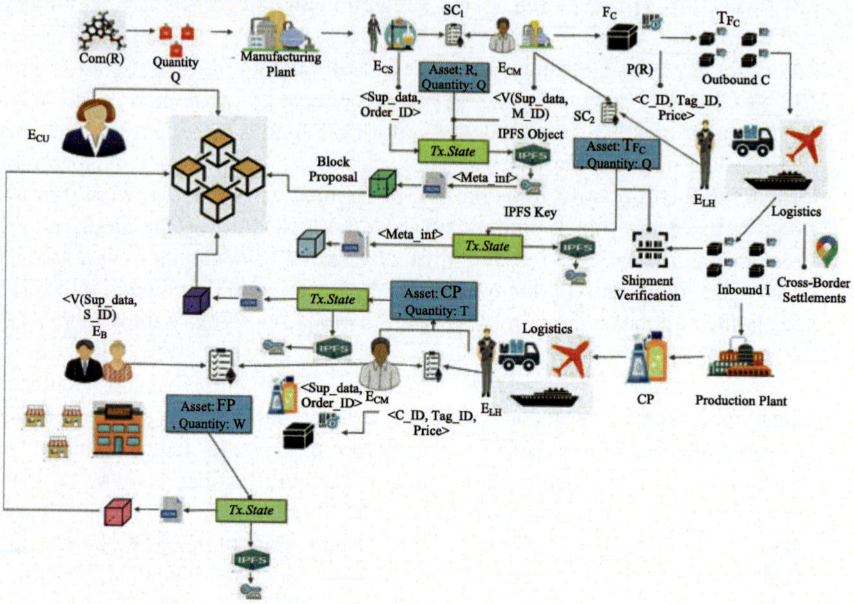

Fig. 4 *ChemBlocks*: the operational flow of the propose case study

4 *ChemBlocks*: A Case Study on RFID-Assisted Smart-Tagging in BC-Leveraged Chemical Supply Chains

In this section, we present an industry-ready practical deployment case study scenario of BC-based chemical supply-chain distribution and management operations [9]. The deployment, named as *Makerchain*, is designed to propose a cyber-driven payment model lifecycle, with SCs auditable trail history. We have modified the proposed model and have included the smart tagging based objects, and have proposed a model, named as *ChemBlocks*, that integrates RFID-assisted smart tagging at every supply-chain point, with the transactional meta-information stored in BC-ledgers. Figure 4 presents the detailed flow of the proposed case study. The components of the model are discussed in detail as follows.

4.1 *Ordering of Raw Material*

As discussed in Sect. 2, we use the same entity notations in the case study for better readability. We consider the raw chemical composition as $\text{Com}(R)$, decided by E_{CS}, and Q units of R are ordered by E_{CM}. At this point, we maintain a SC_1 transactional asset information as <asset:R, quantity:Q >. The SC ledger of E_{CS} includes <Sup_data, Order_ID>, which is orderID, and supplier information, and E_{CM} verifies the SC ledger, and appends < M_id> state to the world-ledger. R is then packaged into containers $P(R)$, and is tagged via RFID-tokens, denoted as T_{F_C}, and is sent to outbound process C. The SC information is recorded as *Tx.State* in Ethereum virtual machine (EVM) ledger as byte-code, and *meta-inf* is stored in IPFS, and block is proposed.

4.2 Logistics: Outbound and Inbound

Outbound C state is maintained as asset $< T_{F_C}, Q >$, and $P(R)$ is attached with < C_ID, Tag_ID, Price>, and is sent to E_{LH} that ships and manages the cross-border settlements through cryptocurrency exchange wallets. At inbound I, the shipment verification is done through a barcode reader, and SC_2 transactional ledger state is maintained and recorded in BC between E_{CM} and E_{LH}.

4.3 Final Chemical Packaging and Outbound to Open Markets

E_{CM} prepares the final chemical product CP from F_C in T quantities, which is RFID-tagged, and is outbound to E_{B}, with logistic intermediary E_{LH}. The cost of the CP, plus the cost of logistics, is included and tagged into the packaged product, with information < C_ID, Tag_ID, Price>. The supplier E_{CM} data *Sup_data* is included in SC_3 asset information, and recorded in BC.

4.4 Inbound at Open Markets and Delivery to End-Customers

E_{LH} delivers the final packaged product in W quantities, with the constraint $W < T$, as we assume $T - W$ items are damaged in transit. An asset $<$ FP, $W >$ is created between E_{CM} and E_{B}, and E_{B} updates its inventory stock information as <Sup_data, S_ID, E_B>, and stores the chemical products at markets. The packaged product is then bought by the end-customer, E_{CU}. As all SC *Tx.States* is recorded as block proposals in public chain, E_{CU} has a view right of the entire supply-chain chronology, and thus, it ensures complete fairness and eliminates any sort of illegal marketing, middlemen, and hoarding of chemical products that inflate the prices of chemical product.

5 Conclusion

In chemical supply-chain ecosystems, BC is gaining widespread prominence to improve the transparency of supply-chain endpoints. Coupled with smart IoT-assisted tagging, it streamlines and automates the entire record and inventory maintenance, with ease in data payment mechanisms, owing to SCs programmable logic. This improves the efficacy of supply chains and lowers production costs. Motivated by these facts, in the paper, we presented a reference architecture of possible BC deployment in chemical supply chains and discussed the key points in the ecosystem. The open issues and challenges are further discussed, which presents the readers with emerging gaps to leverage research in a similar direction. Finally, we presented a supporting case study, *ChemBlocks*, that discusses the integration of BC and RFID-smart tagging in the ecosystem.

As part of future work, we would propose security models to support the data protection against possible attack boundaries, with the inclusion of anti-counterfeiting and copyright protection models. It would allow an end-to-end holistic vision toward the design of formal BC-based chemical supply chains in the chemical sector.

References

1. Al-Jaroodi J, Mohamed N (2019) Blockchain in industries: a survey. IEEE Access 7:36500–36515
2. Bodkhe U, Mehta D, Tanwar S, Bhattacharya P, Singh PK, Hong WC (2020) A survey on decentralized consensus mechanisms for cyber physical systems. IEEE Access 8:54371–54401. https://doi.org/10.1109/ACCESS.2020.2981415
3. Brown R (2021) Connecting energy management systems to enterprise business systems using soap and the xml web services architecture. In: Web based energy information and control systems. River Publishers, pp 447–458
4. Cerny T, Donahoo MJ, Trnka M (2018) Contextual understanding of microservice architecture: current and future directions. SIGAPP Appl Comput Rev 17(4):29–45. https://doi.org/10.1145/3183628.3183631
5. Farsi M, Latsou C, Erkoyuncu JA, Morris G (2020) RFID application in a multi-agent cyber physical manufacturing system. J Manuf Mater Process 4(4). https://doi.org/10.3390/jmmp4040103, https://www.mdpi.com/2504-4494/4/4/103
6. Khalid M, Mujahid U, Najam-ul Islam M, Choi H, Alam I, Sarwar S (2021) Ultralightweight resilient mutual authentication protocol for IoT based edge networks. J Ambient Intell Humanized Comput 1–12
7. Khan SAR, Zkik K, Belhadi A, Kamble SS (2021) Evaluating barriers and solutions for social sustainability adoption in multi-tier supply chains. Int J Prod Res 59(11):3378–3397. https://doi.org/10.1080/00207543.2021.1876271
8. Kleinekorte J, Fleitmann L, Bachmann M, Kätelhön A, Barbosa-Póvoa A, von der Assen N, Bardow A (2020) Life cycle assessment for the design of chemical processes, products, and supply chains. Annu Rev Chem Biomol Eng 11(1):203–233. https://doi.org/10.1146/annurev-chembioeng-011520-075844, pMID: 32216728
9. Leng J, Jiang P, Xu K, Liu Q, Zhao JL, Bian Y, Shi R (2019) Makerchain: a blockchain with chemical signature for self-organizing process in social manufacturing. J Cleaner Prod 234:767–778. https://doi.org/10.1016/j.jclepro.2019.06.265, https://www.sciencedirect.com/science/article/pii/S0959652619322309

10. Patel NS, Bhattacharya P, Patel SB, Tanwar S, Kumar N, Song H (2021) Blockchain-envisioned trusted random oracles for IoT-enabled probabilistic smart contracts. IEEE Internet Things J 8(19):14797–14809. https://doi.org/10.1109/JIOT.2021.3072293
11. Patel SB, Bhattacharya P, Tanwar S, Kumar N (2021) Kirti: a blockchain-based credit recommender system for financial institutions. IEEE Trans Network Sci Eng 8(2):1044–1054. https://doi.org/10.1109/TNSE.2020.3005678
12. Riaz U, Hussain S, Patel H (2021) A comparative study of rest with soap. In: Fu W, Xu Y, Wang SH, Zhang Y (eds) Multimedia technology and enhanced learning. Springer International Publishing, Cham, pp 485–491
13. Shukla A, Bhattacharya P, Tanwar S, Kumar N, Guizani M (2020) Dwara: a deep learning-based dynamic toll pricing scheme for intelligent transportation systems. IEEE Trans Veh Technol 69(11):12510–12520. https://doi.org/10.1109/TVT.2020.3022168
14. Yuvaraj S, Sangeetha M (2016) Smart supply chain management using internet of things (IoT) and low power wireless communication systems. In: 2016 International conference on wireless communications, signal processing and networking (WiSPNET), pp 555–558. https://doi.org/10.1109/WiSPNET.2016.7566196

Correction to: Impact of Green Communication and Technology System

Sumit Dhariwal and Avinash Raipuria

Correction to:
Chapter "Impact of Green Communication and Technology System" in: P. K. Singh et al. (eds.), *Emerging Technologies for Computing, Communication and Smart Cities*, Lecture Notes in Electrical Engineering 875,
https://doi.org/10.1007/978-981-19-0284-0_44

In the original version of the book, the author Sumit Dhariwal's affiliation has been changed from "Manipal University, Jaipur, India" to "Manipal University Jaipur, Rajasthan, India" in the Chapter "Impact of Green Communication and Technology System".

The chapter and book have been updated with the changes.

The updated original version of this chapter can be found at
https://doi.org/10.1007/978-981-19-0284-0_44

P. K. Singh et al. (eds.), *Emerging Technologies for Computing, Communication and Smart Cities*, Lecture Notes in Electrical Engineering 875,
https://doi.org/10.1007/978-981-19-0284-0_58

Author Index

P. K. Singh et al. (eds.), *Emerging Technologies for Computing, Communication and Smart Cities*, Lecture Notes in Electrical Engineering 875,
https://doi.org/10.1007/978-981-19-0284-0